SI unit prefixes

Multiplication factor	Prefix	Symbol	Pronunciation	Term
$1\ 000\ 000\ 000\ 000 = 10^{12}$	tera	T	as in *terra*ce	one trillion
$1\ 000\ 000\ 000 = 10^{9}$	giga	G	jig'a (*a* as in *a*bout)	one billion
$1\ 000\ 000 = 10^{6}$	mega	M	as in *mega*phone	one million
$1\ 000 = 10^{3}$	kilo	k	as in *kilo*watt	one thousand
$100 = 10^{2}$	hecto	h	heck'toe	one hundred
$10 = 10$	deka	da	deck'a (*a* as in *a*bout)	ten
$0.1 = 10^{-1}$	deci	d	as in *deci*mal	one-tenth
$0.01 = 10^{-2}$	centi	c	as in *senti*ment	one-hundredth
$0.001 = 10^{-3}$	milli	m	as in *mili*tary	one-thousandth
$0.000\ 001 = 10^{-6}$	micro	μ	as in *micro*phone	one-millionth
$0.000\ 000\ 001 = 10^{-9}$	nano	n	nan'oh (*an* as in *an*t)	one-billionth
$0.000\ 000\ 000\ 001 = 10^{-12}$	pico	p	peek'oh	one-trillionth

MECHANICS OF FLUIDS

McGraw-Hill Series in Mechanical Engineering

Consulting Editors

MECHANICS
OF FLUIDS

Third Edition

Irving H. Shames

Faculty Professor and
Distinguished Teaching Professor
Faculty of Engineering and Applied Science
State University of New York at Buffalo

McGraw-Hill, Inc.

New York   St. Louis   San Francisco   Auckland   Bogotá
Caracas   Lisbon   London   Madrid   Mexico   Milan   Montreal
New Delhi   Paris   San Juan   Singapore   Sydney   Tokyo   Toronto

MECHANICS OF FLUIDS

International Edition 1992

5 6 7 8 9 0 KKP PMP 9 8 7 6

Library of Congress Cataloging-in-Publication Data

Shames, Irving Hermain, (date).
  Mechanics of fluids/ Irving H. Shames.—3rd ed.
    p.   cm.—(McGraw-Hill series in mechanical engineering)
  Includes bibliographical references and index.
  ISBN 0-07-056387-X
  1. Fluid mechanics.   I. Title.   II. Series.
TA357.S44   1992   91-30217
  620.1'06--dc20

This book was set in Times Roman by Science Typographers, Inc.
The editors were John J. Corrigan and John M. Morriss;
the production supervisor was Richard A. Ausburn.
The cover was designed by Rafael Hernandez.
Project supervision was done by Science Typographers, Inc.
New drawings were done by Science Typographers, Inc.

Cover Photo: Coast Guard vessel being tested in a towing tank for heavy sea conditions. Courtesy of the David Taylor Research Facility, Carderock, Md.

When ordering this title, use ISBN 0-07-112815-8

Printed in Singapore

ABOUT THE AUTHOR

Irving H. Shames holds the position of Faculty Professor with the School of Engineering and Applied Science at the State University of New York at Buffalo and has been awarded the title of Distinguished Teaching Professor within the State University of New York system. The former permits Professor Shames to teach in various engineering departments, and the latter affords resources that are important for his writing efforts.

These writing efforts have extended over a period of 35 years, during which time 10 books have been published. Most of these books have been translated into other languages such as Spanish, Portuguese, Japanese, Korean, Chinese, and Arabic. His first book, *Engineering Mechanics Statics and Dynamics*, published in 1958, was the first widely used mechanics book based on vector principles. It ushered in subsequent books and the almost universal use of the vector approach. The first edition of *Mechanics of Fluids* was the very first textbook to use the Reynolds transport equation for the efficient development of the basic laws and to use the noninertial control volume. Indeed, an examination of this 1962 edition will reveal that most present-day fluid textbooks now closely resemble this ground-breaking text. Other of his texts similarly present innovative approaches and/or viewpoints.

Professor Shames teaches the sophomore sequence—statics, dynamics, and solid mechanics—to virtually the entire engineering student body at Buffalo in one large class without recitation sections. He similarly teaches a course in fluid mechanics to the entire junior class of mechanical and aerospace engineers. In alternating years he teaches a senior-graduate course in variational methods and finite elements and a senior-graduate course in inelastic stress analysis. These senior-graduate classes are electives having enrollments far above the average for this university and involving a wide spectrum of student capabilities.

Professor Shames believes that one reason for the durability of his books stems from the fact that each book is written in conjunction with a course having large and diverse enrollment. This is so because the text must be written to play a key role in such classes, and this represents the most severe test of comprehensibility. Also, he has been a chairman for 18 years of programs in aerospace, engineering science, bioengineering, and nuclear engineering, requiring a close involvement in the curriculum development of these programs. This brings to his writing an exceptionally broad background that allows continuity in his books from earlier course work, while at the same time permitting open-endedness toward later courses.

Professor Shames spent two years as a Visiting Professor at the Technion Institute of Technology, Haifa, Israel, once in mechanical engineering and once in materials engineering. While at SUNY/Buffalo, Professor Shames worked with the famous biologist Dr. James Danielli, on the double molecular layer membrane theory and with him was co-principal investigator on membrane research.

In recent years Professor Shames has expanded his teaching activities and has instituted two summer faculty development workshops sponsored by New York state. In 1991, this has been expanded to a national workshop program sponsored by the National Science Foundation (NSF). The program involves the integration, both conceptually and pedagogically, of mechanics from the sophomore year on through graduate school.

CONTENTS

11   One-Dimensional Compressible Flow   473

Part III   Analysis of Important External Flow

12   Potential Flow   549

PREFACE

With the publication of the third edition this text begins the fourth decade of its existence. In retrospect, it has had three stages of development. The first edition was a radical departure from contemporary fluids texts of its time. For example, it was the very first textbook to use the Reynolds transport equation to formulate the integral forms of the basic laws via control volumes. Additionally, it was the first textbook to introduce and use the noninertial control volume. Furthermore, it featured the development of Stokes' viscosity law and the formulation and use the Navier-Stokes equations. These innovations proved fortunate because the first edition was widely used; it went through 22 printings in 20 years before giving way to the second edition.

The second edition concentrated on topic *coverage*. Chapters on turbomachinery, computational fluid mechanics, and a sizeable appendix on instrumentation data were added. Also, other chapters, particularly the boundary layer chapter, were fleshed out.

The development of the third edition was facilitated by a great opportunity. As Faculty Professor my teaching is not restricted to one department. Hence, although I reside in the Civil Department, I was invited by our Mechanical and Aerospace Department in 1979 to teach the junior fluids course for the entire class and was given a free hand in the manner of presentation and content. There have been 160 to 180 students each year in this class. What was particularly valuable to me was the fact that half the class consisted of transfer students from a wide variety of programs ranging from large university engineering programs to small junior college pre-engineering programs. It has long been my experience that teaching a large class of students with somewhat diverse backgrounds is for me the best way of developing a text. Thus, I was afforded the singular opportunity to work on the third edition. Furthermore, to live up to the extraordinary trust given me by my colleagues in the Mechanical and Aerospace Engineering Department, there was strong motivation for me to improve the book, particularly from a pedagogical standpoint. The third edition is the result of a continuing effort, primarily in this direction, over a decade of time. I will now list some of the changes instituted during this period.

As a writer I have always included material in a book going beyond what might be covered formally in class. This includes what may be considered "advanced material". In my class I have always given a thumbnail sketch of much of this material for orientation purposes. Also, I wish to encourage students to go into the material themselves during the course, and particularly at a later time, in conjunction with other more advanced courses. My students tell me this is a very beneficial practice, so I have incorporated this practice in the third edition. Thus, directly before an advanced (starred or fine print) section there will be a simple cursory rundown explaining what is to be done more carefully and rigorously immediately after.

I have found that students in recent years have trouble projecting curved surfaces that are complex but that have simple openings such as the outside surface of a branched pipe system. In the chapter on hydrostatics, I have presented discussions and problems for such surfaces. This is particularly beneficial later in Chapter 5 when studying momentum flow through a control volume extending over some internal flow through a device. One can include forces that consist of *internal* forces and compute, for example, a thrust by a jet engine on a test stand. Under certain conditions, by simply using *gage pressure* in this calculation we show that the force on the outside surface (not part of the control surface) due to atmospheric pressure will automatically be included as well. I want students to consider control volumes and the associated equations with the same care and precision as was expected of them in using a *free-body diagram* in sophomore mechanics. Specifically, I want them at least initially to consider the control volume and the accompanying momentum equation as a separate calculation from that of the computation of force on a curved (outside) surface (often a complex curved surface not part of the control surface with simple openings and exposed to uniform atmospheric pressure). When this has been made patently clear, I allow the use of gage pressures to simplify the calculations. (However, I have included homework problems where this *cannot* be done.) After this careful beginning, the text generally uses the option of the shorter calculation employing gage pressure when such an approach is permissible.

Some chapters, such as the pipe-flow and boundary-layer chapters, have numerous definitions and formulas with limited ranges of application. I have repackaged these chapters so they are easier to read and use. Furthermore, at the end of these chapters I have presented carefully spaced summary sheets giving essential results of the chapter. Parenthetically, I let students use copies of these summary sheets during exams (closed book).

A number of instructors have asked that I give up some of my pet notations in favor of more widely used notations. Thus, for example, with some sorrow, I have replaced such symbols as $R_{ey, x}$ by Re_x, as well as others. Also, at the request of some instructors I have made far greater use of the pound mass and the "head" measurements of energy. Another notation change has to do with velocity components. It is unfortunately the case that u, v, w represent displacement components in solid mechanics and also velocity components in fluid mechanics. When there can be no confusion, I have used u, v, w as velocity components in this text, as for instance in the Navier-Stokes equations. When the material has more universal use, I have used V_x, V_y, V_z for velocity components. For cylindrical coordinates, I have generally used v_r, v_θ, v_z. Additionally, I have dropped the rather elegant "equivalent height" method in hydrostatics in favor of a more simple direct approach.

Because of long years in teaching and writing in the sophomore mechanics sequence of statics, dynamics, and solid mechanics, I have been very sensitive in this text to continuity between fluid mechanics and the basics of the earlier mechanics courses. Striving for an optimum overall use of velocity component notation previously described is just one instance of this sensitivity.

Two of my reviewers strongly urged the regrouping of certain chapters for the third edition into two major groups involving, respectively, internal flows and external flows. After careful consideration I realized that this would be an improvement, and so I have instituted this change in this edition.

I have found that my students show much interest and curiosity about certain devices, both modern and historic, such as autogyros, dirigibles, forward-swept wings, and so forth. Accordingly, I have assembled photos of some of these devices and have shown them, along with interesting pertinent captions, at the beginning of each chapter. In some cases, examples and problems have been presented based on these devices.

I have retained the rigor and generality of the earlier editions and have avoided recent trends of developing the Reynolds transport equation using a special simple control volume or of developing the first law of thermodynamics for control volumes by using the special case having simple one-dimensional flows in and out. My students don't seem to have any difficulty with my general developments, and I believe they ultimately have a better understanding of these formulations.

There are no new chapters in the third edition, but there are a number of new sections and many new examples. For instance, I have added starred sections in Chap. 7 wherein the differential forms of the four basic laws are all derived via identical procedures, much used in other fields, from the integral forms of these laws. This is done with the aid of Cauchy's formula and Gauss' theorem, whose developments are part of the discussion. The student will be introduced to index notation should he/she decide to read or study this material. I have found my more advanced juniors quite capable of handling this material on their own with only minimal assistance. In addition to new sections and explanations, there are 40 computer projects which, for lack of space, have been presented in the instructor's manual. The instructor is encouraged to reproduce any of these projects for students' use. I assign two or three of these a semester. This is over and above the regular load of reading and problem-solving. Finally, I have added over 300 problems, mostly in S.I. units.

Another feature I have built into the text is *flexibility* in that at various key places there are multiple routes pointed out in the footnotes that the reader may choose to follow. As an example, the Poiseuille flow in pipes is developed from very first principles in Chap. 7. At that time the reader is informed that he/she may opt rather to go to certain sections of Chap. 9 where the Navier-Stokes equations are formulated and where this flow is developed from these equations directly. The reader may then go back to Chap. 7 to continue. As another example, in the chapter on pipe flow the reader has the choice of using Prandtl's mixing length theory, with proper precautions pointed out for its limitations, or a shorter dimensional analysis approach. (The reader may also lose his/her head and do both!)

In examples in the text involving linear and angular momentum, I have used the approach presented in the solutions manuals of the first and second editions where the assumptions made to simplify the general equations to a working equation are prominently displayed and referred to by number. I now require students to do the same in homework assignments and even on tests.

It has long been my practice to assign problems for each class and then to go over every problem in the very next period using carefully prepared transparencies.* This is done after a quick review is made of essentials pertaining to the assignment. Directly after class the masters of these transparencies are posted for 48 hours. In this edition, I am making these masters, which cover all problems, available to users in the instructor's manual. Most of the solutions are on separate pages. They are typed using WordPerfect and printed by a laser printer. The diagrams are either blown up versions of actual text diagrams or are computer executed with a MacIntosh computer. Also, printouts of programs for the 40 computer projects, as well as printouts of results, will be included in the manual. Finally, a disk will be included in the instructor's manual that contain all computer material suitable for an IBM-compatible P.C. Fortran is used.

In summary, I believe the third edition, while maintaining the rigor and level of earlier editions, is easier for the instructor to use and easier and more effective for the student to study. Furthermore, with the enthusiastic approval of many of my students, I have labored to try to make this book one that will be of use long after the course is over. Indeed, my hope is that this book will become a familiar and trusted member of the student's technical library of use throughout his/her career.

I wish to thank Professor Amitabha Gosh from nearby Rochester Institute of Technology, Professor Duen-Ren Jeng of Toledo University, and Professor James Leith of the University of New Mexico. They were excellent reviewers and have offered many helpful suggestions and criticisms. In particular, Professor Leith and Professor Gosh convinced me about the wisdom of internal and external flow groupings. Professor Goodarz Ahmadi of Clarkson University has been a constant source of advice and wisdom over the years going back to the

---

*I have found that the results of this practice are superior in a large class to the use of weekly recitation sections. In particular, I have found this practice keeps students current and helps students to apply the theory to problems while the theory is still fresh in their minds.

first edition. He has given me a careful detailed review of the entire manuscript for the third edition and has made many valuable and insightful observations, all of which resulted in changes in the text. I can't thank him enough! Next I wish to thank my Buffalo colleague and friend, Professor Joseph Atkinson, for a careful, useful review of the chapter on free surface flow and for the use of several photos from his laboratories. Dr. Steve Ma, while a doctoral student at Buffalo, worked on the computer projects and reviewed the final manuscript. I thank him for his skillful contributions. My gratitude goes to Dr. Anoop Dhingra, who, while a student at Buffalo, also worked on the computer projects. Miss Marca Lam and Mr. Jon Luntz, students at Buffalo, checked my solutions for the new problems to be inserted in the third edition. I am grateful to these fine students. Finally, I thank my secretary, Mrs. Debra Kinda, for her tireless and excellent typing efforts.

Irving H. Shames

MECHANICS OF FLUIDS

PART
I

BASIC PRINCIPLES
OF FLUID MECHANICS

X-29 advanced technology demonstrator. (*Courtesy Grumman Corporation, Bethpage, N.Y.*)

German engineers began experimenting with forward swept wings during the Second World War. Grumman Aviation began experimenting with forward swept wings in 1981. The research program has shown that a forward swept wing will produce approximately 20 percent better performance in the transonic regime than an equivalent aft swept wing. The advantage of a lower drag across its entire operational envelope, particularly at speeds around Mach 1, permits the use of a smaller engine.

Compared to an aft swept wing, a forward swept wing offers higher maneuverability, improved slow-speed handling, and lower stall speeds with good post-stall characteristics. And since forward swept wings are placed further back on the fuselage, more flexibility in fuselage design is possible.

However unfavorable aeroelastic effects are prevalent for forward swept metal wings requiring stiffer and hence heavier wings, thus negating the above potential gains. The advent of advanced composite materials provides a solution. Aeroelastic tailoring of graphite epoxy composites allows the forward swept wing to twist leading edge down in order to counteract the upward bending motion that the wing experiences due to flight loads.

Finally, we point out there are control problems for subsonic flight conditions for this kind of aircraft. The instability is controlled by an advanced digital flight control system, which adjusts the control surfaces up to 40 times each second. The system is driven by three computers.

CHAPTER 1

FUNDAMENTAL NOTIONS

1.1 HISTORICAL NOTE

Until the turn of this century the study of fluids was undertaken essentially by two groups—hydraulicians and mathematicians. Hydraulicians worked along empirical lines, while mathematicians concentrated on analytical lines. The vast and often ingenious experimentation of the former group yielded much information of indispensable value to the practicing engineer of the day. However, lacking the generalizing benefits of workable theory, these results were of restricted and limited value in novel situations. Mathematicians, meanwhile, by not availing themselves of experimental information, were forced to make assumptions so simplified as to render their results very often completely at odds with reality.

It became clear to such eminent investigators as Reynolds, Froude, Prandtl, and von Kármán that the study of fluids must be a blend of theory and experimentation. Such was the beginning of the science of fluid mechanics as it is known today. Our modern research and test facilities employ mathematicians, physicists, engineers, and skilled technicians, who, working in teams, bring both viewpoints in varying degrees to their work.

1.2 FLUIDS AND THE CONTINUUM

We define a fluid as a substance which must continue to change shape as long as there is a shear stress, however small, present. By contrast a solid undergoes a definite displacement (or breaks completely) when subjected to a shear stress. For instance, the solid block shown on the left in Fig. 1.1 changes shape in a manner conveniently characterized by the angle $\Delta \alpha$ when subjected to a shear

3

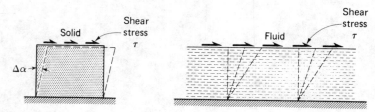

FIGURE 1.1
Shear stress on a solid and on a fluid.

stress τ. If this were an element of fluid (as shown on the right in Fig. 1.1), there would be no fixed $\Delta\alpha$ even for an infinitesimal shear stress. Instead, a continual deformation persists as long as a shear stress τ is applied. In materials which we sometimes call plastic, such as paraffin, either type of shear deformation may be found depending on the shear-stress magnitude. Shear stress below a certain magnitude will induce definite displacements akin to those of a solid body, whereas shear stress above this value causes continuous deformation similar to that of the fluid. This dividing shear-stress magnitude is dependent on the type and state of the material. Certain of these materials are called Bingham materials as will be discussed in Section 1.7.

In considering various types of fluids under *static* conditions, we find that certain fluids undergo very little change in density despite the existence of large pressures. These fluids are invariably in the liquid state for such behavior. Under such circumstances, the fluid is termed *incompressible*, and it is assumed during computations that the density is constant. The study of incompressible fluids under static conditions is called *hydrostatics*. Where the density cannot be considered constant under static conditions, as in a gas, the fluid is termed *compressible* and we sometimes use the name *aerostatics* to identify this class of problems.

The classifications of compressibility given above are reserved for statics. In fluid *dynamics*, the question of when the density may be treated as constant involves more than just the nature of the fluid. Actually, it depends mainly on a certain flow parameter (the Mach number). We then speak of incompressible and compressible *flows*, rather than incompressible or compressible *fluids*. Whenever density variations in a problem are inconsequential, gases and liquids submit to the same manner of analysis. For instance, for flow about fully submerged bodies the basic formulations for low-speed aerodynamics (under about 300 mi/h) are the same as for hydrodynamics. In fact it is entirely possible to examine certain performance characteristics of low-speed airfoils in a water tunnel.

Fluids are composed of molecules in constant motion and collision. To be exact in an analysis, one would have to account for the action of each molecule or group of molecules in a flow. Such procedures are adopted in the kinetic

FIGURE 1.2
Noncontinuum effect on area element.

theory of gases and statistical mechanics but are, in general, too cumbersome for use in engineering applications. In most engineering computations, we are interested in average, measurable manifestations of the many molecules—as, for example, density, pressure, and temperature. These manifestations can be conveniently assumed to arise from a *hypothetical continuous distribution of matter*, called the *continuum*, instead of the actual complex conglomeration of discrete molecules. The continuum concept affords great simplification in analysis and already has been used as an idealization in earlier mechanics courses in the form of a rigid body or a perfectly elastic body.

The continuum approach must be used only where it may yield reasonably correct results. For instance, the continuum approach breaks down when the mean free path[1] of the molecules is of the same order of magnitude as the smallest significant length in the problem. Under such circumstances we may no longer detect meaningful, gross manifestations of molecules. The action of each molecule or group of molecules is then of significance and must be treated accordingly.

To illustrate this, examine the action of a gas on an inside circular element of area of a closed container. With even relatively small amounts of enclosed fluid the innumerable collisions of molecules on the surface result in the gross, non-time-dependent manifestation of force. A truly continuous substance would simulate such action quite well. If only a very tiny amount of gas is now permitted in the container so that the mean free path is of the same order of magnitude as the diameter of the area element, an erratic activity is experienced, as individual molecules or groups of molecules bombard the surface. We can no longer speak of a constant force but must cope with an erratic force variation, as indicated graphically in Fig. 1.2. This action is not what is expected of a continuous distribution of mass. Thus, it is seen that in the first situation the continuum approach would be applicable but in the second case the continuum approach, ignoring as it does individual molecular effects, would be of questionable value.

We may reach the same situation for any amount of enclosed gas by decreasing the size of the area element until irregular molecular effects become

---

[1]Mean free path is the average distance traversed by the molecules between collisions.

significant. Since the continuum approach takes no cognizance of action "in the small," it can yield no accurate information "in the small."

1.3 DIMENSIONS AND UNITS

To study mechanics, we must establish abstractions to describe those manifestations of the body that interest us. These abstractions are called *dimensions*. The dimensions that we pick, which are independent of all other dimensions, are termed *primary*, or *basic*, *dimensions*; the ones that are then developed in terms of the basic dimensions are called *secondary dimensions*. Of the many possible sets of basic dimensions that we could use, we will confine ourselves at present to the set that includes the dimensions of length, time, mass, and temperature. Also we can use force in place of mass in the list of basic dimensions. For quantitative purposes, units have been established for these basic dimensions by various groups and countries. The U.S. Customary System (USCS) employs the pound-force, foot, second, and degree Rankine as the units for the basic dimensions. The International System of Units (SI) uses the newton, meter, second, and degree Kelvin. Table 1.1 lists several common systems of units.

It is convenient to identify these dimensions in the following manner:

Length L
Time T
Force F
Temperature θ

TABLE 1.1
Common systems of units

Metric			
Centimeter-gram-second (cgs)		**SI**	
Mass	gram (g)	Mass	kilogram (kg)
Length	centimeter (cm)	Length	meter (m)
Time	second (s)	Time	second (s)
Force	dyne (dyn)	Force	newton (N)
Temperature	degree Kelvin (K)	Temperature	degree Kelvin (K)

U.S. Customary System			
Type I		**Type II**	
Mass	pound-mass (lbm)	Mass	slug (slug)
Length	foot (ft)	Length	foot (ft)
Time	second (s)	Time	second (s)
Force	pound-force (lbf)	Force	pound-force (lbf)
Temperature	degree Rankine (°R)	Temperature	degree Rankine (°R)

These formal expressions of identification for basic dimensions and the more complicated groupings to be presented for secondary dimensions are called *dimensional representations*.

Secondary dimensions are related by law or by definition to the basic dimensions. Accordingly, the dimensional representation of such quantities will be in terms of the basic dimensions. For instance, the dimensional representation of velocity V is

$$V \equiv \frac{L}{T}$$

By this scheme, pressure then has the dimensions F/L^2 and acceleration is expressed dimensionally as L/T^2.

A change to a new system of units generally entails a change in the scale of measure for the secondary dimensions. The use of the dimensional representation given above permits a simple evaluation of the change of scale. For example, the handbook tells us that one scale unit of pressure in USCS, 1 lb of force per 1 ft^2, is equivalent to 47.9 scale units of pressure in the SI system, or 47.9 N/m^2 (= 47.9 Pa). The unit N/m^2 is called a pascal (Pa) in the SI system. We may arrive at this conclusion by writing the pressure dimensionally, substituting basic units of USCS, and then changing these units to equivalent SI units, as follows:

$$p \equiv \frac{F}{L^2} \equiv \frac{\text{lbf}}{\text{ft}^2} \equiv \frac{4.45 \text{ N}}{0.0929 \text{ m}^2} = 47.9 \text{ N/m}^2$$

Hence,

$$1 \text{ lbf/ft}^2 \equiv 47.9 \text{ N/m}^2 = 47.9 \text{ Pa}$$

On the inside covers of this book we list the physical equivalences of some of the common units of fluid mechanics.

Another technique is to form the ratio of a unit of a basic or secondary dimension and the proper number or fraction of another unit for the basic or secondary dimension such that there is physical equivalence between the quantities. The ratio is then considered as unity because of the one-to-one relation between numerator and denominator from this viewpoint. Thus,

$$\left(\frac{1 \text{ ft}}{12 \text{ in}}\right) \equiv 1$$

Or for another unit, we could say

$$\left(\frac{12 \text{ in}}{305 \text{ mm}}\right) \equiv 1$$

These are to be taken as statements of equivalence, not as algebraic relations in the ordinary sense. Multiplying an expression by such a ratio does not change the measure of the physical quantity represented by the expression. Hence, to change a unit in an expression, we then multiply this unit by a ratio physically equivalent to unity in such a way that the old unit is canceled out, leaving the desired unit. We can then perform a change of units on the previous case in a more convenient way, using the formalism given above on the expressions in the numerator and denominator. Thus,

$$p \equiv \frac{\text{lbf}}{\text{ft}^2} \equiv \frac{\cancel{\text{lbf}}\left(\dfrac{4.45\ \text{N}}{1\ \cancel{\text{lbf}}}\right)}{\left[\cancel{\text{ft}}\left(\dfrac{0.305\ \text{m}}{1\ \cancel{\text{ft}}}\right)\right]^2} \equiv 47.9\ \frac{\text{N}}{\text{m}^2} = 47.9\ \text{Pa}$$

You are urged to employ the latter technique in your work, for the use of less-formal intuitive methods is an invitation to error.

In fluid mechanics, as noted earlier, we deal with secondary dimensions which stem from gross, measurable, molecular manifestations such as pressure and density. Manifestations which are primarily characteristic of a particular fluid and not the manner of flow are called *fluid properties*. Viscosity and surface tension are examples of fluid properties, whereas pressure and density of gases are primarily flow-dependent and hence are not considered fluid properties.

1.4 LAW OF DIMENSIONAL HOMOGENEITY

In order to determine the dimensions of properties established by laws, we must first discuss the law of dimensional homogeneity. This states that *an analytically derived equation representing a physical phenomenon must be valid for all systems of units*. Thus, the equation for the frequency of a pendulum, $f = (1/2\pi)\sqrt{g/L}$, is properly stated for any system of units. A plausible explanation for the law of dimensional homogeneity is that natural phenomena proceed completely oblivious to man-made units, and hence fundamental equations representing such phenomena should have a validity for *any* system of units. For this reason, the fundamental equations of physics are dimensionally homogeneous, so all relations derived from these equations must also be dimensionally homogeneous.

What restriction does this independence of units place on the equation? To answer this, examine the following arbitrary equation:

$$x = y\eta\zeta^3 + \alpha^{3/2}$$

For this equation to be dimensionally homogeneous, the numerical equality between both sides of the equation must be maintained for all systems of units. To accomplish this, the change of scale for each expression must be the same during changes of units. That is, if one expression such as $y\eta\zeta^3$ is doubled in numerical measure for a new systems of units, so must be the expressions x and $\alpha^{3/2}$. *For this to occur under all systems of units, it is necessary that each grouping in the equation have the same dimensional representation*.

As a further illustration, consider the following dimensional representation of an equation which is not dimensionally homogeneous:

$$L = T^2 + T$$

Changing the units from feet to meters will change the value of the left side while not affecting the right side, thus invalidating the equation in the new system of units. We are concerned almost entirely with dimensionally homogeneous equations in this text.

With this in mind, examine a common form of Newton's law, which states that the force on a body is proportional to the resulting acceleration. Thus

$$\mathbf{F} \propto \mathbf{a}$$

We may call the proportionality factor the mass (M). From the law of dimensional homogeneity the dimensions of mass must be

$$M \equiv \frac{FT^2}{L}$$

Mass may be considered as that property of matter which resists acceleration. Hence, it is entirely possible to choose mass as a basic dimension. Force would then be a dependent entity given dimensionally from Newton's law as

$$F \equiv \frac{ML}{T^2}$$

and our basic system of dimensions would then be mass (M), length (L), time (T), and temperature (θ).

1.5 A NOTE ON FORCE AND MASS

In USCS units, we define the amount of mass which accelerates at the rate of 1 ft/s^2 under the action of 1 lbf in accordance with Newton's law as the *slug*. The pound-force could be defined in terms of the deformation of an elastic body such as a spring at prescribed conditions of temperature. Unfortunately a unit of mass stipulated independently of Newton's law is also in common usage. This stems from the law of gravitational attraction, wherein it is posited that the force of attraction between two bodies is proportional to the masses of the bodies—the very same property of a material that enters into Newton's law. Hence, the *pound-mass* (lbm) has been defined as the amount of matter which at the earth's surface is drawn by gravity toward the earth by 1 lbf.

We have thus formulated two units of mass by two different actions, and to relate these units we must subject them to the same action. Thus we can take the pound-mass and see what fraction or multiple of it will accelerate at 1 ft/s^2 under the action of 1 lb of force. This fraction, or multiple, will then represent the number of units of pound-mass that are physically equivalent to 1 slug. It turns out that this coefficient is g_0, where g_0 has the value corresponding to the acceleration of gravity at a position on the earth's surface where the pound-mass was standardized.[2] The value of g_0 is 32.2, to three significant figures. We may

---

[2] The notation g_c is also extensively used for this constant.

then make the statement of equivalence that

$$1 \text{ slug} \equiv 32.2 \text{ lbm} \tag{1.1}$$

How does weight fit into this picture? *Weight is defined as the force of gravity on a body*. Its value will depend on the position of the body relative to the earth's surface. At a location on the earth's surface where the pound-mass is standardized, a mass of 1 lbm has the weight of 1 lbf; but with increasing altitude, the weight will become smaller than 1 lbf. The mass remains at all times a pound-mass, however. If the altitude is not exceedingly high, the measure of weight, in pound-force, will practically equal the measure of mass, in pound-mass. Therefore, it is an unfortunate practice in engineering to think erroneously of weight at positions other than on the earth's surface as the measure of mass and consequently to use the same symbol W to represent pound-mass and pound-force. In this age of rockets and missiles, it behooves us to be careful about the proper usage of units of mass and weight throughout the entire text.

If we know the weight of a body at some point, we can determine its mass very easily, provided that we know the acceleration of gravity g at that point. Thus, according to Newton's law,

$$W(\text{lbf}) = M(\text{slugs}) \, g \left(\text{ft}/\text{s}^2 \right)$$

Therefore,

$$M(\text{slugs}) = \frac{W(\text{lbf})}{g \left(\text{ft}/\text{s}^2 \right)} \tag{1.2}$$

In USCS there are two units of *mass*, namely, the slug and the lbm. In contrast, SI units, as used by many people, involve two units of *force*, as we shall soon see. The basic unit for mass in SI is the *kilogram*, which is the amount of mass that will accelerate 1 m/s^2 under the action of 1 N force. Unfortunately, the kilogram is also used as a measure of force. That is, one often comes across such statements as "body C weighs 5 kg." A kilogram of force is the weight measured at the earth's surface of a body A having a mass of 1 kg. Note that at positions appreciably above the earth's surface, the weight of the body A will decrease; but the mass remains at all times 1 kg. Therefore, the weight in kilograms equals numerically the mass in kilograms *only* at the earth's surface where the acceleration of gravity is 9.806 m/s^2. Care must be taken accordingly in using the kilogram as a measure of weight. In this text we use only the newton, the kilonewton, etc, as the unit for force.

What is the relation between the kilogram force and the newton force? This is easily established when one makes the following observation:

1 N accelerates 1 kg mass 1 m/s^2
1 kg force accelerates 1 kg mass 9.806 m/s^2

Clearly 1 kg force is equivalent to 9.806 N. Furthermore, a newton is about one-fifth of a pound.

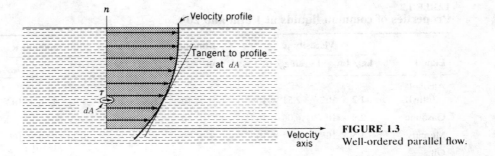

FIGURE 1.3
Well-ordered parallel flow.

What is the mass M of a body weighing W newtons at a location where the acceleration of gravity is g meters per second squared? For this we need only use Newton's law. Thus,

$$W = Mg$$

$$\therefore \ M(\text{kg}) = \frac{W(\text{N})}{g(\text{m/s}^2)} \tag{1.3}$$

In this text we use both systems of units, but with greater emphasis on SI units.

1.6 NEWTON'S VISCOSITY LAW: THE COEFFICIENT OF VISCOSITY

A very important property will now be introduced as a consequence of Newton's viscosity law. For a well-ordered flow[3] whereby fluid particles move in *straight, parallel* lines (parallel flow), the law states that for certain fluids, called *newtonian fluids*, the shear stress on an interface tangent to the direction of flow is proportional to the distance rate of change of velocity, wherein the differentiation is taken in a direction normal to the interface. Mathematically this is stated as

$$\tau \propto \frac{\partial V}{\partial n}$$

Figure 1.3 may further explain this relationship. An infinitesimal area in the flow is chosen parallel to the horizontal velocity axis, as shown. The normal n to this area is drawn. The fluid velocities at points along this normal are plotted, thus forming a velocity profile. The slope of the profile toward the n axis at the position corresponding to the area element is the value $\partial V/\partial n$, which is related, as stated above, to the shear stress τ shown on the interface.

---

[3]Such a flow, called laminar, is free of macroscopic velocity fluctuations. This will be discussed in detail in Chap. 9.

TABLE 1.2
Properties of common liquids at 1 atm and 20°C

Liquid	Viscosity μ		Kinematic viscosity ν		Bulk modulus κ		Surface tension σ	
	kg / (m · s)	slug / (ft · s)	m² / s	ft² / s	GPa	lb / in²	N / m	lb / ft
Alcohol (ethyl)	1.2×10^{-3}	2.51×10^{-5}	1.51×10^{-6}	1.62×10^{-5}	1.21	1.76×10^{5}	0.0223	1.53×10^{-3}
Gasoline	2.9×10^{-4}	6.06×10^{-6}	4.27×10^{-7}	4.59×10^{-6}				
Mercury	1.5×10^{-3}	3.14×10^{-5}	1.16×10^{-7}	1.25×10^{-6}	26.20	3.80×10^{6}	0.514	3.52×10^{-2}
Oil (lubricant)	0.26	5.43×10^{-3}	2.79×10^{-4}	3.00×10^{-3}	⋯	⋯	0.036	2.47×10^{-3}
Water	1.005×10^{-3}	1.67×10^{-5}	0.804×10^{-6}	8.65×10^{-6}	2.23	3.23×10^{5}	0.0730	4.92×10^{-3}

Inserting the coefficient of proportionality into Newton's viscosity law leads to the result

$$\tau = \mu \frac{\partial V}{\partial n} \tag{1.4}$$

where μ is called the *coefficient of viscosity*, having the dimensions $(F/L^2)T$, or M/LT. In the cgs system of units, the unit for viscosity is the *poise*, corresponding to 1 g/cm · s. The *centipoise* is $\frac{1}{100}$ of a poise. The SI unit for viscosity is 1 kg/m · s. It has no particular name. It is 10 times the size of the poise, as is clear from the basic units. In USCS, the coefficient of viscosity has the unit 1 slug/ft · s and like the SI system has no name. Viscosity coefficients for common liquids at 1 atm and 20°C temperature are given in Table 1.2.

The characteristics of viscosity are also given in Fig. B.1 in the appendix for a number of significant fluids. We point out first that viscosity does *not* depend appreciably on *pressure*. However, note from Fig. B.1 that the viscosity of a liquid *decreases* with an *increase* in temperature, whereas a gas, curiously, does quite the opposite. The explanation for these tendencies is as follows. In a *liquid*, the molecules have limited mobility with large cohesive forces present between the molecules. This manifests itself in the property of the fluid which we have called viscosity. An increase in temperature decreases this cohesion between molecules (they are on the average farther apart) and there is a decrease of "stickiness" of the fluid—i.e., a decrease in the viscosity. In a *gas*, the molecules have great mobility and are in general far apart. In contrast to a liquid, there is little cohesion between the molecules. However, the molecules do interact by *colliding* with each other during their rapid movements. The property of viscosity results from these collisions. To illustrate this, consider two small but finite adjacent chunks of fluid A and B at a time t in a simple, parallel flow of a gas of the kind discussed at the outset of this section. This is shown in Fig. 1.4. As seen from the diagram, chunk A is moving faster than chunk B. This means that, *on the average*, molecules in chunk A move faster to

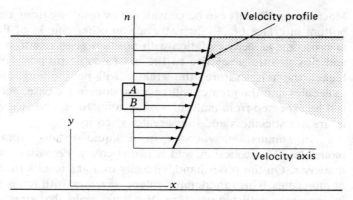

FIGURE 1.4
Parallel flow of a gas at time t.

the right than do molecules in chunk B. But in addition to the aforestated average movement of the molecules, there is also a random migration of molecules from chunk A into chunk B across their interface, and vice versa. Let us first consider the migration from A to B. When the A molecules move to B, there will be some collisions between A molecules and B molecules. Because A molecules are on the average faster in the x direction than are B molecules, there will be a tendency to speed B molecules up in the x direction. This means that there will be a tendency for chunk B macroscopically to speed up. From a continuum point of view, it would appear that there is a shear stress τ at the upper face of B acting to speed B up. This is shown in Fig. 1.5. By a similar action, slow molecules migrating from B into A tend to slow chunk A.

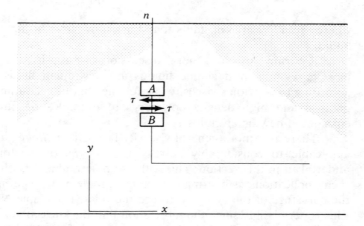

FIGURE 1.5
Shear stress on chunks A and B.

Macroscopically this can be considered as resulting from a shear stress τ on the bottom interface of A. Such stresses on other chunks of fluid where there is a macroscopic velocity variation with position gives rise to a stickiness of the gas and this in turn gives rise to the macroscopic property of viscosity. Now the higher the temperature, the greater will be the migration tendency of the molecules and the greater will be τ for our simple case, because more collisions will be expected from molecules of A going to B, and vice versa. This will result in greater stickiness and thus greater viscosity.

In summation, viscosity in a liquid results from cohesion between molecules. This cohesion, and hence viscosity, *decreases* when the temperature increases. On the other hand, viscosity in a gas results from random movement of molecules. This random movement *increases* with temperature, so the viscosity increases with temperature. We note again that pressure has only a small effect on viscosity—an effect that is usually neglected.

The variation of viscosity of gases with temperature can be approximated by either of two laws called, respectively, the *Sutherland law* and the *power law* and given as follows:

$$\mu = \frac{\mu_0(T/T_0)^{3/2}(T_0 + S)}{T + S} \quad \textit{Sutherland law} \tag{1.5}$$

$$\mu = \mu_0\left(\frac{T}{T_0}\right)^n \quad \textit{power law} \tag{1.6}$$

where μ_0 is some known viscosity at absolute temperature T_0 and where S and n are constants determined by curve fitting. Note that T is the absolute temperature at which μ is being evaluated.

As for liquids, the following simple formula is used to determine viscosity:

$$\mu = Ae^{-BT} \tag{1.7}$$

where A and B are constants found again by curve fitting data for a particular liquid.

Returning to our general discussion of viscosity, we can point out that most gases and most simple liquids are newtonian fluids and hence behave according to Newton's viscosity law for the conditions outlined. Pastes, slurries, greases, and high density polymers are examples of fluids that cannot be considered newtonian fluids.

There is a more general viscosity law, called *Stokes' viscosity law*, that is applicable to considerably more general flows of newtonian fluids than is undertaken in this section. This will also be examined in Chap. 10. However, in such applications as bearing-lubrication problems it is permissible to disregard the curvature of the flow and to use the relatively simple Newton viscosity law. This is permissible since the lubrication film thickness is very small compared to the bearing radius. Therefore domains of such flows having dimensions compa-

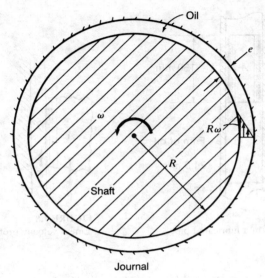

FIGURE 1.6
Shaft rotating in a lubricated journal.

rable to the film thickness involve very little change of direction of flow and may be thought of in such domains as *parallel* flow[4] with the attending permissible use of Newton's viscosity law (for Newtonian fluids). Furthermore, in flows of *real* fluids (which always have some measure of viscosity), in contrast to *hypothetical frictionless*, or as we say *inviscid* flows, the fluid touching a solid boundary must "stick" to such boundaries and thus must have the same velocity as the boundary.[5]

For instance, we consider a shaft A shown in Fig. 1.6 as a cross section rotating with speed ω rad/s inside of a bearing journal with a thin oil film of thickness e separating the bodies. We can approximate the velocity profile here, because e is small compared to the radius, as a *linear profile*, as has been shown in the diagram. The shear stress on all interfaces of oil normal to radial lines can then be given as follows:

$$\tau = \mu \frac{\partial V}{\partial n} = \mu \left[\frac{(R\omega - 0)}{e} \right]$$

---

[4]An intuitive explanation for this can be achieved by noting that looking at a small region around you, where the dimension of this region is much smaller than the radius of the earth, you are not aware of the overall curvature of the earth where you stand.

[5]In very high speed flows 5 or more times the speed of sound there can take place slip of real fluids relative to solid boundaries. We call such flows *slip flows*.

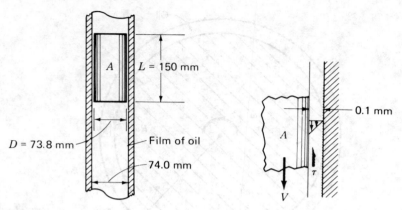

FIGURE 1.7
Cylinder A slides in a lubricated pipe.

FIGURE 1.8
Linear velocity profile in film.

We will examine such problems as homework. Now we consider an oil film problem with actual parallel flow.

Example 1.1. A solid cylinder A of mass 2.5 kg is sliding downward inside a pipe, as shown in Fig. 1.7. The cylinder is perfectly concentric with the centerline of the pipe with a film of oil between the cylinder and inside pipe surface. The coefficient of viscosity of the oil is 7×10^{-3} N · s/m². What is the *terminal* speed V_T of the cylinder—i.e., the final constant speed of the cylinder? Neglect the effects of air pressure.

   We assume a *linear* velocity profile in the oil film, as shown in Fig. 1.8. The value of $\partial V / \partial n$ that we will need for Newton's viscosity law then becomes

$$\frac{\partial V}{\partial n} = \frac{V - 0}{0.0001} = 10,000V \quad \text{s}^{-1} \qquad (a)$$

The shear stress τ on the cylinder wall is then

$$\tau = \mu \frac{\partial V}{\partial n} = (7 \times 10^{-3})(10,000V) = 70V \quad \text{Pa} \qquad (b)$$

We may now equate the weight of the cylinder with the viscous force for the condition of equilibrium which obtains when the cylinder has reached its terminal velocity V_T. Thus,

$$W = (\tau)(\pi D)(L)$$

$$\therefore (2.5)(9.81) = (70 V_T)(\pi)(0.0738)(0.150) \qquad (c)$$

We get for V_T:

$$V_T = 10.07 \text{ m/s} \qquad (d)$$

If we divide μ by ρ, the mass density, we form what is called the *kinematic viscosity*.[6] This property is denoted as ν and has the dimensions L^2/t, as you may yourself verify. In the cgs system, the unit is called the *stoke* (1 cm^2/s). In the SI system, the unit is 1 m^2/s. Clearly the SI unit is 10^4 times the stoke. In USCS, the basic unit is 1 ft^2/s. The kinematic viscosity is independent of pressure for liquids. However, for gases, ν will depend on the pressure. The dependence of ν on temperature for atmospheric pressure is shown in Fig. B.2, Appendix B.

*1.7 A NOTE ON NON-NEWTONIAN MATERIALS

The study of the response of materials to stress is called *rheology*. The Newtonian fluid in this context is a viscous material. Non-Newtonian fluids are also viscous materials wherein the shear stress is related to the shear rate, dV/dy, in a more complicated way. The *power law* is one way to describe the behavior of a viscous material. This would be given for parallel flow as follows:

$$\tau = k \left(\frac{dV}{dy} \right)^n \tag{1.8}$$

For a Newtonian fluid $k = \mu$ and $n = 1$. For other values of n we have a non-Newtonian fluid.

A non-Newtonian fluid whose behavior is described by Eq. (1.8) with $n < 1$ is called a *pseudoplastic*. This name obtains because with increasing shear rate, (dV/dy), there is curiously a decrease in effective viscosity. That is, on increasing shear rate the fluid is "thinning."[7] This stress, shear-rate curve is shown in Fig. 1.9. Many non-Newtonian slurries are pseudoplastic. If, on the other hand, $n > 1$, the fluid is called *dilatant*. Here clearly the fluid "thickens" with increasing shear rate.

Next we have the so-called *linear Bingham material* wherein, as described in Sec. 1.2, there is only a fixed displacement for shear stress less than a value τ_1 and for which there is a Newtonian viscous behavior when the shear stress exceeds τ_1. This behavior is shown in Fig. 1.9. The equation for this behavior is

$$\tau = \tau_1 + \mu_B \frac{dV}{dy} \tag{1.9}$$

---

[6] The viscosity itself is often called the *absolute*, or *dynamic*, viscosity to distinguish it more clearly from the kinematic viscosity.

[7] That is, the pseudoplastic curve falls continually further below that of the Newtonian fluid as can be seen in Fig. 1.9.

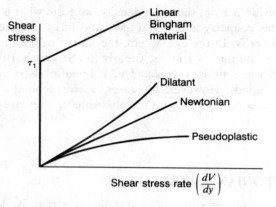

FIGURE 1.9
Rheological behavior of some viscous materials.

Finally we wish to point out that many materials possess a combination of viscous and elastic characteristics. These materials are called *viscoelastic* materials.[8] For instance, plastics at room temperature under load are viscoelastic.

Even with this abbreviated list of material behavior, it should be clear that there is a vast range of possible material characteristics that extend beyond the simple familiar cases generally studied in beginning fluids and solids courses.

In this text, as indicated above, we shall restrict ourselves to Newtonian fluids and shall employ for the definition of a fluid (as opposed to a solid) that given in Section 1.2.

1.8 THE PERFECT GAS: EQUATION OF STATE

If molecules of a fluid are presumed to have a mutual effect arising solely from perfectly elastic collisions, then the kinetic theory of gases indicates that for such a fluid, called a *perfect gas*, there exists a simple formulation relating pressure, specific volume, and absolute temperature. This relation, called the *equation of state*, has the following form for a perfect gas at equilibrium:

$$pv = RT \tag{1.10}$$

where R, the *gas constant*, depends only on the molecular weight of the fluid, v is the specific volume (volume per unit mass), and T is the absolute temperature. Values of R for various gases at low pressure are given in Appendix Table B.3.

In actuality, the behavior of many gases such as air, oxygen, and helium very closely approximates the perfect gas under most conditions and may, with

---

[8]See I. H. Shames and F. Cozzarelli, *Elastic and Inelastic Stress Analysis*, Chap. 7, Prentice-Hall, Englewood Cliffs, N.J., 1992.

good accuracy, be represented by the above equation of state.[9] Since the essence of the perfect gas is the complete lack of intermolecular attraction, gases near condensation conditions depart greatly from perfect-gas behavior. For this reason, steam, ammonia, and Freon at atmospheric pressure and room temperature and, in addition, oxygen and helium at very high pressures cannot properly be considered as perfect gases in many computations.

There are equations of state for other than perfect gases, but these lack the simplicity and range of the relation above. It should be emphasized that all such relations are developed for fluids which are under macroscopic mechanical and thermal equilibrium. This essentially means that the fluid bulk is undergoing no accelerative motion relative to an inertial reference and is free of heat transfer. The term p in the equation of state above and other such equations is usually referred to as the pressure. However, because of the equilibrium nature of this property as used in equations of state, we use the designation *thermodynamic pressure* to differentiate it from quantities involved in dynamic situations. The relations between thermodynamic pressure and nonequilibrium concepts will be discussed in Chaps. 2 and 9. And a more complete discussion of the perfect gas is given in Sec. 11.2.

Example 1.2. Air is kept at a pressure of 200 kPa and a temperature of 30°C in a 500-L container. What is the mass of the air?

We may use the equation of state here with the gas constant R being taken as 287 N · m/(kg · K) and solve for the specific volume v. Thus,

$$pv = RT$$

$$[(200)(1000)]v = (287)(273 + 30)$$

$$\therefore v = 0.435 \text{ m}^3/\text{kg}$$

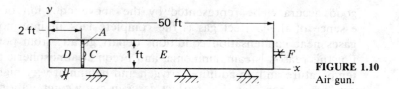

FIGURE 1.10
Air gun.

We may now compute the mass M of the air in the following manner:

$$M = \frac{V}{v} = \frac{[500/1000]}{0.435} = 1.149 \text{ kg}$$

The following starred example is that of an interesting, albeit challenging, problem involving a device used for many years by the U.S. Navy.

*Example 1.3. An *air gun* is used to test the ability of small devices to withstand high accelerations. A "floating" piston A, on which the device to be tested is mounted, is held at position C, and region D is filled with highly compressed air (Fig. 1.10). Region E is initially at atmospheric pressure, but is entirely sealed from the outside. When "fired," a quick-release mechanism releases the piston and it accelerates rapidly toward the other end of the gun, where the trapped air in E "cushions" the motion so that the piston will begin eventually to return. However, as it starts back, the high pressure developed in E is released through valve F and the piston returns only a short distance.

Suppose that the piston and its test specimen have a mass of 2 lbm and the pressure initially in the chamber D is 1000 lb/in^2 gage. Compute the speed of the piston at the halfway point of the air gun if we make the simple assumption that the air in D expands according to $pv = \text{const}$ (i.e., *isothermal* expansion for which we use Boyle's law) and the air in E is compressed also according to $pv = \text{const}$.[10] Take v of this fluid at D to be initially 0.207 ft^3/lbm and v in E to be initially 13.10 ft^3/lbm. Neglect the inertia of the air and friction.

The force on the piston results from the pressures on each face, and we can show that this force is a function of x. Thus, examining the pressure p_D first, we have from initial conditions

$$(p_D v_D)_0 = [(1000 + 14.7)(144)](0.207) = 30,300 \text{ ft} \cdot \text{lbf/lbm} \qquad (a)$$

Furthermore, the mass of fluid D given as M_D is determined from initial data as

$$M_D = \frac{(V_D)_0}{(v_D)_0} = \frac{(2)\left(\pi \frac{1^2}{4}\right)}{0.207} = 7.58 \text{ lbm} \qquad (b)$$

where $(V_D)_0$ indicates the volume of the gas D initially. Returning to Eq. (a), we

---

[10]In Prob. 1.30 we ask you to assume no heat transfer during expansion and compression (a more realistic model). As we will later learn, for this case the equation of state becomes $pv^k = \text{const}$, where k is a constant.

now have for p_D at *any* position x of the piston

$$p_D = \frac{30{,}300}{v_D} = \frac{30{,}300}{V_D/M_D} = \frac{30{,}300}{\pi\left(\frac{1^2}{4}\right)x/M_D}$$

$$\therefore p_D = \frac{293{,}000}{x} \text{ lbf/ft}^2$$

We can similarly get p_E as a function of x. Thus,

$$(p_E v_E)_0 = (14.7)(144)(13.10) = 27{,}700 \text{ ft} \cdot \text{lbf/lbm}$$

and

$$M_E = \frac{(V_E)_0}{(v_E)_0} = \frac{(48)\left(\pi\frac{1^2}{4}\right)}{13.10} = 2.88 \text{ lbm}$$

Hence,

$$p_E = \frac{27{,}700}{v_E} = \frac{27{,}700}{V_E/M_E} = \frac{27{,}700}{\left(\pi\frac{1^2}{4}\right)(50-x)/2.88}$$

$$\therefore p_E = \frac{101{,}600}{50-x} \text{ lbf/ft}^2 \tag{c}$$

Now we can write Newton's law for this case. We get, using V without a subscript as velocity,

$$M\frac{d^2V}{dt^2} = MV\frac{dV}{dx} = \left(\frac{\pi}{4}\right)\left(\frac{293{,}000}{x} - \frac{101{,}600}{50-x}\right) \tag{d}$$

where M is the mass of piston and load. We leave it to the reader to separate variables and integrate the differential equation. The constant of integration is determined by noting that when $x = 2$ ft, $V = 0$. At $x = 25$ ft, we then find that

$$V = 4120 \text{ ft/s}$$

*1.9   COMPRESSIBILITY OF LIQUIDS; SURFACE TENSION

In the preceding section we discussed the compressibility of a perfect gas via the perfect-gas law. We pointed out earlier that liquids undergo only slight compression under pressure. Small as this compressibility might be in liquids, there are times when it is important. For instance, the compressibility of a liquid could be of significance for very high pressures. Also, in underwater acoustics (sonar), the compressibility of water is important even though pressure variation and hence compression may be quite small.

To measure the compressibility of a liquid, we present two quantities. The *coefficient of compressibility* β is defined, using V for volume, as

$$\beta = -\frac{1}{V}\left(\frac{\partial V}{\partial p}\right)_T \tag{1.11}$$

where the subscript T indicates that the compression of the liquid takes place at constant temperature (*isothermal compression*). The reciprocal to β is called the

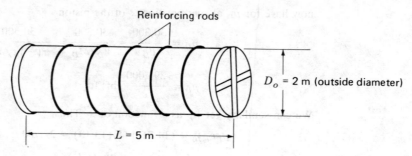

FIGURE 1.11
Cylindrical tank.

bulk modulus, denoted as κ. Thus,

$$\kappa = -V\left(\frac{\partial p}{\partial V}\right)_T \tag{1.12}$$

For water at room temperature and atmospheric pressure, the value of κ is 2068 MPa (300,000 lb/in^2). (M indicates mega $\equiv 10^6$, as indicated in the inside cover sheets.) Note that κ increases with pressure. At a pressure of 3000 atm, water has doubled its bulk modulus. To get a feel for the compressibility of water, consider an isothermal compression of water by a pressure of 10 atm acting on 1 m^3 of water. From Eq. (1.12) we can say, expressing the equation in a *finite difference* mode, that

$$\Delta V = -\frac{V}{\kappa}\,\Delta p \tag{1.13}$$

Assuming κ and V are constant on the right side of the equation, we find that

$$\Delta V = -\frac{1}{2068 \times 10^6}\left[(10)(101,325)\right]$$
$$= -4.90 \times 10^{-4}\ \text{m}^3$$

There is then a percentage decrease in volume of

$$\frac{4.90 \times 10^{-4}}{1} \times 100 = 0.0490\%$$

Example 1.4. A reinforced steel tank (Fig. 1.11) is to contain air at a pressure p of 7.00 MPa gage.[11] To test the tank, we fill it with water and raise the pressure by forcing more water in until a test pressure p_T is reached higher than the operating pressure p. The value of p_T is tied to the safety factor n that will be used. The reason for using water rather than air is to ensure safety during the test. You will

---

[11] Since 0.1 MPa $\approx$ 1 atm, the pressure here is about 70 atm.

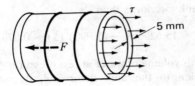

FIGURE 1.12
Free body exposing τ.

learn later that the energy added to a given mass of fluid during compression is $\int p\,dV$ where V is volume. With air there is a large change in volume with pressure, so clearly there is a much larger amount of stored energy in compressed air than in water compressed to the same pressure and volume. A tank fracture with compressed air would accordingly be much more dangerous (resulting in what could be characterized as an explosion) than a tank fracture with compressed water. In the latter case, a small amount of water would shoot out of the failing tank over a short time with little attending risk.

The tank is to be tested using a safety factor n of 1.5. The tank wall has a thickness of 5 mm and is reinforced so as not to permit appreciable change in diameter. Also, the end plates are heavily reinforced to prevent appreciable deformation. However, the longitudinal length of the tank will change as a result of the inside pressure. If the modulus of elasticity E for the tank is 2.07×10^{11} Pa, what *additional* volume of water will be added to an *initially full tank of water at atmospheric pressure* to reach the test pressure p_T? Compute this volume at the test pressure.

For simplicity, let us first compute the deformed volume V' inside the tank under the assumptions made. For this purpose, we first calculate the longitudinal stress τ in the cylinder wall (see Fig. 1.12). Noting that D_i is the inside diameter, we have as a result of *equilibrium* for the free body shown

$$-\tau \frac{\pi}{4}\left(D_o^2 - D_i^2\right) + np\left(\frac{\pi}{4}D_i^2\right) = 0$$

$$-\tau \frac{\pi}{4}(2^2 - 1.990^2) + (1.5)(7.00)\left[\frac{\pi}{4}(1.990)^2\right] = 0$$

$$\tau = 1042 \text{ MPa}$$

The longitudinal strain ϵ according to *Hooke's law* is

$$\epsilon = \frac{\tau}{E} = \frac{1042 \times 10^6}{2.07 \times 10^{11}} = 5.034 \times 10^{-3}$$

The new inside length L' is then

$$L' = L + \epsilon L = 5 + 5(5.034 \times 10^{-3})$$

$$\therefore L' = 5.0252 \text{ m}$$

The deformed inside volume is then

$$V' = \frac{\pi}{4}(1.990^2)(5.0252) = 15.6296 \text{ m}^3$$

The expansion of the tank interior is then

$$V' - V = 15.6296 - \frac{\pi}{4}(1.990)^2(5) = 0.07832 \text{ m}^3$$

We must, in part, add this volume of water at test pressure. Also we must take into account that, in addition to the tank deforming, the water is deformed by compression.

Accordingly we turn our attention to the water. Consider first, for simplicity, that a volume $V' = 15.6296$ m^3 of water enters the deformed tank at atmospheric pressure. Next, compress the water to the final test pressure of $1.5 \times 7.00 = 10.5$ MPa gage and compute the change in volume $(\Delta V)_w$ of the water as a result of this action. This will give us another volume of water to be added. Using the value 2068 MPa for κ, we have for $(\Delta V)_w$ from Eq. (1.13) while noting that the change in pressure is 10.5 MPa:

$$(\Delta V)_w = -\frac{V'}{\kappa}\Delta p = -\frac{15.6296}{2068 \times 10^6}(10.5 \times 10^6)$$

$$= -0.0794 \text{ m}^3$$

With this decrease $(\Delta V)_w$ in volume of the water which originally filled the deformed volume V' of the tank, we can now add an equal volume $(\Delta V)_w$ of water at *test pressure* to fill the tank at test pressure. Thus, the total volume $(\Delta V)_T$ of water added at test pressure to a tank full of water initially at atmospheric pressure is

$$\overset{\substack{\text{Expansion}\\\text{of tank}}}{(\Delta V)_T = (V' - V)} + \overset{\substack{\text{Compression}\\\text{of water}}}{|(\Delta V)_w|}$$

$$= 0.07832 + 0.0794$$

$$= 0.1577 \text{ m}^3$$

In Prob. 1.32 we ask you to compute this excess volume of water added at a pressure corresponding to atmospheric pressure rather than at test pressure. We will then see that only a small amount of water will shoot out should there be a tank fracture at the test pressure.

A second property that we will discuss is *surface tension* at the interface of a liquid and a gas. This phenomenon which is a tensile force distributed along the surface is due primarily to molecular attraction between *like* molecules (*cohesion*) and molecular attraction between *unlike* molecules (adhesion). In the interior of a liquid (see Fig. 1.13), the cohesive forces cancel, but at the free surface the liquid cohesive forces from below exceed the adhesive forces from the gas above resulting in surface tension. It is for this reason that a droplet of water will assume a spherical shape. And it is for this reason likewise that small bugs and insects can alight on the free surface of a pond and not sink. The surface tension is measured as a *line-loading* intensity σ *tangential* to the surface and given per unit length of a line drawn on the free surface.[12]

---

[12] This loading is similar to the line loading $w(x)$ on beams used in strength of materials.

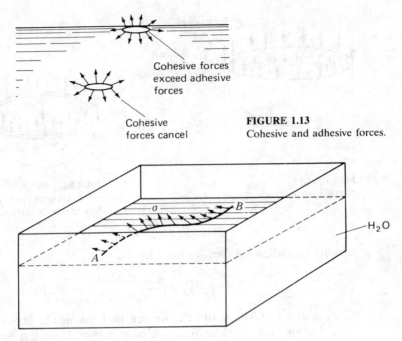

FIGURE 1.13
Cohesive and adhesive forces.

Cohesive forces
exceed adhesive
forces

Cohesive
forces cancel

FIGURE 1.14
Surface tension σ.

Furthermore, the loading is normal to the line, as is shown in Fig. 1.14, where the AB is on the free surface. σ is called the *surface tension coefficient* and is the force per unit length transmitted from the fluid surface just to the left of AB to the fluid surface just to the right of AB with a direction normal to line AB. Thus, the vertical force distribution on the edge of the free body of a half water droplet (see Fig. 1.15) is the surface tension σ on the droplet surface. On the interior cross section, we have shown the distribution of force coming from the pressure p_i inside the droplet. For a droplet of liquid in equilibrium, we can then say that

$$-(p_i)_g(\pi R^2) + (\sigma)(2\pi R) = 0$$

where $(p_i)_g$ is the inside pressure in the droplet above that of the atmosphere. (We are assuming here that the weight of gravity has been counteracted by an outside agent.) Solving for $(p_i)_g$ we get

$$(p_i)_g = \frac{2\sigma}{R} \tag{1.14}$$

At room temperature, σ for water exposed to air is 0.0730 N/m.[13] Hence, for a

---

[13]The coefficient of surface tension for a *water-air* interface is $\sigma = 0.0730$ N/m $\equiv 0.0050$ lb/ft. For a *mercury-air* interface, the coefficient of surface tension is $\sigma = 0.514$ N/m $\equiv 0.0352$ lb/ft.

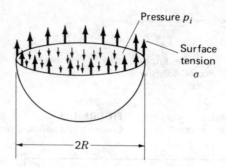

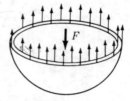

FIGURE 1.15
Surface tension on half water droplet.

FIGURE 1.16
Free body of half a bubble showing two surfaces with surface tensions.

droplet of radius 0.5 mm we have for $(p_i)_g$

$$(p_i)_g = \frac{(2)(0.0730)}{0.0005} = 292 \text{ Pa}$$

Since 1 atm is 1.013×10^5 Pa, we see that the inside pressure is 0.00288 atm.

Let us next consider the case of a *bubble*. If we cut the bubble in half to form a free body (see Fig. 1.16), we see that surface tension exists on two surfaces—the inner and the outer. Taking the radius to be approximately the same for inner and outer surfaces, we can say from equilibrium for the inside gage pressure $(p_i)_g$ that

$$-(p_i)_g \pi R^2 + 2[\sigma(2\pi R)] = 0$$

$$\therefore (p_i)_g = \frac{4\sigma}{R} \qquad (1.15)$$

Consider now the situation where a liquid is in contact with a solid such as liquid in a glass tube. If the adhesion of the liquid to the solid exceeds the cohesion in the liquid, then the liquid will *rise* in the tube and form a *meniscus* curving upward toward the solid, as shown in Fig. 1.17a for water and glass. This curvature toward the solid is measured by the angle θ. The capillary rise, h, depends for a given fluid and solid on θ which in turn depends on the inside diameter of the tube. The capillary rise will increase with decrease in inside diameter of the tube. If the adhesion to the glass is less than the cohesion in the liquid, then we get a meniscus curving downward as measured by θ toward the solid, as shown for the mercury and glass in Fig. 1.17b. Note in this case that the mercury column is *depressed* a distance h. Again, h will increase with a decrease in inside diameter of the tube. These effects are called *capillary* effects.

We will now state certain simple facts that you have most likely covered in previous physics and mechanics courses dealing with hydrostatics. If not, we will cover them in great detail in Chap. 3. First, note that the gage pressure in a

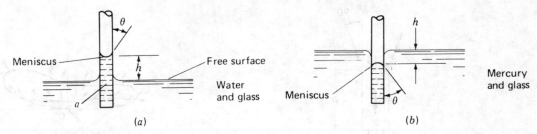

FIGURE 1.17
Capillary effects of cohesion and adhesion.

liquid is computed by the product γd, where d is the depth below the free surface and γ is the specific weight of the liquid. Note however in Fig. 1.17a that the free surface corresponds to the liquid surface away from capillary effects. Thus, the free surface is *not* at the meniscus in the capillary tube. Accordingly, the pressure at a is p_{atm} since it is at the *same elevation* as the free surface where the pressure is p_{atm}. Also note that if you move up a distance l in a liquid, the pressure will decrease by the amount γl. Second, note that if a uniform pressure p acts on a curved surface, the resultant force in a given direction from this pressure is found by simply multiplying p times the area projected in the direction of the desired force. Thus, in Fig. 1.17 the vertical resultant force from the atmospheric pressure on the meniscus is simply $p_{atm}(\pi D^2/4)$, where $\pi D^2/4$ is a circular area—the projection of the meniscus seen as one looks down on it.

Example 1.5. Consider in Fig. 1.17a that the inside diameter of the capillary is 2 mm. If θ is 20° and σ for the water in the presence of air is 0.0730 N/m, determine the height h of the water rise in the capillary.[14]

In Fig. 1.18 we have shown an enlarged free-body diagram of the water standing above the outside level of the free surface. Note that we have atmospheric pressure at the bottom of the column.

Summing forces in the vertical direction, we have, on neglecting the weight of water directly above elevation h and under the meniscus,

$$\sigma(\pi D)\cos\theta - W = 0$$

Hence,

$$(0.0730)(\pi)(0.002)\cos 20° - (9806)\frac{(\pi)(0.002)^2}{4}h = 0$$

$$\therefore h = 13.99 \text{ mm}$$

Thus, knowing θ and σ, we can compute h, the rise or depression of a liquid in a capillary tube.

---

[14] For a very clean glass wall, the angle θ for water is close to 0°. For mercury and clean glass, the angle θ is 40°.

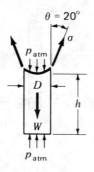

FIGURE 1.18
Free-body diagram of water raised by capillary action.

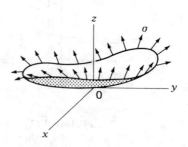

FIGURE 1.19
An arbitrary surface of a liquid.

As a final consideration of surface tension, let us consider an arbitrary surface of a liquid in Fig. 1.19, where the reference is chosen so that at 0, the liquid surface is tangent to the xy plane. If this surface is of infinitesimal dimension about point 0, one can show that the vertical force dF_z resulting from the surface tension along the edge of the surface is given as

$$dF_z = \sigma \left[\left(\frac{\partial^2 w}{\partial x^2} \right)_0 + \left(\frac{\partial^2 w}{\partial y^2} \right)_0 \right] dA \qquad (1.16)$$

We can replace the derivatives by $1/R_1$, and $1/R_2$, where R_1 and R_2 are, respectively, the *radii* of *curvature* at 0 of the intercepts formed with the interface by the xz plane and the yz plane (Fig. 1.20). Dividing by dA, we get a

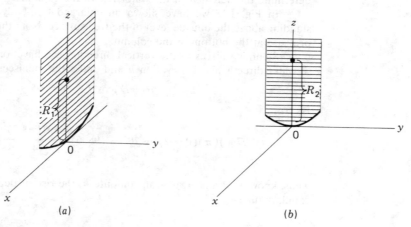

(a)   (b)

FIGURE 1.20
Radii of curvature for liquid surface intercepts.

pressure jump Δp across the interface given as:

$$\Delta p = \sigma \left(\frac{1}{R_1} + \frac{1}{R_2} \right) \qquad (1.17)$$

The case of the droplet of water is a special case of the formulation above where radii of curvature are just R, the radius of the droplet. The pressure jump across the surface—and thus the pressure inside the droplet—is then $2\sigma/R$, as given earlier.

1.10 CLOSURE

In this opening chapter, we have defined a fluid from a mechanics viewpoint and set forth means of describing this substance and its actions in a quantitative manner, using dimensions and their units. There are other attributes of a physical quantity beyond its dimensional representation that are significant in analysis. For instance, you learned in mechanics that some quantities must also have directional descriptions to be meaningful. In the next chapter, we investigate certain of these additional considerations for quantities which form the foundation for the description of fluid phenomena.

PROBLEMS

Problem Categories

Definitions 1.1–1.2
Dimensions and Units 1.3–1.11
Viscosity and Shear Stress 1.12–1.22
Gases (Definitions and Equations of State) 1.23–1.30
Liquids under Compression 1.31–1.38
Surface Tension and Capillary Action 1.39–1.46

1.1. Does the definition of a fluid as used in mechanics differ from what you learned in physics? In chemistry? If so, explain the various definitions and why you think they are different or the same.

1.2. In strength of materials, we used the concept of an elastic, perfectly plastic stress-strain diagram, as shown. Does such a material satisfy the definition of a fluid? Explain.

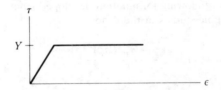

FIGURE P1.2

1.3. What is the dimensional representation of:
   (a) Power
   (b) Modulus of elasticity
   (c) Specific weight
   (d) Angular velocity
   (e) Energy
   (f) Moment of a force
   (g) Poisson's ratio
   (h) Strain

1.4. What is the relation between a scale unit of acceleration in USCS (pound-mass-foot-second) and SI (kilogram-meter-second)?

1.5. How many scale units of power in SI using newtons, meters, and seconds are there to a scale unit in USCS using pounds of force, feet, and seconds?

1.6. Is the following equation a dimensionally homogeneous equation:

$$a = \frac{2d}{t^2} - \frac{2V_0}{t}$$

where $a \equiv$ acceleration
$d \equiv$ distance
$V_0 \equiv$ velocity
$t \equiv$ time

1.7. The following equation is dimensionally homogeneous:

$$F = \frac{4Ey}{(1 - \nu^2)(Rd^2)}\left[(h - y)\left(h - \frac{y}{2}\right)A - A^3\right]$$

where $E \equiv$ Young's modulus
$\nu \equiv$ Poisson's ratio
$d, y, h \equiv$ distances
$R \equiv$ ratio of distances
$F \equiv$ force

What are the dimensions of A?

1.8. The shape of a hanging drop of liquid is expressible by the following formulation developed from photographic studies of the drop:

$$T = \frac{(\gamma - \gamma_0)(d_e)^2}{H}$$

where $\gamma =$ specific weight of liquid drop
$\gamma_0 =$ specific weight of vapor around it
$d_e =$ diameter of drop at its equator
$T =$ surface tension, i.e., force per unit length
$H =$ a function determined by experiment

For the equation above to be dimensionally homogeneous, what dimensions must H possess?

1.9. In the study of elastic solids we must solve the following partial differential equation for the case of a plate where body forces are conservative:

$$\frac{\partial^4\phi}{\partial x^4} + 2\frac{\partial^4\phi}{\partial x^2 \partial y^2} + \frac{\partial^4\phi}{\partial y^4} = -(1 - \nu)\left(\frac{\partial^2 V}{\partial x^2} + \frac{\partial^2 V}{\partial y^2}\right)$$

where $\phi =$ stress function
$\nu =$ Poisson's ratio
$V =$ scalar function whose gradient $[(\partial V/\partial x)\mathbf{i} + (\partial V/\partial y)\mathbf{j}]$ is the body-force distribution where the body force is given per unit volume

What would the dimensions have to be of the stress function?

1.10. Convert the coefficient of viscosity μ from units of dynes, seconds, and centimeters (i.e., *poises*) to units of pound-force, seconds, and feet.

1.11. What are the dimensions of kinematic viscosity? If the viscosity of water at 68°F is 2.11×10^{-5} lb · s/ft^2, what is the kinematic viscosity at these conditions? How many stokes of kinematic viscosity does the water have?

1.12. Water is moving through a pipe. The velocity profile at some section is shown and is given mathematically as

$$V = \frac{\beta}{4\mu}\left(\frac{D^2}{4} - r^2\right)$$

where $\beta =$ a constant
$r =$ radial distance from centerline
$V =$ velocity at any position r

What is the shear stress at the wall of the pipe from the water? What is the shear stress at a position $r = D/4$? If the profile above persists a distance L along the pipe, what drag is induced on the pipe by the water in the direction of flow over this distance?

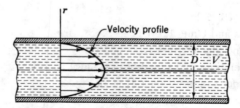

FIGURE P1.12

1.13. A large plate moves with speed V_0 over a stationary plate on a layer of oil. If the velocity profile is that of a parabola, with the oil at the plates having the same velocity as the plates, what is the shear stress on the moving plate from the oil? If a linear profile is assumed, what is then the shear stress on the upper plate?

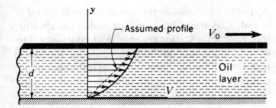

FIGURE P1.13

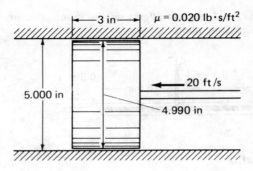

FIGURE P1.16

1.14. A block weighing 1 kN and having dimensions 200 mm on an edge is allowed to slide down an incline on a film of oil having a thickness of 0.0050 mm. If we use a linear velocity profile in the oil, what is the terminal speed of the block? The viscosity of the oil is 7×10^{-2} P.

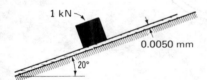

FIGURE P1.14

1.15. A cylinder of weight 20 lb slides in a lubricated pipe. The clearance between cylinder and pipe is 0.001 in. If the cylinder is observed to decelerate at a rate of 2 ft/s² when the speed is 20 ft/s, what is the viscosity of the oil? The diameter of the cylinder D is 6.00 in and the length L is 5.00 in.

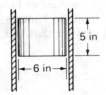

FIGURE P1.15

1.16. A plunger is moving through a cylinder at a speed of 20 ft/s. The film of oil separating the plunger from the cylinder has a viscosity of 0.020 lb · s/ft². What is the force required to maintain this motion?

1.17. A vertical shaft rotates in a bearing. It is assumed that the shaft is concentric with the bearing journal. A film of oil of thickness e and viscosity μ separates the shaft from the bearing journal. If the shaft rotates at a speed of ω radians per second and has a diameter D, what is the frictional torque to be overcome at this speed? Neglect centrifugal effects at the bearing ends and assume a linear velocity profile. What is the power dissipated?

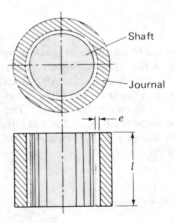

FIGURE P1.17

1.18. In some electric measuring devices, the motion of the pointer mechanism is dampened by having a circular disc turn (with the pointer) in a container of oil. In this way, extraneous rotations are damped out. What is the damping torque for $\omega = 0.2$ rad/s if the oil has a viscosity of 8×10^{-3}

N · s/m²? Neglect effects on the outer edge of the rotating plate.

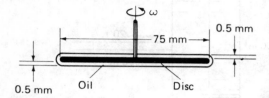

FIGURE P1.18

1.19. For the apparatus in Prob. 1.18, develop an expression giving the damping torque as a function of x (the distance that the midplane of the rotating plate is from its center position). Do this for an angular rotation $\omega = 0.2$ rad/s.

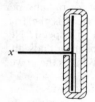

FIGURE P1.19

1.20. A conical body is made to rotate at a constant speed of 10 rad/s. A film of oil having a viscosity of 4.5×10^{-5} lb · s/ft² separates the cone from the container. The film thickness is 0.01 in. What torque is required to maintain this motion? The cone has a 2-in radius at the base and is 4 in tall. Use the straight-line-profile assumption and Newton's viscosity law.

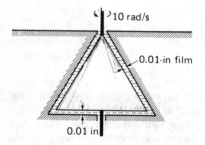

FIGURE P1.20

1.21. A sphere of radius R rotates at constant speed of ω rad/s. A thin film of oil separates the rotating sphere from a stationary spherical container. Develop an expression for the resisting torque in terms of R, ω, μ, and e. Spherical coordinates are shown.

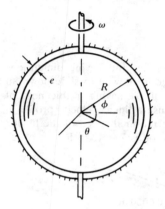

FIGURE P1.21

1.22. An African hunter is operating a blow gun with a poison dart. He maintains a constant pressure of 5 kPa gage behind the poison dart, which has a weight of $\frac{1}{2}$ N and a peripheral area directly adjacent to the inside surface of the blow gun of 1500 mm². The average clearance of this 1500-mm² peripheral area of the dart with the inside surface of the gun is 0.01 mm when shooting directly upward (at a bird in a tree). What is the speed of the dart on leaving the blow gun when fired directly upward? The inside surface of the gun is dry with air and vapor from the hunter's breath as the lubricating fluid between dart and

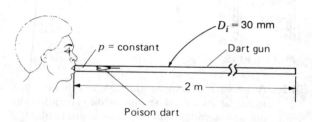

FIGURE P1.22

gun. This mixture has a viscosity of 3×10^{-5} N · s/m². *Hint:* Express dV/dt as $V(dV/dx)$ in Newton's law.

1.23. If specific volume v is given in units of volume per unit mass, and density ρ is given in terms of mass per unit volume, how are they related? Also, if specific weight γ is given in units of weight per unit volume, how is it related to the other quantities?

1.24. What are the dimensions of R, the gas constant, in Eq. (1.10)? Using for air the value 53.3 for R for units degrees Rankine, pound-mass, pound-force, and feet, determine the specific volume of air at a pressure of 50 lb/in² absolute and a temperature of 100°F.

1.25. A perfect gas undergoes a process whereby its pressure is doubled and its specific volume is decreased by two-thirds. If the initial temperature is 100°F, what is the final temperature in degrees Fahrenheit?

1.26. In order to reduce gasoline consumption in city driving, the Department of Energy of the federal government is studying the so-called "inertial transmission" system. In this system, when drivers want to slow up, the wheels are made to drive pumps which pump oil into the compressor tank so as to increase the pressure of the trapped air in the tank. The pumps thus act as brakes. As long as the pressure in the tank stays above a certain minimum value, the tank can supply energy to the aforestated pumps, which then act as motors to drive the wheels when a driver wishes to accelerate. If sufficient braking does not take place to keep the air pressure up, a conventional gas engine cuts in to build up the pressure in the tank. It is expected that a doubling of mileage per gallon can take place in city driving by this system.

Suppose that the volume of air initially in the tank is 80 L and the temperature is 30°C with a pressure of 200 kPa gage. As a result of braking on going down a long hill, the volume decreases to 40 L and the air reaches a pressure of 500 kPa gage. What is the final temperature of the air if there is a loss of air due to a leak of 0.003 kg?

1.27. For Prob. 1.26 suppose that the initial volume of air in the tank is 80 L at a pressure of 120 kPa at $T = 20°C$. The gasoline engine cuts in to double the pressure in the tank while the volume is decreased to 50 L. What is the final temperature and density of the air?

1.28. As you may recall from chemistry, a *pound · mole* of a gas is the number of pounds-mass of the gas equal to its molecular weight M. For 2 lb-mol of air with a molecular weight of 29, a temperature of 100°F, and a pressure of 2 atm, what is the volume V? Show that $pv = RT$ can be expressed as $pV = nMRT$, where n is the number of moles.

1.29. You may recall from chemistry that the gas constant R for a particular gas can be determined from a universal gas constant R_u, having a constant value for all perfect gases, and the molecular weight M of the particular gas. That is, $R = R_u/M$.

The value of R_u in USCS is $R_u = 49{,}700$ ft²/(s²)(°R).

Show that for SI units, we get, $R_u = 8310$ m²/(s²)(K). What is the gas constant R for helium in SI units?

1.30. For Example 1.3, assume that there are *adiabatic* expansions and compressions of the gases, i.e., that $pv^k = $ const, with $k = 1.4$. This assumes no heat transfer to the outside. Compare the results for the speed of the piston. Explain why your result should be higher or lower than that for the isothermal case.

1.31. Anyone who has belly-flopped into water from a high diving board will tell you that the water "feels like concrete." Explain why this is so in terms of the high density of the water and the high bulk modulus of elasticity of the water.

1.32. For Example 1.4, compute the volume of water at atmospheric pressure that would come out under pressure from a crack at the test pressure of 10.5 MPa gage in the tank.

1.33. A high-pressure steel container is partially full of a liquid at a pressure of 10 atm. The volume of the liquid is 1.23200 L. At a pressure of 25 atm, the volume of the liquid equals 1.23100 L. What is the average bulk modulus of elasticity of the liquid over the given range of pressure if the temperature after compression is allowed to return to the original temperature? What is the coefficient of compressibility?

1.34. A heavy tank contains oil (A) and water (B) over which air pressure is varied. The dimensions shown in Fig. P1.34 correspond to the atmospheric pressure of the air. If air is slowly added from a pump to bring the pressure p of the air up to 1 MPa gage, what will be the total downward movement of the free surface of oil and air? Take the average values of bulk moduli of elasticity of the liquids to be, for the pressure range, 2050 MN/m² for oil and 2075 MN/m² for water. Assume that the container does not change volume. Neglect hydrostatic pressures.

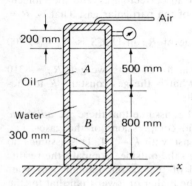

FIGURE P1.34

1.35. For Prob. 1.34 assume that there develops a longitudinal strain $\epsilon_{zz} = 2.34 \times 10^{-5}$ in the cylinder wall and that the diameter changes by 0.01 percent as a result of the pressure increase. What is then the movement of the top free surface relative to a circle on the tank wall coinciding originally with the edge of the top free surface. The change in volume for the fluids from Prob. 1.34 is -4.45×10^{-5} m³.

1.36. Find the bulk modulus of elasticity for steel having a Young's modulus E of 30×10^6 lb/in² and a Poisson's ratio ν of 0.3. To do this, consider an infinitesimal rectangular parallelepiped of steel $dx\, dy\, dz$ under uniform pressure Δp with no shear stress on the faces. Accordingly, there will be equal normal strain ϵ for all edges of the rectangular parallelepiped. Neglecting products of strains compared to strain itself, show that the change in volume per unit volume, $\Delta V / V$, is 3ϵ. From Hooke's law, show that $\Delta p = -E\epsilon/(1 - 2\nu)$. Now you can compute κ for the steel. *Hint:* $(a + b)^3 = a^3 + 3a^2b + 3ab^2 + b^3$.

1.37. A thin-walled steel spherical tank of outside diameter 1 ft and wall thickness of $\frac{1}{4}$ in is full of water, with atmospheric pressure at the top. If the yield stress of the steel is 50,000 lb/in², what volume of water can be forced into the sphere before yielding takes place? Compute this volume at the highest pressure reached in the tank. The modulus of elasticity for the steel is 30×10^6 lb/in². Consider the volume change inside the sphere as a result of the deformation of the container. Use for an average value of κ for the water over the range in pressure involved the value of 305,000 lb/in². Neglect gravitational effects. *Hint:* The force on a curved surface from a uniform pressure equals the pressure times the projected *area* onto a plane *normal* to the direction of the force.

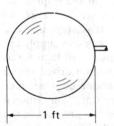

FIGURE P1.37

1.38. For Prob. 1.37 water at a pressure of 4442 lb/in² gage is forced into a tank. If the water is then released from the tank, what is the total mass of water that will be collected at atmospheric pressure? The deformed inside volume in the tank is 800.4069 in³ when the pressure is 4442 lb/in² gage.

1.39. A thin, circular wire is being lifted from contact with water. What force F is required for this action over and above the weight of the wire? The water forms a zero degree contact angle with the outermost and innermost peripheries of the wire for certain metals such as platinum. Compute F for a platinum wire. Explain how you could use this system to measure σ.

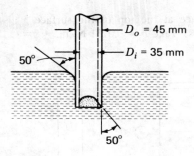

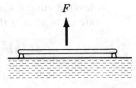

FIGURE P1.41

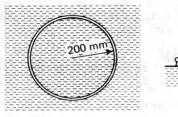

FIGURE P1.39

1.40. Two parallel, wide, clean, glass plates separated by a distance d of 1 mm are placed in water. How far does the water rise due to capillary action away from the ends of the plates? *Hint:* See footnote 14.

1.42. Compute an approximate distance d for mercury in a glass capillary tube. The surface tension σ for mercury and air here is 0.514 N/m, and the angle θ is 40°. The specific gravity of mercury is 13.6. *Hint:* The pressure p_{gage} below the main free surface is the specific weight times the depth below the free surface. Do your assumptions render the actual d larger or smaller than the computed d?

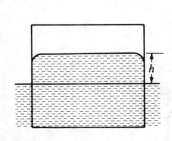

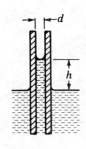

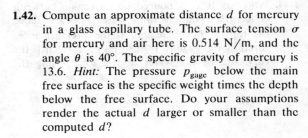

FIGURE P1.42

FIGURE P1.40

1.41. A glass tube is inserted in mercury. What is the upward force on the glass as a result of surface effects? Note that the contact angle is 50° inside and outside. Temperature is 20°C.

1.43. A narrow tank with one end open is filled with water at 45°C carefully and slowly to get the maximum amount of water in without spilling any water. If the pressure gage measures a gage pressure of 2943.7 Pa, what is the radius of curvature of the water surface at the top of the surface away from the ends? Take $\sigma = 0.0731$ N/m.

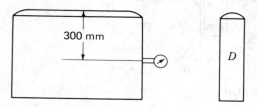

FIGURE P1.43

1.44. Water at 10°C is poured into a region between concentric cylinders until water appears above the top of the open end. If the pressure measured by the gage is 3970.80 Pa gage, what is the curvature of the water at the top? Using the

Taylor series, estimate the height h of the water above the edge of the cylinders. Assume that the highest point of the water is at the midradius of the cylinders.

1.45. In structural mechanics, we can determine the rate of twist α of a shaft of any shape by using *Prandtl's soap-film* analogy. A soap film is attached to a sharp edge having the shape of the outside boundary of the shaft cross section (a rectangle here). Air pressure (Δp) gage is increased under the film so that it forms an elevated curved surface above the boundary. The rate of twist α is then given as:

$$\alpha = \frac{M_x \, \Delta p}{4\sigma G V} \quad \text{radians per unit length} \quad (a)$$

where M_x = torque transmitted by actual shaft
G = shear modulus of actual shaft
V = volume of air under the soap film and above the cross section formed by the sharp edge

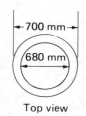

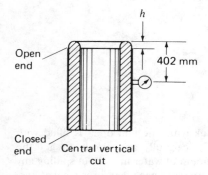

Top view

FIGURE P1.44

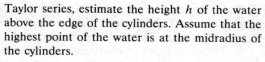

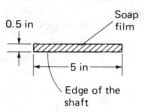

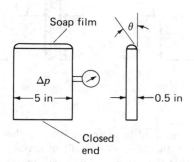

FIGURE P1.45

For the case at hand, the Δp used is 0.4 lb/ft^2 gage. Volume V equals the volume of air forced in to raise the membrane and is measured during the experiment to be 0.5 in^3. The angle θ along the long edge of the cross section is measured optically to be 30°. For a torque of 500 ft · lb on a shaft having $G = 10 \times 10^6$ lb/in^2, what angle of twist does this analogy predict? See hint in Prob. 1.37 concerning pressures on curved surfaces. Note also that like the bubble, there are two surface tensions at the edge of the bubble.

1.46. In using Prandtl's soap-film analogy (see Prob. 1.45), we wish to check the mechanism for measuring the pressure Δp under the soap film. Accordingly, we use a circular cross section for which we have an accurate theory for determining the rate of twist α. The surface tension for the soap film is 0.1460 N/m and volume V under the film is measured to be 0.001120 m^3. Compute Δp from consideration of the soap film and from solid mechanics using Eq. (a) of Prob. 1.45 and the well-known formula from strength of materials $\alpha = M_x/GJ$, where J, the polar moment of area, is $\pi r^4/2$. Compare results.

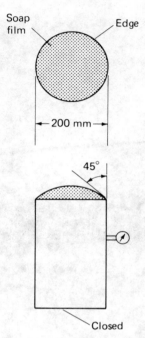

FIGURE P1.46

Closeup of an autogyro. (*Courtesy U.S. Navy.*)

The forerunner of the helicopter was the autogyro invented and developed by a Spaniard, Juan de la Cierva, during the early 1920s. Simply put, much of the lift comes from a set of freely-rotating blades. A conventional propeller provides forward thrust while aerodynamic forces cause the blades to autorotate. To achieve uniform lift for a blade as it moves into and away from the approaching air because of rotation, it is hinged at its base (flapping hinge) so as to allow vertical motion of the blade and thus provide a self-adjustment for uniform lift. There is also a second hinge to counteract change of moment of inertia about a vertical axis caused by the flapping hinge movement. A near vertical take off was achieved by powering the blades to cause a "jump" of about 30 feet in the air. The propeller would by then have created forward motion. A near vertical descent was also possible. However, the autogyro could not hover, and this deficiency caused its demise in favor of the helicopter.

CHAPTER
2

STRESS AT A POINT

2.1 INTRODUCTION

In this chapter, we will consider certain important concepts concerning the stress tensor. In particular, we will see that certain concepts and rules about stress that you learned in your earlier course in strength of materials apply unchanged in the study of fluids.

2.2 SCALAR, VECTOR, AND TENSOR QUANTITIES: FIELDS

Before beginning our discussion, we should classify certain types of quantities which appear in fluid mechanics. A *scalar* quantity requires only the specification of magnitude for a complete description. Temperature, for example, is a scalar quantity. A *vector* quantity, on the other hand, requires a complete directional specification in addition to magnitude, and must add according to the parallelogram law. Usually three values associated with convenient orthogonal directions are employed to specify a vector quantity. These values are called scalar components of the vector. There are more complicated quantities which require the specification of nine or more scalar components for a complete designation. Among these are stress, strain, and mass moment of inertia. These

39

particular quantities, called *tensors*, transform (i.e., change values) in a certain manner under a rotation of the reference axes at a point.[1]

A field is a continuous distribution of a scalar, vector, or tensor quantity described by continuous functions of space coordinates and time. For example, we may describe the temperature at all points in a body at any time by the scalar field expressed as $T(x, y, z, t)$. A vector field, such as the velocity field, may be designated mathematically as $\mathbf{V}(x, y, z, t)$. Usually, however, three scalar fields are employed, each field yielding the value of the velocity component in one of three orthogonal directions. Thus,

$$V_x = f(x, y, z, t)$$

$$V_y = g(x, y, z, t)$$

$$V_z = h(x, y, z, t)$$

This technique of employing a number of scalar fields may be extended to the tensor field, where there may be nine scalar fields.

Since the science of fluid mechanics deals with distributed quantities, there will be considerable opportunity to use the field approach advantageously. Scalar, vector, and tensor fields will all appear in the study of various aspects of fluid phenomena.

2.3 SURFACE AND BODY FORCES; STRESS

In the study of continua, we distinguish between two types of force distributions. Those force distributions which act on matter without the requirement of direct contact are called *body-force distributions*. Gravitational force on a body is the most common body-force distribution. Also, the magnetic-force distribution on a magnetized material in a magnetic field is a body-force distribution. We denote body forces as $\mathbf{B}(x, y, z, t)$ and give them on the basis of per unit mass of the material acted on. The second kind of force distribution on a body arises from direct contact of this body with other surrounding media; it is called a *surface-force distribution* or a *surface-traction distribution*. We denote surface tractions as $\mathbf{T}(x, y, z, t)$ and give them on the basis of per unit area of the material acted on. Surface tractions exist on the physical boundaries of a body or they occur when a "mathematical cut" is made of a body in order to "expose" a surface.

---

[1]You may have also learned in sophomore mechanics that a vector may be defined as having three scalar components associated with a reference *xyz*. These components must change in a certain prescribed manner to become components of a reference *x'y'z'* rotated relative to *xyz*. The transformation equation yielding the new components is very similar but simpler to that which defines tensors. A scalar on the other hand, is invariant with respect to a rotation of axes. From these definitions we can consider all such quantities as tensors of different rank with a vector as a first-order tensor and a scalar as a zero-order tensor.

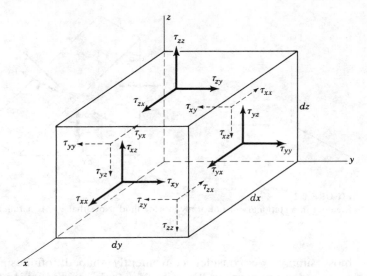

FIGURE 2.1
Stresses on faces of an infinitesimal rectangular parallelepiped taken from inside a body.

In your course in strength of materials or solid mechanics[2] you learned that traction forces on mathematically isolated internal elements (free bodies) gave rise to shear and normal stresses. In Fig. 2.1 we show an infinitesimal rectangular parallelepiped taken from a body with nine stresses acting on the outer faces. A double-index scheme has been utilized to identify the stresses. The first subscript indicates the direction of the normal to the plane associated with the stress, while the second subscript denotes the coordinate direction of the stress itself. The normal stresses have a repeated index, since the stress direction and the normal to the plane on which the stress acts are collinear. The shear stresses will then have mixed indices. For example τ_{yx} is the value of the shear stress acting on a plane whose normal is parallel to the y direction, while the stress itself is parallel to the x direction. The concept of stresses applied to solids holds also for fluids—indeed, it holds for *any* continuum.

2.4 STRESS AT A POINT FOR A STATIONARY FLUID AND FOR NONVISCOUS FLOWS

We will now investigate the relation between the stress on any interface at a point with the stresses on a set of orthogonal interfaces at the point. To do this

---

[2]See the author's *Introduction to Solid Mechanics*, 2nd ed., Prentice-Hall, Englewood Cliffs, N.J., 1988.

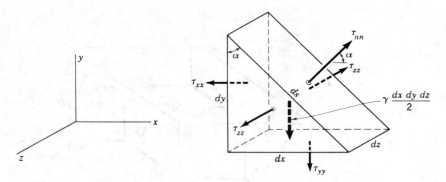

FIGURE 2.2
Element in a stationary or uniformly moving fluid. Note that τ_{nn} is parallel to the yx plane.

most simply, we consider conveniently shaped infinitesimal free bodies of elements of the medium. By using Newton's law and taking the limit as the size of the element goes to zero, we can arrive at the relations that must exist at a point in the medium. We consider at first certain special cases of fluids and then set forth in the next section the general case for any continuous medium.

Case 1. The stationary or uniformly moving fluid. Since a fluid cannot withstand a shear stress without moving, a stationary fluid must necessarily be completely free of shear stress. A uniformly moving fluid, i.e., a flow where all elements have the same velocity, is also devoid of shear stress since the variation of velocity in all directions for uniform flow must be zero ($\partial V/\partial n = 0$), and hence, by virtue of Newton's viscosity law, all the shear stresses are zero.

Assuming that the only body force is that of gravity, we consider an infinitesimal prismatic element of fluid under these conditions as shown in Fig. 2.2. Newton's law in the x direction is

$$-\tau_{xx}\, dy\, dz + \tau_{nn}\, ds\, dz \cos \alpha = 0$$

Since $\cos \alpha = dy/ds$, the equation becomes

$$\tau_{xx} = \tau_{nn}$$

In the y direction Newton's law yields

$$-\tau_{yy}\, dx\, dz + \tau_{nn}\, dz\, ds \sin \alpha - \gamma \frac{dx\, dy\, dz}{2} = 0$$

Again, by recognizing $\sin \alpha$ to be dx/ds and dividing through by $dx\, dz$, we get

$$-\tau_{yy} + \tau_{nn} - \gamma \frac{dy}{2} = 0$$

Now, letting the size of the element shrink to zero, we see that the body force of gravity drops out, so

$$\tau_{yy} = \tau_{nn}$$

Thus we can conclude that in a stationary or uniformly moving fluid the stress at a point is independent of direction and is hence a scalar quantity.[3] Because of the equilibrium nature of the case this quantity may be considered identical to the negative of the thermodynamic pressure, as discussed in Sec. 1.8.

Case 2. Nonviscous fluid in motion. A fluid with theoretically zero viscosity is called a nonviscous, or inviscid, fluid. Since portions of many flows exhibit negligibly small viscous effects, this idealization, with its resulting simplifications, can often be used to good advantage. We will employ Newton's law of motion for an infinitesimal prismatic mass of fluid in the flow. Since there can be no shear stress, we may use the diagram of case 1, remembering that in the present analysis there may be acceleration. In the y direction, Newton's law becomes

$$-\tau_{yy}\, dx\, dz + \tau_{nn}\, ds\, dz \sin \alpha - \gamma \frac{dx\, dy\, dz}{2} = \rho \frac{dx\, dy\, dz}{2} a_y$$

where a_y is the acceleration component. Note that the gravity force and the inertia term vanish in the limiting process, since both these terms are composed of the product of three infinitesimals, as compared with two for the other two terms. Upon replacing $\sin \alpha$ by dx/ds the equation becomes, on dividing through by $dx\, dz$,

$$\tau_{yy} = \tau_{nn}$$

A similar equation in the other direction leads to the conclusion that

$$\tau_{nn} = \tau_{xx} = \tau_{yy}$$

Hence, it may be concluded that for a nonviscous fluid in motion, just as in the case of the stationary or uniformly moving viscous fluid, the stress at a point is a scalar quantity. In Sec. 2.6 it will be pointed out that this quantity is also equivalent to the negative of the thermodynamic pressure.

*2.5 VISCOUS-FLUID MOTION

We now proceed to the general case where viscous effects are taken into account—which means, of course, that shear stresses will be present. That is, there may be nine nonzero stresses on three orthogonal interfaces at a point. In order to discuss the stress at a point, it will be convenient to examine an infinitesimal tetrahedron of fluid, as shown in Fig. 2.3. The nine stresses are on the back faces of the tetrahedron. Employing Newton's law of motion in the direction of the normal to the inclined surface of the tetrahedron, we may solve for the stress τ_{nn} in terms of the nine stresses indicated on the reference planes. The direction cosines of τ_{nn} (and consequently of the normal to ABC) are

---

[3]This is called *Pascal's* law. Also, we point out that this kind of stress distribution is called *hydrostatic stress*.

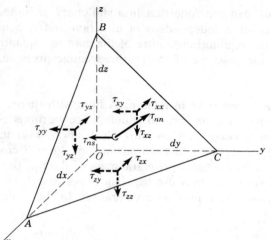

FIGURE 2.3
Tetrahedron of fluid. ABC has arbitrary orientation.

usually denoted as l, m, and n, but we find it expedient to use the letter a with two subscripts. The first subscript identifies the axis whose direction cosines we are measuring—in this case, the n axis—and the second subscript indicates the coordinate axis for the particular direction cosine of interest. Thus we have

$$l \equiv a_{nx}$$

$$m \equiv a_{ny}$$

$$n \equiv a_{nz}$$

We will now use the direction cosines for the normal direction **n** to relate the areas on the faces of the tetrahedron in the following manner[4]:

$$\overline{OCB} = -(\overline{ABC})a_{nx}$$

$$\overline{OAB} = -(\overline{ABC})a_{ny} \qquad (2.1)$$

$$\overline{OCA} = -(\overline{ABC})a_{nz}$$

Expressing Newton's law in the direction of τ_{nn} and dropping gravity and inertia terms as

---

[4]To understand these results more easily, consider that you are *projecting* $\overline{ABC}$ onto the *planes* of the xyz reference. The angle between the $\overline{ABC}$ plane and a reference plane, such as plane yz, is then the angle between the *normals* to these planes—in this case **n** and **i**. The cosine of the angle between **n** and **i** is simply a_{nx}. Thus, the projection of $\overline{ABC}$ onto the yz plane is $(\overline{ABC})a_{nx}$ but, according to the convention of areas pointing outward from a volume, the area $\overline{OCB}$ is negative so we have $\overline{OCB} = -\overline{ABC}a_{nx}$. Hence $|\overline{OCB}| = \overline{ABC}a_{nx}$.

second-order expressions as in the previous cases, we then have

$$\tau_{nn}\overline{ABC} \quad -\tau_{xx}|\overline{OCB}|a_{nx} - \tau_{xy}|\overline{OCB}|a_{ny} - \tau_{xz}|\overline{OCB}|a_{nz}$$
$$-\tau_{yy}|\overline{OAB}|a_{ny} - \tau_{yz}|\overline{OAB}|a_{nz} - \tau_{yx}|\overline{OAB}|a_{nx} - \tau_{zz}|\overline{OCA}|a_{nz}$$
$$-\tau_{zx}|\overline{OCA}|a_{nx} - \tau_{zy}|\overline{OCA}|a_{ny} = 0$$

We replace $|\overline{OCB}|$, $|\overline{OAB}|$, and $|\overline{OCA}|$ using Eq. (2.1). Taking $\tau_{xy} = \tau_{yx}$, $\tau_{xz} = \tau_{zx}$, and $\tau_{yz} = \tau_{zy}$, as will shortly be explained, we get, on cancelling $\overline{ABC}$ and

$$
\begin{aligned}
\tau_{nn} = \quad & \tau_{xx}a_{nx}^2 \quad + \tau_{xy}a_{nx}a_{ny} + \tau_{xz}a_{nx}a_{nz} \\
& + \tau_{yx}a_{ny}a_{nx} + \tau_{yy}a_{ny}^2 \quad + \tau_{yz}a_{ny}a_{nz} \\
& + \tau_{zx}a_{nz}a_{nx} + \tau_{zy}a_{nz}a_{ny} + \tau_{zz}a_{nz}^2
\end{aligned}
\tag{2.2}
$$

Note that we have arranged the right side of the equation as a matrix array of the stresses for ease in remembering the formulation. Let us now imagine that the n axis is the x' axis of a reference $x'y'z'$ rotated relative to xyz. Then we can find $\tau_{x'x'}$ on an interface whose normal is x' at O using the equation above with n replaced by x'. Thus we have

$$
\begin{aligned}
\tau_{x'x'} = \quad & \tau_{xx}a_{x'x}^2 \quad + \tau_{xy}a_{x'x}a_{x'y} + \tau_{xz}a_{x'x}a_{x'z} \\
& + \tau_{yx}a_{x'y}a_{x'x} + \tau_{yy}a_{x'y}^2 \quad + \tau_{yz}a_{x'y}a_{x'z} \\
& + \tau_{zx}a_{x'z}a_{x'x} + \tau_{zy}a_{x'z}a_{x'y} + \tau_{zz}a_{x'z}^2
\end{aligned}
\tag{2.3}
$$

Similarly, for the other two interfaces which are, respectively, perpendicular to the y' and z' axes, we can say

$$
\begin{aligned}
\tau_{y'y'} = \quad & \tau_{xx}a_{y'x}^2 \quad + \tau_{xy}a_{y'x}a_{y'y} + \tau_{xz}a_{y'x}a_{y'z} \\
& + \tau_{yx}a_{y'y}a_{y'x} + \tau_{yy}a_{y'y}^2 \quad + \tau_{yz}a_{y'y}a_{y'z} \\
& + \tau_{zx}a_{y'z}a_{y'x} + \tau_{zy}a_{y'z}a_{y'y} + \tau_{zz}a_{y'z}^2
\end{aligned}
\tag{2.4}
$$

$$
\begin{aligned}
\tau_{z'z'} = \quad & \tau_{xx}a_{z'x}^2 \quad + \tau_{xy}a_{z'x}a_{z'y} + \tau_{xz}a_{z'x}a_{z'z} \\
& + \tau_{yx}a_{z'y}a_{z'x} + \tau_{yy}a_{z'y}^2 \quad + \tau_{yz}a_{z'y}a_{z'z} \\
& + \tau_{zx}a_{z'z}a_{z'x} + \tau_{zy}a_{z'z}a_{z'y} + \tau_{zz}a_{z'z}^2
\end{aligned}
\tag{2.5}
$$

Knowing the nine stresses on orthogonal interfaces at a point, we can thus get the normal stress on any interface inclined to these faces; in particular, we can get the normal stresses for interfaces associated with a reference $x'y'z'$ arbitrarily rotated relative to the xyz axes.

In a similar manner we can get the shear stress having a direction **s** on an interface having normal direction **n** (see Fig. 2.3). Thus, with a_{sx}, a_{sy}, and a_{sz} the direction cosines for the **s** direction, we can say from Newton's law that

$$
\begin{aligned}
\tau_{ns} = \quad & \tau_{xx}a_{nx}a_{sx} + \tau_{xy}a_{nx}a_{sy} + \tau_{xz}a_{nx}a_{sz} \\
& + \tau_{yx}a_{ny}a_{sx} + \tau_{yy}a_{ny}a_{sy} + \tau_{yz}a_{ny}a_{sz} \\
& + \tau_{zx}a_{nz}a_{sx} + \tau_{zy}a_{nz}a_{sy} + \tau_{zz}a_{nz}a_{sz}
\end{aligned}
\tag{2.6}
$$

Again, we have expressed the right side of the equation above in a matrix arrangement of the stresses. As before, we may consider the **n** direction of the interface to be that of an x' axis and the **s** direction tangent to the interface to be that of a z' axis. Hence, for $\tau_{x'z'}$ associated with reference $x'y'z'$ which is rotated relative to xyz we can say on replacing n by x' and s by z' in Eq. (2.6) that

$$
\begin{aligned}
\tau_{x'z'} = \quad & \tau_{xx}a_{x'x}a_{z'x} + \tau_{xy}a_{x'x}a_{z'y} + \tau_{xz}a_{x'x}a_{z'z} \\
+ & \tau_{yx}a_{x'y}a_{z'x} + \tau_{yy}a_{x'y}a_{z'y} + \tau_{yz}a_{x'y}a_{z'z} \\
+ & \tau_{zx}a_{x'z}a_{z'x} + \tau_{zy}a_{x'z}a_{z'y} + \tau_{zz}a_{x'z}a_{z'z}
\end{aligned}
\tag{2.7}
$$

Similar relations can be formed for $\tau_{x'y'}$ and $\tau_{z'y'}$. We thus conclude that once the stresses associated with reference xyz have been established, we can then find any shear stress on any interface at the point; in particular, we can get shear stresses corresponding to reference $x'y'z'$ at the point.

Knowing nine stress components for a reference xyz we can thus compute nine stress components for a reference $x'y'z'$ rotated arbitrarily relative to xyz by making use of the preceding transformation equations [(2.2) and (2.6)]. We can now say that any set of nine components that transform according to the preceding transformation equations, when there is a rotation of axes, is a *second-order tensor*.

The nine scalar components of a stress tensor are usually indicated by the following arrangement, where the first subscript is common for a given row and the second subscript is common for a given column[5]:

$$
\tau_{ij} = \begin{bmatrix} \tau_{xx} & \tau_{xy} & \tau_{xz} \\ \tau_{yx} & \tau_{yy} & \tau_{yz} \\ \tau_{zx} & \tau_{zy} & \tau_{zz} \end{bmatrix}
\tag{2.8}
$$

Whenever it is possible to neglect viscous effects, there is thus the clear advantage of dealing with a single scalar pressure distribution instead of with an obviously more complicated tensor field. Analytical solutions for viscous flow are feasible in relatively simple problems. Particularly in this branch of fluid mechanics, one must rely heavily on experimental information and numerical methods. Equations (2.2) and (2.6) for a viscous fluid are the same as those for the stresses in any continuous media.

2.6 PROPERTIES OF STRESS

Consider a vertical plate loaded by forces in the plane of the plate in Fig. 2.4. The plate is in equilibrium. This is a case of *plane stress*, you may recall, where only stresses τ_{xx}, τ_{yy}, τ_{yx}, and τ_{xy} are nonzero. We have formed a free body in Fig. 2.5 of the element of plate shown in Fig. 2.4. Now set the sum of moments about corners of the element equal to zero. The normal stresses and gravity force give second-order contribution to moments and drop out in the limit. We

---

[5]The subscripts i and j here are called *free* indices having all possible permutations of x, y, and z as represented in the bracketed expression.

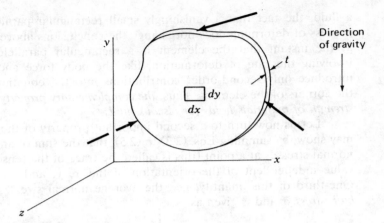

FIGURE 2.4
Plate in equilibrium; plane stress.

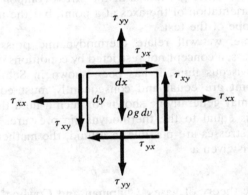

FIGURE 2.5
Infinitesimal rectangular parallelepiped.

conclude first that

$$\tau_{xy} = \tau_{yx}$$

and second that the shear stresses on the infinitesimal element must either point *toward* a corner or *away* from a corner. This is the *complementary* property of shear developed in your strength of materials course. If a state of equilibrium does not exist, then the inertia terms, like the body force, contribute only second-order terms which go out in the limit. The conclusions above still hold. One can extrapolate them to three dimensions[6] wherein $\tau_{yz} = \tau_{zy}$ and $\tau_{xz} = \tau_{zx}$. Again these stresses, as shown in Fig. 2.1, must point toward or away from the edges of a vanishingly small rectangular parallelepiped. In the case of

---

[6] A rigorous proof may be achieved later in the text by doing Prob. 7.46.

a fluid, the fact that a vanishingly small rectangular parallelepiped is in the process of deforming does not change the conclusions above about shear stress at the instant that the element is a rectangular parallelepiped. The terms involving the rate of deformation, like the body force and the inertia term, introduce only second-order contributions into the equation of motion about the corners of the element. Thus, *the complementary property of shear learned in strength of materials holds for viscous fluids.*[7]

Let us now turn to a second important property of the stress tensor. You may show, by summing Eqs. (2.3) to (2.5), that the sum of any set of orthogonal normal stresses at a point (this is called the trace of the tensor) has but a single value independent of the orientation of the x, y, and z axes at that point. One-third of this quantity, i.e., the average normal stress, is often called the *bulk stress* $\bar{\sigma}$ and is given as

$$\bar{\sigma} = \tfrac{1}{3}\left(\tau_{xx} + \tau_{yy} + \tau_{zz}\right) \tag{2.9}$$

Since the bulk stress has no directional properties, it is properly a scalar quantity. There are other groupings of the stress components which are independent of the orientation of the axes at a point, but the use of these relations is beyond the scope of the text.

At this time, we will relate thermodynamic pressure, an equilibrium concept, with stress, a concept not restricted by conditions of equilibrium. In the case of the nonviscous fluid, it has been shown in Sec. 2.4 that all normal stresses at a point are equal and consequently must equal the bulk stress. Furthermore, it may generally be shown for such a fluid that the magnitude of the bulk stress is equal to the thermodynamic pressure. Since normally only negative normal stresses are possible in a fluid, the mathematical statement of this equivalence is given as

$$-\bar{\sigma} = p$$

Using the kinetic theory of gases, Chapman and Cowling[8] have demonstrated this relation to be good for a perfect gas. For the case of real gases, the relation above is not valid when the gas approaches the critical point. Since the vast majority of real-gas problems do not approach this extreme condition, the simple relation given above will be used in this text for all real gases. Finally, experience indicates that this relation may be used with confidence for liquids except when very close to the critical point.

2.7   THE GRADIENT

We have shown how static and frictionless fluids have stress distributions given by the scalar field p. We will now show how a scalar field can give rise to a

---

[7] See footnote 2.

[8] S. Chapman and T. G. Cowling, *The Mathematical Theory of Non-Uniform Gases*, Cambridge University Press, New York, 1953.

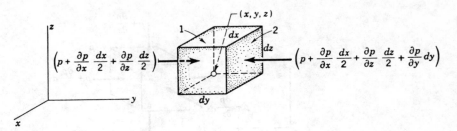

FIGURE 2.6
Pressure variation in the x direction.

vector field of physical significance. To do this, consider an infinitesimal rectangular parallelepiped of fluid at time t, in a frictionless or static fluid as shown in Fig. 2.6. We wish first to compute the resultant force per unit volume on this element from the pressure distribution. For this purpose, a reference with planes parallel to the fluid element has also been shown. The corner of the element nearest the origin is taken as position x, y, z. The pressure at this point is given as p. On face 1 of the element, we have a pressure that may be represented as

$$p_1 = p + \frac{\partial p}{\partial x} \frac{dx}{2} + \frac{\partial p}{\partial z} \frac{dz}{2}$$

This is reached by considering linear variations of pressure in all directions in the immediate vicinity of point x, y, z and computing the pressure in this way at the center of face 1.[9] Face 2 is positioned a distance dy from face 1 so that the pressure there can be considered equal to the pressure on face 1 plus an increment due to this shift in position. We may then say

$$p_2 = p + \frac{\partial p}{\partial x} \frac{dx}{2} + \frac{\partial p}{\partial z} \frac{dz}{2} + \frac{\partial p}{\partial y} dy$$

Note that we could express the increment of pressure from the shift more accurately but this would bring in terms of higher order that would vanish when going to the limit. The net force in the y direction may now be computed from the pressures above. Having chosen the rectangular parallelepiped as the free body, note how we can cancel out the first-order terms of the equation p and leave only the second-order terms of the equation, which give the variation "in the small" of the pressure distribution. Thus

$$dF_y = - \frac{\partial p}{\partial y} dx\, dy\, dz$$

---

[9] What we are doing is actually expressing pressure p as a Taylor series about point x, y, z and retaining terms involving only first-order differentials.

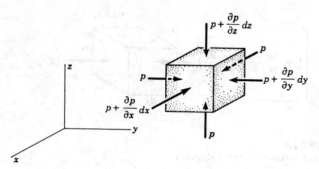

FIGURE 2.7
Pressure variation in *xyz* directions.

Similarly in the x and z directions we get

$$dF_x = -\frac{\partial p}{\partial x} dx\, dy\, dz \qquad dF_z = -\frac{\partial p}{\partial z} dx\, dy\, dz$$

Before proceeding further, we should point out that the above forces on the element could have been achieved had we taken the pressures on the adjacent surfaces nearest the reference to be equal to p and added first-order variations to this value for the outer faces, as is illustrated in Fig. 2.7. It is this formulation which is usually taken in such situations.

The force on the element can then be given as

$$\mathbf{dF} = -\left(\frac{\partial p}{\partial x}\mathbf{i} + \frac{\partial p}{\partial y}\mathbf{j} + \frac{\partial p}{\partial z}\mathbf{k}\right) dx\, dy\, dz$$

The force per unit volume is then

$$\frac{\mathbf{dF}}{dx\, dy\, dz} = \mathbf{f} = -\left(\frac{\partial p}{\partial x}\mathbf{i} + \frac{\partial p}{\partial y}\mathbf{j} + \frac{\partial p}{\partial z}\mathbf{k}\right) \qquad (2.10)$$

Had we used a different element of a shape suitable for computations in a different coordinate system, as, for example, cylindrical coordinates, we would have arrived at a different form of **f** from the one given above. (You will be asked to work out the case of cylindrical coordinates as an exercise.) However, all such formulations have identically the *same physical meaning*, namely force per unit volume at a point, which accordingly is quite *independent* of the coordinate systems used for evaluation purposes. For this reason, we formulate a vector operator,[10] called the *gradient*, which relates scalar and vector fields in such a way that, for the case at hand, we go from a pressure distribution p to

---

[10]A vector operator per se is divorced from coordinate systems until such time as when components are desired.

the vector field **f**, yielding the force per unit volume at a point from surface traction. Thus we can say

$$\mathbf{f} = -\mathbf{grad}\ p \tag{2.11}$$

where, if the gradient operator is referred to a particular coordinate system, it will take on a form dependent on the coordinate system used.[11] For Cartesian coordinates, we thus have for the gradient operator

$$\mathbf{grad} = \mathbf{i}\frac{\partial}{\partial x} + \mathbf{j}\frac{\partial}{\partial y} + \mathbf{k}\frac{\partial}{\partial z} \tag{2.12}$$

In heat transfer, the negative of the gradient of a temperature distribution T yields a vector field **q** which is the *heat-flux* field. Thus the gradient of a scalar gives rise to a *driving action* per unit volume. In particular, $-\mathbf{grad}\ T$ gives rise to a driving action causing flow of heat. And $-\mathbf{grad}\ p$ gives rise to a driving action causing flow of fluid.

In later chapters, we set forth other vector operators such as the *divergence* and *curl* operators that are very useful in that they can describe analytically certain actions occurring commonly in nature without the need for a reference. These operators are used extensively in such fields of study as electricity and magnetism, heat transfer, and theory of elasticity; and although they take on different meanings in these different disciplines, there is still a considerable carryover of meaning from one subject to the other. In the study of fluid mechanics, there is a very vivid picture to be associated with these operators, as they are ordinarily used, so we will employ them throughout the text.

2.8 CLOSURE

In this chapter, we have made some introductory remarks concerning the stress field and its properties. One of the primary goals in fluid mechanics will be to evaluate the stress distribution and the velocity field for certain flows. From this, one can compute forces on bodies, such as airfoils, in the flow and then ascertain probable performance of machines. We generally require use of several basic laws for such undertakings. However, in the next chapter we will be able, by using only Newton's law, to determine the stress field for a static fluid. In other words, static fluids are generally statically determinate, to use the language of rigid-body mechanics.

---

[11]The gradient operator is also expressed by the symbol ∇. Therefore

$$\mathbf{f} = -\nabla p$$

PROBLEMS

Problem Categories

Problems Involving Fields 2.1–2.4
Traction Forces and Stresses 2.5–2.11
Index Notation (starred) 2.12–2.13
Stress Notation 2.14–2.17
Gradient Operator 2.18–2.20

Starred Problems

2.12, 2.13

Derivations and Justifications

2.10, 2.20

2.1. Given the velocity field

$$\mathbf{V}(x, y, z, t) = (6xy^2 + t)\mathbf{i}$$
$$+ (3z + 10)\mathbf{j} + 20\,\mathbf{k}\quad \text{m/s}$$

with x, y, z in meters and t in seconds, what is the velocity vector at position $x = 10$ m, $y = -1$ m, and $z = 2$ m when $t = 5$ s? What is the magnitude of this velocity?

2.2. The velocity components in a flow of fluid are known to be

$$V_x = 6xt + y^2 z + 15 \quad \text{m/s}$$
$$V_y = 3xy^2 + t^2 + y \quad \text{m/s}$$
$$V_z = 2 + 3ty \quad \text{m/s}$$

where x, y, and z are given in meters and t is given in seconds. What is the velocity vector at position $(3, 2, 4)$ m and at time $t = 3$ s? What is the magnitude of the velocity at this point and time?

2.3. A body-force distribution is given as

$$\mathbf{B} = 16x\mathbf{i} + 10\mathbf{j} \quad \text{N/kg}$$

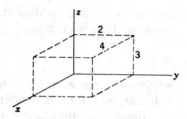

FIGURE P2.3

per unit mass of the material acted on. If the density of the material is given as

$$\rho = x^2 + 2z \quad \text{kg/m}^3$$

what is the resultant body force on material in the region shown in the diagram?

2.4. Oil is moving over a flat surface. We are observing this flow from above in the diagram. A traction force field $\mathbf{T}$ is developed on the flat surface given as

$$\mathbf{T} = (6y + 3)\mathbf{i} + (3x^2 + y)\mathbf{j} + (5 + x^2)\mathbf{k} \quad \text{lb/ft}^2$$

What is the total force on the 3×3 square of area shown in the diagram.

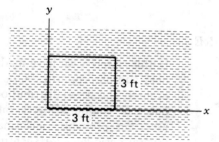

FIGURE P2.4

2.5. The stresses on face A of an infinitesimal rectangular parallelepiped of fluid in a flow are shown in Fig. P2.5 at time t. What is the traction vector for this face at the instant shown? What can you say about shear stresses on faces B and C at this instant?

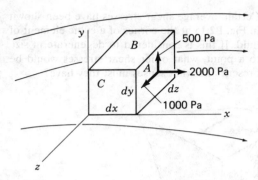

FIGURE P2.5

2.6. Explain why in hydrostatics the traction force exerted on an area element by the fluid is always normal to the area element of the boundary.

2.7. Find the following direction cosines (see Fig. P2.7) between the primed and unprimed axes

$$\begin{bmatrix} a_{x'x} & a_{x'y} & a_{x'z} \\ a_{y'x} & a_{y'y} & a_{y'z} \\ a_{z'x} & a_{z'y} & a_{z'z} \end{bmatrix}$$

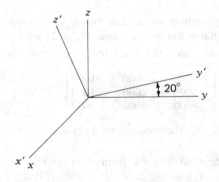

FIGURE P2.7

2.8. A unit vector $\boldsymbol{\epsilon}$ is given as

$$\boldsymbol{\epsilon} = a\mathbf{i} + b\mathbf{j} + c\mathbf{k}$$

Show that a, b, and c are the direction cosines $a_{\epsilon x}$, $a_{\epsilon y}$, and $a_{\epsilon z}$.

2.9. At a point in a flow, the stresses at time t are given as follows:

$$\tau_{ij} = \begin{bmatrix} 2000 & 500 & -1000 \\ 500 & 3000 & 3000 \\ -1000 & 3000 & -4000 \end{bmatrix} \times 10^3 \text{ Pa}$$

What is the normal stress at the point on an interface whose normal unit vector is $\mathbf{n} = 0.6\mathbf{i} + 0.8\mathbf{j} + 0\mathbf{k}$? *Hint:* See Prob. 2.8.

2.10. Express the stress tensor $\tau_{y'x'}$ in terms of the stress tensor τ_{ij} for reference xyz at a point. Then propose a simple set of steps you could follow so that you can easily write the transformation equation for the stress $\tau_{y'x'}$ without having to consult Eq. (2.6). *Hint:* Start by writing τ_{ij} for the unprimed reference [Eq. (2.8)] and proceed from there.

2.11. The stress tensor at point 0 in a flow in Fig. P2.11 is given as

$$\tau_{ij} = \begin{bmatrix} 2000 & 500 & 0 \\ 500 & -4000 & 1000 \\ 0 & 1000 & 1000 \end{bmatrix} \text{ lb/in}^2$$

What is the stress $\tau_{x'z'}$ for reference $x'y'z'$ rotated about the z axis by an angle of 30°?

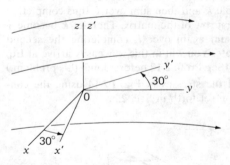

FIGURE P2.11

*2.12. As noted in footnote 5, the subscripts of τ_{ij} are called *free indices* and represent all possible permutations of $i = x, y, z$ and $j = x, y, z$ giving nine different expressions that constitute the stress tensor. If there is a *repetition* of any of the following indices i, j, k, l, m, then we must *sum* over these indices with the letter taking on the notation x, then y, and then z. Thus

$$A_i B_i = A_x B_x + A_y B_y + A_z B_z$$

The indices are called *dummy* indices in this case. This notation is called index or tensor notation.[12] If now we have the following data:

$$A = 6\mathbf{i} + 3\mathbf{j} + 2\mathbf{k}$$
$$B = 10\mathbf{i} + 2\mathbf{j} - 3\mathbf{k}$$
$$C = 3\mathbf{i} + \mathbf{j} - \mathbf{k}$$

what are all the terms for (a) $A_i B_j C_j$ and (b) $A_k B_n C_k$.

*2.13. Equations (2.3), (2.4), (2.5), and (2.7) can all be easily derived from the following equation representing the transformation of the stress terms when there is a rotation of axes:

$$\tau_{ij} = a_{ik} a_{jl} \tau_{kl} \qquad (a)$$

Note there are two free indices and two sets of dummy indices. This equation *defines second-order tensors* in a compact manner. From this equation, formulate $\tau_{x'z'}$. *Suggestion:* Form a 3×3 matrix composed of expressions $a\,a\,\tau$, leaving room for subscripts. Insert for the free indices of Eq. (a) above the values x' and z' in each expression. Finally, assign x to dummy variable k and then sum over l, thus completing the first row of the matrix. Then let k become y and sum again over l, completing the second row, etc. You will in this way easily arrive at Eq. (2.7). (See Prob. 2.12 before doing this problem.)

2.14. Label the stresses in Fig. P2.14 using the convention set forth in Sec. 2.3.

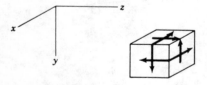

FIGURE P2.14

2.15. Certain average shear stresses have been shown in Fig. P2.15 at the corner of a cubic element of fluid. If this is considered to degenerate in size to a point, what other shear stresses would be present and what senses must they have?

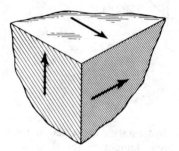

FIGURE P2.15

2.16. We are given the following stress field in megapascals:

$$\tau_{xx} = 16x + 10 \qquad \tau_{zz} = \tau_{xz} = \tau_{yz} = 0$$

$$\tau_{yy} = 10y^2 + 6xy$$

$$\tau_{xy} = -5x^2$$

Express the bulk stress distribution as a scalar field. What is the bulk stress at $(0, 10, 2)$ m?

2.17. In a viscous flow, the stress tensor at a point is

$$\tau_{ij} = \begin{bmatrix} -4000 & 3000 & 1000 \\ 3000 & 2000 & -1000 \\ 1000 & -1000 & -5000 \end{bmatrix} \text{ lb/in}^2$$

What is the thermodynamic pressure at this point?

2.18. A vector field may be formed by taking the gradient of a scalar field. If $\phi = xy + 16t^2 + yz^3$, what is the field **grad** ϕ? What is the magnitude of the vector **grad** ϕ at position $(0, 3, 2)$ when $t = 0$?

2.19. If we have a pressure distribution in a fluid given as

$$p = xy + (x + z^2) + 10 \quad \text{kPa}$$

what is the force per unit volume on an element

of the medium in the direction

$$\mathbf{e} = 0.95\mathbf{i} + 0.32\mathbf{j} \quad \mathrm{m}$$

at position $x = 10$ m, $y = 3$ m, $z = 4$ m?

2.20. Derive the gradient of pressure for cylindrical coordinates in the manner in which we developed the gradient of pressure for Cartesian coordinates. What is the gradient operator in cylindrical coordinates? Use the element shown in Fig. P2.20. *Hint:* Replace $\sin(d\theta/2)$ by $(d\theta/2)$ and $\cos(d\theta/2)$ by 1.

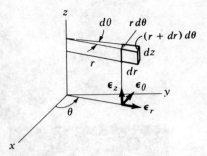

FIGURE P2.20

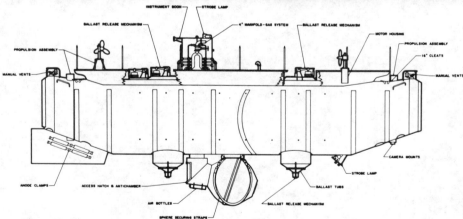

Photo of the deep sea submersible *Trieste* along with a sketch of details. (*Courtesy U.S. Navy.*)

The original bathyscaphe *Trieste* was built by the Swiss physicist Piccard to explore the ocean floor at its lowest depth of 11.3 km. The U.S. Navy purchased this bathyscaphe and has further developed it into a larger, more useful system. This is shown above in an actual scene and also a detailed sketch. An example involving the original bathyscaphe *Trieste* is presented in this chapter (Example 3.9).

CHAPTER 3

FLUID STATICS

3.1 INTRODUCTION

A fluid will be considered static if all particles either are motionless or have identically the same constant velocity relative to an inertial reference. Hence, the conditions for case 1, Sec. 2.4, would now be properly classed as static. For such a case, it was learned, there is no shear stress, so one must deal with a scalar pressure distribution. This chapter evaluates pressure distributions in static fluids and examines some important effects attributable to such pressure distributions.

3.2 PRESSURE VARIATION IN AN INCOMPRESSIBLE STATIC FLUID

In order to ascertain pressure distribution in static fluids, we will consider the equilibrium of forces on an infinitesimal fluid element as shown in Fig. 3.1.[1] The forces acting on the element stem from pressure from the surroundings and the

---

[1]For simplicity we have shown a liquid with a free surface but the resulting equation (3.2) is valid for a gas or a liquid.

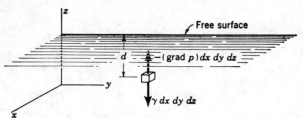

FIGURE 3.1
Free body of element in a static fluid.

gravity force. For equilibrium, we have

$$-\gamma\, dx\, dy\, dz\, \mathbf{k} + (-\mathbf{grad}\, p)\, dx\, dy\, dz = \mathbf{0}$$

where γ is the specific weight. The resulting scalar equations are

$$\frac{\partial p}{\partial x} = 0 \qquad (3.1a)$$

$$\frac{\partial p}{\partial y} = 0 \qquad (3.1b)$$

$$\frac{\partial p}{\partial z} = -\gamma \qquad (3.1c)$$

From this, we see that the pressure can vary only in the z direction, which has been selected as opposite the direction of gravity. (It will be left for you in Prob. 3.4 to deduce from the preceding formulations that the free surface of a liquid at rest must be at right angles to the direction of gravity.)

Since p varies only in the z direction and is not a function of x and y, we may use an ordinary derivative in Eq. (3.1c). Thus

$$\frac{dp}{dz} = -\gamma \qquad (3.2)$$

This differential equation applies to *any static compressible or incompressible fluid* in a gravity field. In order to evaluate the pressure distribution itself, we must integrate between conveniently chosen limits. Choosing the subscript 0 to represent conditions at the free surface, we integrate from any position z, where pressure is p, to position z_0, where the pressure is atmospheric and denoted as p_{atm}. Thus

$$\int_p^{p_{\text{atm}}} dp = \int_z^{z_0} -\gamma\, dz$$

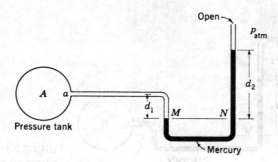

FIGURE 3.2
Simple manometer or U tube.

Taking γ as constant,[2] we may readily integrate. We then get

or

$$p_{atm} - p = -\gamma(z_0 - z)$$

$$p - p_{atm} = \gamma(z_0 - z) = \gamma d \qquad (3.3)$$

where d is the distance below the free surface (see Fig. 3.1). We usually term $p - p_{atm}$, that is, the pressure difference from atmospheric pressure, as the *gage* pressure, with the symbol p_g or p gage. Hence

$$p_g = \gamma d \qquad (3.4)$$

Many pressure-measuring devices yield the pressure above or below that of the atmosphere. As a consequence, gage pressures are used quite often in engineering work.[3] In contrast to pressure p, note that p_g can be negative, with a maximum possible negative value equal to $-p_{atm}$. Note also from the preceding equations that, for any one static fluid, *pressure at a given depth below the free surface remains constant for that depth anywhere in the fluid.*

Example 3.1. *Manometry* is a pressure-measuring technique, the formulations of which follow directly from the preceding discussion. Manometers will be used frequently in the discussions and problems to follow.

The simplest type of manometer is the U tube. This is shown in Fig. 3.2 connected to a tank containing fluid A whose pressure at point a is to be measured. Note that the fluid in the tank extends into the U tube, making contact with the column of mercury. The fluids attain an equilibrium configuration from which it is relatively simple to deduce the tank pressure. Because of its high specific weight, mercury is usually employed as the second fluid when appreciable pressures are expected, since shifts demanded by equilibrium will then be reasonably small.

---

[2] We consider here that g is constant in the range of interest, which is a step that can be taken in most hydrostatic engineering problems. In addition, we consider that the fluid is incompressible so that $\gamma = \rho g$ is constant. We are now restricted to liquids.

[3] We often term the pressure p as *absolute* pressure to distinguish it from gage pressure p_g.

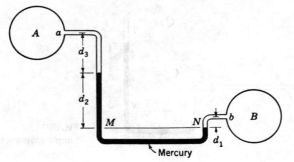

FIGURE 3.3
Differential manometer.

The procedure in deducing p_a is to locate points of equal pressure on the U-tube legs and then to compute these pressures by pressure relations along each leg. This will relate the unknown pressure with the known pressures. Since points N and M of the U tube are at the same elevation and are *joined by the same fluid* (mercury), their corresponding pressures will be equal. Thus

$$p_M = p_a + \gamma_A d_1 = p_N = p_{atm} + \gamma_{Hg} d_2$$
$$\therefore \; p_a = p_{atm} + \gamma_{Hg} d_2 - \gamma_A d_1$$

If fluid A has a very small specific weight compared with mercury, we may, under most circumstances, neglect the term $\gamma_A d_1$. Hence

$$p_a \approx p_{atm} + \gamma_{Hg} d_2$$

Example 3.2. The *differential manometer* will yield the difference in pressures between two regions. In the case of the tanks in Fig. 3.3 containing fluids A and B, note the equality between pressures p_M and p_N. Computing these pressures along each leg of the manometer leads to the desired result.

$$p_b - p_a = \gamma_{Hg} d_2 + \gamma_A d_3 - \gamma_B d_1$$

Example 3.3. In Example 3.2, determine $p_b - p_a$ for the following data:

$$d_1 = 10 \text{ mm}$$
$$d_2 = 80 \text{ mm}$$
$$d_3 = 60 \text{ mm}$$

In tank A and in its capillary arm, we have water at a temperature of 20°C and in B there is air. The specific gravity of mercury is 13.6.

From Appendix Table B.1, we have for γ of water 9788 N/m^3 and for mercury for γ_{Hg} we use the value (13.6)(9788). We then have from the previous example

$$p_b - p_a = \gamma_{Hg} d_2 + \gamma_{H_2O} d_3 - \gamma_B d_1$$

Neglecting $\gamma_B d_1$, we then have

$$p_b - p_a = \left[(13.6)(9788) \text{ N/m}^3\right](0.080 \text{ m}) + (9788 \text{ N/m}^3)(0.060 \text{ m})$$
$$= 11{,}240 \text{ Pa}$$

NOTE: When no temperature is stated, use $\gamma_{H_2O} = 62.4 \text{ lb/ft}^3 \equiv 9806 \text{ N/m}^3$.

3.3 PRESSURE VARIATION WITH ELEVATION FOR A STATIC COMPRESSIBLE FLUID

The vertical distances for the gases in the manometry problems were small and as a consequence we neglected pressure variation with height for such cases. However, in computations involving large vertical distances as may occur in considering atmospheres of planets, we must often consider pressure variation of gas with height. We shall examine two useful cases in this section.

Returning to the differential equation [Eq. (3.2)], relating pressure, specific weight, and elevation for all static fluids, we now assume that γ is a *variable* and thus allow for *compressibility* effects. We restrict ourselves to the perfect gas, which is valid for air or most of its components for relatively large ranges of pressure and temperature. The *equation of state*, containing v, helps us evaluate the required functional variation of the specific weight, γ, since $1/v$ and γ are simply related by their definitions, which are, respectively, the mass and weight of a body per unit volume of the body.[4] Thus, using slugs or kilograms for mass as required by Newton's law, we have from Newton's law

$$\gamma = \frac{1}{v}g = \rho g \tag{3.5}$$

If the mass unit pound-mass is used, the relation above then becomes

$$\gamma = \frac{1}{v}\frac{g}{g_0} \tag{3.6}$$

and since g and g_0 can be considered to have equal values in most practical fluid applications, we often find the relation $1/v = \gamma$ employed under these circumstances. We will formulate our results in terms of slugs or kilograms and make proper conversions when necessary while solving problems.

We will now compute the pressure-elevation relation for two cases, namely, the *isothermal* (constant-temperature) fluid and the case where the temperature of the fluid varies linearly with elevation. These cases occur in certain regions of our atmosphere.

Case 1. Isothermal perfect gas. For this case the equation of state [Eq. (1.10)] indicates that the product pv is constant. Thus, at any position in the fluid, we may say, using the subscript 1 to indicate known data,

$$pv = p_1v_1 = C \tag{3.7}$$

where C is a constant. Solving for v in Eq. (3.5) and substituting into the

---

[4]Note that the specific volume v is the reciprocal of the mass density ρ. That is $\rho = 1/v$.

equation above, we get

$$p\frac{g}{\gamma} = p_1\frac{g_1}{\gamma_1} = C \tag{3.8}$$

We assume that the elevation range is not excessively large so that we can take g as constant. Thus dividing by g,

$$\frac{p}{\gamma} = \frac{p_1}{\gamma_1} = \frac{C}{g} = C' \tag{3.9}$$

Using the relation above, we may express the basic differential equation (3.2) as follows:

$$\frac{dp}{dz} = -\gamma = -\frac{p}{C'}$$

Separating variables and integrating from p_1 to p and z_1 to z, we have

$$\int_{p_1}^{p}\frac{dp}{p} = -\int_{z_1}^{z}\frac{dz}{C'}$$

Carrying out the integration, we get

$$\ln p\big|_{p_1}^{p} = -\frac{z}{C'}\bigg|_{z_1}^{z}$$

Putting in the limits, we have

$$\ln\frac{p}{p_1} = -\frac{1}{C'}(z - z_1)$$

Now we use $p_1/\gamma_1 = C'$ from Eq. (3.9) and solve for p:

$$p = p_1\exp\left[-\frac{\gamma_1}{p_1}(z - z_1)\right] \tag{3.10}$$

This gives us the desired relation between elevation and pressure in terms of the known conditions p_1, γ_1 at elevation z_1. If the datum ($z = 0$) is placed at the position of given data, then z_1, in the equation above, can be set equal to zero. Note that the pressure decreases *exponentially* with elevation.

Case 2. Temperature varies linearly with elevation. The temperature variation for this case is given by

$$T = T_1 + Kz \tag{3.11}$$

where T_1 is the temperature at the datum ($z = 0$). K is often called the *lapse rate* and is a constant. For terrestrial problems, K will be negative. In order to be able to separate the variables of Eq. (3.2), we must solve for γ from the equation of state and, in addition, determine dz from Eq. (3.11). These results are

$$\gamma = \frac{pg}{RT} \qquad (3.12a)$$

$$dz = \frac{dT}{K} \qquad (3.12b)$$

Substituting into the basic equation of statics [Eq. (3.2)], we get, on separating the variables

$$\frac{dp}{p} = -\frac{g}{KR}\frac{dT}{T} \qquad (3.13)$$

Integrating from the datum ($z = 0$) where p_1, T_1, etc., are known, we have

$$\ln\frac{p}{p_1} = \frac{g}{KR}\ln\frac{T_1}{T} = \ln\left(\frac{T_1}{T}\right)^{g/KR}$$

Solving for p and replacing the temperature T by $T_1 + Kz$, we have for the final expression

$$p = p_1\left(\frac{T_1}{T_1 + Kz}\right)^{g/KR} \qquad (3.14)$$

where it should be noted that T_1 must be in degrees absolute.

In concluding this section on compressible static fluids, we should point out that if we know the manner in which the specific weight varies, we can usually separate variables in the basic equation [Eq. (3.2)] and integrate to an algebraic equation between pressure and elevation.

Example 3.4. An atmosphere on a planet has a temperature of 15°C at sea level and drops 1°C per 500 m of elevation. The gas constant R for this atmosphere is 220 N · m/(kg)(K). At what elevation is the pressure 30 percent that of sea level? Take $g = 9.00$ m/s^2.

We must first find K (the lapse rate) for this atmosphere. Note that

$$T = T_1 + Kz$$

$$\therefore T - T_1 = Kz \qquad (a)$$

For $T - T_1 = -1$°C, $z = 500$ m. We get, on applying this condition to Eq. (a),

$$-1 = 500K$$

$$\therefore K = -\frac{1}{500}$$

Now go to Eq. (3.14) in the following form:

$$\frac{p}{p_1} = 0.30 = \left(\frac{T_1}{T_1 + Kz}\right)^{g/KR}$$

Noting that $T_1 = 15 + 273 = 288$ K, we have

$$0.30 = \left(\frac{288}{288 - z/500} \right)^{9.00/[220(-1/500)]}$$

Solving for z, we get

$$z = 8231 \text{ m}$$

This is the desired altitude.

3.4 THE STANDARD ATMOSPHERE

In order to compare the performances of airplanes, missiles, and rockets, a standard atmosphere has been established which approximately resembles the actual atmosphere as it is found in many parts of the world. At sea level, the U.S. Standard Atmosphere conditions are

$p = 29.92$ in Hg $= 2116.2$ lb/ft^2 $= 760$ mm Hg $= 101.325$ kPa

$T = 59°F = 519°R = 15°C = 288$ K

$\gamma = 0.07651$ lb/ft^3 $= 11.99$ N/m^3

$\rho = 0.002378$ slug/ft^3 $= 1.2232$ kg/m^3

$\mu = 3.719 \times 10^{-7}$ lb $\cdot$ s/ft^2 $= 1.777 \times 10^{-8}$ kN $\cdot$ s/m^2

The temperature in the U.S. Standard Atmosphere decreases *linearly with height* according to the relation

$$T = (519 - 0.00357z) \quad °R \ (z \text{ in ft})$$
$$T = (288 - 0.006507z) \quad K \ (z \text{ in m})$$

(3.15)

where z is the elevation above sea level. This region is called the *troposphere*. When a height of about 36,000 ft (or about 11,000 m) is reached, the U.S. Standard Atmosphere becomes isothermal at a temperature of $-69.7°F$ (or $-56.5°C$). This isothermal region is called the *stratosphere*. At about 65,000 ft (or 20,100 m), the temperature starts to increase in value. Appendix Table B.4 gives standard air properties as a function of elevation.

In Fig. 3.4 we have shown a plot of the temperature variation with height for the U.S. Standard Atmosphere.

3.5 EFFECT OF SURFACE FORCE ON A FLUID CONFINED SO AS TO REMAIN STATIC

If external pressure is exerted on a portion of the boundary of a confined fluid compressible or incompressible, this pressure, once all fluid motion has subsided, will extend undiminished throughout the fluid. Two examples are illustrated in Fig. 3.5. The truth of this statement can be demonstrated by examining cylindrical elements of fluid projecting from the pressurized boundary, as is shown in Fig. 3.5. Equilibrium demands that the pressure increase on the

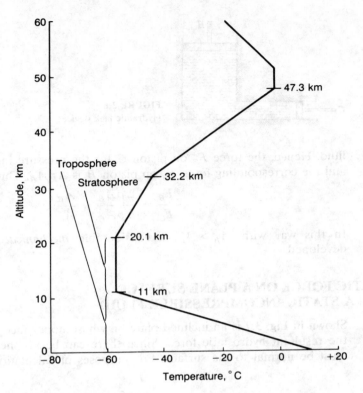

FIGURE 3.4
Temperature variation with height in the U.S. Standard Atmosphere.

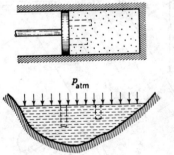

FIGURE 3.5
Pressure on confined fluids.

interior end of the element must keep pace with the pressure applied at the boundary. Since the element may be chosen of any length and at any position, it should be clear that a pressure p developed on the boundary must extend uniformly throughout the fluid.

   This principle underlies the action of the *hydraulic jack* and the *hydraulic brake*. A pressure Δp developed by piston C (Fig. 3.6) is felt throughout the

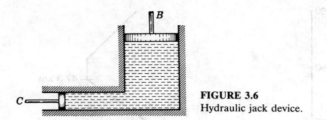

FIGURE 3.6
Hydraulic jack device.

fluid. Hence, the force F_C on piston C for the pressure increase Δp is $\Delta p A_C$ and the corresponding force F_B on piston B is $\Delta p A_B$. Thus,

$$\frac{F_B}{F_C} = \frac{\Delta p A_B}{\Delta p A_C} = \frac{A_B}{A_C}$$

In this way with $A_B > A_C$, a considerable *mechanical advantage* may be developed.

3.6 HYDROSTATIC FORCE ON A PLANE SURFACE SUBMERGED IN A STATIC INCOMPRESSIBLE FLUID

Shown in Fig. 3.7 is an inclined plate on whose upper face we wish to evaluate the resultant hydrostatic force. Since there can be no shear stress, this force must be normal to the surface. For purposes of calculation, the plane of the

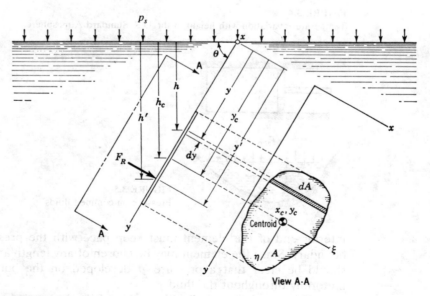

FIGURE 3.7
Submerged plane surface in a liquid.

submerged surface is extended so as to intersect with the plane of the free surface at an angle θ. The trace of the intersection is shown as the x axis in Fig. 3.7. Note that the y axis is coplanar with the top surface of the plate. From previous discussion, we know that there will be superimposed on the plate a uniform pressure p_s, from atmospheric pressure on the free surface, and a uniformly increasing pressure due to the action of gravity on the liquid. It is left for the student to show that the resultant force arising from the uniform pressure p_s has the value $p_s A$ and acts at the *centroid* of the area. Let us proceed then to the analysis of the resultant force from the uniformly increasing pressure.

Note that strip dA has been selected in Fig. 3.7 at uniform depth and consequently is subject to a constant pressure. The *magnitude* of the force on this element is then $\gamma h \, dA$. Integrating over the plate area will then give the value of the resultant force

$$F_R = \int_A \gamma h \, dA = \int_A \gamma y \sin \theta \, dA = \gamma \sin \theta \int_A y \, dA$$

Noting that $\int_A y \, dA$ is the first moment of the plate area about the x axis, we may use in its stead the term $A y_c$, where y_c is the y coordinate of the centroid of this surface. Thus

$$F_R = \gamma \sin \theta \, y_c A = \gamma h_c A = p_c A \tag{3.16}$$

It may be concluded from the formula above that the value of the resultant force due to a uniformly increasing pressure can be most simply evaluated by imagining the pressure at the centroid to extend uniformly over the whole area and computing accordingly.

Clearly the *total* force F_R from the uniform pressure p_s acting at the free surface *and* the uniformly increasing pressure from gravity on the liquid can then be given as

$$F_R = (p_s + \gamma h_c) A = p_c A \tag{3.17}$$

where now p_c is the *total* pressure at the centroid.

The *inclined position* y' of the resultant force from pressure p_s (see Fig. 3.7) on the free surface, as well as from the uniformly increasing pressure of the liquid, will now be evaluated. To do this we will equate the moment of the resultant force F_R about the x axis with the corresponding moment developed by the pressure p_s over the area plus the moment about the x axis of the uniformly increasing pressure of the liquid over the area. Thus

$$F_R y' = \int_A y(p_s + \gamma h) \, dA$$

Replacing F_R and h we have

$$p_c A y' = \int_A y[p_s + \gamma(y \sin \theta)] \, dA$$

where we remind you p_c is the total absolute pressure at y_c from both p_s and the uniformly increasing pressure of the liquid. We now rewrite the above equation:

$$p_c Ay' = p_s \int_A y \, dA + \gamma \sin \theta \int y^2 \, dA$$

$$= p_s Ay_c + \gamma \sin \theta I_{xx}$$

where I_{xx} is the second moment of area about the x axis. Now use the transfer theorem to replace I_{xx} by $I_{\xi\xi} + Ay_c^2$ where $I_{\xi\xi}$ is the second moment of area about the centroidal axis ξ parallel to the x axis (see Fig. 3.7). Hence

$$p_c Ay' = p_s Ay_c + \gamma \sin \theta \left(I_{\xi\xi} + Ay_c^2 \right)$$

Noting that $\gamma y_c \sin \theta + p_s = p_c$ we may rewrite the right side of the above equation as follows:

$$p_c Ay' = p_c Ay_c + \gamma \sin \theta I_{\xi\xi}$$

Rearranging terms we arrive at the desired formulation, namely:

$$\boxed{y' - y_c = \frac{\gamma \sin \theta I_{\xi\xi}}{p_c A}} \tag{3.18}$$

where again p_c is the total absolute pressure at the centroid of the area.

The position of the point of application of the resultant force on the submerged surface is called the *center of pressure*. Since the terms of the right side of the above equation are positive, we see that the center of pressure will always be at a *lower* depth than the centroid.

Next, we investigate the *lateral position* of the resultant. For clarity, the normal view A-A of Fig. 3.7 is shown again in Fig. 3.8. The center of pressure is shown at position y', as determined previously, and at an unknown distance x'

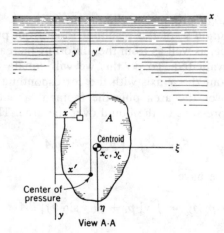

View A-A

FIGURE 3.8
Normal view of plane surface.

from the y axis. Equating the moment about the y axis of the resultant force with the corresponding moment from the pressure distributions, we get

$$(F_R)x' = \int_A x[p_s + \gamma h]\,dA = p_s \int_A x\,dA + \gamma \sin\theta \int_A xy\,dA$$

Replacing F_R, and noting the appearance of the first moment about the y axis and the product of area about the xy axes, we get

$$p_c Ax' = p_s Ax_c + \gamma \sin\theta I_{xy} \tag{3.19}$$

Next, consider the centroidal reference $\xi\eta$ parallel to the xy reference. The transfer formula for products of area between the xy and $\xi\eta$ axes is

$$I_{xy} = I_{\xi\eta} + Ax_c y_c \tag{3.20}$$

and, introducing this into Eq. (3.19), we get

$$p_c Ax' = p_s Ax_c + \gamma \sin\theta(I_{\xi\eta} + Ax_c y_c)$$
$$\therefore \; p_c Ax' = p_s Ax_c + \gamma \sin\theta I_{\xi\eta} + \gamma \sin\theta Ax_c y_c \tag{3.21}$$

Noting once again that $(\gamma \sin\theta y_c + p_s) = p_c$ we can rewrite the above equation as follows:

$$p_c Ax' = p_c Ax_c + \gamma \sin\theta I_{\xi\eta}$$

We can now give the desired result

$$\boxed{x' - x_c = \frac{\gamma \sin\theta I_{\xi\eta}}{p_c A}} \tag{3.22}$$

It must be remembered that $I_{\xi\eta}$ is the product of area about those centroidal axes which are parallel and perpendicular, respectively, to the trace of the plane of the area with the free surface.

In many problems the pressure on the free surface of the liquid in contact with the submerged surface is that of the atmosphere, i.e., p_{atm} and, furthermore, the reverse side of the plate under consideration is exposed only to the atmospheric pressure as shown in Fig. 3.9. For those cases, the combined force on *both* faces of the plate will then be solely due to the action of the uniformly increasing pressure.

There is also the possibility that there may be several layers of immiscible liquids resting on top of the liquid which is in contact with the submerged

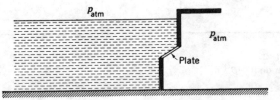

FIGURE 3.9
Atmospheric pressure on the dry side of the plate and at the free surface.

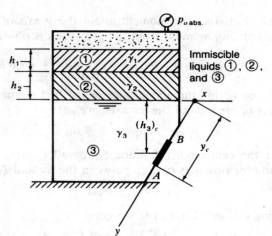

FIGURE 3.10

Liquids ① and ② rest on liquid ③ which wets the surface of door *AB*.

surface of interest to us (see Fig. 3.10). To get the force on door *AB* from the liquids, we recommend that you get the pressure at the centroid of the door as follows:

$$p_c = p_o + \gamma_1 h_1 + \gamma_2 h_2 + \gamma_3 (h_3)_c$$

where $(h_3)_c$ is the height of the liquid that wets *AB* above the centroid of *AB*. We can then use Eqs. (3.17) and (3.18) to get the force and its line of action acting on the wetted part of the door. To get the *total* force inside and outside, use *gage* pressure for p_o.

Example 3.5. A tank is shown in Fig. 3.11 containing water above which the pressure p_A may be established. We wish to determine the force on the door *AB* from internal and external pressures. Note that we have shown axes *xy* and *ξη*.

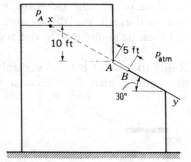 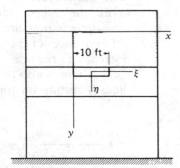

FIGURE 3.11
Find force on door *AB*.

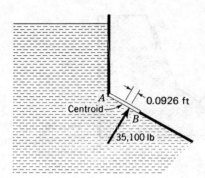

FIGURE 3.12
Resultant force for case $p_A = p_{atm}$.

Case 1. $p_A = p_{atm}$. In this case, the effect of pressure p_A on the inside surface is counteracted by the atmosphere on the outside. The resultant force may then be evaluated by considering only the uniformly increasing pressure distribution from gravitational influence on the water. Thus

$$F_R = p_c A = (62.4)(10 + 2.5 \sin 30°)(50) = (702)(50) = 35,100 \text{ lb}$$

The center of pressure is located at a distance $y' - y_c$ beyond the centroid. Thus

$$y' - y_c = \frac{\gamma \sin \theta I_{\xi\xi}}{p_c A} = \frac{(62.4)(0.5)(\frac{1}{12})(10)(5^3)}{(702)(50)} = 0.0926 \text{ ft}$$

wherein you will note that $\theta = 150°$.

In considering the lateral position, we see that $I_{\xi\eta}$ about the centroidal axes is zero (owing to symmetry), so $x' = x_c$. This means that the lateral position lies along the inclined axis of symmetry of the door. Figure 3.12 shows the resultant.

Case 2. $p_A > p_{atm}$ ($p_A = 18.20$ psi abs). For this case to get the total force on both sides of the door we use the *gage* pressure of p_A namely $(18.20 - 14.7) = 3.50$ psi. Thus:

$$F_R = [(3.50)(144) + (10 + 2.5 \sin 30°)(62.4)]50$$
$$= (1206)(50) = 60,300 \text{ lb}$$

Also

$$y' - y_c = \frac{\gamma \sin \theta I_{\xi\xi}}{p_c A} = \frac{(62.4)(0.5)(\frac{1}{12})(10)(5^3)}{(1206)(50)} = 0.0539 \text{ ft}$$

As a further aid in solving problems we wish to encourage the student not to let geometric complexity obscure the basic simplicity of hydrostatics. For instance, consider the rather ominous geometry in Fig. 3.13 where a high gage pressure p_1 of air is shown acting on the free surface. Assuming the geometry is known, how do you proceed to simplify the problem so that the simple rules of hydrostatics can be used to find the force on door AB?

In short, redraw the diagram as has been shown in Fig. 3.14. Note that it is p_1 and the vertical distance *from the free surface* that dictate pressure *anywhere*

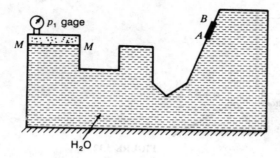

FIGURE 3.13
Complex geometry involving a hydrostatic problem.

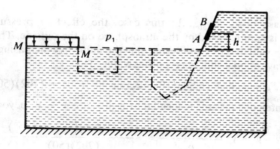

FIGURE 3.14
Simplified geometry noting the key factors p_1 gage and h.

in the liquid[5] and this diagram shows the principal factors, namely p_1, the free surface, and the vertical distance h of the centroid of the door relative to this free surface. The rest of the geometry, part of which is shown dashed, is of no significance here and can be disregarded however complicated it may be. Now we have the simple case of a plane surface AB submerged in a liquid with free surface MM. All we need do is find the hydrostatic force on the right-hand face of the door from p_1 gage on the free surface and from gravity on the water measuring distance h *normal* to the free surface to the centroid of the submerged area. This is a most simple problem. You are encouraged to reduce your problems so as to make solutions accessible to the simple rules of hydrostatics keeping always in mind that *vertical distance* from the free surface and the pressure p_1 on the free surface are the controlling factors and *not boundary geometry* when the liquid is freely connected in the domain of interest.

As a final aid in attacking hydrostatic problems, consider the tank with two different liquids touching a door as shown in Fig. 3.15. The top of the tank is closed to the atmosphere and we wish to find the *total* force on this door. You must observe caution in this undertaking.

---

[5]The liquid must be in free contact throughout the domain where we use this rule—i.e., one part cannot be "sealed" from another part by some rigid plate or body.

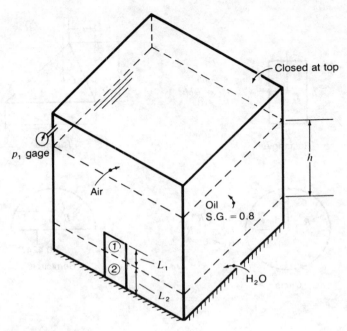

FIGURE 3.15
Two liquids act on door.

First look at portion ① of the door. This is handled by our simple force formulas using p_1 gage at the free surface of the oil, as well as gravity on the oil, to find a force F_1 at the center of pressure of area ①. Now go to area ② of the door. Employing p_1 gage and the weight of the oil, find the gage pressure acting on the free surface of the water. With this pressure and the uniformly increasing pressure from gravity on the water, find the force F_2 and the center of pressure for F_2. Finally combine F_1 and F_2 vectorially giving the line of action of the combined force.

In Fig. 3.16 we have shown some common areas along with useful properties for the students convenience in working problems.

3.7 HYDROSTATIC FORCE ON CURVED SUBMERGED SURFACES

We will now show that forces on curved surfaces submerged in any static fluid can be partially determined by methods used on plane surfaces, as presented in the previous section.

A curved surface is shown in Fig. 3.17 submerged in a static fluid. The force on any area element dA of this surface is directed along the normal to the

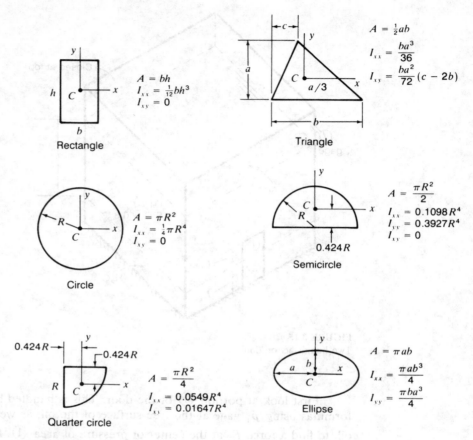

FIGURE 3.16
Properties of some common areas.

area element and is given as

$$dF = -p \, dA$$

where the convention of taking dA pointing outward from the surface has been observed. Taking the dot product of each side of the equation above with the unit vector i, we get the component dF_x on the left side. That is,

$$dF_x = -p \, dA \cdot i$$

But $dA \cdot i$ is actually the projection of the area element onto plane yz yielding dA_x (see Fig. 3.17). To get F_x, we have

$$F_x = -\int_{A_x} p \, dA_x$$

where in the limit of integration A_x is the projection of the curved surface onto the zy plane (or any other plane perpendicular to the x axis). The problem of

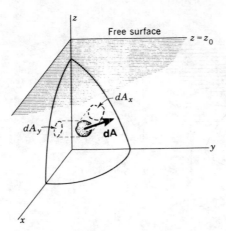

FIGURE 3.17
Area **dA** on curved submerged surface.

finding F_x now becomes the problem of finding the force on a plane submerged surface oriented *perpendicular* to the free surface. We can accordingly use all the techniques set forth earlier for this problem. Similarly, we have for F_y

$$F_y = - \int_{S_y} p \, dA_y$$

where S_y is the projection of the curved surface onto the zx plane (or any plane perpendicular to the y axis). Therefore, two orthogonal components of the resultant force can be determined by methods of submerged plane surfaces. Note that these components are *parallel* to the free surface.

Now let us consider the component *normal* to the free surface. We note that pressure p from *gravitational action* on the fluid at a point on the curved surface is $\int \gamma \, dz$, with limits between z' on the curved surface and z_0 at the free surface (see Fig. 3.18). We can then say that

$$d\mathbf{F} = -p \, d\mathbf{A}$$

$$\therefore dF_z = -p \, d\mathbf{A} \cdot \mathbf{k} = -p \, dA_z$$

$$= - \left(\int_{z'}^{z_0} \gamma \, dz \right) dA_z$$

$$= - \int_{z'}^{z_0} \gamma \, dz \, dA_z$$

Note next in Fig. 3.18 that $\gamma \, dz \, dA_z$ is the weight of an infinitesimal element of fluid in the prismatic column of fluid directly above dA of the curved surface. This column extends to the free surface above. Integrating this quantity from z' to z_0 as we have done above, we obtain from dF_z the *weight* of the column of fluid directly above dA. Clearly, when integrating dF_z over the entire curved surface we get for F_z simply the *weight of the total prismatic column of fluid directly above the curved surface*. The minus sign indicates that a curved surface with a positive dA_z projection (top side of an object) is subjected to a negative force in the z direction (down). It may be shown (see Prob. 3.53) that this force

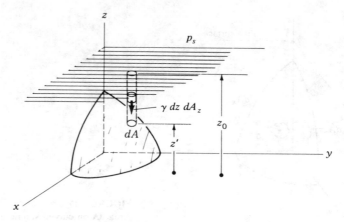

FIGURE 3.18
Note column of fluid above curved surface.

component has a line of action through the center of gravity of the prism of fluid "resting" on the surface.

To account for a pressure p_s on the free surface, we need only multiply the projected area of the curved surface, as seen from above, by p_s. Then add this force to the weight of the column of liquid above the submerged free surface on up to the free surface. We have now formulated the means to determine orthogonal components of the resultant force on the submerged curved surface. These force components give the equivalent action in these directions of the entire surface-force distribution from the fluid on the curved surface. Their lines of action will not necessarily coincide (which means that the simplest resultant system of a curved submerged surface may not be a single force). However, in practical problems it is the components of force in directions parallel and normal to the free surface that are of greatest use.

Note that the conclusions of this section are in no way restricted to incompressible fluids. They are valid for any fluid.

*3.8 A NOTE ON COMPLEX CURVED SURFACES[6]

Let us consider further the *projection process* for curved surfaces. In Fig. 3.19 we see that $(dA_p)_x$ is given as

$$(dA_p)_x = \mathbf{dA} \cdot \mathbf{i}$$

---

[6]In Chapter 5 we shall consider momentum flow to determine force from an internal flow of a fluid through a reducing elbow (Example 5.2). We can, at the same time, include the force from atmospheric pressure on the outside of the elbow in this and most other such problems by simply using *gage pressures* rather than absolute pressures. For those readers ready to accept this approach *intuitively*, there is no need to read this section. In Example 5.3 we shall demonstrate the validity of this approach. Those readers interested in examining this verification, and those wishing to do problems where this simplification is not valid, should study this section.

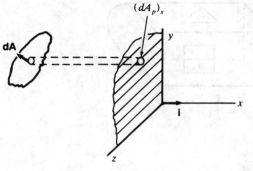

FIGURE 3.19
Projection of **dA** in x direction.

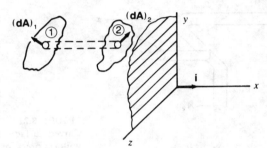

FIGURE 3.20
Projections of $(\mathbf{dA})_1$ and $(\mathbf{dA})_2$ in x direction.

It should be clear that the sign will be negative, denoting that the normal to $(dA_p)_x$ is in the minus x direction. Furthermore, the area has a magnitude which corresponds to that cross section of the prismatic cylinder formed by **dA** in the x direction and shown dashed. Now consider portions of the surface of interest involved in the projection process to be that shown in Fig. 3.20 consisting of the outside of two surfaces ① and ②. Examine the areas $(\mathbf{dA})_1$ and $(\mathbf{dA})_2$ forming the end surfaces of an infinitesimal prismatic cylinder in the x direction. We will now project areas $(\mathbf{dA})_1$ and $(\mathbf{dA})_2$ in the x direction as before. The total projected area from these areas is then given as

$$\left[(dA)_p\right]_{\text{total}} = (\mathbf{dA}_1) \cdot \mathbf{i} + (\mathbf{dA})_2 \cdot \mathbf{i}$$

As indicated earlier, the magnitude of each expression on the right side of the above equation will equal the cross-sectional area of the connecting infinitesimal prismatic cylinder. However, it should be patently clear that the signs will be *opposite*, rendering the right side of the above equation equal to zero. Thus $\mathbf{dA}_1$ "shadows" $\mathbf{dA}_2$ and vice versa.[7] Thus the net projected area in the x direction is zero when the surface is such as to be composed of the end surfaces

---

[7] Your author has used the word "shadows" from terminology used in racing of sailboats wherein when one boat is at a position relative to a competitor so as to block the competitor's wind; we say that the latter has been *shadowed* and is thus devoid of wind for propulsion. This is a very effective tactic against the competitor in racing.

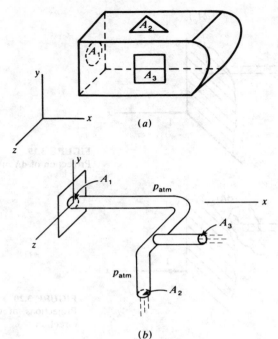

FIGURE 3.21
Curved surfaces submerged in a zone of uniform pressure.

of a continuum of contiguous prismatic cylinders in the x direction. Thus it follows that a *completely closed* surface yields a zero projection in *any* direction.

To illustrate these ideas further consider Fig. 3.21a where we show a closed container with areas A_1, A_2, and A_3 cut out of the faces of the container. What are the projected areas of the *outside* container surface in the plus x, y, and z directions? You should have no difficulty in saying that

$$(A_p)_x = A_1$$
$$(A_p)_y = -A_2$$
$$(A_p)_z = -A_3$$

This follows by considering contiguous prismatic cylinders inside the container separately in the x, y, and z directions as we did in Fig. 3.20 and noting that where we have cut out material from the container we have no end areas of prisms. Or the other way to contemplate the problem is to realize that the cut area A_1 cannot "shadow" an equal plane area of the container in the x direction, leaving a projected area $A_1\mathbf{i}$. Similarly for cut areas A_2 and A_3 in the y and z directions, respectively, giving rise to projected areas $-A_2\mathbf{j}$ and $-A_3\mathbf{k}$. Further we ask what is the force vector from the atmospheric pressure p_{atm} acting on the outer surfaces of the container? Again you should have no difficulty in saying that

$$\mathbf{F} = -p_{atm}A_1\mathbf{i} + p_{atm}A_2\mathbf{j} + p_{atm}A_3\mathbf{k}$$

Next look at Fig. 3.21*b* where we show a bent tube conducting a flow of water. On the outside surface of the tube we have air acting with pressure p_{atm}. What is the force from the air on this outside surface? We have here a curved surface "submerged" in a zone of uniform pressure. Hence using projected areas we can say (here it is easiest to think of the "shadowing" concept)

$$\mathbf{F} = p_{atm}(-A_1 + A_3)\mathbf{i} - p_{atm}(A_2)\mathbf{j}$$

We can use these ideas in parts of Chap. 5.

3.9 EXAMPLES OF HYDROSTATIC FORCE ON CURVED SUBMERGED SURFACES

We will now consider examples that illustrate the formulations of the earlier sections. For simplicity, the fluid in the problems is incompressible.

Example 3.6. Find the vertical force and horizontal force on the quarter circular door *AB*, shown in Fig. 3.22 from water on one side and air on the other. The door is 2 m in width.

For the horizontal force F_x, we project the surface of the curved door onto a plane parallel to the *yz* plane. This projected area is a 1 by 2 m rectangle shown on edge as *OB* in Fig. 3.22. We may now use the formulation for a plane surface to determine F_x. The atmospheric pressure on the free surface clearly develops a horizontal force on the left side of the door *AB* that is canceled completely by the horizontal force from the atmosphere on the right side of the door. We need worry

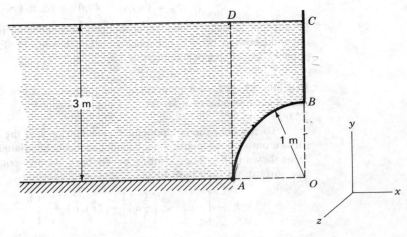

FIGURE 3.22
Find hydrostatic force components on door *AB*.

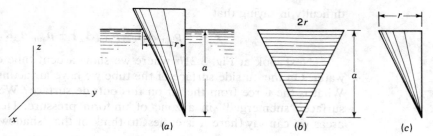

FIGURE 3.23
(a) Submerged semicone. (b) Projection of wetted surface. (c) Water above hypothetical conical shell.

only about the gravitational effect on the water. We then have

$$F_x = p_c A = [(9806)(2.5)](1)(2) = 49,030 \text{ N}$$
$$= 49.0 \text{ kN}$$

Clearly the force component in the z direction is zero because the projected area in this direction is zero.

As for the vertical component, we need only consider the weight of the column of water directly above the door AB. We thus have (see Fig. 3.22)

$$F_y = (9806) \underbrace{[(3)(1)(2)}_{\substack{\text{Volume hav-} \\ \text{ing face } AOCD}} - \underbrace{\tfrac{1}{4}\pi(1^2)(2)]}_{\substack{\text{Volume hav-} \\ \text{ing quarter-} \\ \text{circle face}}}$$

$$\therefore F_y = 43,400 \text{ N} = 43.4 \text{ kN}$$

The resultant force is then

$$F_R = \sqrt{43.4^2 + 49.0^2} = 65.46 \text{ kN}$$

Because the pressure forces are at all times perpendicular to the circular door AB, it should be clear that the simplest resultant has a line of action passing through point O, the center of the circle.

Example 3.7. We wish to evaluate the hydrostatic force components on the *left curved surface* only of a semiconical object inserted in a liquid, as shown in Fig. 3.23a.

To compute the component F_y, we first project the *submerged* semiconical surface onto the xz plane, thus forming an isosceles triangle of height a and base $2r$ as shown in Fig. 3.23b. Noting the we have atmospheric pressure acting on the free surface, we then have for F_y

$$F_y = \left[p_{\text{atm}} + \gamma\left(\frac{1}{3}a\right) \right] \frac{1}{2}(2r)(a) = \gamma\left[\frac{p_{\text{atm}}}{\gamma} + \frac{a}{3} \right](ra)$$

The component F_x is zero, since the net projected surface area onto the yz plane is zero.

For computing the vertical force component, first make believe that the semicone is completely *hollow*, forming only a semiconical surface in the water. The downward force directly from the weight of water on the *inside* surface of the hollow semicone can now be readily determined: It is the weight of the water directly above this surface up to the free surface. Because the semiconical surface we are using here has zero thickness, it should be clear that on the *outside* part of this surface we will have a portion of vertical force which is *equal* and *opposite* to the weight of the column of water supported by the *inside* part of our hypothetical surface. We therefore have a simple means of calculating the vertical *upward* force from gravity acting on the water—a computation we shall make after one more consideration.

Next consider the *atmospheric pressure* on the free surface. Clearly, this pressure extends uniformly throughout the liquid and hence develops an *upward* force component on the wetted semiconical surface. This force is computed by multiplying the atmospheric pressure by the upward projection of the wetted semiconical surface. The *total* vertical upward force component is then

$$F_z = \gamma \left(\frac{1}{2}\right)\left(\frac{1}{3}\right)(\pi r^2)(a) + p_{\text{atm}}\left(\frac{1}{2}\right)\pi r^2$$

$$\therefore F_z = \gamma \left[\frac{1}{6}\pi r^2 a + \frac{p_{\text{atm}}}{2\gamma}\pi r^2\right] \tag{a}$$

wherein the first expression in Eq. (*a*) above is the aforementioned weight of the column of water supported by the hypothetical inside surface of the semicone and the second expression in (*a*), obviously, stems from the atmospheric pressure on the free surface.

The following problem may remind you of the importance of being able to show a proper free body diagram as learned in sophomore mechanics. You are urged to examine this example carefully.

Example 3.8. A large conduit supported from above conducts water and oil under pressure (see Fig. 3.24). It is made of two semicylindrical sections bolted together and weighing 4 kN/m each. If, in a 6-m length of the conduit, 100 bolts clamp the two sections together, what is the total force per bolt to hold the section together with a total force of 6 kN between the flanges to prevent leakage.

The first step is to choose a *free body diagram* that *exposes* the forces from the bolts for equilibrium considerations. This is a job that your studies in statics of rigid bodies should have prepared you for. We have chosen the bottom half of the conduit (Fig. 3.25) for this purpose. This free body includes the water in the lower half cylinder. Our immediate task is to compute the gage pressure $(p_a)_{\text{gage}}$ on surface AA.

For simple *hydrostatics* we have on starting from the top of the mercury column in Fig. 3.24:

$$(p_a)_{\text{gage}} = (13.6)(9806)(0.250) + (0.6)(9806)(0.2 + 1.5 - 0.650)$$

$$+ (9806)(0.650) = 45{,}892 \text{ Pa}$$

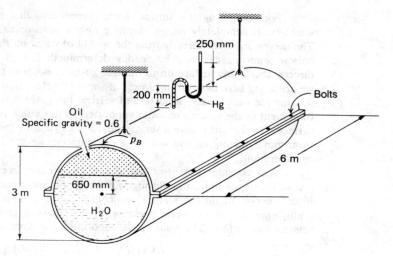

FIGURE 3.24
Large conduit containing water and oil.

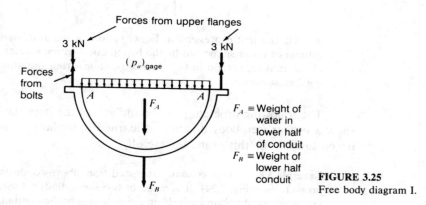

FIGURE 3.25
Free body diagram I.

Now summing forces in the vertical direction, we have from *equilibrium*

$$(100)F_{\text{bolt}} - (45,892)(3)(6) - (4000)(6) - 6000 - \frac{1}{2}\frac{\pi(3^2)}{4}(6)(9806) = 0$$

$$\therefore F_{\text{bolt}} = 10.64 \text{ kN}$$

There is temptation for students not to use pressure $(p_a)_{\text{gage}}$ but to first compute the pressure p_B at the *top* of the tank at B (see Fig. 3.24). It is then proposed that the force acting on the water surface AA in Fig. 3.25 is computed as $p_B(3)(8)$ plus the weight of water and oil inside the upper half of the conduit. The reason given for this approach is that $(p_B)_{\text{gage}}$ acts on a trapped liquid and extends undiminished throughout (this is correct), and, furthermore, that the lower half of the conduit simply supports the liquids "above it." The last assertion is justified by

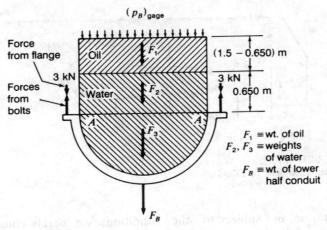

FIGURE 3.26
Free body diagram II using the concept of a submerged surface AA supporting a prismatic column of fluid above it up to the free surface.

referring to Section 3.7. If this is your assessment, you must go back to 3.7 to note that the submerged surface must support a *prismatic column of fluids above it up to the free surface*. Clearly the second assertion above is thus incorrect! If one wanted to use the concept of supporting a column above the surface AA it would be a prismatic column as shown in Fig. 3.26. Now we can properly use equilibrium for this free body. Thus

$$100F_{bolt} = [(13.6)(9806)(0.250) + (0.200)(0.6)(9806)](3)(6)$$

$$+ (0.6)(9806)(1.5 - 0.650)(3)(6)$$

$$+ (9806)(0.650)(3)(6) + \frac{1}{2}\frac{\pi(3^2)}{4}(6)(9806) + (4000)(6) + 6000$$

$$\therefore F_{bolt} = 10.64 \text{ kN}$$

What was really wrong with just getting the weights of the fluids inside the upper cylinder as perhaps some of you may have wanted to do? What is wrong goes back to rigid-body statics wherein the *cardinal error* has been made of *deleting an external force on a free body*! This force is the force stemming from the pressure distribution from the upper half cylinder directly on the surface of oil and water in contact with the upper half cylinder.

3.10 LAWS OF BUOYANCY

The buoyant force on a body is defined as the net vertical force that stems from the fluid or fluids in contact with the body. A body in *flotation* is in contact only with fluids, and the surface force from the fluids is in equilibrium with the force of gravity on the body. To ascertain the buoyant force on bodies both in

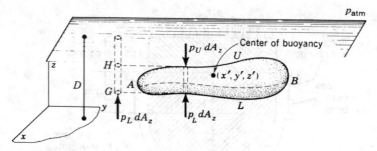

FIGURE 3.27
Submerged body in a fluid.

flotation and subject to other conditions, we merely compute the net vertical force on the surface of the body by methods we have already discussed. Thus no new formulations are involved in buoyancy problems. We consider the cases of:

1. A body completely submerged in one fluid
2. A body at the interface of two immiscible fluids

A totally submerged body is shown in Fig. 3.27 for case 1. The body surface has been divided into an upper portion AUB and a lower portion ALB along a dividing path, shown dotted, forming the outermost periphery of the body as seen by an observer looking down along the direction of gravity. The buoyant force is then the net vertical force exerted by the fluid on these surfaces. Assume for now that we have a liquid surrounding the body and a free surface.

Note first that a pressure p_{atm} on the *free surface* will yield a uniform pressure p_{atm} throughout the fluid below the free surface. Clearly this will yield a zero contribution to the buoyant force and hence we will disregard pressure p_{atm} on the free surface for the computation of buoyant forces. To determine the buoyant force, consider an infinitesimal vertical column *in the body* with cross-sectional area dA_z.

At the top of this column the vertical force, shown as $p_U\, dA_z$, equals the weight of the column of fluid above the upper boundary reaching up to the free surface and having the same cross-sectional area dA_z. At the lower end, the pressure p_L will be the same as that at the bottom of a column of fluid shown to the left where the bottom of this column is at the *same elevation* as the bottom of the column in the submerged body. If dA_z is the same for both columns, the vertical force at the bottom of the one at the left must then be the same as that on the bottom of the column in the body to the right. But the vertical force at the bottom of the left column is simply the weight of the column of fluid up to the free surface. Hence the difference between the upper force $p_U\, dA_z$ and the lower force $p_L\, dA_z$ on the body is the weight of a column of fluid GH having the same size and elevation as the column in the solid body. It is clear that considering all columns in the submerged body the *net upward force is then the weight of fluid displaced*—the familiar *Archimedes principle*. Note that there is no restriction on compressibility involved in the development of this principle and so it is valid for liquids and gases.

We now consider the submerged body in Fig. 3.27 to be composed of infinitesimal vertical prisms, one of which is shown in the diagram. The net force on the prism is

$$dF_B = (p_L - p_U)\, dA_z$$

If we restrict ourselves to an *incompressible* fluid, we can say using the height D of the free surface that

$$dF_B = [(D - z_L)\gamma - (D - z_U)\gamma]\, dA_z = \gamma(z_U - z_L)\, dA_z$$

where z_L and z_U are the elevations, respectively, of the lower end of the prism and the upper end of the prism. Integrating throughout the entire body, we then get the buoyant force

$$F_B = \gamma \int (z_U - z_L)\, dA_z = \gamma V$$

where V is the volume of the submerged body. We thus verify, for the incompressible fluid, the general Archimedes principle which was presented earlier for any fluid.

We next determine the *center of buoyancy*, which is the position in space where the buoyant force may be considered to act. To find the center of buoyancy for this case, we equate the moment of the resultant force F_B about the y axis with that of the pressure distribution of the enveloping fluid. Thus

$$F_B x' = \gamma \int x(z_U - z_L)\, dA_z = \gamma \int_V x\, dv$$

where dv represents the volume of the elemental prism. Replacing F_B by γV and solving for x', we get

$$x' = \frac{\int_V x\, dv}{V} \tag{3.23}$$

We see that x' is the x component of the position vector from xyz to the *centroid* of the volume displaced by the body. We can then conclude by this argument and by similarly taking moments about the x axis that the buoyant force from an incompressible fluid goes through the centroid of the volume of liquid displaced by the body. Clearly this will not be true for a compressible fluid where γ will be some function of z.

Next, examine the case of a body in flotation at the *interface of two immiscible fluids* (Fig. 3.28). This, of course, is the case of every floating vessel, the fluids being water and air. A vertical prism of infinitesimal cross section has

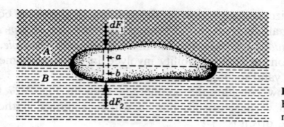

FIGURE 3.28
Body in flotation at interface of two immiscible fluids.

been designated in the floating body. The vertical force components on the upper and lower extremities of the prism are denoted dF_1 and dF_2. It is clear that the net vertical force on the prism from the fluids equals the weight of column a of fluid A plus the weight of column b of fluid B. And by integrating these forces so as to encompass the entire body, we see that the buoyant force equals the sum of the weights of the fluids displaced by the body. Note that with different values of γ present we *cannot* extend our previous argument to state that the buoyant force goes through the centroid of the total volume displaced by the body. However, in nautical work we generally neglect the specific weight of air, and we can consider then that the center of buoyancy is at the centroid of the volume of water displaced by the body.

Example 3.9. Shown in Fig. 3.29 is a highly idealized diagram of the *bathyscaph Trieste*, a device developed by the Swiss physicist Piccard to explore the ocean floor at its greatest depth (11.3 km). A cylindrical tank contains gasoline, which gives the system a buoyant force. A ballast of gravel is attached to the cockpit, which is a steel sphere large enough to house observer and instruments. The vertical motion of the ship is controlled by dropping ballast to get upward force or by releasing gasoline at A and, at the same time, admitting seawater to B to replace the lost gasoline to get a downward force.

If the cockpit and total remaining structure minus gravel and gasoline weigh 15.50 kN, how much should the gravel ballast weigh for flotation at a depth of 3 km below the free surface? The ballast tank of gravel displaces a volume of 2.85

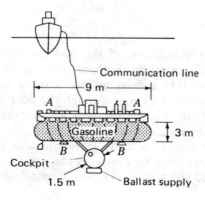

FIGURE 3.29
The bathyscaphe *Trieste* built by Piccard.

m^3. Take the specific weight γ of seawater to be 10.150 kN/m^3 at the location of interest and the specific weight of the gasoline to be 0.65 that of the seawater.

We want to compute the buoyant force for the system. The volume of the gasoline tank and cockpit are, respectively (see Fig. 3.29),

$$V_{gas} = \frac{\pi(3^2)}{4}(9) = 63.6 \text{ m}^3$$

$$V_{cockpit} = \frac{4}{3}\pi\left(\frac{1.5}{2}\right)^3 = 1.767 \text{ m}^3$$

The total volume of displaced seawater then is

$$V_{total} = V_{gas} + V_{cockpit} + V_{ballast} = 63.6 + 1.767 + 2.85 = 68.2 \text{ m}^3 \qquad (a)$$

where we have neglected volume displaced by supporting structural elements. The buoyant force according to the *Archimedes principle* is then

$$F_{buoy} = (V_{total})(\gamma) = (68.2)(10.150) = 692 \text{ kN}$$

If W_B is the weight of the gravel ballast, W_{gas} the weight of the gasoline, and 15.50 kN the weight of the entire structure without gravel or gasoline, then *equilibrium* requires for a full tank of gasoline that

$$-W_B - W_{gas} - 15.50 + F_{buoy} = 0$$

$$\therefore W_B = 692 - 15.50 - (63.6)(0.65)(10.150)$$

$$= 257 \text{ kN}$$

The gravel ballast weight must be 257 kN to maintain neutral buoyancy.

Example 3.10. An empty bucket, with wall thickness and weight considered negligible, is forced open end first into water to a depth E (Fig. 3.30). What is the force F required to maintain this position, assuming the trapped air remains constant in temperature during the entire action?

It is clear that the buoyant force must equal the weight of water displaced whose volume is that of the entrapped air. Therefore,

$$F_B = \gamma e \frac{\pi D^2}{4} \qquad (a)$$

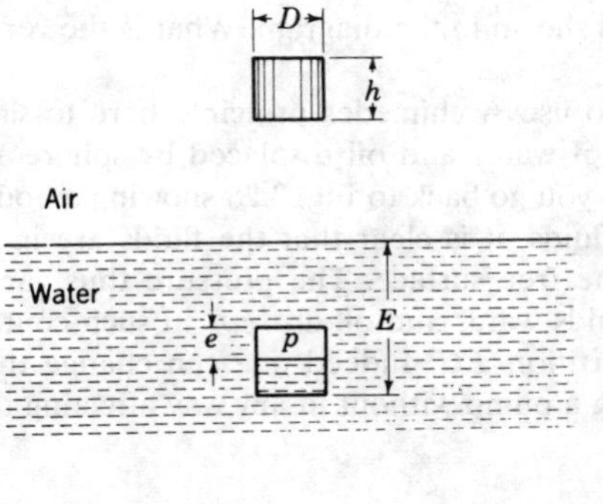

FIGURE 3.30
Empty bucket forced into water.

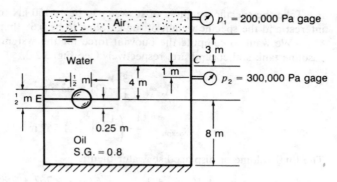

FIGURE 3.31
A partitioned tank.

We do not know e, so we must examine the action of the air and water in the bucket. By using the isothermal-compression formulation for a *perfect gas* to relate the initial state (atmospheric) and the final state of the entrapped air, the following equation holds:

$$p_{atm}\frac{\pi D^2}{4}h = p\frac{\pi D^2}{4}e \qquad (b)$$

This equation introduces another unknown, p. However, p must also be the pressure of the water at the free surface in the bucket, so considering the water now, we can say from *hydrostatics* that

$$p = p_{atm} + \gamma[E - (h - e)] \qquad (c)$$

This gives us a third independent equation by which we can solve for the unknowns F_B, e, and p, a task which we leave to the student. The desired force F is equal and opposite to F_B.

The following example illustrates a case where the student may unwittingly attempt to use Archimedes principle improperly.

Example 3.11. A tank is shown in Fig. 3.31. It is hermetically partitioned into two parts containing water and air above and oil below. A closed sphere D is welded to the thin reinforced partition plate EC and extends equally into the water above and the oil below as shown in the diagram. What is the vertical force on the sphere from the fluids?

If you want to use Archimedes principle here to say that the force desired equals the weights of water and oil displaced by sphere D you will be making a serious error. For if you go back to Fig. 3.28 showing a body sitting at the interface of two immiscible fluids, it is clear that the fluids are in *free direct* contact with each other along the free surface. The pressure thus varies *continuously*. In this example the two fluids water and oil are *not* in such contact with each other as a result of a rigid partition EC. Indeed one can change the pressure in the lower part having oil using a pump without in any way affecting the pressure of the water in the upper part and vice versa. Clearly we must go back to *fundamental considerations* of *submerged curved surfaces*.

First consider the surface of the sphere in the water. The downward force F_1 on this surface is the weight of a column of water directly above it to the free surface plus the gage pressure p_1 times the projected area from above. Thus

$$F_1 = (9806)\left[(3 + 4)\frac{\pi(\frac{1}{2})^2}{4} - \left(\frac{1}{2}\right)\left(\frac{4}{3}\right)(\pi)\left(\frac{1}{4}\right)^3\right] + (200,000)\frac{\pi(\frac{1}{2})^2}{4}$$

$$\therefore F_1 = 52.43 \text{ kN}$$

Now we go to the portion of the sphere in the oil. Let us find the pressure just below the partition and directly above pressure gage 2. Calling this pressure $(p_C)_{\text{gage}}$ we get

$$(p_C)_{\text{gage}} = 300,000 - (0.8)(9806)(1) \doteq 292,155 \text{ Pa}$$

Next we will simplify the problem to allow elementary hydrostatics to be used. Thus, imagine there is no partition *EC* and no water or compressed air above EC. Now raise the level of oil above *EC* so as to give the above pressure at C. The required height of oil above C for this purpose is

$$h = \frac{292,155}{(0.8)(9806)} = 37.24 \text{ m}$$

We next show in Fig. 3.32 a simplified diagram for the lower semispherical surface of interest to us here. We will imagine this body to be a semispherical "cup" with zero wall thickness completely submerged in oil having the free surface shown in the diagram. We will compute the downward force on the inside surface of this cup by computing the weight of oil supported by this cup. Thus we have

$$(F_2)_{\text{down}} = (0.8)(9806)[37.24 + 4]\left(\frac{\pi}{4}\right)\left(\frac{1}{2}\right)^2 + \left(\frac{1}{2}\right)\left(\frac{4}{3}\right)(\pi)\left(\frac{1}{4}\right)^3(0.8)(9806)$$

$$= 63.78 \text{ kN}$$

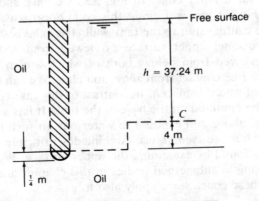

FIGURE 3.32
Simplified diagram for considering semi-sphere portion in oil.

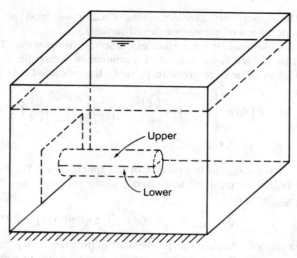

FIGURE 3.33
A closed tube forming a cantilever with the tank wall.

Clearly then the *upward* force on the *outside* surface of the cup will equal this value. Thus the net upward force on the sphere from water *and* oil is then

$$F_{net} = 63.78 - 52.43 = 11.35 \text{ kN}$$

Note that we have used only gage pressures since the atmospheric pressure of 101,325 Pa yields equal and opposite forces on the sphere in the vertical direction and hence plays no role.

Before embarking on assigned problems, we wish to point out how Archimedes principle may further be used. Looking back at Fig. 3.27 notice that the proof of the principle depended on the existence of a distinct upper surface as viewed from above and a distinct lower surface as viewed from below, both having the same outer edge. In Fig. 3.33 we have shown a case where with a moments thought you will agree this condition exists. Thus we have shown a closed tube cantilevered to the tank wall. The tube is submerged in water. Note there is a distinct upper surface as viewed from above and a distinct lower surface as viewed from below, both of which are in contact only with water extending in free contact from above to below. We can find the buoyant force as the weight of water displaced. In contrast to this case consider Fig. 3.34 showing a closed tube mounted on the base of the tank. It has a distinct upper surface as viewed from above and exposed to water but no such lower surface exposed to the water. Here we *cannot* use Archimedes principle. The vertical force from the water is found by examining the upper surface. We will be able to use this way of looking at submerged bodies in this chapter and indeed other chapters. Note that these conclusions apply also to gases.

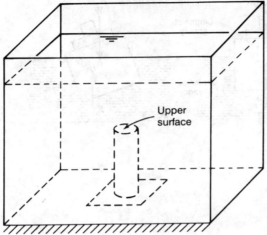

FIGURE 3.34
A closed tube forming a cantilever with
the tank base.

*3.11 STABILITY CONSIDERATIONS FOR BODIES IN FLOTATION

If the imposition of a small displacement on a body in equilibrium brings into action forces tending to restore the body to its original position, the system is said to be in *stable* equilibrium. For instance, in the balloon and basket shown in Fig. 3.35 note that a displacement from the normal position (1) brings into action couple *Wa*, tending to restore the system to the original configuration. The system is thus stable. In general, for completely submerged bodies, as in this example, *stability demands only that the center of gravity of the body be below the center of buoyancy in the normal configuration*. For bodies in flotation at the interface of fluids this requirement is *not* necessary for stability. To illustrate this, observe the vessel shown in Fig. 3.36. Here the weight of the body acts at a point above the center of buoyancy. However, on a "roll," the center of buoyancy *shifts* far enough to develop a righting couple. This explains why a wide rectangular cross section provides a highly stable shape, since a roll causes much fluid to be displaced at

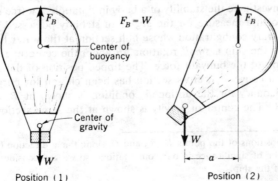

Position (1) Position (2)

FIGURE 3.35
Stability of a balloon.

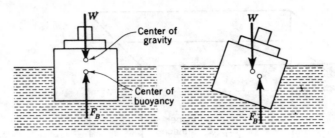

FIGURE 3.36
Stability of ship.

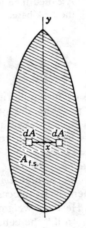

FIGURE 3.37
Cross section of ship at water line.

one extremity at the expense of the other with the result that a large shift for the center of buoyancy takes place toward the tipped end. The weight does not shift, so there is then a comparatively large restoring torque for such shapes.

We investigate the stability of a body in flotation at a free surface with the purpose of setting forth a measure for the *degree* of stability possessed by the body. Consider a ship of arbitrary configuration whose hull section at the water line is shown in Fig. 3.37. We will give the ship a *small* rotation $\Delta\theta$ about the centerline y, and study the shift in line of action of the buoyant force. The tipped position of the ship is illustrated in Fig. 3.38 where a convenient cross section has been chosen for purposes of discussion. The center of buoancy for the untipped condition is shown at B, and the new position is shown at B'. The center of gravity is shown at the same section at G.[8] In rotating the

---

[8] The precise positions of the points B, B', and G along the y direction (i.e., along the axis of the boat) will not be of significance in our computations, so we will consider them to be in the same cross section of the ship.

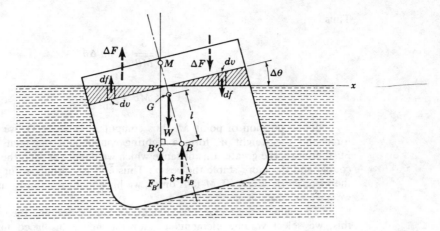

FIGURE 3.38
Ship rotated slightly to show metacentric height MG.

ship about the axis, note that we displace an additional amount of water on the left side of and relinquish an equal amount on the right side. The sections of these volumes have been crosshatched in Fig. 3.38. We will consider, for purposes of calculation, that an upward force ΔF is developed on the left side of the ship as a result of the increased displacement there and that a downward force of equal value is developed on the right side to take care of the loss of displaced water there. These forces then form a couple moment C in the y direction. Thus the total buoyant force system for the tipped configuration can be considered the superposition of the force F_B at B and the couple moment C from forces ΔF. This force system has been shown as dashed arrows in the diagram and is statically equivalent to the single force $F_{B'}$ at B'. We can easily express the distance δ, representing the shift in line of action of the buoyant force, b equating moments of the two systems of forces about an axis parallel to y and going through B'. Thus

$$-F_B\delta + C = 0$$

Hence
$$\delta = \frac{C}{F_B} = \frac{C}{W} \tag{3.24}$$

and so, knowing the couple moment C and the weight of the ship, we can compute the distance δ. Noting in Fig. 3.38 that point M is the intersection of the line of action of $F_{B'}$ and the centerline of the cross section, we can next compute the distance $\overline{MB}$, using δ.

Thus

$$\frac{\delta}{\overline{MB}} = \sin \Delta\theta$$

$$\therefore \overline{MB} = \frac{\delta}{\sin \Delta\theta} \tag{3.25}$$

If the position of point M, thus computed, is *above* G, we see that the buoyant force and the weight W form a righting couple and the boat is said to be stable. Furthermore, the greater this distance which we denote as $\overline{MG}$, the greater this restoring couple and the more stable the vessel. Thus $\overline{MG}$ is a criterion for stability and is called the *metacentric height*. If M falls on G, we have neutral stability, and if it falls below G, we have an unstable condition.

To evaluate the metacentric height, we must determine the couple moment C. For this, we select volume elements dv for the newly displaced fluid and also for the displaced space given up by the vessel as a result of rotation. These are shown in Figs. 3.37 and 3.38 and from these diagrams we can see that

$$dv = x \, \Delta\theta \, dA$$

For each dv we can associate a force df which is the weight of the column of water from dA to the free surface and thus has the value $\gamma x \, \Delta\theta \, dA$. Force df points up for volume elements to the left of y and points down for volume elements to the right of y, as explained earlier. The couple moment C can be determined by taking the moment about y of this force distribution extending over the entire hull section of the ship at the water line (corresponding to the level of the free surface). Denoting the area of this section as $A_{\text{f.s.}}$, we thus have for C

$$C = \int_{A_{\text{f.s.}}} \gamma x^2 \, \Delta\theta \, dA = \gamma \, \Delta\theta \int_{A_{\text{f.s.}}} x^2 \, dA = \gamma \, \Delta\theta \, I_{yy} \tag{3.26}$$

where I_{yy} is the second moment of area $A_{\text{f.s.}}$ about the y axis. Now replace C in Eq. (3.24), using the above result.

$$\delta = \frac{\gamma \, \Delta\theta \, I_{yy}}{W} \tag{3.27}$$

The distance $\overline{MB}$ in Eq. (3.25) can then be written as

$$\overline{MB} = \frac{\gamma \, \Delta\theta \, I_{yy}}{W \sin \Delta\theta} \tag{3.28}$$

From L'Hôpital's rule

$$\lim_{\Delta\theta \to 0} \frac{\Delta\theta}{\sin \Delta\theta} = 1$$

Hence, Eq. (3.28) becomes, in the limit as $\Delta\theta \to 0$,

$$\overline{MB} = \frac{\gamma I_{yy}}{W} \tag{3.29}$$

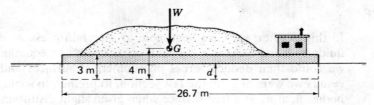

FIGURE 3.39
Loaded barge.

Upon denoting the distance between G and $\dot{B}$ as l (see Fig. 3.38) the metacentric height $\overline{MG}$ then becomes

$$\overline{MG} = (\overline{MB} - l) = \frac{\gamma I_{yy}}{W} - l \qquad (3.30)$$

From this formulation we see that a negative value of $\overline{MG}$ means $\overline{MB} < l$ and hence instability, while a positive value means stability. From our assumptions, both tacit and explicit, we see that the stability criterion presented becomes less meaningful the greater the roll. (We all know that even the most stable vessels can capsize if the disturbance is great enough.) The technique of limiting oneself to small disturbances in order to facilitate computations is common for all engineering sciences. It is important at all times to be cognizant of the limitations associated with the results stemming from such formulations.

Example 3.12. A barge in Fig. 3.39 has the form of a rectangular parallelepiped having dimensions 10 m by 26.7 m by 3 m. The barge weighs 4450 kN when loaded and has a center of gravity 4 m from the bottom. Find the metacentric height for a rotation about its longest centerline, and determine whether or not the barge is stable. If the barge is rotated by 10° about this axis, what is the restoring torque?

We must first find the center of buoyancy of the barge. The barge displaces a volume of fluid having a rectangular cross section 10 m by 26.7 m and a depth d which we determine by using Archimedes principle. Thus

$$[(10)(26.7)d](9806) = W = 4450 \times 10^3$$

$$d = 1.700 \text{ m}$$

The center of buoyancy is at a distance $1.700/2$ m above the bottom of the barge. The distance l, needed in Eq. (3.30), is then $l = 4 - (1.700/2) = 3.15$ m. The metacentric height $\overline{MG}$ is then

$$\overline{MG} = \frac{(9806)\left[\left(\frac{1}{12}\right)(26.7)(10^3)\right]}{(4450)(1000)} - 3.15 = 1.753 \text{ m}$$

Thus the barge is stable.

The restoring couple for a rotation of 10° is given by Eq. (3.26). Thus

$$C = \gamma \, \Delta\theta \, I_{yy} = (9806)\left(\frac{10}{360}\right)(2\pi)\frac{(26.7)(10^3)}{12} = 3808 \text{ kN} \cdot \text{m}$$

3.12 CLOSURE

In this chapter, we have been able to ascertain pressure distributions in static fluids by using Newton's law and, occasionally, an equation of state. From this we could then deduce forces on submerged surfaces and bodies. With these results we were able, in the last section, to predict, to some degree, the action of bodies floating at a free surface when given slight disturbances. In the studies of dynamic fluids in Parts II and III of the text we will follow essentially the same general procedure. That is, we will first determine the velocity field (in this chapter we knew initially by inspection that the velocity was zero relative to an inertial reference); we then get the stress field or that part of it which is of interest and compute certain practical items of interest, as, for example, the lift or drag on some object in the flow.

However, we need more sophisticated methods for quantitatively describing the motion of deformable media beyond those required in the study of particle and rigid-body dynamics. Also, additional laws other than Newton's law are required, and a new way of applying these laws will be helpful. We consider these requirements in Chapter 4.

PROBLEMS

Problem Categories

Pressure Variation in Liquids 3.1–3.7
Manometry Problems 3.8–3.18
Barometers 3.19–3.21
Atmosphere Problems 3.22–3.32
Hydrostatic Forces on Plane Surfaces 3.33–3.49
Forces on Curved Surfaces 3.50–3.73
Buoyancy Problems 3.74–3.95
Stability of Floating Bodies 3.96–3.101

Starred Problems

3.50, 3.51, 3.52, 3.59, 3.73, 3.92

When no stated temperature is given, use $\gamma_{H_2O} = 62.4$ lb/ft^3 = 9806 N/m^3.

3.1. What is meant by an inertial reference?

3.2. If the acceleration of gravity were to vary as K/z^2, where K is a constant, how would the density have to vary if Eq. (3.4) were to be valid?

3.3. The deepest point under water is the Mariana Trench east of Japan where the depth is 11.3 km. What is the pressure there
(a) in absolute pressure?
(b) in gage pressure?
The average specific gravity of seawater there we estimate as 1.300.

3.4. Prove that the free surface of a static liquid must be normal to the direction of gravity.

3.5. Two identical containers, each open to the atmosphere, are initially filled with the same liquid ($\rho = 700$ kg/m^3) to the same level H (Fig. P3.5). The two containers are connected by a pipe in which a frictionless piston of cross section $A = 0.05$ m^2 is made to slide slowly. How much work is done by water on the piston in moving a distance of $L = 0.1$ m? The cross section of each container is *twice* that of the pipe.

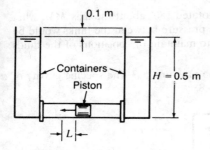

FIGURE P3.5

3.6. Do Prob. 3.5 for the case where the containers are closed and the air above the free surface is at a pressure p_0 of 200 kPa gage. The air expands isentropically in the container on the right side and is compressed isothermally for the container on the left side. $R = 287$ N $\cdot$ m/ kg K.

3.7. A cylindrical tank contains water at a height of 50 mm. Inside is a smaller open cylindrical tank containing kerosene at height h having a specific gravity of 0.8. The following pressures are known from the indicated gages:

$$p_B = 13.80 \text{ kPa gage}$$

$$p_C = 13.82 \text{ kPa gage}$$

What are the gage pressure p_A and the height h of the kerosene? Assume that the kerosene is prevented from moving to the top of the tank.

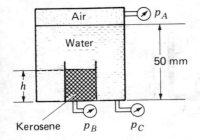

FIGURE P3.7

3.8. Find the difference in pressure between tanks A and B if $d_1 = 300$ mm, $d_2 = 150$ mm, $d_3 = 460$ mm, $d_4 = 200$ mm, and $S_{Hg} = 13.6$.

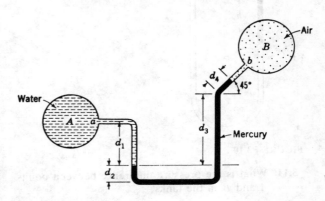

FIGURE P3.8

3.9. An open tube is connected to a tank. The water rises to a height of 900 mm in the tube. A tube used in this way is called a *piezometer*. What are the pressures p_A and p_B of the air above the water? Neglect capillary effects in the tube.

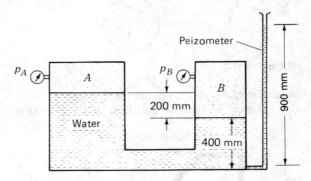

FIGURE P3.9

3.10. Consider the U tube with one end closed and the other end having a funnel of height 2 in. Mercury is poured into the funnel to trap the air in the tube, which is 0.1 in in inside diameter and 3 ft in total length. Assuming that the trapped air is compressed isothermally, what is h when the funnel starts to run over? Neglect capillary effects for this problem.

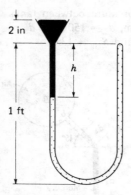

FIGURE P3.10

3.11. What is the pressure difference between points *A* and *B* in the tanks?

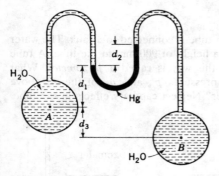

FIGURE P3.11

3.12. Calculate the difference in pressure between centers of tank *A* and tank *B*. If the entire

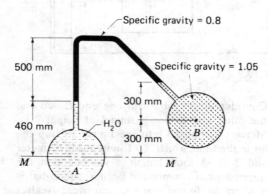

FIGURE P3.12

system is rotated 180° about the axis *MM*, what changes in pressure between the tanks would be necessary to maintain the positions of the fluids intact?

3.13. What is the pressure p_A? The specific gravity of the oil is 0.8.

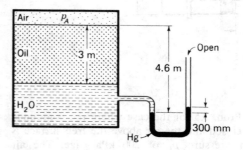

FIGURE P3.13

3.14. What is the specific gravity of fluid *A*?

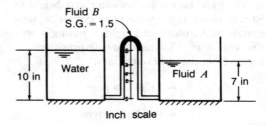

FIGURE P3.14

3.15. Find distance *d* for the U tube.

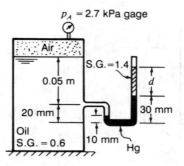

FIGURE P3.15

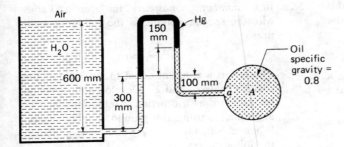

FIGURE P3.16

3.16. What is the absolute pressure in drum A at position a?

3.17. What is the gage pressure in the tank? The tank contains air.

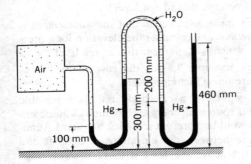

FIGURE P3.17

3.18. When greater precision is required for a pressure measurement, we use a *micromanometer*. Two immiscible liquids having specific weights γ_1 and γ_2, respectively, are used in this system. We assume that the fluids in tanks E and B whose pressure difference we are measuring are gases with negligible specific weight. Compute the pressure difference $p_E - p_B$ in terms of δ, d, γ_1, and γ_2. If the area of the micromanometer tube is a and the cross-sectional areas of the containers C and D are A, determine δ in terms of d, by geometrical considerations. Explain how by having a/A very small and γ_1 almost equal to γ_2, a small pressure difference $p_E - p_B$ will cause a large displacement d, thus making for a sensitive instrument.

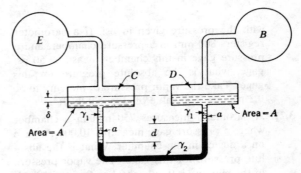

FIGURE P3.18

3.19. A barometer is a device for measuring atmospheric pressure. If we use a liquid having spe-

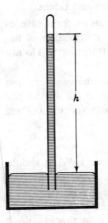

FIGURE P3.19

cific weight γ and invert a tube full of this material as shown, find formulas for h if the absolute vapor pressure of the liquid is p_{vap}
(a) in SI units
(b) in USCS units using psi
Show dimensions in your formulas. If we use a fluid having a specific weight of 850 lb/ft^3 and a vapor pressure of 0.2 psi abs, find h.

3.20. From the preceding problem the height h in a barometer is

$$h = \frac{1}{\gamma}\left[p_{atm} - (p_{vap})\right](144) \text{ ft} \qquad (a)$$

with the pressure given in psi. If a barometer registers 800 mm in a pressure chamber, and a pressure gage in this chamber measures 50 psi gage, what is the absolute pressure for this gage? Take the vapor pressure of mercury to be 0.3 psi abs. The value of γ is 850 lb/ft^3.

3.21. A barometer measures 750 mm in a chamber where a pressure gage measures 10,000 Pa gage on a device in the chamber. What is the absolute pressure for this gage? The vapor pressure of the mercury is 0.5 Pa abs. See Probs. 3.19 and 3.20. What conclusion can be drawn concerning the inclusion of vapor pressure of mercury in most problems?

3.22. The Eiffel Tower in Paris is 984 ft tall with its base about 500 ft above sea level. What is the pressure and temperature at the top in a U.S. Standard Atmosphere? Do not use tables.

3.23. Outside a hot air balloon a barometer measures 690 mm mercury. What is the elevation of the balloon in a U.S. Standard Atmosphere? Do not use tables.

3.24. In the U.S. Standard Atmosphere where the temperature varies according to Eq. (3.15), the equation relating pressure and specific volume is

$$pv^n = \text{const}$$

This is called a *polytropic process*. What should the value of n be?

3.25. At what elevation in feet is the pressure in a standard atmosphere 0.92 that at sea level? Do this *without* tables. What is v at this position? Use sea level data given in Sec. 3.4.

3.26. In an *adiabatic atmosphere*, the pressure varies with the specific volume in the following manner:

$$pv^k = \text{const}$$

where k is a constant equal to the ratio of the specific heats c_p and c_v. Develop an expression for pressure as a function of elevation for this atmosphere, using the ground as a reference. When $z = 0$, take $p = p_0$ and $\gamma = \gamma_0$. Reach the following result:

$$p = \frac{1-k}{k}\gamma z + p_0\frac{\gamma}{\gamma_0}$$

3.27. An atmosphere has a temperature of 27°C at sea level and drops 0.56°C for every 152.5 m. If the gas constant is 287 N · m/(kg)(K), what is the elevation above sea level where the pressure is 70 percent that of sea level?

3.28. In Example 3.4, assume that the atmosphere is *isothermal* and compute the elevation for a pressure which is 30 percent that at sea level.

3.29. Work Example 3.4 for the case of the atmosphere being incompressible.

3.30. The wind has been considered as a possible useful source of energy. How much kinetic energy would be present in a U.S. Standard Atmosphere between the elevations of 5000 ft and 6000 ft above sea level if there is an average wind speed of 5 mi/h? Use an average density. The radius of the earth is 3960 mi. What is the kinetic energy per unit volume of air? Comment on the practical use of wind power. Area of a sphere is πD^2.

3.31. A light rubber balloon containing helium is released in a U.S. Standard Atmosphere. The stretched rubber transmits a membrane force σ proportional to the diameter and given as $5D$ lb/ft with D given in feet. What is the inside pressure in the balloon at an elevation of 5000 ft in a U.S. Standard Atmosphere? The balloon is rising slowly at a constant speed. *Hint:* The force on a curved surface from a uniform pressure equals the pressure times the projected area of the surface onto a plane normal to the direction of the force.

3.32. In a light airplane the cabin pressure is to be maintained at 80 percent that of atmospheric

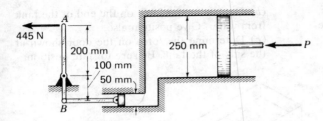

FIGURE P3.33

pressure on the ground which is 10,000 ft above sea level. If for structural reasons the outside-to-inside ambient pressure ratio is not to get smaller than 0.6, what is the maximum height h_{max} that the plane may fly in a U.S. Standard Atmosphere?

3.33. A force of 445 N is exerted on lever AB. End B is connected to a piston which fits into a cylinder having a diameter of 50 mm. What force P must be exerted on the larger piston to prevent it from moving in its cylinder which has a 250 mm diameter?

3.34. Prove that the resultant force from a uniform pressure distribution on an area acts at the centroid of the area.

3.35. Find total force on door AB and the moment of this force about the bottom of the door.

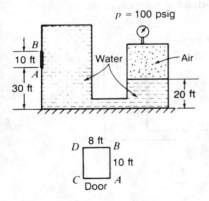

FIGURE P3.35

3.36. A plate is submerged vertically into the water. What is the radius r of a hole to be cut from the center of $ABCD$ to make the hydrostatic force on surface $ABCD$ equal to the hydrostatic force on surface CDE? What is the moment of the total force about AB? Delete p_{atm}.

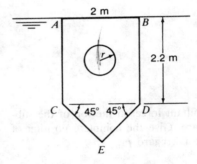

FIGURE P3.36

3.37. A rectangular plate shown as ABC can rotate about hinge B. What length l should BC be so that there is zero torque about B from water and plate weight? Take the weight as 1000 N/m of length. The width is 1 m.

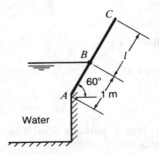

FIGURE P3.37

3.38. Find the total force on door AB from fluids. Take $S_{oil} = 0.6$. Find the position of this force from the bottom of the door.

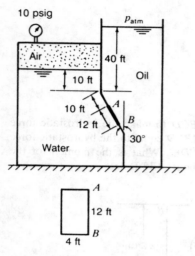

FIGURE P3.38

3.39. Find the resultant force on the top of the submerged surface. Give the complete position of the resultant. Disregard p_{atm}.

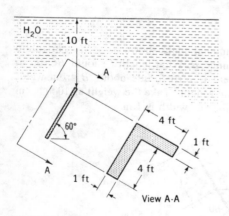

FIGURE P3.39

3.40. An open rectangular tank is partially filled with water. The dimensions are shown.
(a) Determine the force on the bottom of the tank from the water.

(b) Determine the force on the end of the tank from water. Give position also.
(c) Determine the force on the door shown at the side of the tank. Be sure to state position.

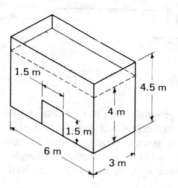

FIGURE P3.40

3.41. A gate AB is hinged at A. When closed, it is inclined at an angle of 60°. It is rectangular and has a length of 0.6 m and a width of 1 m. There is water on both sides of the gate. Furthermore, compressed air exerts a pressure of 20 kPa gage on the surface of the water on the left side of the gate, while the water on the right side is exposed to atmospheric pressure. What is the moment about the hinge A exerted by the water on the gate? *Hint:* With a little thought, you can greatly shorten the solution to the problem.

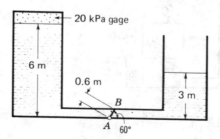

FIGURE P3.41

3.42. In Prob. 3.41, a 1.2-m layer of oil, having specific gravity of 0.8, is added to the top of the water on the right side of the gate. What is the total moment about A from the water on the gate? Hint of Prob. 3.41 applies here.

3.43. Find the resultant force from all fluids acting on the door. Specific gravity of the oil is 0.8.

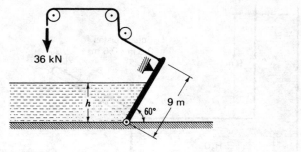

FIGURE P3.45

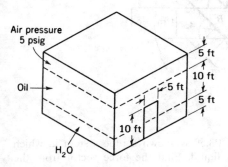

FIGURE P3.43

3.44. Determine the force and its position from fluids acting on the door in Fig. P3.44.

3.45. At what height h will the water cause the door to rotate clockwise (Fig. P3.45)? The door is 3 m wide. Neglect friction and the weight of the door.

3.46. Find F_R on door AB from inside and outside fluids. Give distance d below B for the position of F_R. See Fig. P3.46.

3.47. At what pressure in the air tank will the square piston be in equilibrium if one neglects friction and leakage? See Fig. P3.47.

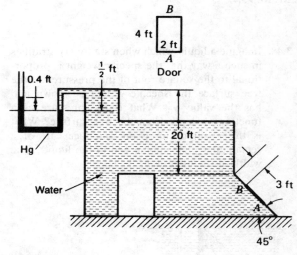

FIGURE P3.46

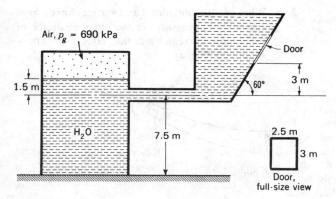

FIGURE P3.44

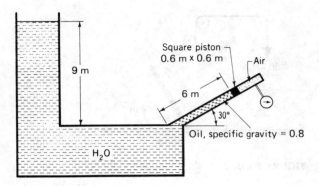

FIGURE P3.47

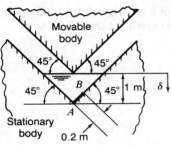

FIGURE P3.49

3.48. Imagine a liquid which when stationary stratifies in such a way that the specific weight is proportional to the square root of the pressure. At the free surface the specific weight is known and has the value γ_0. What is the pressure as a function of depth from the free surface? What is the resultant force on one face AB of a rectangular plate submerged in the liquid? The width of the plate is b.

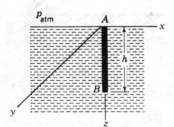

FIGURE P3.48

3.49. A trough of unit length contains water. A solid of identical shape is directly touching the free surface. It is moved directly downward a distance δ relative to the ground. What is the force on door AB of unit width as a function of δ? What happens when $\delta \to 1$ m? Consider only the gravitational force from water. *Hint:* What is the area of a parallelogram?

*3.50. In Fig. P3.50 is shown a pipe system in which flows a liquid. Find the force vector from the atmospheric pressure of 101,325 Pa on the *outside* surface of the pipe system.

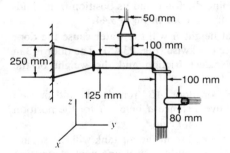

FIGURE P3.50

*3.51. What are the horizontal and vertical forces from atmospheric pressure on the outside surface of the elbow disregarding the effects of atmosphere on flanges.

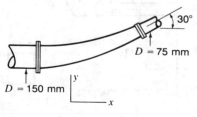

FIGURE P3.51

*3.52. A thin-walled reducing elbow of Prob. 3.51 is shown in Fig. P3.52 inside a pressure tank. Find the horizontal force on the outside surface of the reducing elbow. Neglect pressure of flanges.

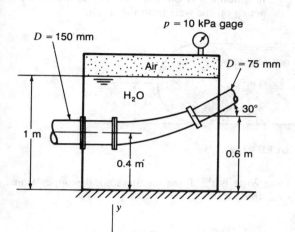

FIGURE P3.52

3.53. Show that the hydrostatic vertical force on a curved submerged surface acts at the center of gravity of the column of liquid above the curved surface and extends to the free surface. *Hint:* Start with Fig. 3.18 and replace $dz\,dA_z$ by dv. Use V as the volume of the prismatic column.

3.54. Determine the magnitude of the resultant force acting on the spherical surface and explain why the line of action goes through the center O.

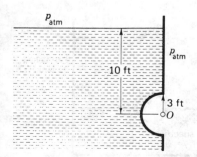

FIGURE P3.54

3.55. What is the resultant force from fluids acting on the door AB which is a quarter circle? The width of the door is 1.3 m. Give the elevation above the ground of the center of pressure.

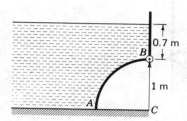

FIGURE P3.55

3.56. What is the horizontal force on the semi-spherical door AB from all fluids inside and out? The specific gravity of oil is 0.8.

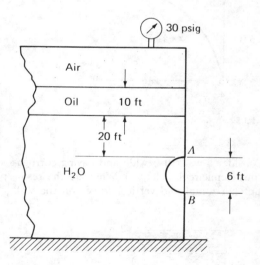

FIGURE P3.56

3.57. Find the horizontal force from the fluids acting on the plug in Fig. P3.57.

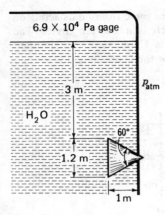

FIGURE P3.57

3.58. A parabolic gate AB is hinged at A and latched at B. If the gate is 10 ft wide, determine the force components on the gate from the water.

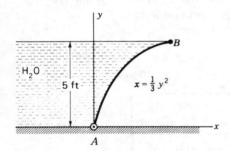

FIGURE P3.58

*3.59. Consider a wall 10 ft wide and having corrugations (semicircular shapes). What are the resultant horizontal and vertical forces on the wall

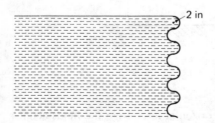

FIGURE P3.59

from the air and water? Give the result per unit width of the wall and for n corrugations.

3.60. A cylindrical control weir is shown. It has a diameter of 3 m and a length of 6 m. Give the magnitude and direction of the resultant force acting on the weir from the fluids.

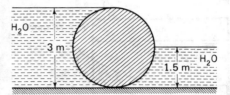

FIGURE P3.60

3.61. What is the force on the conical stopper from the water?

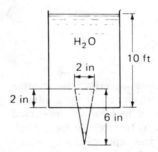

FIGURE P3.61

3.62. What is the vertical force on the sphere if both sections of the tank are completely sealed from each other?

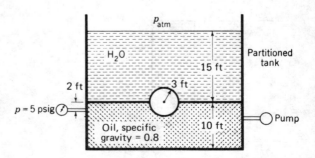

FIGURE P3.62

3.63. The tank is divided into two independent chambers. Air pressure is present in both sections. A manometer measures the difference between these pressures. A sphere of wood (specific gravity is 0.6) is fastened into the wall as shown. (*a*) Compute the vertical force on the sphere. (*b*) Compute the magnitude (only) of the resultant horizontal force on the sphere from the fluids.

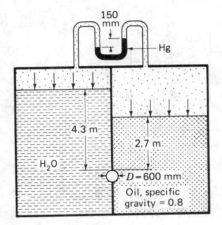

FIGURE P3.63

3.64. A 500-N tank *A* is full of water and is connected to open tank *B* through a pipe. If the tank *A* wall is 2 mm in thickness, determine the

tensile stresses τ_{xx} and τ_{yy} from air and water in the tank wall at a point at $y = 3$ m. Also for 40 bolts at the base, compute the force per bolt holding the end plate of the tank. *Hint:* For the stresses, consider two free-body diagrams including a half circle of a horizontal unit strip.

3.65. Water fills a spherical tank supported from below with a pressure $p_1 = 300$ kPa gage. Fifty bolts hold the upper half of the tank to the lower half with a force between flanges of 5000 N. What is the force per bolt? Each half of the sphere weighs 2000 N.

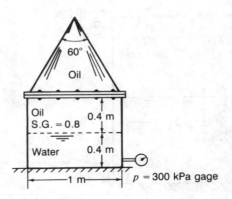

FIGURE P3.65

3.66. An open-ended 60° conical container is bolted to a cylinder. The cylinder contains oil and water with oil extending into the conical container filling the latter. Find the force on each of 30 bolts connecting the cone and cylinder so that there is a force of 6000 N between flanges of the two containers. The volume of a cone is

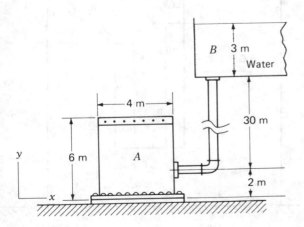

FIGURE P3.64

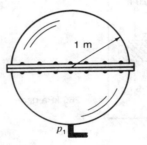

FIGURE P3.66

$\frac{1}{3}Ah$ where A is the area of the base and h is the height. The cone weighs 1000 N and the cylinder weighs 1600 N.

3.67. A tank is hermetically sealed into two compartments by plate AB. A cylinder of diameter 0.3 m protrudes above and below the seal AB and is welded to the seal AB. What is the vertical force on the cylinder?

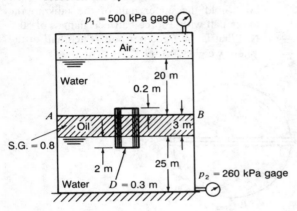

FIGURE P3.67

3.68. Do Prob. 3.67 for when a hemisphere of diameter 0.3 m is added to the top and bottom of the cylinder and $p_2 = 360$ kPa g.

3.69. A tank is separated into two distinct parts by a stiff plate EF. A block A fits in the top part and block B fits in the lower part. If A and B are 3 ft long, find:

(*a*) horizontal force on the blocks from fluids.

(*b*) the total vertical force on the blocks from fluids.

3.70. A tank in Fig. P3.70 is made up of three compartments ①, ②, and ③ separated from each other. Triangle ABC is 3 ft in length and separates the three compartments. Find the net vertical force on ABC from the fluids touching it.

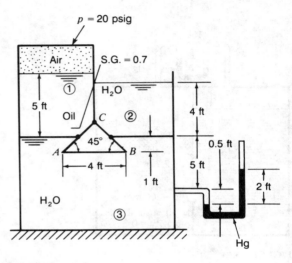

FIGURE P3.70

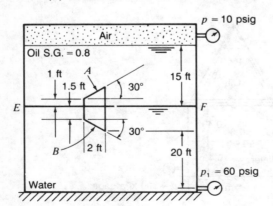

FIGURE P3.69

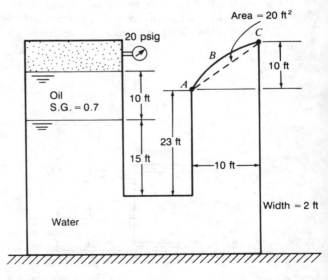

FIGURE P3.71

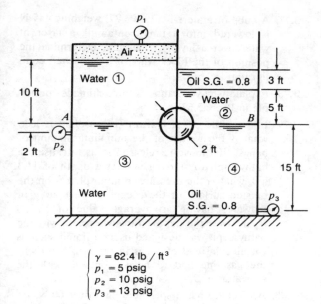

FIGURE P3.72

3.71. A tank containing water and air under pressure is shown in Fig. P3.71. What are the vertical and horizontal forces on *ABC* from water inside and air outside? Note that water completely fills part of the tank on the right and hence wets *ABC*.

3.72. There are four compartments (Fig. P3.72) completely separated from each other. One-quarter of the sphere shown resides in each compartment. Find the:
(*a*) total vertical force from fluids.
(*b*) total horizontal force from fluids.

*3.73. Find the shear force and bending moment on the gate *AB* at *A* in Fig. P3.73. The gate has a width of 1 m. *Hint: ds*(along door) $= \sqrt{dx^2 + dy^2} = [1 + (dy/dx)^2]^{1/2}\, dx$.

3.74. What is the total weight of barge and load (Fig. P3.74)? The barge is 6 m in width.

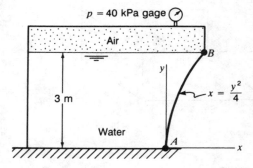

FIGURE P3.73

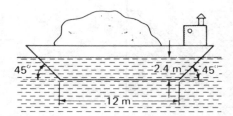

FIGURE P3.74

3.75. A wedge of wood having a specific gravity of 0.6 is forced into water by a 150 lb force. The wedge is 2 ft in width. What is the depth d?

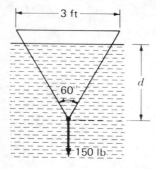

FIGURE P3.75

3.76. A tank is filled to the edge with water. If a cube 600 mm on an edge and weighing 445 N is lowered slowly into the water until it floats, how much water flows over the edge of the tank if no appreciable waves are formed during the action? Neglect effects of adhesion at the edge of the tank.

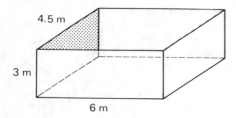

FIGURE P3.76

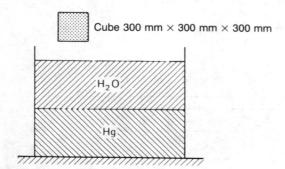

FIGURE P3.77

3.77. A cube of material (Fig. P3.77) weighing 445 N is lowered into a tank containing a layer of water over a layer of mercury. Determine the position of the block when it has reached equilibrium.

3.78. Explain why you cannot use Archimedes principle in Prob. 3.62.

3.79. In Example 3.9 if 0.28 m³ of gasoline is lost, what is the weight of the minimum amount of ballast that must be released so as to cause the bathyscaph to start rising? At a depth of 11.3 km, what is the pressure in atmospheres on the outside surface of the cockpit if we take γ of seawater having an average value of 10,150 N/m³ over the depth? Finally, explain why the bathyscaph was designed using a liquid such as gasoline instead of a gas in the tank, and why the gasoline had to have "contact" with the seawater at B.

3.80. An iceberg has a specific weight of 9000 N/m³ in ocean water, which has a specific weight of 10^4 N/m³. If we observe a volume of 2.8×10^3 m³ of the iceberg protruding above the free surface, what is the volume of the iceberg below the free surface of the ocean?

3.81. A *hydrometer* is a device that uses the principle of buoyancy to determine specific gravity S of a liquid. It is a device weighted by tiny metal spheres to have a total weight W. It has a stem of constant cross section which protrudes through the free surface. It is calibrated by marking the position of the free surface when floating in distilled water ($S = 1$) and by determining its submerged volume V_0. When floated in another liquid, the stem may sit lower or higher at the free surface from this position by distance Δh, as shown to the right in Fig. P3.81.

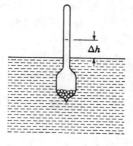

FIGURE P3.81

Show that

$$\Delta h = \frac{V_0}{A_s} \frac{S-1}{S}$$

where A_s is the cross section of the stem and S is the specific gravity of the liquid. We can thus calibrate the stem to read specific gravity directly.

3.82. A rectangular tank of internal width 6 m is partitioned as shown in Fig. P3.82 and contains oil and water. If the specific gravity of oil is 0.82, what must h be? Next, if a 1000-N block of wood is placed in flotation in the oil, what is the rise of the free surface of the water in contact with the air?

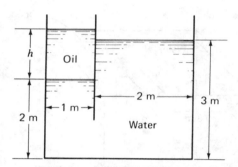

FIGURE P3.82

3.83. A balloon of 2.8×10^3 m^3 is filled with hydrogen having a specific weight of 1.1 N/m^3.
(*a*) What lift is the balloon capable of at the earth's surface if the balloon weighs 1335 N? The temperature is 15°C.
(*b*) What lift is the balloon capable of at 9150 m U.S. Standard Atmosphere, assuming that the volume has increased 5 percent?

3.84. A wooden rod weighing 5 lb is mounted on a hinge below the free surface. The rod is 10 ft

long and uniform in cross section, and the support is 5 ft below the free surface. At what angle α will it come to rest when allowed to drop from a vertical position? The cross section of the rod is $\frac{3}{2}$ in^2 in area.

3.85. A block of material having a volume of 0.028 m^3 and weighing 290 N is allowed to sink in the water. A wooden rod of length 3.3 m and a cross section of 1935 mm^2 is attached to the weight and also to the wall. If the rod weighs 13 N, what will the angle θ be for equilibrium?

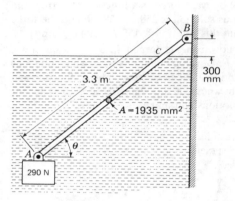

FIGURE P3.85

3.86. An object having the shape of a rectangular parallelepiped is being pushed slowly down an incline on narrow rails into water. The object

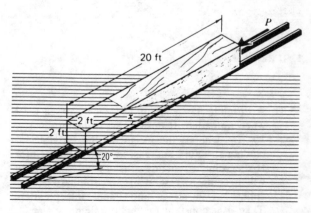

FIGURE P3.86

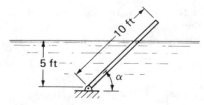

FIGURE P3.84

weighs 4000 lb, and the coefficient of dynamic friction between the object and incline is 0.4. If hydrostatic pressure is assumed to exist all over the submerged surface of the object, express the force P as a function of x, the distance along the bottom surface submerged in the water, to keep the body moving at a constant slow speed along the incline. Begin calculations when water just touches the *top* surface of the object.

3.87. In Prob. 3.86 is there a position x for which there is impending rotation of the object as a result of buoyancy? If so, compute this value of x. The buoyant force as a function of x from the previous solution is $250x - 686$ lb and the force P from this solution is $159.2 - 8.4x$ pounds.

3.88. A hollow cone is forced into the water by a force F. Develop equations from which one may determine e. Neglect the weight of the cone and the thickness of the wall. Be sure to state any assumptions you make.

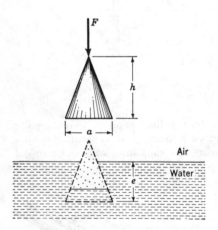

FIGURE P3.88

3.89. A dirigible has a lift of 130,000 lb at sea level when unloaded. If the volume of helium is 3×10^6 ft^3, what is the weight of the dirigible including structure and gases within the dirigible? If the volume remains constant, at what height will it come to rest in a U.S. Standard Atmosphere? Use tables and linear interpolation. Take g as constant for this problem.

3.90. A small balloon has a constant volume of 15 m^3 and has a total weight on earth of 35.5 N. On a planet having $g = 5.02$ m/s^2 and an *isothermal* atmosphere with $\rho = 0.250$ kg/m^3 and $p = 10,000$ Pa at sea level, what is the maximum load capacity at sea level? If released without this load, at what elevation will it come to rest in this atmosphere? Take g as constant for this problem.

3.91. The outside diameter of the pipe is 250 mm. It is submerged in water in the tank. Find the total force on the pipe from the water in the tank.

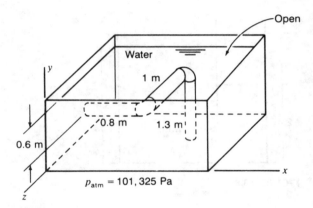

FIGURE P3.91

*3.92. A pipe system goes through a tank of water. The tank is closed on the top with air above it at a pressure $p_1 = 200$ kPa gage. Inside the pipe is a static gas with a uniform pressure $p_2 = 500$ kPa gage.
(*a*) Find the force on the pipe from the static gas on the inside of the pipe.
(*b*) Find the force on the outside surface of the pipe from the water.
Hint: The volume of the frustrum of a cone is

$$\tfrac{1}{3}\left[A_{\text{base}} + A_{\text{top}} + \sqrt{A_{\text{base}} A_{\text{top}}} \right] (\text{height})$$

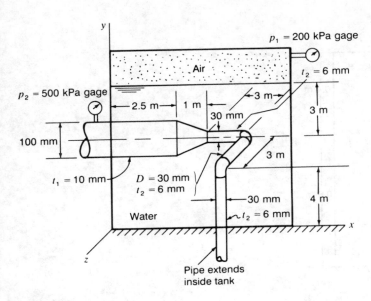

FIGURE P3.92

3.93. Shown is a rectangular tank having a square cross section. A cubic block having dimensions 1 m × 1 m × 1 m and a specific gravity of 0.9 is inserted into the tank. What will then be the

force on the door *A* from all fluids contacting it? The oil has a specific gravity of 0.65. How far below the centroid of the door is the center of pressure?

3.94. A pail open at the bottom and weighing 10 N is slowly made to enter the water open end first until fully submerged. At what depth will the cylinder no longer return to the free surface from buoyant forces? Explain what happens after this elevation has been exceeded. Water is at 20°C. Air is initially at 20°C. The metal thick-

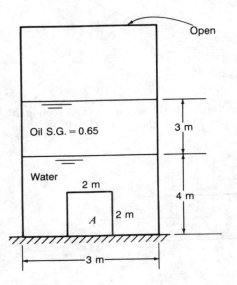

FIGURE P3.93

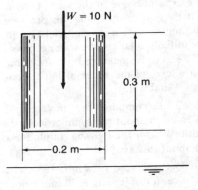

FIGURE P3.94

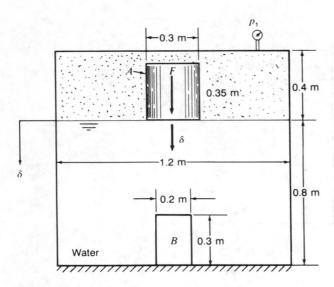

FIGURE P3.95

ness of the cylinder is 2 mm. Assume air compresses isothermally in the cylinder. Account for buoyancy on the metal.

3.95. A cylindrical tank of diameter 1.2 m contains water, air, and a solid cylinder A which initially just touches the free surface. Find the force F to move the cylinder a distance δ downward into the water. Keep δ small enough so that A does not get completely submerged in the water. Then find the force P of the door B as a function of δ. The pressure of the air initially is $p_1 = 200,000$ Pa gage. Any change in pressure of the air during this action is adiabatic. The air temperature initially is 60°C. The water temperature is 60°C. δ is to be measured *relative to the ground* from a level of the water at initial contact between A and water.

3.96. In Example 3.12, compute the metacentric height for a rotation about the symmetrical axis along its width. What is the righting couple for a 10° rotation about this axis?

3.97. A wooden object is placed in water. It weighs 4.5 N, and the center of gravity is 50 mm below the top surface. Is the object stable?

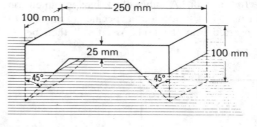

FIGURE P3.97

3.98. A ship weighs 18 MN and has a cross section at the water line as shown in Fig. P3.98. The

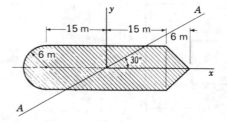

FIGURE P3.98

center of buoyancy is 1.5 m below the free surface, and the center of gravity is 600 mm above the free surface. Compute the metacentric heights for the x and y axes. Also determine the metacentric height for axis AA at an angle of 30° as shown.

3.99. A wooden cylinder of length 2 ft and diameter 1 in and specific weight 20 lb/ft^3 is fastened to a cylinder of metal having a diameter of $\frac{1}{2}$ in, length of 1 ft, and specific weight of 200 lb/ft^3. Is this object stable in water for the orientation shown in Fig. P3.99?

3.100. In Prob. 3.99, is there a specific gravity for which the object attains neutral stability? If so, compute this specific gravity.

3.101. A wooden block having a specific gravity of 0.7 is floating in water. A light rod at the center of the block supports a cylinder A whose weight is 20 N. At what height h will there be neutral stability?

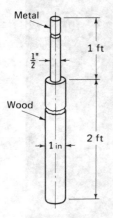

FIGURE P3.99

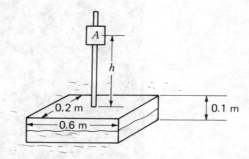

FIGURE P3.101

Shown is part of towing tank facility of the Naval Ship Research and Development Center at Carderock, Maryland. The basin comprises three adjoining sections: (1) A deep water section 22 feet deep, 50.20 feet wide, and 889 feet long. (2) A shallow water section 10 feet deep, 50.96 feet wide, and 303 feet long. The depth of water can be varied. A 32 feet by 5 feet fitting out dock is located here. Above photograph is taken for the shallow water section. (3) A turning basin in the form of a J in which self-propelled models can be allowed to maneuver. The carriage speed can move models up to speeds of 18 knots. In subsequent chapters, towing tanks will be referred to on a number of occasions.

CHAPTER
4

FOUNDATIONS OF FLOW ANALYSIS

4.1 THE VELOCITY FIELD

In particle and rigid-body dynamics we are able to describe the motion of each body in a separate and discrete manner. For instance, the velocity of the nth particle of an aggregate of particles moving in space can be specified by the scalar equations

$$(V_x)_n = f_n(t)$$
$$(V_y)_n = g_n(t) \qquad (4.1)$$
$$(V_z)_n = h_n(t)$$

Note that the identification of a particle is easily facilitated with the use of a subscript. However, in a deformable continuum such as a fluid, there are for practical purposes an infinite number of particles whose motions are to be described, which makes this approach unmanageable; so we employ spatial coordinates to help identify particles in a flow. The velocity of all particles in a flow can therefore be expressed in the following manner:

$$V_x = f(x, y, z, t)$$
$$V_y = g(x, y, z, t) \qquad (4.2)$$
$$V_z = h(x, y, z, t)$$

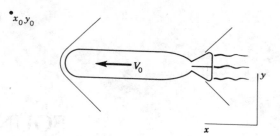

FIGURE 4.1
Unsteady-flow field relative to xy.

Specifying coordinates xyz and the time t and using these values in functions f, g, and h in Eqs. (4.2), we can directly determine the velocity components of a fluid element at the particular position and time specified. The spatial coordinates thus take the place of the subscript n of the discrete systems studied in mechanics. This is called the *field approach*. If properties and flow characteristics at each position in space remain invariant with time, the flow is called *steady flow*. A time-dependent flow, on the other hand, is designated *unsteady flow*. The steady-flow velocity field would then be given as

$$V_x = f(x, y, z)$$
$$V_y = g(x, y, z) \qquad\qquad (4.3)$$
$$V_z = h(x, y, z)$$

It is often the case that a steady flow may be derived from an unsteady-flow field by simply changing the space reference. To illustrate this, examine the flow pattern created by a torpedo moving near the free surface through initially undisturbed water at constant speed V_0 relative to the stationary reference xy, as shown in Fig. 4.1. It can be seen that this is an unsteady-flow field, as seen from xyz. Thus, the velocity at position x_0, y_0 in the field, for instance, will at one instant be zero and later, owing to the oncoming waves and wake of the torpedo, will be subjected to a complicated time variation. To establish a steady-flow field, we now consider a reference $\xi\eta$ *fixed* to the torpedo. The flow field relative to such a moving reference is shown in Fig. 4.2. The velocity at position ξ_0, η_0 clearly must be constant with time, since it is fixed relative to an unchanging flow pattern. Note that the water upstream of the torpedo has now a velocity $-V_0$ relative to the $\xi\eta$ axes, and you can see that this transition from

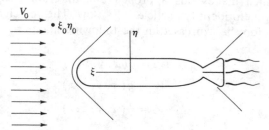

FIGURE 4.2
Steady-flow field relative to $\xi\eta$.

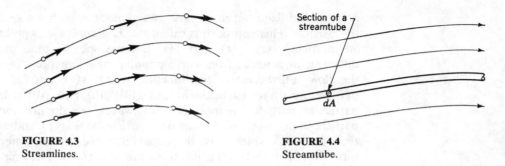

FIGURE 4.3
Streamlines.

FIGURE 4.4
Streamtube.

unsteady to steady flow could have been accomplished by superposing a velocity $-V_0$ on the entire flow field of Fig. 4.1 to arrive at the steady field of Fig. 4.2. *This may be done any time a body is moving with constant speed through an initially undisturbed fluid.*

Flows are usually depicted graphically with the aid of *streamlines*. These lines are drawn so as to be always tangent to the velocity vectors of the fluid particles in a flow. This is illustrated in Fig. 4.3. For a steady flow the orientation of the streamlines will be fixed. Fluid particles, in this case, will proceed along paths coincident with the streamlines. In unsteady flow, however, an indicated streamline pattern yields only an instantaneous flow representation, and for such flow there will no longer be a simple correspondence between path lines and streamlines.

Streamlines proceeding through the periphery of an infinitesimal area at some time t will form a tube, which is useful in discussions of fluid phenomena. This is called the *streamtube*, which is illustrated in Fig. 4.4. From considerations of the definition of the streamline, it is obvious that there can be no flow through the lateral surface of the streamtube. In short, the streamtube acts like an impervious container of zero wall thickness and infinitesimal cross section. A continuum of adjacent streamtubes arranged to form a finite cross section is often called a *bundle of streamtubes*.

4.2 TWO VIEWPOINTS

In the preceding section we discussed various general aspects of the velocity field $\mathbf{V}(x, y, z, t)$. Two procedures will now be set forth by which the field may be utilized in computations involving the motion of fluid particles making up the flow. For instance, by stipulating fixed coordinates x_1, y_1, z_1, in the velocity-field functions and letting time pass, we can express the velocity of particles moving by this position at any time. Mathematically, this may be given by the formulation $\mathbf{V}(x_1, y_1, z_1, t)$. Hence, by this technique we express, at a fixed position in space, the velocities of a continuous "string" of fluid particles moving by this position. This viewpoint is sometimes called the *Eulerian* viewpoint.

On the other hand, to study "any one" particle in the flow one must "follow the particle." This means that x, y, z in the expression $\mathbf{V}(x, y, z, t)$

must not be fixed but must vary continuously in such a way as always to locate the particle. This approach is called the *Lagrangian* viewpoint. For any *particular* particle, $x(t)$, $y(t)$, and $z(t)$ become specific time functions which are different, in general, from corresponding time functions for other particles in the flow. Furthermore, the functions $x(t)$, $y(t)$, and $z(t)$ for a particular particle must have particular values $x(0)$, $y(0)$, and $z(0)$ at time $t = 0$ for that particular particle. In most cases, however, we do *not* identify a particular particle in our work, so for any one particle, $x(t)$, $y(t)$, and $z(t)$ are *unspecified* time functions which have the capability nevertheless, when the form of the time functions and initial positions are chosen, of focusing on any particular particle. Thus we say in this case that

$$V_x = f[x(t), y(t), z(t), t]$$

$$V_y = g[x(t), y(t), z(t), t] \qquad (4.4)$$

$$V_z = h[x(t), y(t), z(t), t]$$

In fluid dynamics there is ample occasion to employ both techniques.[1]

These considerations do not depend on whether the field is steady or unsteady and should not be confused with the conclusions of the previous section. You may note that the Eulerian viewpoint was utilized in that section in both the steady and unsteady flows about the torpedo.

4.3 ACCELERATION OF A FLOW PARTICLE

We will soon use Newton's law for any one particle in a flow, and we will need the time rate of change of velocity of any one particle in a flow. In using the velocity field we will then have to use the Lagrangian viewpoint. Thus, noting that x, y, z are functions of time, we may establish the acceleration field by employing the chain rule of differentiation in the following way:

$$\mathbf{a} = \frac{d}{dt}\mathbf{V}(x, y, z, t) = \left(\frac{\partial \mathbf{V}}{\partial x}\frac{dx}{dt} + \frac{\partial \mathbf{V}}{\partial y}\frac{dy}{dt} + \frac{\partial \mathbf{V}}{\partial z}\frac{dz}{dt}\right) + \left(\frac{\partial \mathbf{V}}{\partial t}\right) \quad (4.5)$$

Since x, y, z are coordinates of any one particle, it is clear that dx/dt, dy/dt,

---

[1] A simple-minded way of thinking of the two viewpoints is to consider a golf tournament where the players are the "particles." If you station yourself as the observer at any particular tee in order to observe the various players coming by this location, you are using the Eulerian viewpoint. On the other hand, if you select your favorite player and move around the course with him/her for purposes of observation, you are using the Lagrangian viewpoint.

and dz/dt must then be the scalar velocity components of any one particle and can be denoted as V_x, V_y, and V_z, respectively. Hence

$$\mathbf{a} = \left(V_x \frac{\partial \mathbf{V}}{\partial x} + V_y \frac{\partial \mathbf{V}}{\partial y} + V_z \frac{\partial \mathbf{V}}{\partial z} \right) + \left(\frac{\partial \mathbf{V}}{\partial t} \right) \qquad (4.6)$$

The three scalar equations corresponding to Eq. (4.6) in the three cartesian-coordinate directions are

$$a_x = \left(V_x \frac{\partial V_x}{\partial x} + V_y \frac{\partial V_x}{\partial y} + V_z \frac{\partial V_x}{\partial z} \right) + \left(\frac{\partial V_x}{\partial t} \right)$$

$$a_y = \left(V_x \frac{\partial V_y}{\partial x} + V_y \frac{\partial V_y}{\partial y} + V_z \frac{\partial V_y}{\partial z} \right) + \left(\frac{\partial V_y}{\partial t} \right) \qquad (4.7)$$

$$a_z = \left(V_x \frac{\partial V_z}{\partial x} + V_y \frac{\partial V_z}{\partial y} + V_z \frac{\partial V_z}{\partial z} \right) + \left(\frac{\partial V_z}{\partial t} \right)$$

Now the acceleration **a** of any one particle is given in terms of the velocity field and partial spatial derivatives and the partial time derivative of **V**. But **V** is a function of x, y, z, and t. Hence the acceleration **a** is then given in terms of x, y, z and t and is thus also a field variable.

The acceleration of fluid particles in a flow field may be imagined as the superposition of two effects:

1. In expressions in the first parentheses on the right-hand sides of Eqs. (4.6) and (4.7), the *explicit* time variable t is held constant. Hence, in these expressions at a given time t, the field is assumed to become and remain steady. The particle, under such circumstances, is in the process of changing position in this steady field. It is thus undergoing a change in velocity because the velocity at various positions in this field will, in general, be different at any time t. This time rate of change of velocity due to changing position in the field is aptly called the *acceleration of transport*, or *convective acceleration*.

2. The term within the second parentheses in the acceleration equations does not arise from the change of particle position, but rather from the rate of change of the velocity field itself at the position occupied by the particle at time t. It is sometimes called the *local acceleration*.

The differentiation carried out in Eq. (4.6) is called the *substantial*, or *total*, *derivative*. In order to emphasize the fact that the time derivative is carried out as one follows the particle, the notation D/Dt is often used in place of d/dt. Hence, the substantial derivative of the velocity is given by $D\mathbf{V}/Dt$. The increased complexity over what we experienced in mechanics of discrete particles is the price we pay for having, by necessity, brought in spatial coordinates to identify particles in a deformable continuous medium. It should be understood that the substantial derivative is by no means restricted to the velocity field vector. Thus for any vector field **H** associated with a flow we can say:

$$\frac{D\mathbf{H}}{Dt} = \left(V_x \frac{\partial \mathbf{H}}{\partial x} + V_y \frac{\partial \mathbf{H}}{\partial y} + V_z \frac{\partial \mathbf{H}}{\partial z} \right) + \frac{\partial \mathbf{H}}{\partial t}$$

Note we have in effect two vector fields involved here. There is first the field **H** undergoing the substantial derivative and for any such vector field **H** associated with the flow there is always the fluid velocity field **V** whose components in the above equation facilitate *following any one particle* as one computes the rate of change of **H** for the particle. We have offered several problems with different **H** fields at the end of the chapter.

In many analyses, it is useful to think of a set of streamlines as part of a coordinate system. In such cases the letter s indicates the position of the particle along a particular streamline, and accordingly $\mathbf{V} = \mathbf{V}(s, t)$. Hence, for the acceleration of transport we have $(\partial \mathbf{V}/\partial s)(ds/dt)$, which gives the acceleration that results from the action of the particle's changing position along a streamline. The complete acceleration is then given as

$$\mathbf{a} = V \frac{\partial \mathbf{V}}{\partial s} + \frac{\partial \mathbf{V}}{\partial t} \tag{4.8}$$

Let us consider the case of steady flow, where, as we pointed out earlier, there is a fixed streamline pattern and streamlines are the same as path lines. We can decompose the acceleration of transport vector for such flow into two scalar components by choosing one component a_T tangent to the path and the other component a_N normal to the path in the osculating plane.[2] You will recall from earlier mechanics courses that the acceleration component a_T can be given as

$$a_T = V \frac{dV}{ds} = \frac{1}{2} \frac{dV^2}{ds} \tag{4.9}$$

and, taking the direction *toward* the center of curvature in the osculating plane as positive, that the other component of acceleration a_N can be given as

$$a_N = \frac{V^2}{R} \tag{4.10}$$

where R is the radius of curvature. We will have occasion to use the acceleration components above in the ensuing chapters.

---

[2] The osculating plane at a particular point on a path is the limiting plane formed by the point and two additional points, on the path, ahead and behind as they are brought ever closer to the particular point. See I. H. Shames, *Engineering Mechanics; Statics and Dynamics*, 3d ed., Prentice-Hall, Englewood Cliffs, N.J., Chap. 11.

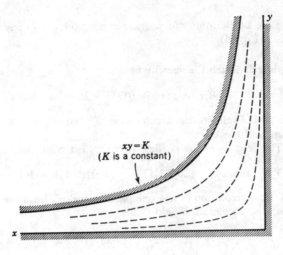

$xy = K$
(K is a constant)

FIGURE 4.5
Two-dimensional flow showing streamlines.

Example 4.1. To illustrate some of the definitions and ideas of the preceding section, examine a simple two-dimensional flow (see Fig. 4.5) with the upper boundary that of a rectangular hyperbola, given by the equation $xy = K$. Assume that the scalar components of the velocity field are known to be

$$V_x = -Ax$$
$$V_y = Ay \qquad A = \text{const} \qquad\qquad (a)$$
$$V_z = 0$$

(Note that the flow is steady.)

The equations of the streamlines will first be ascertained. By definition, they must have the same slope as the velocity vector at all points. Equating these slopes gives

$$\left(\frac{dy}{dx}\right)_{\text{str}} = \frac{V_y}{V_x} = -\frac{y}{x} \qquad\qquad (b)$$

Separating the variables and integrating, we have

$$\ln y = -\ln x + \ln C$$

Hence,
$$xy = C$$

Note that the streamlines form a family of rectangular hyperbolas. The wetted boundaries are part of the family, as is to be expected.

The components of acceleration may now easily be determined. Since this is steady flow, there will be only the acceleration of transport. Employing Eqs. (4.7) under these conditions, we get

$$a_x = (-Ax)(-A) + (Ay)(0) + (0)(0) = A^2x$$
$$a_y = (-Ax)(0) + (Ay)(A) + (0)(0) = A^2y \qquad\qquad (c)$$
$$a_z = 0$$

Hence
$$\mathbf{a} = A^2x\mathbf{i} + A^2y\mathbf{j} \qquad\qquad (d)$$

To give the acceleration of a particle at position $x'y'$ at any time, merely substitute x', y' into Eq. (d).

Example 4.2. Given the velocity field

$$\mathbf{V}(x, y, z, t) = 10x^2\mathbf{i} - 20yx\mathbf{j} + 100t\mathbf{k} \quad \text{m/s}$$

determine the velocity and acceleration of a particle at position $x = 1$ m, $y = 2$ m, $z = 5$ m, and $t = 0.1$ s.

The velocity of the particle is readily established as

$$\mathbf{V} = (10)(1)\mathbf{i} - (20)(2)(1)\mathbf{j} + (100)(0.1)\mathbf{k} = 10\mathbf{i} - 40\mathbf{j} + 10\mathbf{k} \quad \text{m/s}$$

To get the acceleration of any one particle, we must use the Lagrange viewpoint to establish the acceleration field. Thus

$$\mathbf{a}(x, y, z, t) = \left(V_x \frac{\partial \mathbf{V}}{\partial x} + V_y \frac{\partial \mathbf{V}}{\partial y} + V_z \frac{\partial \mathbf{V}}{\partial z} \right) + \left(\frac{\partial \mathbf{V}}{\partial t} \right)$$

$$= \left[(10x^2)(20x\mathbf{i} - 20y\mathbf{j}) + (-20yx)(-20x\mathbf{j}) \right] + 100\mathbf{k}$$

$$= 200x^3\mathbf{i} + (-200x^2y + 400yx^2)\mathbf{j} + 100\mathbf{k} \quad \text{m/s}^2$$

For the particle of interest, the acceleration is

$$\mathbf{a} = (200)(1^3)\mathbf{i} + \left[-200(1^2)(2) + 400(2)(1^2) \right]\mathbf{j} + 100\mathbf{k}$$

$$= 200\mathbf{i} + 400\mathbf{j} + 100\mathbf{k} \quad \text{m/s}^2$$

4.4   IRROTATIONAL FLOW

Earlier we presented the velocity flow $\mathbf{V}(x, y, z, t)$, permitting us to give the velocity of a particle of fluid anywhere in the flow field. We learned in physics that it is the *relative movement* between *adjacent* atoms and molecules which is related to bonding forces between atoms and molecules. Similarly in fluid flow, it is the *relative movement* between *adjacent* flow particles that is related most simply to stresses. We now examine this relative movement.

We wish to point out first that the word "adjacent" will connote for us particles infinitesimally apart. Accordingly, we have shown in Fig. 4.6 two adjacent particles A and B a distance $\mathbf{dr} = dx\,\mathbf{i} + dy\,\mathbf{j} + dz\,\mathbf{k}$ apart at time t. To aid in the consideration of the relative movement between A and B, we have shown in Fig. 4.7 a rectangular parallelepiped for which $\overline{AB}$ is the diagonal. Now if we can effectively describe the deformation and rotation rates of this rectangular parallelepiped, we can in some way give the relative motion between A and B in terms of these rates. To accomplish this, we have shown three additional particles C, D, and E at corners of the rectangular parallelepiped along axes xyz. If we know the relative motion between C and A, between D and A, and between E and A, we then know the deformation and

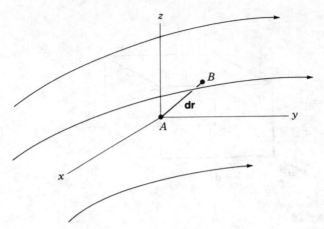

FIGURE 4.6
Adjacent particles A and B.

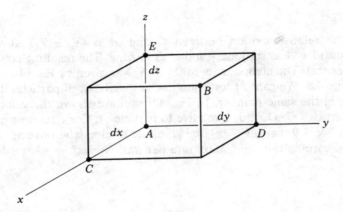

FIGURE 4.7
Adjacent particles along reference axes.

rotation rates of the rectangular parallelepiped, and we can then express the relative motion between B and A in terms of the aforementioned relative motions.

Hence, we start with particle C. The velocity $\mathbf{V}_C$ of this particle can be given in terms of the velocity of particle A, namely $\mathbf{V}_A$, plus an infinitesimal increment, since C is a distance dx apart from A. Thus we have

$$\mathbf{V}_C = \mathbf{V}_A + \left(\frac{\partial \mathbf{V}}{\partial x}\right) dx$$

$$\therefore (\mathbf{V}_C - \mathbf{V}_A) = \left(\frac{\partial V_x}{\partial x}\right) dx\, \mathbf{i} + \left(\frac{\partial V_y}{\partial x}\right) dx\, \mathbf{j} + \left(\frac{\partial V_z}{\partial x}\right) dx\, \mathbf{k}$$

(4.11)

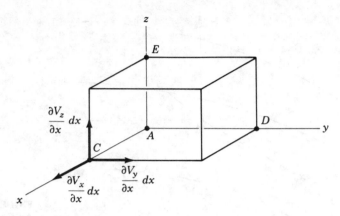

FIGURE 4.8
Components of $(\mathbf{V}_C - \mathbf{V}_A)$.

The relative motion between C and A is $(\mathbf{V}_C - \mathbf{V}_A)$. It will be simplest to consider A as stationary and C as moving. The resulting conclusions will still be general. The components of $(\mathbf{V}_C - \mathbf{V}_A)$ as given by Eq. (4.11) are then shown in Fig. 4.8. We can set forth motion, respectively, of particles D and E relative to A in the same manner. In Fig. 4.9 we have shown the velocity components for particles C, D, and E relative to particle A. Consider now particle C. It is clear in Fig. 4.9 that $(\partial V_x/\partial x)\, dx$ is the rate of elongation of line segment AC. And if we express this elongation rate per *unit original length*, we have simply $\partial V_x/\partial x$.

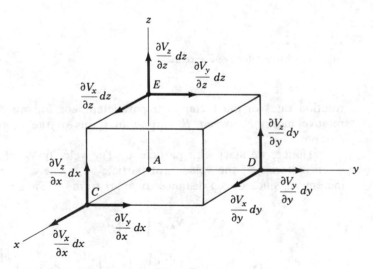

FIGURE 4.9
Relative velocity components for adjacent particles C, D, and E.

But from your course in strength of materials you will recall that the elongation of an infinitesimal line segment in the x direction per unit original length is the normal strain ϵ_{xx}. Thus we may conclude that

$$\frac{\partial V_x}{\partial x} = \dot{\epsilon}_{xx}$$

where the dot represents a time rate of change. Similarly we can say that

$$\frac{\partial V_y}{\partial y} = \dot{\epsilon}_{yy}$$

$$\frac{\partial V_z}{\partial z} = \dot{\epsilon}_{zz}$$

Thus we have depicted the time rates of *elongation* per unit original length (normal strain rates) of the sides of the rectangular parallelepiped. Next, we investigate the rate of *angular* change of the sides of the rectangular parallelepiped. Note in this regard that the velocity $(\partial V_y/\partial x)\,dx$ divided by dx is the angular velocity of AC about the z axis. Similarly at D, $(-\partial V_x/\partial y)\,dy$ divided by dy is the angular velocity of AD about the z axis. We can make two conclusions at this juncture:

1. The *average* rate of rotation about the z axis of the orthogonal line segments AC and AD is

$$\frac{1}{2}\left(\frac{\partial V_y}{\partial x} - \frac{\partial V_x}{\partial y} \right) \tag{4.12}$$

2. *The rate of change of the angle CAD* (a right angle at time t) becomes

$$\frac{\partial V_y}{\partial x} + \frac{\partial V_x}{\partial y} \tag{4.13}$$

The result (2), you may recall from strength of materials, is the time *rate of change* of the *shear angle* γ_{xy} so that

$$\dot{\gamma}_{xy} = \dot{\gamma}_{yx} = \left(\frac{\partial V_y}{\partial x} + \frac{\partial V_x}{\partial y} \right)$$

Similarly,

$$\dot{\gamma}_{xz} = \dot{\gamma}_{zx} = \left(\frac{\partial V_x}{\partial z} + \frac{\partial V_z}{\partial x} \right)$$

$$\dot{\gamma}_{yz} = \dot{\gamma}_{zy} = \left(\frac{\partial V_y}{\partial z} + \frac{\partial V_z}{\partial y} \right)$$

Accordingly, we have available to describe the deformation rate of the rectangular parallelepiped the strain rate terms which we now set forth as follows[3]:

$$
\begin{bmatrix}
\dot{\epsilon}_{xx} & \dfrac{\dot{\gamma}_{xy}}{2} & \dfrac{\dot{\gamma}_{xz}}{2} \\[2ex]
\dfrac{\dot{\gamma}_{yx}}{2} & \dot{\epsilon}_{yy} & \dfrac{\dot{\gamma}_{yz}}{2} \\[2ex]
\dfrac{\dot{\gamma}_{zx}}{2} & \dfrac{\dot{\gamma}_{zy}}{2} & \dot{\epsilon}_{zz}
\end{bmatrix} = \text{strain rate tensor} \tag{4.14}
$$

Now experience from solid mechanics and intuition indicate that it is the strain rate tensor part of relative motion that is most simply related to the stress tensor.

We have thus far described two kinds of relative movement between the adjacent particles along coordinate axes. The normal strain rates give the rate of stretching or shrinking of the sides of the associated rectangular parallelepiped, while the shear-strain rates give rate of change of angularity of the edges of the rectangular parallelepiped. What's left of the relative movement must then be rigid-body *rotation*.[4] Thus, the expression

$$
\frac{1}{2}\left(\frac{\partial V_y}{\partial x} - \frac{\partial V_x}{\partial y} \right)
$$

is actually more than just the average rotation of line segments dx and dy about the z axis—it represents for a deformable medium what may be considered as

---

[3]A note to the advanced reader: By using $\gamma/2$ instead of γ, you may have learned in strength of materials that the nine strain terms without dots form a symmetric second-order tensor. Taking time derivatives of each quantity and thereby forming an array of strain rates does not in any way alter the tensor character of the terms. The strain rate components for a reference $x'y'z'$ are thus found in terms of the strain rate components for xyz using the same transformation formula set forth for stress in Chap. 2. For instance, the normal strain rate in the x' direction, namely $\dot{\epsilon}_{x'x'}$, giving the time rate of change per unit length of a line segment in the x' direction is

$$
\dot{\epsilon}_{x'x'} = \dot{\epsilon}_{xx} a_{x'x}^2 \quad + \frac{\dot{\gamma}_{xy}}{2} a_{x'x} a_{x'y} + \frac{\dot{\gamma}_{xz}}{2} a_{x'x} a_{x'z} +
$$

$$
\frac{\dot{\gamma}_{yx}}{2} a_{x'y} a_{x'x} + \dot{\epsilon}_{yy} a_{x'y}^2 \quad + \frac{\dot{\gamma}_{yz}}{2} a_{x'y} a_{x'z} +
$$

$$
\frac{\dot{\gamma}_{zx}}{2} a_{x'z} a_{x'x} + \frac{\dot{\gamma}_{zy}}{2} a_{x'z} a_{x'y} + \dot{\epsilon}_{zz} a_{x'z}^2
$$

[4]There is clearly no relative motion between adjacent particles for rigid-body *translation*.

the rigid-body angular velocity ω_z about the z axis.[5] This is,

$$\omega_z = \frac{1}{2}\left(\frac{\partial V_y}{\partial x} - \frac{\partial V_x}{\partial y}\right) \tag{4.15}$$

Similarly for the other axes we have, by permuting indices,

$$\omega_x = \frac{1}{2}\left(\frac{\partial V_z}{\partial y} - \frac{\partial V_y}{\partial z}\right) \tag{4.16}$$

$$\omega_y = \frac{1}{2}\left(\frac{\partial V_x}{\partial z} - \frac{\partial V_z}{\partial x}\right) \tag{4.17}$$

$$\therefore \boldsymbol{\omega} = \frac{1}{2}\left(\frac{\partial V_z}{\partial y} - \frac{\partial V_y}{\partial z}\right)\mathbf{i} + \frac{1}{2}\left(\frac{\partial V_x}{\partial z} - \frac{\partial V_z}{\partial x}\right)\mathbf{j} + \frac{1}{2}\left(\frac{\partial V_y}{\partial x} - \frac{\partial V_x}{\partial y}\right)\mathbf{k} \tag{4.18}$$

Had we used a different coordinate system, we would have arrived at formulations which have a different form from Eqs. (4.15) to (4.18), but they would all pertain to the angular motion of fluid elements. Since the angular motion of fluid elements is a physical action not dependent on man-made coordinate systems, we have devised a vector operator called the *curl*[6] which when operating on a vector field **V** portrays twice the angular velocity. Thus Eq. (4.18) becomes

$$\boldsymbol{\omega} = \tfrac{1}{2}(\mathbf{curl\,V}) \equiv \tfrac{1}{2}\boldsymbol{\nabla} \times \mathbf{V} \tag{4.19}$$

Note that Eq. (4.19) alludes to no particular coordinate system. Like the divergence operator and the gradient operator, the curl operator takes on a particular form when carried out in a particular coordinate system.[7] For

---

[5]The expression given by Eq. (4.15) is the *average* angular velocity of two orthogonal vanishingly small line segments dx and dy about the z axis. One can show that it is *also* the average angular velocity about the z axis of *all line segments* in the vanishingly small region dv. The "rigid body" interpretation obtains from the conclusion that if the fluid element in dv were imagined to become frozen at time t with the surrounding fluid made to simultaneously disappear, the frozen element would have the above angular velocity ω_z about the z axis at time t.

[6]The mathematical definition of the curl operator is given as

$$\mathbf{curl\,B} = -\lim_{\Delta V \to 0}\left[\frac{1}{\Delta V}\iint_S \mathbf{B} \times \mathbf{dA}\right]$$

where ΔV is any volume in space and S is the surface enclosing the volume.

[7]It is to be pointed out that there are straightforward general methods for forming the various vector operators for orthogonal coordinate systems. These may be found in mathematics books dealing with vector analysis.

instance, for cartesian coordinates we see from Eq. (4.18) that

$$\text{curl } \mathbf{A} \equiv \nabla \times \mathbf{A} = \left(\frac{\partial A_z}{\partial y} - \frac{\partial A_y}{\partial z} \right)\mathbf{i} + \left(\frac{\partial A_x}{\partial z} - \frac{\partial A_z}{\partial x} \right)\mathbf{j} + \left(\frac{\partial A_y}{\partial x} - \frac{\partial A_x}{\partial y} \right)\mathbf{k}$$

(4.20)

We will not at this time evaluate the curl operator on other coordinate systems. It should be pointed out the the curl can be used on any continuous vector field and the physical interpretation of the resulting curl vector so formed will depend on the particular field operated on. The physical picture of rotation of an element is thus restricted to the curl of the velocity field, but understanding this particular case will help you interpret the curl of other fields.

At this time, we define *irrotational* flows as those for which $\boldsymbol{\omega} = \mathbf{0}$ at each point in the flow. *Rotational* flows are those where $\boldsymbol{\omega} \neq \mathbf{0}$ are points in the flow. For irrotational flow, we require that

$$\begin{aligned} \frac{\partial V_z}{\partial y} - \frac{\partial V_y}{\partial z} &= 0 \\ \frac{\partial V_x}{\partial z} - \frac{\partial V_z}{\partial x} &= 0 \\ \frac{\partial V_y}{\partial x} - \frac{\partial V_x}{\partial y} &= 0 \end{aligned}$$

(4.21)

From Eq. (4.19) it becomes clear that another criterion for irrotationality, and the one we will use, is

$$\boxed{\text{curl } \mathbf{V} = \mathbf{0}}$$

(4.22)

Finally we point out that $2\boldsymbol{\omega}$ is often called the *vorticity* vector.

4.5 RELATION BETWEEN IRROTATIONAL FLOW AND VISCOSITY

We now discuss some conditions under which we can expect rotational and irrotational types of flows. A development of rotation in a fluid particle in an initially irrotational flow would require shear stress to be present on the particle surface. It will be recalled that shear stress on a surface may be evaluated for parallel flows by the relation $\tau = \mu(\partial V/\partial n)$. Thus the shear stress in such flows and in more general flows will depend on the viscosity of the fluid and the manner of spatial variation of velocity (or the so-called velocity gradient) in the region. For fluids of small viscosity, such as air, irrotational flow will then persist in regions where large velocity gradients are not encountered. This may very often be over a great part of the flow. For instance, for an airfoil section moving through initially undisturbed air (Fig. 4.10), the fluid motion relative to the

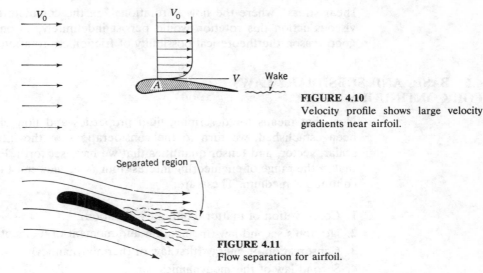

FIGURE 4.10
Velocity profile shows large velocity gradients near airfoil.

FIGURE 4.11
Flow separation for airfoil.

airfoil is that of an irrotational flow over most of the field. However, it is known that no matter how small the viscosity, real fluids "stick" to the surface of a solid body. Thus at point A on the airfoil the fluid velocity must be zero relative to the airfoil, and at a comparatively short distance away it is almost equal to the free-stream velocity V_0. This is illustrated in the velocity profile of the diagram. Thus one sees that there is a thin region adjacent to the boundary where sizable velocity gradients must be present. Here, despite low viscosity, shear stresses of consequential magnitude are present, and the flow becomes rotational. This region adjacent to the solid boundary is called the *boundary layer*. It is fortunate, however, that much of the main flow is very often little affected by the flow conditions in the boundary layer, so that irrotational analysis may be carried out over a large part of the problem.

Another rotational-flow region may be found behind the trailing edge of the airfoil, where flows of different velocities from the upper and lower surfaces come into contact. Here again, large velocity gradients are present and consequently a rotational flow is present over a region behind the airfoil. This region is often called the *wake*.

Finally, we examine a condition called *separation*,[8] where the fluid flow cannot follow the boundary smoothly, as illustrated in Fig. 4.11 in the case of the airfoil at high angle of attack. Inside the separated regions we can again expect rotational flow.

In the flow shown in Fig. 4.11, it may be that the flow downstream of the separation point has regions of relatively small velocity gradients (hence small

---

[8]The boundary layer and the separation process will be discussed at length in Chap. 13.

shear stress), where the flow is rotational. In the complete absence of further viscous action this rotation would persist indefinitely, so one may admit with good reason the theoretical possibility of frictionless rotational flow.

4.6 BASIC AND SUBSIDIARY LAWS FOR CONTINUOUS MEDIA

Now that means for describing fluid properties and flow characteristics have been established, we turn to the considerations of the interrelations among scalar, vector, and tensor quantities that we have set forth. Experience dictates that in the range of engineering interest four *basic laws* must be satisfied for any continuous medium. These are:

1. Conservation of matter (continuity equation)
2. Newton's second law (momentum and moment-of-momentum equations)
3. Conservation of energy (first law of thermodynamics)
4. Second law of thermodynamics

In addition to these general laws, there are numerous *subsidiary laws*, sometimes called *constitutive* relations, that apply to specific types of media. We have already discussed two subsidiary laws, namely, the equation of state for the perfect gas and Newton's viscosity law for certain viscous fluids. Furthermore, for elastic solids there is the well-known Hooke's law, which you studied in strength of materials.

4.7 SYSTEMS AND CONTROL VOLUMES

In employing the basic and subsidiary laws, either one of the following modes of application may be adopted:

1. The activities of each and every given mass must be such as to satisfy the basic laws and the pertinent subsidiary laws.
2. The activities in each and every volume in space must be such that the basic laws and the pertinent subsidiary laws are satisfied.

In the first instance the laws are applied to an identified quantity of matter called the *system*. A system may change shape, position, and thermal condition but must *always entail the same matter*. For example, one may choose the steam

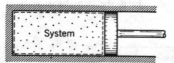

FIGURE 4.12
A system.

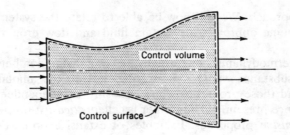

FIGURE 4.13
Control volume for the inside of a nozzle.

in an engine cylinder (Fig. 4.12) after the cutoff[9] to be the system. As the piston moves, the volume of the system changes but there is no change in the quantity and identity of mass.

For the second case, a definite volume, called the *control volume*, is designated in space, and the boundary of this volume is known as the *control surface*.[10] The amount and identity of the matter in the control volume may change with time, but the shape of the control volume is fixed.[11] For instance, in order to study flow through a nozzle, one could choose, as a control volume, the interior of the nozzle as shown in Fig. 4.13.

In rigid-body mechanics it was the system approach (as that time called the free-body diagram) that was invariably used, since it was easy and direct to identify the rigid body, or portions thereof, in the problem and to work with each body as a discrete entity. However, since infinite numbers of particles having complicated motion relative to each other must be dealt with in fluid mechanics, it will often be advantageous to use control volumes in certain computations.

4.8 A RELATION BETWEEN THE SYSTEM APPROACH AND THE CONTROL-VOLUME APPROACH

In Sec. 4.2 we presented two viewpoints involving vector fields associated with a velocity field. These viewpoints allow us either to observe particles moving by a fixed position in space or to follow any one particle. We will now consider these viewpoints for *aggregates* of fluid elements constituting a finite mass where, in following the aggregate as per the Lagrange viewpoint, we are using the system approach. On the other hand, in stationing ourselves and observing in a finite region of space as per the Eulerian viewpoint, we are adopting the control-

---

[9] No further addition of steam takes place after cutoff during the expansion stroke of the steam engine.

[10] In some thermodynamics texts the term *closed system* corresponds to our *system* and *open system* corresponds to our *control volume*.

[11] Some problems can be solved by employing a control volume of variable shape. However, in this text the control volume will always have a fixed shape.

volume approach. We will now be able to relate the system approach and the control-volume approach for certain fluid and flow properties which we next describe.

In thermodynamics one usually makes a distinction between those properties of a substance whose measure depends on the amount of the substance present and those properties whose measure is independent of the amount of the substance present.[12] The former is called *extensive* properties; the latter are called *intensive* properties. Examples of extensive properties are weight, momentum, volume, and energy. Clearly, changing the amount of mass directly changes the measure of these properties, and it is for this reason that we think of extensive properties as directly associated with the material itself. For each extensive variable such as volume V and energy E, one can introduce by *distributive measurements* the corresponding intensive properties, namely, volume per unit mass v and energy per unit mass e, respectively. Thus we have $V = \iiint v\rho \, dv$ and $E = \iiint e\rho \, dv$. Clearly, v and e do not depend on the amount of matter present and are hence the intensive quantities related to the extensive properties V and E by distributive measure. Also, such quantities are termed *specific*, i.e., specific volume and specific energy, and are generally denoted by lowercase letters. Furthermore, such properties as temperature and pressure are by their *mass-independent nature* already in the category of the intensive property. Thus any portion of a metal bar at uniform temperature T_0 also has the same temperature T_0. Nor does the pressure of 1 ft^3 of air in a 10-ft^3 tank at uniform pressure p_0 differ from the pressure of 3 ft^3 of air in the tank. It is with *extensive* properties that we will now relate the system approach with the control-volume approach.

Consider an arbitrary flow field $\mathbf{V}(x, y, z, t)$ as seen from some reference xyz wherein we observe a system of fluid of finite mass at times t and $t + \Delta t$, as shown in a highly idealized manner in Fig. 4.14 by the full line curve and the dashed line curve, respectively. The streamlines correspond to those at time t. In addition to this system, we will consider that the volume in space occupied by the system at time t is a *control volume fixed* in xyz. Hence, at time t our system is identical to the fluid inside our control volume, shown by the full line curve. Let us consider some arbitrary extensive property N of the fluid for the purpose of relating the rate of change of this property for the system with the variations of this property associated with the control volume. The distribution of N per unit mass will be given as η, such that $N = \iiint \eta\rho \, dv$ with dv representing an element of volume.[13]

To do this, we have divided up our system at time $t + \Delta t$ and the system at time t into three regions, as you will note in Fig. 4.14. Region II is common

---

[12]The "amount" of a substance is measured by its mass.

[13]In this text we use v to represent specific volume and dv to represent the volume of a fluid element. Although the same letter is used in both terms, there should be no confusion if the terms are taken in context.

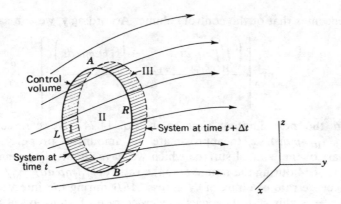

FIGURE 4.14
Simplified view of a moving system.

to the system at both times t and $t + \Delta t$. Let us compute the rate of change of N with respect to time for the system by the following limiting process:

$$\left(\frac{dN}{dt}\right)_{\text{system}} = \frac{DN}{Dt}$$

$$= \lim_{\Delta t \to 0} \left[\frac{\left(\iiint_{\text{III}} \eta\rho \, dv + \iiint_{\text{II}} \eta\rho \, dv \right)_{t+\Delta t} - \left(\iiint_{\text{I}} \eta\rho \, dv + \iiint_{\text{II}} \eta\rho \, dv \right)_{t}}{\Delta t} \right]$$

(4.23)

We may use the rule that the sum of the limits equals the limit of the sums to rearrange the equation above in the following manner:

$$\frac{DN}{Dt} = \lim_{\Delta t \to 0} \left[\frac{\left(\iiint_{\text{II}} \eta\rho \, dv \right)_{t+\Delta t} - \left(\iiint_{\text{II}} \eta\rho \, dv \right)_{t}}{\Delta t} \right]$$

$$+ \lim_{\Delta t \to 0} \left[\frac{\left(\iiint_{\text{III}} \eta\rho \, dv \right)_{t+\Delta t}}{\Delta t} \right] - \lim_{\Delta t \to 0} \left[\frac{\left(\iiint_{\text{I}} \eta\rho \, dv \right)_{t}}{\Delta t} \right] \quad (4.24)$$

Each one of the limiting processes above will now be considered separately. In the first one, we see on noting that $(\iiint_{\text{II}} \eta e \, dv)$ is a function of time that we have here by definition a partial derivative. And as $\Delta t \to 0$, the volume II becomes that of the control volume. Accordingly, we can say that

$$\lim_{\Delta t \to 0} \left[\frac{\left(\iiint_{II} \eta\rho \, dv \right)_{t+\Delta t} - \left(\iiint_{II} \eta\rho \, dv \right)_{t}}{\Delta t} \right] = \frac{\partial}{\partial t} \iiint_{CV} \eta\rho \, dv \qquad (4.25)$$

In the next limiting process of Eq. (4.24), we can consider the integral $(\iiint_{III} \eta\rho \, dv)_{t+\Delta t}$ to approximate the amount of property N that has crossed part of the control surface which we have shown diagrammatically as ARB in Fig. 4.14 during the time Δt, so the ratio $(\iiint_{III} \eta\rho \, dv)_{t+\Delta t}/\Delta t$ approximates the average rate of efflux of N across ARB during the interval Δt. In the limit as $\Delta t \to 0$, this ratio becomes the *exact* rate of *efflux* of N through the control surface. Similarly, in considering the last limiting process of Eq. (4.24), we can consider for flows with continuous-flow characteristics and properties that the integral $(\iiint_{I} \eta\rho \, dv)_{t}$ approximates the amount of N that has passed *into* the control volume during Δt through the remaining portion of the control surface, which we have shown diagrammatically in Fig. 4.14 as ALB. In the limit, the ratio $(\iiint_{I} \eta\rho \, dv)_{t}/\Delta t$ then becomes the *exact* rate of *influx* of N into the control volume at time t. Hence, the last two integrals of Eq. (4.24) give the *net* rate of *efflux* of N from the control volume at time t as

$$\lim_{\Delta t \to 0} \left[\frac{\left(\iiint_{III} \eta\rho \, dv \right)_{t+\Delta t}}{\Delta t} \right]$$

$$- \lim_{\Delta t \to 0} \left[\frac{\left(\iiint_{I} \eta\rho \, dv \right)_{t}}{\Delta t} \right] = \text{Net efflux rate of } N \text{ from } CV \qquad (4.26)$$

We thus see that by these limiting processes, we have equated the rate of change of N for a *system* at time t with the sum of two things:

1. The rate of change of N inside the control volume having the shape of the system at time t [Eq. (4.25)]
2. The rate of efflux of N through the control surface at time t [Eq. (4.26)]

We can express Eq. (4.26) in a more compact, useful form. In this regard, consider Fig. 4.15, where we have a steady-flow velocity field and a portion of a control surface. An area $\mathbf{dA}$ on this surface has been shown. Now this area is also the interface of fluid that is just touching the control surface at the time t shown in the diagram. In Fig. 4.16 we have shown that interface of fluid at time $t + dt$. Note that the interface has moved a distance $V \, dt$ along a direction

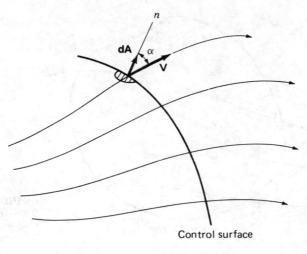

FIGURE 4.15
Interface dA at control surface at time t.

Control surface

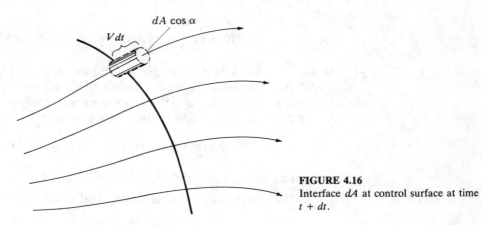

FIGURE 4.16
Interface dA at control surface at time $t + dt$.

tangent to the streamline at that point. The volume of fluid that occupies the region swept out by dA in time dt thus forming a streamtube is

$$dv = (V\,dt)(dA\cos\alpha)$$

Using the definition of the dot product, this becomes

$$dv = \mathbf{V} \cdot \mathbf{dA}\,dt$$

It should be apparent that dv is the volume of fluid that has crossed dA of the control surface in time dt. Multiplying by ρ and dividing by dt then gives the instantaneous rate of mass flow of fluid, $\rho\mathbf{V} \cdot \mathbf{dA}$, leaving the control volume through the indicated area dA.

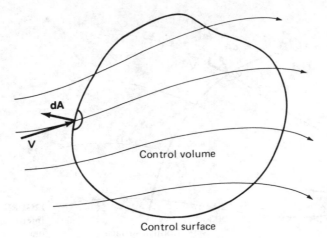

FIGURE 4.17
Control surface showing influx of mass.

The efflux rate of N through the control surface can be given approximately as[14]

$$\text{Efflux rate through CS} \approx \iint\limits_{ARB} \eta(\rho\mathbf{V}\cdot\mathbf{dA})$$

Note next that for fluid *entering* the control volume (see Fig. 4.17) the expression $\rho\mathbf{V}\cdot\mathbf{dA}$ must be negative because of the dot product. Hence, the *influx* rate expression of N through the control surface requires a negative sign to make the result the positive value that we know must exist. Hence, we have

$$\text{Influx rate through CS} \approx -\iint\limits_{ALB} \eta(\rho\mathbf{V}\cdot\mathbf{dA})$$

The approximate *net efflux* rate of N is then

$$\text{Net efflux rate} \approx \text{efflux rate on } ARB - \text{influx rate on } ALB$$

$$= \iint\limits_{ARB} \eta(\rho\mathbf{V}\cdot\mathbf{dA}) - \left[-\iint\limits_{ALB} \eta(\rho\mathbf{V}\cdot\mathbf{dA}) \right]$$

In the limit as $\Delta t \to 0$, the approximations become exact, so we can express the right side of the equation above as $\oiint_{CS}\eta(\rho\mathbf{V}\cdot\mathbf{dA})$, where the integral is a closed surface integral over the entire control surface. Thus Eq. (4.26) can now be given as

$$\text{Net efflux rate of } N \text{ from CV} = \oiint\limits_{CS} \eta(\rho\mathbf{V}\cdot\mathbf{dA}) \qquad (4.27)$$

---

[14]Considering the units of the expression $\eta(\rho\mathbf{V}\cdot\mathbf{dA})$, we get

$$\eta\left(\frac{\text{units of } N}{\text{mass}}\right)\rho\mathbf{V}\cdot\mathbf{dA}\left(\frac{\text{mass}}{\text{unit time}}\right)$$

which is the efflux of N per unit time through $\mathbf{dA}$.

It is to be pointed out that the development of Eq. (4.27) was made for simplicity for a steady-flow velocity field. However, it also holds for unsteady flow, since unsteady effects are of second order for this development. Now using Eqs. (4.27) and (4.25) for the various limiting processes, we can go back to Eq. (4.23) and state that

$$\frac{DN}{Dt} = \oiint_{CS} \eta(\rho\mathbf{V}\cdot\mathbf{dA}) + \frac{\partial}{\partial t}\iiint_{CV}\eta\rho\,dv \tag{4.28}$$

This is called the *Reynolds transport* equation.[15] This equation permits us to change from a system approach to a control-volume approach.

You will note in the development that the velocity field was measured relative to some reference *xyz* and the control volume was *fixed* in this reference. This makes it clear that the fluid velocity **V** in the equation above *is in effect measured relative to the control volume.* Furthermore, you will recall from mechanics that the time rate of change of a vector quantity depends on the reference from which the change is observed. This is an important consideration for us here, since N (and η) can be a vector quantity (as, for example, momentum). Since the system moves in accordance with the velocity field given relative to *xyz* in our development, we see that the time rate of change of N is observed also from the *xyz* reference. Or, a more important conclusion, *the time rate of change of N is in effect observed from the control volume. Thus all velocities and time rates of change of* (4.28) *are those seen from the control volume.* Since we could have used a reference *xyz* having any arbitrary motion in the development above, it means that our control volume can have any motion whatever. Equation (4.28) will then instantaneously be correct if we measure the time derivatives and velocities relative to the control volume, no matter what the motion of the control volume may be. Finally, it can be shown that for an *infinitesimal control volume*, and an *infinitesimal system*, Eq. (4.28) reduces to an identity. This will explain why the system and control-volume equations as developed in subsequent chapters become redundant for infinitesimal considerations.

In Chaps. 5 and 6 we formulate the control-volume approach for the basic laws mentioned earlier by starting in each case with the familiar system formulation and extending it with the aid of the Reynolds transport equation to

---

[15]Although the Reynolds transport equation has been carefully developed from a mathematical point of view, it does have a rather straightforward physical interpretation. We can illustrate this most simply by considering your classroom as the control volume and the system consisting of all the students in the classroom at any time t. Let N be the mass of the system. After the bell has rung for the end of the class period, there will be, at time t, an efflux rate of mass through the doorways (part of the control surface) with a resulting rate of change of mass inside the classroom. The Reynolds transport equation requires that $dN/dt = 0$ at any time t since we are not destroying (physically, that is) students nor are we creating students. Thus, the efflux rate of mass plus the rate of change of mass inside at this time t clearly should be zero. (Would you have it any other way?)

the control-volume formulation. As you do this several times in the next chapter, you will develop a greater physical feel for the Reynolds transport equation, which may seem at this time to be "artificial." Perhaps the realization that all human efforts to explain nature analytically are artificial may be of some comfort. Two additional "artificialities" will now be presented to permit us to use the basic laws, soon to be developed, with greater effects.

4.9 ONE- AND TWO-DIMENSIONAL FLOWS

In every analysis a hypothetical substance or process is set forth which lends itself to mathematical treatment while still yielding results of practical value. We have already discussed the continuum concept. Now, simplified flows are set forth, which, when used with discretion, will permit the use of highly developed theory on problems of engineering interest.

One-dimensional flow is a simplification where all properties and flow characteristics are assumed to be expressible as functions of *one* space coordinate and *time*. The position is usually the location along some path or conduit. For instance, a one-dimensional flow in the pipe shown in Fig. 4.18 would require that the velocity, pressure, etc., be constant over any given cross section at any given time, and vary only with s at this time.

In reality, flow in pipes and conduits is never truly one dimensional, since the velocity will vary over the cross section. Shown in Fig. 4.19 are the respective velocity profiles of a truly one-dimensional flow and that of an actual case. Nevertheless, if the departure is not too great or if average effects at a cross section are of interest, one-dimensional flow may be assumed to exist. For instance, in pipes and ducts this assumption is often acceptable where

1. Variation of cross section of the container is not too excessive.
2. Curvature of the streamlines is not excessive.
3. Velocity profile is known not to change appreciably along the duct.

Two-dimensional flow is distinguished by the condition that all properties and flow characteristics are functions of two cartesian coordinates, say, x, y, and time, and hence do not change along the z direction at a given instant. All planes normal to the z direction will, at the given instant, have the same

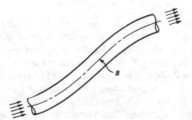

FIGURE 4.18
One-dimensional (1-D) flow.

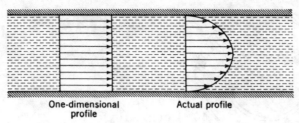

FIGURE 4.19
Comparison of 1-D flow and actual flow.

One-dimensional profile

Actual profile

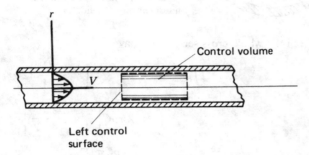

Control volume

Left control surface

FIGURE 4.20
Steady viscous flow in a pipe.

streamline pattern. The flow past an airfoil of infinite aspect ratio[16] or the flow over a dam of infinite length and uniform cross section are mathematical examples of two-dimensional flows. Actually, in a real problem a two-dimensional flow is assumed over most of the airfoil or dam, and "end corrections" are made to modify the results properly.

Example 4.3. Consider a viscous, steady flow through a pipe (Fig. 4.20). We will learn in Chap. 10 that the velocity profile forms a paraboloid about the pipe centerline, given as

$$V = -C\left(r^2 - \frac{D^2}{4}\right) \quad \text{m/s} \qquad (a)$$

where C is a constant.

(a) What is the flow of mass through the left end of the control surface shown dashed?

(b) What is the flow of kinetic energy through the left end of the control surface? Assume that the velocity profile does not change along the pipe.

In Fig. 4.21, we have shown a cross section of the pipe. For an infinitesimal circular strip, we can say noting that **V** and **dA** are collinear but of opposite sense:

$$\rho\mathbf{V} \cdot \mathbf{dA} = \rho\left[C\left(r^2 - \frac{D^2}{4}\right)\right]2\pi r\,dr$$

---

[16]A wing of constant cross section and infinite length.

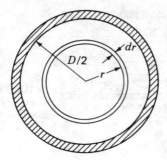

FIGURE 4.21
Cross section of pipe with infinitesimal ring of fluid.

From the whole cross section, we have

$$\iint \rho \mathbf{V} \cdot d\mathbf{A} = \rho \int_0^{D/2} C\left(r^2 - \frac{D^2}{4}\right) 2\pi r\, dr$$

$$= 2\pi \rho C \left[\frac{r^4}{4} - \frac{D^2}{4}\frac{r^2}{2}\right]_0^{D/2}$$

$$= -\frac{\rho C \pi D^4}{32} \quad \text{kg/s} \tag{b}$$

We now turn to the flow of kinetic energy through the left end of the control surface. The kinetic energy for an element of fluid is $\frac{1}{2}dm\,V^2$. This corresponds to an infinitesimal amount of an extensive property. To get η, the corresponding intensive property, we divide by dm to get

$$\eta = \tfrac{1}{2}V^2 \tag{c}$$

We accordingly wish to compute

$$\iint \eta(\rho \mathbf{V} \cdot d\mathbf{A}) = \iint \left(\tfrac{1}{2}V^2\right)\{\rho \mathbf{V} \cdot d\mathbf{A}\}$$

Employing Eq. (a) for V, and noting again that $\mathbf{V}$ and $d\mathbf{A}$ are collinear but of opposite sense, we get

$$\iint \eta(\rho \mathbf{V} \cdot d\mathbf{A}) = \int_0^{D/2} \frac{1}{2}C^2\left(r^2 - \frac{D^2}{4}\right)^2 \left\{\rho\left[C\left(r^2 - \frac{D^2}{4}\right)2\pi r\, dr\right]\right\}$$

$$= \rho C^3 \pi \int_0^{D/2}\left(r^2 - \frac{D^2}{4}\right)^3 r\, dr$$

$$= \frac{\rho C^3 \pi D^8}{2048} \quad \text{N} \cdot \text{m/s} \tag{d}$$

where we can facilitate the integration by making a change of variable for $(r^2 - \frac{D^2}{4})$ to a single variable—say ξ.

Example 4.4. For Example 4.3, assume a *one-dimensional* model with the same mass flow. Compute the kinetic energy flow through a section of the pipe for this model. That is, compute kinetic energy flow with an average constant velocity. What is the ratio of the actual kinetic energy to the kinetic energy flow for the one-dimensional model flow?

We first compute the constant velocity at a section for the one-dimensional model. Hence, using Eq. (*b*) of Example 4.3

$$-(V_{av})\left(\frac{\rho\pi D^2}{4}\right) = -\frac{\rho C D^4 \pi}{32}$$

$$\therefore V_{av} = \frac{CD^2}{8} \quad \text{m/s} \tag{a}$$

The kinetic energy flow for the one-dimensional model is then

$$\iint \frac{V^2}{2}(\rho \mathbf{V} \cdot \mathbf{dA}) = -\frac{\rho}{2}\left(\frac{CD^2}{8}\right)^3\left(\frac{\pi D^2}{4}\right)$$

$$= -\frac{\rho C^3 D^8 \pi}{4096} \quad \text{N} \cdot \text{m/s} \tag{b}$$

We now define the *kinetic-energy correction* factor α as the ratio of the actual flow of kinetic energy through a cross section to the flow of kinetic energy for a one-dimensional model for the same mass flow. That is

$$\alpha = \frac{\text{KE flow for section}}{\text{KE flow for 1-D model}} \tag{c}$$

For the case at hand, we have from Eq. (*b*) of this example and Eq. (*d*) of Example 4.3

$$\alpha = \frac{-\rho C^3 \pi D^8/2048}{-\rho C^3 \pi D^8/4096} = 2 \tag{d}$$

The factor α exceeds unity, so there is an underestimation of kinetic energy flow for a one-dimensional model. We will have more to say about this point later in the text.

4.10   CLOSURE

In this chapter we have laid the foundation for the handling of fluid flow. Specifically, we have presented (1) kinematical procedures and concepts which enable us to describe the motion of fluids including the concept of irrotationality; (2) the four basic laws which will form the basis for our calculating the motion and flow characteristics of fluids; (3) the system and control-volume viewpoints by which we can apply these laws effectively to physical problems, and (4) the Reynolds transport equation relating the system approach to the control-volume approach or, in other words, relating the Eulerian and the Lagrangian viewpoints. In the next two chapters, we will develop these basic laws for both finite systems and finite-control volumes in a very general form. And, in solving problems in those chapters, we make liberal usage of the one- and two-dimensional flow models.

PROBLEMS

Problem Categories

4.1. A flow field is given as

$$\mathbf{V} = 6x\mathbf{i} + 6y\mathbf{j} - 7t\mathbf{k} \quad \text{m/s}$$

What is the velocity at position $x = 10$ m and $y = 6$ m when $t = 10$ s? What is the slope of the streamlines for this flow at $t = 0$ s? What is the equation of the streamlines at $t = 0$ up to an arbitrary constant? Finally, sketch streamlines at $t = 0$.

4.2. We will later learn that the two-dimensional flow around an infinite stationary cylinder is given as

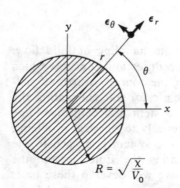

FIGURE P4.2

follows, using cylindrical coordinates:

$$V_r = V_0 \cos\theta - \frac{\chi \cos\theta}{r^2}$$

$$V_\theta = -V_0 \sin\theta - \frac{\chi \sin\theta}{r^2}$$

where V_0 and χ are constants. (Note that there is no flow in the z direction.) What is the slope (dy/dx) of a streamline at $r = 2$ m and $\theta = 30°$? Take $V_0 = 5$ m/s and $\chi = \frac{5}{4}$ m³/s. Show that at $r = \sqrt{\chi/V_0}$ (i.e., on the boundary of the cylinder) the streamline must be tangent to the cylinder wall. *Hint*: What does this imply about normal component V_N at the boundary?

4.3. Given the following *unsteady*-flow field,

$$\mathbf{V} = 3(x - 2t)(y - 3t)^2\mathbf{i}$$
$$+ (6 + z + 4t)\mathbf{j} + 25\mathbf{k} \quad \text{ft/s}$$

can you specify by inspection a reference $x'y'z'$ moving at constant speed relative to xyz so that $\mathbf{V}$ relative to $x'y'z'$ is *steady*? What is $\mathbf{V}$ for this reference? What is the speed of translation of $x'y'z'$ relative to xyz? *Hint*: For the last step, imagine a point fixed in $x'y'z'$. How must $x'y'z'$ then move relative to xyz to get correct relations between x' and x, y' and y, and z' and z?

4.4. Using data from Prob. 4.1, determine the acceleration field for the flow. What is the acceleration of the particle at the position and time designated in Prob. 4.1?

4.5. Given the velocity field

$$\mathbf{V} = 10\mathbf{i} + (x^2 + y^2)\mathbf{j} - 2yx\mathbf{k} \quad \text{ft/s}$$

what is the acceleration of a particle at position $(3, 1, 0)$ ft?

4.6. Given the velocity field

$$\mathbf{V} = (6 + 2xy + t^2)\mathbf{i} - (xy^2 + 10t)\mathbf{j}$$
$$+ 25\mathbf{k} \quad \text{m/s}$$

what is the acceleration of a particle at $(3, 0, 2)$ m at time $t = 1$ s?

4.7. A flow of charged particles (a plasma) is moving through an electric field $\mathbf{E}$ given as

$$\mathbf{E} = (x^2 + 3t)\mathbf{i} + yz^2\mathbf{j} + (x^2 + z^2)\mathbf{k} \quad \text{N/C}$$

The velocity field of the particles is given as

$$\mathbf{V} = 10x^2\mathbf{i} + \left(5t + \sqrt{y}\right)\mathbf{j} + t^3\mathbf{k} \quad \text{m/s}$$

If the charge per particle is 10^{-5} C, what is the time rate of change of force on any one particle as it moves through the field?

4.8. The force $\mathbf{F}$ on a particle with electric charge q moving through a magnetic field $\mathbf{B}$ is given as

$$\mathbf{F} = q\mathbf{V} \times \mathbf{B}$$

Consider a flow of charged particles moving through a magnetic field $\mathbf{B}$ given as

$$\mathbf{B} = (10 + t^2)\mathbf{i} + (z^2 + y^2)\mathbf{k} \quad \text{W/m}^2$$

where the velocity field is given as

$$\mathbf{V} = (20x + t^2)\mathbf{i} + (18 + zy)\mathbf{j} \quad \text{m/s}$$

What is the time rate of change of $\mathbf{F}$ for a flow particle with charge 10^{-5} C? Do not take time to multiply out terms in final computation.

4.9. The equation for streamlines corresponding to a two-dimensional doublet (to be studied in Chap. 12) is given in meters as

$$x^2 + y^2 - \frac{\chi}{C}y = 0 \quad\quad (a)$$

where χ is a constant for the flow and C is a constant for a streamline. What is the direction of the velocity of a particle at position $x = 5$ m and $y = 10$ m? If $V_x = 5$ m/s, what is V_y at the point of interest?

4.10. In Prob. 4.9, it should be apparent from analytic geometry that the streamlines represent circles. For a given value of χ and for different values of C, along what axis do the centers of the aforementioned circles lie? Show that all circles go through the origin. Sketch a system of streamlines.

4.11. In Example 4.1, what is the equation of the streamline passing through position $x = 2$, $y = 4$? Remembering that the radius of curvature of a curve is

$$R = \frac{\left[1 + (dy/dx)^2\right]^{3/2}}{|d^2y/dx^2|}$$

determine the acceleration of a particle in a direction normal to the streamline and toward the center of curvature at the aforementioned position.

4.12. We are given the following family of curves representing streamlines for a two-dimensional source (Chap. 12):

$$y = Cx \quad\quad (1)$$

where C is a constant for each streamline. Also we know that

$$|\mathbf{V}| = \frac{K}{\sqrt{x^2 + y^2}} \quad\quad (2)$$

where K is a constant for the flow. What is the velocity field $\mathbf{V}(x, y, z)$ for the flow? That is, show that

$$V_x = \frac{Kx}{x^2 + y^2} \quad\quad V_y = \frac{Ky}{x^2 + y^2}$$

Suggestion: Start by showing that

$$|\mathbf{V}| = V_x\sqrt{1 + \left(\frac{V_y}{V_x}\right)^2} \quad \text{and} \quad \frac{V_y}{V_x} = C = \frac{y}{x}$$

4.13. A *path line* is the curve traversed by any one particle in the flow and corresponds to the *trajectory* as employed in your earlier course in particle mechanics. Given the velocity field

$$\mathbf{V} = (6x)\mathbf{i} + (16y + 10)\mathbf{j} + (20t^2)\mathbf{k} \quad \text{m/s}$$

what is the path line of a particle which is at $(2, 4, 6)$ m at time $t = 2$ s? *Suggestion:* Form dx/dt, dy/dt, and dz/dt. Integrate: solve for constants of integration; then eliminate the time t to relate xyz in a single equation.

4.14. Consider a velocity field $\mathbf{V}(x, y, z, t)$ as measured from reference xyz. The reference xyz is moving relative to another reference XYZ with an angular velocity $\boldsymbol{\omega}$ and a translational velocity $\dot{\mathbf{R}}$ and has, in addition, an angular acceleration $\dot{\boldsymbol{\omega}}$ and a translational acceleration $\ddot{\mathbf{R}}$. From your earlier dynamics course, you may have learned that the acceleration of a particle relative to XYZ (that is, $\mathbf{a}_{XYZ}$) is given as

$$\mathbf{a}_{XYZ} = \mathbf{a}_{xyz} + \ddot{\mathbf{R}} + 2\boldsymbol{\omega} \times \mathbf{V}_{xyz} + \dot{\boldsymbol{\omega}} \times \boldsymbol{\rho}$$
$$+ \boldsymbol{\omega} \times (\boldsymbol{\omega} \times \boldsymbol{\rho})$$

where $\mathbf{a}_{xyz}$ and $\mathbf{V}_{xyz}$ are taken relative to xyz.

We have the following data at an instant:

$$\mathbf{V} = 10x\mathbf{i} + 30xy\mathbf{j} + (3x^2z + 10)\mathbf{k}\quad m/s$$
$$\boldsymbol{\omega} = 10\mathbf{i}\quad rad/s$$
$$\dot{\mathbf{R}} = \mathbf{0}\quad m/s$$
$$\ddot{\mathbf{R}} = 16\mathbf{k}\quad m/s^2$$
$$\dot{\boldsymbol{\omega}} = 5\mathbf{k}\quad rad/s^2$$

What is the acceleration relative to xyz and XYZ, respectively, of a particle at

$$\boldsymbol{\rho} = 3\mathbf{i} + 3\mathbf{k}\quad m$$

at the instant of interest?

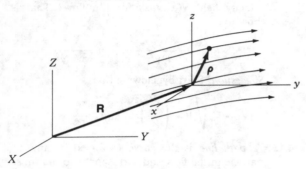

FIGURE P4.14

4.15. Think up and discuss a few situations where the formulations developed in Prob. 4.14 would be of use.

4.16. Consider a steady two-dimensional inviscid flow about a cylinder of radius a. Using cylindrical coordinates, we can express the velocity field of a nonviscous incompressible flow in the following manner,

$$\mathbf{V}(r,\theta) = -\left(V_0\cos\theta - \frac{a^2V_0}{r^2}\cos\theta\right)\boldsymbol{\epsilon}_r$$
$$+ \left(V_0\sin\theta + \frac{a^2V_0}{r^2}\sin\theta\right)\boldsymbol{\epsilon}_\theta$$

where V_0 is a constant and $\boldsymbol{\epsilon}_r$ and $\boldsymbol{\epsilon}_\theta$ are unit vectors in the radial and transverse directions, respectively, as shown in the diagram. What is the acceleration of a fluid particle at $\theta = \theta_0$ at the *boundary* of the cylinder whose radius is a? *Suggestion*: Use path coordinates. *Hint*: What must V_r be at the boundary?

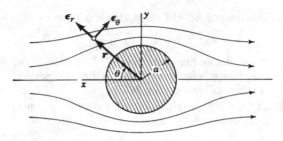

FIGURE P4.16

4.17. Given the following velocity field

$$\mathbf{V} = 10x^2y\mathbf{i} + 20(yz + x)\mathbf{j} + 13\mathbf{k}\quad m/s$$

what is the strain rate tensor at $(6, 1, 2)$ m.

4.18. In Prob. 4.17, what is the total angular velocity of a fluid particle at $(1, 4, 3)$ m?

4.19. Given the velocity field

$$\mathbf{V} = 5x^2y\mathbf{i} - (3x - 3z)\mathbf{j} + 10z^2\mathbf{k}\quad m/s$$

compute the angular velocity field $\boldsymbol{\omega}(x, y, z)$.

4.20. A flow has the following velocity field:

$$\mathbf{V} = (10t + x)\mathbf{i} + yz\mathbf{j} + 5t^2\mathbf{k}\quad ft/s$$

What is the angular velocity of a fluid element at $x = 10$ ft, $y = 3$ ft, and $z = 5$ ft? Along what surface is the flow always irrotational?

4.21. Show that any velocity field $\mathbf{V}$ expressible as the gradient of a scalar ϕ must be an irrotational field.

4.22. If $\mathbf{V} = \mathbf{grad}\,\phi$, what irrotational flow is associated with the function

$$\phi = 3x^2y - 3x + 3y^2 + 16t^3 + 12zt$$

Read Prob. 4.21 before proceeding.

4.23. Is the following flow field irrotational or not?

$$\mathbf{V} = 6x^2y\mathbf{i} + 2x^3\mathbf{j} + 10\mathbf{k}\quad ft/s$$

4.24. Explain why in a capillary tube the flow is virtually always rotational.

4.25. What were the basic laws and subsidiary laws that you used in your course in strength of materials?

4.26. In the studies of rigid-body mechanics, how was conservation of mass ensured? Also, was conservation of energy a law independent and apart from Newton's laws? Explain the reason for your answer.

4.27. Have we placed any restrictions on the motion of a control volume? Can it have material other than fluid inside or passing through?

4.28. A fluid is moving along a curved circular pipe such that the pressure, velocity, etc., are uniform at each section of the pipe and are functions of the position s along the centerline of the pipe and time. How would we classify this flow in the light of our discussion in this chapter? If the flow properties were also functions at a section of the radial distance r from the centerline in addition to s and t, would this then be a two-dimensional flow? Why?

4.29. In Example 4.3, compute the linear momentum flow through a cross section of the control volume. Recall that the linear momentum of a particle is $m\mathbf{V}$.

4.30. In Prob. 4.29 find a momentum correction factor which would be the ratio for the actual momentum flow to that of the one-dimensional model of the flow for the same mass flow. In the previous problem, we got the result

$$\iint V(\rho \mathbf{V} \cdot d\mathbf{A}) = -\frac{\rho C^2 \pi D^6}{192}$$

Do not consult Example 4.3 while doing this problem.

4.31. In Example 4.3, compute the kinetic energy flow through one face of the control surface if it is moving to the left at a speed of V_0 relative to the ground.

4.32. In Chap. 12, we discuss the simple *vortex* where in cylindrical coordinates

$$V_r = 0 \qquad V_z = 0$$

$$V_\theta = \frac{\Lambda}{2\pi r}$$

Λ is a constant called the *strength* of the vortex. Draw the streamlines for the simple vortex. What is the mass flow and kinetic energy flow through the surface shown in the diagram?

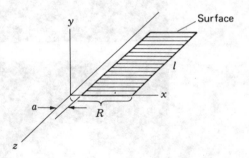

FIGURE P4.32

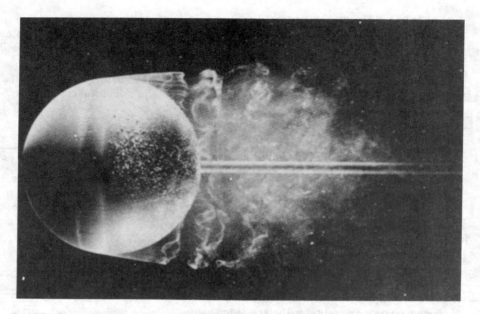

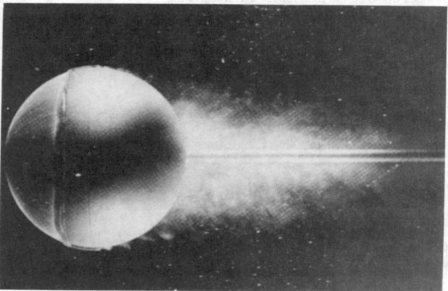

Flows over spheres illustrating separation from laminar and from turbulent boundary layers. (*Courtesy Dr. Henry Werlé, Onera, France.*)

The top photo shows a flow of water with dye inserted in the boundary layer. There is a laminar boundary layer showing a separation ahead of the equator and remaining laminar for almost one radius before becoming turbulent. In the second photo a wire hoop is placed ahead of the equator to *trip* the boundary layer into turbulent flow. It now separates further rearward than if it were laminar. Drag is dramatically reduced. It is for this reason that for some airfoils, a string of small vortex generator blades is to be found. This last photo was made with air bubbles in water. A discussion on this process is described in Section 13.12.

CHAPTER
5

BASIC LAWS FOR FINITE SYSTEMS AND FINITE CONTROL VOLUMES I: CONTINUITY AND MOMENTUM

5.1 INTRODUCTION

In the previous chapters, fluid properties and flow characteristics have been described with the use of the field concept. We will now develop two of the basic laws relating these quantities in various forms. In each development the more familiar finite system approach will first be undertaken, and then the formulations will be extended to the case of the finite control volume. The chapter will be presented in three parts, dealing with the topics of:

A. Conservation of mass
B. Linear momentum
C. Moment of momentum

In this chapter and the following chapter we will develop many of the basic equations which form the basis for much of the analytical work in the text. Consequently, these are vital chapters.

PART A
CONSERVATION OF MASS

5.2 CONTINUITY EQUATION

Recall that a *system*, by our definition, always entails the same quantity of matter. Thus, by employing the definition of a system properly, i.e., by keeping M, the mass, constant, we find the conservation of mass is in this instance automatically ensured.

In using the *control volume* (Fig. 5.1) for the handling of flow problems, it is clear that matter is *not* identified and that there is not the simple and direct manner for ensuring conservation of mass as was the case in previous studies of discrete particles and rigid bodies where a systems approach was used.

To go from the systems approach to the control-volume approach here, we employ the Reynolds transport equation (4.28), where

1. The extensive property N is for our case M, the mass of a fluid system.

2. The quantity η is unity for our case, since $M = \iiint_V \rho \, dv$.

Since the mass M of any system is constant as noted above, we can then say at any time t on using the Reynolds transport equation that

$$\frac{DM}{Dt} = 0 = \oiint_{CS} (\rho\mathbf{V} \cdot \mathbf{dA}) + \frac{\partial}{\partial t} \iiint_{CV} \rho \, dv$$

Since we can choose a system of any shape at time t, the relation above is then valid for any control volume at time t. Furthermore, as we pointed out in the last chapter, this control volume may have any motion whatever provided that the velocity $\mathbf{V}$ and the time derivative $\partial/\partial t$ are measured relative to the control volume. To interpret this equation most simply, we can rewrite it as follows:

$$\boxed{\oiint_{CS} (\rho\mathbf{V} \cdot \mathbf{dA}) = -\frac{\partial}{\partial t} \iiint_{CV} \rho \, dv} \qquad (5.1)$$

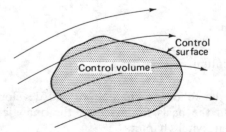

FIGURE 5.1
Flow through a control volume.

That is, *the net efflux rate of mass through the control surface equals the rate of decrease of mass inside the control volume.* In this way we account for the mass entering or leaving any volume that we may choose in the flow at any time. Equation (5.1) and its simplified forms are called *equations of continuity.*

If the flow is steady relative to a reference fixed to the control volume, all fluid properties, including the density at any fixed position in the reference, must remain invariant with time. Since we are dealing with control volumes of fixed shape, the right side of Eq. (5.1) can be written in the form $\iiint (\partial \rho / \partial t)\, dv$ and it is clear that this integral is zero. Hence we can state that *any steady flow* involving one or many species of fluids must satisfy the equation

$$\oiint_{CS} (\rho \mathbf{V} \cdot \mathbf{dA}) = 0 \qquad (5.2)$$

Next, consider the case of *incompressible* flow involving only a *single species* of fluid in the domain of our control volume. In this case, ρ is constant at all positions in the domain and for all time even if the velocity field is unsteady. The right side of Eq. (5.1) vanishes then, and on the left side of this equation we can extract ρ from under the integral sign. We may then say that

$$\oiint_{CS} (\mathbf{V} \cdot \mathbf{dA}) = 0 \qquad (5.3)$$

Thus, for any incompressible flow involving a single species of fluid, conservation of mass reduces to conservation of volume.

You should not be intimidated by the rather forbidding integration operations given in the preceding equations. These equations should be thought of as precise mathematical language for conservation of mass equivalent to, but more precise than, the statement made after Eq. (5.1). From these general formulations, we can now develop a useful special equation.

Let us, for example, consider the very common situation in which fluid enters some device through a pipe and leaves the device through a second pipe, as shown diagrammatically in Fig. 5.2. The control surface we have chosen is indicated by a dashed line. We assume that the flow is *steady* relative to the control volume and that the inlet and outlet flows are *one-dimensional*. Applying Eq. (5.2) for this case, we get

$$\oiint_{CS} (\rho \mathbf{V} \cdot \mathbf{dA}) = \iint_{A_1} (\rho \mathbf{V} \cdot \mathbf{dA}) + \iint_{A_2} (\rho \mathbf{V} \cdot \mathbf{dA}) = 0$$

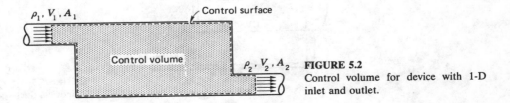

ρ_1, V_1, A_1

Control surface

Control volume

ρ_2, V_2, A_2

FIGURE 5.2
Control volume for device with 1-D inlet and outlet.

where A_1 and A_2 are, respectively, the entrance and exit areas. Upon noting that the velocities are normal to the control surfaces at these areas and employing the previously discussed sign convention of an outward directed normal for the representation of area, the equation above becomes

$$\oint_{CS} (\rho \mathbf{V} \cdot \mathbf{dA}) = - \iint_{A_1} \rho V \, dA + \iint_{A_2} \rho V \, dA = 0$$

With ρ and V constant at each section as a result of the one-dimensional restriction for the inlet and outlet flows, we get for this equation

$$-\rho_1 V_1 \iint_{A_1} dA + \rho_2 V_2 \iint_{A_2} dA = 0$$

Integrating, we get

$$\rho_1 V_1 A_1 = \rho_2 V_2 A_2 \tag{5.4}$$

which is a simple relation known by every high-school physics student. The main purpose in going through the development above was to see how a particular algebraic equation for a simple case is fashioned from the general formulation. Note that $\rho_1 V_1 A_1$ is the mass flow and can be given as ρQ where Q is the volume flow.

In the following example we shall illustrate the use of the continuity equation as well as a format that we encourage the student to follow.

Example 5.1. Water is flowing through a large tank having a diameter of 5 ft (see Fig. 5.3a). The velocity of water relative to the tank is given as

$$V = 6.25 - r^2 \text{ ft/s}$$

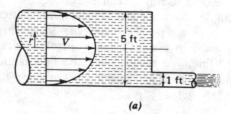

(a)

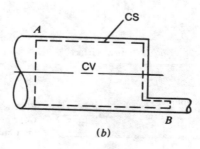

(b)

FIGURE 5.3
Cylindrical duct system.

What is the *average* velocity of the water leaving by the smaller pipe having an inside diameter of 1 ft?

The first step is to choose a control volume, part of whose surface involves the desired unknown information. This is shown in Fig. 5.3b where at B the control surface is normal to and cuts through the flow in the pipe away from the larger cylinder.

Next we state the general *continuity* equation for this control volume. Thus

$$\oiint_{CS} (\rho \mathbf{V} \cdot \mathbf{dA}) = -\frac{\partial}{\partial t} \iiint_{CV} \rho \, dv \qquad (a)$$

Now we shall state certain assumptions that will enable us to simplify the above equation for direct use. These assumptions are a vital part of the analysis and for this reason we box them in.

---

1. The flow is steady.
2. The flow is incompressible.
3. The flow at B is one dimensional.

How do these assumptions affect Eq. (a)? Note that assumption 1 permits us to delete the time derivative on the right side of equation (a). Assumption 2 allows us to take ρ as constant and this term can be canceled out of Eq. (a). Finally, assumption 3 permits us to express the volume flow through portion B of the control surface simply as $V_B A_B$ with V_B as a constant.

We can now write Eq. (a) as follows:

$$-\int_0^{2.5} (6.25 - r^2)(2\pi r) \, dr + V_B \frac{\pi(1^2)}{4} = 0 \qquad (b)$$

Note that the first minus sign above obtains because the velocity vector and the area vector at A are colinear and of *opposite* direction. Integrating and solving for V_B we get

$$-(2\pi)\left(6.25\frac{r^2}{2} - \frac{r^4}{4}\right)\Bigg|_0^{2.5} + V_B \frac{\pi}{4} = 0 \qquad (c)$$

$$\therefore V_B = 78.125 \text{ ft/s}$$

The one-dimensional model at B then gives us the desired average velocity in the small pipe.

Later you will find that the conservation of mass consideration will occur in problems *after* other basic laws have been formulated. In such cases assumptions of the flow will have already been made. We then suggest that if a simple form of continuity such as $\rho_1 V_1 A_1 = \rho_2 V_2 A_2$ is applicable, you use it directly without going through the detailed discussion such as that of the previous example for which only conservation of mass was the prime consideration. We shall illustrate this in subsequent problems.

5.3   SYSTEM ANALYSIS

Let us now consider a finite fluid system moving in a flow. Newton's law says that

$$\mathbf{F}_R = \frac{d}{dt}_{\text{system}} \left[\iiint_M \mathbf{V}\,dm \right] = \frac{d\mathbf{P}}{dt}_{\text{system}} \tag{5.5}$$

where $\mathbf{F}_R$ is the resultant external force acting on the system and $\mathbf{V}$ and the time derivative are taken for an *inertial reference*. You will recall that $\mathbf{P}$ is the linear momentum vector. Since we are following the system, we may express Eq. (5.5) as follows:

$$\mathbf{F}_R = \frac{D}{Dt} \iiint_M \mathbf{V}\,dm = \frac{D\mathbf{P}}{Dt} \tag{5.6}$$

We will distinguish between two types of forces which combine to give the resultant force $\mathbf{F}_R$. Recall from Chap. 2 that force distributions acting on the boundary of the system are called *surface-force* distributions or *surface tractions*, denoted as $\mathbf{T}(x, y, z, t)$, given as the force per unit area on the boundary surfaces. Force distributions acting on the material inside the boundary are called *body-force* distributions, denoted as $\mathbf{B}(x, y, z, t)$, and given as force per unit mass at a point. For example, gravity is the most common body-force distribution and for gravity, $\mathbf{B} = -g\mathbf{k}$. We can now rewrite Eq. (5.6) as follows:

$$\oiint_S \mathbf{T}\,dA + \iiint_V \mathbf{B}\rho\,dv = \frac{D\mathbf{P}}{Dt} \tag{5.7}$$

We have thus expressed Newton's law for a finite system. Of greater interest to us in fluid mechanics is the control-volume approach, which can now be readily set forth with the help of the Reynolds transport equation (4.28).

5.4   CONTROL VOLUMES FIXED IN INERTIAL SPACE

We will consider linear momentum $\mathbf{P}$ as the extensive property to be considered in the Reynolds transport equation (4.28). The quantity η to be used in this equation now becomes momentum per unit mass, which is simply $\mathbf{V}$, the velocity of fluid elements. This is readily verified by noting that $\mathbf{P} = \iiint_V \mathbf{V}(\rho\,dv)$. We then have

$$\frac{D\mathbf{P}}{Dt} = \oiint_{CS} \mathbf{V}(\rho\mathbf{V} \cdot d\mathbf{A}) + \frac{\partial}{\partial t} \iiint_{CV} \mathbf{V}(\rho\,dv) \tag{5.8}$$

It was pointed out in Chap. 4 that the velocities and time derivatives must be those seen from the control volume. If the control volume is *fixed in inertial space*, then the derivative on the left side is taken for an inertial reference and we may then use Newton's law, Eq. (5.7), to replace this term to form the desired linear momentum equation.[1]

$$\oiint_{CS} \mathbf{T}\, dA + \iiint_{CV} \mathbf{B}\rho\, dv = \oiint_{CS} \mathbf{V}(\rho\mathbf{V} \cdot \mathbf{dA}) + \frac{\partial}{\partial t} \iiint_{CV} \mathbf{V}(\rho\, dv) \qquad (5.9)$$

Since the system and control volume have the same shape at time t, the surface-force distribution $\mathbf{T}$ is now the total-force distribution acting on the control surface, and the body-force distribution $\mathbf{B}$ is now the total-force distribution acting on the fluid inside the control volume. *This equation then equates the sum of these force distributions with the rate of efflux of linear momentum across the control surface plus the rate of increase of linear momentum inside the control volume.* For steady flow and negligible body forces, as is often the case, the equation above becomes

$$\oiint_{CS} \mathbf{T}\, dA = \oiint_{CS} \mathbf{V}(\rho\mathbf{V} \cdot \mathbf{dA}) \qquad (5.10)$$

It should be kept in mind that the momentum equation is a vector equation. The scalar-component equations in the orthogonal x, y, and z directions may then be written by simply taking the components of the vectors $\mathbf{V}$, $\mathbf{T}$, and $\mathbf{B}$. Thus

$$\oiint_{CS} T_x\, dA + \iiint_{CV} B_x\rho\, dv = \oiint_{CS} V_x(\rho\mathbf{V} \cdot \mathbf{dA}) + \frac{\partial}{\partial t}\left(\iiint_{CV} V_x\rho\, dv \right)$$

$$\oiint_{CS} T_y\, dA + \iiint_{CV} B_y\rho\, dv = \oiint_{CS} V_y(\rho\mathbf{V} \cdot \mathbf{dA}) + \frac{\partial}{\partial t}\left(\iiint_{CV} V_y\rho\, dv \right) \quad (5.11)$$

$$\oiint_{CS} T_z\, dA + \iiint_{CV} B_z\rho\, dv = \oiint_{CS} V_z(\rho\mathbf{V} \cdot \mathbf{dA}) + \frac{\partial}{\partial t}\left(\iiint_{CV} V_z\rho\, dv \right)$$

In using Eqs. (5.11) one selects directions for the inertial reference axes xyz, so that positive directions of the velocities V_x, V_y, and V_z as well as the surface and body forces T_x and B_x, etc., are established. It must be kept in mind that this sign consideration is independent of the sign of $\mathbf{V} \cdot \mathbf{dA}$. To exemplify

---

[1] As with the Reynolds transport equation, we must not let the imposing mathematical appearance of Eq. (5.9) obscure its rather simple physical interpretation. In short, we can say here that the resultant force on the material in a given domain "drives" linear momentum through the control surface and, in addition, changes linear momentum inside the control surface. The imposing mathematical formulation of this simple physical picture allows us to precisely apply data to properly carry out the formulation in given circumstances.

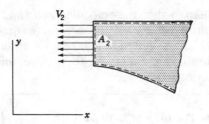

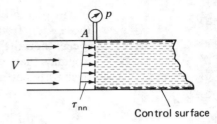

FIGURE 5.4
1-D flow out of control volume.

FIGURE 5.5
Parallel flow.

this, examine a portion of a device shown in Fig. 5.4, where the control surface extends over the outlet area. A one-dimensional flow normal to the outlet area is indicated. The mass flow for the control surface at the outlet is simply $+\rho_2 V_2 A_2$, since V_2 and A_2 are oriented in the *same* direction. On the other hand, the velocity component V_x in the surface integral is given by the term $-V_2$, since it is directed in the negative direction of the selected coordinate system. Thus the portion of the surface integration over the outlet area is $(-V_2)(+\rho_2 V_2 A_2)$.

We should also like to point out that for a *parallel flow* (i.e., the streamlines are straight and parallel as shown in Fig. 5.5) we will prove later in the text that the normal stress τ_{nn} on a control surface perpendicular to the flow will have a value equal to the pressure p measured by the gage shown plus, a *hydrostatic* increase in pressure as one goes down into the flow from the inlet A of the pressure gage. For such flows in small cross sections, as in pipes, where p may be reasonably high, we can usually neglect the hydrostatic increase with depth. We then have a uniform traction force distribution of magnitude p over this surface.

The momentum equations developed in this section are very general and are of great importance in fluid mechanics. In the following section a number of problems have been undertaken in detail to help explain the meaning of the terms and the manner of use of these equations.

5.5 USE OF THE LINEAR MOMENTUM EQUATION FOR THE CONTROL VOLUME

We have thus far developed very general formulations for the law of conservation of mass and Newton's law as applied to control volumes. From the general continuity equation, we developed simpler specialized equations, one of which was probably quite familiar to you. For most problems, it is advisable to go directly to the proper continuity equation unless for instructional purposes you wish to begin with the general case. However, in the case of linear momentum we have not developed any of the common specialized forms of this equation such as the thrust formula of a nozzle, since we feel that in view of the complexity of the momentum equation, you should develop the simpler equations yourself as they are needed for particular problems. Doing this will give you a greater awareness of the limitations of your results that are imposed by the simplifications and idealizations employed in reaching the working equations. It is the experience of the author that overreliance on specialized formulas in this area, coupled with unclear specifications of control volumes, is often the source of serious errors by engineers both in and out of school.

Since the linear momentum equation is primarily a relation between forces and velocities, you should choose a control volume that will involve forces and velocities in an economical manner that will contribute to the solution of the problem. Generally, you will expose for consideration in the problems the *reaction* to the force you are seeking. That is, you may want the force *on* a pipe or vane *from water*, but using your fluid mechanics you will actually first seek the force *from* the pipe or vane *on the water*. Just as when you selected several free-body diagrams in your statics course, you may be required to select several different control volumes to have enough independent equations to carry through to the solution. It is extremely important to designate each control volume carefully and to denote clearly the particular control volume for which each equation is written. Furthermore and most importantly, as in free body diagrams you must include *all* force systems involved with materials in the control volume. Not to do this is a cardinal error.

Finally, one must be reminded that the linear momentum equation (5.9) was developed from $\mathbf{F} = m\mathbf{a}$, so it has the limitation that the control volume, relative to which the velocities in the equation are measured, must be *fixed* in inertial space. (In the next section we consider the general case of noninertial control volumes.)

We will now examine the use of the linear momentum and continuity equations in the following examples, which you are urged to study carefully.

In Example 5.2 we shall use *absolute pressures* and thereby compute the force from the *internal flow* on the reducing elbow of the problem. We shall then point out that by using *gage pressures* instead of absolute pressure, we can get the *combined force* on the elbow from internal flow as well as from the force from the atmosphere on the outside surface of the elbow. Those that accept this approach intuitively need not go to Example 5.3 where the aforementioned procedure is justified. Those readers who wish this justification and who wish to be able to solve special cases where this approach is not valid, should then read starred Sec. 3.8 and go to Example 5.3.

Example 5.2. We wish to evaluate the force coming onto the reducing elbow shown in Fig. 5.6 as a result of an internal steady flow of liquid. The average values of the flow characteristics at the inlet and outlet are known, as is the geometry of the reducer.

A control volume constituting the interior of the reducer will enable us to relate known quantities at the inlet and outlet with the force $\mathbf{R}$ on the fluid *from* the reducer wall. (The reaction to this latter force is the quantity desired.) This is shown in Fig. 5.6. *All* the forces acting on the fluid in the control volume at any time t have been designated. The surface forces include the effects of *absolute* pressures p_1 and p_2 at the entrance and exit of the reducer as well as distributions of normal and shear stresses p_W and τ_W, whose resultant force is $\mathbf{R}$ coming from

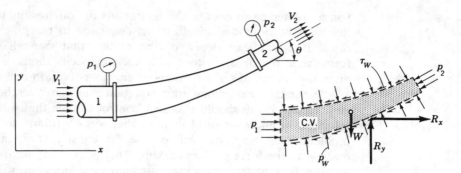

FIGURE 5.6
Flow through reducing elbow.

the wall of the elbow onto the fluid. The body force is simply the weight of the fluid inside the control volume at time t and is indicated in Fig. 5.6 as W.

The following are our assumptions for the flow in this control volume:

1. Flow is steady.
2. Flow is incompressible.
3. One-dimensional, parallel flow entering at 1 and leaving at 2.

We begin writing the linear momentum equation in its general form:

$$\oiint_{CS} \mathbf{T}\,dA + \iiint_{CV} \mathbf{B}\rho\,dv = \oiint_{CS} \mathbf{V}(\rho\mathbf{V}\cdot d\mathbf{A}) + \frac{\partial}{\partial t}\iiint_{CV}\mathbf{V}(\rho\,dv)$$

Next we shall simplify the equation in light of the flow model we have proposed above embodied in our assumptions. First we note that the last expression is zero by virtue of assumptions 1 or 2. Now we examine the other expressions using separate horizontal and vertical components of the equation. In doing so we will keep the control volume and its surface under close scrutiny. Thus, considering forces first, note that we neglect hydrostatic pressure variation at the entrance and exit of the control surface and have taken uniform absolute pressures p_1 and p_2 over these sections, respectively, in accordance with assumption 3.

The x and y components of the resultant force on the fluid may be expressed as

$$\oiint_{CS} T_x\,dA + \iiint_{CV} B_x\rho\,dv = p_1A_1 - p_2A_2\cos\theta + R_x$$

$$\oiint_{CS} T_y\,dA + \iiint_{CV} B_y\rho\,dv = -p_2A_2\sin\theta - W + R_y$$

(a)

where R_x and R_y are the net force components of the reducer wall on the fluid. R_x and R_y, being unknown, have been selected positive.

Examine the linear momentum flow through the control surface. The surface integration need be carried out only at the inlet and outlet surfaces of the control volume since $\mathbf{V} \cdot d\mathbf{A}$ is zero at the walls. (Why?) The normal components of velocity at the inlet and outlet surfaces are seen to equal V_1 and V_2, respectively. By virtue of the one-dimensionality assumption 3, the efflux rate of linear momentum may then be expressed as

$$\oint_{CS} \mathbf{V}(\rho \mathbf{V} \cdot d\mathbf{A}) = \mathbf{V}_1(-\rho_1 V_1 A_1) + \mathbf{V}_2(\rho_2 V_2 A_2) \qquad (b)$$

The scalar components of Eq. (b) in the x and y directions are given as

$$\oint_{CS} V_x(\rho \mathbf{V} \cdot d\mathbf{A}) = -V_1(\rho_1 V_1 A_1) + (V_2 \cos\theta)(\rho_2 V_2 A_2)$$

$$\oint_{CS} V_y(\rho \mathbf{V} \cdot d\mathbf{A}) = (V_2 \sin\theta)(\rho_2 V_2 A_2) \qquad (c)$$

The continuity equation for this control volume meanwhile using assumptions 1 and 3 may be stated in a simple form. Thus

$$\rho_1 V_1 A_1 = \rho_2 V_2 A_2 \qquad (d)$$

Now substitute the preceding results into the linear momentum equations in the x and y directions. Thus we get

$$p_1 A_1 - p_2 A_2 \cos\theta + R_x = (V_2 \cos\theta - V_1)\rho_1 V_1 A_1$$

$$- p_2 A_2 \sin\theta - W + R_y = (V_2 \sin\theta)\rho_1 V_1 A_1 \qquad (e)$$

One may now solve for R_x and R_y. Changing the sign of these results will then give the force components *on the elbow* from the fluid. Using the symbols K_x and K_y for these components, we get

$$K_x = p_1 A_1 - p_2 A_2 \cos\theta - \rho_1 V_1 A_1(V_2 \cos\theta - V_1)$$

$$K_y = -p_2 A_2 \sin\theta - W - \rho_1 V_1 A_1(V_2 \sin\theta) \qquad (f)$$

Now for thin-walled devices such as the elbow of this example, using gage pressures for p_1 and p_2, we will then get *both* the force on the elbow from internal flow as well as the force on the elbow from air outside. In the next example we shall justify this assertion.

Example 5.3. For Example 5.2 we wish to include the atmospheric force on the *outside* of the elbow. It will be shown that if *gage* pressures are used for p_1 and p_2 in the preceding problem, the total force on the elbow from internal *and* external fluids will automatically be determined. This will be valid for elbows with *thin* walls and will be approximate where the walls are thick compared with the inlet and outlet diameters[2] and where pressure on the flanges is neglected. More

---

[2] If the wall is not thin and we wish an accurate result, we will have to consider the *actual outside* surface geometry in a manner to be described below.

important is the requirement that uniform atmospheric pressure acts on the outside surface. This need not be the case for all problems as you will see in the homework set.

Accordingly, let us now examine the surface forces in the x direction on the *outer* surface of the elbow from atmospheric pressure. This is a *static* problem involving a curved surface submerged in a uniform pressure taking us back to Chap. 3.

Taking the wall to be thin, we use inside geometry, and, from our work in Sec. 3.8 for projecting curved surfaces, you should be able to verify that the force in the x direction from p_{atm} and denoted as $(K_x)_{air}$ is

$$(K_x)_{air} = -p_{atm}A_1 + p_{atm}A_2 \cos \theta \tag{a}$$

Similarly in the y direction we have

$$(K_y)_{air} = p_{atm}A_2 \sin \theta \tag{b}$$

Using Eqs. (a) and (b), and Eqs. (f) of Example 5.2 we compute the *total* force *on* the elbow from the water inside and the air outside. Collecting terms we then get

$$(K_x)_{total} = (p_1 - p_{atm})A_1 - (p_2 - p_{atm})A_2 \cos \theta - \rho_1 V_1 A_1 (V_2 \cos \theta - V_1)$$

$$(K_y)_{total} = -(p_2 - p_{atm})A_2 \sin \theta - W - \rho_1 V_1 A_1 (V_2 \sin \theta) \tag{c}$$

Now compare Eqs. (f) of Example 5.2 and Eqs. (c) above. It is clear that you could have arrived at Eq. (c) directly had you used *gage pressures* at the outset of Example 5.2 when considering the water flowing through the reducing elbow. We use this artifice when we want the effects of inside fluid as well as the *outside atmosphere*.

Thus we see from this example which is an internal flow that if the wall of the reducing elbow is *thin* and if there is *uniform atmospheric pressure* outside we can get the force from inside water *and* outside air using the inside control volume while simply using *gage pressures* when computing K_x and K_y. When similar conditions prevail in other problems one can shorten and simplify calculations via this procedure. However, when these conditions do *not* prevail one must make *separate* calculations for the flow inside the device and for conditions outside the device. The outside could conceivably be a hydrostatic pressure field or even a different outside flow! For the former we would then deal with a *curved surface* submerged in the *hydrostatic pressure field*—a problem we have dealt with in Chap. 3. We shall present problems requiring this two-pronged approach in the homework problems.

Having considered an internal flow in Examples 5.2 and 5.3, we now consider an *external* flow. We shall use *absolute pressures* to get the force from the *water* on a trough. We then point out that by using gage pressures, we would then get for K_x the combined force from water as well as from the atmospheric pressure impinging on the non-wetted surface of the trough. The

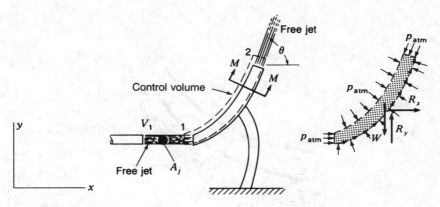

FIGURE 5.7
Stationary trough.

justification of this step is simple to present and we have done just that for the interested reader at the end of the example.

> **Example 5.4.** A stream of water is shown in Fig. 5.7 directed against a stationary curved trough. We will determine the horizontal force on the trough from the water. A convenient control volume is chosen so as to include the surface of contact between the water and the trough (i.e., the wetted surface). This control volume will expose the force on the water from the trough. In this way we can introduce the reaction R_x to the desired force in a *linear momentum* equation. The surface and body forces for the control volume are shown in Fig. 5.7. Note that the "free-jet"[3] part of the flow has the stress field corresponding to the surrounding pressure p_{atm}. Observe in Figs. 5.8a and 5.8b that we have identified for a cross-section such as *MM* the *wetted* portion of the trough part of the control surface as *ABC*. It is through this surface that the force from the trough is imposed on the water. Over the remaining control surface a uniform pressure equal to atmospheric pressure acts. It should be clear from Fig. 5.8 that the net projected area in the x direction on which we have p_{atm} equals the projected area of the above-mentioned *wetted* portion of the control surface *ABC* (see Fig. 5.8c).[4] We shall indicate this projected area as $(A_t)_x$.

---

[3]A more detailed discussion of free jets will be given in Chap. 11. Remember now that p_{atm} is simply a measure of pressure and is the measure of the pressure inside the free jets which go respectively in and out of the control volume.

[4]Note that the x projection of the wetted surface of the trough (see Fig. 5.9) is *not* available for *shadowing* that part of the control surface having atmospheric pressure acting on it. Thus there is cancellation everywhere of the x projection of that part of the control surface on which we have p_{atm} *except* for the aforementioned *wetted area* of the *trough*. The *net* projection in the x direction of the control surface having p_{atm} acting on it thus equals that of the wetted area of the trough.

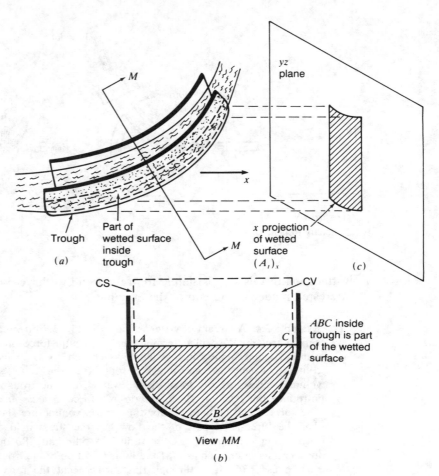

FIGURE 5.8
Details of flow through the trough.

The *assumptions* for this flow are:

1. Neglect action of friction and gravity on the speed of the flow.
2. Steady flow.
3. Incompressible flow.
4. 1-D flow at entrance and exit of the control volume.

As a result of assumption 1 the speed of flow is constant. If this is not entirely clear now, you will be able to prove this when you have studied Bernoulli's equation in the next chapter. The constant speed of flow we denote as V_1. The *linear momentum equation* in the x direction is as follows:

$$\iiint_{CV} B_x \rho \, dv + \oiint_{CS} T_x \, dA = \oiint_{CS} V_x (\rho \mathbf{V} \cdot \mathbf{dA}) + \frac{\partial}{\partial t} \iiint_{CV} V_x \rho \, dv$$

Assumption 2 permits us to delete the last integral while assumption 4 allows for simple integration at the inlet and outlet. Finally, assumption 3 allows us to use a constant ρ throughout. Since $B_x = 0$, we have using the aforementioned constant V_1

$$p_{atm}(A_t)_x + R_x = -(V_1)(\rho V_1 A_j) + (V_1 \cos \theta)(\rho V_1 A_2) \qquad (a)$$

where $(A_t)_x$ is the net projected area in the x direction of that part of the control surface exposed to atmospheric pressure as discussed earlier.

From *continuity* we can say for the chosen control volume that

$$\rho V_1 A_j = \rho V_1 A_2 \qquad (b)$$

Substituting from (b) into (a) and solving for R_x, we get

$$R_x = -p_{atm}(A_t)_x - \left(\rho V_1^2 A_j \right)(1 - \cos \theta) \qquad (c)$$

Taking the reaction to this force gives us the desired force component *on the trough* from the *water*. Thus

$$K_x = p_{atm}(A_t)_x + \rho V_1^2 A_j (1 - \cos \theta) \qquad (d)$$

Now that we have the force from the water acting on the trough, we next look at the rest of the surface of the trough. Consider again Fig. 5.9 showing a cross section of the trough including the water in the trough. Note now that atmospheric pressure acts over the entire surface of the trough except along ABC where water impinges. The x component of force from p_{atm} on the nonwetted surface is the static problem of a curved surface submerged in a uniform pressure, p_{atm}. The projected area in the x direction of this nonwetted surface is the same $(A_t)_x$ we used earlier as should now be clear to you. Thus the force from p_{atm} on the trough is $-p_{atm}(A_t)_x$ and so the total force from water and air is

$$(K_x)_T = \rho V_1^2 A_j (1 - \cos \theta) \qquad (e)$$

Note that like the reducing elbow problem, we would have reached this result by using gage pressures when we get to K_x. We have accordingly shown for an internal flow and now for an external flow, how gage pressures can be

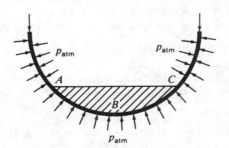

FIGURE 5.9
Cross section of trough.

used to include the effects of a uniform pressure on the outside surface of a device when getting the K forces. However, we shall present problems in the homework where separate analyses will be required on the one hand for the control volume using absolute pressures, and on the other hand for the force on an outside surface which is not part of the control surface.

A *correct* quick procedure accordingly is as follows. Use absolute pressures in the momentum equation since in accordance with the mechanics involved we must *only* use forces *on* the material inside or on the control surface and not on some other regions outside the control surface such as the outside surface of the reducing elbow in Example 5.2 or parts of the nonwetted surfaces of the trough in Example 5.4. But when we get to the K_T forces we can switch to gage pressure to include the *total* force on the device from the flow being considered in the control volume *and* the uniform pressure acting on the nonwetted portions of the device outside the control volume. A *formal* procedure which is even quicker is to use *gage pressures directly* in the momentum equation with the full understanding that this is not a correct procedure from a mechanics point of view (hence, a "formal" procedure) but will nevertheless lead to a correct formulation of K_T on the device. We shall generally follow the first approach in the examples but, if you are fully cognizant of what you are doing, we suggest you use the aforementioned "formal" approach in the homework problems.

Example 5.5. We will now calculate the results of Example 5.4 more directly by choosing a *different* control volume such as is shown in Fig. 5.10. This control volume surrounds the trough entirely and *cuts through* the support in a plane perpendicular to the y axis. The surface stresses on the control surface include atmospheric pressure over the entire control surface except at the area where it cuts the support.[5] There we have indicated F_x and F_y and M as the components of the resultant force system transmitted through the cut support surface. The body force is W, the *total weight* of all the material inside the control volume (including water, trough, and part of support). We may write the horizontal component of the *linear momentum equation* in the following way (remembering the assumptions set forth earlier):

$$F_x = -\left(\rho V_1^2 A_j\right)(1 - \cos \theta) \qquad (a)$$

Note that the effects of atmospheric pressure cancel in the x direction for the control volume chosen. The result F_x is the total force in the x direction that the ground must exert on the trough. Consider next the trough plus its cut support as a *free body* (or a *system*) as shown in Fig. 5.11, where $(K_x)_T$ are force components from the enveloping fluids acting on the free body.[6] From *equilibrium* of this free

---

[5] Note that in the free jets going into and out of the control volume we have a pressure equal to p_{atm} of the surrounding air and this is the pressure that the control surface experiences from the water going through.

[6] The atmospheric pressure forces acting on the support arm of the trough clearly is *zero* in the x direction except directly at the point of attachment to the trough itself. Here atmospheric force has already been accounted for in $(K_x)_T$ formulated in Prob. 5.4.

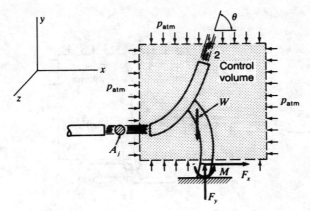

FIGURE 5.10
Control volume cuts through support.

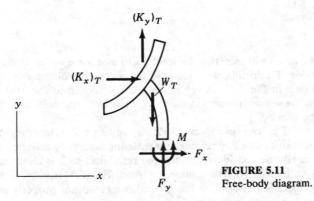

FIGURE 5.11
Free-body diagram.

body we have in the x directions

$$F_x + (K_x)_T = 0$$

$$\therefore (K_x)_T = -F_x$$

Thus:

$$(K_x)_T = \rho V_1^2 A_j (1 - \cos \theta) \qquad (b)$$

which is the same result as Eq. (e) of Example 5.4.

Note that in Examples 5.4 and 5.5 we used

(a) a control volume for linear momentum and continuity
(b) a curved surface submerged in a uniform pressure for which we used hydrostatics
(c) a free body diagram for which we used rigid body equations of equilibrium.

Example 5.6. Consider next the case where the curved trough of Example 5.5 is *moving* with a constant speed V_0 relative to the ground, as is shown in Fig. 5.12.

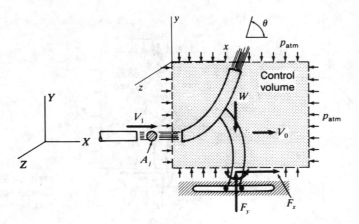

FIGURE 5.12
Moving trough.

We again will ascertain the total horizontal force on the trough from the enveloping fluids. To do this, we have selected a control volume moving with the trough, as shown in Fig. 5.12. A reference xyz is attached to the control volume. Reference xyz moves with constant speed V_0 relative to the reference XYZ, which is fixed to the ground.

For our purposes, XYZ is an inertial reference. You will recall from mechanics that any reference translating uniformly relative to an inertial reference can also be considered an inertial reference, so it is clear that our control volume is an inertial control volume for which Eq. (5.9) is valid. Thus, if the trough were accelerating relative to the ground, we could not properly use this equation. (We shall examine such cases in the next section.)

Remembering to use velocities relative to the control volume and carrying over the assumptions of the previous examples, we then have for the component of the *linear momentum* equation in the x direction

$$F_x = (V_1 - V_0)\left[-(V_1 - V_0)(\rho)(A_j)\right] + (V_1 - V_0)\cos\theta\left[(V_1 - V_0)(\rho)(A_2)\right] \tag{a}$$

The *continuity* equation for the control volume is

$$(V_1 - V_0)(\rho)(A_j) = (V_1 - V_0)(\rho)(A_2) \tag{b}$$

Upon using Eqs. (*b*) and (*a*), the solution for F_x can be stated as

$$F_x = -(V_1 - V_0)^2(\rho A_j)(1 - \cos\theta) \tag{c}$$

If we take a free body of the trough and support which is inside the control volume (as in Fig. 5.12), it should be clear that

$$(K_x)_T = -F_x = (V_1 - V_0)^2\rho A_j(1 - \cos\theta) \tag{d}$$

where $(K_x)_T$ is the total thrust in the x direction stemming from the water and the air enveloping the trough.

In the preceding example we considered a device which we will call a simple *turbomachine*. Note for this device that the motion of an *unconfined* fluid is altered in such a way that the propulsive thrust is created on the device. As it moves, power is developed on the device from energy supplied by the jet of water. In more complex turbomachines, such as turbojets and ram jets, fluid is made to undergo certain processes so as to achieve a flow pattern which gives rise to a propulsive thrust. For these devices, burning fuel supplies the requisite energy for maintaining the necessary flow to accomplish this task. Other turbomachines have different missions from that of propulsion. Steam turbines, for example, are devices in which the flow pattern is arranged so as to develop a torque on a rotor and, in this way, to drive a generator for electric power. In addition we have rotary pumps, torque converters, fluid couplings, centrifugal compressors, etc. All these devices are considered apart from the familiar *reciprocating machine* in that during the process that the fluid undergoes in a turbomachine it is at no time "trapped," or confined, by the machine, as is the case with reciprocating machines such as a diesel engine, where the fluid is confined in cylinders during most of the action.

In Example 5.7 we will consider the evaluation of thrust by a turbojet aircraft engine. Later in Part C we will further consider turbomachines after we have considered the moment-of-momentum equation for control volumes.

Example 5.7. A *turbojet engine* is shown in Fig. 5.13 moving through the air at a constant velocity V_1. Air enters at speed V_1 relative to the engine and is slowed down while gaining pressure in the diffuser section. A compressor further increases the pressure of the air, and in the combustion chambers, fuel is burned in the airstream to maintain this high pressure. The fluid then expands through the turbine which drives the compressor, and, in so doing, the fluid gives up some of its energy. Finally, the fluid expands in the nozzle section so that on leaving it has a pressure close to that of the atmosphere. The buildup and decay of pressure in the flow inside the machine result in traction force distributions on the interior surfaces of the machine that combine to give a thrust to the device. It is this thrust which we will compute for this machine where the exit velocity of the fluid is known from other calculations to be V_2 relative to the machine and where it is

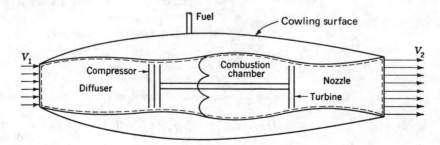

FIGURE 5.13
Simple sketch of jet engine.

FIGURE 5.14
Cutaway view of an early jet engine. (*Courtesy Curtiss Wright Corp.*)

known that $(1/N)$ kilogram of fuel is burned per kilogram of incoming air. (See Fig. 5.14 for a cutaway section of a real jet engine.)

We select as a control volume the interior region of the jet engine (Fig. 5.13) such that the control surface *cuts through* the supports of the compressor, combustion chamber, etc., in the manner of the control volume discussed in Example 5.6. Such a control volume clearly "exposes" for calculations the force distribution exerted on the fluid by the wall of the jet engine as well as the forces transmitted from the wall to the elements interior to the wall (and thus inside the control volume), such as compressors, combustion chambers, etc. The reaction to the total of these forces is the thrust that the machine feels from internal flow.

The assumptions for this flow are as follows:

---

1. Steady flow.
2. 1-D flow of air into the control surface and 1-D flow of combustion products out of the control surface.
3. Inlet and exhaust pressures are at atmospheric pressure.

With ρ_1 and N known, the rate of efflux of mass from the control volume is given by the *continuity* equation using the above assumptions:

$$\rho_1 V_1 A_1 + \frac{1}{N}(\rho_1 V_1 A_1) = \rho_2 V_2 A_2$$

$$\therefore \rho_2 V_2 A_2 = \left(1 + \frac{1}{N}\right)\rho_1 V_1 A_1 \tag{a}$$

The *linear momentum equation* in the x direction is next examined:

$$\iiint_{CV} B_x \rho \, dv + \oiint_{CS} T_x \, dA = \oiint_{CS} V_x(\rho \mathbf{V} \cdot \mathbf{dA}) + \frac{\partial}{\partial t} \iiint V_x(\rho \, dv)$$

Assumption 1 eliminates the time derivative while 2 simplifies the surface integrations. Using p_{atm} in the traction force calculations as per assumption 3, we

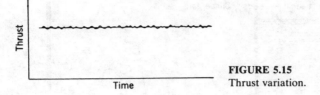

FIGURE 5.15
Thrust variation.

then have with $B_x = 0$,

$$p_{\text{atm}} A_1 - p_{\text{atm}} A_2 + R_x = \left(\rho_2 V_2^2 A_2 \right) - \left(\rho_1 V_1^2 A_1 \right) \qquad (b)$$

Solving for R_x, replacing $\rho_2 V_2 A_2$ using Eq. (a), and taking the reaction to R_x gives the thrust from the working fluid:

$$K_x = \rho_1 V_1 A_1 \left[V_1 - \left(1 + \frac{1}{N} \right) V_2 \right] + p_{\text{atm}} A_1 - p_{\text{atm}} A_2 \qquad (c)$$

Note that the linear momentum flow of the entering fuel has no component in the x direction and hence does not appear in Eqs. (b) and (c).

If the pressure on the cowling surface is close to atmospheric pressure, the use of statics on the cowling surface gives the following force due to air:

$$(K_x)_{\text{air}} = -p_{\text{atm}} A_1 + p_{\text{atm}} A_2 \qquad (d)$$

Adding Eqs. (c) and (d) will include inside and outside forces with the exception of outside shear tractions in accordance with our conclusions in Example 5.3. Since the shear drag is not usually "charged" against the engine, it may be stated approximately that[7]

$$\text{Thrust} = \rho_1 V_1 A_1 \left[V_1 - \left(1 + \frac{1}{N} \right) V_2 \right] \qquad (e)$$

Let us consider the analysis in more detail.

There was the tacit assumption in the analysis that the flow is steady. Actually, this is not the case in regions around the compressor blades and turbine blades. However, these departures from steady flow if taken into account accurately would superpose on the constant thrust computed earlier a high-frequency small-amplitude variation having a mean time average variation of zero over time intervals which are large compared with the periods of these variations. This is shown in Fig. 5.15. In computing the thrust for purposes of evaluating performance of an aircraft, these variations are insignificant, so the assumption of steady flow throughout yields with greatest ease the desired average thrust. However, in vibration studies, it is the variation that is important, since even small disturbances can cause large stresses in parts if resonance is reached for these parts. Clearly, a more precise study would be needed for such problems.

---

[7]Here is an instance where care must be used in examining flows inside the control volume and flow on the outside of the body. Overly glib use of gage pressures must be guarded against at this stage of your studies.

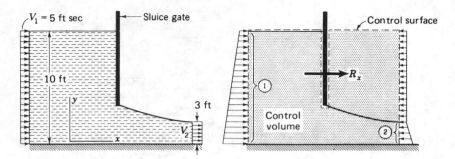

FIGURE 5.16
Flow by a sluice gate.

Example 5.8. Figure 5.16 shows the cross section of a *sluice gate*, which is a device used for controlling the flow of water in channels. Determine the force on the gate per unit width of the gate.

It is clear that the linear momentum equation should be used, since a force is desired, which in this case results from an easily simplified fluid flow. We must choose a control volume which exposes the desired force for use in the linear momentum equation. Such a control volume has been shown in Fig. 5.16, where the control surface envelops that part of the gate over which there does not exist a cancellation of pressure effects on both sides of the gate. The force R_x from the gate on the control surface is then the reaction to the desired force for this problem.

The following are the assumptions for this flow.

---

1. At the inlet and outlet we assume a uniform pressure p_{atm} on which is superposed a hydrostatic pressure variation.[8]

2. Steady flow.

3. Incompressible flow.

4. 1-D velocity at entrance and exit.

5. Zero shear stress at bed of channel.

The *linear momentum* equation in the x direction is first stated:

$$\iiint_{CS} B_x \rho \, dv + \oiint_{CS} T_x \, dA \equiv \oiint_{CS} V_x (\rho \mathbf{V} \cdot d\mathbf{A}) + \frac{\partial}{\partial t} \iiint_{CV} V_x (\rho \, dv)$$

The condition $B_x = 0$ and assumption 2 kill the first and last expressions, while assumption 1 permits a simple calculation of the traction integral at inlet and

---

[8]Experience and studies of free-surface flows (flows where one boundary is a free surface), which are also parallel flows, indicate that this step will not incur serious error.

outlet. Assumption 3 allows for constant ρ and γ in the calculation, while assumption 5 eliminates the x component of the traction force on the bed surface. Assumption 4 allows simple momentum flow calculation. Finally, noting that the atmospheric effects cancel, we have for the above equation on using hydrostatics from Chap. 3 for computing the force on a plane submerged surface,

$$\gamma(y_1)_c A_1 - \gamma(y_2)_c A_2 + R_x = \rho V_2^2 A_2 - \rho V_1^2 A_1 \qquad (a)$$

Substituting numerical values and noting that V_2 can be determined as $V_1(A_1/A_2)$ from *continuity*, we have

$$(62.4)(5)(10) - (62.4)(1.5)(3) + R_x = (1.938)(16.67)^2(3) - (1.938)(5^2)(10)$$

$$R_x = -1708 \text{ lb/ft of width} \qquad (b)$$

Therefore, the force on the gate is then 1708 lb/ft of width.

All the examples up to this time in this chapter have been steady flow problems. Unsteady flow problems require the inclusion of the expression $\partial/\partial t[\iiint \mathbf{V}(\rho \, dv)]$. This requires knowledge of the velocity field inside the entire control volume and not just at the control surface as is the case for steady flow problems. This negates one of the key advantages of the integral approach which allowed up to now the calculation of useful information with a minimum knowledge of the flow details. In the next example we look at an interesting unsteady flow problem which as you will see submits to a rather simple solution.

Example 5.9. A fighter plane is being refueled in flight (see Fig. 5.17) at the rate of 150 gal/min of fuel having a specific gravity of 0.68. What additional thrust does the plane need to develop to maintain the constant velocity it had just before the hookup? The inside diameter of the flexible pipe is 5 in. The fluid pressure in the pipe at the entrance to the plane is 4 lb/in^2 gage. Do not consider the mechanical forces on the plane directly from the flexible pipe itself.

In Fig. 5.18 we show the plane with a hypothetical control volume which represents the gas tank and duct from the inlet to the gas tank. A reference *xyz* has been attached to the plane. The plane we can assume is moving with constant speed and so the reference and the control volume are inertial. The flow in the tank is *unsteady* since gasoline is constantly accumulating inside the tank. The

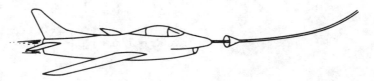

FIGURE 5.17
Fighter plane being refueled in flight.

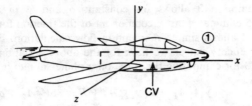

FIGURE 5.18
Fighter plane showing a hypothetical gasoline tank system.

assumptions for the flow in the control volume are as follows:

1. Flow is 1-D at entrance.
2. Incompressible flow.
3. *Average* velocity in the tank in the x direction is *zero* relative to *xyz* and is constant in the duct leading to the tank.

The general linear momentum equation in the x direction is next stated:

$$\oiint_{CS} T_x\, dA + \iiint_{CV} B_x \rho\, dv = \oiint_{CS} V_x(\rho \mathbf{V} \cdot d\mathbf{A}) + \frac{\partial}{\partial t} \iiint_{CV} V_x(\rho\, dv)$$

Since V_x has been assumed to average zero value in the tank (assumption 3), we can delete the last expression even though the flow technically is unsteady.[9] There is no body force contribution in the x direction. The velocity entering at (1) is given as follows:

$$V_1 = \frac{150(\text{gal/min})\left(0.002228 \, \dfrac{\text{ft}^3/\text{s}}{\text{gal/min}}\right)}{\left[\dfrac{\pi}{4}\left(\dfrac{5}{12}\right)^2 (\text{ft}^2)\right]}$$

$$\therefore V_1 = 2.45 \text{ ft/s}$$

Hence entering numerical data into the linear momentum equation we get, noting assumptions 1 and 2,

$$R_x - (p_{\text{atm}} + 4)\left(\frac{\pi 5^2}{4}\right) = -(2.45)\left[-\frac{62.4}{g}(0.68)(2.45)\frac{\pi 5^2}{(4)(144)}\right]$$

$$\therefore R_x = 79.6 + p_{\text{atm}}\left(\frac{\pi 5^2}{4}\right) \text{ lb}$$

---

[9]In the vertical direction there will be an unsteady contribution as the gasoline rises in the tank.

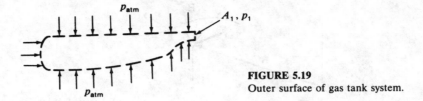

FIGURE 5.19
Outer surface of gas tank system.

The force on the plane from the gasoline is then

$$K_x = -79.6 - p_{atm}\left(\frac{\pi 5^2}{4}\right) \text{ lb} \qquad (a)$$

Next we look at the *outside surface* of the tank system (see Fig. 5.19). Atmospheric pressure is *assumed* to act over this curved surface. The net projected area of this surface in the x direction is clearly the area A_1 through which gasoline is passing. From *hydrostatics* we can say for the force in the x direction:

$$(F_x)_{atm} = p_{atm}\left[\frac{\pi(5^2)}{4}\right] \text{ lb} \qquad (b)$$

The total force from gasoline and air is then the sum of results (a) and (b). Thus,

$$(K_x)_{total} = -79.6 \text{ lb} \qquad (c)$$

The student may reach this directly by using gage pressure in Eq. (a) as discussed earlier.

You will note in thinking back over these sample problems that what we have been doing is simplifying the flows through carefully selected control volumes so that with the integral forms of the linear momentum and continuity equations we could solve for certain resultant forces or certain average velocities. This procedure is no different from what you followed in strength of materials, where you assumed certain behavior of the material in beams and shafts, namely, that plane sections remain plane, in order to compute bending and torsional stresses. To analyze the behavior of the material in beams and shafts more accurately, we would have to consider the basic laws and pertinent subsidiary laws in *differential* form with the view to integrating them to fit the boundary condition of the problem (this is done in the theory of elasticity). You were perhaps not distressed in your strength of material course, since the assumption of plane sections was made only once by the text. From there on very few additional assumptions had to be made to handle the problems, whereas in the work of the present chapter you are required to make assumptions for each problem. Lacking experience, you may be ill at ease in doing this. Also it may seem "unscientific" to you.

The thing to do is to make the most reasonable assumptions you can that permit you to reach answers that according to your judgment will be reasonably close to reality. The scientific aspect to your analysis will lie in whether there is

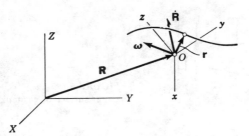

FIGURE 5.20
xyz moves arbitrarily relative to *XYZ*.

consistency between your computations and the flow models you have chosen, according to the basic and subsidiary laws. (This is why you will be requested to state your assumptions prominently.) On perhaps a much finer scale, this is how all analytic studies proceed. Thus you see that the analytical investigator can use imagination and intuition in such undertakings. The quality and success of the work can usually be assessed by observing how the results, predicted by the use of fundamental laws as applied to the models, check with what is observed and measured in the physical world for reasonably similar conditions.

You are therefore encouraged to proceed boldly with the problems and to try each time to gage the success of your analysis, using your instructor as a guide. If you consider your results studiously each time, you will develop more confidence and precision in carrying out problems.

In Parts 2 and 3 of the text we consider *differential* forms of the laws, and using finer and more realistic models, we are able to learn more accurately what the flows are really like in certain situations.

*5.6 NONINERTIAL CONTROL VOLUME

Equation (5.7) was developed essentially by using Newton's law in the form $\mathbf{F} = (D/Dt)$ $(m\mathbf{V}) = m\mathbf{a}$. You will recall from earlier courses in mechanics that this requires that the acceleration be measured relative to an inertial reference. Since the motion of the fluid in the previous development of the linear momentum equation is taken relative to the control volume, it is patent that the control volumes for which Eq. (5.9) is valid must then be fixed in, or translating at constant velocity relative to, an inertial reference. In this section we extend the linear momentum equation so as to be applicable to a control volume having *any* motion.

Consider an inertial reference *XYZ* and a reference *xyz* moving in an arbitrary manner relative to *XYZ*, as is shown in Fig. 5.20. The origin *O* of *xyz* is located in *XYZ* by the position vector **R**, and the motion of *xyz* relative to *XYZ* can then be described as the superposition of a translatory velocity $\dot{\mathbf{R}}$, corresponding to the actual motion of the origin *O* plus a pure rotation with angular velocity **ω** about an axis of rotation through the origin. (This is the result of Chasles' theorem.[10])

---

[10]See I. H. Shames, *Engineering Mechanics*, 3rd ed., Prentice-Hall, Englewood Cliffs, N.J., 1980 (Dynamics, Chap. 15).

We have learned in mechanics that the acceleration of a particle relative to the *XYZ* reference, $\mathbf{a}_{XYZ}$, is related to the acceleration of the particle relative to the *xyz* reference, $\mathbf{a}_{xyz}$, in the following manner:

$$\mathbf{a}_{XYZ} = \mathbf{a}_{xyz} + \ddot{\mathbf{R}} + 2\boldsymbol{\omega} \times \mathbf{V}_{xyz} + \dot{\boldsymbol{\omega}} \times \mathbf{r} + \boldsymbol{\omega} \times (\boldsymbol{\omega} \times \mathbf{r}) \qquad (5.12)$$

where $\mathbf{r}$ is the position vector of the particle in the *xyz* reference and $\mathbf{V}_{xyz}$ is the velocity of the particle relative to the *xyz* reference. To write Newton's law properly for an infinitesimal particle in terms of motion relative to *xyz*, we employ the relation above to form

$$d\mathbf{F} = dm\, \mathbf{a}_{XYZ} = dm\Big[\mathbf{a}_{xyz} + \ddot{\mathbf{R}} + 2\boldsymbol{\omega} \times \mathbf{V}_{xyz} + \dot{\boldsymbol{\omega}} \times \mathbf{r} + \boldsymbol{\omega} \times (\boldsymbol{\omega} \times \mathbf{r})\Big] \qquad (5.13)$$

We next rearrange this equation in the following manner:

$$d\mathbf{F} - dm\Big[\ddot{\mathbf{R}} + 2\boldsymbol{\omega} \times \mathbf{V}_{xyz} + \dot{\boldsymbol{\omega}} \times \mathbf{r} + \boldsymbol{\omega} \times (\boldsymbol{\omega} \times \mathbf{r})\Big] = dm\, \mathbf{a}_{xyz} = \frac{D}{Dt_{xyz}}(dm\,\mathbf{V}_{xyz})$$

$$(5.14)$$

Here D/Dt_{xyz} indicates a time derivative as seen from the *xyz* reference and performed as one follows the infinitesimal system. We now have the acceleration expression sitting alone on the right-hand side of the equation, and, by thinking of the second set of terms on the left-hand side of the equation as hypothetical forces, we have expressed Newton's law in a form which is familiar. You may recognize some of the hypothetical forces. For instance, $-(dm)(2\boldsymbol{\omega} \times \mathbf{V}_{xyz})$ is the famous Coriolis force, and $-(dm)[\boldsymbol{\omega} \times (\boldsymbol{\omega} \times \mathbf{r})]$ is the centrifugal force. We now go to the corresponding equation for a finite system by integrating the equation above for all elements in the system.

$$\oiint_S \mathbf{T}\, dA + \iiint_V \mathbf{B}\rho\, dv - \iiint_V \Big[\ddot{\mathbf{R}} + 2\boldsymbol{\omega} \times \mathbf{V}_{xyz} + \dot{\boldsymbol{\omega}} \times \mathbf{r} + \boldsymbol{\omega} \times (\boldsymbol{\omega} \times \mathbf{r})\Big]\rho\, dv$$

$$= \frac{D}{Dt_{xyz}} \iiint_V \mathbf{V}_{xyz}(\rho\, dv) = \frac{D}{Dt_{xyz}}(\mathbf{P}_{xyz}) \qquad (5.15)$$

Note that the hypothetical forces become body-force distribution in the case of a finite system. The expression on the right side of the equation above may be interpreted as the time variation as seen from the reference *xyz* of the linear momentum of the system relative to the reference *xyz*.

Let us now return to the Reynolds transport equation (4.28) and apply it to the extensive property $\mathbf{P}$, the linear momentum of the same finite system considered in Eq. (5.15). The reference to be used in Eq. (4.28) is the noninertial *xyz* reference of the preceding discussion, so the linear momentum is measured relative to this reference (we will continue to use the notation $\mathbf{P}_{xyz}$ to indicate this) as are the velocity $\mathbf{V}_{xyz}$ and the time variations. You will recall that the control volume generated by this formulation [Eq. (4.28)] will be fixed in the *xyz* reference and hence will be a so-called *noninertial* control volume. Noting that η becomes $\mathbf{V}_{xyz}$ for our case, we then have for Eq. (4.28)

$$\frac{D\mathbf{P}_{xyz}}{Dt_{xyz}} = \oiint_{CS} \mathbf{V}_{xyz}(\rho\mathbf{V}_{xyz} \cdot d\mathbf{A}) + \frac{\partial}{\partial t_{xyz}} \iiint_{CV} \mathbf{V}_{xyz}(\rho\, dv) \qquad (5.16)$$

We may now substitute Eq. (5.16) into Eq. (5.15). Note that the system over which we integrate the body force and hypothetical forces in Eq. (5.15) has at time *t* the same volume as the control volume of Eq. (5.16), so when we combine the equations we can

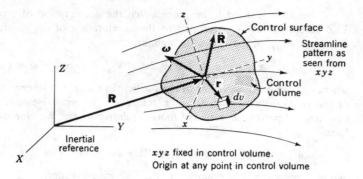

FIGURE 5.21
Noninertial control volume.

make the limits of the integrations in Eq. (5.15) the familiar control-volume limits. We then get the desired linear momentum equation for a noninertial control volume.

$$
\oint_{CS} \mathbf{T}\, da + \iiint_{CV} \mathbf{B}\rho\, dv - \iiint_{CV} \left[\ddot{\mathbf{R}} + 2\boldsymbol{\omega} \times \mathbf{V}_{xyz} + \dot{\boldsymbol{\omega}} \times \mathbf{r} + \boldsymbol{\omega} \times (\boldsymbol{\omega} \times \mathbf{r}) \right] \rho\, dv
$$

$$
= \oint_{CS} \mathbf{V}_{xyz}(\rho \mathbf{V}_{xyz} \cdot d\mathbf{A}) + \frac{\partial}{\partial t_{xyz}} \iiint_{CV} \mathbf{V}_{xyz}(\rho\, dv)
$$

(5.17)

Since the control volume is fixed in xyz, we usually find the control volume more convenient to use as the observation post for measuring velocities and time derivatives when we actually use Eq. (5.17) or its simplified forms. Thus $\ddot{\mathbf{R}}$ is the acceleration relative to the inertial reference XYZ of some convenient point O in the control volume, $\boldsymbol{\omega}$ is the angular velocity of the control volume measured from XYZ with an axis of rotation going through point O, and $\dot{\boldsymbol{\omega}}$ is the angular acceleration of the control volume relative to the inertial reference XYZ. Equation (5.17) then states that the total of the surface and body forces acting on the fluid inside the control surface minus the total of the various hypothetical body-force distributions, stemming from the fact that the control volume is noninertial, equals what an observer on the control volume sees as the rate of efflux of linear momentum through the control surface plus the rate of increase of linear momentum inside the control volume. Figure 5.21 shows the various elements involved in this equation in a general diagrammatic way.

Equation (5.17) indeed is a formidable equation and may intimidate all but the hardiest of students. Actually we rarely use it at full strength, and we will now consider the special case where the control volume translates along a straight line relative to the inertial reference (Fig. 5.22). (You will be asked as an exercise to formulate the case of a control volume undergoing pure rotation relative to an inertial reference.) Here we have $\boldsymbol{\omega} = 0$, $\dot{\boldsymbol{\omega}} = 0$, so the hypothetical forces can be written as

$$
- \iiint_{CV} \ddot{\mathbf{R}}\rho\, dv = -\ddot{\mathbf{R}} \iiint_{CV} \rho\, dv = -M\ddot{\mathbf{R}} \tag{5.18}
$$

since $\ddot{\mathbf{R}}$ is not a function of the coordinates xyz. M is the total mass of material within

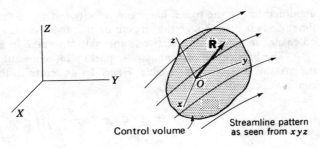

FIGURE 5.22
Translating noninertial volume.

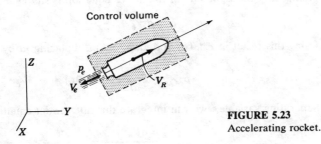

FIGURE 5.23
Accelerating rocket.

the control surface at any time. The linear momentum equation then becomes

$$\oiint_{CS} \mathbf{T}\, dA + \iiint_{CV} \mathbf{B}\rho\, dv - M\ddot{\mathbf{R}} = \oiint_{CS} \mathbf{V}_{xyz}(\rho\mathbf{V}_{xyz} \cdot \mathbf{dA}) + \frac{\partial}{\partial t_{xyz}} \iiint_{CV} \mathbf{V}_{xyz}(\rho\, dv)$$

(5.19)

We will now consider the problem of the rocket, for which this form of the linear momentum equation is extremely useful.

Example 5.10. A rocket (Fig. 5.23) is fired from rest along a straight line in outer space ($p_{\text{atm}} = 0$), where we can neglect air friction and gravitational influence. The rocket burns β kilograms of fuel per unit time and has initially a total mass m_0. The mass at any time t after firing is then $m = m_0 - \beta t$, and for this reason problems such as this are sometimes called "variable-mass problems." The exhaust velocity relative to the rocket V_e is constant and is computed from interior ballistics of the rocket engine, as are the exit pressure p_e and density ρ_e. The velocity of the rocket relative to an inertial reference is denoted as V_R. We are to determine the motion of the rocket relative to the inertial reference.

  We choose a control volume that moves *with* the rocket, as shown in Fig. 5.23, so that $\ddot{\mathbf{R}}$, the acceleration of the control volume, is then $\dot{V}_R$, the acceleration of the rocket relative to XYZ. The only real force is a surface force $p_e A_e$. Thus we get for the component of Eq. (5.19) in the direction of flight

$$p_e A_e - (\dot{V}_R)(m) = -\rho_e V_e^2 A_e \qquad (a)$$

The last volume integral in Eq. (5.19) is zero for the following reason: The amount of combustion gases in the control volume remains approximately constant, and since the velocity V_e is constant, there is then no change of momentum as seen from the control volume for this material. The structure of the rocket and the unburned fuel in the rocket have zero velocity relative to the control volume at all times, so that even though the amount of unburned fuel is a variable, there can be no change in momentum relative to the control volume from these materials.

The application of *continuity* requires a bit more attention than usual since the mass of the rocket is variable. Hence let us return to the general form of the continuity equation [Eq. (5.1)].

$$\oint_{CS} \rho \mathbf{V}_{xyz} \cdot d\mathbf{A} = -\frac{\partial}{\partial t_{xyz}} \iiint_{CV} \rho \, dv$$

Noting that the right side of the above equation is simply $-\dot{m}$, we then have

$$\rho_e V_e A_e = -\dot{m} = \beta \qquad (b)$$

Using this result in Eq. (a) and, in addition, replacing m by $m_0 - \beta t$, we get

$$p_e A_e - \frac{dV_R}{dt}(m_0 - \beta t) = -\beta V_e \qquad (c)$$

Separating variables, we can integrate this differential equation. Thus

$$\frac{dV_R}{p_e A_e + \beta V_e} = \frac{dt}{m_0 - \beta t}$$

Therefore

$$V_R = -\frac{p_e A_e + \beta V_e}{\beta}\big[\ln(m_0 - \beta t) - \ln m_0\big] \qquad (d)$$

where we have put in for $t = 0$, the condition $V_R = 0$. Rearranging the terms, we get

$$V_R = \left(V_e + \frac{p_e A_e}{\beta}\right)\ln\frac{m_0}{m_0 - \beta t} \qquad (e)$$

*PART C
MOMENT OF MOMENTUM

5.7   MOMENT OF MOMENTUM FOR A SYSTEM

Consider a finite system of fluid as shown in Fig. 5.24. An element dm of the system is acted on by a force $d\mathbf{F}$ and has a linear momentum $dm\,\mathbf{V}$. From Newton's law, we can say that

$$d\mathbf{F} = \frac{D}{Dt}(\mathbf{V}\,dm) \qquad (5.20)$$

Now take the cross product of each side of the equation using the position

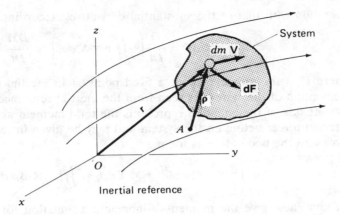

FIGURE 5.24
Mass *dm* in a finite system.

vector **r**. Thus,

$$\mathbf{r} \times d\mathbf{F} = \mathbf{r} \times \frac{D}{Dt}(dm\,\mathbf{V}) \qquad (5.21)$$

Consider next the following operation:

$$\frac{D}{Dt}(\mathbf{r} \times dm\,\mathbf{V}) = \frac{D\mathbf{r}}{Dt} \times dm\,\mathbf{V} + \mathbf{r} \times \frac{D}{Dt}(dm\,\mathbf{V}) \qquad (5.22)$$

Note that $D\mathbf{r}/Dt = \mathbf{V}$, so that the first expression on the right side is zero, since $\mathbf{V} \times \mathbf{V} = \mathbf{0}$. The second expression on the right side of the equation is identical to the right side of Eq. (5.21). Accordingly, we can now write Eq. (5.21) as follows:

$$\mathbf{r} \times d\mathbf{F} = \frac{D}{Dt}(\mathbf{r} \times dm\,\mathbf{V}) \qquad (5.23)$$

This equation equates the moment of the total force on an element *dm* about the origin of an inertial reference with the time rate of the moment about the origin of the linear momentum viewed from the inertial reference. It is a trivial matter to prove that instead of the origin we could have chosen any fixed point *A* in the inertial reference about which to take moments. We would have arrived at Eq. (5.23) with $\boldsymbol{\rho}$, the position vector from *A* to *dm*, replacing **r**.

We now integrate the expressions in Eq. (5.23) over the entire system using *O* as the fixed point. Thus,

$$\int \mathbf{r} \times d\mathbf{F} = \iiint_M \frac{D}{Dt}(\mathbf{r} \times \mathbf{V})\,dm \qquad (5.24)$$

The mass of the system is fixed so that the limits of the integration on the right side of Eq. (5.24) are fixed, permitting us to interchange the integration operation with that of the substantial derivative. Accordingly, we may say that

$$\int \mathbf{r} \times \mathbf{dF} = \frac{D}{Dt}\left(\iiint_M \mathbf{r} \times \mathbf{V}\,dm\right) = \frac{D\mathbf{H}}{Dt} \tag{5.25}$$

where $\mathbf{H}$ is the moment about a fixed point A in inertial space of the linear momentum of the system as seen from the inertial reference.[11] The integral on the left side of the equation represents the total moment about point A of the external forces acting on the system and may be given in terms of the traction force and the body force as follows.[12]

$$\int \mathbf{r} \times \mathbf{dF} = \oiint_S \mathbf{r} \times \mathbf{T}\,dA + \iiint_V \mathbf{r} \times \mathbf{B}\rho\,dv$$

We may now give the moment-of-momentum equation for a finite system as follows:

$$\boxed{\oiint_S \mathbf{r} \times \mathbf{T}\,dA + \iiint_V \mathbf{r} \times \mathbf{B}\rho\,dv = \frac{D\mathbf{H}}{Dt}} \tag{5.26}$$

As in the case of linear momentum we find the finite-control-volume approach extremely useful, so we now use Eq. (5.26) to formulate the equation for moment of momentum of a finite control volume.

5.8 CONTROL-VOLUME APPROACH FOR THE MOMENT-OF-MOMENTUM EQUATION FOR INERTIAL CONTROL VOLUMES

We may easily express the moment-of-momentum equation by considering $\mathbf{H}$ to be the extensive property in the Reynolds transport equation. Since $\mathbf{H} = \iiint_{\text{sys.}}(\mathbf{r} \times \mathbf{V})\rho\,dv$, the quantity η then becomes $\mathbf{r} \times \mathbf{V}$ for this case. Thus

$$\frac{D\mathbf{H}}{Dt} = \oiint_{CS}(\mathbf{r} \times \mathbf{V})(\rho\mathbf{V}\cdot\mathbf{dA}) + \frac{\partial}{\partial t}\iiint_{CV}(\mathbf{r} \times \mathbf{V})(\rho\,dv) \tag{5.27}$$

By substituting Eq. (5.27) into Eq. (5.26), we are limiting the reference for which the resulting equation is valid to that of an inertial reference XYZ, so the control volume is inertial for this equation. Also, since the system and the control volume occupy the same space at time t we can interpret

$$\iint_S \mathbf{r} \times \mathbf{T}\,dA \quad \text{and} \quad \iiint_V \mathbf{r} \times \mathbf{B}\rho\,dv$$

---

[11]Equation (5.25) is the same as $\mathbf{M} = \dot{\mathbf{H}}$ derived in particle mechanics for any system of particles. It is to be pointed out that $\mathbf{H}$ is also termed the *angular* momentum.

[12]Note that the moments of the internal forces cancel out because of Newton's third law.

in the resulting equation to be, respectively, the total moment about some point A in XYZ of the surface-force distribution on the control surface and the total moment about point A in XYZ of the body-force distribution throughout the fluid inside the control volume. We then have the desired moment-of-momentum equation for an inertial control volume.

$$\oiint_{CS} \mathbf{r} \times \mathbf{T}\, dA + \iiint_{CV} \mathbf{r} \times \mathbf{B}\rho\, dv = \oiint_{CS} (\mathbf{r} \times \mathbf{V})(\rho \mathbf{V} \cdot \mathbf{dA})$$
$$+ \frac{\partial}{\partial t} \iiint_{CV} (\mathbf{r} \times \mathbf{V})(\rho\, dv) \tag{5.28}$$

The terms on the right side represent the efflux of moment of momentum through the control surface plus the rate of increase of moment of momentum inside the control volume where both quantities are observed from the control volume.

Again we ask the student not to be intimidated by this rather impressive-looking equation. It has a simple physical connotation whereby we have the total moment vector of all body and traction forces for a control volume, wherein this moment is taken about some convenient point A fixed in inertial space driving angular momentum about A through the control surface and changing angular momentum about A inside the control volume. The complex mathematical apparatus employed allows us to apply numerical data precisely and directly for a specific problem involving the moment-of-momentum principle.

Thus for use of Eq. (5.28) we choose a useful control volume (see Fig. 5.25) and a useful fixed point A with both A and the control volume fixed in inertial space. As indicated in the diagram we must then find the moment $\mathbf{M}_A$ of traction forces and body forces and equate their sum with the efflux rate of angular momentum about A flowing through the control surface plus the rate of change of angular momentum about A inside the control volume.

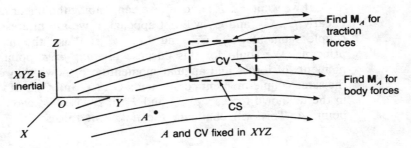

FIGURE 5.25
Elements entering moment-of-momentum equation.

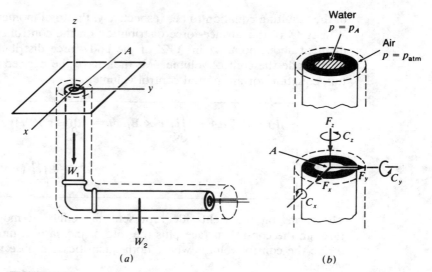

FIGURE 5.26
Cantilevered pipe with outside control volume.

A more specific situation is shown in Fig. 5.26a showing a cantilevered pipe with water flowing through it. We desire the stresses in the pipe cross section at the base of the pipe system. For this purpose a control volume has been chosen enveloping the pipe system on the outside and cutting through the pipe and incoming water at the base as well as cutting through the exiting free jet of water. A reference xyz is shown with the origin at point A which is the center of the pipe section at the base. Now examine the traction forces on the control surface. Where we cut the pipe we expose a general force system from the wall onto the pipe section given by force components F_x, F_y, and F_z and couple-moment components C_x, C_y, and C_z. Also on the control surface cutting the incoming water there is pressure p_A. Over the rest of the control surface including the exiting free jet there is atmospheric pressure. The body forces are the total weights W_1 and W_2 of pipe and water in the two pipe lengths.

If we want F_x, F_y, and F_z, we can employ the linear momentum equations (clearly C_x, C_y, and C_z will not appear). If we are interested in computing the couple moments C_x, C_y, and C_z, we can use the moment-of-momentum principle. We could for this purpose use *any fixed* point in space for taking moments of forces and angular momentum of flows. Any such point will yield the same couple-moment components C_x, C_y, and C_z.[13] If we are not interested in the unknown forces F_x, F_y, and F_z, it is wisest to choose the origin as fixed point A. This will eliminate the unknown forces F_x, F_y, and F_z from the

---

[13] Remember from mechanics that the couple moment is a free vector.

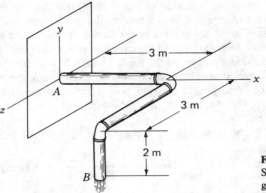

FIGURE 5.27
Steady flow through a pipe with right-angled elbows.

moment-of-momentum equation since they will all go through point A as will the force from pressure p_0 on the flow at the base. Finally, the force from p_{atm} will act over a projected area which equals the area enclosed by the pipe section at the base. The force from p_{atm} also has a line of action through point A.

The following example will reiterate some of the comments and illustrate the use of the moment-of-momentum equation for a control volume.

Example 5.11. Water is flowing at a constant rate Q of 0.01 m^3/s through a pipe having two right-angle bends as shown in Fig. 5.27. If the pipe has an interior cross-sectional area of 2580 mm^2 and weighs 300 N/m, what are the bending moment components at A?

A control volume is shown in Fig. 5.28 cutting the pipe at A. The stresses in the pipe *at the cut section* give rise to moments about the x, y, and z axes at A as

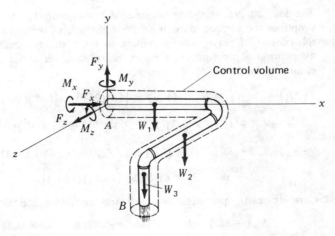

FIGURE 5.28
Pipe showing control volume.

well as forces, as indicated in Fig. 5.28. As for the rest of the control surface, we note that the pressure from water on the inside cross section of the pipe at A has a resultant force *going through* the origin A and hence gives *zero* moment about A. The *net* area exposed to atmospheric pressure is that of a circle corresponding to the outside diameter of the pipe. This circle is about the x axis, and gives rise to a force from atmospheric pressure also going along the x axis through the origin A. Hence atmospheric pressure also yields no moment about A. The body force resultants W_1, W_2, and W_3 are the weights of water and pipe for each section of the system.

We note first that the average velocity V in the pipe is

$$V = \frac{Q}{A} = \frac{0.01}{2580 \times 10^{-6}} = 3.88 \text{ m/s} \tag{a}$$

The moment-of-momentum equation for point A is next stated as

$$\oiint_{\text{CS}} \mathbf{r} \times \mathbf{T}\, dA + \iiint_{\text{CV}} \mathbf{r} \times \mathbf{B}\rho\, dv = \oiint_{\text{CS}} (\mathbf{r} \times \mathbf{V})(\rho \mathbf{V} \cdot d\mathbf{A})$$

$$+ \frac{\partial}{\partial t} \iiint_{\text{CV}} (\mathbf{r} \times \mathbf{V})(\rho\, dv)$$

To simplify this equation we next make the following assumptions.

1. Steady flow.
2. Incompressible flow.
3. Sheet of exiting water is at p_{atm}.
4. 1-D flow entering the flow.

The last integral vanishes because of assumption 1. The use of assumption 3 simplifies the surface integral on the right side of the equation. Using weights at the centers of gravity corresponding to the geometric centers of the pipes (see assumption 4) we formulate our working equation about the origin A so as to eliminate the forces F_x, F_y, and F_z:

$$(M_x\mathbf{i} + M_y\mathbf{j} + M_z\mathbf{k}) + (1.5\mathbf{i}) \times \left[(3)(2580 \times 10^{-6})(9806) + (3)(300)\right](-\mathbf{j})$$

$$+ (3\mathbf{i} + 1.5\mathbf{k}) \times \left[(3)(2580 \times 10^{-6})(9806) + (3)(300)\right](-\mathbf{j})$$

$$+ (3\mathbf{i} + 3\mathbf{k} - 1\mathbf{j}) \times \left[(2)(2580 \times 10^{-6})(9806) + (2)(300)\right](-\mathbf{j})$$

$$= (3\mathbf{i} + 3\mathbf{k} - 2\mathbf{j}) \times (-3.88\mathbf{j})[(0.01)(1000)] \tag{b}$$

Carrying out the products, we may reach the following equation:

$$M_x\mathbf{i} + M_y\mathbf{j} + M_z\mathbf{k} - 1464\mathbf{k} - 2928\mathbf{k} + 1464\mathbf{i} - 1952\mathbf{k} + 1952\mathbf{i}$$

$$= -116.4\mathbf{k} + 116.4\mathbf{i} \tag{c}$$

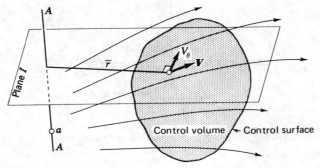

FIGURE 5.29
For computation of moment of momentum about axis AA.

The torques are then calculated as

$$M_x = -3300 \ \text{N} \cdot \text{m}$$
$$M_y = 0 \ \text{N} \cdot \text{m}$$
$$M_z = 6230 \ \text{N} \cdot \text{m}$$

These are the torques from the wall onto the pipe at A, which is useful information for the structural engineer.

If we wanted the moment due only to water and air, we would use a control volume covering the *interior* of the pipe. The moments *on* the water *from* the pipe about A would then be the *reactions* for the desired moments at A.

In many practical problems we usually employ only a single scalar component of Eq. (5.28) at any time, which means that we are taking moments of forces and momenta about an *axis* rather than a point. It will then be easiest to use cylindrical coordinates (Fig. 5.29) with the z direction along the axis shown as AA.[14] The term $\mathbf{r} \times \mathbf{V}$ may be replaced by $\bar{r}V_\theta$, where $\bar{r}$ is the radial distance from the axis to a particle and V_θ is the transverse component of the velocity of the particle. You will remember from mechanics that V_θ is so directed that it is at right angles to $\bar{r}$ and with $\bar{r}$ forms a plane normal to the axis (plane I in Fig. 5.29). Similarly $\mathbf{r} \times \mathbf{T}$ is replaced by $\bar{r}T_\theta$ and $\mathbf{r} \times \mathbf{B}$ is replaced by $\bar{r}B_\theta$. This scalar component of the moment-of-momentum equation then becomes

$$\oiint_{CS} \bar{r}T_\theta \, dA + \iiint_{CV} \bar{r}B_\theta \rho \, dv$$

$$= \oiint_{CS} (\bar{r}V_\theta)(\rho\mathbf{V} \cdot \mathbf{dA}) + \frac{\partial}{\partial t} \iiint_{CV} (\bar{r}V_\theta)(\rho \, dv) \qquad (5.29)$$

We will now illustrate the use of this equation in a simple example.

Example 5.12. A lawn sprinkler shown in Fig. 5.30 is held stationary by an external agent acting on the outside rotor surface while water is flowing through it. What is

---

[14]Although for simplicity we have shown the axis to be vertical it can have any orientation. Flows

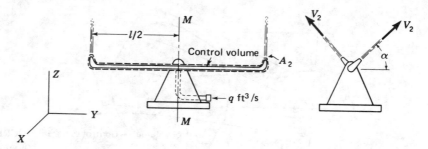

FIGURE 5.30
Lawn sprinkler held stationary.

the torque on the rotor from the outside agent if q cubic feet per second flows through the apparatus?

We choose as the control volume the exterior of the rotating arm of the sprinkler, and for the axis we select MM, the axis of rotation.

We make the following assumptions

---

1. Flow is steady.
2. Flow is incompressible
3. 1-D flow at exits.

The *moment-of-momentum* equation about an axis is

$$\oint_{CS} \bar{r} T_\theta \, dA + \iiint_{CV} \bar{r} B_\theta \rho \, dv = \oint_{CS} \bar{r} V_\theta (\rho \mathbf{V} \cdot \mathbf{dA}) + \frac{\partial}{\partial t} \iiint_{CV} \bar{r} V_\theta (\rho \, dv)$$

The volume integrals vanish because of assumption 1 and because gravity is parallel to the axis. From *continuity* and assumption 3 the efflux velocity is $V_2 = q/2A_2$. The rate of efflux of mass at each nozzle is $\rho q/2$. Also, the transverse component V_θ of the efflux velocity can be seen to be $V_2 \cos \alpha$ and the corresponding arm $\bar{r} = l/2$. The *moment-of-momentum* equation then gives us

$$M_\theta = 2 \left[\underbrace{\frac{l}{2}}_{\bar{r}} \underbrace{\left(\frac{q}{2A_2} \cos \alpha \right)}_{V_\theta} \right] \underbrace{\left(\rho \frac{q}{2} \right)}_{\substack{\text{mass} \\ \text{efflux}}} = \frac{\rho l q^2 \cos \alpha}{4 A_2} \qquad (a)$$

where M_θ is the torque from an outside agent to hold the rotor stationary. Note that the entering water having zero arm about the rotation axis does not appear in Eq. (a).

We now examine a problem wherein one of the important flows at the control surface is not a one-dimensional flow.

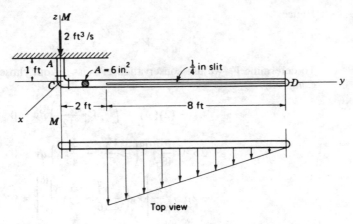

FIGURE 5.31
Pipe flow with efflux not a one-dimensional flow.

Example 5.13. A slit of thickness $\frac{1}{4}$ in. in pipe CD is so shaped on the inside that a sheet of water of uniform thickness $\frac{1}{4}$ in. issues out radially from the pipe (see Fig. 5.31). The velocity varies linearly along the pipe as shown, and 2 ft^3/s of water enters at the top. Find the moment about MM from the flow of water inside the pipe system and the air outside.

  We choose as a control volume the interior of the pipe. Next we make the following assumptions:

---

  1. Steady flow.
  2. Incompressible flow.
  3. Sheet of exiting water is at p_{atm}.
  4. 1-D flow entering the flow.

  As a first step we shall determine the velocity of the exiting flow as a function of y. We start with the equation of a straight line for V. Thus

$$V = my + b$$

We subject this to the conditions that

$$\text{at} \quad y = 10 \text{ ft} \quad V = 0 \text{ ft/s}$$

$$\text{at} \quad y = 2 \text{ ft} \quad V = V_0 \text{ ft/s}$$

where V_0 is as yet undetermined. We find that

$$m = -\frac{V_0}{8} \qquad b = \frac{5}{4}V_0$$

Hence

$$V = -\frac{V_0}{8}y + \frac{5}{4}V_0 \qquad (a)$$

To determine V_0 we use conservation of mass for our chosen control volume. *Continuity:*

$$-(2)(\rho) + \int_2^{10} \rho V \left(\frac{1/4}{12}\right) dy = 0$$

$$-2\rho + \frac{\rho}{48} \int_2^{10} \left(-\frac{V_0}{8}y + \frac{5}{4}V_0\right) dy = 0 \qquad (b)$$

$$-2\rho + \frac{\rho}{48}\left[V_0\left(-\frac{y^2}{16} + \frac{5}{4}y\right)\right]\Bigg|_2^{10} = 0$$

$$\therefore V_0 = 24 \text{ ft/s}$$

Hence Eq. (a) becomes

$$V = -3y + 30 \text{ ft/s} \qquad (c)$$

Next we go to the *moment-of-momentum* equation about axis *MM*:

$$\oint_{CS} \bar{r}T_\theta \, dA + \iiint_{CV} \bar{r}B_\theta \rho \, dv = \oint_{CS} \bar{r}V_\theta(\rho\mathbf{V} \cdot \mathbf{dA}) + \frac{\partial}{\partial t}\iiint_{CV} \bar{r}V_\theta(\rho \, dv)$$

Clearly the expressions involving volume integrals are zero since gravity is parallel to the axis *MM* and the fact that the flow is steady (1) and/or incompressible (2). We then have

$$\underbrace{(p_{\text{atm}})(8)\left(\frac{1/4}{12}\right)(6)}_{\text{at exit slot}} + \underbrace{T_{MM}}_{\substack{\text{from pipe} \\ \text{wall}}} = -\int_2^{10}[y(-3y+30)](\rho)(-3y+30)\left(\frac{1/4}{12}\right) dy$$

The minus sign on the right side obtains because V_θ here for cylindrical coordinates is negative. Note that both the traction force at the entrance and the entering velocity are parallel to *MM*, thus yielding zero moment about axis *MM* for that part of the control surface. Solving for T_{MM} we have

$$T_{MM} = -\frac{1.938}{48}\int_2^{10}(30^2 y - 180y^2 + 9y^3) \, dy - p_{\text{atm}}\left[\frac{1/4}{12}8\right](6)$$

$$= -248 - p_{\text{atm}} \text{ ft-lb}$$

Taking the reaction we get the torque from the water on the pipe. That is,

$$(T_{\text{pipe}})_{MM} = 248 + p_{\text{atm}} \text{ ft-lb} \qquad (d)$$

At this point we can include the torque on the pipe from *both water inside* and *air outside* by simply using gage pressures in the above formulation as was discussed earlier for linear momentum. We thus get for the *total* torque

$$(T_{MM})_{\text{total}} = 248 \text{ ft-lb}$$

5.9 MOMENT-OF-MOMENTUM EQUATION APPLIED TO PUMPS AND TURBINES

In Sec. 5.5 we defined a turbomachine as a device which alters the motion of an unconfined fluid in such a way as to transmit a torque on a rotor or to develop a propulsive thrust. At that time, we discussed the turbojet engine briefly in an example. This turbomachine provides propulsive thrust directly. Other such turbomachines are rocket motors and ram jets. At this time, we turn our attention to those pumps and turbines which we are able to classify as turbomachines. A more detailed study will be found in Chap. 15. We will introduce the subject here primarily to see the use of momentum and moment-of-momentum equations in this area of study.

Turbines extract useful work from fluid energy, while pumps, just oppositely, add energy to fluids. This is accomplished by directing the fluid-flow field with vanes fixed rigidly to a shaft. If, in its passage through the turbine or pump, the fluid is at no time confined, we can consider the device as a turbomachine. Otherwise, it is called a *positive-displacement* machine. In the case of a turbine the assembly of blades that is attached to the shaft is often called a *runner*, while in pumps this assembly is called an *impeller*.

Turbines are generally classified either as *impulse turbines* or as *reaction turbines*. Impulse turbines are driven by one or several high-speed free jets of fluid very much like the simple device considered in Example 5.6. The kinetic energy of the jet is developed outside the runner in a nozzle wherein the fluid goes from a high to a low pressure in attaining its kinetic energy. If friction and gravity are neglected, the jet undergoes no change in speed relative to the vane in the runner over which it moves and the fluid emerging from the runner is at the same pressure as when it entered. Hence, the fundamental characteristic of an impulse turbine is that the expansion in pressure takes place *in the nozzle* and not in the blades. In reaction turbines, fluid enters the runner over the *entire periphery* of the runner and at a high pressure. This pressure is reduced as the fluid flows through the runner and at exit is at a lower pressure. Thus, even if friction and gravity are disregarded, the speed of the fluid relative to the runner vanes is not constant, as was the case of the impulse turbine. Instead, the speed will increase along the runner vanes, so one sees that the runner acts partly as a nozzle. The flow pattern required for the fluid to enter the runner properly is developed with the use of stationary vanes called *guide vanes* (Fig. 5.32) which are placed outside the periphery of the runner. Because the runner in a reaction turbine is always full of fluid, in contrast to an impulse turbine, where only a few vanes direct fluid at any one time, it is clear that for the same output, the reaction turbine has a smaller rotor than has the impulse turbine.

Example 5.14. Pelton water wheel. A simple impulse type of turbomachine is shown in Fig. 5.33 wherein a single jet of water issues out of a nozzle and impinges on the system of buckets attached to a wheel. The runner, which is the assembly of buckets and wheel, has a radius of r to the center of the buckets. The shape of the bucket is also shown in Fig. 5.33 where a horizontal midsection of the bucket has

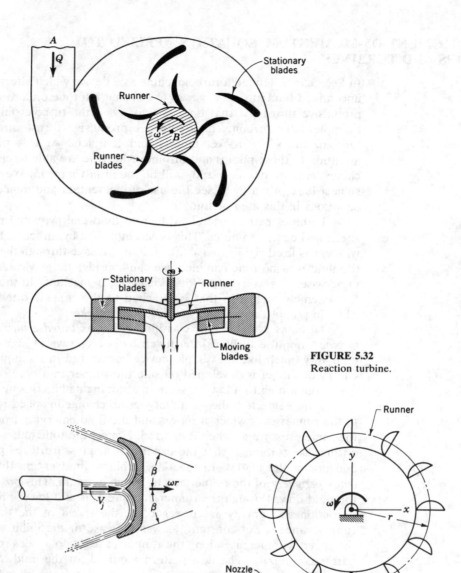

FIGURE 5.32
Reaction turbine.

FIGURE 5.33
Pelton water wheel.

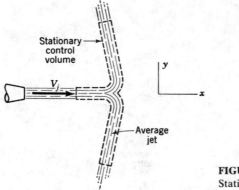

FIGURE 5.34
Stationary average jet.

been shown. Note that the jet is split in two parts by the bucket and is rotated almost 180° relative to the bucket in the horizontal plane. If we neglect gravity and friction, the speed of the water relative to the bucket is unchanged during the action. If Q cubic feet per second of water issues out of the nozzle at a velocity V_j and the runner is loaded by a generator so as to rotate at a constant angular speed of ω radians per second, what is the torque and power developed by the water on the water wheel?

We shall do this problem in two ways, namely, by the method of linear momentum and by the method of moment of momentum.

Method 1. Linear momentum. Note first that the flow of water is not quite steady relative to the ground, since the orientation of each bucket is continually changing relative to the impinging jet. Furthermore, there may be differing numbers of buckets in various degrees of contact with the jet from instant to instant. However, the variations in torque arising from these departures from steady flow are small and not of interest here. To compute the *average* torque, we must consider that there is an *average* steady flow of the jet through the runner and that this steady flow is *identical* to the *instantaneous* flow of a full jet when it is impinging on a bucket at its lowermost position, i.e., when the bucket is moving horizontally with a constant speed of ωr. We have shown this average jet in Fig. 5.34 and have selected a control volume around a portion of this jet so as to expose that part of the bucket surface. A_w, wetted by the jet. This control volume is *stationary*. The velocity of the fluid entering is understood to be V_j, but we must be careful in ascertaining the exit velocity of the fluid relative to this stationary control volume. To do this, we consider the flow along the representative bucket which we have decided gives rise to the average jet. The velocity of the fluid relative to this bucket is $V_j - \omega r$ in Fig. 5.35, where we have shown on the left the view of the jet as seen from an observer on the bucket. In the absence of friction and gravity this relative velocity is unchanged in magnitude, as discussed in earlier problems, so that the velocity of the fluid leaving the bucket has a magnitude $V_j - \omega r$ relative to the bucket, as has been shown in Fig. 5.35. To ascertain the exit velocity V_2 relative to the *ground*, we must *add* to the exit velocity relative to the bucket the velocity ωr of the bucket relative to the ground. This has been shown graphically in the vector diagram on the right side of Fig. 5.35, for half of the

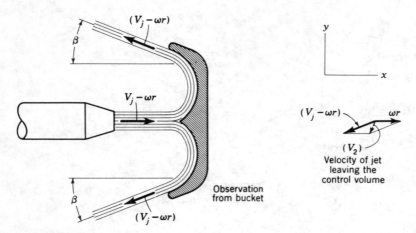

FIGURE 5.35
Moving bucket.

exiting fluid and it should now be clear why the stationary control volume enveloping the average jet has the shape shown in Fig. 5.34.

We can now employ the linear momentum equation for our stationary control volume. In the x direction we then have

$$p_{\text{atm}}(A_w)_x + R_x = -Q\rho V_j + Q\rho\left[-(V_j - \omega r)\cos\beta + \omega r\right] \qquad (a)$$

where, in computing the x component of the linear momentum leaving the control volume, we employed the components of the exit velocity with the aid of the vector diagram in Fig. 5.35. Taking the reaction to R_x gives us the average force from the jet on the buckets at an average position. Thus, collecting terms, we get

$$K_x = Q\rho(V_j - \omega r)(1 + \cos\beta) + p_{\text{atm}}(A_w)_x \qquad (b)$$

Now, upon considering the affect of atmospheric pressure on the nonwetted part of the bucket, we simply use gage pressure in Eq. (b) so that the total average force $(K_T)_x$ is computed. Thus

$$(K_T)_x = Q\rho(V_j - \omega r)(1 + \cos\beta) \qquad (c)$$

The average torque on the water wheel is then

$$\text{Torque} = Q\rho r(V_j - \omega r)(1 + \cos\beta) \qquad (d)$$

and the power developed is

$$\text{Power} = (\text{torque})(\omega) = Q\rho r\omega(V_j - \omega r)(1 + \cos\beta) \qquad (e)$$

Of importance is the question of what relation must exist between ω and V_j for

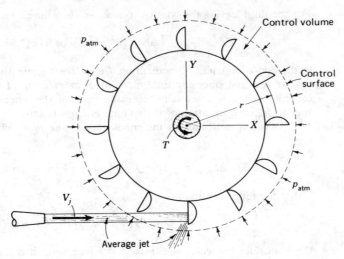

FIGURE 5.36
Control volume for moment of momentum.

maximum power. You will be asked to show in an exercise that $V_j = 2\omega r$ for maximum power.

Method 2. Moment of momentum. We will choose as our control volume a stationary volume that completely encloses the water wheel as shown in Fig. 5.36. The control surface cuts the shaft of the water wheel on both sides of the wheel. The average jet, as discussed in method 1, has been shown. We take the component of the moment-of-momentum equation along the centerline of the water wheel, so we can use Eq. (5.29) which we rewrite for convenience.

$$\oint_{CS} \bar{r}T_\theta \, dA + \iiint_{CV} \bar{r}B_\theta \rho \, dv = \oint_{CS} (\bar{r}V_\theta)(\rho \mathbf{V} \cdot \mathbf{dA}) + \frac{\partial}{\partial t} \iiint_{CV} (\bar{r}V_\theta)(\rho \, dv) \quad (a)$$

We have atmospheric pressure everywhere along the control surface except where it cuts the shaft, so the only torque from the force distribution on the control surface stems from the force distribution at these two cut surfaces of the shaft. This torque, which we denote simply as M_{shaft}, is clearly the torque transmitted to the wheel from the shaft. With axial symmetry for the wheel, and neglecting the torque from the gravitational action on that part of the average jet inside the control volume, we see that $\iiint rB_\theta \rho \, dv$ is zero. M_{shaft} is thus the only torque involved in Eq. (a). We next consider the flow of moment of momentum. For fluid entering the control volume, this is easily computed as $-rV_j\rho Q$. To ascertain the corresponding quantity for flow leaving the control volume, we must use the velocity of the outgoing jet *relative to the ground*. Observing Fig. 5.35 we get for cylindrical coordinates, using the components of the exit velocity,

$$V_\theta = \omega r - (V_j - \omega r)\cos \beta$$

For the fluid leaving the control volume we then have

$$r\big[\omega r - (V_j - \omega r)\cos\beta\big]\rho Q$$

as the flow of moment of momentum. Considering finally the last expression in Eq. (*a*), we see that since the total moment of momentum of the material inside the control volume is essentially the same value if the wheel rotates with constant speed, the time variation of the volume integral is zero.

We can now express the *moment-of-momentum* equation in the following way:

$$M_{\text{shaft}} = -rV_j\rho Q + r\big[\omega r - (V_j - \omega r)\cos\beta\big]\rho Q \qquad (b)$$

Collecting terms, we have

$$M_{\text{shaft}} = -r(V_j - \omega r)(1 + \cos\beta)\rho Q \qquad (c)$$

If we consider the water wheel itself as a free body, it is seen that for constant ω, the torque from the shaft onto the wheel (i.e., the above value M_{shaft}) is equal and opposite to the torque developed by the action of the jet on the water wheel. So we can give the desired torque from the jet as

$$\text{Torque} = r(V_j - \omega r)(1 + \cos\beta)\rho Q \qquad (d)$$

which is the same result achieved by the method of linear momentum.

While the method of linear momentum can be used in simpler turbomachine problems, it becomes intractable in the more complicated problems, so the method of moment of momentum is generally to be preferred.

We have introduced the turbomachine in this chapter to illustrate the use of momentum equations as well as to give the reader at least a nodding acquaintance with this important class of machinery. As pointed out earlier, in Chap. 15 a much more detailed study of turbomachinery is presented for those readers desiring a more complete study of this subject.

Pumps and turbines on space vehicles can achieve very complex motions relative to inertial space, and the rather complex formulations we next develop will unavoidably become more useful in the future.

*5.10 MOMENT OF MOMENTUM FOR NONINERTIAL CONTROL VOLUMES

In the development in Sec. 5.8, the control volume and point A about which moments of forces and momenta are taken, are fixed in an inertial reference XYZ. Suppose now that we wish to consider the situation where the control volume and point (or the axis through A) are *not* fixed in inertial space but are fixed in a reference xyz whose origin O has an acceleration $\ddot{\mathbf{R}}$ relative to the inertial reference XYZ and whose axes have an

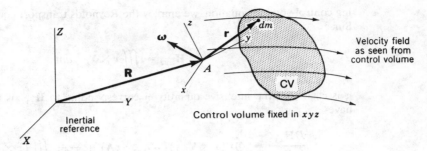

FIGURE 5.37
Noninertial control volume.

angular velocity $\boldsymbol{\omega}$ relative to XYZ about an axis of rotation going through O with possibly an angular acceleration $\dot{\boldsymbol{\omega}}$ as seen from the reference XYZ. We take the origin of xyz at A for simplicity (see Fig. 5.37). To express the moment-of-momentum equation in terms of observations from the noninertial control volume, we start with Newton's law for an element of fluid using xyz as the reference axes. We can accordingly say that

$$d\mathbf{F} - dm\left[\ddot{\mathbf{R}} + 2\boldsymbol{\omega} \times \mathbf{V}_{xyz} + \dot{\boldsymbol{\omega}} \times \mathbf{r} + \boldsymbol{\omega} \times (\boldsymbol{\omega} \times \mathbf{r})\right] = \frac{D}{Dt_{xyz}}\left(dm\,\mathbf{V}_{xyz}\right) \quad (5.30)$$

Taking the cross product of each of the above terms by $\mathbf{r}$, we get

$$\mathbf{r} \times d\mathbf{F} - dm\left\{\mathbf{r} \times \left[\ddot{\mathbf{R}} + 2\dot{\boldsymbol{\omega}} \times \mathbf{V}_{xyz} + \dot{\boldsymbol{\omega}} \times \mathbf{r} + \boldsymbol{\omega} \times (\boldsymbol{\omega} \times \mathbf{r})\right]\right\}$$

$$= \mathbf{r} \times \frac{D}{Dt_{xyz}}\left(dm\,\mathbf{V}_{xyz}\right)$$

$$= \frac{D}{Dt_{xyz}}\left(\mathbf{r} \times dm\,\mathbf{V}_{xyz}\right) \quad (5.31)$$

We have shown in Sec. 5.7 the validity of moving the differentiation operator D/Dt ahead of the $\mathbf{r}$, as we have done in the right side of the Eq. (5.31). Integrating these terms for all infinitesimal elements comprising the system at time t, we get

$$\mathbf{M}_S + \mathbf{M}_B - \iiint\limits_V \left\{\mathbf{r} \times \left[\ddot{\mathbf{R}} + 2\boldsymbol{\omega} \times \mathbf{V}_{xyz} + \dot{\boldsymbol{\omega}} \times \mathbf{r} + \boldsymbol{\omega} \times (\boldsymbol{\omega} \times \mathbf{r})\right]\right\}\rho\,dv = \frac{D\mathbf{H}_{xyz}}{Dt_{xyz}}$$

$$(5.32)$$

where $\mathbf{M}_S$ and $\mathbf{M}_B$ are, respectively, the moment about A of the traction force distribution and the body-force distribution and where $\mathbf{H}_{xyz}$ is the total moment of momentum of the system about point A as seen from xyz. To formulate the corresponding control-volume equation, we employ the Reynolds transport equation for DH_{xyz}/Dt. Since

$$\mathbf{H}_{xyz} = \iiint_V \mathbf{r} \times \mathbf{V}_{xyz}\, dm$$

it is clear that the intensive quantity η corresponding to $\mathbf{H}_{xyz}$ is $(\mathbf{r} \times \mathbf{V}_{xyz})$. Hence we have

$$\frac{D\mathbf{H}_{xyz}}{Dt} = \oiint_S (\mathbf{r} \times \mathbf{V}_{xyz})(\rho \mathbf{V}_{xyz} \cdot d\mathbf{A}) + \frac{\partial}{\partial t_{xyz}} \iiint_V (\mathbf{r} \times \mathbf{V}_{xyz})(\rho\, dv)$$

Replacing the right side of Eq. (5.32) using the above result, we get

$$\boxed{\begin{aligned}
\mathbf{M}_S + \mathbf{M}_B &- \iiint_{CV} \left\{ \mathbf{r} \times \left[\ddot{\mathbf{R}} + 2\boldsymbol{\omega} \times \mathbf{V}_{xyz} + \dot{\boldsymbol{\omega}} \times \mathbf{r} + \boldsymbol{\omega} \times (\boldsymbol{\omega} \times \mathbf{r}) \right] \right\} \rho\, dv \\
&= \iint_{CS} (\mathbf{r} \times \mathbf{V}_{xyz})(\rho \mathbf{V}_{xyz} \cdot d\mathbf{A}) + \frac{\partial}{\partial t_{xyz}} \iiint_{CV} (\mathbf{r} \times \mathbf{V}_{xyz})(\rho\, dv)
\end{aligned}}$$

(5.33)

Except for the vectors $\ddot{\mathbf{R}}$, $\boldsymbol{\omega}$, and $\dot{\boldsymbol{\omega}}$ all quantities and time derivatives are measured from the control volume in Eq. (5.33).

We will now simplify this equation for the case where $\boldsymbol{\omega}$ and $\dot{\boldsymbol{\omega}}$ have a *fixed* direction in inertial space. An example of such a case would be a control volume fixed to the compressor section of an aircraft jet engine where the aircraft is only translating relative to the ground, which is considered an inertial reference (see Fig. 5.38). Clearly, $|\dot{\boldsymbol{\omega}}|$ will be simply the rate of change of angular speed of the compressor. For such

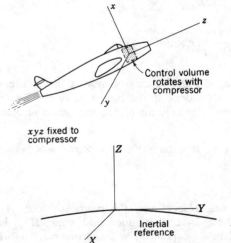

FIGURE 5.38
Accelerating jet plane.

problems it is convenient to choose A at a point whose acceleration is easily ascertained relative to XYZ. In the case of the aircraft shown in Fig. 5.38, A is chosen along the centerline of the compressor and hence has an acceleration corresponding to that of the plane itself. Furthermore, it is simplest to choose the x axes to coincide with the direction of $\boldsymbol{\omega}$ and $\dot{\boldsymbol{\omega}}$. We will use cylindrical coordinates to allow for more simple calculations and more meaningful results. We therefore decompose the vectors $\mathbf{r}$, $\mathbf{V}_{xyz}$, $\boldsymbol{\omega}$, and $\dot{\boldsymbol{\omega}}$ as follows:

$$\mathbf{r} = \bar{r}\boldsymbol{\epsilon}_{\bar{r}} + z\boldsymbol{\epsilon}_z \tag{5.34a}$$

$$\mathbf{V}_{xyz} = (V_{\bar{r}})_{xyz}\boldsymbol{\epsilon}_{\bar{r}} + (V_{\theta})_{xyz}\boldsymbol{\epsilon}_{\bar{r}} + (V_z)_{xyz}\boldsymbol{\epsilon}_z \tag{5.34b}$$

$$\boldsymbol{\omega} = \omega\boldsymbol{\epsilon}_z \tag{5.34c}$$

$$\dot{\boldsymbol{\omega}} = \dot{\omega}\boldsymbol{\epsilon}_z \tag{5.34d}$$

Before substituting these terms into Eq. (5.33), let us first consider the integral

$$- \iiint\limits_{\text{CV}} (\mathbf{r} \times \ddot{\mathbf{R}})(\rho \, dv)$$

This may be expressed as

$$\ddot{\mathbf{R}} \times \iiint\limits_{M} \mathbf{r} \, dm \tag{5.35}$$

since $\ddot{\mathbf{R}}$ is constant as far as the integration over the control volume is concerned. But

$$\iiint\limits_{M} \mathbf{r} \, dm$$

is the first moment of mass about point A and with the use of the center of mass can be replaced by $M\mathbf{r}_c$. Clearly for symmetrical mass distributions about A in the control volume this term is zero. We will employ the formulation (5.35) in rewriting Eq. (5.33), and we will use cylindrical components for $\mathbf{r}$, $\mathbf{V}_{xyz}$, $\boldsymbol{\omega}$, and $\dot{\boldsymbol{\omega}}$. To avoid cluttering the equation, we omit the subscript xyz on the velocities since by this time these subscripts should be superfluous. We get

$$\mathbf{M}_S + \mathbf{M}_B + \ddot{\mathbf{R}} \times M\mathbf{r}_c - \iiint\limits_{\text{CV}} (\bar{r}\boldsymbol{\epsilon}_{\bar{r}} + z\boldsymbol{\epsilon}_z) \times \left\{ (2\omega\boldsymbol{\epsilon}_z) \times (V_{\bar{r}}\boldsymbol{\epsilon}_{\bar{r}} + V_{\theta}\boldsymbol{\epsilon}_{\theta} + V_z\boldsymbol{\epsilon}_z) \right.$$

$$\left. + \dot{\omega}\boldsymbol{\epsilon}_z \times (\bar{r}\boldsymbol{\epsilon}_{\bar{r}} + z\boldsymbol{\epsilon}_z) + \omega\boldsymbol{\epsilon}_z \times [\omega\boldsymbol{\epsilon}_z \times (\bar{r}\boldsymbol{\epsilon}_{\bar{r}} + z\boldsymbol{\epsilon}_z)] \right\} (\rho \, dv)$$

$$= \oiint\limits_{\text{CS}} \left[(\bar{r}\boldsymbol{\epsilon}_{\bar{r}} + z\boldsymbol{\epsilon}_z) \times (V_{\bar{r}}\boldsymbol{\epsilon}_{\bar{r}} + V_{\theta}\boldsymbol{\epsilon}_{\theta} + V_z\boldsymbol{\epsilon}_z) \right] (\rho\mathbf{V} \cdot d\mathbf{A})$$

$$+ \frac{\partial}{\partial t} \iiint\limits_{\text{CV}} \left[(\bar{r}\boldsymbol{\epsilon}_{\bar{r}} + z\boldsymbol{\epsilon}_z) \times (V_{\bar{r}}\boldsymbol{\epsilon}_{\bar{r}} + V_{\theta}\boldsymbol{\epsilon}_{\theta} + V_z\boldsymbol{\epsilon}_z) \right] (\rho \, dv) \tag{5.36}$$

The components of Eq. (5.36) in the radial, transverse, and axial directions then become

Radial component:

$$(\mathbf{M}_S + \mathbf{M}_B + \ddot{\mathbf{R}} \times M\mathbf{r}_c)_{\bar{r}} + \iiint\limits_{\text{CV}} (2\omega z V_{\bar{r}} + \dot{\omega}\bar{r}z)(\rho \, dv)$$

$$= - \oiint\limits_{\text{CS}} (zV_{\theta})(\rho\mathbf{V} \cdot d\mathbf{A}) - \frac{\partial}{\partial t} \iiint\limits_{\text{CV}} (zV_{\theta})(\rho \, dv) \tag{5.37a}$$

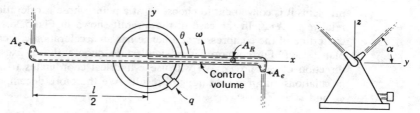

FIGURE 5.39
Rotating sprinkler.

Transverse component:

$$\left(\mathbf{M}_S + \mathbf{M}_B + \ddot{\mathbf{R}} \times M\mathbf{r}_c\right)_\theta + \iiint_{CV} \left(2\omega z V_\theta + \omega^2 \bar{r} z\right)(\rho \, dv)$$

$$= \oiint_{CS} \left(z V_{\bar{r}} - \bar{r} V_z\right)(\rho \mathbf{V} \cdot \mathbf{dA}) + \frac{\partial}{\partial t} \iiint_{CV} \left(z V_{\bar{r}} - \bar{r} V_z\right)(\rho \, dv) \qquad (5.37b)$$

Axial component:

$$\left(\mathbf{M}_S + \mathbf{M}_B + \ddot{\mathbf{R}} \times M\mathbf{r}_c\right)_z - \iiint_{CV} \left(2\omega \bar{r} V_{\bar{r}} + \dot{\omega} \bar{r}^2\right)(\rho \, dv)$$

$$= \oiint_{CS} \left(\bar{r} V_\theta\right)(\rho \mathbf{V} \cdot \mathbf{dA}) + \frac{\partial}{\partial t} \iiint_{CV} \left(\bar{r} V_\theta\right)(\rho \, dv) \qquad (5.37c)$$

We have thus simplified Eq. (5.36) to the case where $\boldsymbol{\omega}$ and $\dot{\boldsymbol{\omega}}$ are of constant direction in inertial space. These equations permit us to evaluate the torques coming onto a device from the supports. Note that even if the flow is steady relative to the control volume, one still requires knowledge of the flow of all particles inside the control volume in order to evaluate the volume integrals on the left side of the equations.

We will now illustrate the use of these equations in the following example.

Example 5.15. Let a sprinkler similar to the one in Example 5.12 rotate with constant angular velocity ω (Fig. 5.39). Compute the torque on the rotor about the axis of rotation from the water moving through inside and the air pressure outside.

The control volume consists of the interior of the rotor and is consequently noninertial. We choose the z axis to be the axis of rotation and the x axis to be the centerline of the rotor as has been shown in Fig. 5.39. The equation we must use, clearly, is Eq. (5.37c). The flow is steady relative to the rotor; so the last expression in this equation is zero. Furthermore, $\ddot{\mathbf{R}}$ and $\dot{\boldsymbol{\omega}}$ are both zero vectors. The body force is gravity, and it obviously cannot exert a torque about the z axis, leaving as the only real torque the torque from surface forces, which we denote as M_S. Keeping in mind that velocities are measured relative to the control volume, we can then say

$$M_S - \iiint_{CV} (2\omega \bar{r} V_{\bar{r}})(\rho \, dv) = \oiint_{CS} (\bar{r} V_\theta)(\rho \mathbf{V} \cdot \mathbf{dA}) \qquad (a)$$

The torque acting on the water in the control volume consists of the torque from the rotor wall, which we denote as T_{rotor}, and the torque from the atmospheric pressure in contact with the fluid at the rotor outlets. Denoting the latter as T_{atm}, we have for the total real torque

$$M_S = T_{\text{rotor}} + T_{\text{atm}} \qquad (b)$$

The torque from the hypothetical force, neglecting the water in the two small nozzles, is easily evaluated in the following manner:

$$\iiint_{CV} (2\omega \bar{r} V_{\bar{r}})(\rho \, dv) = 2\omega \iiint_{CV} \bar{r} \left(\frac{q}{2A_R} \right)(\rho \, dv) = \frac{\omega q \rho}{A_R} \iiint_{CV} \bar{r} \, dv \qquad (c)$$

Upon noting that $dv = A_R \, d\bar{r}$, the above integral can be evaluated as

$$\frac{\omega q \rho}{A_R} \iiint_{CV} \bar{r} \, dv = 2\omega q \rho \int_0^{l/2} \bar{r} \, d\bar{r} = 2\omega q \rho \frac{(\bar{r})^2}{2} \Bigg]_0^{l/2} = \frac{\omega q \rho l^2}{4} \qquad (d)$$

Finally we have for the flow of moment of momentum through the control surfaces, with $\bar{r} = l/2$ and $V_\theta = -(q/2A_e)\cos \alpha$ at the nozzles,

$$\oiint_{CS} (\bar{r} V_\theta)(\rho \mathbf{V} \cdot \mathbf{dA}) = -\frac{l}{2} \frac{q}{2A_e}(\cos \alpha) \frac{\rho q}{2} - \frac{l}{2} \frac{q}{2A_e}(\cos \alpha) \frac{\rho q}{2}$$

$$= -\frac{\rho q^2 l \cos \alpha}{4A_e} \qquad (e)$$

We have used continuity several times in Eq. (e) to evaluate the velocities. Equation (a) can now be given as

$$T_{\text{rotor}} + T_{\text{atm}} - \frac{\omega q \rho l^2}{4} = -\frac{\rho q^2 l}{4A_e} \cos \alpha \qquad (f)$$

The torque from the water on the rotor is then the reaction to T_{rotor}. Thus

$$T_{\text{water}} = -T_{\text{rotor}} = T_{\text{atm}} - \frac{\omega q \rho l^2}{4} + \frac{\rho q^2 l}{4A_e} \cos \alpha \qquad (g)$$

Now using gage pressures, $T_{\text{atm}} = 0$, and we then get the total torque on the rotor from water and air

$$T_{\text{air, water}} = -\frac{\omega q \rho l^2}{4} + \frac{\rho q^2 l}{4A_e} \cos \alpha \qquad (h)$$

5.11   CLOSURE

In this important chapter we presented two of the basic laws for finite systems and finite control volumes. And just as you were required to draw carefully considered free body diagrams in your sophomore mechanics courses, it should be patently clear that the same care has to be exercised in drawing appropriate control volumes. Now in addition you should list assumptions made in constructing the flow model you are using to represent the actual problem. All told we presented a rather formal attack on each problem. You are urged at least until full mastery has been achieved to follow a similar formal, methodical approach.

In the next chapter we will present the first law of thermodynamics and from it the very useful Bernoulli equation. We will merely introduce the second law of thermodynamics and leave for its use to a later chapter when it will be more directly needed.

The Reynolds transport equation will once again be featured in the development of the first law of thermodynamics. You should emerge from Chaps. 5 and 6 with a good feel and sound appreciation of the Reynolds transport equation.

PROBLEMS

Problem Categories

Conservation of Mass 5.1–5.18

Linear Momentum 5.15–5.71 (Probs. 5.39–5.44, 5.52–5.53, 5.69–5.70 deal with discrete particles)

Noninertial Control Volumes (Linear Momentum) 5.71–5.97

Moment of Momentum 5.78–5.96

Noninertial Control Volume (Moment of Momentum) 5.98–5.103

Starred Problems

5.70

5.1. A flow of 0.3 m³/s of water enters a rectangular duct. Two of the faces of the duct are porous. On the upper face water is added at a rate shown by the parabolic curve; on the front face a portion of the water leaves at a rate deter-

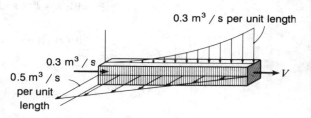

FIGURE P5.1

mined linearly by the distance from the end. The maximum values of both rates are given in cubic meters per second per unit length along the duct. What is the average velocity V of the water leaving the end of the duct if it is 0.3 m long and has a cross section of 0.01 m²?

5.2. In Prob. 5.1, determine the position along the duct where the average velocity of flow along the duct is a maximum.

5.3. Hot steel is being rolled at a rolling mill. The steel emerging from the rollers is 10 percent more dense than before entering. If the steel is being fed at the rate of 0.2 m/s, what is the

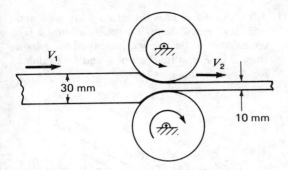

FIGURE P5.3

speed of the rolled material? There is a 9 percent increase in the width of the steel.

5.4. Shown is a device into which 0.3 m³/s of water is admitted at the axis of rotation and directed out radially through three identical channels whose exit areas are each 0.05 m² perpendicular to the direction of flow relative to the device. The water leaves at an angle of 30° relative to the device as measured from a radial direction, as is shown in the diagram. If the device rotates clockwise with a speed of 10 rad/s relative to the ground, what is the magnitude of the average velocity of the fluid leaving the vane as seen from the ground?

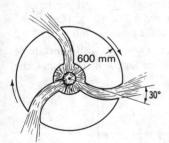

FIGURE P5.4

5.5. If the device in Prob. 5.4 has a clockwise angular acceleration of 5 rad/s² at the instant that the other data are given and the rate of increase of influx is 0.03 m³/s², what is the magnitude of the acceleration of the water leaving the device relative to the ground? Take the fluid as completely incompressible for the calculations.

5.6. Water is forced into the device at the rate of 0.1 m³/s through pipe *A*, while oil of specific gravity 0.8 is forced in at the rate of 0.03 m³/s through pipe *B*. If the liquids are incompressible and form a homogeneous mixture of oil globules in water, what is the average velocity and density of the mixture leaving through pipe *C* having a 0.3-m diameter?

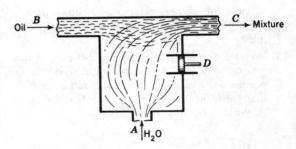

FIGURE P5.6

5.7. In Prob. 5.6 the piston at *D* having a 150-mm diameter moves at the rate of 0.3 m/s to the left. What is the average velocity of the fluid leaving at *C*?

5.8. Water flows into a cylindrical tank through pipe ① at the rate of 20 ft/s and leaves through pipes ② and ③ at the rates of 8 ft/s and 10 ft/s, respectively. At ④, we have an open air

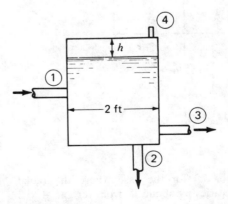

FIGURE P5.8

vent. Using the entire inside volume of the tank as a control volume, what is dh/dt? The following are the inside diameters of the pipes:

$$D_1 = 3 \text{ in}$$
$$D_2 = 2 \text{ in}$$
$$D_3 = 2\frac{1}{2} \text{ in}$$
$$D_4 = 2 \text{ in}$$

What is the average velocity of airflow through vent ④, assuming that the flow is incompressible?

5.9. Water is flowing downward in a 2-ft pipe at the rate of 50 ft³/s. It enters a conical section with porous walls such that there is a radial outflow that varies linearly from zero at A to 3 ft/s at B. What is the flow at B?

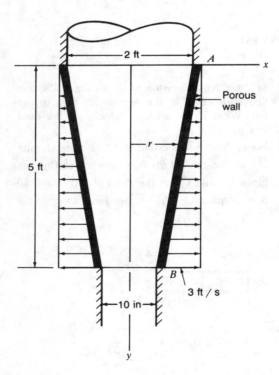

FIGURE P5.9

5.10. Do Prob. 5.9 for the case where the radial velocity varies parabolically from zero at A to 3 ft/s at B.

5.11. 5 ft³/s of water goes through a 12-in pipe and then is directed through a region around a 60° cone. What is the average velocity in the region from C to E as a function of η and δ? Evaluate for $\delta = 2$ in and $\eta = 16$ in.

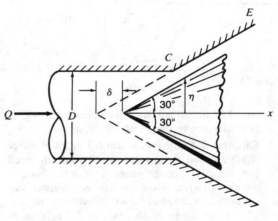

FIGURE P5.11

5.12. Water is flowing in at ① into a rectangular tank A with a length of 5 ft and a width of 5 ft. The rate of flow Q_1 at ① is 5 ft³/s. At the instant of interest, $h_1 = 15$ ft and water is flowing into tank B through ③ at the rate of 4 ft³/s. At this instant, h_2 is 12 ft. If the free surface in tank B is dropping at the rate of 0.2 ft/s, what is the flow Q_2 at ② at the instant of interest? Tank B is of length 8 ft and has the

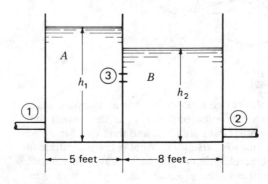

FIGURE P5.12

same width as tank A. Also, what is the rate at which h_1 is changing value?

5.13. A rectangular ditch of width 10 m has a sloping bottom as shown. Water is added at the rate Q of 100 L/s. What is dh/dt when $h = 1$ m? How long does it take for the free surface to go from $h = 1$ m to $h = 1.2$ m?

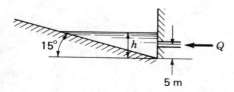

FIGURE P5.13

5.14. Suppose in Prob. 5.13 that there is influx q normal to the sloping bottom of 2 L/m^2 of the bottom. What is dh/dt and the time Δt for h to go from 1 to 1.2 m? *Hint:*

$$\int \frac{x\,dx}{a + bx} = \frac{1}{b^2}\left[a + bx - a\ln(a + bx)\right]$$

5.15. A rectangular cube a meters on an edge is moving in space at a very high velocity V m/s. The box is designed to capture solar dust particles inside the box as they strike the box. Because the speed of the box is much faster than the speed of the dust particles, we can assume that the latter are stationary. If there are n dust particles per unit volume, and if N represents the number of dust particles inside the box,

what is the rate of accumulation of dust particles in the box (dN/dt)? What is the total number of dust particles ΔN collected during a time interval Δt seconds?

5.16. Do Prob. 5.15 using a sphere of radius a as the dust collector. Start by considering a strip of length $a\,d\theta$, then integrate to cover surface of impact. What general simple rule can you now state in words for a collector of any shape? *Hint:*

$$\int \sin\theta\cos\,d\theta = \frac{\sin^2\theta}{2}$$

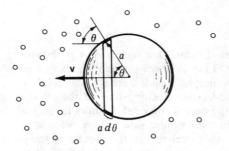

FIGURE P5.16

5.17. A Kingsbury thrust bearing consists of a number of small pads whose bottom surface is inclined

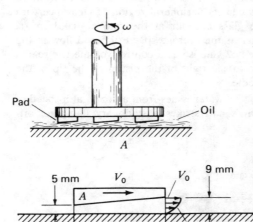

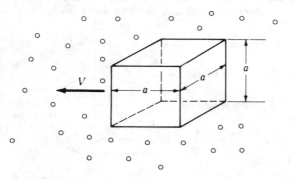

FIGURE P5.15

FIGURE P5.17

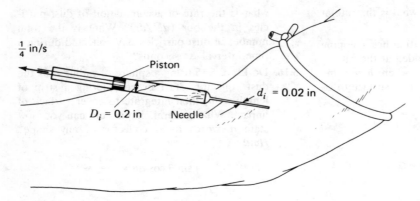

FIGURE P5.18

to the horizontal. One such pad A is shown in detail. Oil moves under the pad; and because of the flow under the pad, there is a vertical thrust developed. The velocity profile of the oil relative to the base is shown at the inlet to the pad. If we do not consider side leakage, what is the average speed relative to the ground of the oil on leaving the region under the pad? V_0 is the average velocity of the pad relative to the stationary base.

5.18. A nurse is withdrawing blood from a patient (Fig. P5.18). The piston is being withdrawn at a speed of $\frac{1}{4}$ in/s. The piston allows air to move through its peripheral region of clearance with the glass cylinder at the rate of 0.001 in^3/s. What is the average speed of blood flow in the needle? Choose as a control volume the region just to the right of the piston to the tip of the needle.

5.19. A jet of water issues from a nozzle at a speed of 6 m/s and strikes a stationary flat plate ori-

ented normal to the jet. The exit area of the nozzle is 645 mm^2. What is the total horizontal force on the plate from the fluids in contact with it? Solve this problem using two different control volumes.

5.20. In Prob. 5.19 the nozzle is moving with a speed of 1.5 m/s to the left relative to the ground.
(*a*) If the water issues out at 6 m/s relative to the nozzle, what is the horizontal force on the plate from all fluids?
(*b*) If, in addition, the plate is moving at a uniform speed of 3 m/s to the right relative to the ground, what is the horizontal force on the plate from the fluids?
Use only one control volume in each case.

5.21. For example 5.6, determine the value of V_0 and θ for maximum power.

5.22. What is the force on the elbow-nozzle assembly from the water and air? The water issues out as

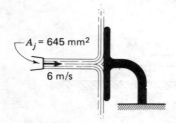

FIGURE P5.19

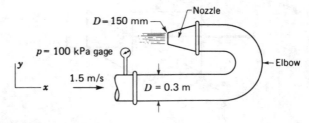

FIGURE P5.22

a free jet from the nozzle. The interior volume of the nozzle elbow assembly is 0.1 m³.

5.23. Find the horizontal force on the device of Prob. 5.6 if the oil enters at a pressure of 1.4×10^5 Pa gage, the water at 1.2×10^5 Pa gage, and the mixture leaves at a pressure of 1.0×10^5 Pa gage. Pipe B has a diameter of 0.5 m, pipe A has a diameter of 0.3 m, and pipe C has a diameter of 0.3 m. From Prob. 5.6 we have $\rho_c = 955.5$ kg/m³ and $V_c = 1.914$ m/s.

5.24. A dredging operation delivers 5000 gal/min of a mixture of mud and water having a specific gravity of 3 into a stationary barge. What is the force on the barge which tends to separate the barge from the dredger? The area of the nozzle exit is 1 ft².

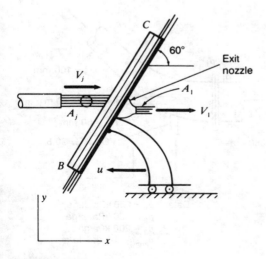

FIGURE P5.24

5.25. A trough in Fig. P5.25 moves at constant speed $u = 2$ m/s. A jet of water having a speed of $V_j = 6$ m/s impinges on the trough as shown. The water leaves the trough in three places. At the exit nozzle, the speed of water V_1 is 10 m/s relative to the trough. The area $A_1 = 0.02$ m² while the area $A_j = 0.08$ m². Twice as much water leaves at B then leaves at C. Compute thrust on the trough. Use a control volume that does not cut the trough support. Assume no friction and no affect of gravity on the *unconfined* flow in the trough itself. However, the exit nozzle flow results in a different fluid exit velocity because in the nozzle the flow is *confined* and squirts out at a higher velocity relative to the trough.

5.26. Do Prob. 5.25 using a control volume that cuts the trough support.

5.27. We are looking down from above on a large tank of water which is connected to a 12-in horizontal pipe. The water, once in the pipe, has a speed of 5 ft/s before reaching the end of a second thin pipe, AB, through which water is pumped at a speed of 25 ft/s. The pressure p_1 in the main stream at the position shown is 5 psig and at A the high speed jet emerges as a free jet. At about 3 ft from A the two flows are thoroughly mixed. If we neglect friction at the walls of the 12-in pipe, what is the pressure p_2?

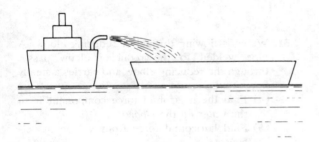

FIGURE P5.25

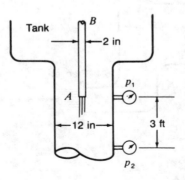

FIGURE P5.27

5.28. What is the dynamic force (i.e., excluding gravity) on the flat plate from the water on one side and air on the other? Water is at 10°C?

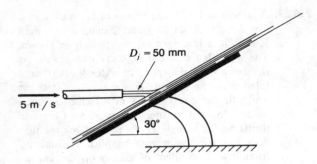

FIGURE P5.28

5.29. Water flows at a steady rate through the device shown. The following data apply:

$$p_1 = 20 \text{ psig}$$
$$V_1 = 10 \text{ ft/s}$$
$$D_1 = 15 \text{ in}$$
$$D_2 = 8 \text{ in}$$
$$D_3 = 4 \text{ in}$$
$$V_2 = 20 \text{ ft/s}$$

What is the horizontal thrust from water and air?

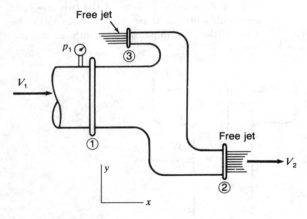

FIGURE P5.29

5.30. Water is moving steadily through a double-exit elbow for which $V_1 = 5$ m/s. The inside volume of the elbow is 1 m³. Find the vertical and horizontal forces from air and water on the elbow. Take $V_2 = 10$ m/s.

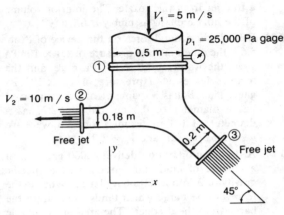

FIGURE P5.30

5.31. Water is flowing through a reducing elbow. A pipe welded to the reducing elbow passes through the reducing elbow and carries a steady flow of oil.
   (a) Find the horizontal force component from the *water* on the *elbow*.
   (b) Find horizontal force from *air* on the elbow.

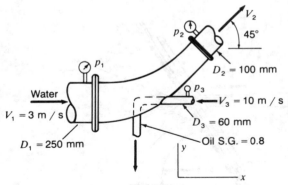

Data: $p_1 = 250$ kPa gage
$p_2 = 180$ kPa gage
$p_3 = 200$ kPa gage
$p_{atm} = 101.325$ kPa

FIGURE P5.31

(*c*) Find horizontal force from *oil* on the elbow.

(*d*) Give total horizontal force on the elbow from water, air, and oil.

5.32. A dustcropper is spraying a field with an insecticide at the rate of 0.01 m³/s. The fluid is coming out as *free jets* from six openings each of diameter 20 mm. If the coefficient of drag C_D for the horizontal part of the device extending from the plane is 0.45, compute:

(*a*) the moment at the base at *A* from the fluid flow inside the pipe. Do not use gauge pressure.

(*b*) the moment at the base at *A* from the air flow over the base portion of the system outside.

(*c*) the total moment at *A*.

Note: We will later learn that for the drag force F_D we have $F_D = \frac{1}{2}C_D A \rho V^2$ where *A* is the frontal projected area of the surface of the base portion in the direction of the velocity and ρ is the density of the air.

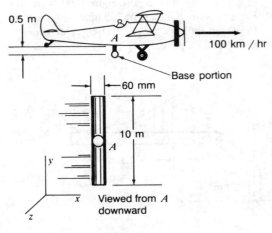

$\rho_{insecticide} = 900\ kg\,/\,m^3$

$\text{Air:}\begin{cases} T = 30°\text{C} \\ R = 287\ \text{N}\cdot\text{m}\,/\,\text{kg K} \\ p = 101,325\ \text{Pa} \end{cases}$

FIGURE P5.32

5.33. 30 L/s of water at 10°C enter a jet pump at *A* at a pressure $p_1 = 300,000$ Pa gage. Oil is sucked

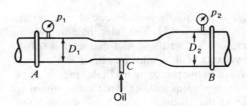

FIGURE P5.33

in at *C* at the rate of 1 L/s. The oil has a specific gravity of 0.65. A thoroughly mixed flow of water and oil leave at *B* at a pressure p_2 of 150,000 Pa gage. The dimensions of D_1 and D_2 are 200 and 250 mm, respectively. What is the horizontal thrust on the pump from water, oil, and air? Density of the water is 999.7 kg/in³.

5.34. A *vertical* system in Fig. P5.34 conducts water from a large reservoir at the rate of 1 m³/s. At the tee at *B*, $\frac{1}{3}$ m³/s goes to the left and $\frac{2}{3}$ m³/s goes to the right. Pipe *EB* weighs 1 kN/m, pipe *AB* weighs 0.6 kN/m, and pipe *BC* weighs 0.8 kN/m. Find the *total* vertical and horizontal forces on the pipes from fluid flow and air as well from gravity on water and pipe. Free jets are at *A* and *C*. $p_E = 390.4$ kPa gage.

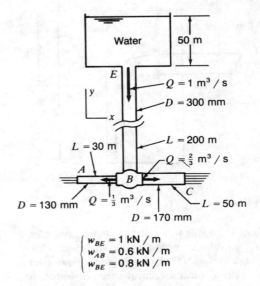

$\begin{cases} w_{BE} = 1\ \text{kN}\,/\,\text{m} \\ w_{AB} = 0.6\ \text{kN}\,/\,\text{m} \\ w_{BE} = 0.8\ \text{kN}\,/\,\text{m} \end{cases}$

FIGURE P5.34

5.35. Water is flowing through a cone-shaped nozzle which weighs 500 N. The inlet diameter to the nozzle is 1 m and the outlet diameter is 0.3 m. The velocity of flow into the nozzle is 1 m/s. If the force between nozzle and inlet pipe is to be 1000 N to prevent leakage, what is the minimum strain in the 10 bolts holding the system? The diameter of each bolt is 25 mm and the modulus of elasticity is 2×10^{11} Pa.

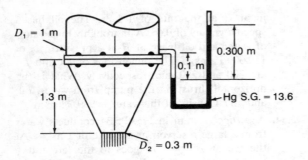

$D_1 = 1$ m
0.300 m
0.1 m
1.3 m
Hg S.G. = 13.6
$D_2 = 0.3$ m

FIGURE P5.35

***5.36.** Find the vertical force on the elbow.

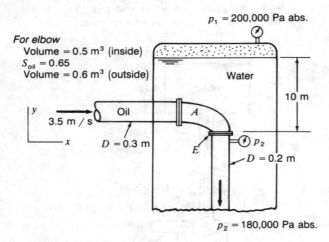

$p_1 = 200{,}000$ Pa abs.

For elbow
Volume $= 0.5$ m³ (inside)
$S_{oil} = 0.65$
Volume $= 0.6$ m³ (outside)

Water

10 m

y
Oil
3.5 m/s
x
$D = 0.3$ m A E p_2
$D = 0.2$ m
$p_2 = 180{,}000$ Pa abs.

FIGURE P5.36

5.37. Water is flowing in a circular duct of diameter 1 m. The velocity profile is paraboloidal with a volume flow of 5000 L/s. The water flows over a 90° cone to form a conical sheet of water that then exits to the atmosphere. If $\delta = 0.1$ m, what

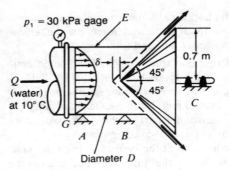

$p_1 = 30$ kPa gage E
0.7 m
δ
45°
Q
(water)
at 10° C
45°
C
G
A B
Diameter D

FIGURE P5.37

is the total horizontal force on the ground supports A, B, and C from the cylinder E and cone F? Neglect hydrostatic pressure at G.

5.38. Five-hundred liters of water per second flow through the pipe shown in Fig. P5.38. The flow exits through a rectangular area of length 0.8 m and width of 40 mm. The velocity profile is that of a parabola. The pipe weighs 1000 N/m and has an inside diameter of 250 mm. What are the forces on the pipe at A? The entering pressure p_1 is 100 kPa gage.

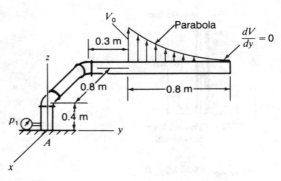

V_0
Parabola
0.3 m
$\frac{dV}{dy} = 0$
z
0.8 m
0.8 m
p_1
0.4 m
y
A
x

FIGURE P5.38

5.39. A control volume in Fig. P5.39 is moving at constant speed V_0. There are n stationary particles per unit volume uniformly distributed, each of mass m. Compute the flow of mass through left face A_1 realizing that the formulation $\iint_{A_1}(\rho\mathbf{V} \cdot \mathbf{dA})$ must be adjusted from a continuum approach to a flow of distinct separate

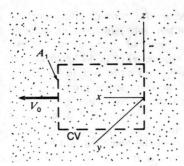

FIGURE P5.39

particles. Do same for linear momentum and kinetic energy. *Hint:* Use an approach similar to Fig. 4.16. Show that $\iint \rho \mathbf{V} \cdot \mathbf{dA} \equiv V_0 mnA_1$.

5.40. A light plane is moving through a hail storm at a speed of 200 km/h. The windshield shown has an area of 0.25 m² and its normal unit vector $\boldsymbol{\epsilon}_n$ is

$$\boldsymbol{\epsilon}_n = 0.2\mathbf{i} + 0.25\mathbf{j} + 0.947\mathbf{k}$$

The hail element has a mass of 1 mg and slides off the windshield after *plastic* collision with negligible friction. There are $n = 500$ hail particles per unit volume—that is, $n = 500$ m³. The hail particles have vertical speed of 10 km/h downward. What is the average force on the windshield from the collisions with the hail? Do not include body force contributions. *Hint:* Choose a rectangular parallelepiped for a con-

trol volume of small thickness with the windshield as one face of the control volume.

5.41. A double-wedge airfoil is moving at very high speed V_0 through stationary dust particles in outer space, each particle of mass m. The particles are uniformly distributed with n particles per unit volume. The collisions are elastic with angle of incidence $\beta = (\pi/2 - \alpha)$ equal to the angle of reflection β as shown for a colliding particle. What is the drag on the airfoil per unit length assuming a two-dimensional approach? See Prob. 5.39 before doing this problem and use a control volume such as is shown in Fig. P5.41.

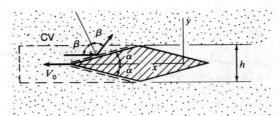

FIGURE P5.41

5.42. You learned in particle mechanics that for an inelastic collision of a particle with a large rigid body, the coefficient of restitution ϵ is related to the velocity of approach component normal to the surface of a rigid body to the velocity of rebound normal to the surface by the following simple relation:

$$\epsilon = -\frac{(V_N)_{\text{rebound}}}{(V_N)_{\text{approach}}}$$

Do the preceding problem with a coefficient of restitution ϵ for all collisions.

5.43. A spherical communications satellite is moving in outer space with a speed 20 or more times the speed of sound (i.e., a high Mach number). Molecules of mass m move at the speed of sound. Because of the speed disparity between satellite and molecule, we will assume molecules are stationary and that the satellite moves with speed V_0 hitting molecules in front of it. We will assume that the collisions are elastic so that

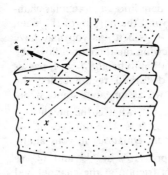

FIGURE P5.40

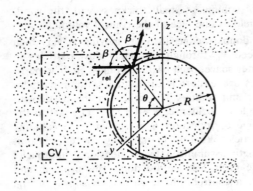

FIGURE P5.43

the angle of incidence α equals the angle of reflection β (remember optics?). Using a control volume fixed to the satellite as shown, compute the *drag*. The molecules each have a mass m. There are n molecules per unit volume. Make use of strips shown, integrating from $\theta = 0$ to $\theta = \pi/2$. The radius of the sphere is R. Because we do not have continuum, we have no pressure as such; only the collision of discrete molecules with the satellite surface. See Prob. 5.39 before proceeding.

5.44. A light attack boat is leaving an engagement at full speed. To help in the process, a battery of four 50-calibre machine guns is fired to the rear continuously. The muzzle velocity of the guns is 1000 m/s and the rate of firing for each gun is 3000 rounds per minute. Each bullet weighs 0.5 N. The ship weighs 440 kN. What is the additional thrust from the guns when the boat has attained constant speed of 40 knots? Neglect the change of total mass of the boat and its contents.

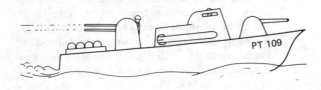

FIGURE P5.44

5.45. A jet plane is on the runway after touching down. The pilot puts into play movable vanes so as to achieve a reverse thrust from his two engines. Each engine takes in 40 kg of air per second. The fuel-to-air ratio is 1 to 40. If the exit velocity of the combustion products is 800 m/s relative to the plane, what is the total reverse thrust of the airplane if it is moving at the instant of interest at a speed of 150 km/h? The deceleration of the plane will not be great, so that little error is incurred if you consider your control volume to be an inertial control volume. The exit jets are close to atmospheric pressure.

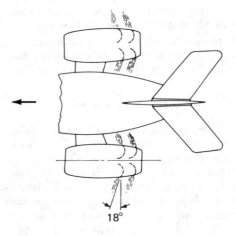

FIGURE P5.45

5.46. Water flows over a dam into a rectangular channel of width b. Beyond the bottom of the dam, the depth of flow h_1 is 3 ft. As you may have seen yourself on viewing rapid channel flow, the water changes elevation to height h_2 as it passes through a highly disturbed region called the *hydraulic jump*. If we assume one-dimensional flows at ① and at ②, what is the height h_2? The velocity V_1 is 25 ft/s. Use only continuity and linear momentum equations in your analysis, and take the pressures at ① and ② as hydrostatic. Neglect friction at the channel bed and walls.

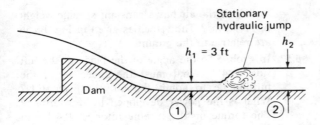

FIGURE P5.46

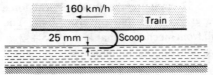

FIGURE P5.50

5.47. An airplane is moving at a constant speed of 600 m/h. Each of its four jet engines has an inlet area of 10 ft^2 and an exit area of 3 ft^2. The fuel-to-air ratio is 1 to 40. The exit velocity of the combustion products is 6000 ft/s relative to the plane at a local gage pressure of 4 lb/in^2. The plane is at 40,000 ft where the air density is 0.000594 slugs/ft^3. What is the drag of the airplane?

5.48. In Prob. 5.47, consider the drag D of the plane to be proportional to the velocity squared. What would the speed of flight be for the plane if two engines conk out and the fuel-to-air ratio is kept the same as in the previous problem? The computed thrust per engine in the previous problem was found to be 29,300 lb. Assume that the exit pressure for combustion products remains at 4 lb/in^2 gage.

5.49. A fighter plane is climbing at an angle θ of 60° at constant speed of 950 km/h. The plane takes in air at a rate of 450 kg/s. The fuel-to-air ratio is 1 to 40. The exit speed of the combustion products is 1825 m/s relative to the plane. If the plane changes to an inclination θ of 20°, what will be the speed of the plane when it reaches uniform speed? The new engine settings are such that the same amount of air is taken in and the exhaust speed is the same relative to the plane. The plane weighs 130 kN. The drag force is proportional to the speed squared of the plane. The exhaust jet is at the outside pressure.

5.50. A train is to take on water on the run by scooping up water from a trough. The scoop is 1 m wide and skims off a 25-mm layer of water. If the train is moving at 160 km/h, how much

water does it take on per second and what is the drag on the train due to this action?

5.51. Consider Example 5.1. Determine the horizontal force on the device from inside and outside fluids if on entering, the water has a uniform pressure of 25 lb/in^2 gage (we are neglecting hydrostatic variations) and on leaving has a uniform pressure of 13 lb/in^2 gage. The entering water has a paraboloidal velocity profile, but owing to the action in the device it has on exit a velocity with almost a uniform profile. Do not use a one-dimensional assumption on the inlet flow.

5.52. Two cubic meters per minute of gravel is being dropped on a conveyor belt which moves at the speed of 5 m/s. The gravel has a specific weight of 20 kN/m^3. The gravel leaves the hopper at a speed of 1 m/s and then has an average free fall of height $h = 2$ m. What torque T is needed by the conveyor to do the job? Neglect friction of rollers. *Hint:* Does the vertical motion of the gravel enter into your calculations?

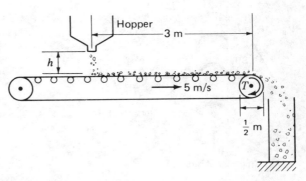

FIGURE P5.52

5.53. In Prob. 5.52 compute the total vertical force from the gravel onto the conveyor. *Hint:* You will recall from mechanics that $V = \sqrt{V_0^2 + 2gh}$ for a free fall from V_0 initial speed.

5.54. Water issues from a large tank through a 1300-mm^2 nozzle at a velocity of 3 m/s relative to the cart to which the tank is attached. The jet then strikes a trough which turns the direction of flow by an angle of 30°, as is shown in Fig. P5.54. Assuming steady flow, determine the thrust on the cart which is held stationary relative to the ground by the cord.

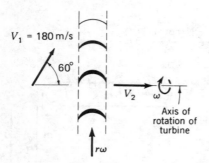

FIGURE P5.54

5.55. (*a*) If the trough in Prob. 5.54 is moving at a uniform speed of 1.5 m/s to the right relative of the cart, what is the thrust on the cart?
(*b*) If, furthermore, the cart is moving to the left at a uniform speed of 9 m/s, what is the thrust on the cart?

5.56. A jet of air from a 50-mm nozzle impinges on a series of vanes on a turbine rotor. The turbine has an average radius r of 0.6 m to the vanes and rotates at a constant angular speed ω. What are the transverse force and the torque on the

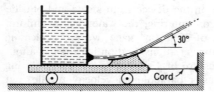

Top view of turbine

FIGURE P5.56

turbine if the air has a constant specific weight of 12 N/m^3? The velocities given in Fig. P5.56 are relative to the ground.

5.57. In Prob. 5.56 the angle of the blade at the left side is 45°. What must the speed ω of the turbine be to admit the air most smoothly? What is the power developed by the turbine? The torque on the turbine rotor is 40.4 N · m from the previous problem.

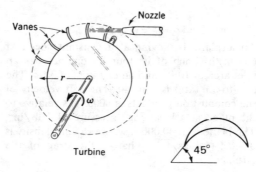

Turbine

FIGURE P5.57

5.58. A *rocket engine* is a turbomachine which differs from the turbojet engine in that it carries its own oxidizer, i.e., it is not air-breathing. An oxidizer such as nitric acid and a fuel such as aniline are burned in the combustion chamber to attain a pressure of about 2×10^3 kPa. This then expands out to a lower pressure, which is usually close to the atmosphere on leaving the nozzle. (Since the flow will be supersonic on exit, it does not have necessarily the same pressure as the surroundings, as was the case of the subsonic free jets we have been using in this chapter.) The thrust of the rocket motor is attributed to the force developed by the fluid in the rocket thrust chamber above that of the surrounding atmosphere of the rocket.

If a rocket (see Fig. P5.58) using nitric acid and aniline on a test stand has an oxidizer flow rate of 2.60 kg/s and a fuel flow of 0.945 kg/s (hence a propellant flow rate of 3.545 kg/s) and if the flow leaves the nozzle at 1900 m/s through an area of 0.0119 m^2 with a pressure of 110 kPa abs., what is the thrust of the rocket motor? Neglect momentum of entering fluids.

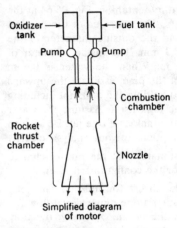

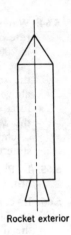

Oxidizer tank — Fuel tank

Pump — Pump

Combustion chamber

Rocket thrust chamber

Nozzle

Simplified diagram of motor

Rocket exterior

FIGURE P5.58

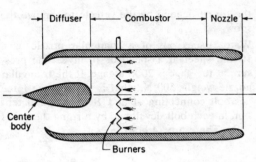

Diffuser — Combustor — Nozzle

Center body

Burners

FIGURE P5.60

5.59. A rocket has a propellant flow rate of 11.40 kg/s. The exit area is 0.0335 m², and the exhaust velocity is 2000 m/s relative to the nozzle at the pressure of 101.4 kPa. What is the teststand thrust of the motor (*a*) at sea level and (*b*) in an atmosphere equivalent to standard atmosphere at 9150 m? (Read the first paragraph of Prob. 5.58 before doing this problem.)

5.60. Figure P5.60 shows a supersonic *ram-jet engine*. It performs the same function as the turbojet discussed in Example 5.7. However, it is more efficient than the turbojet when operated at high supersonic speeds, and it is deceptively more simple in appearance. The diffuser section slows up the flow while compressing it. Fuel is burned in the stream in the combustion zone, and the products of combustion then expand down to some pressure p_e coming out of the nozzle. Because the fluid leaves at supersonic speed, the pressure p_e is not necessarily that of the ambient pressure around the jet.

Assume that the ram jet is moving at a speed V_1 and that the effective inlet area is A_1. Assume that w_f pound-mass of fuel is burned per unit time by the system and that the exit velocity is V_2 relative to the nozzle. If we disregard skin friction on the outside surface of the engine cowling, what is the thrust developed by the ram jet?

5.61. A jet of water of area A_j of 2 in² and speed V_j of 60 ft/s impinges on a trough which is moving at a speed u of 10 ft/s. If the water divides so that two-thirds goes up and one-third goes down, what is the force on the vane?

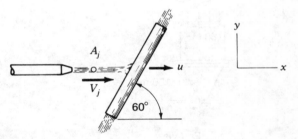

A_j

V_j

u

y

x

60°

FIGURE P5.61

5.62. A locomotive snow remover is clearing snow from a track as seen from the top view. The snow is 2 ft above the top of the tracks and has an average density of 20 lbm/ft³. If the locomotive is going 30 ft/s at a steady rate, estimate

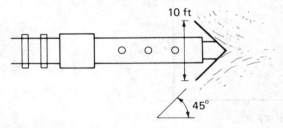

10 ft

45°

FIGURE P5.62

the thrust needed by the vehicle to remove the snow.

5.63. Water issues out of a triangular nozzle as a 10-mm sheet at a speed of 10 m/s. The pressure at the gage is 20 kPa gage. If the triangular nozzle weighs 500 N, what is the average force per bolt connecting A and B? The initial tension in each bolt developed by turning the nut is 50 N. There are 14 bolts.

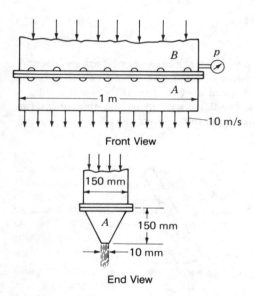

Front View

End View

FIGURE P5.63

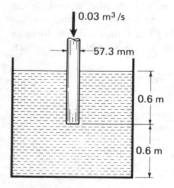

FIGURE P5.64

5.64. Water is pumped into a tank in Fig. P5.64 at the rate of 0.03 m^3/s. The exit area of the pipe jet is 2000 mm^2, and the outside diameter of the pipe itself is 57.3 mm. The inside diameter of the tank is 1.2 m. When the water is 0.6 m above the exit of the pipe, estimate the upward force required to hold up the tank not including the weight of the tank itself. Assume the water discharges into the tank as a free jet. Be sure to state other assumptions of your analysis clearly.

5.65. In Prob. 5.64 estimate what the rate of change of the force is for the configuration given.

5.66. In computing the thrust for rockets, ram jets, etc., we have assumed one-dimensional flow for the fluid leaving the nozzle. Actually, the flow issues out of the nozzle in a somewhat radial manner. If the exit speed is of constant magnitude V_e relative to the nozzle, what is the proper expression for the linear momentum flow in the x direction across the exit of the nozzle, using this flow model?

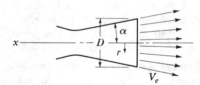

FIGURE P5.66

5.67. Using the model you have established in Prob. 5.66 for exit flow in a nozzle, recompute the thrust of the rocket at sea level in Prob. 5.59 for the angle $\alpha = 20°$. Retain the assumption that the exit pressure p_e is uniform across the jet. *Hint:* Use the result $r = (D \tan \theta)/(2 \tan \alpha)$ in your calculations for ρ.

5.68. Water is flowing over a dam (Fig. P5.68). Upstream the flow has an elevation of 12 m and has an average speed of 0.3 m/s, while at a position downstream the water has a fairly uniform elevation measured as 1 m. If the width of the dam is 9 m, find the horizontal force on the dam.

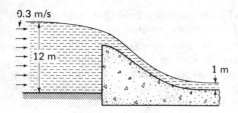

FIGURE P5.68

5.69. A row of identical blocks are lined up as shown. Each block weighs 5 lb and has a coefficient of dynamic friction with the ground of 0.3. A bulldozer moving at a constant speed V_0 of 10 ft/s is going to move these blocks to the right. If the impacts are completely inelastic (completely plastic), calculate the *average* force developed by the bulldozer as a function of time after the first block is touched. *Hint:* Consider a stationary control volume encompassing all the blocks. Then consider the average change in linear momentum per second inside the control volume. Consider linear momentum change and friction separately and then combine.

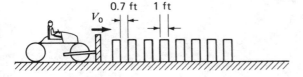

FIGURE P5.69

*5.70.** Wrought-iron chain is held above a wooden board so as to just touch at $t = 0$. The chain is then released and simultaneously a force F is applied to the board to accelerate at a constant rate of 32.2 ft/s^2. If we have plastic impact between chain and wood, what function of time must F be to do the task? The coefficient of dynamic friction between the board and the floor is 0.3. The chain weighs 10 lb/ft. The board weighs 5 lb. Use stationary control volume for wrought iron.

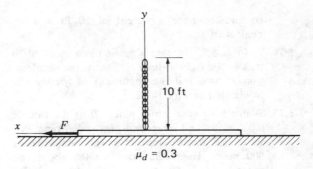

FIGURE P5.70

The following seven problems require the use of a noninertial control volume.

5.71. Express the general linear momentum equation for a control volume which is spinning arbitrarily but is not translating in inertial space.

5.72. In Prob. 5.58 we developed the thrust of a rocket motor on a test stand where clearly the control volume used was inertial. Describe what additional term should properly be included in the formulation of thrust of the rocket motor if the rocket were accelerating along the direction of the centerline of the rocket, and explain why we can neglect this term in practical problems.

5.73. Consider a tank of water in a container which rests on a sled. A high pressure is maintained by a compressor so that the water leaving the tank through the orifice does so at a constant speed of 30 ft/s relative to the tank. If there is 2 ft^3 of water in the tank at time t and the area of the jet is $\frac{1}{2}$ in^2, what will be the acceleration of the sled at time t if it and the empty tank and

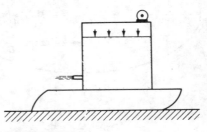

FIGURE P5.73

compressor have a weight of 50 lb and we neglect friction?

5.74. In Prob. 5.73 the sled is moving at a speed of 20 ft/s to the right at time t. What is the acceleration at time t if the coefficient of friction between sled and ice is 0.2?

5.75. Water is flowing into a pipe AB at the rate of 0.006 m³/s and issues out of a radial pipe BC. Pipes AB and BC have cross-sectional areas of 600 mm². The radial arm rotates about the centerline of AB, having a constant angular acceleration of 10 rad/s², and when the arm is vertical, it has an angular velocity of 3 rad/s. What is the force at this instant on CB (include

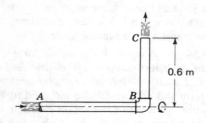

FIGURE P5.75

elbow) from the water flowing through the pipe and the atmosphere outside?

5.76. Oil (Fig. P5.76) of specific gravity 0.8 is flowing at a steady rate of 3×10^{-3} m³/s through a pipe section AB of area $A = 300$ mm² and length 0.6 m. The pressure p_1 at one end of the section is 7×10^4 Pa gage, and at the other end it is 6×10^4 Pa gage according to instruments. This section is mounted on a space vehicle for which, relative to inertial space, point O has an acceleration of $-20g\mathbf{k}$ m/s² and has an angular velocity $10\mathbf{i} + 6\mathbf{j}$ rad/s, with an angular acceleration $2\mathbf{k}$ rad/s². The pipe section is oriented, at the instant of interest, to be a distance $d = 1.5$ m from O along the direction Y, and the pipe section at that instant is parallel to X. What is the force from the oil on the pipe section including the elbows at that instant?

5.77. Compute the horizontal force on the trough (Fig. P5.77) on which a jet of water impinges. The trough accelerates uniformly in the horizontal direction at the rate of 10 ft/s² and at the instant of interest has a speed of 20 ft/s. The length of the trough is 18 in, and the jet has a speed of 50 ft/s.

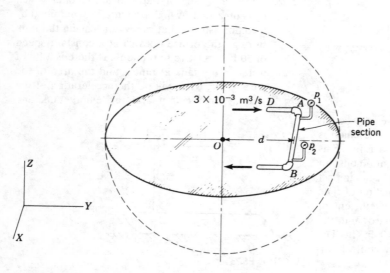

FIGURE P5.76

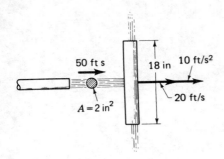

FIGURE P5.77

5.78. Water is flowing through a pipe having an inside diameter of 6 in. Find the total moment on the pipe at the base A from water, air, and the weight of the pipe. The pipe weighs 10 lb/ft. The pressure at A is 10 lb/in² gage. The flow is steady. Use a control volume as was used in example 5.11.

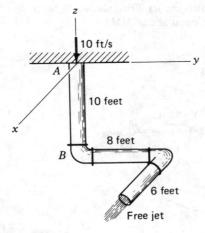

FIGURE P5.78

5.79. Do Prob. 5.78 by using first a control volume covering the interior volume of the pipe and then, to take care of the weight of the pipe, a free-body diagram of the pipe.

5.80. Compute the bending moment from the water at point E of the pipe system containing water

using the method of moment of momentum. The flow is steady.

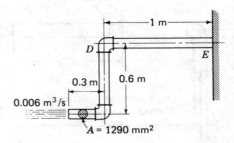

FIGURE P5.80

5.81. For a steady flow of water, compute the bending moment from the water at section A of the pipe by the method of moment of momentum.

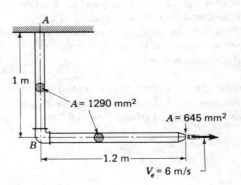

FIGURE P5.81

5.82. Water is flowing steadily through the 200-mm pipe (Fig. P5.82).

(a) Find the moment components on the base of the pipe at A from water, air, and pipe weight.

(b) Find the force components at the base of the pipe at A from water, air, and pipe weight.

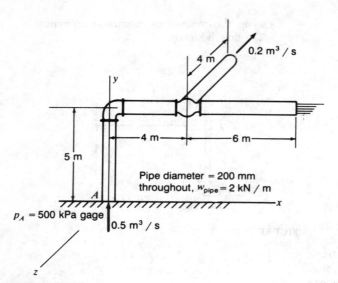

FIGURE P5.82

5.83. Water is steadily flowing through a pipe at the rate Q of 0.2 m³/s. The pipe weighs 0.3 kN/m. The pressure at A is 10 kPa gage.

(a) What are the total bending moments and twisting moments in the pipe at the base A of the pipe system?

(b) What are the shear forces and axial force in the pipe at A due *only* to the *water* and *air*. The diameter of the pipe is 0.2 m.

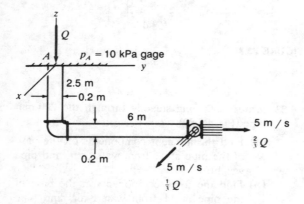

FIGURE P5.83

5.84. A platform is shown which can rotate about axis MM. A jet of water is directed out from the center of the platform while it is stationary and strikes a vane at the periphery of the platform. The vane turns the jet 90° as shown. What is the torque developed about MM?

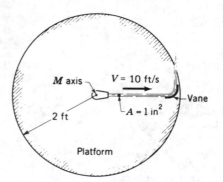

FIGURE P5.84

5.85. A jet of water (Fig. P5.95) 645 mm² in cross section is directed out at a speed of 3 m/s at a 90° vane positioned 0.6 m from the center of the platform from which the jet issues as shown. It then strikes a 90° vane to attain a direction of motion parallel to its original direction. If the platform is stationary, what is the torque about MM as a result of this action?

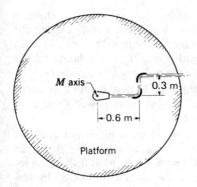

FIGURE P5.85

5.86. In Example 5.12 the following data apply: $q = 5$ L/s, $\alpha = 20°$, $l = 300$ mm, $A_e = 600$ mm^2. Find the angular speed ω of the arm for zero frictional torque.

5.87. In Prob. 5.86 find ω for steady rotation if there is a resisting torque T_f due to bearing friction and windage given as $0.08\omega^2$ newton-meters, with ω in radians per second.

5.88. Consider Prob. 5.4. What is the steady-state rotational speed if there is a constant resisting torque of 30 N · m?

5.89. If the flow of water divides up equally at the tee at D, what is the total force system (force and

torque) transmitted through section C owing to the water and air? $A_e = 1290$ mm^2 and $p_C = 70$ kPa gage.

5.90. If in Prob. 5.89 the flow is reversed at the tee at B by putting the plug on the other side, what is the total force system (force and torque) transmitted through section C owing to the water?

5.91. A rotor is held stationary while 5 L/s of water enters at C and flows out through three channels, each of which has a cross-sectional area of 1800 mm^2. What is the angular speed ω of the rotor 2 s after its release? Let us assume that there is no frictional resistance to rotation about a vertical axis z coming out of the paper to you at C. The moment of inertia about z, I_{zz}, for the rotor plus water is 10 kg · m^2 (that is, $\iiint \bar{r}^2 \rho \, dv = 10$ kg · m^2). Use a stationary control volume.

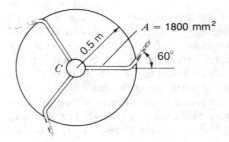

FIGURE P5.91

5.92. If in Prob. 5.91, there is a resisting torque T_f for rotation of rotor given as $T_f = 1.5 - 4\omega$ newton-meters, what is the speed of the rotor after 2 s? From your general solution of ω, indicate what is the final steady rotation speed ω.

5.93. A rotor in Fig. P5.93 with four channels is held stationary and, with exits at B blocked, is filled with water through inlet at C. Now at $t = 0$, the outlets are opened at B, the rotor is released, and a flow q is started at the inlet such that q varies as $q = 0.05t$ m^3/s, with t in seconds. What is the differential equation for ω of the rotor if there is no resistance to rotation about the axis of the rotor at C? The area of each of the channels is 1500 mm^2. Use a stationary control volume. The moment of inertia I_{zz} of rotor and water ($\iiint \bar{r}^2 \rho \, dv$) is 10 kg · m^2.

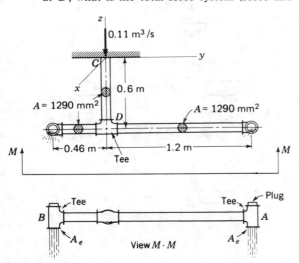

FIGURE P5.89

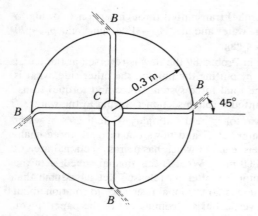

FIGURE P5.93

5.94. A horizontal disc A in Fig. P5.94 has a torque T applied about axis C that causes a constant angular acceleration $\dot{\omega} = \kappa$. A chain of length l and weight per unit length w lies on a frictionless horizontal surface. At time $t = 0$, $\omega = 0$, and at this instant the chain is connected to the disc at D. What is the torque T needed? The moment of inertia about the axis C of the disc is I. Use a stationary control volume that includes the entire chain and disc. *Hint:* For the disc, $\iiint (rV_\theta)\rho \, dv = \iiint r(r\omega) \, dm = \omega \iiint r^2 \, dm = I\omega$. Is the torque T that you have computed valid after H touches the disc?

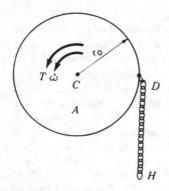

FIGURE P5.94

5.95. Do Prob. 5.94 for the case where the chain that has not come into contact with the disc rests on

a horizontal surface having with the chain a dynamic coefficient of friction of μ_d. Compute the torque for the time interval before H comes into contact with the disc. The solution to the preceding problem is $T_s = r_0^2 \kappa(w/gl) + I\kappa$.

5.96. Show in Example 5.14, that the Pelton water wheel develops its maximum power when $\beta = 180°$ and $V_j = 2\omega r$.

5.97. A Pelton water wheel has a mean radius of 1.2 m, and each bucket has an angle $\pi - \beta$ of 172° as shown. If the bucket velocity is 0.48 of the jet velocity, what is the torque developed by the wheel if a 75-mm-diameter jet having a speed of 30 m/s is employed? How many kilowatts are being developed?

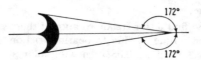

FIGURE P5.97

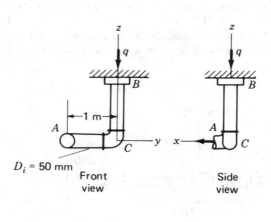

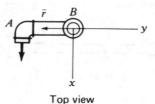

FIGURE P5.98

The following six problems involve noninertial control volumes.

5.98. Pipe system *ABC* in Fig. P5.98 is free to rotate about the indicated *z* axis. The system is full of water and the end *A* is capped closed. At time $t = 0$, the cap is opened and a constant flow of 10 L/s is forced through the pipe. If there is no resisting torque at *B*, what is the angular speed ω of *ABC* at $t = 1$ s? The moment of inertia I_{zz} of *ABC* with water about *z* is 15 kg · m². The inside diameter of all pipes is 50 mm. Use a nonstationary control volume. *Hint:* Recall from mechanics that $I_{zz} = \iiint r^2 \rho \, dv$.

5.99. In Prob. 5.98 suppose that the device is rotating at constant speed $\omega = 1$ rad/s about axis *CB* in a clockwise direction as seen looking down from the support. What torque is required about *CB*? For the stationary case, we got $\mathbf{T} = -7.77\mathbf{i} + 2814\mathbf{j} + 1735\mathbf{k}$ newton-meters.

5.100. Suppose in addition to the angular velocity of 1 rad/s at time *t* in Prob. 5.99 that there is an angular acceleration of 2 rad/s² in a clockwise direction as one looks down from the support. What is the additional torque required at that instant resulting from the angular acceleration?

5.101. In Prob. 5.81 the system is rotating at constant angular speed of 2 rad/s about *AB* with a clockwise sense as you look from *A* to *B*, and the support at *A* is accelerating at a rate of 3 m/s² in the horizontal direction to the right. What is the total bending moment at *A* in the position shown, as a result of fluid in the pipe? The result from Prob. 5.81 is $\mathbf{T}_A = 32.3\mathbf{j}$ newton-meters.

5.102. In Prob. 5.80 the pipe is rotating about axis *ED* at a rate of 2 rad/s with a clockwise sense as you look from *E* to *D*. The support *E* has an acceleration relative to the inertial reference given as

$$\ddot{\mathbf{R}} = 3\mathbf{i} + 1.8\mathbf{k} \text{ meters per second squared}$$

At the configuration shown, what is the total bending moment at the base of the pipe *E*, from the fluid flow? The result for Prob. 5.80 is $\mathbf{T} = 35.01\mathbf{k}$ newton-meters.

5.103. In Example 5.13 the system rotates about *MM* at a speed of 3 rad/s in a clockwise sense as you look down from the support. What torque about *MM* is required to maintain this rotation?

Air cushion vehicle. (Courtesy Bell Aerospace Textron.)

Bell Aerospace Textron's LACV-30 (Lighter, Amphibian Air Cushion Vehicle - 30 ton payload). This is the first **production** contract for an air cushion vehicle awarded to an American company.

The LACV-30 is a high-speed amphibious cargo carrier. It can run at speeds up to 56 miles per hour carrying 60 tons of gross weight over water, land, snow, ice, even marshes and swamps. It can haul a wide variety of wheeled and tracked equipment and containerized cargo, and it needs no dock or berth facilities.

The bow has a crane for unloading.

CHAPTER
6

BASIC LAWS FOR FINITE
SYSTEMS AND FINITE
CONTROL VOLUMES II:
THERMODYNAMICS

6.1 INTRODUCTION

We have thus far considered conservation of mass and Newton's law, the latter in the forms of linear momentum and moment of momentum, for both finite systems and finite control volumes. We started with the familiar system approach and then using the Reynolds transport equation quickly went to the corresponding control volume formulations. In this chapter we shall do the same for the first and second laws of thermodynamics. We shall defer, however, detailed discussion of the second law of thermodynamics until a later chapter where its use will be more directly needed.

6.2 PRELIMINARY NOTE

The first law of thermodynamics is a statement of macroscopic experience which stipulates that energy must at all times be conserved. Hence, the first law accounts for energy entering, leaving, and accumulating in either a system or a control volume.

It will be convenient to classify energy under two main categories, *stored energy* and *energy in transition*. Energy primarily associated with a given mass will be considered stored energy. On the other hand, energy which is going from one system to another is called energy in transition. We will consider only stored energy as extensive from the viewpoint that this energy is directly identified with and "resides in" the matter involved in a discussion. One may list the following types of stored energy of an element of mass:

1. *Kinetic energy E_K:* energy associated with the motion of the mass
2. *Potential energy E_P:* energy associated with the position of the mass in conservative external fields
3. *Internal energy U:* molecular and atomic energy associated with the internal fields of the mass[1]

Two types of energy in transition are listed, heat and work. Heat is the energy in transition from one mass to another as a result of a temperature difference. On the other hand, work, as we learned in mechanics, is the energy in transition to or from a system which occurs when external forces, acting on the system, move through a distance. In thermodynamics, we further generalize the concept of work to include energy transferred from or to a system by any action such that the total external effect of the given action can be reduced by hypothetical frictionless mechanisms entirely to that of raising a mass in the gravitational field.[2] Thus electric current can be arranged to lift a weight by using an electric motor, and if there is no friction or electric resistance, this can be the only effect of the current. Thus it represents a flow of energy which we classify as work. Heat, however, even with frictionless machines, cannot raise a weight and have no other effect. There must be rejected heat to a sink.

Considering stored energy again, we note that since the kinetic energy of an infinitesimal particle is by definition equal to $\frac{1}{2} dm V^2$, the change in kinetic energy during a process is obviously dependent only on the final and initial *velocity* of the infinitesimal system for the process. Change in the potential energy is defined only for *conservative* force fields and is by definition equal to minus the work done *by* these conservative force fields on the infinitesimal system during a process. As you will recall from your studies in mechanics and electrostatics, this work depends only on the final and initial *positions* of the infinitesimal system for the process. Finally, the internal atomic and molecular energy of a fluid stems from force fields which are approximately conservative. Hence, it should be noted that *stored energy* is a *point function*; i.e., all changes during a process are expressible in terms of values at the end points. On the

---

[1] By considering stored energy as the sum of these quantities, we impose the restrictions of classical mechanics on the resulting equations. This means that the control volumes later to be used for the first law of thermodynamics will of necessity be inertial control volumes.

[2] The reader is referred to current textbooks on thermodynamics for a more detailed discussion of work and heat.

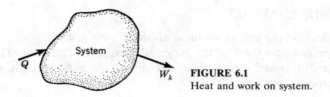

FIGURE 6.1
Heat and work on system.

other hand, energy in transition is a *path function*; i.e., changes are dependent not only on the end points but also on the actual path between the end points.

6.3 SYSTEM ANALYSIS

An arbitrary system is shown in Fig. 6.1 which, by definition, may move and deform without restriction but may not transfer mass across its boundary. The net heat *added* to the system and the net work *done* by the system on the surroundings during the time interval Δt are designated as Q and W_K, respectively.

If E is used to represent the total stored energy of a system at any time t and its property as a point function is employed, conservation of energy demands that for a process occurring during interval t_1 to t_2,

$$Q - W_K = \Delta E = E_2 - E_1 = (E_K + E_P + U)_2 - (E_K + E_P + U)_1 \quad (6.1)$$

The differential form of Eq. (6.1) may be written in the following manner[3]:

$$dE = dQ - dW_K \quad (6.2)$$

We have now listed the usual form of the first law of thermodynamics applied to systems. Since Q and W_K are not point functions, they are then representable as explicit functions of time. Accordingly, we can employ the usual derivative notation dQ/dt and dW_K/dt for time derivatives. However, E is a point function and expressible in terms of spatial variables and time. To indicate that we are following the system, we use the substantial derivative. Thus we have for the time variations of stored energy and energy in transition for a system

$$\boxed{\dfrac{DE}{Dt} = \dfrac{dQ}{dt} - \dfrac{dW_K}{dt}} \quad (6.3)$$

---

[3] In thermodynamics texts the differentials of Q and W are frequently denoted as δQ and δW_K or as $đQ$ and $đW_K$, respectively, to remind the reader that since both Q and W depend on the path these are not perfect differentials.

6.4   CONTROL-VOLUME ANALYSIS

To develop the control-volume approach, we will consider E to be the extensive property to be used in the Reynolds transport equation. The term e will then represent stored energy per unit mass. We can then say using the Reynolds transport equation

$$\frac{DE}{Dt} = \oiint_{CS} (e)(\rho \mathbf{V} \cdot \mathbf{dA}) + \frac{\partial}{\partial t} \iiint_{CV} (e)(\rho \, dv) \tag{6.4}$$

Using Eq. (6.3) in the left side of Eq. (6.4), we get

$$\boxed{\frac{dQ}{dt} - \frac{dW_K}{dt} = \oiint_{CS} (e)(\rho \mathbf{V} \cdot \mathbf{dA}) + \frac{\partial}{\partial t} \iiint_{CV} (e)(\rho \, dv)} \tag{6.5}$$

Equation (6.5) then states that the net rate of energy transferred into the control volume by heat and work[4] equals the rate of efflux of stored energy from the control volume plus the rate of increase of stored energy inside the control volume.

According to the discussion of Sec. 6.2, it is seen that the term e may be given as the sum of the following specific types of stored energy per unit mass:

1. *Kinetic energy e_K.* The kinetic energy of an infinitesimal particle is $dm \, V^2/2$, where dm is in units of slugs or kilograms in this text. Per unit mass this then becomes $V^2/2$.

2. *Potential energy e_P.* Assuming that the only external field is the earth's gravitational field, we have for the potential energy of an infinitesimal particle at an elevation z above some datum, the quantity $\int_0^z dm \, g \, dz$. Considering g as constant, we then have as the potential energy per unit mass the quantity gz.

3. *Internal energy u.* If certain properties of a fluid are known, the internal energy per unit mass relative to some datum state may usually be evaluated or found in experimentally contrived tables.

Hence we give e as

$$e = \frac{V^2}{2} + gz + u \tag{6.6}$$

---

[4]In the present development, dW_K/dt does *not* include work done by gravitational body forces, since this effect is contained in e as potential energy.

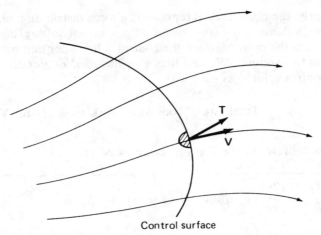

Control surface

FIGURE 6.2
Flow work at control surface.

Next let us discuss the term dW_K/dt in Eq. (6.5). It will be convenient to form three classifications of W_K. They are:

1. Net work done *on the surroundings* as a result of *tractions* on that part of the control surface across which there is a *flow of fluid*. We call this work *flow work*.

2. Any other work transferred at the rest of the control surface *to the surroundings* by direct contact between inside and outside nonfluid elements. For example, the work transferred through a control surface by shafts or by electric currents would fall into this category. We call this work *shaft work* and denote it as W_S.

3. Classifications 1 and 2 take care of the total work transferred at the control surface by direct contact. Inside the control surface, we may have work on the surroundings resulting from the reactions to *body forces*. Note carefully, in this regard, that the effects of gravity have already been taken into account as the potential energy (in e), so the body force **B** must *not* include gravity; it may include, for instance, contributions from magnetic and electric force distributions.

Let us examine flow work carefully (see Fig. 6.2). First, note that **T** by definition is the traction force from the *surroundings* acting *on* the control surface. Hence, $\mathbf{T} \cdot \mathbf{V}$ is the time rate of work (power) done *by* the surroundings at the control surface per unit area of the control surface. It thus represents power per unit area *entering* the control volume. Hence, the time rate of work *leaving* the control volume—the total rate of *flow work*—is given as

$$\text{Total rate of flow work} = -\oiint_{CS} \mathbf{T} \cdot \mathbf{V}\, dA \qquad (6.7)$$

Similarly, the body force **B** represents a force distribution on the material *in* the control volume *from* the surroundings not requiring direct contact. Hence $-\mathbf{B} \cdot \mathbf{V}$ is the power *leaving* the control volume per unit mass of material inside the control volume. We can thus give the *total rate of body force work leaving* the control volume as

$$\text{Total rate of body force work} = -\iiint_{CV} \mathbf{B} \cdot \mathbf{V}\rho \, dv \tag{6.8}$$

A general form of the first law can now be given as

$$
\boxed{
\begin{aligned}
\frac{dQ}{dt} &- \frac{dW_S}{dt} + \oiint_{CS} \mathbf{T} \cdot \mathbf{V}\, dA + \iiint_{CV} \mathbf{B} \cdot \mathbf{V}\rho \, dv \\
&= \oiint_{CS} \left(\frac{V^2}{2} + gz + u \right)(\rho\mathbf{V} \cdot \mathbf{dA}) + \frac{\partial}{\partial t} \iiint_{CV} \left(\frac{V^2}{2} + gz + u \right)(\rho \, dv)
\end{aligned}
}
$$

$$\tag{6.9}$$

We will next consider two cases for which Eq. (6.9) reduces to a more familiar form. First, if **T** is normal to the control surface (i.e., for a *frictionless flow*), then we can express **T** as follows:

$$\mathbf{T} = \tau_{nn} \frac{\mathbf{dA}}{dA}$$

where τ_{nn} is the normal stress. And since $\tau_{nn} = -p$ for inviscid flow, the rate of efflux of flow work can then be given as follows:

$$\text{Rate of flow work} = -\oiint_{CS} \mathbf{T} \cdot \mathbf{V}\, dA = \oiint_{CS} \left(p \frac{\mathbf{dA}}{dA} \right) \cdot \mathbf{V}\, dA$$

$$= \oiint_{CS} p\mathbf{V} \cdot \mathbf{dA}$$

Since the product of v, the specific volume, and ρ, the mass density, is unity, we may introduce ρv into the integrand on the right side of the equation above to form the following that goes into Eq. (6.5) for part of dW_k/dt:

$$\text{Rate of flow work} = \oiint_{CS} p v (\rho\mathbf{V} \cdot \mathbf{dA}) \tag{6.10}$$

We can arrive at the above formulation also for the case of a viscous flow wherein *the velocity* **V** *of fluid passing through the control surface is everywhere normal to the control surface*. The flow through the cross section of a pipe is an example. Thus, returning to Eq. (6.7), we replace **V** by **n**V (see Fig. 6.3), where **n**

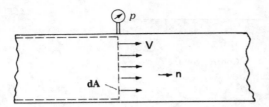

FIGURE 6.3
Flow perpendicular to control surface.

is the outward normal unit vector of the area element, and we form the equation

$$\text{Rate of flow work} = - \oiint_{CS} \mathbf{T} \cdot \mathbf{V} \, dA = - \oiint_{CS} (\mathbf{T} \cdot \mathbf{n}V) \, dA = - \oiint_{CS} \tau_{nn} \mathbf{V} \cdot \mathbf{dA}$$

wherein $\mathbf{T} \cdot \mathbf{n}$ has been set equal to τ_{nn}. Also, since $\mathbf{V}$ and $\mathbf{A}$ are collinear for this case, $V \, dA$ has been set equal to $\mathbf{V} \cdot \mathbf{dA}$. Insert next ρv in the integrand of the last expression above. Also, we have pointed out that for *parallel* flow, such as Fig. 6.3, we can replace τ_{nn} by $-p$ if we neglect hydrostatic pressure variations over the cross section. This can usually be done in pipe flows, so the rate of flow work for such cases is again expressed as in Eq. (6.10) for part of dW_k/dt in Eq. (6.5). The first law of thermodynamics can then be written for inviscid flow and for flows with one-dimensional inlets and outlets[5] as

$$\frac{dQ}{dt} - \frac{dW_S}{dt} + \iiint_{CV} \mathbf{B} \cdot \mathbf{V} \rho \, dv = \oiint_{CS} \left(\frac{V^2}{2} + gz + u + pv \right)(\rho \mathbf{V} \cdot \mathbf{dA})$$
$$+ \frac{\partial}{\partial t} \iiint_{CV} \left(\frac{V^2}{2} + gz + u \right)(\rho \, dv)$$

(6.11)

The internal energy u and the flow work pv are often combined, that is, $(u + pv)$, to form the property called *specific enthalpy*, h, in such situations. Equation (6.1) then becomes

$$\frac{dQ}{dt} - \frac{dW_S}{dt} + \iiint_{CV} \mathbf{B} \cdot \mathbf{V} \rho \, dv = \oiint_{CS} \left(\frac{V^2}{2} + h + gz \right)(\rho \mathbf{V} \cdot \mathbf{dA})$$
$$+ \frac{\partial}{\partial t} \iiint_{CV} e\rho \, dv$$

(6.12)

A very important simplification commonly encountered is the case of a steady flow where inlet and outlet flows to and from a device, respectively, are

---

[5]Note that when we use v in the differential dv, it simply represents volume, but when v is isolated as in Eq. (6.11), it is meant to represent the specific volume. If this is understood, there should be no confusion about the meaning of v in our discussions.

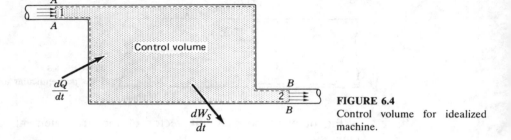

FIGURE 6.4
Control volume for idealized machine.

considered one dimensional. Such an example is shown in Fig. 6.4. This may represent for instance, a steam turbine where the control volume has been selected to represent the inside volume of the turbine casing and sections AA and BB of the control surface have been established in the inlet and outlet pipes of the turbine.

Since all properties are taken constant over cross sections AA and BB, and since the inlet and outlet velocities are normal to the control surface, one may carry out the surface integration of Eq. (6.12) with ease. Furthermore, because this is steady flow, the stored energy inside the control volume remains constant with time and the last term on the right side of Eq. (6.12) is zero. It being noted that

$$\iint_{A_1} z_1 \, dA_1 = (z_c)_1 A_1$$

where $(z_c)_1$ is the z coordinate of the centroid of the inlet area, and similarly for the exit area, the resulting equation is

$$\frac{dQ}{dt} - \frac{dW_S}{dt} = -\left[\frac{V_1^2}{2} + g(z_c)_1 + h_1\right]\rho_1 V_1 A_1 + \left[\frac{V_2^2}{2} + g(z_c)_2 + h_2\right]\rho_2 V_2 A_2$$

Rearranging the equation, we then have a form of the first law of thermodynamics which has much direct use and may well be familiar to you from your course in thermodynamics:

$$\boxed{\begin{aligned}&\left[\frac{V_1^2}{2} + gz_1 + u_1 + \frac{p_1}{\rho_1}\right]\rho_1 V_1 A_1 + \frac{dQ}{dt} \\ &= \left[\frac{V_2^2}{2} + gz_2 + u_2 + \frac{p_2}{\rho_2}\right]\rho_2 V_2 A_2 + \frac{dW_S}{dt}\end{aligned}} \qquad (6.13)$$

We shall call this form of the first law as well as other simple variations of it to be presented later in this section the *simplified first law*. Continuity conditions for the control volume of Fig. 6.4 yield the relation $\rho_1 V_1 A_1 = \rho_2 V_2 A_2 = dm/dt$, where dm/dt is the mass-flow rate. Dividing the previous equation by dm/dt

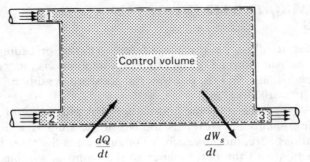

FIGURE 6.5
Control volume with two inlets.

and rearranging terms gives us

$$\left[\frac{V_1^2}{2} + g(z_c)_1 + h_1 \right] + \frac{dQ/dt}{dm/dt} = \left[\frac{V_2^2}{2} + g(z_c)_2 + h_2 \right] + \frac{dW_S/dt}{dm/dt} \quad (6.14)$$

The expression $(dQ/dt)/(dm/dt)$ becomes dQ/dm and is simply the net heat added per unit mass of flow, while $(dW_S/dt)/(dm/dt)$ becomes dW_S/dm, the net shaft work done per unit mass of flow. Thus the final form is a familiar one, given as

$$\left[\frac{V_1^2}{2} + g(z_c)_1 + h_1 \right] + \frac{dQ}{dm} = \left[\frac{V_2^2}{2} + g(z_c)_2 + h_2 \right] + \frac{dW_S}{dm} \quad (6.15)$$

For two entrances and one exit of essentially one-dimensional flows (see Fig. 6.5), we may proceed by applying Eq. (6.13) and continuity in the preceding manner for steady flow to reach the following equation:

$$\left[\frac{V_1^2}{2} + g(z_c)_1 + h_1 \right] \frac{dm_1}{dt} + \left[\frac{V_2^2}{2} + g(z_c)_2 + h_2 \right] \frac{dm_2}{dt} + \frac{dQ}{dt}$$

$$= \left[\frac{V_3^2}{2} + g(z_c)_3 + h_3 \right] \left(\frac{dm_1}{dt} + \frac{dm_2}{dt} \right) + \frac{dW_S}{dt} \quad (6.16)$$

Dividing through by dm_1/dt gives

$$\left[\frac{V_1^2}{2} + g(z_c)_1 + h_1 \right] + \left[\frac{V_2^2}{2} + g(z_c)_2 + h_2 \right] \frac{dm_2/dt}{dm_1/dt} + \frac{dQ}{dm_1}$$

$$= \left[\frac{V_3^2}{2} + g(z_c)_3 + h_3 \right] \left(1 + \frac{dm_2/dt}{dm_1/dt} \right) + \frac{dW_S}{dm_1} \quad (6.17)$$

where the net work done and heat transferred are based on a unit mass of fluid entering inlet 1.

6.5   PROBLEMS INVOLVING THE FIRST LAW OF THERMODYNAMICS

In order to illustrate the use of some of the preceding equations, several examples will now be given. In the computations, it will be helpful if we remember that $v = 1/\rho$. For water at standard conditions, $\rho = 62.4/g = 1.938$ slugs/ft$^3 \equiv 1000$ kg/m^3.

We now present some information that will be useful in working problems involving pumps and turbines for which the fluid is a liquid and hence, for calculations here, incompressible. For such cases the term *head* is often used in conjunction with the performance of the pump or turbine. Using the notation H_D, head is defined as

$$H_D = \left(\frac{V^2}{2g} + z + \frac{p}{\gamma} \right) \quad \text{(dimension is length)} \tag{6.18}$$

and may be considered to be the "mechanical energy" per unit *weight* of flow. The *change* in head ΔH_D for a pump or turbine is then

$$\Delta H_D = \Delta \left(\frac{V^2}{2} \right) + \Delta(z) + \Delta \left(\frac{p}{\gamma} \right) \tag{6.19}$$

where Δ indicates change between outlet and inlet conditions of the device.

How does ΔH_D relate to dW_S/dm? Since dW_S/dm is energy per unit mass we will now consider $g\,\Delta H_D$ which also has these dimensions. For a *pump*, $g\,\Delta H_D$ is a *positive* number because of the increase of mechanical energy of the fluid, while clearly for a *turbine*, $g\,\Delta H_D$ will be *negative*. But dW_S/dm for a pump is negative (energy going into the control volume), while for a turbine it will be positive (energy leaving the control volume). Thus we see that

$$\frac{dW_S}{dm} = -g\,\Delta H_D \tag{6.20}$$

Hence Eq. (6.15) can be written for constant value of u as

$$\frac{dQ}{dm} + g\,\Delta H_D + \left(\frac{V_1^2}{2} + gz_1 + \frac{p_1}{\rho} \right) = \left(\frac{V_2^2}{2} + gz_2 + \frac{p_2}{\rho} \right) \tag{6.21}$$

with the proper sign applied to ΔH_D (positive for pumps and negative for turbines).

Finally we wish to point out that for cases involving liquid, we may use the gage pressure on both sides of the simplified first law equation since the portion p_{atm}/γ of the absolute pressures will cancel, leaving only the gage pressures.

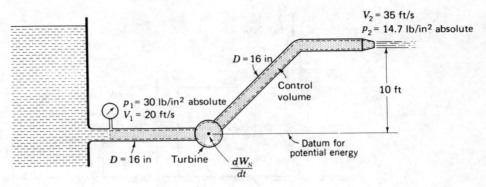

FIGURE 6.6
Water flows through pipe system.

Example 6.1. Shown in Fig. 6.6 is a large-diameter pipe through which water flows. The inlet and outlet conditions are specified in the diagram. What is the power of the turbine if we neglect friction?[6]

The control volume is chosen to include the interior of the pipe and turbine and to terminate at places where known information is available. This is indicated by the dashed line in Fig. 6.6. Employ the centerline of the lower pipe as the elevation datum.

We make the following assumptions:

1. Steady flow.
2. Incompressible flow.
3. 1-D flow in and out of CV.
4. Neglect friction (short large-diameter pipes).
5. Take $dQ/dt = 0$ as a result of condition 4.
6. Consider u is constant because of conditions 2 and 4.

These conditions allow us to use the *simplified first law*. Thus noting assumptions 5 and 6 in particular we have

$$\frac{V_1^2}{2} + g(z_c)_1 + p_1 v_1 = \frac{V_2^2}{2} + g(z_c)_2 + p_2 v_2 + \frac{dW_S}{dm} \qquad (a)$$

---

[6]For other than short large-diameter pipes, friction will result in considerable heat transfer from pipe to surroundings and must be accounted for. We will consider this problem in Chap. 9 on pipe flow.

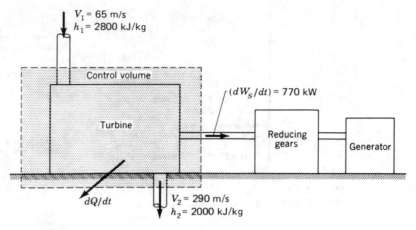

FIGURE 6.7
Power plant.

Substituting numerical values, we get

$$\frac{20^2}{2} + 0 + \frac{(30)(144)}{1.938} = \frac{35^2}{2} + 10g + \frac{(14.7)(144)}{1.938} + \frac{dW_S}{dm}$$

Solving for the work per slug of flow gives us $dW_S/dm = 402$ ft · lb/slug. To find the power, multiply dW_S/dm by dm/dt. Thus noting that $dm/dt = \rho V A$, we have

$$\text{Power} = 402\left[1.938\frac{\pi(16^2/4)}{144}20\right] = 21{,}773 \text{ ft · lb/s}$$

Dividing by 550 produces the desired result—39.6 hp.

Example 6.2. A steam turbine (Fig. 6.7) uses 4500 kg/h of steam while delivering 770 kW of power at the turbine shaft. The inlet and outlet velocities of the steam are 65 m/s and 290 m/s, respectively. Measurements indicate that the inlet and outlet enthalpies of the steam are 2800 kJ/kg and 2000 kJ/kg, respectively. Calculate the rate at which heat is lost from the turbine casing and bearings.

A control volume has been drawn around the steam turbine and bearings as indicated in Fig. 6.7. We make the following assumptions:

1. Steady flow.
2. Neglect potential energy change of steam in and out of the control volume.
3. 1-D flow in and out.

We can go directly to the *simplified first law of thermodynamics* which stipulates that

$$\left(\frac{V_1^2}{2} + h_1\right)\left(\frac{dm}{dt}\right) + \frac{dQ}{dt} = \left(\frac{V_2^2}{2} + h_2\right)\left(\frac{dm}{dt}\right) + \frac{dW_S}{dt}$$

$$\left(\frac{65^2}{2} + 2800 \times 10^3\right)\frac{4500}{3600} + \frac{dQ}{dt} = \left(\frac{290^2}{2} + 2000 \times 10^3\right)\frac{4500}{3600} + 770 \times 10^3$$

Solving for dQ/dt, we get -180.1 kJ/s, the negative sign indicating that the heat is leaving the control volume. It will be left for the student to show that the inclusion of the potential energies would not noticeably affect the result for changes in height of 3 m between the inlet and outlet pipe sections in our control volume.

Example 6.3. A water turbine is operational at an efficiency of 58% and is developing 200 kW of power. The flow is 1000 L/s at a temperature of 10°C. What is ΔH_p across the turbine?

The power going through the turbine from the water is

$$\frac{dW_S}{dt} = \frac{200{,}000}{0.58} = 3.448 \times 10^5 \text{ W}$$

The power per unit mass of flow is then

$$\frac{dW_S}{dt} = \frac{dW_S}{dm/dt}\frac{dt}{} = \frac{3.448 \times 10^5}{(1)(999.7)} = 344.9 \text{ N-m/kg}$$

Hence we can say

$$g\,\Delta H_D = -\frac{dW_S}{dm} = -344.9 \text{ N-m/kg}$$

$$\therefore \Delta H_D = -\frac{344.9}{9.81} = -35.16 \text{ m}$$

Another situation that bears mentioning is the first law involving enthalpy h expressed in British thermal units (Btu) per *pound mass* flow while mass flow $\dot{m}$ is given as pounds mass per second. For such a situation we would write the first law with Q in Btu as

$$778\frac{dQ}{dt} + \left(\frac{V_1^2}{2g} + z_1 + 778h_1\right)\dot{m} = \left(\frac{V_2^2}{2g} + z_2 + 778h_2\right)\dot{m} + \frac{dW_S}{dt}$$

Note we have divided $V^2/2$ and gz by g. We explain this step in the following way. The terms $V^2/2$ and gz come from Newton's law and so are based on the slug. Thus they are energy per slug of flow. But $\dot{m}$ is based on pounds mass. Hence on dividing by g, the kinetic energy and potential energy terms then are given per unit pound mass of flow as is h in this case. We then get an equation involving energy per unit time. Please be alert in the homework problems for the nomenclature and units described in the preceding paragraphs.

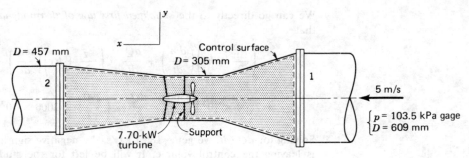

FIGURE 6.8
Water turbine.

In the last problem of this series we will be required to use linear momentum, the first law, and continuity simultaneously.

Example 6.4. A small water turbine is shown in Fig. 6.8 absorbing 7.70 kW from the flow of water proceeding through a water tunnel. What horizontal force is developed on the tunnel from the flow of water inside and the atmospheric pressure outside?

We choose a control volume that includes the entire interior space of the water tunnel shown in the diagram as the dotted region. The control surface thus cuts the supports of the water turbine in the manner described in Example 5.7. The desired force is the reaction to the force from the walls of the water tunnel onto the turbine as transmitted through the cut supports, plus the reaction to the force from the walls onto the water at the wetted area. By using the linear momentum equation for the chosen control volume, we can relate the reaction to the desired force with known quantities and with other quantities which may be evaluated by the aid of additional considerations.

Assumptions:

1. Steady flow.
2. Incompressible flow.
3. 1-D flow in and out.

The general *linear momentum* equation in the horizontal direction is now stated:

$$\iint_{\text{CS}} T_x \, dA + \iiint_{\text{CV}} B_x \rho \, dv = \iint_{\text{CS}} V_x(\rho \mathbf{V} \cdot \mathbf{dA}) + \frac{\partial}{\partial t} \iiint_{\text{CV}} V_x(\rho \, dv)$$

The body force integral vanishes for the horizontal direction and the last expression is zero because of assumption 1. Observing assumptions 2 and 3 we then have as the working equation,

Momentum equation in horizontal direction:

$$p_1 A_1 - p_2 A_2 + R_x = \rho_2 V_2^2 A_2 - \rho_1 V_1^2 A$$

Putting in numerical values, we get using gage pressures

$$(103.5 \times 10^3)\frac{(\pi)(0.609)^2}{4} - (p_2)(\pi)\frac{0.457^2}{4} + R_x$$

$$= (1000)(V_2^2)\frac{(\pi)(0.457^2)}{4} - (1000)(5^2)\frac{(\pi)(0.609^2)}{4}$$

Carrying out the algebraic operations and collecting terms, we get

$$-0.1640 p_2 + R_x = 164.0 V_2^2 - 3.743 \times 10^4 \qquad (a)$$

We have here as additional unknowns the quantities V_2 and p_2. The quantity V_2 is easily determined by using the chosen control volume for continuity considerations. Thus

Continuity equation:

$$\rho_1 V_1 A_1 = \rho_2 V_2 A_2$$

Therefore,

$$V_2 = \frac{A_1}{A_2} V_1 = \frac{(\pi)(0.609^2)/4}{(\pi)(0.457^2)/4} 5 = 8.88 \text{ m/s} \qquad (b)$$

We next consider the first law of thermodynamics for our control volume. We can argue that the totality of the local fluctuations of stored energy around the turbine gives rise to a mean time average of stored energy which is constant with time and so by considering the flow to be steady inside this control volume we will be able to ascertain the average pressure p_2 unknown up to this point. For a device of this kind, the flow will be reasonably close to adiabatic and since we are considering the flow as incompressible, we can take the internal energy as constant. These assumptions and others are summarized below.

Assumptions

---

1. Steady flow.
2. Incompressible flow.
3. Internal energy is constant.
4. Negligible heat transfer.
5. 1-D flow at conditions 1 and 2.

We can thus go directly to the *simplified first law*. Canceling u_1 and u_2 and deleting dQ/dm we then have in accordance with our assumed flow conditions

First law of thermodynamics:

$$\frac{V_1^2}{2} + p_1 v_1 = \frac{dW_S}{dm} + \frac{V_2^2}{2} + p_2 v_2 \qquad (c)$$

Noting that $\dot{m} = \rho V_1 A_1 = (1000)(5)(\pi)(0.609^2)/4 = 1.456 \times 10^3$ kg/s, we have using gage pressures

$$\left(\frac{5^2}{2}\right) + (103.5 \times 10^3)\left(\frac{1}{1000}\right) = \frac{7.70 \times 10^3}{1.456 \times 10^3} + \left(\frac{8.88^2}{2}\right) + p_2\left(\frac{1}{1000}\right)$$

Solving for p_2, we get

$$p_2 = 71.28 \text{ kPa gage} \qquad (d)$$

Substituting the solved value of p_2 and V_2 into Eq. (a) and solving for R_x, we get

$$R_x = -12.81 \text{ kN}$$

Hence,

$$K_x = 12.81 \text{ kN}$$

Because we used gage pressures in the momentum equation K_x is the total horizontal force from water inside and air outside. We could use gage pressure in the first law equation because ρ is constant in this problem.

Before you go to problems we point out that if your stated assumptions warrant, you are encouraged to go directly to the simplified forms of the first law [Eqs. (6.13) and (6.15)] as we have here. This is the same procedure proposed for conservation of mass, and is in contrast to the linear and moment-of-momentum equations wherein, because of the complexity of these equations, we urged you to start with the general equations first and then carefully to fashion your working equations.

6.6 BERNOULLI'S EQUATION FROM THE FIRST LAW OF THERMODYNAMICS

Let us consider that a portion of a streamtube in a *steady*, *incompressible*, *nonviscous* flow as shown in Fig. 6.9 as our control volume. In applying the first law of thermodynamics for this control volume we note that Eq. (6.15) is valid. There is obviously no work other than flow work, and hence the term dW_s/dm is zero. Furthermore, since the cross sections of the control volume are infinitesimal, we can replace $(z_c)_1$ and $(z_c)_2$ by z_1 and z_2, respectively. We then obtain after some rearrangement

$$\left(\frac{V_1^2}{2} + p_1 v + gz_1 \right) = \left(\frac{V_2^2}{2} + p_2 v + gz_2 \right) + \left[(u_2 - u_1) - \frac{dQ}{dm} \right]$$

For frictionless flow involving mechanical energy only, i.e., no heat transfer or change in internal energy, the last bracketed expression in the equation above vanishes and we obtain

$$\boxed{\frac{V_1^2}{2} + p_1 v + gz_1 = \frac{V_2^2}{2} + p_2 v + gz_2} \qquad (6.22)$$

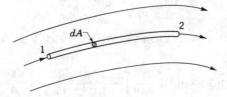

FIGURE 6.9
Streamtube in a steady, incompressible, nonviscous flow.

This equation is called *Bernoulli's equation*. Shrinking the streamtube cross-section without limit, Bernoulli then states that along a streamline the *mechanical energy* per unit mass is conserved. Or, along any one streamline:

$$\frac{V^2}{2} + pv + gz = \text{const} \tag{6.23}$$

The constant may have a different value for each streamline. However, in many problems one can deduce that somewhere in the flow the streamlines have the same mechanical energy per unit mass, so the mechanical energy per unit mass is constant *everywhere* in the flow. We can present Bernoulli's equation in a different form by dividing Eqs. (6.22) and (6.23) by g and replacing v/g by $1/\gamma$. We then get

$$\frac{V_1^2}{2g} + \frac{p_1}{\gamma} + z_1 = \frac{V_2^2}{2g} + \frac{p_2}{\gamma} + z_2 \tag{6.24}$$

Or, one can say that

$$\frac{V^2}{2g} + \frac{p}{\gamma} + z = \text{const} \tag{6.25}$$

Note that the dimension for each expression is L (length). The terms are accordingly called *heads*. Recall that we introduced the head H_D in conjunction with pumps and turbines when the fluid was a liquid. We can say for Eq. (6.25) that the sum of the velocity head, the pressure head, and the elevation head is constant along a streamline.

We now illustrate the use of Bernoulli's equation in the following examples. We often will use gage pressures in Bernoulli's equation since the $p_{\text{atm}}v$ part of the absolute pressure terms will cancel out.

Example 6.5. A large tank with a well-rounded, small opening as an outlet is shown in Fig. 6.10. What is the velocity of a jet of water issuing from the tank?

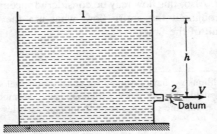

FIGURE 6.10
Efflux from a large tank through a well-rounded opening.

This is not strictly steady flow, since the elevation of the water surface h is decreasing. However, since h changes slowly, no serious error will be incurred if it is assumed at time t that the corresponding height h is constant in computing the jet velocity. The flow may thus be considered *quasi-steady*. It can be assumed, furthermore, that density is constant and friction can be neglected. However, corrections may later be made to account for the latter. Under these conditions and in light of the fact *that all streamlines have the same total energy per unit mass at the free surface*, we may use Bernoulli's equation at all positions in the flow.

By equating mechanical heads between point 1, at the free surface, and point 2, at the free jet, known quantities are related to the desired velocity. A position datum is established at the level of the jet. Hence, neglecting the kinetic energy at the free surface and taking the *pressure in the free jet to be atmospheric pressure*, we have[7]

$$h + \frac{p_{\text{atm}}}{\gamma} = \frac{V^2}{2g} + \frac{p_{\text{atm}}}{\gamma}$$

$$\therefore V = \sqrt{2gh}$$

For more accurate results one may account for friction by utilizing an experimentally determined coefficient called the *velocity coefficient* c_v to multiply $\sqrt{2gh}$. This coefficient depends on the size and shape of the opening as well as the elevation h of the free surface. The value of c_v usually is no smaller than 0.98 for well-rounded openings.

For other than a well-rounded opening there will be a contraction of the jet stream as it leaves the container. The smallest section of the jet is called the *vena contracta* (Fig. 6.11) and the area at this section is determined experimentally. The *coefficient of contraction* c_c, defined by the expression $A_c = c_c A$, is used for this purpose. This coefficient depends on the shape and size of the opening as well as the elevation of the free surface above the jet. Contraction coefficients usually run from 0.60, for a sharp-edged outlet, on up to 1, for the well-rounded outlet.

Therefore, to determine the rate of efflux of fluid, q, we have

$$q = c_v \sqrt{2gh} \, c_c A = c_d \sqrt{2gh} \, A$$

where $c_d = c_v c_c$ is called the *coefficient of discharge*. Tables and charts of the various aforementioned coefficients can be found in hydraulic handbooks and in Appendix A.I, where there is a more detailed discussion of flow-measuring devices.

Example 6.6. The cross section of an infinitely long cylinder is shown in Fig. 6.12 oriented at right angles to a steady flow which may be assumed to be uniform far from the cylinder. Also, this flow may be considered incompressible. If friction is to be neglected throughout the entire flow, what are the forces coming onto the cylinder as a result of the flow?

---

[7]Note for the simple conditions we have assumed we get the same velocity here as that of a freely falling particle in a gravity field when we neglect friction. Thus, thermodynamics and Newton's law give identical results here as was the case in the mechanics of particles and rigid bodies studied in your sophomore classes.

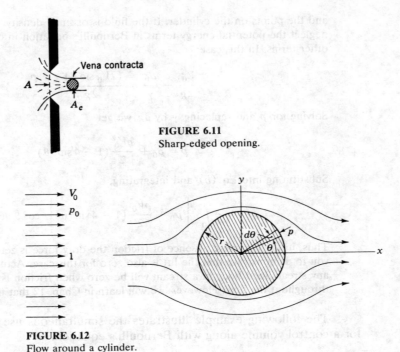

FIGURE 6.11
Sharp-edged opening.

FIGURE 6.12
Flow around a cylinder.

In aerodynamic work, engineers are concerned with two components of the force from the fluid flow. The component parallel to the free-stream velocity is called the *drag* component, while the component normal to the free stream is termed the *lift* component. Since this is a two-dimensional flow, evaluations will be made per unit length of the cylinder.

The free-stream velocity and pressure are given as V_0 and p_0, respectively. One may evaluate the fluid velocity at the boundary from two-dimensional incompressible nonviscous theory to be[8]

$$V = 2V_0 \sin \theta \qquad (a)$$

In the absence of shear stresses the force on the strip $r\,d\theta$ equals $pr\,d\theta$ and is normal to the boundary of the cylinder. The component parallel to the free stream is then $-pr\,d\theta \cos \theta$. Integrating over the surface gives us the drag per unit length

$$D = -\int_0^{2\pi} pr \cos \theta \, d\theta \qquad (b)$$

We must now find the pressure at each point on the cylinder surface in terms of the known free-stream characteristics and the variable θ. This can be done by using Bernoulli's equation between the free-stream conditions far from the cylinder

---

[8]This will be done in Chap. 12.

and the points on the cylinder. If the fluid is of small density, such as air, we may neglect the potential-energy terms in Bernoulli's equation in comparison with the other terms. In this case,

$$\frac{V_0^2}{2g} + \frac{p_0}{\gamma} = \frac{(2V_0 \sin \theta)^2}{2g} + \frac{p}{\gamma} \qquad (c)$$

Solving for p and replacing γ by ρg we get

$$p = p_0 + \frac{\rho V_0^2}{2}(1 - 4\sin^2 \theta) \qquad (d)$$

Substituting into Eq. (b), and integrating,

$$D = -\int_0^{2\pi} \left[p_0 + \frac{\rho V_0^2}{2}(1 - 4\sin^2 \theta) \right] r \cos \theta \, d\theta = 0 \qquad (e)$$

Thus, in the complete absence of friction the drag force is zero. It will be left for you to demonstrate that the lift is also zero for this case. Actually, the drag about any streamlined body in a stream will be zero when friction is completely ignored throughout the flow. However, we will learn in Chap. 12 that this is not so for lift.

The following example illustrates the simultaneous use of the basic laws for a control volume along with Bernoulli's equation.

Example 6.7. Water flows from a very large reservoir and drives a turbine, as is shown in Fig. 6.13. If we neglect friction through the pipes (they are short with large diameters), determine the horsepower developed by the flow on the turbine for the data given in the diagram.

First law of thermodynamics. We select a control volume which surrounds the turbine as shown in Fig. 6.13. We make the following assumptions for this control

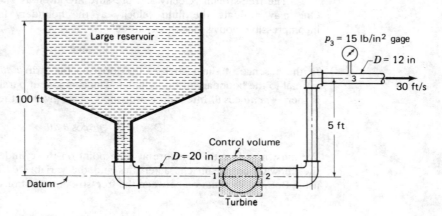

FIGURE 6.13
Flow from reservoir drives turbine.

volume:

---

1. Flow is steady.
2. No heat transfer.
3. Incompressible flow.
4. 1-D flow in and out.
5. No change in specific internal energy.

We can use the *simplified first law* for this case. Thus we have

$$\frac{V_1^2}{2} + p_1 v = \frac{V_2^2}{2} + p_2 v + \frac{dW_S}{dm} \tag{a}$$

The term dW_S/dm may be expressed as

$$\frac{dW_S}{dm} = \frac{P}{\rho V_1 A_1} \tag{b}$$

where P, the power, is the desired unknown. The unknowns are V_1, V_2, p_1, p_2, and P.

We now make additional assumptions for flow domains outside the control volume:

---

1. Neglect friction in reservoir and (as already noted) in the short large diameter pipes and elbows.
2. Flows are incompressible.
3. Flows are steady.
4. Flows are 1-D in pipes.
5. Neglect velocity at free surface of reservoir.

Bernoulli's equation. We may accordingly apply Bernoulli's equation between any point on the free surface and point 1 at the control surface at the centerline of the pipe just ahead of the turbine. Using gage pressures, we then get

$$100g = p_1 v + \frac{V_1^2}{2} \tag{c}$$

Similarly between points 2 and 3 we get from Bernoulli's equation

$$p_2 v + \frac{V_2^2}{2} = p_3 v + \frac{V_3^2}{2} + 5g \tag{d}$$

Continuity equations. We get for the control volume about the turbine

$$V_1 = V_2 \tag{e}$$

and for a control volume consisting of the interior of the pipe from point 2 to point

3 we get

$$V_2 = \left(\frac{12}{20}\right)^2 V_3 = 0.36V_3 \qquad (f)$$

We now have enough equations for all the unknowns, and we may solve for P in units of horsepower. Thus

$$P = 124.1 \text{ hp}$$

6.7 A NOTE ON THE SECOND LAW OF THERMODYNAMICS

It will be recalled from Chap. 1 that a property is a measurable characteristic of a material. The *state* of a substance is determined when enough properties are specified to establish uniquely the thermodynamic condition of the substance. A change of state takes place during a *process*. In a process, some or all of the properties will then have changing values.

If a system is involved in a process, there will usually be an exchange of energy between it and its surroundings, as well as a possible change in form of the stored energy. The first law of thermodynamics stipulates that during any and all processes all the energies are to be accounted for. The second law of thermodynamics now places restrictions on the *direction of energy transfer* as well as *the direction in which a real process may proceed*. For instance, heat must always proceed from a higher temperature to a lower temperature if no external influence is exerted on the process. Also, it is known that friction will always tend to retard the relative motion between two solid bodies in contact.

In frictionless incompressible flow, the second law of thermodynamics will be intrinsically satisfied as a result of the basic simplicity of the flow. In such flows there will only be exchanges of kinetic and potential energies between fluid elements, with the complete absence of friction and heat transfer; and under such circumstances there will be no restrictions on the manner of interchange of these energies. If friction without heat transfer is now assumed to be present, only correct directional effects need be prescribed to satisfy the second law. However, the inclusion of heat transfer dictates a more careful procedure, and it is when we reach this part of our study that we actively employ the second law of thermodynamics.

*6.8 THE SECOND LAW OF THERMODYNAMICS

You learned in your thermodynamics course that for a reversible process

$$\oint \frac{dQ}{T} = 0 \qquad (6.26)$$

This in turn means that between any two states (1 and 2), the integral $\int_1^2 dQ/T$ is independent of the path. To see this, examine Fig. 6.14, where two states, a and b are shown. Paths 1 and 2 are shown, representing arbitrary reversible

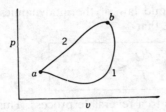

FIGURE 6.14
Paths 1 and 2 form cycle.

processes. The paths may be considered to form a cycle, so we may say, according to Eq. (6.26),

$$\int_{a \atop 1}^{b} \frac{dQ}{T} + \int_{b \atop 2}^{a} \frac{dQ}{T} = 0$$

Hence, on inverting the limits of the second integral, we can say

$$\int_{a \atop 1}^{b} \frac{dQ}{T} = \int_{a \atop 2}^{b} \frac{dQ}{T}$$

Since the paths were arbitrary, this integration would yield the same result for *any* path between a and b. Hence, it is possible to associate a quantity at each state of a substance such that the difference of these quantities at two points represents the integral $\int dQ/T$ between the points. That is,

$$S_2 - S_1 = \Delta S = \left(\int_1^2 \frac{dQ}{T} \right)_{\text{rev}} \tag{6.27}$$

where S represents the *entropy* of a substance, a property defined through considerations of the second law of thermodynamics and the idealization of reversibility. We will not concern ourselves here with the absolute values of entropy. The change of entropy will be the quantity of interest, thus making the datum of measure unnecessary.

For a process resulting in an infinitesimal change of state, Eq. (6.27) becomes

$$dS = \frac{dQ}{T} \tag{6.28}$$

$$\therefore dQ = T\,dS$$

We now pause to go back to Eq. (6.2) and consider dW_K to be from a gas at uniform pressure so that $dW_K = p\,dv$ for a unit mass. Then using Eq. (6.28) we get a relation called the *combined first and second law of thermodynamics* given as follows using s as the entropy per unit mass:

$$T\,ds = du + p\,dv \tag{6.29}$$

Continuing now, the second law of thermodynamics states that for *any* process between states 1 and 2 that

$$\boxed{\int_1^2 \frac{dQ}{T} \leq (S_2 - S_1)}$$ (6.30)

where the equality sign applies to a reversible process. This is a familiar way of expressing the second law of thermodynamics mathematically. Thus, a process for which the integral $\int dQ/T$ exceeds the change of entropy between two states is not possible. If a system undergoes an *adiabatic* process, the left side of the equation is always zero and we then have the result

$$\Delta S \geq 0$$ (6.31)

where the equality sign applies to the additional requirement that the process be reversible. We see then that for a reversible adiabatic process there is no change in entropy; such processes as pointed out earlier are termed *isentropic*.

We have thus far considered changes of the property entropy between equilibrium states. We have indicated how to compute the change in entropy between equilibrium states and have indicated that a reversible process between two equilibrium states results in a larger value of $\int dQ/T$ than an irreversible process between the states. What *do* we do with *nonequilibrium* states and the actions during a real process for which at any time t we do not have equilibrium? This is the province of *nonequilibrium* thermodynamics rather than thermostatics that we have been dealing with in more elementary courses. We shall make the following assumptions in regard to nonequilibrium thermodynamics—the correctness of which can only be deduced by the ability to describe real processes quantitatively properly. First, we shall assume that the entropy for any set of state parameters as measured by instruments during a real process is the same as for an equilibrium system having those parameters. Thus, we will take the measure of entropy at any time to be something independent of whether the system is in equilibrium or not at that time. The second assumption is that the system need not have uniform properties but can be nonuniform. Thus we can say at any time that:

$$\Delta S = \Delta\left[\iiint s\rho \, dv \right]$$ (6.32)

where s is the specific value of S and measured locally rather than in some finite uniform system as presupposed earlier.

We can now write the second law of thermodynamics for a system as

$$\frac{DS}{Dt} \geq \frac{1}{T}\frac{dQ}{dt}$$ (6.33)

Using the Reynolds transport equation we then have for a control volume:

$$\boxed{\oiint_{CS} s(\rho \mathbf{V} \cdot \mathbf{dA}) + \frac{\partial}{\partial t}\iiint_{CV} s(\rho \, dv) \geq \frac{1}{T}\frac{dQ}{dt}}$$ (6.34)

6.9   CLOSURE

We have developed the basic laws for systems and control volumes in this and the preceding chapter and have solved a number of problems. You will note that we used essentially the integral forms of these basic laws and solved for certain flow parameters at a certain part of a carefully chosen control surface. Little had to be known of the details of the flow inside the control volume, particularly for steady flow, and we invariably made certain simplifying assumptions concerning flow crossing the control surface (very often the one-dimensional-flow assumptions). In these problems, we required little information, and we reached limited results in the form of resultant forces, average velocities, etc. In other words, we did not determine the velocity field or the stress distribution throughout a region of flow by the use we made of the integral forms of the basic laws. To do this, and thus to be able to make more detailed deductions, we must use the *differential forms* of the basic laws at a point and integrate them in such a way that the given boundary conditions of the problem are satisfied. This is a very difficult undertaking—so difficult that we cannot present general integration procedures applicable for any fluid under any circumstance. Instead, we must present idealizations for certain classifications of flow, and introducing concepts useful in discussing these flows, we can then consider the basic laws in *differential* form tailored for these cases. Sometimes empirical and experimental results have to be included to reach meaningful results. In Parts II and III of this text, we will examine, in this manner, a number of important flow classifications.

In the following chapter, we will consider certain key differential equations that will be of much use to us as we proceed through Parts II and III of the text.

PROBLEMS

Problem Categories

First Law Problems 6.1–6.15
Bernoulli's Equation 6.16–6.37
First Law plus Bernoulli 6.38–6.50

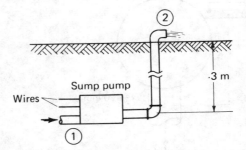

FIGURE P6.1

6.1. A sump pump is a sealed pump usually underground that pumps water from its inlet at ① to above ground at ②. The inlet has an inside diameter of 75 mm and the outlet at ② has a diameter of 50 mm. A current of 10 amp is flowing at a voltage of 220 V to the pump. What is the maximum possible capacity of the pump? Neglect friction in the pipes and heat transfer. Take p_1 as p_{atm}.

6.2. Air at an absolute pressure of 500 kPa and at a temperature of 35°C enters a highly insulated air motor and leaves as a free jet into the atmosphere at a temperature of −5°C. The inlet velocity is 25 m/s and the exit velocity is 70 m/s. If 3 kg of air flows per minute and if we take the internal energy, u, as $c_v T$, with c_v as a constant giving the specific heat at constant volume, what power is developed by the air motor? Take the specific heat as 4.08×10^{-5} N · m/(kg)(K). The atmospheric pressure is 101.4 kPa.

6.3. Steam enters a condenser at the rate of 600 kg/h with an enthalpy h of 2.70×10^6 N · m/(kg). To condense the steam, water at 15°C is brought in at the ratio of 7 kg of water per kilogram of steam. The water enters through a pipe with a 75-mm inside diameter and mixes directly with the steam. The velocity of the entering steam is 120 m/s. What is the temperature of the water leaving the condenser at the same elevation as the water inlet in a pipe having an inside diameter of 100 mm? We may take the enthalpy of a liquid to be $c_p T$ where c_p, the specific heat at constant pressure, is given as 4210 N · m/(kg)(K) for water. Neglect heat transfer from the condenser to the surroundings.

6.4. Water moves steadily through the turbine shown at the rate of 220 L/s. The pressures at 1 and 2 are 170 kPa gage and −20 kPa gage, respectively. If we neglect heat transfer, what is the horsepower delivered to the turbine from the water?

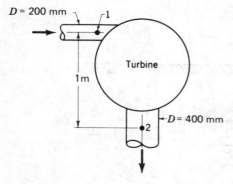

FIGURE P6.4

6.5. The flow rate through a turbine is 9000 kg/h, and the heat loss through the casing is 100,000 kJ/h. The inlet and exit enthalpies are 2300 kJ/kg and 1800 kJ/kg, respectively, while the

inlet and exit velocities are 25 m/s and 115 m/s, respectively. Compute the shaft horsepower of the turbine.

6.6. It takes 50 hp to drive the centrifugal water pump. The pressure of the water at 2 is 30 lb/in^2 gage, and at 1, where the water enters, it is at 10 lb/in^2 gage. How much water is the pump delivering?

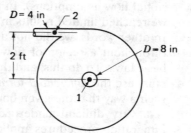

FIGURE P6.6

6.7. Shown in Fig. P6.7 is a system of highly insulated pipes through which water is flowing. In the upper pipe, the water leaving the pipe shows an increase of internal energy of 23 kJ/kg over the water entering at A; and the water leaving the lower pipe has an increase in internal energy of 116 kJ/kg (these increases are a result of friction in the flow). Compute the velocity V_3 for the data given in the diagram. Take the water as

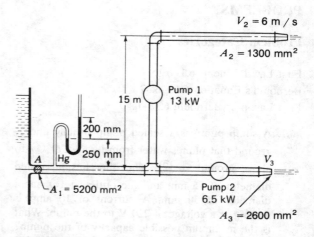

FIGURE P6.7

incompressible with an internal energy entering the pipe of 140 kJ/kg. (Set up to simultaneous equations, but do not solve.)

6.8. A gas undergoes steady flow through a porous plug in a *well insulated* pipe as shown in Fig. P6.8. Show that if we have no change in kinetic energy and no heat transfer that the enthalpy h is conserved on going through the plug. This is an example of what is called a *throttling process* which mimics what occurs as a gas passes through a partially opened valve.

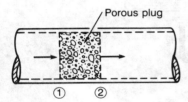

FIGURE P6.8

6.9. A jet condenser condenses steam into water by mixing a spray of water with exhaust steam from some device inside of a well-insulated tank. Water then leaves the tank. If entering steam has an enthalpy of 1200 Btu/lbm and enters at the rate of 300 lbm/h and if 4000 lb of water is injected per hour, what must the enthalpy of the incoming water be? The enthalpy of the water leaving the condenser is 120 Btu/lbm. Neglect kinetic and potential energy changes.

6.10. In the preceding problem, if the enthalpy of the entering water is 41 Btu/lbm, what must be the heat loss per hour from the condenser?

6.11. A *heat exchanger* shown in Fig. P6.11 has water entering at A, going through a set of horizontal pipes, and leaving at B. The purpose of this flow is to heat a flow of kerosene entering the heat exchanger at C and leaving at D after passing over the horizontal pipes. Water comes in at A at a temperature of 200°F and leaves at a temperature of 100°F. The kerosene is to be heated from 40°F to 120°F. If we are to heat 3 lbm/s of kerosene, what is the mass flow of water required? Diameters of pipes at A, B, C, and D are equal. The heat exchanger is well insulated. Use the following formulations for specific en-

thalpy per pound mass of the fluids (where t is in degrees Fahrenheit):

$$h_{water} = t - 32 \ \text{Btu/lbm}$$

$$h_{kerosene} = 0.5t + 0.0003t^2 \ \text{Btu/lbm}$$

Neglect kinetic energy.

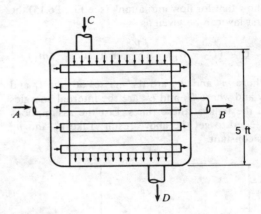

FIGURE P6.11

6.12. In Prob. 6.11 what is the heat transfer from the heat exchanger to the atmosphere if the amount of water used is 5000 lbm/h for the conditions given in this problem, and the exit temperature of the kerosene is 100°F?

6.13. A gas turbine is idling at steady state incurring very little heat transfer with the surroundings. Preheated air at a temperature of 400°F enters the *combustion chamber* of the gas turbine at the rate of 40 lbm/s with a velocity of 340 ft/s. Liquid fuel is brought in at the rate of 68 parts by weight of air to fuel. The liquid fuel is at 60°F. The combustion products leave the combustion chamber at a temperature of 1400°F, a velocity of 680 ft/s, and an enthalpy of 360 Btu/lbm. What is the enthalpy of the entering fuel? The enthalpy of the preheated air is given as

$$h = 124.3 + \int_{60}^{T} c_p \, dT \ \text{Btu/lbm}$$

where the reference enthalpy is taken at 60°F and T is in degrees Fahrenheit. Also for air we have at low pressure

$$c_p = 0.219 + \frac{0.342T}{10^4} - \frac{0.293T^2}{10^8} \text{ Btu/lbm °R}$$

where T is in degrees Rankine.

6.14. If in Prob. 6.13 the enthalpy of the liquid fuel is 12,000 Btu/lbm, what is the heat loss per second from the combustion chamber?

6.15. Show that for flow into a tank (see Fig. P6.15) the first law can be given as

$$Q = (U_2 - U_1) - \left(\frac{V_p^2}{2} + h_p\right)(m_2 - m_1)$$

where m_1 and m_2 are the masses at time t_1 and t_2 and where U_1 and U_2 are the internal energies in the tank at these times. List the assumptions needed to get the above result. Take V_p and h_p as constant.

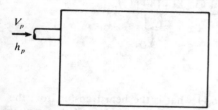

FIGURE P6.15

6.16. If friction is neglected, what is the velocity of the water issuing from the tank as a free jet? What is the discharge rate?

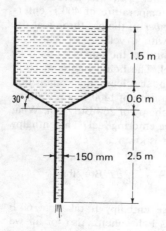

FIGURE P6.16

6.17. One end of a U tube is oriented directly into the flow so that the velocity of the stream is zero at this point. The pressure at a point in the flow which has been stopped in this way is called the *stagnation pressure*. The other end of the U tube measures the "undisturbed" pressure at a section in the flow. Neglecting friction, determine the volume flow of water in the pipe.

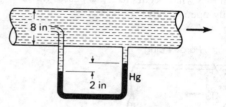

FIGURE P6.17

6.18. Compute the ideal flow rate through the pipe system shown in Fig. P6.18 *Hint:* Read Prob. 6.17.

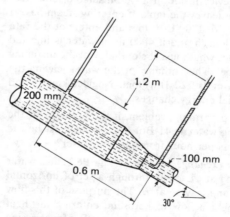

FIGURE P6.18

6.19. A cylindrical tank contains air, oil, and water (Fig. P6.19). On the oil a pressure p of 5 lb/in^2 gage is maintained. What is the velocity of the water leaving if we neglect both friction everywhere and the kinetic energy of the fluid above elevation A? The jet of water leaving has a diameter of 1 ft.

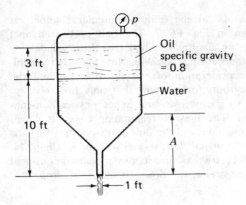

FIGURE P6.19

6.20. A large tank contains compressed air, gasoline at specific gravity 0.68, light oil at specific gravity 0.80, and water. The pressure p of the air is 150 kPa gage. If we neglect friction, what is the mass flow $\dot{m}$ of oil from a 20-mm diameter jet?

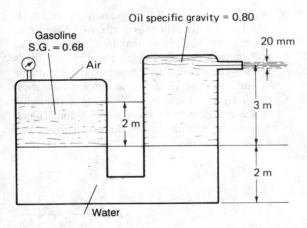

FIGURE P6.20

6.21. A *venturi meter* is a device which is inserted into a pipe line to measure incompressible-flow rates. It consists of a convergent section which reduces the diameter to between one-half and one-fourth the pipe diameter. This is followed by a divergent section. The pressure difference between the position just before the venturi and at the throat of the venturi is measured by a differential

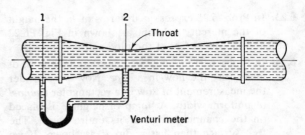

FIGURE P6.21

manometer as shown. Show that

$$q = c_d \left[\frac{A_2}{\sqrt{1 - (A_2/A_1)^2}} \sqrt{2g \frac{p_1 - p_2}{\gamma}} \right]$$

where c_d is the *coefficient of discharge*, which takes into account frictional effects and is determined experimentally.

6.22. Another way of measuring flow rates is to use the *flow nozzle*, which is a device inserted into the pipe as shown in Fig. P6.22. If A_2 is the exit area of the flow nozzle, show that for incompressible flow we get for q

$$q = c_d \left[\frac{A_2}{\sqrt{1 - (A_2/A_1)^2}} \sqrt{2g \frac{p_1 - p_2}{\gamma}} \right]$$

where c_d is the *coefficient of discharge*, which takes into account frictional effects and is determined experimentally.

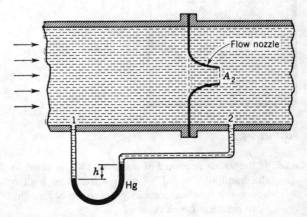

FIGURE P6.22

6.23. In Prob. 6.22 express q in terms of h, the height of the mercury column, as shown in Fig. P6.22 and the diameters of the pipe and flow nozzle.

6.24. In Probs. 6.21 and 6.22 we considered methods of measuring the flow in a *pipe*. Now we consider the measurement of flow in a rectangular *channel* of uniform width. A hump of height δ is placed on the channel bed over its entire width. The free surface then has a dip d as shown. If we neglect friction we can consider that we have one-dimensional flow. Compute the flow q for the channel per unit width. This system is called a *venturi flume*.

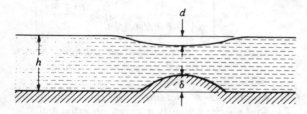

FIGURE P6.24

6.25. A *siphon* is shown in Fig. P6.25. If we neglect friction entirely, what is the velocity of the water leaving at C as a free jet? What are the pressures of the water in the tube at B and at A?

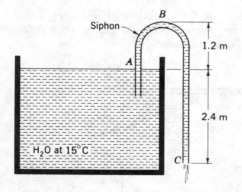

FIGURE P6.25

6.26. If the vapor pressure of water at 15°C is given in the handbook as 0.1799 m of water, how high h above the free surface can point B be before the siphon action breaks down?

6.27. Water is flowing in a rectangular channel, as shown in Fig. P6.27. The bed of the channel drops an amount H. Show that three values of y are theoretically possible. Now by rough graphical consideration of the function yielding the three roots for y, show that only two roots y_1 and y_2 are positive and hence physically meaningful. The flow corresponding to y_1 is called *shooting flow* and the flow corresponding to y_2 is called *tranquil flow*, as you will see in Chap. 14. Neglect friction and consider one-dimensional flow upstream and downstream of the drop.

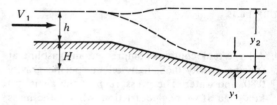

FIGURE P6.27

6.28. Water is moving with high velocity V_1 in a rectangular channel. There is a rise of height H in the channel bed. Show that to the right of this rise there are three depths given by the fluids calculations. By rough graphical considerations of the equation yielding the three roots, show that only two roots y_1 and y_2 are meaningful. Neglect friction and consider one-dimensional flow at the sections shown downstream and upstream of the rise.

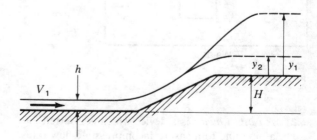

FIGURE P6.28

6.29. Water flows steadily up the vertical pipe and enters the annular region between the circular plates as shown. It then moves out radially, issuing out as a free sheet of water. If we neglect friction entirely, what is the flow of water through the pipe if the pressure at A is 69 kPa gage?

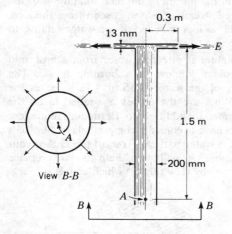

View B-B

FIGURE P6.29

6.30. In Prob. 6.29, compute the upward force on the device from water and air. The volume flow is 0.408 m³/s. Explain why you cannot profitably use Bernoulli's equation here for a force calculation.

6.31. The velocity at point A is 18 m/s. What is the pressure at point B if we neglect friction?

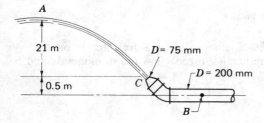

FIGURE P6.31

6.32. A diver is directing a flexible pipe into which is sucked sand and water so as to expose part of a sunken ship. If the pressure at the inlet A is close to the hydrostatic pressure of the surrounding water, what amount of sand will be sucked up per second by a 2-kW pump? The specific gravity

of the sand and water mixture picked up is 1.8. The inside diameter of the pipe is 250 mm. Neglect friction losses in the pipe.

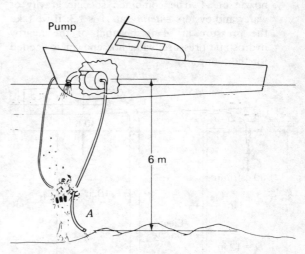

FIGURE P6.32

6.33. Air is made to flow through a well-insulated pipe by a pump. It is desired that 50.0 ft³ of air per second flow by A. The inlet pressure p_A is 10 lb/in² absolute and the temperature is 60°F. What power is required for the blower? Take u equal to $c_v T$, where c_v is the specific heat at constant volume and T is the absolute temperature. Use the value 0.171 Btu/(lbm)(°R) for c_v. The exit temperature of the air is 90°F.

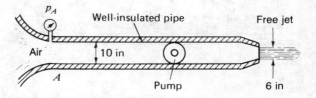

FIGURE P6.33

6.34. A system for collecting oysters from the bottom floor of the ocean has been attempted in the past. A pump on board ship passes 1500 gal/min of water into a 6-in pipe to a so-called ejector nozzle which is housed in a second larger nozzle E open at end A and connected to a 10-in pipe.

The jet of water from the ejector nozzle at *B* entrains water in the larger nozzle *E* and draws 250 gal/min of water and oysters into the larger nozzle at *A*. The combined specific gravity of water and oysters *entering* at *A* is 1.3. If we take the pressure at *A* to be that of the nearby hydrostatic pressure, what horsepower is needed for the pump?

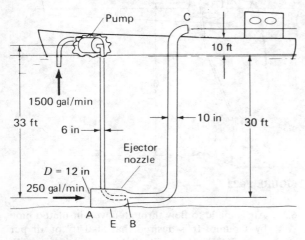

FIGURE P6.34

6.35. A rocket-powered test sled slides over rails. This test sled is used for experimentation on the ability of human beings to undergo large persistent

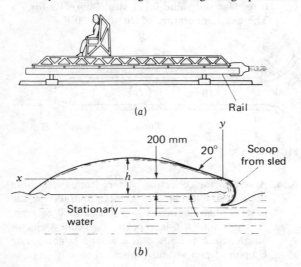

FIGURE P6.35

accelerations. To brake the sled from high speeds, small scoops are lowered to deflect water from a stationary tank of water placed near the end of the run. If the sled is moving at a speed of 100 km/h at the instant of interest, compute *h* of the deflected stream of water as seen from the sled. Assume no loss in speed of the water relative to the scoop.

6.36. A firefighter is directing water from a hose into the broken window of a burning house. The velocity of the water is 15 m/s as it leaves the hose. What are the angles α needed to do the job? *Hint:* In addition to Bernoulli's equation, you will have to consider components of Newton's law for a water particle. Toward the end of your calculations it will also help if you replace $1/(\cos^2 \alpha)$ by $(1 + \tan^2 \alpha)$ which equals $\sec^2 \alpha$.

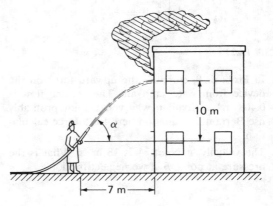

FIGURE P6.36

6.37. The engine room of a freighter is on fire. A fire-fighting tugboat has drawn alongside and is

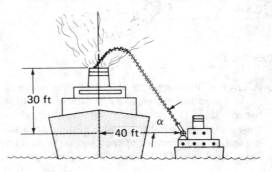

FIGURE P6.37

directing a stream of water to go into the stack of the freighter. If the exit speed of the jet of water is 70 ft/s, what angle α is needed to accomplish the task? *Hint:* Only one α will result in water getting into the stack. To decide the proper α of the two results, locate positions x where y_{max} occurs for the stream and decide which stream can enter stack. See hint in Prob. 6.36.

6.38. A fluid expands through a nozzle from a pressure of 300 lb/in^2 absolute to a pressure of 5 lb/in^2 absolute. The initial and final enthalpies of the fluid are 1187 Btu/lbm and 1041 Btu/lbm, respectively. Calculate the final velocity by neglecting the inlet velocity (called the approach velocity), gravitational effects, and heat transfer out of the casing and along the fluid flow. If the internal energy u of the fluid is known at the exit conditions to be 800 Btu/lbm and the inlet and outlet areas are 3 in^2 and 2 in^2, respectively, compute the thrust of the nozzle.

6.39. Neglecting friction in the pipe shown in Fig. P6.39, compute the power developed on the turbine from the water coming from a large reservoir.

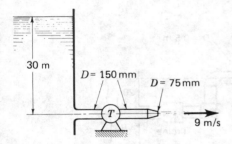

FIGURE P6.39

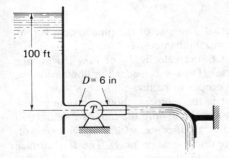

FIGURE P6.40

6.40. Water enters a pipe in Fig. P6.40 from a large reservoir and on issuing out of the pipe strikes a 90° deflector plate as shown. If a horizontal thrust of 200 lb is developed on the deflector, what is the horsepower developed by the turbine?

6.41. Water in a large tank is under a pressure of 35 kPa gage at the free surface. It is pumped through a pipe as shown and issues out of a nozzle to form a free jet. For the data given, what is the power required by the pump?

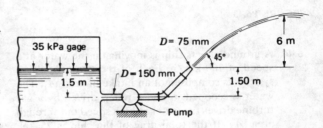

FIGURE P6.41

6.42. A pump draws water out of a reservoir as shown in Fig. P6.42. The pump develops 10 hp on the flow. What is the horizontal force at support D required as a result of the fluid flow?

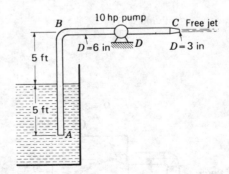

FIGURE P6.42

6.43. If the pump in Fig P6.43 develops 3.75 kW on the flow, what is the flow rate? *Hint:* What is the velocity of flow at the right opening of the U tube?

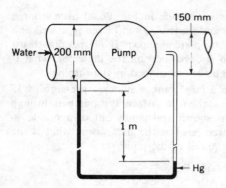

FIGURE P6.43

6.44. A ground effects ship is moving on the water at a speed of 100 km/h. Each of the two propulsion systems is composed of an intake of area A_1. The water is scooped in and a pump driven by a gas turbine drives the water at high speed out through area A_2. If the total drag of the ship is 25 kN, and if there are two drive systems described above, what is the area A_1 for each inlet?

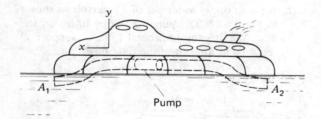

FIGURE P6.44

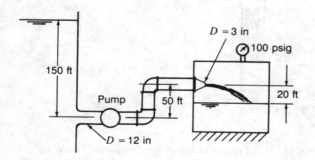

FIGURE P6.45

6.45. If 10 ft^3/s of water is to flow in Fig. P6.45, what must be the horsepower of the pump? Neglect friction.

6.46. What horsepower is needed to cause 30 ft^3/s of water flow in Fig. P6.46? Neglect friction in pipes. The exit diameter of the nozzle is 10 in.

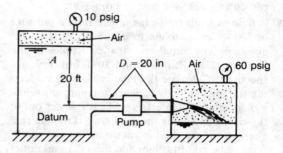

FIGURE P6.46

6.47. Neglecting friction, what is the power developed by the turbine in Fig. P6.47? At B we have a free jet. The mass flow is 500 kg/s.

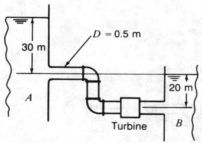

FIGURE P6.47

6.48. The internal diameter of the pipe system is 6 in. The exit nozzle diameter is 3 in.
  (a) What is the velocity V_e of flow leaving the nozzle? (Do *not* consider the flow inside the pipe proper to be inviscid.)
  (b) What is the moment about A coming from the water alone onto the pipe? BC is parallel to the z direction. (Set up moment-of-momentum equation only.) The free surface may be considered at constant height.

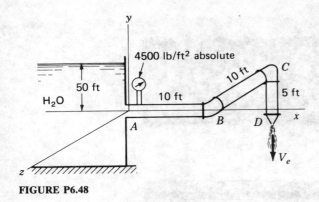

FIGURE P6.48

6.49. A fountain consists of a tank G containing a water pump feeding four pipes out of which

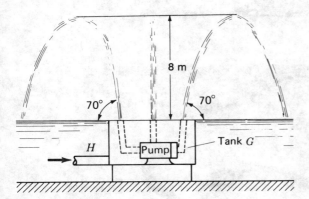

FIGURE P6.49

come water streams. The top of the tank is open. At H, we inject enough water to replace the water taken in by the pump from the tank G to keep the level in the tank the same as outside the tank. If the inside diameter of the four pipes is 75 mm, what total vertical force is developed on the tank supports stemming from the flow of water?

6.50. A water fountain has four identical spouts of water emerging from a tank inside of which (not shown) is a pump driving the flows in the four spouts. Consider spout A. The unit vector ϵ at the centerline of pipe A is given as

$$\epsilon = 0.5\mathbf{i} + 0.4\mathbf{j} + 0.768\mathbf{k}$$

If each pipe has an inside diameter of 3 in, what is the vertical force on the tank and the torque on the tank about the y axis at the center of the tank both as a result of the water flowing? The inlet flow at B is the xy plane.

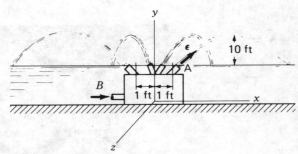

FIGURE P6.50

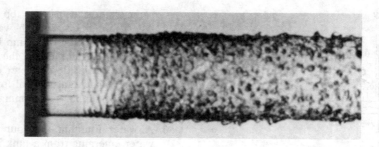

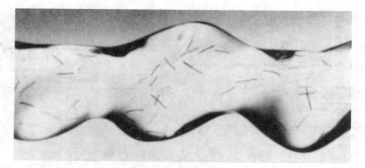

Photographs of free jets showing effects on surface due to turbulance as well as impending breakup of a jet. (Courtesy J. W. Hoyt and J. J. Taylor, Flow Visualization using the "floc" Technique, Flow Visualization II Wolfgang Merzkirch, Hemisphere Publishing Corp., 1982, p. 683.)

A free jet is shown in the top figure at exit illustrating a transition to turbulence on the jet surface. In the second photo, we have flow from a nozzle designed to delay transition to turbulence on the jet surface. The third photo shows a jet about to break up. Notice the random orientation of the "floc" particles. We have only considered in the book the free jet at immediate exit. Obviously the free jet in its entirety represents a complex flow about which books have been written.

CHAPTER
7

DIFFERENTIAL
FORMS OF
THE BASIC LAWS

7.1 INTRODUCTION

In the last two chapters we considered basic laws for finite systems, and using the Reynolds transport equation, we formulated the basic laws for finite control volumes. Particularly for steady flows, we were able to solve for certain resultant force components, average velocities, etc., by utilizing information concerning the flow at the control surface. We did not need detailed information concerning the characteristics of the flow everywhere inside the control volume.[1] At times this is an advantage in that it affords great simplicity in computing certain quantities. The weakness of this approach is that you can get only average values of quantities or only components of resultant forces but cannot learn about the details of the flow. For this information we employ appropriate *differential equations* valid at a point. These equations must then be integrated in order to satisfy the boundary conditions of the problem.

In this chapter, we first present differential equations for the conservation of mass by considering an infinitesimal control volume and then for Newton's

---

[1]This is analogous to the blackbody approach of systems theory.

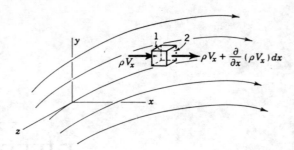

FIGURE 7.1
Infinitesimal fixed control volume.

law by considering an infinitesimal system. To illustrate the use of the differential form of Newton's law, we examine the steady-state configuration of liquids under constant acceleration and under constant rotation about a vertical axis. Furthermore, in starred sections we later present differential forms of all four basic laws using a neat procedure involving index notation. This procedure is used in other fields such as solid mechanics.

It is important to say once again that infinitesimal systems and infinitesimal control volumes yield exactly the same differential equations for any given law. We will use both approaches in the ensuing work.

PART A
ELEMENTARY DEVELOPMENT OF DIFFERENTIAL FORMS OF THE BASIC LAWS

7.2 CONSERVATION OF MASS

In dealing with the differential forms of the basic laws, we generally use the notation u, v, w to respectively replace V_x, V_y, V_z. We shall do this in this and succeeding chapters.

We now examine an infinitesimal control volume in the shape of a rectangular parallelepiped (Fig. 7.1) fixed in xyz for some general flow $\mathbf{V}(x, y, z, t)$ measured relative to xyz. In computing the net efflux rate for this control volume, we first consider flow through the surfaces 1 and 2, which are parallel to the yz plane. Note from Fig. 7.1 that the efflux rate through area 1 is $-\rho u$ per unit area and that for a continuum this varies continuously in the x direction, so the efflux rate per unit area through area 2 can be given as $\rho u + [\partial(\rho u)/\partial x]\,dx$.[2] The net efflux rate through these surfaces is then $[\partial(\rho u)/\partial x]\,dx\,dy\,dz$. Performing similar computations for the other pairs of

---

[2] We are using a two-term Taylor series expansion here.

sides and adding the results, we get the net efflux rate:

$$\text{Net efflux rate} = \left[\frac{\partial(\rho u)}{\partial x} + \frac{\partial(\rho v)}{\partial y} + \frac{\partial(\rho w)}{\partial z}\right] dx\,dy\,dz \qquad (7.1)$$

Equating this to the rate of decrease of mass inside the control volume, $-(\partial\rho/\partial t)\,dx\,dy\,dz$, we get the desired equation for conservation of mass after we cancel $dx\,dy\,dz$.

$$\boxed{\frac{\partial(\rho u)}{\partial x} + \frac{\partial(\rho v)}{\partial y} + \frac{\partial(\rho w)}{\partial z} = -\frac{\partial\rho}{\partial t}} \qquad (7.2)$$

We often call this the *differential continuity equation*. For *steady flow* this equation becomes

$$\frac{\partial(\rho u)}{\partial x} + \frac{\partial(\rho v)}{\partial y} + \frac{\partial(\rho w)}{\partial z} = 0 \qquad (7.3)$$

and for *incompressible flow* we get

$$\frac{\partial u}{\partial x} + \frac{\partial v}{\partial y} + \frac{\partial w}{\partial z} = 0 \qquad (7.4)$$

even if the flow is unsteady.

By using other suitable infinitesimal control volumes you can develop, in a manner analogous to what we have here, corresponding differential equations for cylindrical and spherical coordinates. These equations are all *differential* forms of the continuity equation, in contrast to Eq. (5.1), which is the *integral* form of the continuity equation. You will be asked to work out as an assignment (Prob. 7.1) the case of cylindrical coordinates (see Fig. 7.2) to get

$$\frac{1}{\bar{r}}\frac{\partial}{\partial\bar{r}}(\bar{r}\rho v_{\bar{r}}) + \frac{1}{\bar{r}}\frac{\partial(\rho v_\theta)}{\partial\theta} + \frac{\partial(\rho v_z)}{\partial z} = -\frac{\partial\rho}{\partial t} \qquad (7.5)$$

Although the left sides of Eqs. (7.2) and (7.5) appear quite different, they both convey the same information, namely, the measure of the *rate of efflux of mass per unit volume at a point*. Thus the differences in form are due only to the use of different coordinate systems. To divorce ourselves from the artificially con-

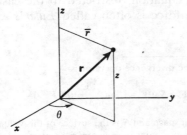

FIGURE 7.2
Cylindrical coordinates.

trived coordinate systems in expressing the efflux rate per unit volume at a point, we introduce the *divergence operator*[3] so that for any and all coordinate systems we can say for the continuity equation that

$$\boxed{\text{div}(\rho\mathbf{V}) = -\frac{\partial\rho}{\partial t}} \quad \text{or} \quad \boxed{\nabla\cdot(\rho\mathbf{V}) = -\frac{\partial\rho}{\partial t}} \tag{7.6}$$

For different coordinate systems the divergence operator (like the gradient operator) takes on different forms. Thus, for Cartesian coordinates the divergence operating on any vector field **A** becomes

$$\boxed{\nabla\cdot\mathbf{A} = \frac{\partial A_x}{\partial x} + \frac{\partial A_y}{\partial y} + \frac{\partial A_z}{\partial z}} \tag{7.7}$$

and in cylindrical coordinates we have

$$\boxed{\nabla\cdot\mathbf{A} = \frac{1}{\bar{r}}\frac{\partial}{\partial\bar{r}}(\bar{r}A_{\bar{r}}) + \frac{1}{\bar{r}}\frac{\partial A_\theta}{\partial\theta} + \frac{\partial A_z}{\partial z}} \tag{7.8}$$

Clearly, for different vector fields **A**, the divergence of **A** takes on a somewhat different meaning from that corresponding to the vector field $\rho\mathbf{V}$.

7.3 NEWTON'S LAW; EULER'S EQUATION

Linear momentum of an element of mass dm is a vector quantity defined as $dm\,\mathbf{V}$. The fundamental statement of Newton's law for an inertial reference is given in terms of linear momentum as

$$\mathbf{dF} = \frac{D}{Dt}(dm\,\mathbf{V})$$

In the case where the infinitesimal particle of mass dm is part of a velocity field $\mathbf{V}(x, y, z, t)$ as seen from the inertial reference, this equation may be given as

$$\mathbf{dF} = \frac{D}{Dt}(dm\,\mathbf{V}) = dm\left(u\frac{\partial\mathbf{V}}{\partial x} + v\frac{\partial\mathbf{V}}{\partial y} + w\frac{\partial\mathbf{V}}{\partial z} + \frac{\partial\mathbf{V}}{\partial t}\right) \tag{7.9}$$

The extreme right-hand quantity in parentheses is the substantial derivative discussed in Chap. 4. This equation, restricted to the case of *no shear* stress with *only gravity* as a body force is often called *Euler's equation*. The surface

---

[3]The mathematical definition of the divergence operator is

$$\text{div}\,\mathbf{F} \equiv \nabla\cdot\mathbf{F} = \lim_{\Delta V\to 0}\frac{1}{\Delta V}\oint_S \mathbf{F}\cdot\mathbf{dA}$$

where ΔV is any volume element in space and S is the surface of this volume element. In electricity, div **E** is is the flux per unit volume at a point in space from an electric field.

force on a fluid element is then due only to pressure p, and in accordance with our discussion in Sec. 2.7, we can express this force in the form $-(\nabla p)\,dv$. We take the negative z direction to correspond to the direction of gravity. We can thus express the gravity force as $-g\rho\,dv\,\mathbf{k}$, which can also be given as $-g(\nabla z)(\rho\,dv)$. Employing the aforesaid forces in Eq. (7.9) and dividing through by $\rho\,dv\,(=dm)$, we then have Euler's equation

$$-\frac{1}{\rho}\,\nabla p - g\,\nabla z = \left(u\frac{\partial \mathbf{V}}{\partial x} + v\frac{\partial \mathbf{V}}{\partial y} + w\frac{\partial \mathbf{V}}{\partial z}\right) + \frac{\partial \mathbf{V}}{\partial t} = \frac{D\mathbf{V}}{Dt} \qquad (7.10)$$

Using familiar vector operations, we can next express the acceleration of transport in the following way:

$$u\frac{\partial \mathbf{V}}{\partial x} + v\frac{\partial \mathbf{V}}{\partial y} + w\frac{\partial \mathbf{V}}{\partial z} = (\mathbf{V}\cdot\nabla)\mathbf{V}$$

You may readily verify this yourself by carrying our the operations on the right side of the equation in terms of Cartesian components, remembering to evaluate the operator in the parentheses first before operating on the last term $\mathbf{V}$. We then may present another form of Euler's equation:

$$-\frac{1}{\rho}\,\nabla p - g\,\nabla z = (\mathbf{V}\cdot\nabla)\mathbf{V} + \frac{\partial \mathbf{V}}{\partial t} \qquad (7.11)$$

The expanded forms of the above equation in rectangular, cylindrical, and streamline coordinates are shown below.

Rectangular Coordinates

$$-\frac{1}{\rho}\frac{\partial p}{\partial x} + B_x = \left(u\frac{\partial u}{\partial x} + v\frac{\partial u}{\partial y} + w\frac{\partial u}{\partial z}\right) + \frac{\partial u}{\partial t} \qquad (7.12a)$$

$$-\frac{1}{\rho}\frac{\partial p}{\partial y} + B_y = \left(u\frac{\partial v}{\partial x} + v\frac{\partial v}{\partial y} + w\frac{\partial v}{\partial z}\right) + \frac{\partial v}{\partial t} \qquad (7.12b)$$

$$-\frac{1}{\rho}\frac{\partial p}{\partial z} + B_z = \left(u\frac{\partial w}{\partial x} + v\frac{\partial w}{\partial y} + w\frac{\partial w}{\partial z}\right) + \frac{\partial w}{\partial t} \qquad (7.12c)$$

Cylindrical Coordinates

$$-\frac{1}{\rho}\frac{\partial p}{\partial \bar{r}} + B_{\bar{r}} = \left(v_{\bar{r}}\frac{\partial v_{\bar{r}}}{\partial \bar{r}} + \frac{v_\theta}{\bar{r}}\frac{\partial v_{\bar{r}}}{\partial \theta} - \frac{v_\theta^2}{\bar{r}} + v_z\frac{\partial v_{\bar{r}}}{\partial z}\right) + \frac{\partial v_{\bar{r}}}{\partial t} \qquad (7.12d)$$

$$-\frac{1}{\rho}\frac{\partial p}{\bar{r}\partial\theta} + B_\theta = \left(v_{\bar{r}}\frac{\partial v_\theta}{\partial \bar{r}} + \frac{v_\theta}{\bar{r}}\frac{\partial v_\theta}{\partial \theta} + \frac{v_{\bar{r}}v_\theta}{\bar{r}} + v_z\frac{\partial v_\theta}{\partial z}\right) + \frac{\partial v_\theta}{\partial t} \qquad (7.12e)$$

$$-\frac{1}{\rho}\frac{\partial p}{\partial z} + B_z = \left(v_{\bar{r}}\frac{\partial v_z}{\partial \bar{r}} + \frac{v_\theta}{\bar{r}}\frac{\partial v_z}{\partial \theta} + v_z\frac{\partial v_z}{\partial z}\right) + \frac{\partial v_z}{\partial t} \qquad (7.12f)$$

Streamline Coordinates in the Osculating Plane[4]

$$-\frac{1}{\rho}\frac{\partial p}{\partial s} + B_s = V\frac{\partial V}{\partial s} + \frac{\partial V}{\partial t} \tag{7.12g}$$

$$-\frac{1}{\rho}\frac{\partial p}{\partial n} + B_n = \frac{V^2}{R} + \frac{\partial V_n}{\partial t} \tag{7.12h}$$

We consider next two applications of Euler's equation.

*7.4 LIQUIDS UNDER CONSTANT RECTILINEAR ACCELERATION OR UNDER CONSTANT ANGULAR SPEED

Case 1. Uniformly accelerating liquid. Consider a liquid in a container undergoing constant acceleration **a** in inertial space. The motion of the container will be along a straight line collinear with **a**. Suppose that we choose the z axis of the inertial reference to be vertical to the earth's surface and with the y axis to form coordinate plane yz parallel to the acceleration and velocity vectors of the container (see Fig. 7.3). Thus $a_x = V_x = 0$. After a period of time, the liquid will reach a fixed orientation relative to the container.[5] Accordingly, all fluid elements will then have the same velocity at any time t. This means that $\partial \mathbf{V}/\partial x = \partial \mathbf{V}/\partial y = \partial \mathbf{V}/\partial z = 0$, and $\partial \mathbf{V}/\partial t = \mathbf{a}$ throughout the liquid. From Newton's viscosity law, furthermore, with $\partial \mathbf{V}/\partial n = 0$ in any direction **n**, we conclude that there is no shear stress present. In that case, the stress field degenerates to a pressure field. We can accordingly apply Euler's equation for this case. Thus, using p_g as the gage pressure we have

$$-\frac{1}{\rho}\boldsymbol{\nabla}(p_g + p_{\text{atm}}) - g\,\boldsymbol{\nabla}z = \mathbf{a}$$

$$\therefore -\frac{1}{\rho}\boldsymbol{\nabla}p_g - g\,\boldsymbol{\nabla}z = \mathbf{a} \tag{7.13}$$

---

[4]The osculating plane at a position on a curve is the limiting plane of three different points on the curve as they come together at the position of interest on the curve.

[5]This is analogous to a linear mass-spring problem of sophomore mechanics, where a sudden constant force is applied to the mass. The mass will oscillate as a result of the disturbance, but this motion will die out as a result of friction, leaving in due course a fixed deflection of the spring. The motion that dies out is the so-called *transient* leaving the *steady-state* fixed deflection. We are here concerned with the steady-state orientation of the liquid.

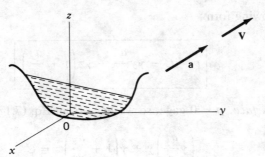

FIGURE 7.3
Container of liquid accelerates uniformly along a straight line parallel to yz plane.

Let us first consider the z component of this equation. We get

$$-\frac{1}{\rho}\frac{\partial p_g}{\partial z} - g = a_z$$

$$\boxed{\therefore \quad \frac{\partial p_g}{\partial z} = -\rho(g + a_z)}$$

(7.14)

The gage pressure p_g thus varies in the z direction like a *hydrostatic* pressure variation, except that we add to the acceleration of gravity g the acceleration component a_z. With $a_z = 0$ the pressure variation is *exactly hydrostatic*, as studied in Chap. 2.

What is the shape of the free surface? To determine this, take the dot product of Eq. (7.13) with the infinitesimal position vector **dr** in *any* direction. Thus dropping the subscript g on gage pressure p for convenience,

$$-\frac{1}{\rho}\nabla p \cdot \mathbf{dr} - g\,\nabla z \cdot \mathbf{dr} = \mathbf{a} \cdot \mathbf{dr}$$

$$\therefore -\frac{1}{\rho}\left(\frac{\partial p}{\partial x}\,dx + \frac{\partial p}{\partial y}\,dy + \frac{\partial p}{\partial z}\,dz\right) - g\,dz = a_z\,dz + a_y\,dy$$

Noting that the expression in parentheses is the total differential dp, we get

$$-\frac{dp}{\rho} - g\,dz = a_z\,dz + a_y\,dy$$

(7.15)

Remember that the term dp is the change in pressure in *any* direction. We now integrate Eq. (7.15). Thus

$$-\frac{p}{\rho} - gz = a_z z + a_y y + C'$$

$$\therefore p = -\rho a_z z - \rho a_y y - \rho gz + C$$

(7.16)

We determine the constant of integration C from knowledge of the pressure at some known location. We may rewrite Eq. (7.16) on noting that $\rho = \gamma/g$ to have the following form:

$$p = C - \gamma y \frac{a_y}{g} - \gamma z \left(1 + \frac{a_z}{g}\right) \qquad (7.17)$$

At the *free surface*, $p = 0$ gage, so we have from Eq. (7.17) for this surface

$$\left(\gamma \frac{a_y}{g}\right) y + \gamma \left(1 + \frac{a_z}{g}\right) z = C \qquad (7.18)$$

This is clearly the equation of a *plane surface*—a not unexpected result. The slope of this surface is

$$\left(\frac{dz}{dy}\right)_{fs} = -\frac{a_y/g}{1 + a_z/g} = -\frac{a_y}{g + a_z} \qquad (7.19)$$

We now consider an example to illustrate the use of the formulations above.

Example 7.1. An open, small, rectangular container of water is placed on a conveyor belt (Fig. 7.4) which is accelerating at a constant rate of 5 m/s². Will the water spill from the tank during the steady-state configuration of the water?

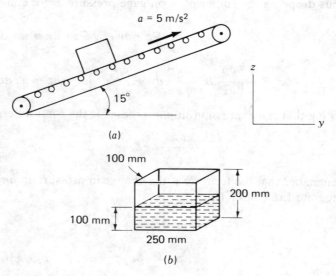

FIGURE 7.4
Tank of water on an accelerating conveyor belt.

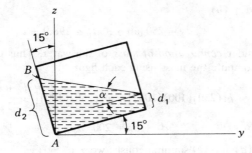

FIGURE 7.5
Water in steady-state configuration.

We first compute the *slope* of the free surface of the water for steady-state operation (Fig. 7.5). From Eq. (7.19), we have

$$\left(\frac{dz}{dy}\right)_{\text{fs}} = -\frac{a_y}{g + a_z} = -\frac{5\cos 15°}{9.81 + 5\sin 15°} = -0.435 \qquad (a)$$

From Fig. 7.6, we can then say for the angle of inclination β (see right end of Fig. 7.6) of the free surface

$$\tan^{-1}\left(\frac{dz}{dy}\right)_{\text{fs}} = \tan^{-1}(-0.435) = 180° - \beta$$

$$\therefore \beta = 23.5°$$

Now, considering Fig. 7.5 again, we can relate d_1 and d_2 for the free surface as follows:

$$\frac{d_2 - d_1}{250} = \tan \alpha \qquad (b)$$

Next consider Fig. 7.6. It should be clear from this diagram that

$$\alpha + 75° + (90° - \beta) = 180°$$

$$\therefore \alpha = 15° + \beta \qquad (c)$$

$$= 15° + 23.5° = 38.5°$$

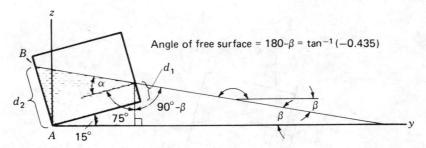

FIGURE 7.6
Orientation of the free surface.

Hence, from Eq. (*b*),

$$d_2 = 250 \tan \alpha + d_1 = 198.9 + d_1 \qquad (d)$$

We now consider *conservation of mass* of the water. Thus considering Figs. 7.4b and 7.5, we compute the mass using each figure.

$$\rho[(250)(100)(100)] = \rho\left[(250)(100)\left(\frac{d_1 + d_2}{2}\right)\right] \qquad (e)$$

$$\therefore d_1 + d_2 = 200$$

Solving Eqs. (*d*) and (*e*) simultaneously, we get

$$d_2 = 199.5 \text{ mm}$$

On consulting Figs. 7.4b and 7.5, we can conclude that the water does *not* spill during steady-state operation.

Example 7.2. In Example 7.1 determine the force on the left side *AB* of the tank from the water during steady-state operation.

The gage pressure in the tank is given as follows in accordance with Eq. (7.16):

$$p = -\rho a_z z - \rho a_y y - \rho g z + C$$

$$= -(1000)(5 \sin 15°)z - (1000)(5 \cos 15°)y - (1000)(9.81)z + C \qquad (a)$$

$$= -4830y - 11,100z + C$$

where C is the constant of integration. In Fig. 7.6, note that $p = 0$ at position B where

$$y = -d_2 \sin 15° = -0.1995 \sin 15° = -0.0516 \text{ m}$$

$$z = d_2 \cos 15° = 0.1995 \cos 15° = 0.1927 \text{ m}$$

Going back to Eq. (*a*), we have at B

$$0 = -(4830)(-0.0516) - (11,100)(0.1927) + C$$

$$\therefore C = 1890 \text{ N/m}^2 \qquad (b)$$

Now consider the side BA in Fig. 7.7. The force df on segment $d\eta$ is found using

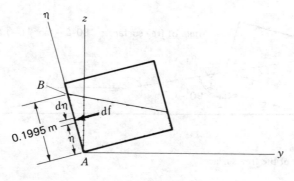

FIGURE 7.7
Force df on wall from pressure.

Eq. (a) as:

$$df = p\,dA = p(0.100)\,d\eta$$

$$= [-4830(-\eta \sin 15°) - (11,100)(\eta \cos 15°) + 1890](0.100)\,d\eta$$

$$= [-947\eta + 189.0]\,d\eta \quad \text{N} \tag{c}$$

Integrating over the entire side AB, we have

$$F = \int_0^{0.1995} (-947\eta + 189.0)\,d\eta = 18.86 \text{ N}$$

Can you find the center of pressure?

Case 2. Uniform rotation about a vertical axis. We now consider a cylindrical container of liquid (see Fig. 7.8) which is maintained at a uniform angular rotation for a long enough period of time to have the liquid reach a steady-state orientation of the liquid relative to the container. We would like the shape of the free surface as well as the pressure distribution below the free surface for this steady state configuration.

Clearly, for steady-state operation, the liquid rotates as if it were a solid (i.e., with no relative deformational motion between particles) and hence there is no shearing action between elements in the flow. Thus, there is no shear stress on the elements and the stress field once more degenerates to a pressure distribution. We can again use Eq. (7.13) as follows:

$$-\frac{1}{\rho}\nabla(p + p_{\text{atm}}) - g\,\nabla z = \mathbf{a} \tag{7.20}$$

where p is the gage pressure. The acceleration vector is toward the axis of

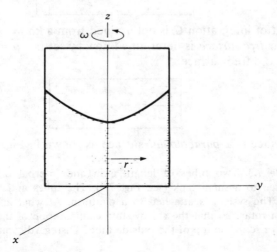

FIGURE 7.8
Cylinder rotating at constant speed. Liquid has reached steady-state configuration.

rotation and is given as $-\bar{r}\omega^2\boldsymbol{\epsilon}_{\bar{r}}$. We then have

$$-\frac{1}{\rho}\nabla p - g\,\nabla z = -\bar{r}\omega^2\boldsymbol{\epsilon}_{\bar{r}} \tag{7.21}$$

First, we consider the component of Eq. (7.21) in the z direction—i.e., the direction of gravity. We get

$$-\frac{1}{\rho}\frac{\partial p}{\partial z} - g = 0$$

$$\therefore \boxed{\frac{\partial p}{\partial z} = -\gamma} \tag{7.22}$$

We see that the pressure p varies *hydrostatically* in the z direction. We next take the dot product of Eq. (7.21) with the arbitrary infinitesimal position vector $\mathbf{dr} = d\bar{r}\,\boldsymbol{\epsilon}_{\bar{r}} + \bar{r}\,d\theta\,\boldsymbol{\epsilon}_{\theta} + dz\,\boldsymbol{\epsilon}_z$, where we have used cylindrical coordinates. Recalling from case 1 of Sec. 7.4 that $\nabla p \cdot \mathbf{dr}$ results in the change in pressure dp, we get

$$-\frac{dp}{\rho} - g\,dz = -\bar{r}\omega^2\,d\bar{r}$$

We now integrate the equation to get

$$-\frac{p}{\rho} - gz = -\frac{\bar{r}^2\omega^2}{2} + C'$$

Rearranging the equation, we have

$$\boxed{p = -\rho g z + \frac{\rho\bar{r}^2\omega^2}{2} + C} \tag{7.23}$$

The constant of integration C is determined from a known pressure at a known location. The *free surface* is determined now by setting $p = 0$ in Eq. (7.23). We then get for the free surface

$$\boxed{\frac{\rho\omega^2}{2}\bar{r}^2 - \rho g z = -C} \tag{7.24}$$

The free surface is a *paraboloidal* surface as shown in Fig. 7.8.

Example 7.3. Two tubes of length 6 in and internal diameter of 0.1 in are connected to a small tank (see Fig. 7.9). The tubes and tank contain water as shown. The system is attached to a platform. At what angular speed must the platform rotate so that the steady-state configuration of the water will cause the water to rise to the top of the outside tube? Disregard capillary effects.

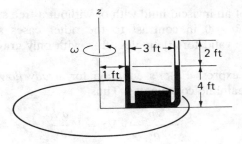

FIGURE 7.9
Initial configuration for $\omega = 0$.

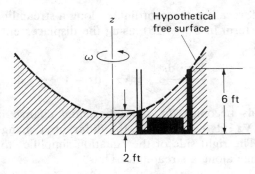

FIGURE 7.10
Desired steady-state configuration.

The desired steady-state configuration is shown in Fig. 7.10. Now the heights reached by the water in the tubes would be the same as the free surface of water in a hypothetical cylindrical tank rotating with the same speed ω provided that the hypothetical tank initially had 4 ft depth of water like the tubes. Accordingly, we can use Eq. (7.24) for this hypothetical free surface. Thus,

$$\frac{\rho \omega^2}{2}\bar{r}^2 - \rho g z = -C \qquad (a)$$

We have two unknowns—ω and C. However, we know (see Fig. 7.10) that when $\bar{r} = 4$ ft, then $z = 6$ ft, and from conservation of mass in the tube-tank system, that when $\bar{r} = 1$ ft, then $z = 2$ ft. Applying these conditions to Eq. (a), we have

$$\frac{1.938\omega^2}{2}4^2 - (1.938)(32.2)(6) = -C$$

$$\frac{1.938\omega^2}{2}1^2 - (1.938)(32.2)(2) = -C$$

Solving simultaneously, we get

$$\omega = 4.14 \text{ rad/s}$$

7.5 INTEGRATION OF THE STEADY-STATE EULER EQUATION; BERNOULLI'S EQUATION

In the previous section, we integrated Euler's equation for the case of uniform acceleration of liquid with a free surface and for steady-state angular rotation of a liquid with free surface. In both cases, $\partial \mathbf{V}/\partial t \neq \mathbf{0}$. We now consider general steady-state flow of an inviscid fluid with or without a free surface. Here, we will require that $\partial \mathbf{V}/\partial t = \mathbf{0}$ in contrast to the other cases studied. Recall that Euler's equation is valid for frictionless flow with only gravity acting as a body force.

Let us then express Euler's equation for *steady flow* with the transport term given for streamline coordinates. Thus

$$-\frac{\nabla p}{\rho} - g \, \nabla z = V \frac{\partial \mathbf{V}}{\partial s} \tag{7.25}$$

where s, you will recall, is the coordinate along a streamline. Now take the dot product of each term in Eq. (7.25), using the displacement vector **ds** *along the streamline*. Thus

$$-\frac{\nabla p \cdot \mathbf{ds}}{\rho} - g \, \nabla z \cdot \mathbf{ds} = V \frac{\partial \mathbf{V}}{\partial s} \cdot \mathbf{ds}$$

The term $\nabla p \cdot \mathbf{ds}$ becomes dp, the differential change in pressure along a streamline; and $\nabla z \cdot \mathbf{ds}$ becomes dz, the differential change in elevation along the streamline. The right side of the equation simplifies to $V\,dV$, where dV is the velocity change along a streamline. Thus[6]

$$-\frac{dp}{\rho} - g \, dz = V \, dV = d\left(\frac{V^2}{2}\right)$$

Taking g as constant and integrating along a streamline, we get

$$\boxed{\int_0^p \frac{dp}{\rho} + gz + \frac{V^2}{2} = \text{const}} \tag{7.26}$$

This equation is often called the *compressible* form of Bernoulli's equation. If ρ is expressible as a function of p only, that is, $\rho = \rho(p)$, the first expression is integrable. Flows having this characteristic are called *barotropic flows*. If, on the other hand, the flow is *incompressible*, we then get[7]

$$\boxed{\frac{p}{\rho} + gz + \frac{V^2}{2} = \text{const}} \tag{7.27}$$

This is Bernoulli's equation, derived in Chap. 6 from the first law of thermody-

---

[6] We have had to restrict **ds** to be along a streamline in order to get the right side of the equation into the form of a differential and thus to permit the integration to give the simple result $V^2/2$.
[7] We may use p/ρ or pv in Bernoulli's equation. In thermodynamics, the tendency is to use pv. We will generally use p/ρ in fluid mechanics.

namics. Between any two points along a streamline, we have

$$\frac{p_1}{\rho} + gz_1 + \frac{V_1^2}{2} = \frac{p_2}{\rho} + gz_2 + \frac{V_2^2}{2} \tag{7.28}$$

Multiplying Eq. (7.27) by $1/g$ and replacing $g\rho$ by γ, we get

$$\boxed{\frac{p}{\gamma} + z + \frac{V^2}{2g} = \text{const}} \tag{7.29}$$

You will recall that the terms in this equation are in units of length and are frequently designated as pressure, elevation, and velocity *heads*, respectively. The analogous equation to Eq. (7.28) between two points in a flow can then be given for the various heads as

$$\frac{p_1}{\gamma} + z_1 + \frac{V_1^2}{2g} = \frac{p_2}{\gamma} + z_2 + \frac{V_2^2}{2g} \tag{7.30}$$

We see that for frictionless, isothermal, flow the first law of thermodynamics is equivalent to Newton's law. This is the same situation that prevailed in particle and rigid-body mechanics, where Newton's law, linear momentum methods, and energy methods were equivalent to each other. That is, one could solve a problem considering any of these three methods. (To be sure, certain problems were more easily solved by certain ones of the three methods.) We point out next that if the flow is compressible or if friction causes changes in the properties of the fluid by causing temperature changes, among other things, the first law of thermodynamics and Newton's law (in the form of momentum equations) are *independent* equations and must be separately satisfied.

7.6 BERNOULLI'S EQUATION APPLIED TO IRROTATIONAL FLOW

The Bernoulli equation developed in Sec. 6.6 may be applied between points along any one streamline when there is frictionless, incompressible, steady flow. If we stipulate further that the flow be *irrotational*, we can show that Bernoulli's equation is valid between *any two* points in a flow.

Thus consider Euler's equation for steady flow in the form given by Eq. (7.13):

$$-\frac{\nabla p}{\rho} - g\,\nabla z = (\mathbf{V} \cdot \nabla)\mathbf{V}$$

You will recall that this equation is valid for frictionless flow with gravity as the only body force. The term $(\mathbf{V} \cdot \nabla)\mathbf{V}$ in this equation may be replaced in the following manner:

$$(\mathbf{V} \cdot \nabla)\mathbf{V} = \nabla\left(\frac{V^2}{2}\right) - \mathbf{V} \times \text{curl } \mathbf{V} \tag{7.31}$$

a step which you may readily verify by carrying out the operations on both sides of the equation, using Cartesian components of the vectors and operators. Euler's equation can then be written as

$$-\frac{\nabla p}{\rho} - g\,\nabla z = \nabla\left(\frac{V^2}{2}\right) - \mathbf{V} \times \text{curl } \mathbf{V} \tag{7.32}$$

If the flow is *irrotational*, **curl V** = **0** and Eq. (7.32) simplifies to

$$\frac{\nabla p}{\rho} + g\,\nabla z + \nabla\left(\frac{V^2}{2}\right) = \mathbf{0} \tag{7.33}$$

We next take the dot product of the terms in the equation with the *arbitrary* infinitesimal displacement vector **dr**. Noting that $\nabla p \cdot \mathbf{dr} = dp$, $\nabla z \cdot \mathbf{dr} = dz$, and $\nabla(V^2/2) \cdot \mathbf{dr} = d(V^2/2)$, where the differentials represent infinitesimal changes of the quantities in the *arbitrary* direction of **dr**, we get

$$\frac{dp}{\rho} + g\,dz + d\left(\frac{V^2}{2}\right) = 0$$

Taking g as a constant, and integrating, we get

$$\int_0^p \frac{dp}{\rho} + gz + \frac{V^2}{2} = \text{const} \tag{7.34}$$

By limiting ourselves to incompressible flow we then get

$$\frac{p}{\rho} + gz + \frac{V^2}{2} = \text{const} \tag{7.35}$$

Multiplying through by $1/g$,

$$\frac{p}{\gamma} + z + \frac{V^2}{2g} = \text{const} \tag{7.36}$$

Since **dr** was arbitrary in direction, there was no directional restriction on the differentials formed, with the result that the integrated formulations are valid *everywhere* in the flow. Thus, by including the restriction of irrotationality, we can throw out the earlier requirement of remaining on a particular streamline for a given constant in Bernoulli's equation. Hence, between *any* two points 1 and 2 in such a flow we have

$$\frac{p_1}{\gamma} + z_1 + \frac{V_1^2}{2g} = \frac{p_2}{\gamma} + z_2 + \frac{V_2^2}{2g} \tag{7.37}$$

We now consider Newton's law for a general flow.

*7.7 NEWTON'S LAW FOR GENERAL FLOWS

In Sec. 7.3, we considered the differential form of Newton's law for inviscid flows in a gravitational field and developed the Euler equations of motion for a fluid. Now we consider a general flow of fluid of any kind under any circumstance. Accordingly, consider an infinitesimal element of fluid which at time t is a rectangular parallelepiped (Fig. 7.11). We show only stresses on the faces of the element which give force increments in the x direction. Note that we have assumed the stresses to vary continuously and have used Taylor series expansions limited to two terms. Thus, on face 1, we have shown τ_{xx} and on face 2, a distance dx apart from face 1, we have used $\tau_{xx} + (\partial \tau_{xx}/\partial x)\, dx$, where τ_{xx} in this expression corresponds to the stress on face 1 and the derivative is taken at a position corresponding to face 1. Now use Newton's law in the x direction relative to inertial reference XYZ. With body-force component B_x, we have

$$
\begin{aligned}
&\left(\tau_{xx} + \frac{\partial \tau_{xx}}{\partial x}\, dx \right) dy\, dz - \tau_{xx}\, dy\, dz + \\[2mm]
&\left(\tau_{yx} + \frac{\partial \tau_{yx}}{\partial y}\, dy \right) dx\, dz - \tau_{yx}\, dx\, dz + \\[2mm]
&\left(\tau_{zx} + \frac{\partial \tau_{zx}}{\partial z}\, dz \right) dx\, dy - \tau_{zx}\, dx\, dy + \\[2mm]
&B_x \rho\, dx\, dy\, dz = \rho (\, dx\, dy\, dz)\, a_x
\end{aligned}
\tag{7.38}
$$

where a_x is the acceleration component in the x direction. Canceling terms and

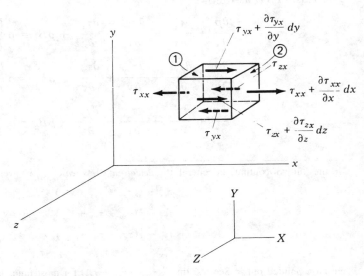

FIGURE 7.11
Stresses contributing force in x direction.

dividing through by $dx\, dy\, dz$, we get

$$\frac{\partial \tau_{xx}}{\partial x} + \frac{\partial \tau_{yx}}{\partial y} + \frac{\partial \tau_{zx}}{\partial z} + \rho B_x = \rho a_x$$

Now using the complementary property of shear ($\tau_{ij} = \tau_{ji}$) and formulating Newton's law in the y and z directions, we get the desired form of Newton's law as follows:

$$
\begin{aligned}
\frac{\partial \tau_{xx}}{\partial x} + \frac{\partial \tau_{xy}}{\partial y} + \frac{\partial \tau_{xz}}{\partial z} + \rho B_x &= \rho \frac{Du}{Dt} \\[2mm]
\frac{\partial \tau_{yx}}{\partial x} + \frac{\partial \tau_{yy}}{\partial y} + \frac{\partial \tau_{yz}}{\partial z} + \rho B_y &= \rho \frac{Dv}{Dt} \\[2mm]
\frac{\partial \tau_{zx}}{\partial x} + \frac{\partial \tau_{zy}}{\partial y} + \frac{\partial \tau_{zz}}{\partial z} + \rho B_z &= \rho \frac{Dw}{Dt}
\end{aligned}
\qquad (7.39)
$$

Note that the equations have a simple format if you simply start off with the matrix array of stresses. These equations may be familiar to some readers, having seen them in strength of materials with the right side of the equation set equal to zero for equilibrium.

If we consider that the stresses are the superposition of a dilatation deformation associated with a scalar pressure field p plus a distortional deformation associated with a stress field τ'_{ij} called the *deviatoric* stress field,[8] we can give Eqs. (7.39) as follows:

$$-\frac{\partial p}{\partial x} + \frac{\partial \tau'_{xx}}{\partial x} + \frac{\partial \tau'_{xy}}{\partial y} + \frac{\partial \tau'_{xz}}{\partial z} + \rho B_x = \rho \frac{Du}{Dt} \qquad (7.40a)$$

$$-\frac{\partial p}{\partial y} + \frac{\partial \tau'_{yx}}{\partial x} + \frac{\partial \tau'_{yy}}{\partial y} + \frac{\partial \tau'_{yz}}{\partial z} + \rho B_y = \rho \frac{Dv}{Dt} \qquad (7.40b)$$

$$-\frac{\partial p}{\partial z} + \frac{\partial \tau'_{zx}}{\partial x} + \frac{\partial \tau'_{zy}}{\partial y} + \frac{\partial \tau'_{zz}}{\partial z} + \rho B_z = \rho \frac{Dw}{Dt} \qquad (7.40c)$$

---

[8] In the study of continua in general, the *deviatoric stress tensor* τ'_{ij} is given as

$$\tau'_{xx} = \tau_{xx} - \tfrac{1}{3}(\tau_{xx} + \tau_{yy} + \tau_{zz})$$

$$\tau'_{yy} = \tau_{yy} - \tfrac{1}{3}(\tau_{xx} + \tau_{yy} + \tau_{zz})$$

$$\tau'_{zz} = \tau_{zz} - \tfrac{1}{3}(\tau_{xx} + \tau_{yy} + \tau_{zz})$$

$$\tau'_{xy} = \tau_{xy} \qquad \tau'_{xy} = \tau_{xz} \qquad \tau'_{yz} = \tau_{yz}$$

We have pointed out in Sec. 2.6 that $\tfrac{1}{3}(\tau_{xx} + \tau_{yy} + \tau_{zz})$ for most fluids is $-p$. Equation (7.40) is thus Newton's law for such fluids involving the deviatoric stress tensor. Substituting for τ'_{ij} in Eq. (7.40) we then get back to Eq. (7.39).

If we have an inviscid flow with only gravity as the body force, Eqs. (7.40) become

$$-\frac{\partial p}{\partial x} = \rho \frac{Du}{Dt}$$

$$-\frac{\partial p}{\partial y} = \rho \frac{Dv}{Dt} \qquad (7.41)$$

$$-\frac{\partial p}{\partial z} - \rho g = \rho \frac{Dw}{Dt}$$

In vector form, we have

$$-\boldsymbol{\nabla} p - \rho g \, \boldsymbol{\nabla} z = \rho \frac{D\mathbf{V}}{Dt} = \rho \left[(\mathbf{V} \cdot \boldsymbol{\nabla})\mathbf{V} + \frac{\partial \mathbf{V}}{\partial t} \right] \qquad (7.42)$$

which you will recognize as Euler's equation.

7.8 PROBLEMS INVOLVING LAMINAR[9] PARALLEL FLOWS

In Part II of the text we shall be concerned with integrating differential forms of basic laws. At this time as a preview we shall integrate certain differential equations of previous sections.[10] This will allow us to use these equations immediately while their development is still fresh. We consider here two *parallel*, *steady*, *incompressible isothermal* flows of a *Newtonian* fluid.

Before we begin, we wish to show from the continuity equation that for incompressible parallel flow the velocity profile *cannot change* in the direction of flow. Thus for such a flow in the x direction, we have $v = w = 0$ and from *continuity* we have at a point

$$\frac{\partial u}{\partial x} + \frac{\partial v}{\partial y} + \frac{\partial w}{\partial z} = 0$$

$$\therefore \frac{\partial u}{\partial x} + 0 + 0 = 0$$

The velocity V which equals u thus cannot change in the direction of flow with the result that the *velocity profile must remain intact*.

We now examine two cases, namely, flows between infinite parallel plates and flows through pipes.

---

[9]You will recall from Chap. 1 that a laminar flow is a well-ordered flow in contrast to turbulent flow which has a random small velocity variation superposed over what otherwise would be well ordered. We shall begin studies of turbulent flows in Chap. 9.

[10]The reader may now opt to go to Chap. 10, Secs. 10.1, 10.3, and 10.6, to first develop the *Navier-Stokes equations* and then to return here once the differential equations of the parallel plate and pipe flows have been reached. The reader need then only consider here the solutions to these differential equations.

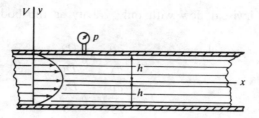

FIGURE 7.12
Flow between two infinite parallel plates of a
Newtonian fluid.

Case 1. Flow between two infinite parallel plates. We now consider steady, laminar, incompressible flow between two infinite parallel plates shown in Fig. 7.12. We shall apply the *linear momentum equation*. Clearly the only nonzero stress from τ'_{ij} is τ'_{xy}. Furthermore, this stress can only be a function of y since the velocity profile does not change in the direction of flow x and also since the problem is two dimensional thus eliminating variations with z. With this in mind, Eqs. (7.40) reduce to

$$-\frac{\partial p}{\partial x} + \frac{\partial \tau'_{xy}}{\partial y} = 0 \qquad (7.43a)$$

$$-\frac{\partial p}{\partial y} - \rho g = 0 \qquad (7.43b)$$

$$-\frac{\partial p}{\partial z} = 0 \qquad (7.43c)$$

Note we have used the fact that $v = w = 0$ and $u = u(y)$, as well as the condition of steady flow to render all substantial derivatives equal to zero. Note also that the pressure *varies hydrostatically* in the y direction [Eq. (7.43b)][11] and will be a function additionally only of x [Eqs. (7.43a) and (7.43c)]. Hence, at any section x, the gage pressure at the upper plate will be that recorded by a pressure gage mounted on the plate at that position (see Fig. 7.12) and will increase hydrostatically as one descends into the flow at that position x.

Now we apply the *Newton viscosity law* and we focus in on Eq. (7.43a). Thus we have

$$\frac{\partial p}{\partial x} = \mu \frac{\partial^2 u}{\partial y^2} \qquad (7.44)$$

If we delete the hydrostatic increase in pressure in the y direction (since it is constant at any level y and clearly has no influence on the velocity field), then p, which is the measurement at the top, will be constant at a section x. Also, $\partial p/\partial x$ is constant at a section x. Hence $\partial p/\partial x$ can only vary with x.

On the other hand, the velocity u has been shown to be independent of x, and in the two-dimensional problem can be a function of only the coordinate y.

---

[11]This will also be shown more generally in Sec. 10.3.

Thus each side of the equation is a function of a separate distinct variable. If an equality is to be maintained for all values of the independent variables x and y, it is necessary that each side always equal the same constant. Otherwise, one could vary one variable, say, x, and alter the value of the left side of the equation independently of the right side and so invalidate the equality. Hence, it may be said that

$$\frac{\partial p}{\partial x} = -\beta \tag{7.45a}$$

$$\mu \frac{\partial^2 u}{\partial y^2} = -\beta \tag{7.45b}$$

where $-\beta$ is the aforementioned constant. Since Eq. (7.45a) involves only the independent variable x and Eq. (7.45b) involves only the independent variable y, it is clear that the preceding equations are in effect two independent ordinary differential equations. They may then be written as

$$\frac{dp}{dx} = -\beta \tag{7.46a}$$

$$\mu \frac{d^2 u}{dy^2} = -\beta \tag{7.46b}$$

In essence, a second-order *partial* differential equation has then been replaced by a couple of *ordinary* differential equations.

It is now a simple matter to integrate both equations and arrive at the desired results of a velocity profile and an expression for pressure variation. Integrating Eq. (7.46b) first, we get

$$u = -\frac{\beta}{\mu}\left(\frac{y^2}{2} + C_1 y + C_2\right) \tag{7.47}$$

The constants of integration may be determined by employing the boundary conditions, which are

$$u = 0 \quad \text{when} \quad y = \pm h \tag{7.48}$$

Applying these conditions to the equation leads to the following relations:

$$0 = -\frac{\beta}{\mu}\left(\frac{h^2}{2} + C_1 h + C_2\right)$$

$$\tag{7.49}$$

$$0 = -\frac{\beta}{\mu}\left(\frac{h^2}{2} - C_1 h + C_2\right)$$

These equations are satisfied if $C_2 = -h^2/2$ and $C_1 = 0$. The velocity profile may then be given as

$$u = \frac{\beta}{2\mu}(h^2 - y^2) \tag{7.50}$$

As has been indicated in Fig. 7.12, the profile is that of a two-dimensional parabolic surface. The maximum velocity occurs at $y = 0$, so that

$$(u)_{max} = \frac{\beta h^2}{2\mu} \tag{7.51}$$

We wish next to express the separation constant β in terms of the flow q. Hence, we have for q per unit width of the flow

$$q = \int_{-h}^{+h} u(1)(dy)$$

$$= \int_{-h}^{+h} \frac{\beta}{2\mu}(h^2 - y^2)\,dy$$

$$= \frac{\beta}{2\mu}\left[h^2 y - \frac{y^3}{3}\right]\Bigg|_{-h}^{+h} = \frac{\beta}{2\mu}\left[\frac{4}{3}h^3\right] \tag{7.52}$$

Solving for β, we get

$$\beta = \tfrac{3}{2}h^{-3}q\mu \tag{7.53}$$

With β established, the velocity profile from Eq. (7.50) is now fully determined.
    Let us now turn our attention to the pressure. Substituting from Eq. (7.53), we may rewrite Eq. (7.46a) as

$$\frac{dp}{dx} = -\frac{3}{2}h^{-3}q\mu \tag{7.54}$$

Integrating from position 1 to position 2, a distance L apart along the direction of flow, we get

$$(p)_1 - (p)_2 = \tfrac{3}{2}qh^{-3}L\mu \tag{7.55}$$

The loss in pressure in the direction of flow represented in Eq. (7.55) is attributable only to the action of friction. By dividing both sides of the equation by ρ, there results an expression which we call *head loss*, h_l. This expression represents the loss of pressure due to friction over distance L per unit of mass flowing.[12] Thus

$$\frac{p_1 - p_2}{\rho} \equiv h_l = \frac{3}{2}\frac{qL\mu}{\rho h^3} \tag{7.56}$$

We now consider axisymmetric laminar incompressible steady flow in a circular duct of pipe.

---

[12] With no friction clearly there would be zero pressure change. The physical interpretation of h_l will be set forth in more detail when we approach it from a thermodynamic viewpoint in Sec. 9.3.

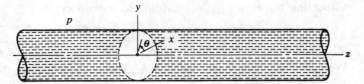

FIGURE 7.13
Flow in a circular pipe or duct.

Case 2. Flow in a circular duct. Consider the flow in a circular duct or pipe in Fig. 7.13. (Please note the coordinates we are using.) This is another case of parallel flow. As will be shown in Sec. 10.3 the pressure p must vary hydrostatically as will be the case of all parallel incompressible flows. We shall delete hydrostatic contributions and will consider pressure p which is that pressure at the top and which is constant over the section but which will vary with z.

We must now develop an equation for cylindrical coordinates corresponding to Eq. (7.40c). To do this we consider an element in the flow suitable for use with cylindrical coordinates. This is shown in Fig. 7.14 wherein the shear stresses contributing force in the z direction are shown. Newton's law gives us

$$-\tau_{zz}\bar{r}\,d\theta\,d\bar{r} + \left(\tau_{zz} + \frac{\partial \tau_{zz}}{\partial z}\,dz\right)\bar{r}\,d\theta\,d\bar{r} - \tau_{\bar{r}z}\bar{r}\,d\theta\,dz$$

$$+ \left(\tau_{\bar{r}z} + \frac{\partial \tau_{\bar{r}z}}{\partial \bar{r}}\,d\bar{r}\right)(\bar{r} + d\bar{r})\,d\theta\,dz + \bar{\rho}B_z\,d\bar{r}\,\bar{r}\,d\theta\,dz$$

$$= \rho\bar{r}\,d\theta\,d\bar{r}\,dz\,\frac{Dv_z}{Dt} \qquad (7.57)$$

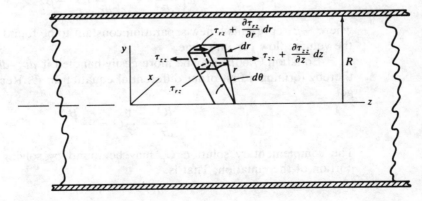

FIGURE 7.14
Infinitesimal element for cylindrical coordinates with stresses shown only in the z direction.

Canceling terms, we get on dividing by $r\,dr\,d\theta\,dx$,

$$\frac{\partial \tau_{zz}}{\partial z} + \frac{\tau_{\bar{r}z}}{\bar{r}} + \frac{\partial \tau_{\bar{r}z}}{\partial \bar{r}} + \frac{\partial \tau_{\bar{r}z}}{\partial \bar{r}}\frac{d\bar{r}}{\bar{r}} + \bar{\rho}B_z = \rho\frac{Dv_z}{Dt} \tag{7.58}$$

We can delete the fourth term as negligible. Now use the Newton viscosity law. We get

$$\frac{\partial \tau_{zz}}{\partial z} + \frac{1}{\bar{r}}\mu\frac{\partial v_z}{\partial \bar{r}} + \mu\frac{\partial^2 v_z}{\partial \bar{r}^2} + \rho B_z = \rho\frac{Dv_z}{Dt}$$

Next, replace τ_{zz} by $(\tau'_{zz} - p)$ as we have done earlier. We thus have

$$-\frac{\partial p}{\partial z} + \frac{\partial \tau'_{zz}}{\partial z} + \mu\left[\frac{1}{\bar{r}}\frac{\partial v_z}{\partial \bar{r}} + \frac{\partial^2 v_z}{\partial \bar{r}^2}\right] + \rho B_z = \rho\frac{Dv_z}{Dt}$$

For the case at hand, $B_z = Dv_z/Dt = 0$. Also τ'_{zz} cannot vary with z because of the fixed profile in the direction of flow, so that we end up with the following equation to solve:

$$\frac{\partial p}{\partial z} = \mu\left(\frac{\partial^2 v_z}{\partial \bar{r}^2} + \frac{1}{\bar{r}}\frac{\partial v_z}{\partial \bar{r}}\right) \tag{7.59}$$

Exactly as in the previous analysis one sees that disregarding hydrostatic pressure variation, the left side of the equation is a function of only the independent variable z. The right side, meanwhile, cannot vary in the direction of flow because of the fixed profile and consequently is a function of only the variable r. Hence, a separation of the independent variables may be carried out as in the previous section. The resulting ordinary differential equations may then be expressed as

$$\frac{dp}{dz} = -\beta \tag{7.60a}$$

$$\mu\left(\frac{d^2 v_z}{d\bar{r}^2} + \frac{1}{\bar{r}}\frac{dv_z}{d\bar{r}}\right) = -\beta \tag{7.60b}$$

where $-\beta$ represents a new separation constant to be found later in terms of q, the volume flow per unit time.

Equation (7.60b) may be more easily handled if $dv_z/d\bar{r}$ is replaced by G, thereby forming a first-order differential equation in G. Rearranging terms, we get

$$\frac{dG}{d\bar{r}} + \frac{1}{\bar{r}}G = -\frac{\beta}{\mu} \tag{7.61}$$

The complementary solution G_c may be found by solving the homogeneous portion of the equation. That is,

$$\frac{dG_c}{d\bar{r}} = -\frac{1}{\bar{r}}G_c$$

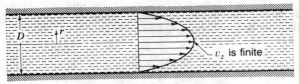

(*a*)   Physically possible case

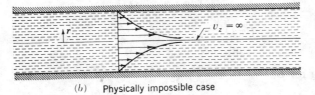

(*b*)   Physically impossible case

FIGURE 7.15
Mathematical velocity profile possibilities.

Separating variables and integrating,

$$\ln G_c = -\ln \bar{r} + \ln C_1$$

where $\ln C_1$ is the constant of integration. Combining the logarithmic expressions and taking the antilogarithm gives the complementary solution as

$$G_c = \frac{C_1}{\bar{r}}$$

A particular solution may easily be found by inspection to be $G_p = -\beta\bar{r}/2\mu$, so that the general solution for G becomes

$$G = -\frac{1}{2}\frac{\beta}{\mu}\bar{r} + \frac{C_1}{\bar{r}} \qquad (7.62)$$

Replacing G by $dv_z/d\bar{r}$, we may perform a second integration. Thus

$$v_z = -\frac{\beta}{4\mu}\bar{r}^2 + C_1 \ln \bar{r} + C_2 \qquad (7.63)$$

When $\bar{r} = 0$, the term $\ln \bar{r}$ "blows up" and becomes infinite (see Fig. 7.15*b*), so that the constant C_1 must be zero to render the equation physically meaningful. Also at the pipe wall—that is, $\bar{r} = D/2$—v_z is zero, so that the other constant C_2 must equal $\beta D^2/16\mu$. The final equation for the velocity profile then becomes

$$v_z = \frac{\beta}{4\mu}\left(\frac{D^2}{4} - \bar{r}^2\right) \qquad (7.64)$$

The reader may recognize from this result that the profile for this case is that of a paraboloidal surface of revolution (see Fig. 7.15*a*).

We wish next to express the separation constant $-\beta$ in terms of the volume flow q. We accordingly have for q

$$q = \int_0^{D/2} v_z (2\pi \bar{r}) \, d\bar{r} = \int_0^{D/2} \left(\frac{\beta}{4\mu} \right) \left(\frac{D^2}{4} - \bar{r}^2 \right) (2\pi \bar{r} \, d\bar{r})$$

$$= \frac{\beta}{4\mu} (2\pi) \left[\frac{D^2}{2} \frac{\bar{r}^2}{2} - \frac{\bar{r}^4}{4} \right]_0^{D/2}$$

$$= \frac{\beta \pi D^4}{128 \mu}$$

Solving for β,

$$\beta = \frac{128 q \mu}{\pi D^4} \tag{7.65}$$

The profile is now fully established in terms of the volume flow.

The differential equation [Eq. (7.60a)] will now be examined to give us the second item of information of this section, namely, the pressure drop. This equation may be written as

$$\frac{dp}{dz} = -\beta = -\frac{128 q \mu}{\pi D^4}$$

Integrating between sections 1 and 2 a distance L apart, we get

$$(p)_1 - (p)_2 = \frac{128 q \mu L}{\pi D^4} \tag{7.66}$$

Since this pressure drop is due only to friction, we can get the head loss h_l by dividing through by ρ.[13] Thus

$$\boxed{\frac{\Delta p}{\rho} = h_l = \frac{128 q \mu L}{\pi D^4 \rho}} \tag{7.67}$$

Solving for q in Eq. (7.66) we may form the following useful formula:

$$q = \frac{\pi (p_1 - p_2) D^4}{128 \mu L} \tag{7.68}$$

Also, going back to Eq. (7.64) and replacing β using Eqs. (7.65) and (7.68) we get another useful formula for the velocity profile which we write as follows:

$$\boxed{v_z = \frac{p_1 - p_2}{4 \mu L} \left(\frac{D^2}{4} - r^2 \right)} \tag{7.69}$$

---

[13]Here again the head loss is the drop in pressure over distance L per unit mass flow.

In the homework problems we will present a number of simple flows of Newtonian liquids for solution by methods presented in this section.

Furthermore, in Chap. 10, we shall come back to Eq. (7.40) and, using a more general viscosity law for Newtonian fluids (Stokes' viscosity law), valid for more general flows, along with equations of continuity, we shall develop very powerful equations called the *Navier-Stokes equations* for incompressible viscous flows in the laminar and turbulent ranges. Because of the complexity of these equations and the many ramifications associated with their development and use, we have reserved a whole starred chapter for the development and study of these equations.

In Part B of this chapter we present for the advanced reader a Cartesian tensor development of the differential forms of the basic laws constructed directly from the integral forms of Chaps. 5 and 6. Much of the presentation goes beyond fluid mechanics in its generality.

Your author urges you not to be intimidated by the complex appearance of the material in Part B. Actually once going into it you will find it not nearly as complex as it appears, particularly if you do not attempt to physically picture the meaning of intermediate formulations needed on the way to a final form of the law under study. You must simply make sure in these steps that you are using correct mathematics for the intermediate steps and reserve the physical interpretation for the final form.

Specifically in Part B we will rederive the continuity equation and Newton's law presented in Part A, this time in a more elegant manner using Cartesian tensor notation plus the Cauchy formula (much used in solid mechanics) and Gauss' law (a powerful and much used integral theorem). Rederiving these familiar equations will give the reader a good grasp of a technique that is used in Part B to *also* derive general differential forms of the first and second laws of thermodynamics.

In fairness to the reader we must point out that we shall not have occasion to use the first and second laws at full strength in this text and so we have starred Part B. However, these formulations will be of great value in your later more advanced courses.

*PART B
DIFFERENTIAL FORM OF THE BASIC LAWS:
A MORE GENERAL APPROACH

7.9 INDEX NOTATION AND CAUCHY'S FORMULA

We now introduce index notation. This notation is widely used in solid and fluid mechanics. Like vector notation, one can become comfortable with the powerful notation in a short time.

For this purpose we introduce the concept of the *free index*. The free index is any letter which appears *only once* as a subscript in a group of terms.

Thus, i is the free index in the following expressions:

$$V_i \qquad A_{ij}V_j$$

When we have an expression with a free index such as V_i, we can consider that we are presenting *any one* component of the array of terms formed by having i become an x, a y, or a z. This array accordingly is

$$\begin{pmatrix} V_x \\ V_y \\ V_z \end{pmatrix}$$

Alternatively, V_i can be interpreted to represent the entire *set* of components given above. Clearly, V_i may then be considered a *vector* since it includes the three rectangular components of the vector. The particular meaning of the free index depends on the *context* of the discussion. Similarly, the expression A_{ij} has two free indices and can represent any one component of the array of terms formed from all possible permutations of the subscripts with i and j taking on the values x, y, and z. This array of terms must then be

$$\begin{pmatrix} A_{xx} & A_{xy} & A_{xz} \\ A_{yx} & A_{yy} & A_{yz} \\ A_{zx} & A_{zy} & A_{zz} \end{pmatrix}$$

Alternatively, the expression A_{ij} can represent the entire set of the above components. Clearly, A_{ij} could possibly in this context represent a second-order tensor.

Next we shall introduce the concept of *dummy indices* by prescribing that when letters i, j, k, l, and m are repeated in an expression we *sum* terms formed by letting the repeated indices take on successively the values x, y, and z. Thus, the expression $A_{ij}B_j$ used earlier may represent a set of three terms each of which is a sum of three expressions. Thus, retaining the free index i we have

$$A_{ij}B_j = A_{ix}B_x + A_{iy}B_y + A_{iz}B_z$$

If we wish to express the set for the free index i, we get

$$A_{ij}V_j \equiv \begin{bmatrix} (A_{xx}B_x + A_{xy}B_y + A_{xz}B_z) \\ (A_{yx}B_x + A_{yy}B_y + A_{yz}B_z) \\ (A_{zx}B_x + A_{zy}B_y + A_{zz}B_z) \end{bmatrix}$$

If we have a double set of repeated indices such as $A_{ij}V_{ij}$, we let i take on the value x and sum terms with j ranging over x, y, and z; then we add to this three terms found by letting i become y while letting j range again over x, y,

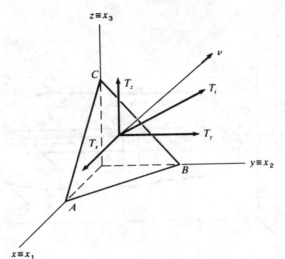

FIGURE 7.16
Tetrahedron showing traction force **T** and its components.

and z, etc. We thus get a *sum* of nine terms. It should be clear that, since a pair of dummy indices always sums out, the particular letters of the set *ijklm* used is immaterial. We can accordingly say that $A_i B_i = A_j B_j = A_m B_m$, etc.[14]

Our first use of index notation will now appear in the development of the very useful *Cauchy formula*. For this purpose we have shown an infinitesimal tetrahedron in Fig. 7.16. On face *ABC* we have shown the traction force intensity **T**, or in index notation T_i, with the orthogonal components of this vector, namely, T_x, T_y, and T_z. If ν is the normal axis for *ABC*, then $a_{\nu x}$, $a_{\nu y}$, and $a_{\nu z}$ are the direction cosines between ν and the x, y, and z axes, respectively. We have asked you to show in Prob. 7.35 that the traction components relate to the stresses on the inside faces of the tetrahedron according to the following formulas:

$$T_x = \tau_{xx} a_{\nu x} + \tau_{xy} a_{\nu y} + \tau_{xz} a_{\nu z}$$

$$T_y = \tau_{yx} a_{\nu x} + \tau_{yy} a_{\nu y} + \tau_{yz} a_{\nu z} \qquad (7.70)$$

$$T_z = \tau_{zx} a_{\nu x} + \tau_{zy} a_{\nu y} + \tau_{zz} a_{\nu z}$$

These formulations are called *Cauchy's formula*. You should now be able to show that in index notation the above can be efficiently given as

$$T_i = \tau_{ij} a_{\nu j} \qquad (7.71)$$

If we replace $a_{\nu j}$ by ν_j representing the jth component of the unit normal

---

[14]Note that $A_i B_i$ is equal to **A** · **B**. Also note that with no free index we have a scalar.

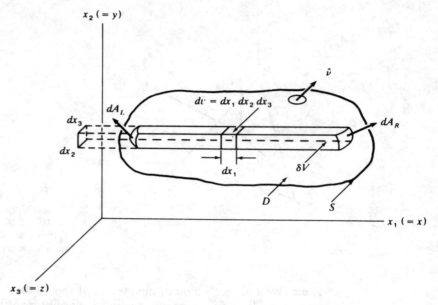

FIGURE 7.17
Body with prismatic element in x_1 direction.

vector $\boldsymbol{\nu}$, we can give the above formula in a convenient form that we will use. Thus

$$\boxed{T_i = \tau_{ij}\nu_j} \tag{7.72}$$

7.10 GAUSS' THEOREM

We have considered stress τ_{ij} with two free indices giving an array of nine terms. Consider next a more general set of terms $T_{ij\ldots}$ with n free indices. This includes scalars with no free indices, vectors with one free index, etc. We will now derive a very important theorem called Gauss' theorem for $T_{ij\ldots}$.

Consider a domain D in space bounded by a closed surface S of such a shape that lines parallel to the x_1 axis can pierce the boundary only twice (see Fig. 7.17). Imagine now that this volume is composed of infinitesimal prisms having sides $dx_2\ (= dy)$ and $dx_3\ (= dx)$ as has been shown in the diagram. Consider one such prism and compute the following integral over the volume δV of this prism:

$$\int_{\delta V} \frac{\partial}{\partial x_1}(T_{jk\ldots})\,dx_1\,dx_2\,dx_3$$

Carrying out integration with respect to x_1, we get

$$\int_{\delta V} \frac{\partial}{\partial x_1} \left(T_{jk} \cdots \right) dx_1\, dx_2\, dx_3 = \left(T_{jk} \cdots\, dx_2\, dx_3 \right)_R - \left(T_{jk} \cdots\, dx_2\, dx_3 \right)_L$$

where the first expression on the right side of the above equation is evaluated at the right end of the prism while the second expression is evaluated at the left end of the prism. Using $\boldsymbol{\nu}$ to represent the unit vector normal to the boundary surface of D, and considering ν_1, ν_2, and ν_3 to be direction cosine components of $\boldsymbol{\nu}$ as we have earlier, we can replace $dx_2\, dx_3$ at the right end of the prism by $(+\nu_1\, dA)_R$ and by $(-\nu_1\, dA)_L$ on the left end of the prism.[15] We thus have

$$\int_{\delta V} \frac{\partial}{\partial x_1} \left(T_{jk} \cdots \right) dx_1\, dx_2\, dx_3 = \left(T_{jk} \cdots\, \nu_1\, dA \right)_R + \left(T_{jk} \cdots\, \nu_1\, dA \right)_L$$

Now integrating over all the prisms comprising the domain of the volume D, we get:

$$\int_{D} \frac{\partial}{\partial x_1} \left(T_{jk} \cdots \right) dx_1\, dx_2\, dx_3 = \iint_{R} \left(T_{jk} \cdots\, \nu_1\, dA \right) + \iint_{L} \left(T_{jk} \cdots\, \nu_1\, dA \right)$$

The right side of the above equation clearly covers the entire surface of domain D and is accordingly replaceable by a closed-surface integral. We have then the following statement:

$$\int_{D} \frac{\partial}{\partial x_1} \left(T_{jk} \cdots \right) dx_1\, dx_2\, dx_3 = \oiint_{S} \left(T_{jk} \cdots \right) \nu_1\, dA \qquad (7.73)$$

We developed Eq. (7.73) for the x_1 direction. We could have proceeded in a similar manner in any direction x_i. Accordingly, the generalization of the above statement is given as follows:

$$\boxed{\int_{D} \frac{\partial T_{jk} \cdots}{\partial x_i}\, dv = \oiint_{S} \left(T_{jk} \cdots \right) \nu_i\, dA} \qquad (7.74)$$

where i is a free index. This is *Gauss'* theorem in a generalized form. Note that

---

[15] The appearance of the minus sign here is a result of the use of the outward-normal convention for area vectors. Thus, ν_1 on the left side of the prism is clearly negative for the kind of domain we have chosen to work with, and a minus sign must be included so the product $(-\nu_1\, dA)$ will be the positive number needed to replace $dx_2\, dx_3$ for that side.

the derivative $\partial/\partial x_i$ represents the derivatives

$$
\begin{bmatrix} \dfrac{\partial}{\partial x} \\[2mm] \dfrac{\partial}{\partial y} \\[2mm] \dfrac{\partial}{\partial z} \end{bmatrix} \quad \text{or} \quad \begin{bmatrix} \dfrac{\partial}{\partial x_1} \\[2mm] \dfrac{\partial}{\partial x_2} \\[2mm] \dfrac{\partial}{\partial x_3} \end{bmatrix}
$$

A simple way to remember Gauss' theorem is to replace ν_i by $\partial/\partial x_i$ so as to act on the integrand of the closed surface integral while at the same time making the closed surface integral into a volume integral. This theorem thus permits us to equate a volume integral with a surface integral and is the source of a number of useful theorems such as the divergence theorem. We will ask you to develop some of these (with some help) in the homework problems.[16]

We now have tools which will allow us to go smoothly from the integral forms of the basic laws to differential forms. Before embarking we wish to advise the reader again that it is not necessary or even useful to try to establish a physical picture of the series of formulations that are formed in the process of deriving the differential equations of interest. All that should concern you is that correct mathematics governs each step of the process. When the final formulation has been reached you can then ponder physical meaning. We start with conservation of mass and rederive the equations presented in Sec. 7.2.

7.11 CONSERVATION OF MASS

The procedure we will follow is to first write the integral form of conservation of mass for a control volume as developed in Chap. 5. Thus

$$
\oiint_{\text{CS}} (\rho \mathbf{V} \cdot \mathbf{dA}) = \frac{\partial}{\partial t} \iiint_{\text{CV}} \rho \, dv
$$

In index notation we have on expressing $\mathbf{dA}$ first as $dA\,\boldsymbol{\nu}$ and then as $dA\,\nu_i$:

$$
\oiint_{\text{CS}} \rho V_i \nu_i \, dA = \iiint_{\text{CV}} \frac{\partial \rho}{\partial t} \, dv
$$

Now go to Gauss' theorem and replace the surface integral by a volume integral

---

[16] For further study of Cartesian tensor notation, see I. H. Shames and C. L. Dym, *Energy and Finite Element Methods in Structural Mechanics*, 1985, Hemisphere Publishing Corp., New York, 1985, Appendix I.

so that we get

$$\iiint\limits_{CV} \frac{\partial}{\partial x_i}(\rho V_i)\, dv + \iiint\limits_{CV} \frac{\partial \rho}{\partial t}\, dv = 0$$

$$\therefore \iiint\limits_{CV} \left[\frac{\partial}{\partial x_i}(\rho V_i) + \frac{\partial \rho}{\partial t}\right] dv = 0$$

The control volume could have been chosen arbitrarily including an infinitesimal control volume. This means that the integrand must be zero. Thus

$$\boxed{\frac{\partial}{\partial x_i}(\rho V_i) + \frac{\partial \rho}{\partial t} = 0} \qquad (7.75)$$

This is the desired form we have been seeking. We leave it for Prob. 7.42 for you to show that this formulation is exactly the same as developed in Sec. 7.2.

7.12 MOMENTUM EQUATIONS

As in the case of conservation of mass, we start with the integral form of the linear momentum equation presented in Chap. 5. Thus

$$\oiint\limits_{CS} \mathbf{T}\, dA + \iiint\limits_{CV} \mathbf{B}\rho\, dv = \oiint\limits_{CS} \mathbf{V}(\rho \mathbf{V} \cdot d\mathbf{A}) + \frac{\partial}{\partial t}\iiint\limits_{CV} \mathbf{V}(\rho\, dv) \quad (7.76)$$

Now go over to index notation and use Cauchy's formula to replace T_i by $\tau_{ij}\nu_j$. Noting that $d\mathbf{A}$ can be replaced by $dA\,\boldsymbol{\nu} \equiv dA\,\nu_i$, we have on bringing the time derivative under the integral sign

$$\oiint\limits_{CS} \tau_{ij}\nu_j\, dA + \iiint\limits_{CV} B_i\rho\, dv = \oiint\limits_{CS} V_i(\rho V_j\nu_j)\, dA + \iiint\limits_{CV} \frac{\partial(\rho V_i)}{\partial t}\, dv$$

Next employ Gauss' theorem to replace the two closed surface integrals by volume integrals. Thus we get

$$\iiint\limits_{CV} \frac{\partial}{\partial x_j}(\tau_{ij})\, dv + \iiint\limits_{CV} B_i\rho\, dv = \iiint\limits_{CV} \frac{\partial}{\partial x_j}(\rho V_i V_j)\, dv + \iiint\limits_{CV} \frac{\partial(\rho V_i)}{\partial t}\, dv$$

$$\therefore \iiint\limits_{CV} \left[\frac{\partial(\tau_{ij})}{\partial x_j} + B_i\rho - \frac{\partial}{\partial x_j}(\rho V_i V_j) - \frac{\partial(\rho V_i)}{\partial t}\right] dv = 0$$

Since the control volume is arbitrary, we can conclude that the above integrand is zero. Carrying out differentiation for the third and fourth expressions using the product rule, we may then write

$$\frac{\partial \tau_{ij}}{\partial x_j} + \rho B_i - V_i\frac{\partial(\rho V_j)}{\partial x_j} - \rho V_j\frac{\partial V_i}{\partial x_j} - \rho\frac{\partial V_i}{\partial t} - V_i\frac{\partial \rho}{\partial t} = 0$$

Rearranging and collecting the third and last terms, we have

$$\frac{\partial \tau_{ij}}{\partial x_j} + \rho B_i - \rho V_j \frac{\partial V_i}{\partial x_j} - \rho \frac{\partial V_i}{\partial t} - V_i \left[\frac{\partial (\rho V_j)}{\partial x_j} - \frac{\partial \rho}{\partial t} \right] = 0 \qquad (7.77)$$

On comparing the bracketed expression with the left side of Eq. (7.75), namely the continuity equation, we see that the bracketed expression is equal to zero. We thus arrive at the desired form of the linear momentum equation

$$\boxed{\frac{\partial \tau_{ij}}{\partial x_j} + \rho B_i = \rho \left[V_j \frac{\partial V_i}{\partial x_j} + \frac{\partial V_i}{\partial t} \right]} \qquad (7.78)$$

We ask you to demonstrate that the bracketed expression is equivalent to $(\mathbf{V} \cdot \boldsymbol{\nabla})\mathbf{V} + \partial \mathbf{V}/\partial t$ which in turn is $D\mathbf{V}/Dt \equiv DV_i/Dt$.

We thus have as an alternate form of the linear momentum equation

$$\boxed{\frac{\partial \tau_{ij}}{\partial x_j} + \rho B_i = \rho \frac{DV_i}{Dt}} \qquad (7.79)$$

You should have no difficulty in establishing the fact that this equation is exactly the same as Eq. (7.39).

In Prob. 7.46 we have asked you via the same general approach to show that the *moment-of-momentum equation* when cast into differential form reveals that the stress tensor is symmetric. That is

$$\tau_{ij} = \tau_{ji}$$

Previously we have argued this fact in a rather simplistic manner.

7.13   FIRST LAW OF THERMODYNAMICS

The integral form of the first law employing the shaft work concept was very useful for fluid flow application. The term dW_S/dt included energy transmitted by shafts and wires crossing the control surface. The shafts and wires represent a discontinuous change in the density of the material in the control volume and because the control volume concept is geared for fluids, their contributions were lumped into the term dW_S/dt. In the differential element of fluid we delete shafts. As for the electrical current, we shall also delete such a possibility. Because a more detailed understanding of this contribution is needed in the differential approach, we must refer such considerations to a study of the electromagnetics of continua, which is beyond the scope of this text. Thus, the first law that we shall use is given as:

$$\frac{dQ}{dt} + \iiint \mathbf{B} \cdot \mathbf{V} \rho \, dv + \oiint \mathbf{T} \cdot \mathbf{V} \, dA$$

$$= \oiint \left(\frac{V^2}{2} + u + gz \right) \rho \mathbf{V} \cdot \mathbf{dA} + \frac{\partial}{\partial t} \iiint \left(\frac{V^2}{2} + u + gz \right) \rho \, dv \qquad (7.80)$$

We will first consider in more detail the term dQ/dt. It will be helpful to express this quantity in terms of a *flux of energy through* the control surface, plus a *source distribution inside* the control volume. The former stems from radiation or conduction of heat, while the latter stems possibly from a chemical or a nuclear reaction. Thus we have:

$$\frac{dQ}{dt} = - \oiint_{CS} \mathbf{q} \cdot \mathbf{dA} + \iiint_{CV} \dot{\Theta} \rho \, dv \qquad (7.81)$$

where $\mathbf{q}$ is the *heat flux vector* and $\dot{\Theta}$ is the *rate of energy production* per unit mass at a point in the continuum.

We now incorporate Eq. (7.81) into Eq. (7.80). Going to index notation and now including gravity in the body force $\mathbf{B}$, we have

$$- \oiint_{CS} q_i \nu_i \, dA + \iiint_{CV} \dot{\Theta} \rho \, dv + \iiint_{CV} B_i V_i \rho \, dv + \oiint_{CS} T_i V_i \, dA$$

$$= \oiint_{CS} \left(\frac{V^2}{2} + u \right)(\rho V_i \nu_i \, dA) + \iiint_{CV} \frac{\partial}{\partial t}\left[\frac{V^2}{2} + u \right]\rho \, dv \qquad (7.82)$$

Now replacing $T_i^{(\nu)}$ by $\tau_{ij}\nu_j$ and employing Gauss' law for the three surface integrals, we get

$$- \iiint_{CV} \frac{\partial q_i}{\partial x_i} \, dv + \iiint_{CV} \dot{\Theta} \rho \, dv + \iiint_{CV} B_i V_i \rho \, dv + \iiint_{CV} \frac{\partial(\tau_{ij}V_i)}{\partial x_j} \, dv$$

$$- \iiint_{CV} \frac{\partial}{\partial x_i}\left[\left(\frac{V^2}{2} + u \right)\rho V_i \right] dv + \iiint_{CV} \frac{\partial}{\partial t}\left[\frac{V^2}{2} + u \right]\rho \, dv = 0$$

Collecting terms under one integration

$$\iiint_{CV} \left\{ -\frac{\partial q_i}{\partial x_i} + \rho\dot{\Theta} + \rho B_i V_i + \frac{\partial}{\partial x_j}(\tau_{ij}V_i) - \frac{\partial}{\partial x_i}\left[\left(\frac{V^2}{2} + u \right)\rho V_i \right] \right.$$

$$\left. + \frac{\partial}{\partial t}\left[\rho\left(\frac{V^2}{2} + u \right) \right] \right\} dv = 0$$

Setting the integrand equal to zero since again the control volume is arbitrary, we arrive at the following preliminary form of the first law of thermodynamics:

$$\frac{\partial}{\partial x_j}(\tau_{ij}V_i) + \rho B_i V_i - \frac{\partial q_i}{\partial x_i} + \rho\dot{\Theta} = \frac{\partial}{\partial x_i}\left[\left(\frac{V^2}{2} + u \right)\rho V_i \right] + \frac{\partial}{\partial t}\left[\left(\frac{V^2}{2} + u \right)\rho \right]$$

$$(7.83)$$

We can reach a very useful form of the above equation by employing the previously developed differential forms of the continuity and linear momentum equations. We must first expand out the differentiation in the above equation as follows:

$$V_i \frac{\partial \tau_{ij}}{\partial x_j} + \tau_{ij} \frac{\partial V_i}{\partial x_j} + \rho B_i V_i - \frac{\partial q_i}{\partial x_i} + \rho \dot{\Theta}$$

$$= \rho V_i \frac{\partial}{\partial x_i} \left(\frac{V^2}{2} \right) + \frac{V^2}{2} \frac{\partial}{\partial x_i} (\rho V_i) + u\rho \frac{\partial V_i}{\partial x_i} + uV_i \frac{\partial \rho}{\partial x_i} + \rho V_i \frac{\partial u}{\partial x_i}$$

$$+ \rho \frac{\partial}{\partial t} \left(\frac{V^2}{2} \right) + \frac{V^2}{2} \frac{\partial \rho}{\partial t} + u \frac{\partial \rho}{\partial t} + \rho \frac{\partial u}{\partial t}$$

Rearranging the terms we have

$$V_i \left[\frac{\partial \tau_{ij}}{\partial x_j} + \rho B_i \right] + \tau_{ij} \frac{\partial V_i}{\partial x_j} - \frac{\partial q_i}{\partial x_i} + \rho \dot{\Theta}$$

$$= \frac{\rho}{2} \left[\frac{\partial V^2}{\partial t} + V_i \frac{\partial V^2}{\partial x_i} \right] + \frac{V^2}{2} \left[\frac{\partial}{\partial x_i} (\rho V_i) + \frac{\partial \rho}{\partial t} \right]$$

$$+ \rho \left[\frac{\partial u}{\partial t} + V_i \frac{\partial u}{\partial x_i} \right] + u \left[\frac{\partial \rho}{\partial t} + V_i \frac{\partial \rho}{\partial x_i} \right] + u\rho \frac{\partial V_i}{\partial x_i}$$

The first bracket on the left side of the above equation is replaceable by $\rho \, DV_i/Dt$ in accordance with the linear momentum equation [see Eq. (7.79)] and accordingly the first expression is $\rho V_i(DV_i/Dt)$ which can be written as $(\rho/2)(DV^2/Dt)$.[17] Using the material derivative the first bracketed expression on the right side of the equation can be expressed as $(\rho/2)(DV^2/Dt)$, which is identical to the aforementioned representation of the first expression on the left side of the equation, causing a cancellation of these terms. The second bracketed expression on the right side of the equation is zero according to the continuity equation [see Eq. (7.75)]. The remaining last two bracketed expressions are seen readily to be Du/Dt and $D\rho/Dt$, respectively. The above equation then simplifies to the following form with these substitutions:

$$\tau_{ij} \frac{\partial V_i}{\partial x_j} - \frac{\partial q_i}{\partial x_i} + \rho \dot{\Theta} = \rho \frac{Du}{Dt} + u \left(\frac{D\rho}{Dt} + \rho \frac{\partial V_i}{\partial x_i} \right) \tag{7.84}$$

where we have collected some terms on the right side of the equation. Looking at the last bracketed expression we see once again from the continuity equation

---

[17]That is,

$$\frac{\rho}{2} \frac{DV^2}{Dt} = \frac{\rho}{2} \frac{D}{Dt} (V_i V_i) = \rho V_i \frac{DV_i}{Dt}$$

[Eq. (7.75) that the term is zero and so we get:

$$\tau_{ij}\frac{\partial V_i}{\partial x_j} - \frac{\partial q_i}{\partial x_i} + \rho\dot{\Theta} = \rho\frac{Du}{Dt} \tag{7.85}$$

We may further develop this equation by using the *Fourier heat conduction law* which we state now in vector form and in Cartesian tensor form:

$$\mathbf{q} = -k\,\nabla T = -k\left[\frac{\partial T}{\partial x}\mathbf{i} + \frac{\partial T}{\partial y}\mathbf{j} + \frac{\partial T}{\partial z}\mathbf{k}\right]$$

$$\therefore q_i = -k\frac{\partial T}{\partial x_i} \tag{7.86}$$

(Can you justify the second form?) We substitute q_i into the previous equation now to get a useful form of the first law of thermodynamics:

$$\tau_{ij}\frac{\partial V_i}{\partial x_j} + \frac{\partial}{\partial x_i}\left(k\frac{\partial T}{\partial x_i}\right) + \rho\dot{\Theta} = \rho\frac{Du}{Dt} \tag{7.87}$$

If we replace τ_{ij} above as follows using the deviatoric stress tensor τ'_{ij},[18]

$$\tau_{ij} = -p\delta_{ij} + \tau'_{ij}$$

we get for Eq. (7.87),

$$-p\frac{\partial V_i}{\partial x_j}\delta_{ij} + \tau'_{ij}\frac{\partial V_i}{\partial x_j} + \frac{\partial}{\partial x_i}\left(k\frac{\partial T}{\partial x_i}\right) + \rho\dot{\Theta} = \rho\frac{Du}{Dt}$$

Noting that δ_{ij} is zero unless $j = i$, the first expression above becomes $p\,\partial V_i/\partial x_i$. Calling $\tau'_{ij}\,\partial V_i/\partial x_j$ the *dissipation function*, Φ, we get

$$-p\frac{\partial V_i}{\partial x_i} + \Phi + \frac{\partial}{\partial x_i}\left(k\frac{\partial T}{\partial x_i}\right) + \rho\dot{\Theta} = \rho\frac{Du}{Dt}$$

Rearranging terms we then have

$$\rho\frac{Du}{Dt} + p\frac{\partial V_i}{\partial x_i} = \frac{\partial}{\partial x_i}\left(k\frac{\partial T}{\partial x_i}\right) + \Phi + \dot{\Theta} \tag{7.88}$$

---

[18] δ_{ij} is called *Kronecker delta* and is defined as

$$\delta_{ij} = \begin{bmatrix} 1 & 0 & 0 \\ 0 & 1 & 0 \\ 0 & 0 & 1 \end{bmatrix}$$

7.14 SECOND LAW OF THERMODYNAMICS

The second law of thermodynamics was given as follows:

$$\oint_{CS} s\rho \mathbf{V} \cdot d\mathbf{A} + \frac{\partial}{\partial t} \iiint_{CV} s\rho \, dv \geq \frac{1}{T} \frac{dQ}{dt}$$

Using Eq. (7.81) for dQ/dt and inserting $\partial/\partial t$ under the integral sign we get:

$$\oint_{CS} s\rho \mathbf{V} \cdot d\mathbf{A} + \iiint_{CV} \frac{\partial}{\partial t}(s\rho) \, dv \geq \frac{1}{T}\left[-\oint_{CS} \mathbf{q} \cdot d\mathbf{A} + \iiint_{CV} \dot{\Theta}\rho \, dv \right]$$

Putting this equation into an index notation format, we get:

$$\oint_{CS} s\rho V_i \nu_i \, dA + \iiint_{CV} \frac{\partial}{\partial t}(s\rho) \, dv \geq \frac{1}{T}\left[-\oint_{CS} q_i \nu_i \, dA + \iiint_{CV} \dot{\Theta}\rho \, dv \right]$$

Now employ Gauss' theorem twice:

$$\iiint_{CV} \frac{\partial}{\partial x_i}(s\rho V_i) \, dA + \iiint_{CV} \frac{\partial}{\partial t}(s\rho) \, dv \geq \frac{1}{T}\left[\iiint_{CV} \left(-\frac{\partial q_i}{\partial x_i} + \rho\dot{\Theta} \right) dv \right]$$

Collecting terms and carrying out differentiation processes, we get on rearranging terms

$$\iiint_{CV} \left[\frac{\partial}{\partial t}(s\rho) + V_i \frac{\partial(s\rho)}{\partial x_i} + \frac{s\rho \, \partial V_i}{\partial x_i} \right] dv \geq \frac{1}{T}\left[\iiint \left[-\frac{\partial q_i}{\partial x_i} + \rho\dot{\Theta} \right] dv \right]$$

We choose the domain vanishingly small around some point *xyz* and so we have from the above

$$\left[\frac{\partial}{\partial t}(s\rho) + V_i \frac{\partial(s\rho)}{\partial x_i} \right] + \frac{s\rho \, \partial V_i}{\partial x_i} \geq \frac{1}{T}\left(-\frac{\partial q_i}{\partial x_i} + \rho\dot{\Theta} \right)$$

The first bracketed expression is clearly $(D(s\rho)/Dt)$ and so we get the desired equation

$$\boxed{\frac{D(s\rho)}{Dt} + s\rho \frac{\partial V_i}{\partial x_i} \geq \frac{1}{T}\left(-\frac{\partial q_i}{\partial x_i} + \rho\dot{\Theta} \right)} \tag{7.89}$$

We have thus set forth basic laws in differential form.

7.15 BASIC LAWS IN CYLINDRICAL COORDINATES

In the previous section we have presented the basic laws in Cartesian tensor notation; they are then limited to rectangular coordinates. The unabridged forms of these equations are available by simply using the rules for subscripts presented in Sec. 7.9 and expanding the equations. We will ask you to do this in the homework problems.

In this section we will merely present the cylindrical component equations for the basic laws.[19] Note for a scalar A that $DA/Dt = v_{\bar{r}}\partial A/\partial\bar{r} + v_\theta(\partial A/\bar{r}\,\partial\theta) + v_z(\partial A/\partial z)) + \partial A/\partial t$.

Continuity equation. We have already presented this equation [see Eq. (7.5)] and we now rewrite it:

$$\frac{1}{\bar{r}}\frac{\partial}{\partial\bar{r}}(\bar{r}\rho v_{\bar{r}}) + \frac{1}{\bar{r}}\frac{\partial(\rho v_\theta)}{\partial\theta} + \frac{\partial(\rho v_z)}{\partial z} = -\frac{\partial\rho}{\partial t} \quad (7.90)$$

Linear momentum equation.

$$\rho\left(\frac{Dv_{\bar{r}}}{Dt} - \frac{v_\theta^2}{\bar{r}}\right) = \rho B_{\bar{r}} + \frac{\partial\tau_{\bar{r}\bar{r}}}{\partial\bar{r}} + \frac{1}{\bar{r}}\frac{\partial\tau_{\bar{r}\theta}}{\partial\theta} + \frac{\partial\tau_{z\bar{r}}}{\partial z} + \frac{\tau_{\bar{r}\bar{r}} - \tau_{\theta\theta}}{\bar{r}} \quad (7.91a)$$

$$\rho\left(\frac{Dv_\theta}{Dt} + \frac{v_{\bar{r}}v_\theta}{\bar{r}}\right) = \rho B_\theta + \frac{\partial\tau_{\bar{r}\theta}}{\partial\bar{r}} + \frac{1}{\bar{r}}\frac{\partial\tau_{\theta\theta}}{\partial\theta} + \frac{\partial\tau_{\theta z}}{\partial z} + \frac{2\tau_{\bar{r}\theta}}{\bar{r}} \quad (7.91b)$$

$$\rho\left(\frac{Dv_z}{Dt}\right) = \rho B_z + \frac{\partial\tau_{z\bar{r}}}{\partial\bar{r}} + \frac{1}{\bar{r}}\frac{\partial\tau_{\theta z}}{\partial\theta} + \frac{\partial\tau_{zz}}{\partial z} + \frac{\tau_{z\bar{r}}}{\bar{r}} \quad (7.91c)$$

where

$$\frac{D}{Dt} = \left(v_{\bar{r}}\frac{\partial}{\partial\bar{r}} + \frac{v_\theta}{\bar{r}}\frac{\partial}{\partial\theta} + v_z\frac{\partial}{\partial z}\right) + \frac{\partial}{\partial t} \quad (7.92)$$

First law of thermodynamics.

$$\frac{\rho\,Dh}{Dt} - \frac{Dp}{Dt} = \frac{1}{\bar{r}}\frac{\partial}{\partial\bar{r}}\left(k\bar{r}\frac{\partial T}{\partial\bar{r}}\right) + \frac{1}{\bar{r}^2}\frac{\partial}{\partial\theta}\left(k\frac{\partial T}{\partial\theta}\right) + \frac{\partial}{\partial z}\left(k\frac{\partial T}{\partial z}\right) + \Phi + \dot{\Theta} \quad (7.93)$$

The dissipation function Φ for viscous flow is given as

$$\Phi = 2\mu\left[\left(\frac{\partial v_{\bar{r}}}{\partial\bar{r}}\right)^2 + \left(\frac{1}{\bar{r}}\frac{\partial v_\theta}{\partial\theta} + \frac{v_{\bar{r}}}{\bar{r}}\right)^2 + \left(\frac{\partial v_z}{\partial z}\right)^2\right.$$

$$+ \frac{1}{2}\left(\frac{\partial v_\theta}{\partial\bar{r}} - \frac{v_\theta}{\bar{r}} + \frac{1}{\bar{r}}\frac{\partial v_{\bar{r}}}{\partial\theta}\right)^2 + \frac{1}{2}\left(\frac{1}{\bar{r}}\frac{\partial v_z}{\partial\theta} + \frac{\partial v_\theta}{\partial z}\right)^2$$

$$\left. + \frac{1}{2}\left(\frac{\partial v_{\bar{r}}}{\partial z} + \frac{\partial v_z}{\partial\bar{r}}\right)^2 - \frac{1}{3}(\nabla\cdot\mathbf{q})^2\right] \quad (7.9.4)$$

7.16 CLOSURE

We have presented in Part A the continuity equation and the linear momentum equation in differential form, and in addition once again Bernoulli's equation. The Bernoulli equation for inviscid flow with only gravity as the body force is

---

[19] For a detailed development of these equations as well as those for spherical coordinates, see S. W. Yuan, *Foundations of Fluid Mechanics*, Prentice-Hall Inc., Englewood Cliffs, N.J., 1967, Chap. 5.

also the first law of thermodynamics for such flows. Furthermore, we have already presented in Chap. 6 a very elementary form of the differential first law [see Eq. (6.3)] as well as the differential second law of thermodynamics [see Eq. (6.31)].

In Part B we introduced Cartesian tensor notation as well as Cauchy's formula and Gauss' theorem. In a very neat way using integral forms as a beginning, we were able then to rederive the differential forms of the continuity equation and the linear momentum equation of Part A. A specific common orderly approach was taken for each law. With this *same* approach we then developed general forms of the first and second laws of thermodynamics. In the homework assignments for Part B (see Probs. 7.45 and 7.46) we have also asked you to use the very same approach as presented for the four aforementioned laws to consider the differential form of the *moment-of-momentum* equation. From this development will come the result that the stress tensor *must* be *symmetric*—that is, $\tau_{ij} = \tau_{ji}$. The material in Part B should be of much use as you proceed deeper into fluid mechanics, heat transfer, and continuum mechanics courses.

We will make much use of differential forms of the basic and subsidiary laws in Part II of the text, wherein we examine in far greater detail certain key flows that are important in engineering.

In addition to these differential formulations, we have also to make considerable use of *experimental* information and data. Accordingly, in the next chapter we examine certain general and vital aspects concerning the use of experimental data as well as the important procedure of model testing. The results of the next chapter will be vital for proper understanding of much of what will follow in later chapters. In particular, you will find that model testing, to be meaningful, requires a sound knowledge of the fundamentals of fluid mechanics.

PROBLEMS

Problem Categories

Starred Problems

7.1. Determine the divergence operator for the vector $\rho\mathbf{V}$ for cylindrical coordinates. Use the infinitesimal control volume in Fig. P7.1.

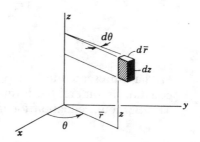

FIGURE P7.1

7.2. Check to see whether or not the velocity fields presented in Probs. 4.1, 4.5, 4.6, and 4.16 satisfy the law of conservation of mass for an incompressible flow.

7.3. For an incompressible flow, $V_x = 30y^2x^3$ m/s and $V_z = 20$ m/s. What is the most information you can give for V_y?

7.4. In steady flows of liquids through a straight pipe wherein $v_{\hat{r}} = v_\theta = 0$, why can we assert that the velocity profiles along sections of the flow must be invariant?

7.5. In the study of two-dimensional potential flow, we express the velocity $\mathbf{V}$ in terms of a function ψ called the *stream function*. That is,

$$V_x = \frac{\partial \psi}{\partial y}$$

$$V_y = -\frac{\partial \psi}{\partial x}$$

Show that we can satisfy conservation of mass by doing this for incompressible flow.

7.6. Given the following hypothetical velocity field

$$\mathbf{V} = x^2\mathbf{i} + yx\mathbf{j} + t^2\mathbf{k} \quad \text{m/s}$$

and a density distribution given as

$$\rho = \rho_0[1 + x \times 10^{-2}] \quad \text{kg/m}^3$$

what is the time rate of change of ρ at a position $(2, 2, 0)$ m at time $t = 2$ s? ρ_0 is a constant.

7.7. One of the famous four Maxwell's equations in electromagnetic theory is given as follows for a vacuum:

$$\text{div } \mathbf{E} = \frac{\rho}{\epsilon_0}$$

where $\mathbf{E}$ = electric field, N/C
ρ = charge density, C/M^3
ϵ_0 = *dielectric* constant
If the following field is given as

$$\mathbf{E} = (y^2 + x^3)\mathbf{i} + (xy + t^2)\mathbf{j}$$
$$+ (3z + 5)\mathbf{k} \quad \text{N/C}$$

with coordinates in meters, what is the charge density at $(2, 5, 3)$ at any time?

7.8. Suppose that pressure distribution in a steady-flow wave is given as

$$p = 6x^2 + (y + z^2) + 10 \quad \text{Pa}$$

If the fluid has a mass density of 1000 kg/m^3, ascertain the acceleration that a fluid particle would have at position

$$\mathbf{r} = 6\mathbf{i} + 2\mathbf{j} + 10\mathbf{k} \quad \text{m}$$

7.9. Show that the operation $(\mathbf{V} \cdot \boldsymbol{\nabla})\mathbf{V}$ gives the acceleration of transport in Euler's equation.

7.10. A nonviscous flow has the following velocity field:

$$\mathbf{V} = (x^2 + y^2)\mathbf{i} + 3xy^2\mathbf{j} + (16t^2 + z)\mathbf{k} \quad \text{m/s}$$

The density ρ is to be considered constant. What is the rate of change of pressure in the x direction at a position $(1, 1, 0)$? Is the pressure variation in any coordinate direction changing with time? What is this time variation at $(1, 0, 2)$ m?

7.11. A tank weighs 80 N and contains 0.25 m^3 of water. A force of 100 N acts on the tank. What is θ when the free surface of the water assumes a fixed orientation relative to the tank?

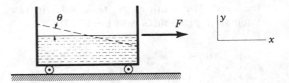

FIGURE P7.11

7.12. A tank of water in Fig. P7.12 is given a constant acceleration a_y. If the water is not to spill out when a fixed configuration of the water is reached relative to the tank, what is the largest acceleration permissible?

7.13. To construct a simple device for measuring acceleration, take a capillary tube in the shape of a U tube (Fig. P7.13) and put in oil to the level shown of 300 mm. If the vehicle to which this U tube is attached accelerates so that the oil assumes the orientation shown, what is the accel-

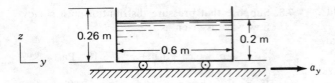

FIGURE P7.12

eration that you would mark on the acceleration scale at position A?

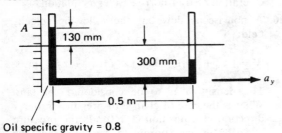

FIGURE P7.13

7.14. A rectangular container is given a constant acceleration a of 0.4 g's. What is the force from fluids on the left wall AB when a fixed configuration of the water has been reached relative to the tank? The width of the tank is 1.5 ft. Use integration procedures with Eq. (7.17).

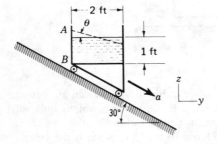

FIGURE P7.14

7.15. Using Eq. (7.17), show that for a *vertical* plane submerged surface in a liquid having constant acceleration components a_y and a_z, the gage

pressure p can be given as

$$p = \bar{\gamma}d$$

where d is the depth *below* the free surface and $\bar{\gamma}$ is given as

$$\bar{\gamma} = \gamma\left(1 + \frac{a_z}{g}\right)$$

This means that using $\bar{\gamma}$ we can compute the force of the liquid on a vertical plane surface as well as the center of pressure exactly as was done in the chapter on hydrostatics.

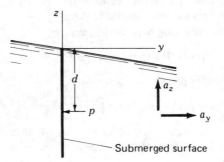

FIGURE P7.15

7.16. In Prob. 7.14, using the results of the preceding problem, find the force on wall AB as well as the center of pressure. The height of water on the wall AB was worked out to be 1.433 ft.

7.17. In Example 7.2, locate the position of the center of pressure from point A.

*7.18. A container has a constant width of 500 mm and contains water as shown in Fig. P7.18. It is accelerated uniformly to the right at a rate of 2 m/s². What is the total force on side AB when the water has assumed a stationary configuration relative to the container?

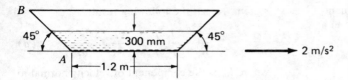

FIGURE P7.18

7.19. Show that the free-surface profile of a rotating liquid once steady state has been achieved is independent of the density ρ.

7.20. The system shown at rest is rotated at a speed of 24 r/min. When steady state has been reached, what height h will the fluid reach in the outer capillary tubes? Do not consider capillary effects.

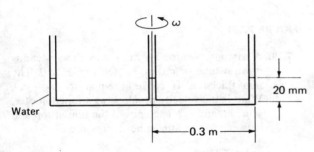

FIGURE P7.20

7.21. On rotating the system at a speed ω of 30 r/min, what height h does the water rise to in the vertical capillary tubes after steady state has been achieved? Do not consider capillary effects.

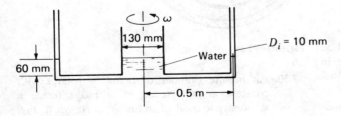

FIGURE P7.21

7.22. A tank of water is to be spun at an angular speed of ω radians per second. At what speed will the water begin to spill out when it reaches a steady-state configuration relative to the tank?

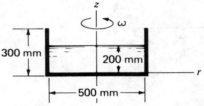

FIGURE P7.22

7.23. In Prob. 7.22, we found that $\omega = 7.92$ rad/s and the equation of the free surface is

$$z = \frac{1}{\rho g}\left(\frac{\rho \omega^2}{2} r^2 + 2942 - 31.25\omega^2\right) \quad \text{m}$$

Find the center of pressure at the bottom of the tank for a semicircular part of the base area.

7.24. What body-force distribution is needed to maintain the following stress field in equilibrium in a solid?

$$\tau_{ij} = \begin{bmatrix} 500x^3 & 0 & (10z^2 + 580) \\ 0 & -800y^2x^2 & -1000zy^2 \\ (10z^2 + 580) & -1000zy^2 & 0 \end{bmatrix} \text{ Pa}$$

*7.25. For cylindrical coordinates, derive the equations of motion using the differential element shown in Fig. P7.25.

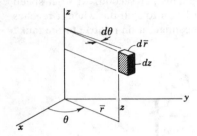

FIGURE P7.25

7.26. A flow field has the stress field given in Prob. 7.24 with only gravity as a body force in the z direction. What is the convective acceleration at position $(1, 2, 0)$m? Take ρ as constant.

7.27. What are the equations of motion for two-dimensional flow parallel to the xy plane with no body forces except gravity in the z direction? Show that if

$$\tau_{xx} = \frac{\partial^2 \Phi}{\partial y^2}$$

$$\tau_{yy} = \frac{\partial^2 \Phi}{\partial x^2}$$

$$\tau_{xy} = -\frac{\partial^2 \Phi}{\partial x \, \partial y}$$

where Φ is a scalar function, then there will be zero acceleration everywhere. This is done in solid mechanics to satisfy equilibrium. The function Φ is then called the *Airy function*.

7.28. Consider a flow as shown in Fig. P7.28. Formulate Newton's law in the direction n normal to the streamline using the indicated infinitesimal

system. Reach the following result:

$$\frac{V^2}{R} - \frac{1}{\rho} \frac{\partial p}{\partial n} - g \frac{\partial z}{\partial n} = \frac{\partial V_n}{\partial t} \qquad (a)$$

where V_n is the component of velocity normal to the streamline.

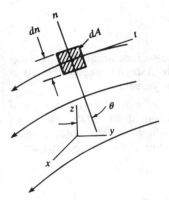

FIGURE P7.28

7.29. A viscous Newtonian fluid film flows steadily down a tube of radius r_2 (see Fig. P7.29). The film thickness is constant equal to $(r_2 - r_1)$. What is the velocity profile V as a function of r, γ, μ, r_1, and r_2? The ends of the film at top and between are at atmospheric pressure.

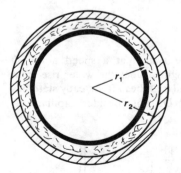

FIGURE P7.29

7.30. An infinite cylinder of radius a rotates with constant angular speed of ω rad/s inside a stationary journal of radius b as shown in Fig. P7.30. Oil of viscosity μ kg/ms separates the cylinder from the journal. The oil is Newtonian.

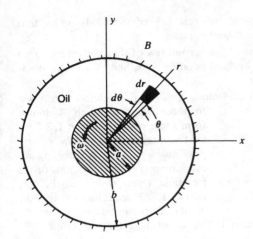

FIGURE P7.30

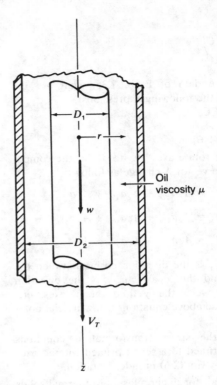

FIGURE P7.33

Find the transverse velocity field v_θ of the oil as a function of r and the pertinent parameters of geometry and fluid properties. Assume steady-state conditions have been reached and that we can use Newton's viscosity law despite the fact that the flow is not a parallel flow. First show that $(d/dr)(r(dv_\theta/dr)) = 0$. Then get

$$v_\theta = \frac{\omega a}{\ln(a/b)} \ln\left(\frac{r}{b}\right)$$

7.31. In the previous problem for the data $\omega = 0.1$ rad/s and $r_b = 0.1$ m, determine the largest value of $(r_b - r_a)$ in order that the linear profile approach presented in Chap. 1 gives a torsional resistance within 10% of the exact resistance for laminar flow.

7.32. In Problem 7.30, develop the differential equation for the pressure as the dependent variable and r as independent variable. Do not try to solve analytically. The equation is nonlinear and must be solved numerically. Use v_θ of Problem 7.30.

7.33. A vertical shaft weighing w per unit length is sliding down concentrically inside a pipe (see Fig. P7.33). Oil separates the two members. Determine the terminal velocity V_T of the shaft without the linear profile assumption made in Chap. 1. Neglect the weight of oil.

7.34. In Prob. 7.33 we got for the terminal velocity the result

$$V_T = \frac{w \ln(D_2/D_1)}{2\pi\mu}$$

If $D_1 = 200$ mm and $D_2 = 210$ mm, what is the error by computing V_T using a thin-film linear profile approach as we did in Chap. 1. Take $w = 100$ N/m.

7.35. Using Fig. 7.16 derive Eqs. (7.70) from Newton's law.

7.36. Given the following information

$$A_x = 8 \qquad B_x = 3 \qquad C_x = 10$$
$$A_y = -2 \qquad B_y = 0 \qquad C_y = 3$$
$$A_z = 3 \qquad B_z = 4 \qquad C_z = -2$$

find

(a) $A_i B_i C_2$

(b) $A_i B_i A_1$

(c) $B_3 C_i A_i$

7.37. Using the data of Prob. 7.36, what are the arrays of the following expressions:

(a) $A_k B_k C_j$

(b) $C_j B_k A_j$

(c) $C_i B_i A_j B_k$

7.38. When we rotate axes xyz to $x'y'z'$ the components of a vector **A** change as follows:

$$A'_x = A_x a_{x'x} + A_y a_{x'y} + A_z a_{x'z}$$

$$A'_y = A_x a_{y'x} + A_y a_{y'y} + A_z a_{y'z}$$

$$A'_z = A_x a_{z'x} + A_y a_{z'y} + A_z A_{z'z}$$

where $a_{x'x}$ is the direction cosine between the x' axis and the x axis, $a_{y'y}$ is the direction cosine between the y' and the y axes, etc. Express the above equations using indicial notation.

7.39. Express the stress transformation equations from unprimed to a set of primed rotated axes as given by Eq. (2.6) in index notation.

7.40. Hooke's law for linear, elastic, isothermal solids (as learned in strength of materials) is given in index notation as

$$\epsilon_{ij} = \frac{1 + \nu}{E} \tau_{ij} - \frac{\nu}{E} \tau_{kk} \delta_{ij} \qquad (a)$$

where δ_{ij} is called *Kronecker delta* and is given as

$$\delta_{ij} = \begin{bmatrix} 1 & 0 & 0 \\ 0 & 1 & 0 \\ 0 & 0 & 1 \end{bmatrix}$$

Expand Eq. (a) into unabridged notation to form six independent equations.

7.41. Use the vector field B_i for $T_{ijk \cdots}$ in Gauss' theorem. Develop the familiar divergence theorem in vector notation which is

$$\iiint_V \text{div } \mathbf{B} \, dv = \oiint_S \mathbf{B} \cdot d\mathbf{A}$$

7.42. Expand the continuity equation and the linear momentum equation into unabridged notation.

7.43. Expand the first law of thermodynamics into unabridged notation.

7.44. Expand the second law of thermodynamics into unabridged notation and then into vector notation.

7.45. The *alternating tensor* ϵ_{ijk} is defined as follows:

$\epsilon_{ijk} = 0$ for those terms for which ijk do not form some permutation of $1, 2, 3$

$\epsilon_{ijk} = 1$ for those terms for which ijk can be put into a $1, 2, 3$ sequence by an even number of permutations of ijk

$\epsilon_{ijk} = -1$ for those terms which can be put into a $1, 2, 3$ sequence by an odd number of permutations.

Indicate the values of the following terms:

$$\epsilon_{112} \qquad \epsilon_{123} \qquad \epsilon_{132} \qquad \epsilon_{321}$$

7.46. Expand $C_i = \epsilon_{ijk} A_j B_k$ for $i = x$ and get the component C_x. Also get C_x for $\mathbf{C} = \mathbf{A} \times \mathbf{B}$.

(a) Show that you get the same x component of **C** and C_i. Thus $\mathbf{A} \times \mathbf{B} \equiv \epsilon_{ijk} A_j B_k$.

(b) Express the integral moment of momentum equation in index notation. Note that we can write

$$\frac{\partial}{\partial t} \iiint_{CV} \mathbf{r} \times \mathbf{V} \rho \, dv \quad \text{as} \quad \iiint_{CV} (\mathbf{r} \times \dot{\mathbf{V}}) \rho \, dv$$

*__7.47.__ Starting with the moment-of-momentum equation in index notation in the form

$$\oiint_{CS} \epsilon_{ijk} x_j T_k \, dA + \iiint_{CV} \epsilon_{ijk} x_j B_k \rho \, dv$$

$$= \oiint_{CS} \epsilon_{ijk} x_j V_k (\rho V_l \nu_l \, dA)$$

$$+ \iiint_{CV} \epsilon_{ijk} x_j \frac{\partial}{\partial t} (\rho V_k) \, dv$$

(1) Replace T_k using Cauchy's formula.

(2) Use Gauss' theorem on surface integrals. Collect under one integral.

(3) Carry out differentiation using the product rule.

(4) Collect terms with x_j in the form $x_j[\cdots]$ and collect terms with V_k in the form $V_k[\cdots]$.

(5) Note that $\partial x_j/\partial x_l = \delta_{jl}$ and that $\delta_{jl}V_l = V_j$, etc.

(6) Use the *linear momentum* equation to get rid of the first bracketed expression and the *continuity* equation to delete all but one term in the second bracket. Reach the following result:

$$\epsilon_{ijk}\left[V_k V_j - \tau_{kj}\right] = 0$$

(7) Now show that $\tau_{jk} = \tau_{kj}$. That is, τ_{kj} is symmetric. Thus moment of momentum in differential form simply stipulates that stress is symmetric.

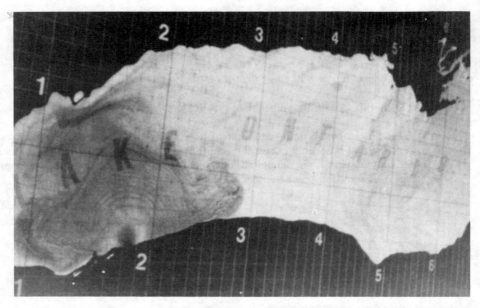

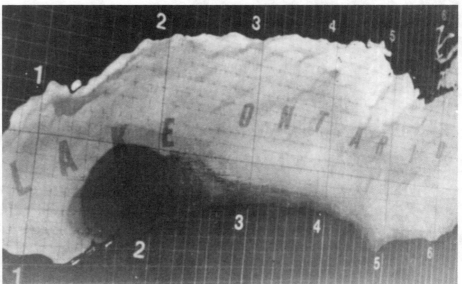

Model of Lake Ontario for environmental studies. (*Courtesy Dr. J. Atkinson, State University of New York at Buffalo.*)

A model of Lake Ontario is shown in the above photos. The horizontal scale of this model is 1/100,000. The model is about 3 m long and 1 m wide. The model is rotated at 1.71 r/min to include the Coriolis force. Dynamic similarity is achieved via duplicating the Froude and Rossby numbers. The effect of Coriolis is illustrated in the two photos: the top one without rotation and the bottom one with rotation. The dark dye is admitted from the Niagara River at the left. Notice that the Coriolis force caused the dye to move to the right hugging the shoreline. Such studies are valuable for environmental considerations.

CHAPTER 8

DIMENSIONAL ANALYSIS AND SIMILITUDE

8.1 DIMENSIONLESS GROUPS

Before proceeding to specialized discussions of selected types of flow, it will be useful to study the dimensional aspects of fluid flow. This will enable us to understand more clearly the differences between the various flows to be considered in the succeeding chapters. Furthermore, fundamental considerations essential for experimental investigation of fluid phenomena will be set forth in this chapter with the aid of dimensional studies.

In mechanics, a formalism was presented for expressing a dependent dimension in terms of a chosen set of independent basic dimensions. Thus, velocity is given dimensionally by the relation $V \equiv L/T$. To give the most simple dimensional representation of a product of quantities, one need only carry out ordinary algebraic operations on the basic dimensions appearing in the dimensional representation of the quantities. For instance, $Vt \equiv (L/T)T \equiv L$, thus indicating that the product of velocity and time most simply dimensionally is a distance. If a group of quantities has a dimensional representation most simply of unity when multiplied together, the group is called a *dimensionless group*. As an example, the product $\rho VD/\mu$ is a dimensionless group, since

$$\frac{\rho VD}{\mu} \equiv \frac{(M/L^3)(L/T)L}{M/LT} \equiv 1$$

Many of these dimensionless products have been given names, the above group being the well-known *Reynolds* number. In a later section the physical significance of the Reynolds number as well as other dimensionless groups will be discussed.

PART A
DIMENSIONAL ANALYSIS

8.2 NATURE OF DIMENSIONAL ANALYSIS

Recall from mechanics that analytically derived equations are correct for any system of units and consequently each group of terms in the equation must have the same dimensional representation. This is the law of *dimensional homogeneity*. Use was made of this law in establishing the dimensions of quantities such as viscosity.

Another important use may be made of this rule in a situation whereby *the variables involved in a physical phenomenon are known*, *while the relationship between the variables is not known*. By a procedure, called dimensional analysis, the phenomenon may be formulated as a relation between a set of dimensionless groups of the variables, the groups numbering less than the variables. The immediate advantage of this procedure is that considerably less experimentation is required to establish a relationship between the variables over a given range. Furthermore, the nature of the experimentation will often be considerably simplified.

To illustrate this, consider the problem of determining the drag F of a smooth sphere of diameter D moving comparatively slowly with velocity V through a viscous fluid. Other variables involved are ρ and μ, the mass density and viscosity, respectively, of the fluid. The drag F may be stated as some unknown function of these variables. That is,

$$F = f(D, V, \rho, \mu)$$

To determine this relationship experimentally would be a considerable undertaking, since only one variable in the parentheses must be allowed to vary at a time, with the resulting accumulation of many plots. A possible representation of the results of such a procedure is indicated in Fig. 8.1, where F is plotted against D for various values of V. Each plot, however, corresponds to a definite value of ρ and μ, so you can see from the diagram that there will be very many

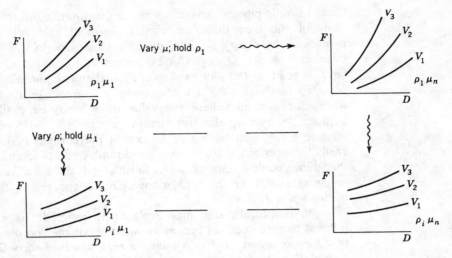

FIGURE 8.1
Many plots needed to get F versus D, ρ, V, μ.

plots required for an effective description of the process. Also, such an approach would mean the use of many spheres of varying diameter and a variety of fluids of varying viscosity and density. Thus, one sees that this could be an extremely time-consuming, as well as expensive, investigation.

Let us now employ dimensional analysis before embarking on an experimental program. As will be shown presently, the drag process above may be formulated as a functional relation between only two dimensionless groups. Each group is called a π (with no relation implied to the number 3.1416...). Thus

$$\frac{F}{\rho V^2 D^2} = g\left(\frac{\rho V D}{\mu}\right)$$

where the nature of the function g is not known. However, by experiment, a *single* curve may be established relating the π's. This is indicated in Fig. 8.2.

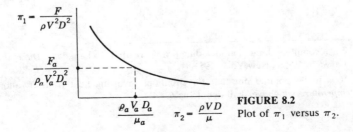

FIGURE 8.2
Plot of π_1 versus π_2.

Such a simple plot can give as complete quantitative information as hundreds of plots of the type discussed earlier. Suppose that the drag is desired for conditions V_a, D_a, ρ_a, μ_a. The dimensionless group $(\pi_2)_a$ can immediately be evaluated as $\rho_a V_a D_a / \mu_a$. Corresponding to this value of $(\pi_2)_a$, the value of $(\pi_1)_a$ is read off the plot as shown. F_a is then computed as $(\rho_a V_a^2 D_a^2)(\pi_1)_a$.

To establish such a useful curve, one may use a wind or water tunnel where for a given sphere the value of π_2 may be easily and continuously adjusted by varying the free-stream velocity V. The force on the sphere is measured for each setting of V so that the corresponding values of π_1 may easily be ascertained. Thus, with considerably less time and expense, a curve of the dimensionless groups is established which, as a result of dimensional analysis, is valid for any fluid or any diameter sphere in a flow within the range of the π's tested.

Philosophically one may explain the rationale of what is to ensue as follows. Nature does not bother knowing about the coordinates and dimensions that we use when we try to mimic a real process. Accordingly, when we find formulations that also do not depend on dimensions and units we quite often have available physically meaningful formulations. Thus the aforementioned dimensionless groups are better able to mimic real processes than the variables themselves. This will become increasingly evident as we proceed through this chapter and indeed through Part II of the text.

The question now arises about the number of dimensionless π's to be formed from a group of variables known to be involved in a physical phenomenon. For this we turn next to the Buckingham π theorem.

8.3   BUCKINGHAM'S π THEOREM[1]

According to this theorem, *the number of independent dimensionless groups that may be employed to describe a phenomenon known to involve n variables is equal to the number n − r, where r is usually the number of basic dimensions needed to express the variables dimensionally.* In the previous example, the variables were $F, V, D, \rho,$ and μ, making n equal to 5. In expressing these quantities dimensionally three basic dimensions M, L, T or F, L, T must be employed so that $n - r$ becomes equal to 2.[2] Also, it is clear that the dimensionless groups employed are independent, i.e., not related to each other through algebraic operations, since F appears only in one group and μ appears only in the other

---

[1] For a complete discussion including proofs, read H. L. Langhaar, *Dimensional Analysis and Theory of Models*, Wiley, New York, 1951.

[2] It is to be pointed out that for situations involving heat transfer, another basic dimension is *temperature* which we shall denote as θ.

group. The theorem above states that there can be no additional independent dimensionless group. Hence, any other dimensionless group that the reader may propose will invariably be one that can be developed by algebraic operations on the groups $F/\rho V^2 D^2$ and $\rho VD/\mu$. For example, $F/\mu VD$ is a dimensionless group formed by the product of the aforementioned groups.

The computation of r in Buckingham's π theorem as the number of basic dimensions needed to express the variables dimensionally is not always correct. For instance, in stress analysis there are problems dealing with force and distance whereby the basic dimensions may number two (F, L) for the FLT system or may number three (M, L, T) if the MLT system is selected. A correct procedure for ascertaining the value of r will now be put forth.

The variables α, β, γ, etc., are listed along a horizontal axis, and the basic dimensions M, L, T, etc., to be used are listed along a vertical axis, as shown below. Under each variable a column of numbers is listed representing the powers to which the basic dimension must be raised in the dimensional representation of the particular variable.

	α	β	γ	δ
M	1	0	3	0
L	-1	-2	1	2
T	2	1	1	1

Thus, in the representation above the variable α must have the dimensions MT^2/L, while the variable $\beta \equiv T/L^2$. The array of numbers so formed is called the *dimensional matrix* of the process and is represented as

$$\begin{pmatrix} 1 & 0 & 3 & 0 \\ -1 & -2 & 1 & 2 \\ 2 & 1 & 1 & 1 \end{pmatrix}$$

It will be remembered from algebra that one may take the determinant of a group of numbers forming an arrangement having equal numbers of rows and columns. The array above may be "squared up" by adding a row of zeros. However, this determinant is obviously zero. An important question then is: What is the size of the next smaller square subgroup that has a nonzero determinant? The number of rows or columns in this determinant then defines the *rank* of the original matrix. In this case any one of the several possibilities is present. For instance, using the first three rows and columns gives

$$\begin{vmatrix} 1 & 0 & 3 \\ -1 & -2 & 1 \\ 2 & 1 & 1 \end{vmatrix} = 6$$

making the rank of the dimensional matrix equal to 3.

The correct value for r in Buckingham's π theorem may now be stated as the rank of the dimensional matrix.

8.4   IMPORTANT DIMENSIONLESS GROUPS IN FLUID MECHANICS

In most fluid phenomena where heat transfer can be neglected, the following variables may be important:

1. Pressure change, Δp
2. Length, L
3. Viscosity, μ
4. Surface tension, σ
5. Velocity of sound, c
6. Acceleration of gravity, g
7. Mass density, ρ
8. Velocity, V

From these variables the following dimensionless groups can be formed:

1. Reynolds number, $\mathrm{Re} = \rho V D / \mu$
2. Froude number, $\mathrm{Fr} = V^2 / Lg$
3. Mach number, $\mathrm{M} = V/c$
4. Weber number, $\mathrm{We} = \rho V^2 L / \sigma$
5. Euler number, $\mathrm{Eu} = \Delta p / \rho V^2$

Note that since three basic dimensions are needed to describe the variables, there are $8 - 3 = 5$ independent dimensionless groups in this listing in accordance with the Buckingham theorem. That these groups are independent is easily seen by noting that the Euler number is the only expression containing the variable Δp; the Weber number is the only one with σ, etc. Hence, no one of these groups could have been formulated by any manner of algebraic combination or operation of the others.

Fortunately, in many engineering problems, only a few of the variables in the previous listing are simultaneously involved to any appreciable degree. For instance, in aeronautical work, surface tension and gravity are not important enough to warrant consideration so that the Froude number and the Weber number would not be involved. Later, the physical significance of the dimensionless groups listed will be discussed so that the reader will better appreciate when certain groups may be deleted.

8.5   CALCULATION OF THE DIMENSIONLESS GROUPS

Now that the correct number of dimensionless groups in a process has been determined, we turn to the problem of how to form the groupings. One way is to establish forms of the correct number of independent groups by trial and error. However, when this is not feasible, the following procedure will be effective.

As an illustration of practical value, let us dimensionally investigate the pressure drop in a viscous incompressible flow through a length L of straight pipe. The variables known to be involved in such a process are pressure drop Δp; average velocity V; viscosity μ; inside diameter of pipe, D; length of pipe section, L; density ρ; and, finally, the roughness of the pipe as represented by the average variation e of inside radius. Functionally, the pressure drop may be expressed as

$$\Delta p = h(\rho, \mu, V, L, D, e) \tag{8.1}$$

The right side of Eq. (8.1) is replaced by an infinite series[3]

$$\Delta p = \left(K_1 \rho^{a_1} \mu^{b_1} V^{c_1} L^{d_1} D^{f_1} e^{g_1} \right) + \left(K_2 \rho^{a_2} \mu^{b_2} V^{c_2} L^{d_2} D^{f_2} e^{g_2} \right) + \cdots \tag{8.2}$$

where $K_1, K_2, \ldots$ are dimensionless coefficients and $a_1, b_1, \ldots, a_2, b_2, \ldots$ are exponents required by the series. Since each grouping in Eq. (8.2) must have the same dimensions by the law of dimensional homogeneity, we need include in the dimensional representation of Eq. (8.2) only the first expression of the series. Hence, dropping the subscript of the exponents and expressing the equation dimensionally, we get

$$\left[\frac{M}{LT^2} \right] \equiv \left[\frac{M}{L^3} \right]^a \left[\frac{M}{LT} \right]^b \left[\frac{L}{T} \right]^c [L]^d [L]^f [L]^g$$

Now the exponents of the basic dimensions M, L, and T on both sides of the equation may be, respectively, equated according to the law of dimensional homogeneity to form the following set of simultaneous algebraic equations:

For M: $\qquad\qquad\qquad 1 = a + b \qquad\qquad\qquad\qquad (1)$

For L: $\qquad\qquad\qquad -1 = -3a - b + c + d + f + g \qquad (2)$

For T: $\qquad\qquad\qquad -2 = -b - c \qquad\qquad\qquad\qquad (3)$

Since there are six quantities related by only three equations, we may solve any three quantities in terms of the remaining three quantities. We choose as the three *dependent* quantities to be *eliminated* those quantities that are associated with three variables that you would want in one of the dimensionless groups.

---

[3]Here again we are attempting to mimic nature. Thus since natural phenomena proceed such that the variables relate continuously to the variable of interest (here Δp), then we should expect that Δp would be expressible as an infinite series which converges uniformly.

Suppose that we wish ρ, V, and D to be in one dimensionless group. Then [see Eq. (8.2)] we will take a, c, and f to be eliminated—i.e., to be expressed in terms of the remaining quantities b, d, and g. Equation (1) accordingly gives us

$$a = 1 - b$$

while Eq. (3) gives us c,

$$c = 2 - b$$

Finally, substituting these results into (2) permits the solution of f in terms of the selected independent variables[4]:

$$f = -b - d - g$$

Returning to Eq. (8.2), we restrict the discussion to the first term of the series and replacing a, c, and f by the previous relations, we get

$$\Delta p = K(\rho^{1-b})(\mu^{b})(V^{2-b})(L^{d})(D^{-b-d-g})(e^{g})$$

Upon grouping those terms with the same exponents together and extending the results to the other members of the series, the equation above may be expressed as

$$\frac{\Delta p}{\rho V^2} = K_1\left(\frac{\mu}{\rho V D}\right)^{b_1}\left(\frac{L}{D}\right)^{d_1}\left(\frac{e}{D}\right)^{g_1} + K_2\left(\frac{\mu}{\rho V D}\right)^{b_2}\left(\frac{L}{D}\right)^{d_2}\left(\frac{e}{D}\right)^{g_2} + \cdots$$

Note that any one of the groupings of variables in the equation above is raised to different powers as one goes from expression 1 onward to expression 2, etc., as is required by the series expansion. That is, the grouping $(\mu/\rho V D)$ is raised to different powers b_1, b_2, etc. Because of dimensional homogeneity, it follows that each of the groupings must perforce be *dimensionless*. Finally, returning to the functional representation of the series, we have

$$\frac{\Delta p}{\rho V^2} = f\left[\left(\frac{\mu}{\rho V D}\right), \left(\frac{L}{D}\right), \left(\frac{e}{D}\right)\right]$$

where f denotes a function. Note that by this procedure the correct number of independent dimensionless groups has been formed. Also, one grouping has $\rho V D$ in it as proposed earlier. Since f is an unknown function, the term $\mu/\rho V D$ may be inverted, thus forming the Reynolds number. Also, note the appearance of the Euler number, as well as two geometrical ratios. The pressure loss through a pipe may then be characterized by the equation

$$\text{Eu} = f\left(\text{Re}, \frac{L}{D}, \frac{e}{D}\right)$$

---

[4]If we had n variables and r basic dimensions, we would get any one chosen set of r exponents in terms of the other $n - r$ exponents.

In a later chapter we consider this relationship further. Meanwhile, it has served to show how, in a direct manner, a set of dimensionless groups can be formed which are independent and of a number consistent with Buckingham's π theorem.

Example 8.1. The end deflection δ of a tip-loaded cantilever beam of length L, as you learned in strength of materials, is given by the formula

$$\delta = \frac{1}{3}\frac{PL^3}{EI} \qquad (a)$$

where E = modulus of elasticity
  P = load
  I = second moment of area of the cross section of the beam about the centroidal axis.

What does dimensional analysis tell you about the relation between δ and the other variables?

First we want to decide carefully about the number of independent π's. For this reason, we look at the dimensional matrix using FLT as our basic dimensions. We have

	P	L	E	I	δ
F	1	0	1	0	0
L	0	1	-2	4	1
T	0	0	0	0	0

The rank of the matrix clearly is 2, so there will be three independent π's. We now proceed to find such π's.

$$\delta = f(P, L, E, I)$$

Hence,

$$\delta = K_1\left[(P)^{a_1}(L)^{b_1}(E)^{c_1}(I)^{d_1}\right] + \cdots$$

Considering the dimensional representation, we can say on dropping the subscript of the exponents:

$$[L] \equiv [F]^a[L]^b\left[\frac{F}{L^2}\right]^c[L^4]^d \qquad (b)$$

From the law of dimensional homogeneity, we can say on equating exponents for the basic dimensions:

For F: $\qquad\qquad 0 = a + c$

For L: $\qquad\qquad 1 = b - 2c + 4d$

For t: $\qquad\qquad 0 = 0$

Hence, selecting a and b to be solved in terms of c and d (this means that P and

L will appear in one dimensionless group), we require that

$$a = -c$$

$$b = 1 + 2c - 4d \qquad (c)$$

We can then conclude on going back to Eq. (b) that:

$$\delta = K_1\left[(P)^{-c}(L)^{1+2c-4d}(E)^c(I)^d\right] + \cdots \qquad (d)$$

Grouping the terms with the same powers,

$$\left(\frac{\delta}{L}\right) = K_1\left[\left(\frac{EL^2}{P}\right)^c\left(\frac{I}{L^4}\right)^d\right] + \cdots$$

Going back to the functional form,

$$\left(\frac{\delta}{L}\right) = f\left[\left(\frac{L^2E}{P}\right),\left(\frac{I}{L^4}\right)\right] \qquad (e)$$

We thus get the expected three dimensionless groups. This is as much as we can get from dimensional analysis. If we multiply the two dimensionless groups in the function we get on inverting the results:

$$\left(\frac{\delta}{L}\right) = g\left[\frac{PL^2}{EI}\right] \qquad (f)$$

Going back to Eq. (a) from strength of materials, we have

$$\left(\frac{\delta}{L}\right) = \frac{1}{3}\left(\frac{PL^2}{EI}\right) \qquad (g)$$

Either the theory of beams or experiments will reveal that the functional relation between δ/L and PL^2/EI in Eq. (f) is that of direct proportionality where the constant of proportionality is $\frac{1}{3}$.

In general, you will get different dimensionless groups when you choose different sets of powers to be solved for in terms of the remaining powers. However, in accordance with Buckingham's π theorem, there will be only $n - r$ *independent* π's. Hence the various possible sets of π's that can be found can be brought into coincidence with each other by simple algebraic manipulations such as multiplying, dividing, and/or raising the π's to powers. You will have a chance to do this in your homework.

As an aid in performing the chapter exercises, we have shown in Table 8.1 a list of commonly used quantities with their dimensional representations for both the $FLT\theta$ and the $MLT\theta$ systems.

TABLE 8.1
Dimensions

Quantity	*MLT*θ system	*FLT*θ system
Force	ML/T^2	F
Area	L^2	L^2
Volume	L^3	L^3
Acceleration	L/t^2	L/t^2
Angular velocity	$1/T$	$1/T$
Angular acceleration	$1/T^2$	$1/T^2$
Linear momentum	ML/T	FT
Moment of momentum	ML^2/T	FLT
Energy	ML^2/T^2	FL
Work	ML^2/T^2	FL
Power	ML^2/T^3	FL/T
Pressure and stress	M/T^2L	F/L^2
Moments and products of area	L^4	L^4
Inertia tensor	ML^2	FT^2L
Moment of torque	ML^2/T^2	FL
Heat	ML^2/T^2	FL
Density	M/L^3	FT^2/L^4
Specific weight	M/T^2L^2	F/L^3
Absolute viscosity	M/LT	FT/L^2
Kinematic viscosity	L^2/T	L^2/T
Enthalpy	L^2/T^2	L^2/T^2
Specific heat	$L^2/T^2\theta$	$L^2/T^2\theta$
Surface tension	M/T^2	F/L
Thermal conductivity	$ML/T^3\theta$	$F/T\theta$

We now present an *alternative* procedure for establishing dimensionless groups. This procedure has the virtue of being quick to execute. We first choose three variables which between them involve the basic dimensions M, L, T. For instance going back to the pressure drop in a pipe (Eq. 8.1) which we now rewrite

$$\Delta p = h(\rho, \mu, V, l, D, e)$$

we may choose D having dimension L, V having a dimension T in it, and ρ having a dimension M in it—thus including all three of our basic dimensions. Now using these variables, form the basic dimensions of L, M, and T in the following way:

$$(L) = (D) \tag{a}$$

$$(T) = (D/V) \tag{b}$$

$$(M) = (\rho D^3) \tag{c}$$

Next take each of the four variables not used above namely Δp, μ, l, and e and divide each one by its dimensional representation. We are thus forming dimensionless expressions. Writing these and calling them π_1, π_2, π_3, and π_4 we have:

$$\pi_1 = \frac{\Delta p}{M/LT^2}$$

$$\pi_2 = \frac{\mu}{M/LT}$$

$$\pi_3 = \frac{l}{L}$$

$$\pi_4 = \frac{e}{L}$$

Finally in each of the above expressions replace the basic dimensions using results (a), (b), and (c) above in place of L, T, and M. Thus we get

$$\pi_1 = \frac{\Delta p}{\dfrac{\rho D^3}{(D)(D/V)^2}} = \frac{\Delta p}{\rho V^2}$$

$$\pi_2 = \frac{\mu}{\dfrac{\rho D^3}{(D)(D/V)}} = \frac{\mu}{\rho V D}$$

$$\pi_3 = \frac{l}{D}$$

$$\pi_4 = \frac{e}{D}$$

We have thus produced the dimensionless groups in a less formal manner.

We now examine similitude, an important consideration for model testing, and in Sec. 8.7 we will relate similitude with dimensional analysis.

**PART B
SIMILITUDE**

8.6 DYNAMIC SIMILARITY

Similitude in a general sense is the indication of a known relationship between two phenomena. In fluid mechanics this is usually the relation between a full-scale flow and a flow involving smaller but geometrically similar boundaries.

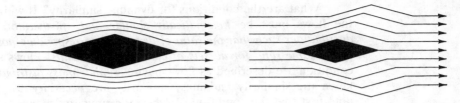

FIGURE 8.3
Kinematically dissimilar flows with geometrically similar boundaries.

However, it must be pointed out that there are similarity laws in common use in fluid mechanics involving flows with dissimilar boundaries. For instance, there is a similarity relation between a subsonic compressible flow (Mach number less than unity) and an incompressible flow about a body distorted in a prescribed manner from that of the compressible flow.[5] Also, in hydrology one uses a model of a river, which is geometrically similar as a plan view, but which is often not similar in depth to the actual river. In this text we restrict the discussion to that of *geometrically similar flows*, i.e., to flows with geometrically similar boundaries.[6]

Two flows consisting of geometrically similar sets of *streamlines* are called *kinematically similar flows*. Since the boundaries will form some of the streamlines, it is clear that kinematically similar flows must also be geometrically similar. However, the converse to this statement is not true, as it is quite easy to arrange kinematically dissimilar flows despite the presence of geometrically similar boundaries. In Fig. 8.3, streamlines are shown about similar double wedges in two-dimensional flows. The one on the left is a low-speed subsonic flow, $M < 1$, while the one on the right is a high-speed supersonic flow, $M > 1$. The lack of similarity between the streamlines is quite apparent.

We define yet a third similarity called *dynamic similarity* whereby the force distribution between the two flows is such that, at corresponding points in the flows, identical types of forces (such as shear, pressure, etc.) are parallel and in addition have a ratio which is the same value at all sets of corresponding points between the two flows. Furthermore, this ratio must be common for the various types of forces present. *For dynamically similar flows there will then be this same ratio between corresponding resultant forces on corresponding boundaries.*

---

[5]Gothert's subsonic similarity rule.
[6]Geometrically similar boundaries can be brought into coincidence by either a uniform dilatation or a uniform shrinkage.

What are the conditions for dynamic similarity? It will now be shown that the flows must be *kinematically similar and*, *furthermore*, *must have mass distributions such that the ratio of density at corresponding points of the flows are of the same ratio for all sets of corresponding points*. Flows satisfying the latter condition are described as flows having *similar mass distributions*. To show that kinematic similarity and mass similarity are *necessary* for dynamic similarity, note first that the condition of *kinematic similarity* means that accelerations

1. Are *parallel* at corresponding points.
2. Have a constant ratio of *magnitude* between all corresponding sets of points.

Item 1 and Newton's law mean that the *resultant* force on each particle must be *parallel* at corresponding points. The condition of *similar mass distributions* and item 2 then mean, also in view of Newton's law, that these *resultant* forces must also have a *constant ratio* of magnitude between all corresponding points in the flow. Since the direction of *each type* of force on a particle is intrinsically tied up with the direction of the streamlines, one may conclude, furthermore, that in kinematically similar flows *identical types of forces* at corresponding points are also *parallel*. Hence we can conclude that since the *resultant* forces on particles have a constant ratio of magnitude between flows, it is necessarily true that *all corresponding components* of the resultant forces (such as shear forces, pressure forces, etc.) have the *same ratio of magnitude* between flows. In short, it is seen that kinematically similar flows with similar mass distributions satisfy all conditions of dynamically similar flows as set forth in our definition at the outset of this paragraph.

Why is the dynamic similarity important in model testing? The reason is very simple and was presented earlier. We now elaborate. If the same ratio exists between corresponding forces at corresponding points, this ratio being the same for the entire flow, then we can say that the *integration of the force distribution giving rise to perhaps lift or drag will also have this ratio between model and prototype flows*. If we do not have flows at least close to having dynamic similarity, then the force ratios between model and prototype flows at different sets of corresponding points will be different, and there is *no simple way* of relating the aforementioned resultants such as drag and lift between the model and prototype. The testing may be useless. For dynamically similar flows, the ratio between corresponding forces at corresponding points and the accompanying ratio between desired resultant forces for model and prototype flows is not hard to establish. One need only multiply the free-stream pressure times the square of a characteristic length for each flow. This gives rise to corresponding forces. The ratio of these forces is the ratio of the desired resultant forces over corresponding boundaries between flows. That is,

$$\frac{[(p_0)(L^2)]_m}{[(p_0)(L^2)]_p} = \frac{(\text{Resultant forces})_m}{(\text{Resultant forces})_p}$$

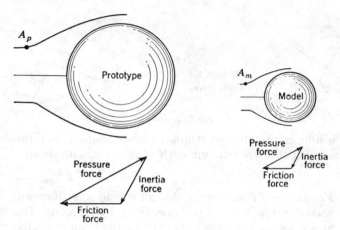

FIGURE 8.4
Prototype and model flows around sphere.

8.7 RELATION BETWEEN DIMENSIONAL ANALYSIS AND SIMILITUDE

Let us examine two dynamically similar incompressible, viscous flows about spheres denoted as model and prototype flows in Fig. 8.4. Neglecting body forces, two types of forces may be distinguished acting on each particle, namely, shear and pressure forces. If the inertia term in Newton's law is now written as a D'Alembert force $-m\mathbf{a}$, you will remember from mechanics that for depicting Newton's law we may consider the sum of the two external forces plus the D'Alembert force as equal to zero and thus in "equilibrium." Hence, one may establish a force triangle at each point of the flow. This has been done in Fig. 8.4 for corresponding points A_p and A_m. By rules of dynamic similarity the force triangles for these points are similar, since the sides of the triangles must be parallel. We may form the following equations, which are true for all corresponding points:

$$\frac{(\text{Pressure force})_m}{(\text{Pressure force})_p} = \frac{(\text{friction force})_m}{(\text{friction force})_p} = \frac{(\text{inertia force})_m}{(\text{inertia force})_p} = \text{const} \quad (8.3)$$

The following relations may be derived from these equations:

$$\frac{(\text{Inertia force})_m}{(\text{Friction force})_m} = \frac{(\text{inertia force})_p}{(\text{friction force})_p} = (\text{const})_1 \quad (8.4)$$

$$\frac{(\text{Inertia force})_m}{(\text{Pressure force})_m} = \frac{(\text{inertia force})_p}{(\text{pressure force})_p} = (\text{const})_2 \quad (8.5)$$

FIGURE 8.5
Shear forces.

FIGURE 8.6
Pressure forces.

It will be informative to evaluate these equations in terms of the flow variables. Let us accordingly examine each of the forces involved:

1. *Viscous or friction force.* An infinitesimal system with rectangular sides is shown in Fig. 8.5 at a position along a streamline. The dimensions are given as ds, dn, and dz, the latter not having been indicated in the diagram. Note from the figure that the net shear force on one pair of sides is $(\partial\tau/\partial n)\, dn\, ds\, dz$. Employing Newton's viscosity law, we may replace τ by $\mu(\partial|\mathbf{V}|/\partial n)$. Using dv as the volume of the system, this shear force is expressed as $\mu(\partial^2|\mathbf{V}|/\partial n^2)\, dv$.

2. *Pressure force.* The net pressure force on the pair of sides shown in Fig. 8.6 is easily evaluated as $(\partial p/\partial n)\, dv$.

3. *Inertia force.* The component selected is along the streamline so that employing Eq. (4.9), we may say for steady flow

$$dm\, a_T = (\rho\, dv)|\mathbf{V}|\frac{\partial|\mathbf{V}|}{\partial s}$$

Although only components of the three types of forces have been developed, it should be clear that the ratios of these components between model and prototype flows will be the same as the ratios of the respective complete forces. Hence, by canceling the term dv, Eqs. (8.4) and (8.5) may be written in the following way:

$$\left[\frac{\rho|\mathbf{V}|\dfrac{\partial|\mathbf{V}|}{\partial s}}{\mu\dfrac{\partial^2|\mathbf{V}|}{\partial n^2}}\right]_m = \left[\frac{\rho|\mathbf{V}|\dfrac{\partial|\mathbf{V}|}{\partial s}}{\mu\dfrac{\partial^2|\mathbf{V}|}{\partial n^2}}\right]_p \tag{8.6}$$

$$\left[\frac{\rho|\mathbf{V}|\dfrac{\partial|\mathbf{V}|}{\partial s}}{\dfrac{\partial p}{\partial n}}\right]_m = \left[\frac{\rho|\mathbf{V}|\dfrac{\partial|\mathbf{V}|}{\partial s}}{\dfrac{\partial p}{\partial n}}\right]_p \tag{8.7}$$

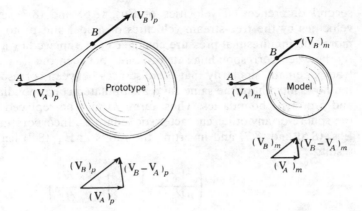

FIGURE 8.7
Kinematically similar flows.

Now the velocities at all corresponding points in the two flows have the same ratio in magnitude. Hence, one could correctly employ the respective model and prototype *free-stream* velocities for the local velocities in the preceding equations. Also, the difference between the velocity vectors at any two points in one flow and the difference between the velocity vectors at corresponding points of the other flow are vectors which are parallel and of a ratio equal in magnitude to the ratios between velocities themselves at corresponding points. This may be understood by observing Fig. 8.7, where the velocities of points A and B of model and prototype have been indicated. The velocity triangles formed by these velocities and their respective differences form a pair of similar triangles, since two adjacent sides of each triangle, $(\mathbf{V}_A)_m$ and $(\mathbf{V}_B)_m$ of model, and $(\mathbf{V}_A)_p$ and $(\mathbf{V}_B)_p$ of prototype, are parallel and of equal ratio. Thus

$$\frac{|\mathbf{V}_A|_p}{|\mathbf{V}_A|_m} = \frac{|\mathbf{V}_B|_p}{|\mathbf{V}_B|_m} = \frac{|\mathbf{V}_B - \mathbf{V}_A|_p}{|\mathbf{V}_B - V_A|_m} \tag{8.8}$$

Since the ratio $|V_A|_p/|V_A|_m$ holds for all corresponding points, it is proper to replace it by $|V_0|_p/|V_0|_m$, the ratio of the free-stream velocities. Furthermore, by taking points A and B infinitesimally close, the equation above may be expressed as

$$\frac{|\mathbf{V}_0|_p}{|\mathbf{V}_0|_m} = \frac{|\mathbf{dV}|_p}{|\mathbf{dV}|_m} \tag{8.9}$$

The same is true for second difference of velocity appearing in the friction forces. From this it is possible to replace all the differential velocities and second differences of velocities of Eqs. (8.6) and (8.7) as well as the local velocities by the free-stream velocities of model and prototype flows. Furthermore, the infinitesimal pressure change can be replaced by a change in pressure Δp between corresponding sets of points between the flows. Finally, one notes from kinematic similarity that the distances between corresponding positions of the flows must have the same ratio as stipulated corresponding lengths of model and prototype boundaries. Thus, term dn may be replaced by the diameter of the sphere or any other characteristic length L. Incorporating these results into Eqs. (8.6) and (8.7) and inverting the ratios in Eq. (8.7) leads to the following results:

$$\left(\frac{\rho V_0^2 / L}{\mu V_0 / L^2} \right)_m = \left(\frac{\rho V_0^2 / L}{\mu V_0 / L^2} \right)_p$$

and therefore on canceling terms

$$\left(\frac{\rho V_0 L}{\mu} \right)_m = \left(\frac{\rho V_0 L}{\mu} \right)_p \tag{8.10a}$$

Also,

$$\left(\frac{\Delta p / L}{\rho V_0^2 / L} \right)_m = \left(\frac{\Delta p / L}{\rho V_0^2 / L} \right)_p$$

and therefore

$$\left(\frac{\Delta p}{\rho V_0^2} \right)_m = \left(\frac{\Delta p}{\rho V_0^2} \right)_p \tag{8.10b}$$

It may be concluded from these results that a *necessary condition for dynamic similarity for the particular flows undertaken is the equality of the Reynolds number and Euler number between the flows, with convenient flow parameters and geometrical measurements being used.* Also a physical interpretation in terms of forces is now available for two of the dimensionless groups introduced in Sec. 8.4. Other groups will presently be discussed.

In Sec. 8.2, a dimensional analysis was carried out for the drag on a sphere moving in an incompressible viscous flow of precisely the same nature as was investigated in the preceding paragraphs. The following result was given:

$$\left(\frac{F}{\rho V^2 D^2} \right) = g \left(\frac{\rho V D}{\mu} \right) \tag{8.11}$$

where g is an unknown function. Now the quantity F/D^2 is proportional to the change in pressure Δp between corresponding sets of points between the flows because of the dynamic similarity between the flows. Hence, we may restate the equation above in the form

$$\left(\frac{\Delta p}{\rho V^2} \right) = h \left(\frac{\rho V D}{\mu} \right) \tag{8.12a}$$

or

$$\mathrm{Eu} = h(\mathrm{Re}) \tag{8.12b}$$

Thus we see that *dimensional analysis produces the dimensionless groups whose values must be duplicated between geometrically similar flows if dynamic similarity is to be attained between the flows.* Furthermore, we see from Eqs. (8.12) that if the Reynolds numbers are duplicated, then the Euler number will be duplicated and therefore, as a result of dimensional analysis, we may relax the requirement for dynamic similarity between the flows under discussion to that of duplicating the Reynolds number only.

Since the shape of the boundary in this discussion was of no consequence beyond the condition of being geometrically similar, we can then say for all viscous, steady, incompressible, geometrically similar flows that conveniently contrived Reynolds numbers must be duplicated between flows to achieve dynamic similarity and so permit data stemming from force distributions on boundaries of models to be meaningful in predicting full-scale results. Experience indicates that this condition with some added corrections is generally sufficient. We will learn more about sufficiency conditions as we study particular flows in more detail in later chapters.

Generalizing once again to *any* flow, we may state that *dimensional analysis will yield the dimensionless groups in a flow for which we must duplicate at least all but one in geometrically similar flows to achieve dynamic similarity.*

8.8 PHYSICAL MEANING OF IMPORTANT DIMENSIONLESS GROUPS OF FLUID MECHANICS

In the previous section, physical connotations were placed on the Reynolds number and the Euler number. Proceeding in a similar manner, we find that physical interpretations can be developed for the remaining dimensionless groups introduced in Sec. 8.4. These are now listed and discussed in a cursory manner:

1. Reynolds number   Ratio of the inertia force to the friction force, usually in terms of convenient flow and geometrical parameters.

$$\frac{\rho V^2/L}{\mu V/L^2} = \frac{\rho V L}{\mu}$$

2. Mach number   Ratio of the square root of the inertia force to the square root of the force stemming from the compressibility of the fluid. This becomes of paramount importance in high-speed flow, where density variations from pressure become significant

$$\sqrt{\frac{\rho V^2/L}{\rho c^2/L}} = \frac{V}{c}$$

3. Froude number   Ratio of the inertia force to the force of gravity. If there is a free surface, such as in the case of a river, the shape of this surface in the form of waves will be directly affected by the force of gravity, so the Froude number in such problems is significant.

$$\frac{\rho V^2/L}{\gamma} = \frac{V^2}{Lg}$$

4. Weber number   Ratio of the inertia force to the force of surface tension. This also requires the presence of a free surface, but where large objects are involved, like boats in a fluid such as water, this effect is very small.

$$\frac{\rho V^2/L}{\sigma/L^2} = \frac{\rho V^2 L}{\sigma}$$

5. Euler number   Ratio of the pressure force to the inertia force. In practical testing work the *pressure coefficient* $\Delta p/(\frac{1}{2}\rho V^2)$ equal to twice the Euler number is ordinarily used.

$$\frac{\Delta p/L}{\rho V^2/L} = \frac{\Delta p}{\rho V^2}$$

With a physical picture of what these dimensionless groups mean, it is much simpler to stipulate which are to be significant and which can be neglected during an investigation. We will continually refer to these throughout the remainder of the text.

Example 8.2. The drag on a submarine moving well below the free surface is to be determined by a test on a model scaled one-twentieth of the prototype. The test is to be carried out in a water tunnel. Determine the ratio of model to prototype drag needed for determining the prototype drag when the speed of the prototype is 5 kn. The kinematic viscosity of seawater is 1.30×10^{-6} m^2/s and the density is 1010 kg/m^3 at the depth of the prototype. The water in the tunnel has a temperature of 50°C.

Since the submarine will be moving well below the free surface, we need not be concerned with wave effects; hence the Froude number will play no role. Because of the slow speed of the submarine, compressibility plays no role, so the Mach number will play no role. Indeed, all we need be concerned about is the Reynolds number and the Euler number for dynamic similarity. Denoting the length of the submarine as L, we have the following Reynolds number for the prototype flow:

$$[Re]_p = \frac{V_p L_p}{\nu_p} = \frac{[5 \text{ kn}][0.5144 \text{ (m/s)/kn}][L(m)]_p}{[1.30 \times 10^{-6} \text{ m}^2/\text{s}]}$$

$$= 1.978 \times 10^6 L_p \qquad\qquad (a)$$

The Reynolds number for the model flow must then equal the value above. Using the Table B.1 in the appendix for ν, we have

$$[\text{Re}]_m = \frac{(V_m)(\frac{1}{20}L)_p}{0.556 \times 10^{-6}} \qquad (b)$$

Equating (a) and (b), we may solve for V_m to be

$$V_m = 22.0 \text{ m/s} \qquad (c)$$

This is the free-stream velocity for the water tunnel.

We will measure a drag F_m in the water-tunnel test. We want the drag F_p on the prototype. For *dynamic similarity* all force components have the same ratio at corresponding points. The drag forces for the submarine must have this *same* ratio. What is this ratio? For this information, we can look at the Euler numbers which we know must be duplicated between the flows.

$$\frac{\Delta p_m}{\rho_m V_m^2} = \frac{\Delta p_p}{\rho_p V_p^2}$$

Replacing Δp_m by F_m/L_m^2, which is proportional to Δp_m, and for the same reason, Δp_p by F_p/L_p^2, we get

$$\frac{F_m/L_m^2}{\rho_m V_m^2} = \frac{F_p/L_p^2}{\rho_p V_p^2}$$

$$\qquad (d)$$

$$\therefore F_p = \left(\frac{\rho_p V_p^2}{\rho_m V_m^2}\right)\left(\frac{L_p}{L_m}\right)^2 F_m$$

Inserting values, we get for F_p

$$F_p = \frac{(1010)[(5)(0.5144)]^2}{(988)(22^2)}\left[\frac{L_p}{(L_p/20)}\right]^2 F_m$$

$$\therefore F_p = 5.59 F_m$$

Therefore we multiply a drag measured in the water tunnel by 5.59 to get the proper drag on the prototype.

Example 8.3. The power P to drive an axial flow pump depends on the following variables:

Density of the fluid, ρ
Angular speed of rotor, N
Diameter of rotor, D
Head, ΔH_D
Volumetric flow, Q

A model scaled to one-third the size of the prototype has the following characteristics:

$$N_m = 900 \text{ r/min}$$
$$D_m = 5 \text{ in}$$
$$(\Delta H_D)_m = 10 \text{ ft}$$
$$Q_m = 3 \text{ ft}^3/\text{s}$$
$$P_m = 2 \text{ hp}$$

If the full-size pump is to run at 300 r/min, what is the power required for this pump? What head will the pump maintain? What will the volumetric flow rate Q be?

We leave it to you to show that the pump process can be described by three π's. That is,

$$\underbrace{\left[\frac{P}{\rho D^5 N^3}\right]}_{\pi_1} = f\left[\underbrace{\left(\frac{\Delta H_D}{D}\right)}_{\pi_2}, \underbrace{\frac{Q}{ND^3}}_{\pi_3}\right]$$

For the model flow we have

$$\pi_2 = \left[\frac{\Delta H_D}{D}\right] = \frac{10}{(5/12)} = 24$$

$$\pi_3 = \left[\frac{Q}{ND^3}\right] = \frac{3}{(900)(2\pi/60)(5/12)^3} = 0.440$$

For dynamic similarity, we must maintain these values of the π's for the full scale pump. Hence,

$$\left[\frac{\Delta H_D}{D}\right]_p = 24$$

$$\therefore \frac{(\Delta H_D)_p}{(5/12)(3)} = 24 \qquad (\Delta H_D)_p = 30 \text{ ft}$$

Also,

$$\left[\frac{Q}{ND^3}\right]_p = 0.440$$

$$\therefore \frac{Q_p}{(300)(2\pi/60)[3(5/12)]^3} = 0.440 \qquad Q_p = 27 \text{ ft}^3/\text{s}$$

Finally we require

$$\left[\frac{P}{\rho D^5 N^3}\right]_m = \left[\frac{P}{\rho D^5 N^3}\right]_p$$

$$\frac{2}{\rho(5^5)(900^3)} = \frac{P_p}{\rho[(3)(5)]^5(300^3)}$$

$$\therefore P_p = 18 \text{ hp}$$

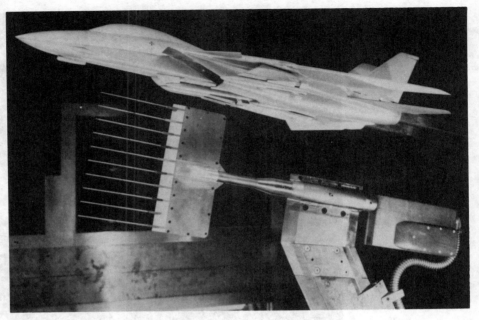

FIGURE 8.8
Grumman F-14 Tomcat model undergoing flow survey in 7 × 10 ft transonic wind tunnel. (*Aviation and Surface Effects Department, David W. Taylor Naval Ship Research and Development Center, Carderock, Maryland.*)

8.9 PRACTICAL USE OF THE DIMENSIONLESS GROUPS

If the variables of a fluid phenomenon are known, we have learned that a dimensional analysis will yield a set of independent dimensionless groups which may usually be put in the form of the various "numbers" discussed previously as well as dimensionless groups in the form of simple geometrical ratios. If at least all but one of the groups are duplicated for geometrically similar flows, we reasoned in Sec. 8.7 that the flows will probably be dynamically similar. This fact introduces the possibility of testing a model of some proposed apparatus in order to study, less expensively, full-scale performance and possible design variations, as we have already indicated. For instance, in the aircraft industry, model testing of some proposed airfoil or entire plane is a very important and significant part of a development program (Fig. 8.8). It must be pointed out that model testing is not inexpensive: the models run into many thousands of dollars and use of test facilities often costs thousands of dollars per hour. In addition to these deterrents, there is the important practical question of how much dynamic similarity can be achieved in a test. Of course, this is a very important criterion with respect to the usefulness of the results.

One of the oldest forms of scientific model testing involving considerations of flow similarity is that of the towing tank (see Fig. 8.9), where models of

FIGURE 8.9
Testing a model of America Cup entry *Heart of America* in a towing tank. Unfotunately this sailboat was not a winner. (*Courtesy Davidson Laboratories, Stevens Institute of Technology.*)

proposed water hulls are moved along a channel of water and drag estimates made with the aid of certain measurements. It will now be shown that even in this apparently straightforward test, true dynamic similarity cannot be achieved for practical testing purposes. A dimensional analysis will reveal that three groups are involved in determining the drag on the hull—the pressure coefficient, the Reynolds number, and the Froude number. Therefore, for dynamic similarity it is necessary that at least the Reynolds number and Froude number be duplicated for prototype flow and geometrically similar model flow. Suppose that the prototype has a length of 100 ft and a velocity of 10 kn and is to be propelled through fresh water with a viscosity of 2.10×10^{-5} lbf-s/ft^2 and a density of 62.4 lbm/ft^3. Then

$$\text{Re} = \frac{\rho V D}{\mu} = \frac{\left[(62.4/32.2) \text{ slugs/ft}^3\right]\left[(10)(1.689) \text{ ft/s}\right]\left[100 \text{ ft}\right]}{\left[2.10 \times 10^{-5} \text{ lbf-s/ft}^2\right]}$$

Replacing pound-force by $(1 \text{ slug})(\text{ft/s}^2)$ from Newton's law, we get for Re,

$$\text{Re} = 1.559 \times 10^8$$

For the Froude number, we get

$$\text{Fr} = \frac{V^2}{Lg} = \frac{[(10)(1.689) \text{ ft/s}]^2}{[100 \text{ ft}][32.2 \text{ ft/s}^2]} = 0.0886$$

For a geometrically similar model of a scale one-twentieth of the prototype, to duplicate Froude number we require a velocity determined as

$$\left(\frac{V^2}{Lg}\right)_m = \frac{V_m^2}{(\frac{1}{20})(100)(32.2)} = 0.0886$$

$$\therefore V_m = 2.24 \text{ kn}$$

To duplicate the Reynolds number we require next

$$\left(\frac{\rho VD}{\mu}\right)_m = \left(\frac{VD}{\nu}\right)_m = \left[\frac{[(2.24)(1.689)](\frac{1}{20})(100)}{\nu_m}\right] = 1.559 \times 10^8$$

$$\therefore \nu_m = 1.2113 \times 10^{-7} \text{ ft/s}$$

Here we reach an impasse, since such a fluid cannot be practically formed. Furthermore, for other tests, there would probably be different requirements on the kinematical viscosity of the fluid. Hence, using ordinary water as is normally the case, dynamic similarity is not achieved in towing-tank tests. However, by duplicating Froude numbers and by additional *theoretical* computations the errors arising from the dissimilar aspects of the flows can be successfully taken into account. Likewise, in wind-tunnel testing, one sometimes must take wind-tunnel corrections into account so as to render results which are meaningful for full-scale operation.

It must always be kept in mind that model testing, although tremendously cheaper than full-scale testing, is nevertheless usually quite expensive. When theoretical computer computations can be made instead of model testing, this is usually the least-expensive avenue of approach and should be thoroughly explored before embarking on a long, expensive model-testing program.

8.10 SIMILITUDE WHEN THE DIFFERENTIAL EQUATION IS KNOWN

In this chapter, we considered a process where the significant variables were known. This permitted us to ascertain the dimensionless groups characterizing the process. You will recall that we *arbitrarily defined* various similarity relations and then, with the aid of Newton's second law and Newton's viscosity law applied to a simple type of flow, we showed that duplicating all but one of the dimensionless groups, we could achieve these similitude relations between flows with geometrically similar boundaries. In some other process of another field of study where we do not know the laws involved, how would we establish the effect of duplicating dimensionless groups? In such a case much experience with the process might lead to an insight about the nature of the similarity laws achieved by duplicating dimensionless groups.

If, at the other extreme, the *differential equation* depicting the process is known, we can *deduce* the dimensionless groups *and* the similarity laws resulting from their duplication, even if the differential equation is not solvable. In Chap. 10 we develop the powerful Navier-Stokes equations governing fluid flow. We will then *deduce* at that time (Sec. 10.8) the kinematic- and dynamic-similarity laws of this chapter in a direct and straightforward fashion. The approach we have followed in this chapter, although not so direct, will give us a greater physical feel for the dimensionless groups and will permit us to think of similitude early in our study of fluid mechanics.

8.11 CLOSURE

In this chapter we have examined the procedures and concepts involved in dimensional analysis and similitude and the relation between these concepts. Although the techniques presented and the concepts discussed are important in themselves, there is an even more important lesson to be inferred from this chapter which has not been stressed in the discussion thus far.

It is no doubt clear from the tremendous cost of many devices such as ships, airplanes, and rockets that careful preliminary design and model testing are a requirement before one can "freeze" plans and start on the construction of such a system. Wind-tunnel tests and towing-tank tests on *models* are a major part of a development program. And in these tests dynamic similarity between model and prototype is *almost never* actually completely achieved. There is then usually the necessity of making corrections and adjustments of the model data to make it more accurate. Such steps can be effectively taken only by engineers who are well grounded in the fundamental theory of fluid mechanics. Thus, the "practical" field of testing requires a thorough understanding of fundamentals for other than the most routine full-scale testing.

In retrospect, you can see that the engineer must first get as close to dynamic similarity as is possible. Even then he or she may have to make additional corrections via theoretical procedures. Therefore, this chapter may perhaps motivate you to examine carefully the fundamental concepts of the various types of flow to be examined in Parts II and III of the text and not to become despondent when for certain situations we cannot employ theory in such a fashion as directly to produce answers to given problems by straightforward analytical methods. In such cases, one must often resort to experiments, particularly with models, and such experiments and tests can be carried out successfully in such areas only when the basic laws of fluid mechanics that we have been studying are well understood.

PROBLEMS

Problem Categories

8.1. Show that the Weber number given as $\rho V^2 L/\sigma$ is dimensionless where
ρ = density of fluid
V = velocity
L = length
σ = surface tension given as force per unit length

8.2. A dimensionless group that is used in studies of heat transfer is the *Prandtl number* given as

$$Pr = \frac{c_p \mu}{k}$$

where c_p = specific heat at constant pressure
μ = coefficient of viscosity
k = thermal conductivity of a fluid
What is the dimensional representation of k in the *MLT* system of basic dimensions and in the *FLT* system of basic dimensions?

8.3. In Chap. 13, we will be introduced to the so-called shear velocity τ_* defined as

$$\tau_* = \sqrt{\tau_w/\rho}$$

where τ_w is the shear stress at the wall. Show that τ_* has the dimensions of a velocity—hence its name.

8.4. In heat transfer the heat *convection coefficient h* is defined as the wall heat flux (energy per unit time per unit area) divided by the difference between the wall temperature and average temperature of the fluid at the wall. The *Stanton* number St is a useful dimensionless group defined as

$$St = \frac{h}{\rho c_p V}$$

Show that it is dimensionless.

8.5. The *Grashof* number Gr is used in bouyancy induced flows where temperature is nonuniform and is defined as

$$Gr = \frac{g \beta L^3 t}{\nu^2}$$

where β is the thermal expansion coefficient defined as the change in volume per unit volume per unit temperature and t is temperature. Show that the Grashof number is dimensionless.

8.6. What is the rank of the following dimensional matrix? What are the dimensions of α, β, γ, and δ?

	α	β	γ	δ
M	1	0	1	0
L	2	1	−1	1
T	−1	2	−1	0

8.7. Consider a mass on a weightless, frictionless spring as in Fig. P8.7. The spring constant is K, and the position of the body of mass M is measured by the displacement x from the position of static equilibrium. In ascertaining the amplitude of vibration A on such a system resulting from a harmonic disturbance, we know that the following variables are involved:
A = amplitude of vibration
M = mass of body
K = spring constant
F_0 = amplitude of disturbance
ω = frequency of disturbance
Assume that this problem cannot be handled theoretically but must be done experimentally.
(*a*) Explain how you would carry out an experimental program without the use of dimensional analysis.
(*b*) Form two independent dimensionless groups by trial and error, and then explain how you would carry out your experiments.

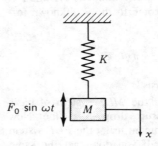

FIGURE P8.7

8.8. The maximum pitching moment that is developed by the water on a flying boat as it lands is noted as C_{max}. The following are the variables that are involved in this action:

  α = angle made by flight path of plane with horizontal
  β = angle defining attitude of plane
  M = mass of plane
  L = length of hull
  ρ = density of water
  g = acceleration of gravity
  R = radius of gyration of plane about axis of pitching

According to Buckingham's π theorem, how many independent dimensionless groups should there be which characterize this problem?

8.9. The power required to drive a propeller is known to depend on

  D = diameter of propeller
  ρ = density of fluid
  c = velocity of sound in fluid
  ω = angular velocity of propeller
  V = free-stream velocity
  μ = viscosity of fluid

According to Buckingham's π theorem, how many dimensionless groups characterize this problem?

8.10. What are the dimensional matrices for Probs. 8.8 and 9.9? What are the ranks of these matrices?

8.11. Consider a freely falling body near the earth's surface. The time t of descent we believe depends on the height h of the fall, the weight w, and the gravitational acceleration g. What is the minimal experimentation to find the time t? Take g as a constant.

8.12. The period τ of oscillation for a pendulum is known to depend on l, the length of the pendulum, its mass m, and gravitational acceleration g. How close can you come to the well-known formula

$$\tau = 2\pi\sqrt{l/g}$$

by dimensional analysis?

8.13. In Sec. 8.5, we used the MLT system of basic dimensions for pressure drop in a pipe. Now carry out the development using the FLT system of basic dimensions. If you don't get the same

π's as in Sec. 8.5, manipulate π's algebraically until you do.

8.14. By a formal procedure, evaluate a set of dimensionless groups for Prob. 8.7.

8.15. Determine a set of dimensionless groups for Prob. 8.8.

8.16. Determine a set of dimensionless groups for Prob. 8.9. Adjust your π's by algebra to get

$$\frac{P}{D^5\rho\omega^3} = f\left(\text{Re}, \text{M}, \frac{D\omega}{c}\right)$$

8.17. In strength of materials, you learned that the shear stress in a rod under torsion is given as

$$\tau = \frac{M_x r}{J} \qquad (a)$$

where M_x is the torque, and J is the polar moment of area of the cross section. We can give the formula above in dimensionless form as follows:

$$\left(\frac{\tau r^3}{M_x}\right) = \left(\frac{r^4}{J}\right) \qquad (b)$$

How close to this formula can you get by dimensional analysis? Use the FLT system of basic dimensions.

8.18. Do Example 8.1 using the MLT system of basic dimensions.

8.19. A disc A having a moment of inertia I_{xx} is held by a light rod of length L (see Fig. 8.19). The formula for free torsional oscillation of the disc is given as:

$$\theta = A\sin\sqrt{\frac{K_t}{I_{xx}}}\,t + B\cos\sqrt{\frac{K_t}{I_{xx}}}\,t$$

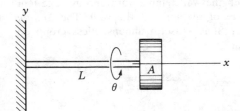

FIGURE P8.19

where K_t is the equivalent torsional spring constant coming from the rod and has a value given as

$$K_t = \frac{GJ}{L}$$

where G = shear modulus of the rod
J = polar moment of area of the rod
L = length of the rod

How close can you come to these results by using dimensional analysis?

8.20. Experience dictates that the head ΔH_D developed by turbomachines depends on the following variables:
Diameter of rotor, D
Rotational speed, N
Volume flow through the machine, Q
Kinematic viscosity, ν
Gravity, g
Show that

$$\frac{\Delta H_D}{D} = f\left(\frac{Q}{ND^3}, \frac{g}{N^2 D}, \frac{ND^2}{\nu} \right)$$

8.21. The velocity of sound c in a perfect gas is given as $c = \sqrt{kRT}$, where k is the ratio of specific heats and hence dimensionless, and T is the temperature. What can you learn about c by using dimensional analysis only?

8.22. In the laminar flow of a viscous fluid through a capillary tube, the pressure drop over length L is a function of the velocity, diameter, viscosity, and length. Determine the π's involved. How close can you get to the formula $\Delta p = 32(V\mu/L)(L/D)^2$ to be derived in Chap. 9?

8.23. A boat moving along the free surface has a drag D which we know depends on the following variables:
$V \equiv$ velocity
$L \equiv$ length of boat
$\mu \equiv$ viscosity
$g \equiv$ gravity
$\rho \equiv$ density
$B \equiv$ beam or width of boat
Formulate the dimensionless groups involved. If you don't get the Reynolds number, Froude number, and Euler number, manipulate your π's algebraically to get them.

8.24. The pressure drop Δp in a compressible one-dimensional flow in a circular duct is a function of the following variables:
Density, ρ
Velocity of sound, c
Viscosity, μ
Velocity of flow, V
Diameter of duct, D
Length of duct, L
What are the dimensionless groups involved? Manipulate your results until you get an Euler number, a Reynolds number, and a Mach number as three of your π's.

8.25. Formally develop the dimensionless groups given in Sec. 8.4 from the variables presented. If you don't get the particular groups presented in this section, manipulate the π's until you do.

8.26. In Fig. 13.33 we have shown vortices shedding from flow past a cylinder. If the frequency of vortex shedding N is a function of ρ, μ, V, and diameter D, what are the dimensionless groups characterizing the process?

8.27. We learned in solid mechanics that the twist of a circular shaft is given by the following formula:

$$\Delta \phi = \frac{M_x L}{GJ} \qquad (1)$$

How close can you come to this result by using dimensional analysis? Proceed as follows:
(1) Using *FLT* system write dimensional matrix.
(2) What is the rank r of this matrix?
(3) Now get as close as possible to Eq. (1) using dimensional analysis.

8.28. A jet of liquid (1) enters liquid (2). The length L from discharge to complete disintegration is to be studied. If the variables known to be involved are the densities and viscosities of the fluids and the jet velocity V_j as well as the jet diameter D_j, determine the dimensionless groups involved.

8.29. Stoke's law [see Eq. (A.I.29) of Appendix A] for a small sphere of radius R has a drag F for steady creeping flow around the sphere given as

$$F = 6\pi\mu VR$$

How would you get this formula experimentally with a minimum of experimentation knowing the variables involved?

8.30. In strength of materials you learned that the buckling load of a pin ended column is given as

$$P_{cr} = \frac{\pi^2 EI}{L^2}$$

How close can you come to this formulation via dimensional analysis?

8.31. The drag D of a diving bell (see Fig. P8.31) depends on the following variables:
Volume of vehicle, V
Density of water, ρ
Viscosity of water, μ
The speed of the vehicle, $\dot{s}$
The roughness of the surfaces, e
Derive a set of independent dimensionless groups. Use the *MLT* system of dimensions. We want ρ, $\dot{s}$, and e in the same group.

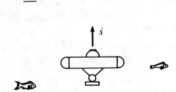

FIGURE P8.31

8.32. The drag on a rectangular $a \times b$ plate at an angle α relative to a wind of velocity V is desired. The drag depends on a, b, α, V, μ, and ρ. What dimensionless groups characterize the process?

8.33. Fourier's law of heat conduction in a solid is known to be

$$q = \frac{kA}{L}(t_2 - t_1)$$

where $q \equiv$ energy flow per unit time
$k \equiv$ is the thermal conductivity and is energy times thickness per unit area, per unit time, per unit temperature
$t \equiv$ temperature
How close can you come to Fourier's conduction law by dimensional analysis?

8.34. In the chapter on boundary layers (Chap. 13, see Fig. 13.1), you will learn that the boundary layer thickness δ depends for a smooth plate on the following items:
μ, viscosity of the fluid
ρ, mass density of the fluid
V_0, the free stream velocity
x, the distance from the leading edge of the plate
Theory indicates for a laminar boundary layer that

$$\frac{\delta}{x} = 4.96 \left/ \sqrt{\frac{\rho V_0 x}{\mu}} \right.$$

How close can you come by dimensional analysis? Notice that $\rho V_0 x/\mu$ is a form of Reynolds number with x as the length dimension.

8.35. The flow through a square-edged circular orifice is given in Appendix A.I by Eq. (A.I.3) as follows for an inviscid liquid:

$$q = A_2 \left\{ \frac{2(p_1 - p_2)/\rho}{\left[1 - (A_2/A_1)^2 \right]} \right\}^{1/2}$$

where A_2 is the area of the orifice and A_1 is the area of the pipe. If we rewrite this formula as

$$q = \frac{\pi d^2}{4} \left\{ \frac{2\Delta p/\rho}{1 - (d/D)^2} \right\}^{1/2} \qquad (a)$$

where d is the orifice diameter, how close can we come to this result by dimensional analysis alone?

8.36. The following variables are known to be involved in a flow:
ρ, mass density
L, characteristic length
c, velocity of sound
μ, viscosity
g, acceleration of gravity
V, average velocity
Δp, pressure change
What are the π's involved? Form the Reynolds number, Froude number, Mach number, and Euler number from your results.

8.37. The rise in a tube due to capillary action is a function of D, θ, σ, g, and ρ.

$$\therefore h = f(D, \theta, \sigma, g, \rho)$$

where σ is the surface tension (F/L). Rewrite

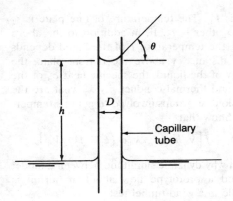

FIGURE P8.37

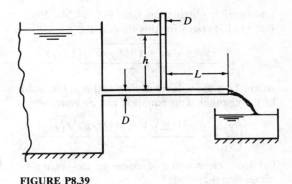

FIGURE P8.39

this relation in terms of dimensionless groups. Have σ, ρ, and g be in the *same* π. Use F, L, and T as basic dimensions. What is r as determined by the dimensional matrix? By algebraic manipulation of π's reach the following result:

$$\left(\frac{h}{D}\right) = G\left[\left(\frac{\sigma}{D^2\rho g}\right), \theta\right]$$

8.38. The thrust from an airplane propeller is a function of the following variables:
V_0 = speed of airplane
D = diameter of propeller
ρ = density of air
μ = viscosity of the air
c = speed of sound
ω = ang. speed of propeller
Hence,

$$T = f(V_0, D, \rho, \mu, c, \omega)$$

Find the dimensionless groups that characterize the process. Manipulate so you get:

$$\frac{T}{\rho\omega^2 D^4} = g\left(\frac{\rho V_0 D}{\mu}, \frac{V_0}{c}, \frac{V_0}{\omega D}\right)$$

8.39. A capillary tube (Fig. P8.39) can be used for measuring viscosity. It is known that for this device the viscosity μ is a function of the following variables:
D, diameter of tube
ρ, mass density of fluid
g, acceleration of gravity
L, length of tube from capillary to exit
h, height of capillary fluid
q, volume flow of fluid

How close can you get to the following solution:

$$\left(\frac{\mu}{\rho g^{1/5} q^{2/5}}\right) = \frac{\pi}{128}\left(\frac{D^4 g^{4/5}}{q^{8/5}}\right)\left(\frac{h}{L}\right)$$

8.40. The viscosity μ in a viscosimeter (see Fig. A.I.18 in Appendix A) depends on the following variables:
T_q, torque on the spring
ω, angular speed or container
r_1, radius of drum
h, height of drum
ϵ, distance between drum base and container
$\alpha = (r_2 - r_1)$, distance between container and drum bottom
Evaluate the dimensionless groups getting μ, ω, and T in one group. How close can you come to the following analytic solution?

$$\left(\frac{\mu\omega r_1^3}{T_q}\right) = \frac{1}{2\pi}\left[\frac{1}{(h/\alpha) + (1/4)(r_1/\epsilon)}\right]$$

8.41. In Section A.I.7 of Appendix A we computed the viscosity of a fluid by observing the terminal speed V_T of a small sphere of radius R and mass density ρ_s in the viscous fluid whose density is ρ_L. We got the following result

$$\mu = \frac{2}{9}\frac{gR^2}{V_T}[\rho_s - \rho_L] \qquad (a)$$

How close can you come using dimensional analysis?

8.42. In the chapter on free surface flow you will learn that a *hydraulic jump* can occur in a rapidly moving flow that must slow down as a result of a downstream obstruction (see Fig. 14.24). Equation (14.87) states for this flow that

$$y_2 = \frac{-y_1 \pm \sqrt{y_1^2 + (8Q^2/gb^2)(1/y_1)}}{2}$$

where Q is the volume of flow and b is the width of the channel. This equation can be rewritten as

$$\left(\frac{y_2}{y_1}\right) = \frac{-1 \pm \sqrt{1 + (8Q^2/gb^2 y_1^3)}}{2}$$

(a) How close can you come to this result by dimensional analysis?
(b) Show that $Q^2/gb^2 y_1^3$ is a Froude number with y_1 as the length dimension.

8.43. In the study of turbomachines (Chap. 15) there are generally six variables involved. They are:
Size of machine diameter D
Rotational speed N
Volume flow through the machine Q
Kinematic viscosity ν
Gravity g
Change in total head ΔH_D
Show that the following describes the performance of turbomachines:

$$f\left(\frac{Q}{ND^3}, \frac{\Delta H_D}{D}, \frac{g}{N^2D}, \frac{ND^2}{\nu}\right) = 0$$

8.44. Consider a flow of fluid past a cylinder involving heat transfer. The heat transfer coefficient h is known for certain conditions to depend on the following variables:
Free-stream velocity V
Fluid density ρ
Fluid viscosity μ
Coefficient of thermal conductivity k
Diameter of cylinder D
Specific heat c_p
What are a set of dimensionless groups for this process? The dimensions for h and k are

$$[h] = \left[\frac{L}{\theta T^3}\right]$$

$$[k] = \left[\frac{F}{T\theta}\right]$$

where recall, θ is the dimensional representitive of temperature. Note we get a Reynolds number and a Prandtl number, $c_p\mu/k$.

8.45. A liquid flows between two parallel plates separated by a distance h. The average velocity of the

liquid is V_0. The temperature of one plate is t_1 and the other is t_2. If, in addition to the above factors, the temperature t of the liquid depends on the distance y above the bottom plate, the viscosity of the liquid, the specific heat c_p of the liquid, and thermal conductivity k, what are the dimensionless groups involved to get this temperature? Show that

$$t/t_1 = f\left[y/h, \mu c_p/k, c_p(t_1 - t_2)/V_0^2\right].$$

8.46. Explain why dynamic similitude between a model flow and a prototype flow about an airfoil is desirable in a wind-tunnel test.

8.47. The drag of a two-man submarine hull is desired when it is moving far below the free surface of water. A model scaled down one-tenth from the prototype is to be tested. What dimensionless group should be duplicated between model and prototype flows? If the drag of the prototype at 1 kn is desired, at what speed should the model be moved to give the drag to be expected by the prototype?

8.48. Oil having a kinematic viscosity of 6.05×10^{-5} ft^2/s is flowing through a 10-in pipe. At what velocity would water at 60°F have to flow through the pipe for dynamically similar flow? What is the ratio of drags for corresponding lengths of pipe from the flows? The specific gravity of the oil is 0.8.

8.49. The wave resistance of an ocean liner scaled down $\frac{1}{100}$ is to be measured. If the drag of the prototype at 20 kn is desired, what must the speed of the model be? Ascertain the ratio of the drag forces between model and prototype.

8.50. A *Venturi meter* is a device for measuring flow in a pipe. It is merely a section of pipe having a reduced diameter. Suppose a model is one-tenth

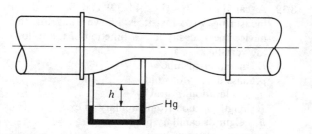

FIGURE P8.50

the size of the prototype. If the diameter of the model is 60 mm and the approach velocity is 5 m/s, what is the discharge in liters per second in the prototype for dynamic similarity? The kinematic viscosity in the model fluid is 0.9 times the kinematic viscosity of the prototype fluid.

8.51. A V-notch *weir* is a vertical plate with a V notch through which fluid in a channel flows. If the shape of the free surface is vital in governing the flow, what π should be duplicated between the flows of a model and a prototype? If a model is one-fiftieth the size of the prototype and the free-stream velocity upstream is 10 ft/s for the prototype, what should the free-stream velocity be for the model? What is the ratio of force on the weir between model and prototype?

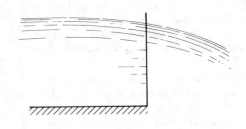

Front View

FIGURE P8.51

8.52. A model of a submarine is scaled down to one-twentieth the size of the prototype and is to be tested in a wind tunnel where at free stream $p = 300$ lb/in² absolute and $t = 120°F$. The speed of the prototype at which we are to estimate the drag is 15 kn. What should the free-stream velocity of the air be in the wind tunnel? What will be the ratio of the drags between model and prototype? Explain why, despite the high pressure in the wind tunnel, we can consider the *flow* to be incompressible. The following is

given:

$$\left(\frac{\mu}{\rho}\right) \text{ for seawater} = 1.121 \times 10^{-5} \text{ ft}^2/\text{s}$$

Explain why you would not have dynamic similarity if the submarine prototype moved near the free surface.

8.53. A long cylinder is immersed in a large tank of liquid. The diameter of the cylinder is D and the viscosity of the liquid is μ. If the cylinder is spun slowly about its centerline at a speed ω rad/s, what dimensionless group or groups represent the torque per unit length T from viscous action? Suppose the data for a model of this system is known to be

$$D_M = 0.02 \text{ m}$$
$$\mu_M = 4.79 \times 10^{-4} \text{ N-s/m}^2$$
$$\omega_M = 3 \text{ r/min}$$
$$T_M = 0.2 \text{ N-m/m}$$

What will the torque per unit length T_P be for a prototype with the following data

$$D_P = 0.6 \text{ m}$$
$$\mu_P = 6 \times 10^{-4} \text{ N-s/m}^2$$
$$\omega_P = 0.2 \text{ r/min}$$

8.54. A centrifugal pump is rated at 50 m³/s at 1750 r/min with a head if 30 m. What is the flow rate and head if the speed is changed to 1250 r/min? See Prob. 8.43.

8.55. A pump on the earth's surface delivers 10 m³/s of water at 60°C while rotating at 1750 r/min. It has a head of 20 m and the diameter of the impeller is 0.4 m. On a space vehicle a geometrically similar pump $\frac{3}{4}$ the size pumps oil of kinematic viscosity 3×10^{-6} m²/s at a rotational speed of 1450 r/min. At what distance d from the earth's surface will there be possible dynamic similarity between space and earth pump flows? Determine the volume flow and head for the space pump. (The radius of the earth is 6372 km.) See Prob. 8.43.

8.56. A barge (Fig. P8.56) is towed at a speed V of 3 m/s. The width is 20 m. In a towing tank a model scaled down $\frac{1}{30}$ is being tested. What should the ratio of drags be if we duplicate wave drag with the idea that we will correct for skin

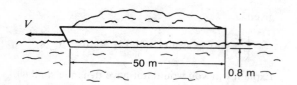

FIGURE P8.56

friction drag later? Take ρ of water in both cases as 1000 kg/m^3 and $\nu = 0.0113 \times 10^{-4}$ m^2/s for prototype and model.

8.57. A transport plane is expected to fly at 550 mi/h at an elevation of 30,000 ft standard atmosphere. A model of this plane scaled down to $\frac{1}{15}$ of the prototype is to be tested in a wind tunnel at a temperature of 70°F. To duplicate both *Reynolds* and *Mach* numbers, what is the tunnel *velocity* and the tunnel *pressure* absolute? Take μ_{air} for prototype as 2×10^{-7} lb · s/ft^2. Take μ_{air} for model as 4.2×10^{-7} lb · s/ft^2. The speed of sound for a perfect gas is $\sqrt{kRT}$.

8.58. A prototype of a boat of length 100 ft is to move at a speed of 10 kn in fresh water where $\mu = 2.10 \times 10^{-5}$ lbf-s/ft^2 and $\rho = 62.4$ lbm/ft^3. A model scaled down $\frac{1}{20}$ of the prototype is to be tested in a towing tank. For dynamic similarity what three dimensionless groups must be duplicated?
(*a*) Find proper V of model
(*b*) Find proper kinematic viscosity of liquid in the towing tank
(*c*) Find ratio of drags
Keep ρ the same for model and prototype.

8.59. The model of a submarine is scaled to $\frac{1}{30}$ of the full scale size of the prototype. The speed of the full scale submarine is to be 20 kn while at the free surface of sea water where

$$(\nu)_{\text{sea water}} = 1.210 \times 10^{-5} \text{ ft}^2/\text{s}$$

$$(\rho)_{\text{sea water}} = 1.940 \text{ lbm/ft}^3$$

In the towing tank, where $\nu = 1.217 \times 10^{-5}$ ft^2/s and $\rho = 1.938$ lbm/ft^3 what should be the free-stream velocity for movement at the free surface? What is the ratio of drags? Next, the submarine is considered to move much below the free surface at a speed of 0.5 kn. In a water tunnel, what

should be the speed and the ratio of drags? Take $\nu = 1.217 \times 10^{-5}$ ft^2/s again.

8.60. Wave motion along a section of coast is to be studied experimentally in the laboratory using a geometrically similar geometry reduced by a factor of 20. The density of ocean water is 1030 kg/m^3 and the laboratory fresh water is 1000 kg/m^3. If we neglect surface tension and friction, what is the wave velocity in the model if the wave velocity in the prototype is 0.15 m/s. What is the ratio of force between prototype and model for these flows?

8.61. A set of blades is used to mix crude oil in a large tank well below the free surface at a temperature of 20°C at an angular speed ω of 0.2 rad/s. A geometrically similar model of this device reduced by a scale factor of $\frac{1}{5}$ is run at a speed ω_M required for dynamic similarity in the large tank of water well below the free surface at a temperature of 60°C. If the model requires a torque of 0.4 N-m, what are the torque and power for the prototype?

8.62. We wish to determine the wind force on a water tower when a wind normal to the centerline of the water tower is 60 km/h. To do this we examine in a water tunnel a geometrically similar model reduced by $\frac{1}{20}$ scale. What should the water tunnel velocity be if the static temperatures of both prototype and model flows are the same, namely 60°C? What is the ratio of bending moments about the base of the building?

8.63. We wish to model an irrigation canal reduced by $\frac{1}{20}$ scale. Water is flowing in the canal at a speed of 1 m/s at a temperature of 30°C. If we are to duplicate *both* Reynolds and Froude numbers, what must be the kinematic viscosity of the model flow?

8.64. The model of an airfoil reduced to one-twentieth of the prototype is to be tested in a wind tunnel where the temperature is 70°F and the pressure is atmospheric. If the prototype is to fly at 500 mi/h at 5000 ft in the standard atmosphere, what should the velocity be in the wind tunnel for dynamic similarity considering only compressibility? What should the ratio of drags be for model to prototype? The velocity of sound in a perfect gas is $\sqrt{kRT}$, where for air $k = 1.4$, $R = 53.3$ ft · lb/(lbm)(°R), and T is the absolute temperature. Take ρ of air in the wind tunnel to be 0.002378 slug/ft^3.

8.65. In Prob. 8.64, consider the prototype to be moving at a speed of 150 mi/h at ground level. If we used the wind tunnel under conditions described above to measure the drag of the model, at what speed should the flow be to have dynamic similarity where viscous effects are significant? Considering the required velocity V_M, why is such a test not meaningful? Explain why for such tests one must have highly compressed air in the wind tunnel or use a water tunnel? Take ρ for model and prototype flows to be equal. Similarly, for μ.

8.66. We wish to use a model of an airfoil which is one-tenth the size of the prototype. The prototype is at a speed of 150 km/h in the process of landing where T of the air is 25°C. Because viscous effects are significant here, we will test the model in a water tunnel. What speed should we have if for the water the temperature is 50°C and the pressure is atmospheric at free stream? What is the ratio of lifts for the prototype to the model? At larger angles of attack, what must you be concerned with in this kind of a test?

8.67. Suppose in Probs. 8.9 and 8.16 that we exclude viscosity from the variables determining the power

required to drive a propeller. A model of a propeller which is 2 ft in length is scaled down to one-fifth of the full-scale propeller. If the model requires 5 hp, what is the power needed for the full-scale propeller rotating at a speed of 150 r/min. The full-size propeller is to operate at 30,000 ft in the standard atmosphere at a free-stream speed of 300 mi/h. What free-stream speed should we use for the model test? What is the angular speed for the model? Take $T_m = 59$°F.

8.68. In Example 8.3, determine the indicated dimensionless groups presented in the example. Consider the following characteristics of the model.

$$P_m = 5 \text{ kW}$$
$$Q_m = 5 \text{ L/s}$$
$$\Delta H_m = 2 \text{ m}$$
$$N_m = 900 \text{ r/min}$$
$$D_m = 800 \text{ mm}$$

If the full-scale pump is to deliver 50 kW of power at a speed of 400 r/min, what should the scale factor for the full-scale pump be? What are the head and the volumetric flow for the full-scale pump?

PART
II

ANALYSIS OF IMPORTANT INTERNAL FLOWS

In Part II of this text we will look into certain specific flows with the purpose of examining the details of each case utilizing the differential approach as well as experimental data. Particularly we will look into *internal flows* starting with incompressible flow through pipes both circular and noncircular. We will have to use certain ideas and concepts that will be shown later to be valid in a rigorous manner in the chapter following pipe flow wherein we develop Navier-Stokes equations and also solve certain internal flows using these equations. Next we go to compressible flow through ducts including separately variable cross-sectional area, friction in constant area ducts, and heat transfer, also in constant area ducts. We will, at times, be discussing in these chapters certain aspects of external flow around bodies, but we'll leave the detailed coverage of these flows to Part III of the text.

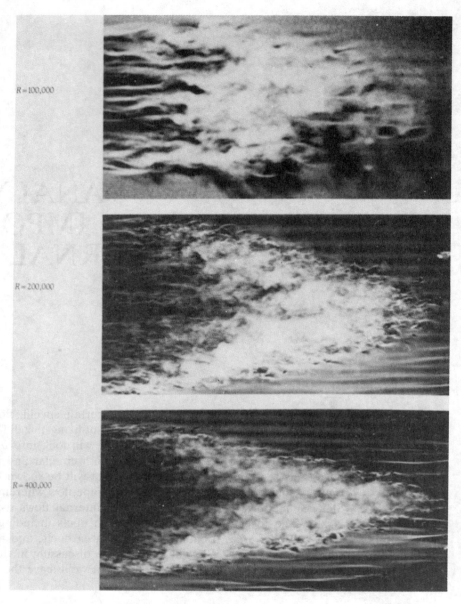

R = 100,000

R = 200,000

R = 400,000

Turbulence spots. (Courtesy Professor Robert Falco, Michigan State University.)

Here are shown pictures of a turbulence spot as seen looking down on a flat plate using smoke in air and flood lighting. The spot is shown at different Reynolds numbers. Transition from laminar to turbulent flow takes place intermittantly with random appearance of spots such as the ones shown above. A spot grows almost linearly with distance and has an arrowhead shape as can be seen above.

CHAPTER 9

INCOMPRESSIBLE VISCOUS FLOW THROUGH PIPES

9.1 INTRODUCTION

An important topic which we must discuss is concerned with the question of when viscous effects are significant to a degree which dictates a shift from irrotational flow to an approach taking friction into account. For this purpose it is useful to distinguish between two broad classes of flow. The flows around bodies such as airfoils, rockets, and surface vessels are termed *external flows* when other boundaries of the flow, such as the earth's surface, are comparatively distant from the body. On the other hand, flows which are enclosed by boundaries of interest will be termed *internal flows*. Examples of internal flow include the flow through pipes, ducts, and nozzles.

We start by considering the case of *external* flow. You will recall that Newton's viscosity law indicates that two primary factors are related to the shear stress in a flow. These are the viscosity of the fluid and the gradient (rate of change with respect to distance) of the velocity field. As a consequence, we pointed out in Chap. 4 that even for small viscosity, there will always be a thin region around a body where owing to the large velocity gradient brought about by the fluid "sticking" to the boundary, there will be significant shear stress. We called this region the boundary layer. On the other hand, if for the same body the flow is that of a fluid of very high viscosity, clearly we must consider viscous flow over much of the flow field and not just in the boundary layer (creeping flow). Thus, we can present only rough guides here, and the decision as to the essential nature of the flow in a region of interest must be made for individual problems based on the details of the problem and the accuracy required. In general, external gas flows such as are encountered in aeronautical engineering and external hydrodynamical flows such as are encountered in marine engineering can be considered frictionless except for the region of the boundary layer.

In considering *internal* flow we must still take into account the behavior of the boundary layer. At the entrance of a duct or pipe the boundary layer is generally very thin, so that in this region the flow can be considered as nonviscous except near the boundary.[1] However, farther along the flow there is a thickening of the boundary layer. In many flows the boundary layer may readily encompass the entire cross section of the flow. And when this occurs early in the flow, we generally consider the flow to be entirely viscous. The growth of the boundary layer is thus an important criterion. At this time, therefore, we can cite only simple extreme cases. For instance, except for extremely minute volume of flow, the flow in a capillary tube is generally considered entirely viscous even for low-viscosity fluids and short lengths of tube. Also the pipe flows involving the transport of oil and water can be considered after a few diameters of length from the entrance to be viscous flows. However, duct flow involving the movement of air over comparatively short distances such as in air-conditioning ducts and wind tunnels can generally be considered as frictionless flow except for the very important boundary-layer region.

In this chapter we consider pipe flows wherein viscous action can be considered to pervade the entire flow. Pipe flow is of great importance in our technology and will always be significant as long as we transport liquids. Also, much valuable experimentation has been performed on pipe flow which has rather general significance, as you will soon see.

Before setting up formulations for describing an action in nature, we must carefully observe the action. Our first step then is to examine flow in which viscous effects are significant.

9.2 LAMINAR AND TURBULENT FLOWS

The concept of laminar flow was introduced in Chap. 1 in connection with Newton's viscosity law. This flow was described as *a well-ordered pattern whereby fluid layers are assumed to slide over one another*. To illustrate this flow, let us examine the classic Reynolds experiment on viscous flow. Water is made to flow through a glass pipe as shown in Fig. 9.1, the velocity being controlled by an outlet valve. At the inlet of the pipe, a dye having the same specific weight as the water is injected into the flow. When the outlet valve is only slightly open, the dye will move through the glass pipe intact, forming a thread as illustrated in Fig. 9.1. The orderly nature of this flow is apparent from this demonstration.

---

[1] We will discuss pipe-entrance conditions in greater detail.

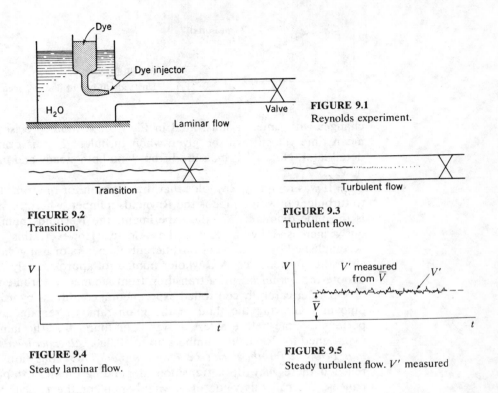

FIGURE 9.1
Reynolds experiment.

FIGURE 9.2
Transition.

FIGURE 9.3
Turbulent flow.

FIGURE 9.4
Steady laminar flow.

FIGURE 9.5
Steady turbulent flow. V' measured

However, as the valve is progressively opened, a condition will be reached whereby the dye assumes a fluctuating motion as it proceeds through the pipe (Fig. 9.2). A *transition* is taking place from the previous well-ordered flow, which may be considered as *laminar flow*, to an unstable type of flow. Further opening of the valve then results in a condition whereby irregular fluctuations are developed in the flow so that the thread of dye is completely dispersed before proceeding very far along the pipe (Fig. 9.3). This irregular flow is called *turbulent flow*.

The experiment brings out the essential difference between laminar and turbulent flow. The former, while having irregular molecular motions, is macroscopically a well-ordered flow. However, in the case of turbulent flow there is the effect of a small but macroscopic fluctuating velocity V' superposed on a well-ordered flow $\bar{V}$. A graph of velocity versus time at a given position in the pipe of the Reynolds apparatus would appear as in Fig. 9.4 for laminar flow and as in Fig. 9.5 for turbulent flow. In the graph for turbulent flow a mean time average velocity denoted as $\bar{V}$ has been indicated. Because this velocity is constant with time, the flow has been designated as steady. An unsteady turbulent flow may be considered one where the mean[2] time velocity field

---

[2]Henceforth, we shall for simplicity at times call the mean time average velocity simply the mean time velocity. And the term average velocity will indicate q/A.

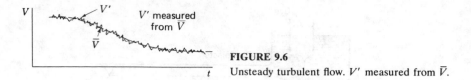

FIGURE 9.6
Unsteady turbulent flow. V' measured from $\bar{V}$.

changes with time, as illustrated in Fig. 9.6. A more precise definition of the mean time velocity will be given when turbulent flow is examined in greater detail later. Note V' is measured from $\bar{V}$ in both steady and unsteady turbulent flow.

It was found by Reynolds that the criterion for the transition from laminar to turbulent flow in a pipe is the Reynolds number, where the length parameter is the pipe diameter.[3] In the experiment, the Reynolds number was continuously increased by increasing the velocity. However, this could have been accomplished by using pipes of different diameters or using fluids with different viscosities or densities. A Reynolds number of approximately 2300 was found to denote the *imminence* of a transition from laminar to turbulent flow.

Under carefully controlled experimental conditions by using a very smooth pipe and allowing the fluid in the main tank to remain quiescent for long periods of time before a test, it was later found that the laminar flow can be maintained for Reynolds numbers up to 40,000. *All experiments thus far indicate that below* 2300 *there can be only laminar flow*. Thus, after 2300 has been reached there *may* be a transition, depending on the extent of local disturbances. We call this value of Reynolds number the *critical Reynolds number*. The flow corresponding to a Reynolds number exceeding 2300 may be considered unstable to an extent where a certain amount of disturbance will cause the random fluctuations of turbulent flow to appear. Below the critical Reynolds number the amount of damping present is sufficient to quell the effects of any local disturbance, so the flow is then always well ordered. In practical engineering pipe problems there is usually enough local disturbance present to cause turbulent flow to appear whenever the critical Reynolds number is exceeded.

The flow in the boundary layer also exhibits the two types of flow discussed above. It should be remembered, however, that the value of the aforementioned critical Reynolds number applies only to pipe flow; we have to make a separate study of transition conditions from laminar to turbulent flow for the boundary layer when we get to Chap. 13 on boundary-layer theory.

In the flows which we consider as frictionless, there may possibly be turbulence present. Thus there are times when free-stream turbulence in flows which are considered frictionless in usual computations must be taken into account, as you will see later when we consider the criteria for transition in the boundary layer in Chap. 13.

---

[3]We will be using various kinds of Reynolds numbers whose main difference will be the distance parameter. Consequently, we shall at times denote Reynolds number as Re_D with the D indicating that the pipe diameter is the length parameter.

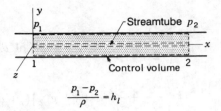

$$\frac{p_1 - p_2}{\rho} = h_l$$

FIGURE 9.7
Horizontal constant-area flow.

**PART B
LAMINAR FLOW**

9.3   FIRST LAW OF THERMODYNAMICS FOR PIPE FLOW; HEAD LOSS

In earlier studies in Chap. 6, we used the one-dimensional model for flow through ducts and pipes. In this section, we consider the thermodynamics of laminar, steady, incompressible flow through pipes with some care. We arrive at a one-dimensional equation as a result, which is consistent with our earlier less formal work.

We begin by considering steady, laminar, incompressible flow through a straight, horizontal pipe, as shown in Fig. 9.7. This is clearly a case of parallel flow. We have stated in Chap. 6 that for parallel, horizontal flow, the pressure will be *uniform* at a section except for a hydrostatic variation.[4] (This will be proved in Chap. 10.) Also we have shown in Sec. 7.8 from the continuity equation that the velocity profile of the flow must *not change* in the direction of flow. Before embarking on the details of this section, we wish to point out that although we are now concerned with laminar flow, we will later with some pertinent comments be able to extend the results of this section to turbulent flow.

We now examine pipe flows between sections 1 and 2 in Fig. 9.7. Consider this flow to be made up entirely of streamtubes one of which is shown. Consider next areas at the ends of a streamtube. These area elements will be referred to as *corresponding* area elements. Because the velocity profile does not change for parallel flow, we see that $V^2/2$ has the same value at corresponding area elements. Also there is zero change in elevation, $y_2 - y_1$, between corresponding elements of area. Finally, it can be assumed for pipe flow that the internal energy u is constant over a cross section. Now the *first law of thermodynamics* for the control volume consisting of the interior region of the pipe between sections 1 and 2, and thus including all the enclosed streamtubes, can be given

---

[4]Since there is no acceleration in the vertical direction for horizontal parallel flow, one may intuitively accept the fact that the pressure varies hydrostatically in the vertical direction.

as

$$\frac{dQ}{dt} = - \iint_{A_1} \left(\frac{V_1^2}{2} + \frac{p_1}{\rho} + u_1 + gy_1 \right) (\rho V \, dA)$$

$$+ \iint_{A_2} \left(\frac{V_2^2}{2} + \frac{p_2}{\rho} + u_2 + gy_2 \right) (\rho V \, dA)$$

where we have used p/ρ instead of pv. This may be rewritten as:

$$\frac{dQ}{dt} = \iint_{A} \left[\left(\frac{V_2^2}{2} - \frac{V_1^2}{2} \right) + \frac{p_2 - p_1}{\rho} + (u_2 - u_1) + g(y_2 - y_1) \right] (\rho V \, dA)$$

$$(9.1)$$

The first and last expressions in parentheses in the integrand are clearly zero on considering corresponding area elements at sections 1 and 2 of each streamtube. Note that the hydrostatic pressure contributions cancel at corresponding area elements of streamtubes, leaving uniform pressures which we will denote simply as p_1 and p_2 at the respective sections. The two expressions that are left are constant for the integration over the cross section of the pipe. Thus we have

$$\frac{dQ}{dt} = \left[\frac{p_2 - p_1}{\rho} + u_2 - u_1 \right] \iint \rho V \, dA$$

$$= \left[\frac{p_2 - p_1}{\rho} + (u_2 - u_1) \right] \frac{dm}{dt}$$

Dividing through by the mass flow, dm/dt, we get on rearranging terms

$$\frac{p_1 - p_2}{\rho} = \frac{\Delta p}{\rho} = \left[-\frac{dQ}{dm} + (u_2 - u_1) \right] \qquad (9.2)$$

The expression on the right side of Eq. (9.2) represents the loss of "mechanical energy" per unit mass of fluid flowing. Why is this energy considered a loss? First note that it represents the increase in internal energy of the fluid and the heat transfer from the fluid inside the control volume to the surroundings. In practical situations where we are transporting fluid, any increase in internal energy is of little use, since it is usually lost in subsequent storage, and the contribution to heating up surroundings, particularly the atmosphere, is certainly usually not economically desirable. We therefore group these terms together, calling the combination the *head loss*, which as in Chap. 7 we denote as h_l. Thus

$$\boxed{h_l = (u_2 - u_1) - \frac{dQ}{dm}} \qquad (9.3)$$

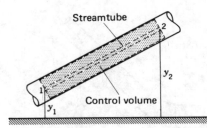

FIGURE 9.8
Inclined constant-area flow.

We then can say on going back to Eq. (9.2),

$$\frac{\Delta p}{\rho} = h_l$$

If there were no friction in this flow, Newton's second law tells us that there would be no change in pressure, so we can interpret the head loss h_l as we did in Chap. 7 as the loss in pressure per unit mass, $\Delta p/\rho$, *due to friction*.

The head loss is *also* defined as h_l/g, which is the loss in pressure due to friction per unit *weight* (rather than per unit mass). The dimension simplifies to length L in this case and it is a form of "head" as discussed in Sec. 6.6. We denote this head loss as H_l. Thus[5]

$$H_l = \frac{h_l}{g} \tag{9.4}$$

The definition H_l has the advantage of simple dimension, which is a length—the kind of measurement given directly by manometers. Many engineers use this formulation along with the *head* terms of kinetic energy head $V^2/2g$, potential energy head y, and pressure head $\Delta p/\gamma$.

However, the head loss H_l defined in this manner is *not* a unique quantity for a particular flow. Since it is based on weight it must depend on the location of the flow in any gravitational field. The formulation h_l on the other hand permits the head loss to be the same for identical flows on earth or on a distant space vehicle. We will use either definition as the occasion merits.

Let us next consider the parallel flow under discussion as not parallel to the earth's surface. We return to Eq. (9.2) for the control volume in the pipe between 1 and 2 as shown in Fig. 9.8. Now $g(y_2 - y_1)$ at corresponding area elements at sections 1 and 2 is *not* zero as before but is instead a nonzero constant for *all* pairs of corresponding area elements (i.e., at the ends of

---

[5]Please do not confuse the head loss H_l, which is a *loss in pressure* per unit weight as a result of friction, with the head H_D of Chap. 6. Recall that $H_D = (V^2/2g + p/\gamma + z)$ and is the *mechanical energy* per unit weight of flow.

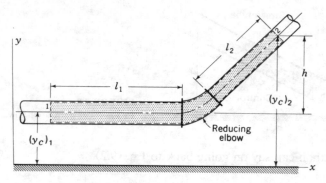

FIGURE 9.9
Pipe flow with area and elevation change.

streamtubes). Equation (9.1) then degenerates to the following equation

$$\frac{\Delta p}{\rho} = g(y_2 - y_1) + \left[(u_2 - u_1) - \frac{dQ}{dm} \right]$$

$$= g(y_2 - y_1) + h_l \qquad (9.5)$$

The pressure drop is due now to an *elevation* head *and* the *head loss* due to friction. It is important to note that the head loss is dependent on the velocity profile, the type of fluid, and sometimes the roughness of the pipe surface. Thus, for head loss the inclination of the pipe is of no consequence. Hence, the head loss may be determined independently of the pipe orientation—i.e., it is still the pressure loss due to friction divided by ρ.

Next, consider the flow in pipes of different diameters connected by a reducing elbow as shown in Fig. 9.9. For the indicated control volume, we may again use Eq. (9.1) for the first law of thermodynamics. However, now the term $g(y_2 - y_1)$ for corresponding points at sections 1 and 2 (ends of a streamtube) is not a constant value for all sets of corresponding points. However, the *average* value of $g(y_2 - y_1)$ for the totality of corresponding points will be equal to the value $g[(y_c)_2 - (y_c)_1]$, where the y_c's are measured to the respective centers of the cross sections. Similarly, $V_2^2/2 - V_1^2/2$ will not be constant for correspond-ing points at sections 1 and 2.[6] It is the practice, nevertheless, to use at each section the velocity $V = $ volume flow$/A = q/A$ for the velocity at each point in the section and consequently to take $V_2^2/2 - V_1^2/2$ as a constant in Eq. (9.1) for all streamtubes.[7] It might appear that appreciable error can be incurred by such an improper averaging process. However, in laminar flow the velocity-head term will be a small quantity when compared with the other terms of the

---

[6]Near the boundary at both sections the velocity will be close to zero so that $(V_2^2/2 - V_1^2/2)$ is a very small number. However, at the centerline the velocity at station 2 will be comparatively large and will exceed that at 1 so that $(V_2^2/2 - V_1^2/2)$ for that streamtube may be appreciably larger than zero.

[7]Actually we should average the *squares* of the velocity at a cross section to be correct.

equation so that the error incurred in the computations of pressure change will be small. In the case of turbulent flow, the velocity profile is much more uniform than that of laminar flow. Here then, even though larger velocity-head terms may be present, the basic error is considerably less, so that it would seem permissible in almost all turbulent pipe flows to proceed in the simpler manner. Also, note that experimental information must often be used to ascertain head loss. The amount of error inherent in these computations often exceeds the errors incurred by an incorrect averaging process as outlined. We then reach the following result analogous to Eq. (9.5):

$$\frac{\Delta p}{\rho} = \left(\frac{V_2^2}{2} - \frac{V_1^2}{2} \right) + g(y_2 - y_1) + (h_l)_T \tag{9.6}$$

where $(h_l)_T$ is the sum of the head losses in each pipe plus that occurring in the elbow. We usually distinguish the various head-loss contributions separately. Those occurring in fittings such as elbows are called *minor losses* and are denoted as $(h_l)_M$. In this case, we would say

$$(h_l)_T = (h_l)_1 + (h_l)_M + (h_l)_2 \tag{9.7}$$

where $(h_l)_1 =$ head loss in the horizontal pipe
$(h_l)_2 =$ head loss in the inclined pipe
$(h_l)_M =$ head loss in the reducing elbow

On rearranging terms of Eq. (9.6), we find that the *first law of thermodynamics* may then have the following form:

$$\left(\frac{V_1^2}{2} + \frac{p_1}{\rho} + gy_1 \right) = \left(\frac{V_2^2}{2} + \frac{p_2}{\rho} + gy_2 \right) + (h_l)_T \tag{9.8}$$

We often call this equation the *modified Bernoulli equation*. This equation can be applied to *any series of straight pipes interconnected by various kinds of connecting fittings*. In this equation we are maintaining the density constant but are accounting for changes in internal energy and heat transfer. When is such a model useful in portraying physical situations? Consider pipes carrying liquids such as water or oil over comparatively long distances. The liquids tend to remain at constant temperatures close to that of the surroundings during such long trips, and consequently, with even high-pressure changes present, the density remains essentially constant. Although the heat-transfer rate may be very small over any small section of pipe, keep in mind that dQ/dm is the *total* heat transfer from a unit mass of fluid as it moves over the *entire* distance between sections 1 and 2 of the pipe. Thus, even though a small temperature difference may be present between the fluid surroundings, the accumulated heat transfer per unit mass over long distances may be considerable. Furthermore, changes in internal energy due to pressure change and whatever small temperature differences there may be present are also taken into account by the head-loss term. All told, the head loss may hence have a very large value. Thus, you see that in such cases there is no inconsistency in talking about *incompressible flows having appreciable head loss*. Chemical engineers, mechanical engineers, and particularly civil engineers must often consider long pipelines transporting liquids, and it is for such problems that the formulations presented in this section are most useful. The general procedure in such problems is to ascertain the head loss by using theory, when possible, as illustrated in the following section, or by using experimental data. This is then inserted in the modified Bernoulli equation (or the first law of thermodynamics), which, when used with continuity equations, is then usually sufficient to solve for the desired unknowns. In the following section we will be able to determine the head loss analytically for pipes.

Note that since we are dealing with incompressible flows the portion p_{atm}/γ of the total pressure head will be present on both sides of the simple energy equation and the modified Bernoulli equation and so can be canceled, thus allowing us to use *gage pressure*.

9.4   LAMINAR FLOW PIPE PROBLEM

In Sec. 7.8 we studied flow in a pipe undergoing fully developed laminar flow and arrived at two key results, the velocity profile and the head-loss formula. We rewrite then here:

or

$$V = \frac{\beta}{4\mu}\left(\frac{D^2}{4} - r^2\right) \qquad \beta = \frac{128q\mu}{\pi D^4}$$

$$V = \frac{p_1 - p_2}{4\mu L}\left(\frac{D^2}{4} - r^2\right)$$

(9.9)

and

$$h_l = \frac{128q\mu L}{\pi D^4 \rho}$$

(9.10)

Before proceeding to problems, we must emphasize the difference between *hydrostatics* and *steady flow*. It will be vital in ensuing calculations that we clearly distinguish between these conditions. For *hydrostatics* the fluid particles must be *stationary* relative to some *inertial reference*. For *steady flow* the fluid particles passing any fixed point in a reference must have all properties and kinematic variables *constant* with respect to time at the point. The particles under this condition may accelerate at a point. But each particle at that point must have the same acceleration at any time. Thus looking at a flow from a large reservoir into a pipe (see Fig. 9.10), the flow is *steady* if the height of the free surface is kept constant. We *do not* have a hydrostatic state here because the fluid particles are *accelerating* as they approach the pipe inlet.

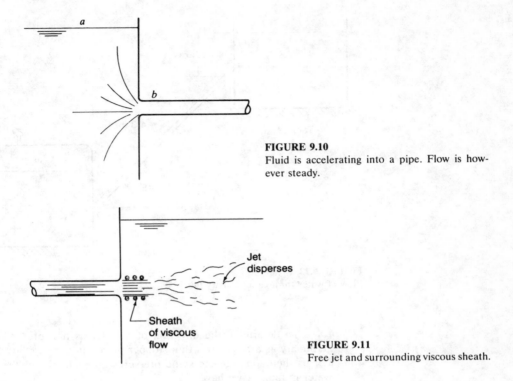

FIGURE 9.10
Fluid is accelerating into a pipe. Flow is however steady.

FIGURE 9.11
Free jet and surrounding viscous sheath.

Next consider a subsonic jet of water issuing into a tank of water as shown in Fig. 9.11. At the entrance to the tank and a short distance beyond, the jet is intact with parallel flow as shown in the diagram. A thin sheath of *viscous* flow having small vortices will be found surrounding the intact jet. This is the darkened region shown in the diagram, and beyond the sheath but near the wall, we have fairly static conditions. Note away from the entrance region the jet disperses into an irregular pattern. Because of the parallel flow in the jet and the static conditions directly outside the sheath, we can use *hydrostatic pressure* of the *surrounding fluid* as the *pressure in the emerging jet*. However, we *cannot* use the *Bernoulli* equation between the emerging jet and the free surface because we would pass through the sheath where there is significant friction. This friction invalidates the use of the Bernoulli equation. We call this emerging fluid into the tank a *free jet*. As we will learn later, a *supersonic* jet need not have the same pressure of the surrounding fluid in contrast to a subsonic free jet.

Example 9.1. A capillary tube of inside diameter 6 mm connects closed tank A and open tank B as is shown in Fig. 9.12. The liquid in A, B, and capillary CD is water having a specific weight of 9780 N/m^3 and a viscosity of 0.0008 kg/m · s. The pressure $p_A = 34.5$ kPa gage. Which direction will the water flow? What is the volume of flow q? Neglect losses at C and D.

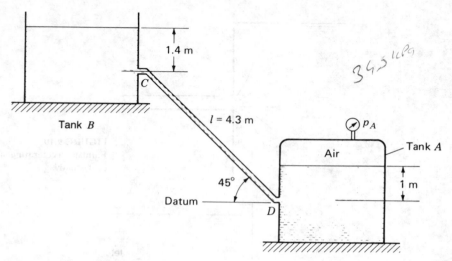

FIGURE 9.12
Flow of water through a capillary tube.

To determine the direction of flow let us first pretend that the fluids are stationary as a result of a thin stopper placed in the capillary tube at position D. Let us then compute the static pressures on each face of the stopper. From the water in tank A we have

$$p_1 = p_A + p_{atm} + (\gamma)(1)$$

$$= 34{,}500 + p_{atm} + (9780)(1)$$

$$= 44{,}280 + p_{atm} \quad \text{Pa} \qquad (a)$$

On the other side of the stopper, we have

$$p_2 = p_{atm} + \gamma[1.4 + (4.3)(0.707)]$$

$$= p_{atm} + (9780)(4.44) = 43{,}420 + p_{atm} \quad \text{Pa} \qquad (b)$$

Thus, the water must flow from tank A to tank B in the absence of the stopper.

We will use the *modified Bernoulli* equation inside the capillary between D and C. Assume that we have laminar flow for the moment and use Eq. (9.10) for the head-loss term in the tube. Thus we have

$$\frac{V_D^2}{2} + \frac{p_D}{\rho} + gy_D = \frac{V_C^2}{2} + \frac{p_C}{\rho} + gy_C + \frac{128\mu Lq}{\pi\rho D^4} \qquad (c)$$

The velocity terms will cancel because of *continuity*. We can compute pressure p_D

by considering frictionless incompressible flow in tank A.[8] Thus, we can use *Bernoulli* between the free surface and the entrance at D. We accordingly have on using gage pressures and using D as a datum:

$$\frac{V_{f.s.}^2}{2} + \frac{p_A}{\rho} + g(1) = \frac{V^2}{2} + \frac{p_D}{\rho} + 0 \tag{d}$$

We neglect the kinetic energy at the free surface, and we have for p_D/ρ using gage pressures

$$\frac{p_D}{\rho} = -\frac{V^2}{2} + \frac{34,500}{9780/g} + (9.81)(1)$$

$$= -\frac{V^2}{2} + 44.4 \quad \text{N} \cdot \text{m/kg} \tag{e}$$

We consider at C that the liquid emerges as a *free jet* and the pressure p_C is taken as the *hydrostatic* pressure in the tank at the depth of the free jet.

$$p_C = (\gamma)(1.4) = (9780)(1.4) = 13,960 \text{ Pa} \tag{f}$$

Now going back to Eq. (c) and replacing q by $[V\pi D^2/4]$, we get on using Eqs. (e) and (f),

$$\left(-\frac{V^2}{2} + 44.4\right) + g(0) = \frac{13,690}{9780/g} + (g)(0.707)(4.3)$$

$$+ \frac{(128)(\mu)(L)(V)(\pi D^2/4)}{\pi\rho D^4}$$

Inserting known numerical values, we get for the equation above,

$$-\frac{V^2}{2} + 44.4 = 13.73 + 29.8 + \frac{(128)(0.0008)(4.3)V(\pi)(0.006)^2/4}{\pi(9780/g)(0.006^4)}$$

We thus have a quadratic equation for V:

$$V^2 + 6.13V - 1.740 = 0$$

Solving for V,

$$V = \frac{-6.13 + \sqrt{6.13^2 + (4)(1)(1.740)}}{2}$$

$$= 0.2718 \text{ m/s}$$

---

[8]This can be done because the walls of the tank are far from the main bulk of flow in the tank and we are neglecting friction losses at the entrance D. Therefore, the velocity gradients are small everywhere except directly at the tank boundary. Also, we remind you that the modified Bernoulli equation is only valid for flows in pipes or tubes connected by fittings with no pumps or turbines in the pipes considered in the equation.

Now let us compute the Reynolds number, Re_D:

$$\text{Re}_D = \frac{\rho V D}{\mu} = \frac{(9780/g)(0.2718)(0.006)}{(0.0008)}$$

$$= 2032$$

Our assumption of laminar flow is thus justified.[9] The flow q can now be given

$$q = \frac{(0.2718)(\pi)(0.006^2)}{4} = 7.685 \times 10^{-6} \text{ m}^3/\text{s}$$

$$= 7.685 \times 10^{-3} \text{ L/s}$$

9.5 PIPE-ENTRANCE CONDITIONS

In the previous discussions on pipe flow, attention was centered on *fully developed* laminar flow ($\text{Re}_D \leq 2300$), a condition that is reached when viscous action has pervaded the entire cross section of flow so as to result in an invariant velocity profile in the direction of flow and straight, parallel streamlines. Directly after a pipe entrance, these conditions are not present, as is illustrated in Fig. 9.13 wherein velocity profiles have been shown downstream of a well-rounded pipe entrance. Note that the flow is initially almost uniform. Further on, this uniform region of the profile, sometimes called the core, shrinks as viscous effects extend deeper into the flow. Finally, a fully developed laminar flow is established assuming $\text{Re}_D \leq 2300$ for which the formulations of Sec. 9.4 are valid. A convenient viewpoint is to consider that a laminar or turbulent boundary layer is formed at the entrance and thickens downstream until the entire cross section is occupied by this layer. This is illustrated in Fig. 9.13 where a laminar boundary layer is shown darkened.

The distance L' from the entrance to the position in the pipe for fully developed laminar flow has been determined to be given by the following formula[10]:

$$L' = 0.058\text{Re}_D D \tag{9.11}$$

---

[9]In Prob. 9.51 we present a problem wherein this reasoning fails; that is, using Moody's curves (to be studied later) a laminar flow and a turbulent flow can be deduced, resulting in two different solutions.

[10]H. L. Langhaar, "Steady Flow in the Transition Length of a Straight Tube," *J. Appl. Mech.*, vol. 9, 1942.

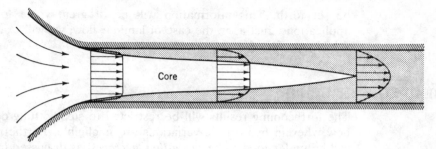

FIGURE 9.13
Boundary-layer growth at pipe entrance.

For turbulent flow ($\text{Re}_D > 2300$) we have for L' the following approximate result:

$$L' = 4.4\text{Re}_D^{1/6}D \tag{9.12}$$

The pipe-entrance region is not one of a parallel flow in the strict sense, since streamlines will have curvature along the interface between the uniform core and the boundary layer. Hence one cannot properly apply the results which stem from a parallel-flow restriction. However, in most problems this region is short compared with the pipe distances wherein fully developed flow is present. Consequently such items as head loss for the pipe are computed on the basis that the entire pipe length has fully developed laminar or turbulent flow. Also a minor loss is used at the entrance. We will consider this later.

**PART C
TURBULENT FLOW:
EXPERIMENTAL CONSIDERATIONS**

9.6 PRELIMINARY NOTE

It was indicated at the outset of this chapter that turbulent flow can be considered as the superposition of a random fluctuating flow over a well-ordered flow. Unfortunately, the nature of the fluctuating component is little understood, so that no adequate theory has yet been formulated for analyzing turbulent flow. Hypotheses have been set forth with some success, but these are useful only in limited areas of application, and all require certain experimentally derived information. More will be said about such undertakings later. At this time, with the help of dimensional analysis, significant experimental results will be set forth. This information will be of great value for engineering pipe applications. Just as in the case of laminar flow, we will examine head loss and velocity profiles.

9.7 HEAD LOSS IN A PIPE

The forthcoming results will be restricted to steady, fully developed, turbulent flow wherein hydrostatic variations are negligible. Furthermore, all quantities will be understood to be *mean-time averages*, as discussed in Sec. 9.2.

It will be shown in Chap. 10 that the mean-time averages[11] of flow parameters and properties in turbulent flow behave as if the flow were laminar except for the presence of additional stresses, called *apparent stresses*, which encompass the effects of turbulence. The apparent stresses in turbulent flow far exceed the ordinary viscous stress present, so the latter is usually neglected in turbulent flows except near the boundary where viscous stress predominates. Apparent stresses do not prevent the extension of the general parallel-flow conclusions of Secs. 9.3 and 9.4 to the mean-time-average quantities in steady parallel turbulent flow, so we will continue to consider that the mean-time-average profile remains fixed in the direction of flow and will continue to use uniform mean-time-average pressure p at sections of the pipe.

It is known that pressure changes Δp along a pipe in turbulent flow depend on the following quantities:

1. D, pipe diameter
2. L, length of pipe over which the pressure change is to be determined
3. μ, the familiar coefficient of viscosity
4. V, the average, over a cross section, of the mean-time-average velocity, which is equivalent to q/A
5. ρ, mass density
6. e, the average variation in pipe radius—a measure of pipe roughness

In functional notation this becomes

$$\Delta p = f(D, L, \mu, V, \rho, e)$$

In Sec. 8.5, a dimensional analysis was carried out for the variables above. The

---

[11]The mean-time-average velocity $\overline{V}$ is given as

$$\overline{V} = \frac{1}{\Delta t} \int_0^{\Delta t} V \, dt$$

where Δt is large enough in steady flow to make $\overline{V}$ be independent of time.

result is given by the following relation involving four dimensionless groups:

$$\frac{\Delta p}{\rho V^2} = G\left(\frac{\rho VD}{\mu}, \frac{L}{D}, \frac{e}{D}\right)$$

It is intuitively clear that the pressure change Δp is directly proportional to the pipe length L. We may then simplify the equation above to include this relationship in the following manner:

$$\frac{\Delta p}{\rho V^2} = \frac{L}{D} H\left(\frac{\rho VD}{\mu}, \frac{e}{D}\right)$$

The unknown function G of three π's has been replaced by an unknown function H of only two π's. Replacing Δp by $h_l\rho$ because we have horizontal, parallel, mean-time-average flow and rearranging terms, we get

$$h_l = \frac{V^2}{2}\frac{L}{D}\left[K\left(\frac{\rho VD}{\mu}, \frac{e}{D}\right)\right]$$

The number 2 has been inserted to form the familiar kinetic-energy term. This is permissible, since there is as yet an undetermined function in the equation. Finally, it is the practice to call the unknown function $K(\rho VD/\mu, e/D)$ the *friction factor*. Upon using the notation f for this term, the final form of the dimensional considerations becomes the *Darcy-Weisbach* formula:

$$\boxed{h_l = f\frac{L}{D}\frac{V^2}{2}} \tag{9.13}$$

The term f is determined by experiment, so that the modified Bernoulli equation, using mean-time averages, is satisfied. In the plot of f versus Re_D shown in Fig. 9.14 for various conditions of roughness, the data due to Nikuradse have been employed. The pipes in these experiments were artificially roughened by gluing sand of various degrees of coarseness and in varying degrees of spacing on the pipe walls. Note that the data cover both the laminar and the turbulent ranges. For Reynolds numbers under 2300 there is indicated a simple relation between the friction factor and Reynolds number *completely independent of roughness*. It is a simple matter to compute the relation between f and Re_D from the theoretical work of the preceding sections for the range of laminar flow and so have a check between theory and experiment. To accomplish this, substitute the theoretical expression for head loss for laminar flow [Eq. (9.10)] into Eq. (9.13). We get

$$\frac{128q\mu L}{\pi D^4 \rho} = \frac{V^2}{2}\frac{L}{D}f$$

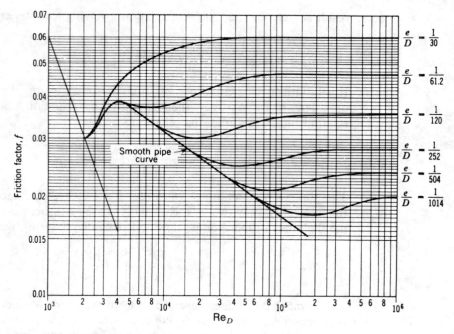

FIGURE 9.14
Nikuradse's data from artificially roughened pipe flows.

Replace q by $V(\pi D^2/4)$, and solve for f.

$$f = \frac{64}{\rho V D / \mu} = \frac{64}{\text{Re}_D} \tag{9.14}$$

Theory indicates that f versus Re_D forms a rectangular hyperbola. On logarithmic paper such as in Fig. 9.14 this becomes a straight line. The excellent agreement between the experimental results and the theoretical curve is seen by checking a couple of points.

We can easily relate f to the shear stress at the wall τ_w by considering as a system a chunk of fluid forming a cylinder of length L and diameter D inside the pipe at time t (see Fig. 9.15).

Since the mean-time-average velocity of all fluid elements of this system is constant, we can conclude that the forces in the direction of flow from the

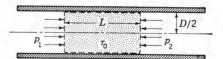

FIGURE 9.15
System of fluid in pipe.

apparent stresses are in equilibrium. Equating these forces, we get

$$\Delta p \frac{\pi D^2}{4} = \tau_w \pi D L \tag{9.15}$$

Using the definition of the head loss, we can replace Δp by $h_l \rho$, and additionally, using the friction factor, we can replace $h_l \rho$ by $f(L/D)(V^2/2)\rho$. Employing this result in the equation above and solving for f, we get

$$f = \frac{8\tau_w}{\rho V^2} \tag{9.16}$$

Solving for τ_w we also have the useful formula

$$\tau_w = \frac{f}{8}(\rho V^2) \tag{9.17}$$

Let us now turn to the turbulent part of the plot. Note that directly after the critical Reynolds number all roughness curves are coincident with the smooth-pipe curve. Later, each curve "peels off" from the smooth-pipe curve in a sequence such that the greater the roughness, the earlier the departure. That part of any curve coincident with the smooth-pipe curve is called the *smooth-pipe zone of flow*. Note, furthermore, that after passing the smooth-pipe zone, each curve eventually flattens out to a straight line parallel to the abscissa. Here the friction factor f is independent of Reynolds number. This region for each curve is called the *rough-pipe zone of flow*, while the region between the two extremes is called the *frictional transition zone*. Thus, each curve except that of the smooth pipe goes through three zones of flow, the position and extent of each zone depending on the roughness of the pipe.

The behavior of the friction curves in the three zones of flow may be explained in a qualitative manner. From a physical viewpoint, it is necessary that the macroscopic flow fluctuations of turbulent flow subside considerably as one approaches a solid boundary. Hence, adjacent to each boundary there is a thin layer of flow called the *viscous sublayer* where viscous effects predominate over turbulence effects.[12] In the case of a pipe, the greater the Reynolds number the thinner the sublayer, since higher Reynolds numbers mean stronger velocity fluctuations and consequently greater penetration of turbulence toward a boundary. In the smooth-pipe zone of a curve in the preceding plot, the sublayer thickness is large enough to exceed e, the average height of sand. It has already been noted that roughness has no effect on the head loss for laminar flow. Since the region of flow to which the sand particles are exposed is *viscous*

---

[12] In the literature, the viscous sublayer is often called the laminar sublayer. This gives the impression that the flow is free of fluctuations in this region. Actually this is not the case—there will be some velocity fluctuations. However, the fluctuations do not contribute appreciably to stress compared with the contribution of viscosity. For this reason, the region is termed the viscous sublayer rather than the laminar sublayer.

rather than *turbulent*, it then becomes clear that all friction curves in the smooth-pipe zone must coincide. At higher Reynolds numbers, the sublayer thickness decreases to expose the grains of sand to the turbulent flow outside the viscous sublayer. Naturally, the pipes of greater roughness undergo this exposure at smaller Reynolds numbers. When the rough-pipe zone is reached, much of the rough surface is exposed to the turbulent flow outside the sublayer. In this region, larger velocities predominate. For such velocities, the resistance to flow is primarily that of *wave drag*.[13] This type of resistance is due primarily to normal stresses distributed along the boundary in such a manner as to resist flow, whereas in smooth-pipe flow the resistance is a result primarily of shear stresses. It is known that wave drag for velocities such as are encountered in pipe flow is proportional to the square of the average velocity q/A. Hence, the pressure drop is proportional to the velocity squared. Since the head-loss formula [Eq. (9.13)] is already in a form proportional to the square of this velocity, it is clear that the friction factor must be a constant. Finally, the transition zone is explained as involving both viscous and wave-drag effects to varying degrees.

It must be remembered that the Nikuradse data have been developed for artificial conditions of roughness. There is the question of how well this type of roughness approximates actual conditions of roughness as found in real situations. Moody has made an extensive study of commercial piping data in order to alter the previous plot to be of use in practical problems. His plot is shown in Fig. 9.16 along with Table 9.1, from which the pertinent roughness coefficient e may readily be found for numerous surfaces. The use of these charts will be demonstrated in Sec. 9.10. Note at this time that for Reynolds numbers from 2000 to 4000 the curves are dashed, signifying that the data here for f are not accurate. This is further emphasized by the shading inserted to cover this portion of the Moody data.

In this chapter, we will use Moody's chart for the solving of pipe problems. We shall also be able to use the Moody diagram for channel flow and for compressible flow in ducts. If we consider numerical methods suitable for use with a digital computer, it is desirable to have a mathematical formulation of f in terms of e/D and Re_D. There are different semiempirical correlation formulas relating f with e/D and Re_D. The best-known formula is the Colebrook formula, which for the *frictional transition* zone is as follows:

$$\frac{1}{\sqrt{f}} = 1.14 - 2.0 \log\left[\frac{e}{D} + \frac{9.35}{\mathrm{Re}_D\sqrt{f}}\right] \qquad (9.18)$$

where you must note that log is to the base 10 (ln refers to base e). This formula has the disadvantage that f does not appear explicitly and for a given e/D and Re_D must be solved by iteration. In the *completely rough zone* of flow, we know

---

[13] Wave drag will be discussed in more detail in Chap. 12.

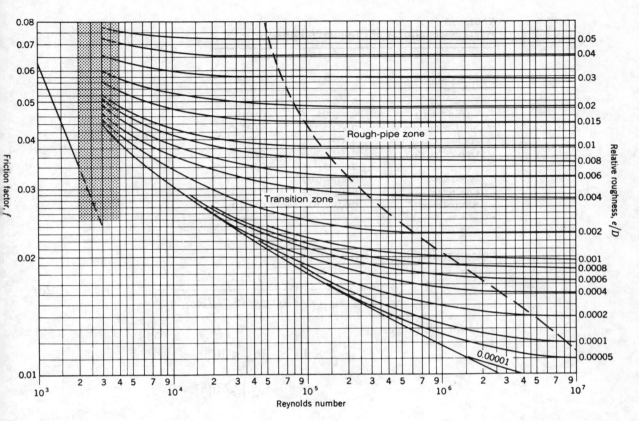

FIGURE 9.16
Friction factors for flow in pipes.

that f does not depend on the Reynolds number, so considering Eq. (9.18), it means that $e/D \gg 9.35/(Re_D\sqrt{f})$. The formula for f can then be given explicitly as follows for the *completely rough* zone of flow:

$$f = \frac{1}{\left[1.14 - 2.0\log(e/D)\right]^2} \tag{9.19}$$

We continue by presenting another more-recent formula[14] valid for certain large ranges of e/D and Re_D which span much of the frictional transition zone as well as the rough-pipe zone. Furthermore, the formula has the virtue of

---

[14] P. K. Swamee and A. K. Jain, "Explicit Equations for Pipe-Flow Problems," *J. Hydraulic Div. Proc. ASCE*, pp. 657–664, May 1976.

TABLE 9.1
Average roughness of commercial pipes

Material (new)	e	
	ft	mm
Glass	0.000001	0.0003
Drawn tubing	0.000005	0.0015
Steel, wrought iron	0.00015	0.046
Asphalted cast iron	0.0004	0.12
Galvanized iron	0.0005	0.15
Cast iron	0.00085	0.26
Wood stave	0.0006–0.003	0.18–0.9
Concrete	0.001–0.01	0.3–3.0
Riveted steel	0.003–0.03	0.9–9.0

being explicit in f. We have

$$f = \frac{0.25}{\left\{ \log\left[(e/3.7D) + (5.74/\mathrm{Re}_D^{0.9}) \right] \right\}^2} \tag{9.20}$$

valid for the ranges

$$5 \times 10^3 \le \mathrm{Re}_D \le 10^8$$

$$10^{-6} \le \left(\frac{e}{D} \right) \le 10^{-2}$$

Finally, for the *hydraulically smooth* zone of turbulent pipe flow we have an empirical formula developed by Blasius and valid for $\mathrm{Re}_D \le 100,000$.

$$f = \frac{0.3164}{\mathrm{Re}_D^{1/4}} \tag{9.21}$$

In a homework assignment (Prob. 9.29), you will be asked to compare formulations for f using these semiempirical formulas as well as the Moody diagram. You will find excellent correlation between the various methods of calculating f.

9.8 VELOCITY PROFILE AND WALL SHEAR STRESS FOR TURBULENT FLOW

Nikuradse has made an extensive experimental study of velocity profiles for turbulent flow at relatively *low* Reynolds numbers. A few profiles from his data are shown in Fig. 9.17 in the case of a *smooth* pipe. Each curve corresponds to a certain Reynolds number. The symbol y represents the radial distance from the pipe wall. It is possible to represent these curves empirically away from the

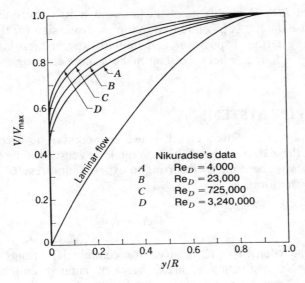

FIGURE 9.17
Velocity profiles for laminar and turbulent flow.

wall by the following relation for Reynolds numbers under 3×10^6 approximately:

$$\frac{\overline{V}}{\overline{V}_{\max}} = \left(\frac{y}{D/2} \right)^{1/n} \tag{9.22}$$

The exponent n varies with the Reynolds number, ranging from 6 to 10 for Reynolds numbers from 4000 to 3.240×10^6. The value of 7 is often employed for n thus giving the name *seventh-root law* to this formulation.

A laminar profile has been included in Fig. 9.17 for purposes of comparison. Note that the turbulent profile has a *steeper slope* near the wall than the laminar profile. This is true also in the case of laminar and turbulent boundary layers. As will be shown in Chap. 13, this is an important factor when it comes to consideration of flow separation for immersed bodies.

In the case of a *rough* pipe, one may approximate the profile in a similar manner for the range of Reynolds numbers of the preceding discussion, although with less accuracy.

It is to be pointed out, before leaving this section, that for smooth pipes at low Reynolds numbers below 3×10^6 Blasius has developed from empirical studies a very useful formula for shear stress at the wall of a pipe. This is called the Blasius friction formula and is given for English units as

$$\tau_W = 0.03325 \rho V^2 \left(\frac{\nu}{RV} \right)^{1/4} \tag{9.23}$$

where $V = q/A$ here. We will have occasion to use this formulation in connection with flow over a plate in Chap. 13.

For *high* Reynolds numbers (above 3×10^6) exceeding the range of the Nikuradse data, it is possible to develop an expression for the velocity profile by employing certain hypotheses and limited experimental information. This will be done in Part E where we shall make a more thorough study of the velocity profile.

9.9 MINOR LOSSES IN PIPE SYSTEMS

With pipe bends, valves, etc., it is usually necessary to account for head losses through these devices, in addition to the losses sustained by the pipes. This must almost always be done by resorting to experimental results. Such information is given in the form

$$h_l = K \frac{V^2}{2} \tag{9.24}$$

where the coefficient K is given for commercial fittings in numerous hand books.[15] No distinction is made between laminar and turbulent flow. The velocity V may be stipulated in the handbook as either the upstream or the downstream average velocity q/A to or from the device. These minor losses are then included into the modified Bernoulli equation (or the first law of thermodynamics) along with the pipe losses, as will be indicated in the following section.

There is an important case whereby the estimation of the head loss may be made by analytical methods. This is the case of the sudden opening as shown in Fig. 9.18. Note that two regions may be specified in the larger pipe—one which appears as a highly irregular flow and a region of reasonably smooth flow. It will be useful to imagine a control volume enclosing the smoother region of flow, with one end 1 at the entrance to the large pipe and the other end 2 far enough downstream so as to be in a region of parallel flow. This control volume is shown separately in Fig. 9.19. With a given average velocity V_1 in the smaller pipe, there is a fixed average velocity V_2 in the larger pipe determined by continuity. For incompressible flow, the linear momentum efflux through the control surface is thus essentially a *fixed quantity* once the velocity in the upstream pipe is specified. This means that the *total force* in the horizontal direction on the control surface is then *fixed*. For such a condition, suppose that the normal stresses on the curved part of the control surface are lowered, while the shear stresses are not changed. It is then necessary that pressure p_2 *decrease* so as to maintain the same linear momentum flow through the control volume (and hence the same total force in the horizontal direction on the control surface). There is then a *greater* head loss between sections 1 and 2. In the actual case, because of the high turbulence and mixing action in the pocket

---

[15]The minor loss is also given as an equivalent length of pipe L_{eq} to be added to the system.

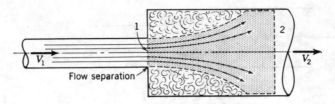

FIGURE 9.18
Sudden opening.

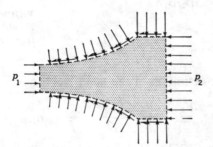

FIGURE 9.19
Control volume around smooth-flow region.

of fluid outside the control volume, there is little recovery of pressure from the kinetic energy of the fluid entering the pocket from the main stream. Instead, the kinetic energy is dissipated into internal energy and heat transfer. The pressure in the pocket thus tends to remain equal to the low pressure p_1 at the point of separation (see Fig. 9.19) resulting in a comparatively high head loss. If a conduit having the shape of the control volume is employed, there will be higher normal stresses along the wall in the absence of the detrimental effects of the low-pressure pocket of the sudden opening and, assuming similar shear action, we see that for given entering conditions the pressure p_2 must be higher for such an arrangement, which explains the lower head loss that one expects from such "internal streamlining."

In the ensuing computations, we assume a zero pressure recovery in the pocket so the p_1 is considered to persist throughout this region. If shear stresses are neglected, the *linear momentum* equation becomes for the aforementioned control volume

$$p_1 A_2 - p_2 A_2 = \rho V_2^2 A_2 - \rho V_1^2 A_1$$

Replacing V_1 by $V_2(A_2/A_1)$ by *continuity* we get, on rearranging terms,

$$\frac{p_1 - p_2}{\rho} = V_2^2\left(1 - \frac{A_2}{A_1}\right) \qquad (9.25)$$

To ascertain the head loss, write the *first law of thermodynamics* for the control

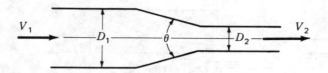

If $\theta \leq 45°$, $K = \dfrac{0.8 \sin (\theta/2)[1 - (D_2/D_1)^2]}{(D_2/D_1)^4}$

If $45° < \theta \leq 180°$, $K = \dfrac{0.5[1 - (D_2/D_1)^2]\sqrt{\sin \dfrac{\theta}{2}}}{(D_2/D_1)^4}$

$h_l = K \dfrac{V_1^2}{2}$

FIGURE 9.20
Head-loss factor for gradual contraction.

volume, using the basic definition of the head loss. Thus

$$\frac{V_1^2}{2} + \frac{p_1}{\rho} = \frac{V_2^2}{2} + \frac{p_2}{\rho} + h_l$$

By substituting from Eq. (9.25) for $(p_1 - p_2)/\rho$, the final result for h_l may then be stated

$$h_l = \frac{V_2^2}{2}\left(1 - \frac{A_2}{A_1}\right)^2 = \frac{V_2^2}{2}\left[1 - \left(\frac{D_2}{D_1}\right)^2\right]^2 \qquad (9.26)$$

For *gradual contraction* or for *gradual enlargement* (see Figs. 9.20 and 9.21, respectively), we give the formulas for the friction-factor coefficient K.[16] Please note that the velocities used in $V^2/2$ for the expression $K(V^2/2)$ in the

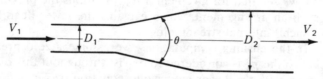

If $\theta \leq 45°$, $K = \dfrac{2.6 \sin (\theta/2)\left[1 - (D_1/D_2)^2\right]^2}{(D_1/D_2)^4}$

If $45° < \theta \leq 180°$, $K = \dfrac{\left[1 - (D_1/D_2)^2\right]^2}{(D_1/D_2)^4}$

$h_l = K \dfrac{V_2^2}{2}$

FIGURE 9.21
Head-loss factor for gradual enlargement.

TABLE 9.2
K factors for fitting*

	Nominal diameter, in											
	$\frac{1}{2}$	$\frac{3}{4}$	1	$1\frac{1}{2}$	2	3	4	5	6	8–10	12–16	18–24
Gate valve (open)	0.22	0.20	0.18	0.16	0.15	0.14	0.14	0.13	0.12	0.11	0.10	0.096
Globe valve (open)	9.2	8.5	7.8	7.1	6.5	6.1	5.8	5.4	5.1	4.8	4.4	4.1
Standard elbow (screwed) 90°	0.80	0.75	0.69	0.63	0.57	0.54	0.51	0.48	0.45	0.42	0.39	0.36
Standard elbow (screwed) 45°	0.43	0.40	0.37	0.34	0.30	0.29	0.27	0.26	0.24	0.22	0.21	0.19
Standard tee (flow through)	0.54	0.50	0.46	0.42	0.38	0.36	0.34	0.32	0.30	0.28	0.26	0.24
Standard tee (flow branched)	1.62	1.50	1.38	1.26	1.14	1.08	1.02	0.96	0.90	0.84	0.78	0.72

*Data from Crane C., "Flow of Fluids," Tech. Paper 410, 1979.

respective diagrams are different. You will note that for the sudden enlargement ($\theta = \pi$), the K factor from Fig. 9.21 becomes

$$K = \frac{\left[1 - (D_1/D_2)^2\right]^2}{(D_1/D_2)^4} = \left[\left(\frac{D_2}{D_1}\right)^2 - 1\right]^2 = \left[1 - \left(\frac{D_2}{D_1}\right)^2\right]^2 \quad (9.27)$$

This is identical to the result for K in Eq. (9.26), which was the analytically derived formulation. Also, for the *sudden contraction* (Fig. 9.20), we get the value

$$K = \frac{0.5\left[1 - (D_2/D_1)^2\right]}{(D_2/D_1)^4} \quad (9.28)$$

where the formula above corresponds to $\theta = \pi$.

In Tables 9.2 and 9.3, we have listed the K factors for a number of important fittings connected in pipes having certain *nominal* diameters. Note that the nominal diameter does *not* correspond exactly to the inside diameter of the pipe. In Table 9.4, we have listed the nominal diameter for carbon and alloy steel pipe as well as stainless steel along with the inside diameter as well as the internal cross-sectional area. Note that the inside diameter up to size 12 in is slightly larger than the nominal diameter whereas after 12 in the inside diameter is less than the nominal diameter.

---

[16]Adapted from Crane Co. "Flow of Fluids," Tech. Paper 410, 1979.

TABLE 9.3
K factors for flanged 90° elbows*

					Nominal pipe size, in						
r/d	$\frac{1}{2}$	$\frac{3}{4}$	1	2	3	4	5	6	8–10	12–16	18–24
1	0.54	0.50	0.46	0.38	0.36	0.34	0.32	0.30	0.28	0.26	0.24
3	0.32	0.30	0.276	0.228	0.216	0.204	0.192	0.018	0.168	0.156	0.144
6	0.459	0.425	0.391	0.32	0.31	0.29	0.27	0.26	0.24	0.22	0.20
10	0.81	0.75	0.69	0.57	0.54	0.51	0.48	0.45	0.42	0.39	0.36
14	1.03	0.95	0.87	0.72	0.68	0.65	0.61	0.57	0.53	0.49	0.46
20	1.35	1.25	1.15	0.95	0.90	0.85	0.80	0.75	0.70	0.65	0.60

*Data from Crane Co., "Flow of Fluids," Tech. Paper 410, 1979.

TABLE 9.4
Nominal pipe sizes for standard pipes*

Nominal pipe diameter, in	Inside diameter, in	Inside cross-sectional area, in^2
$\frac{1}{2}$	0.364	0.1041
$\frac{3}{4}$	0.824	0.533
1	1.049	0.864
$1\frac{1}{2}$	1.610	2.036
2	2.067	3.356
3	3.068	7.393
4	4.026	12.73
5	5.047	20.01
6	6.065	28.89
8	7.981	50.03
10	10.020	78.85
12	12.000	113.10
14	13.000	132.73
16	15.25	182.65
18	17.250	233.74
20	19.250	291.04
22	21.250	354.66
24	23.250	424.56

*Data from Crane Co., "Flow of Fluids," Tech. Paper 410, 1979.

In the examples to follow in Part *D* of this chapter and in many of the homework problems, when we specify a diameter we are referring to an *inside* diameter unless we specify a *nominal* diameter. In the latter case, you will have to use Table 9.4 for the inside diameter.

Finally, in Fig. 9.22, we have shown *K* factors for pipe entrances.

We point out further that in the examples of Part *D* and in many of the homework problems we specify the *K* factors in the interest of saving time.

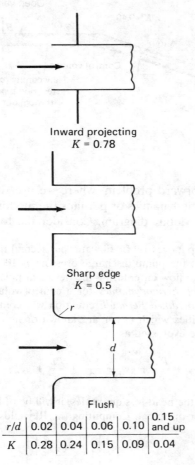

Inward projecting
$K = 0.78$

Sharp edge
$K = 0.5$

Flush

r/d	0.02	0.04	0.06	0.10	0.15 and up
K	0.28	0.24	0.15	0.09	0.04

FIGURE 9.22
Pipe entrances.

PART D
PIPE FLOW PROBLEMS

9.10   SOLUTION OF SINGLE-PATH PIPE PROBLEMS

In the pipe problems to follow we shall not list our assumptions prominently for each case as we did in Chap. 5 since they are now obvious and essentially the same for each problem. We will discuss three cases for the single-path system.

Case 1.

Flow conditions at one section are known, as are the roughness coefficient and the entire pipe geometry. Flow conditions at some other section are desired.

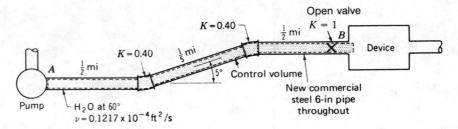

FIGURE 9.23
Flow pumped to device.

This is a straightforward problem, where we use the continuity equation and the first law of thermodynamics or perhaps the modified Bernoulli equation in conjunction with the Moody diagram. Consider the following two examples.

Example 9.2. A pump moves 1.0 ft^3 of water per second through a 6-in pipeline, as shown in Fig. 9.23. If the pump discharge pressure is 100 lb/in^2 gage, what must be the pressure of the flow on entering the device at position B?

The *first law of thermodynamics* for a control volume including the entire pipe interior or the *modified Bernoulli* equation will contain the desired quantity p_B, as well as quantities which either are known or may be directly ascertained. This equation may be expressed as

$$\frac{V_A^2}{2} + \frac{p_A}{\rho} + gy_A = \frac{V_B^2}{2} + \frac{p_B}{\rho} + gy_B + (h_l)_T \qquad (a)$$

In order to compute the head-loss quantities, it will be of interest to know whether the flow is laminar or turbulent. Computing the Reynolds number we get

$$\text{Re}_D = \frac{(q/A)D}{\nu} = \frac{\left[1 \Big/ \dfrac{\pi(\frac{1}{2})^2}{4}\right]\dfrac{1}{2}}{0.1217 \times 10^{-4}} = 209,000$$

It is clear that the flow is well into the turbulent range. Using the Moody roughness chart (Table 9.1) we find for the pipe diameter of 6 in that the relative roughness e/D is 0.0003 in the case of commercial steel pipe. The friction factor may next be read from the Moody diagram for the preceding Reynolds number and relative roughness. We get for f the value 0.017, and the major pipe head loss $(h_l)_p$ then becomes on noting that $V = q/A = 1/[\pi(\frac{1}{2})^2/4] = 5.09$ ft/s,

$$(h_l)_p = f\frac{V^2}{2}\frac{L}{D} = 0.017\frac{5.09^2}{2}\frac{6340}{0.5} = 2792 \frac{\text{ft} \cdot \text{lb}}{\text{slug}}$$

The *minor losses* are given as

$$(h_l)_m = \sum_i \left(K_i \frac{V_i^2}{2} \right) = 0.8\frac{5.09^2}{2} = 10.36 \text{ ft} \cdot \text{lb/slug}$$

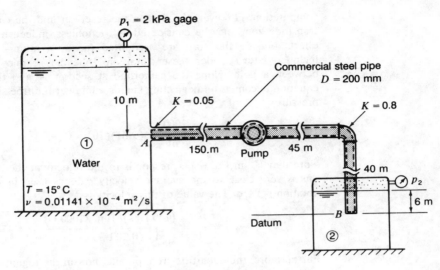

FIGURE 9.24
Water is pumped between two large tanks.

By using the lowest pipe centerline as a datum, we find that the solution for p_B may readily be carried out by substituting the above head losses into Eq. (a) and noting from *continuity* that the velocity terms cancel. Thus

$$\frac{(100)(144)}{1.938} = \frac{p_B(144)}{1.938} + \frac{(32.2)(5280)}{5} \sin 5° + 2790 + 10.36 \qquad (b)$$

Solving for p_B,

$$p_B = 22.4 \text{ lb/in}^2 \text{ gage}$$

Example 9.3. Water is pumped from a large tank ① to a second large tank ② through a pipe system, as shown in Fig. 9.24. The pump is developing 20 kW on the flow. For a steady flow of 140 L/s what must be the pressure p_2 in the air trapped above the water in tank ②?

We shall consider steady incompressible flow. A control volume is chosen to include the entire interior of the pipe as well as the pump. At the reservoir end [tank ①], the control surface is positioned directly after the rounded entrance, as indicated in Fig. 9.24. The *first law of thermodynamics* (we cannot use the modified Bernoulli equation between A and B because of the pump) is given as

$$\frac{V_A^2}{2} + \frac{p_A}{\rho} + gy_A = \frac{V_B^2}{2} + \frac{p_B}{\rho} + gy_B + (h_l)_p + (h_l)_m + \frac{dW_s}{dm} \qquad (a)$$

where $(h_l)_p$ is head loss in pipes. Among the quantities that must first be ascertained is the pressure p_A. You may be tempted to say that $p_A = (2000 + 10\gamma)$ Pa gage., but a moment's reflection will reveal that the equilibrium required by this formulation is not present near the reservoir outlet, so this is not a correct computation. However, the flow in the reservoir and the core region of the pipe transition zone may be considered as frictionless, indeed irrotational. (Frictional effects outside the core are accounted for by a minor-loss expression to be included later.) Under these conditions, one may employ *Bernoulli's* equation between a point along the centerline at section A and the free surface. This equation becomes, on neglecting kinetic energy at the free surface and using gage pressures,

$$\frac{2000}{\rho} + (9.81)(10) = \frac{V_A^2}{2} + \frac{p_a}{\rho} \tag{b}$$

Note that V_A in Eq. (b) corresponds to the velocity at the core center. However, this is very close to the average velocity q/A as used in Eq. (a), so the same notation is used. The value of this velocity is

$$V_A = \frac{q}{A} = \frac{(140 \text{ L/s})(1 \text{ m}^3/1000 \text{ L})}{\left[\pi(0.200^2)/4\right] \text{ m}^2} = 4.456 \text{ m/s}$$

Furthermore, the departure from parallel flow in the region of section A is slight, so that by neglecting hydrostatic variations one may reasonably consider a uniform pressure to exist over the section. It is clear that p_A in Bernoulli's equation can be considered equal to the pressure used in the first law of thermodynamics given by Eq. (a). Solving for p_A/ρ, we get from Eq. (b)

$$\frac{p_A}{\rho} = (9.81)(10) - \frac{4.456^2}{2} + \frac{2000}{999.1} = 90.17 \text{ N} \cdot \text{m/kg} \tag{c}$$

A Reynolds number of 7.81×10^5 is computed for the pipe flow so that the friction factor for head-loss computations is found from the Moody diagram (using $e/D = 0.046/200 = 0.00023$) to be 0.0152. The *head loss* due to friction in the pipes is

$$(h_l)_p = 0.0152\frac{4.456^2}{2}\frac{235}{0.200} = 177.3 \text{ N} \cdot \text{m/kg} \tag{d}$$

The total *minor loss* is evaluated as

$$(h_l)_m = \left(\sum_i K_i\right)\frac{V_A^2}{2} = (0.05 + 0.8)\left(\frac{4.456^2}{2}\right) = 8.44 \text{ N} \cdot \text{m/kg} \tag{e}$$

Next, the work expression dW_s/dm in Eq. (a) must be given in units of work per kilogram of flow. The computation from kilowatts to these units may be carried out as follows:

$$\frac{dW_s}{dm} = \frac{dW_s/dt}{q\rho} = \frac{-(20)(1000)}{(0.140)(999.1)} = -143.0 \text{ N} \cdot \text{m/kg} \tag{f}$$

Finally we consider the flow from the pipe into tank ②. Here we have a subsonic free jet on immediate exit from the pipe and so we can give p_B as follows using hydrostatics of the water immediately surrounding the emerging jet:

$$p_B = p_2 + (6)(9.81) \text{ Pa gage} \tag{g}$$

Going back to Eq. (*a*) we note that $V_A = V_B$ so we cancel the velocity terms. Substituting from Eqs. (*c*)–(*g*) we have, on using the outlet B as our datum as well as using gauge pressures,

$$90.17 + (9.81)(40) = \frac{p_2 + (6)(9.81)}{999.1} + 0 + 177.3 + 8.44 - 143.0$$

$$\therefore p_2 = 439{,}400 \text{ Pa gage}$$

We have thus arrived at the desired information.

Case 2.

> Pressures at several sections are known, as are the roughness coefficient and the entire geometry. The amount of flow q is to be determined.

In this class of problems we are to solve for the velocity V or the volume flow q. In the case of turbulent flow, this will not be done directly, because we will not use the friction formulas presented earlier to express the friction factor f in terms of V mathematically so as to allow for algebraic isolation of V in the first-law equation. Instead, an easy iterative process will be employed in conjunction with Moody's diagram, as will be demonstrated in the following example.

Example 9.4. A pipe system carries water from a reservoir and discharges it as a free jet, as shown in Fig. 9.25. How much flow is to be expected through a 200-mm steel commercial pipe with the fittings shown?

By using the indicated control volume for the *first law of thermodynamics* or the *modified Bernoulli* equation and substituting for the entering pressure by using

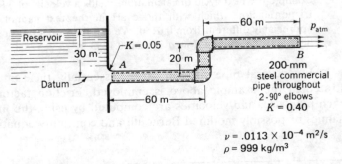

FIGURE 9.25
Water flows from reservoir.

Bernoulli's equation, as in the previous problem, the following equation may be written:

$$\frac{V_A^2}{2} + \left[\frac{p_{atm}}{\rho} + g(30) - \frac{V_A^2}{2} \right] + gy_A = \frac{V_B^2}{2} + \frac{p_B}{\rho} + gy_B + (h_l)_p + (h_l)_m \quad (a)$$

Applying the pressure condition of a subsonic free jet, $p_B = p_{atm}$. Inserting known elevations, and using head-loss formulas, we have

$$\frac{p_{atm}}{\rho} + g(30) = \frac{V_B^2}{2} + \frac{p_{atm}}{\rho} + g(20) + f\frac{V_B^2}{2}\frac{140}{0.200} + (0.05 + 0.4 + 0.4)\frac{V_B^2}{2}$$
$$(b)$$

Collecting and rearranging terms, we get

$$\frac{V_B^2}{2} + f\frac{V_B^2}{2}700 + 0.85\frac{V_B^2}{2} = 10g \quad (c)$$

Since f is not known, we will work simultaneously with the friction diagram. The roughness coefficient e is read from Table 9.1 and for the given pipe diameter and material for this problem, we have $e/D = 0.00023$. Observing the friction diagram, estimate a friction factor for this relative roughness. We choose 0.014 as a first guess.[17] It is now possible to solve Eq. (c) for a velocity $(V_B)_1$ designated as the first velocity estimate. This becomes

$$(V_B)_1 = 4.10 \text{ m/s}$$

To check the estimated friction factor, solve for the Reynolds number for the above velocity.

$$\text{Re}_D = \frac{(4.10)(0.200)}{0.0113 \times 10^{-4}} = 7.26 \times 10^5$$

For this number, the Moody diagram yields a friction factor of 0.0150. With this value of friction factor, the preceding computations are repeated and the cycle of computations is performed until there is no appreciable change in the friction factor. At that time, the correct solution will have been reached. Actually only one or two iterations are ever necessary, as the process is rapidly convergent. In this example the next cycle of calculations yields a velocity of 3.99 m/s; this results in a friction factor, which with the chart in the text cannot be detected as being appreciably different from 0.0150. We can then say that the velocity desired is 3.99 m/s and the volume flow q is then 0.1253 m³/s = 125.3 L/s.

For several pipe sizes in series, essentially the same procedure demonstrated in the example above is employed. Friction factors are estimated for each pipe size, and velocities are computed by using the first law of thermodynamics or possibly modified Bernoulli and continuity equations. Reynolds num-

---

[17]The flow in many pipe flows will generally be in the rough pipe zone. This fact will help you get a good first guess for f.

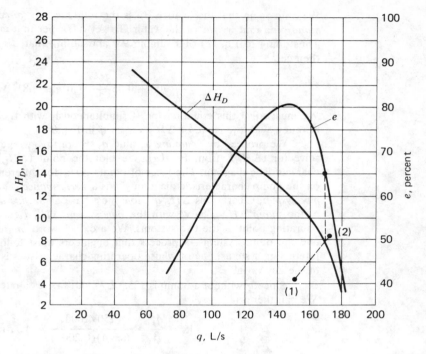

FIGURE 9.26
Plot of total head ΔH_D versus q and efficiency e versus q for a pump.

bers are then computed for each pipe size with these velocities. New friction factors are chosen from the Moody diagram and the set of operations repeated.

In Example 9.3, we knew the power developed on the water for a given volume flow q. In the next example, we look at the very practical problem of hooking up a given pump having certain performance characteristics into a pipe system and ask the question: What will be the volume flow q when the pump is working and what is the power needed by the pump for the computed operation? Here we will have to *match* the operating characteristics of the pump with the flow characteristics of the pipe system. In complex turbomachines and in complex duct flows, such as in wind tunnels, one must match characteristics of various components of the system. Thus in a jet engine, we match the diffuser, compressor, turbine, and nozzle characteristics to determine the overall operation. The next example on a simple scale illustrates this kind of problem.

Example 9.5. In the previous example a pump is inserted somewhere along the pipe. The characteristics of this pump are shown in Fig. 9.26 where the total head (ΔH_D) developed by the pump is plotted against the volumetric flow q. Also, the efficiency e of the pump is plotted versus q.

We use the *first law of thermodynamics* for the control volume of Fig. 9.25. However, now we must include a shaft work term to take the pump into account.

We note again that the head ΔH_D is the *total mechanical energy developed by the pump* on a unit *weight* of the water. To get ΔH_D per unit mass we multiply by g. Thus going to Eq. (c) of Example 9.4 and noting that $dW_s/dm = -g\,\Delta H_D$ we then get[18]

$$\frac{V^2}{2} + f\frac{V^2}{2}700 + 0.85\frac{V^2}{2} = 10g + 9.81(\Delta H_D) \qquad (a)$$

We must solve this equation for V simultaneously with the Moody diagram and also now with the curve of ΔH_D versus q in Fig. 9.26.

We proceed by *assuming* a value q_1. From this, we get V and f, so we can solve for $(\Delta H_D)_1$ from Eq. (a). We plot the point $(\Delta H_D)_1, (q_1)$ on the pump characteristic diagram. The locus of points ΔH_D for different values of q plotted on the pump characteristic diagram yields a *performance characteristic curve of the pipe system*. Clearly this ΔH_D-q curve must have a *positive* slope. The *intersection* of the *pump* ΔH_D-q curve with the *pipe system* ΔH_D-q curve will be the *desired* operating point of the two systems. We are, in a word, *matching* the pipe system and the pump system—a process that engineers must follow when dealing with interacting systems, one of whose operating characteristics is known only as a plot of the key variables.

Since q without a pump is 125.3 L/s, our first estimate q_1, will be 150 L/s.[19] We can then say

$$V_1 = \frac{q_1}{A} = \frac{(150/1000)}{(\pi/4)(0.200^2)} = 4.77 \text{ m/s}$$

$$(\text{Re}_D)_1 = \frac{(4.77)(0.200)}{0.0113 \times 10^{-4}} = 8.45 \times 10^5 \qquad (b)$$

$$f = 0.0149$$

Now going back to Eq. (a) and substituting the above values, we get

$$\frac{4.77^2}{2} + 0.0149\frac{4.77^2}{2}(700) + 0.85\frac{4.77^2}{2} = 98.1 + 98.1(\Delta H_D)_1$$

We plot $[q_1, (\Delta H_D)_1]$ on the plot in Fig. 9.26. Because it is below the pump ΔH_D-q curve, we choose as a second estimate the value $q_2 = 170$ L/s to try to get just above the ΔH_D-q curve. We then can say for this flow in the pipe that

$$V_2 = \frac{170/1000}{(\pi/4)(0.200^2)} = 5.41 \text{ m/s}$$

$$(\text{Re}_D)_2 = \frac{(5.41)(0.200)}{0.0113 \times 10^{-4}} = 9.58 \times 10^5$$

$$f = 0.0148$$

---

[18] If you were using the first law in the form of head terms $V^2/2g$, p/γ, z, etc.—that is, per unit weight—we would simply use ΔH_D without the g.

[19] If the guess is too low, ΔH_D will come out negative in your first-law calculation. Since for a pump $\Delta H_D > 0$, Choose a larger value of q for an estimate.

From Eq. (a) we have for $(\Delta H_D)_2$ the result

$$(\Delta H_D)_2 = 8.21 \text{ m}$$

We plot point 2 in Fig. 9.26 as shown. Points 1 and 2 are close enough to the ΔH_D-q curve of the pump so we can estimate the flow q for the system by connecting the two points by a straight line and noting the intersection with the ΔH_D-q curve for the pump. We get

$$q = 169 \text{ L/s}$$

The estimated power requirements for the pump can now readily be determined using the efficiency curve for the pump. Thus, noting $\Delta H_D = 7.80$ m and $e = 65$ percent from Fig. 9.26 for the operating point,

$$\begin{aligned}
\text{Power} &= \frac{(\Delta H_D)(g)(\rho q)}{e} \\
&= \frac{(7.80)(9.81)[(169/1000)(999)]}{0.65} \\
&= 19.87 \text{ kW}
\end{aligned}$$

Case 4.

The centerline geometry of the pipe is known, as are the flow q and pressures at two sections. The proper pipe diameter for a specified type of pipe material is to be ascertained for these conditions.

In this case, we are to evaluate a proper pipe diameter to perform a given function. As in Example 9.4, iteration will be necessary.

Example 9.6. A pipe system having a given centerline geometry as shown in Fig. 9.27 is to be chosen to transport a maximum of 120 ft^3/s of water from tank A to tank B. What is a pipe size which will do the job? A pump is developing 556 hp on the flow.

A control volume is chosen encompassing the pipe and pump interior and extending a short distance into the receiving tank B. We may consider that the water issues into this tank as a free jet and is therefore at a pressure corresponding to the hydrostatic pressure of the surrounding water. The pressure at the other end of the control volume may be computed by employing *Bernoulli's* equation between points c and a, as is shown in the diagram. Thus, neglecting the velocity head at position c, we have on using gauge pressures

$$\frac{p_a}{\rho} = \left(\frac{p_c}{\rho} - \frac{V_a^2}{2} \right) + gz_c = \frac{(35)(144)}{1.938} - \frac{V_a^2}{2} + (30)(32.2) \tag{a}$$

$$\therefore \frac{p_a}{\rho} = 3567 - \frac{V_a^2}{2}$$

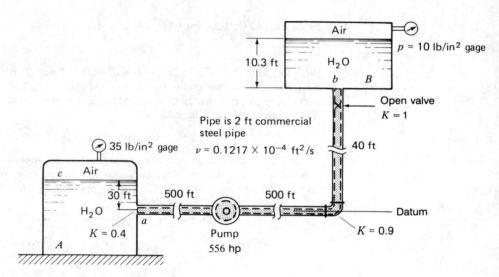

FIGURE 9.27
Design problem.

Next we compute the p_b/ρ of the free jet emerging into tank B. Thus

$$\frac{p_b}{1.938} = [(10)(144) + (10.3)(62.4)]\frac{1}{1.938} = 1075 \text{ ft-lb/slug} \qquad (b)$$

Now we go to the *first law* of thermodynamics for the control volume we have chosen.

$$\frac{p_a}{\rho} + \frac{V_a^2}{2} + gz_a = \frac{p_b}{\rho} + \frac{V_b^2}{2} + gz_b + \frac{dW_s}{dm} + (h_l)_T \qquad (c)$$

Using the datum shown in the diagram as well as Eqs. (a) and (b) we get for Eq. (c), while noting that $V_a = V_b \equiv V$,

$$\left(3567 - \frac{V^2}{2}\right) + \frac{V^2}{2} + 0 = 1075 + \frac{V^2}{2} + (32.2)(40)$$

$$- \frac{(556)(550)}{(1.938)(V)(\pi/4)(D^2)} + f\left(\frac{1040}{D}\right)\frac{V^2}{2} + 2.3\frac{V^2}{2}$$

$$(d)$$

The value of V to be inserted above is

$$V = \frac{q}{A} = \frac{120}{\pi D^2/4} = \frac{152.8}{D^2} \qquad (e)$$

At this time we make a guess for f (remember we are seeking the proper value of D). Our first selection will be $f = 0.015$. Equation (d) then becomes, on multiplying through by D^5,

$$2519D^5 - 3.852 \times 10^4 D^4 - 1.821 \times 10^5 = 0 \qquad (f)$$

To solve this equation is not an easy matter. If the second term were not present, the solution for D could readily be carried out. An examination of this term reveals that it stems from minor losses and the kinetic-energy term in Eq. (d). In most pipe problems, these terms, while not negligible at all times, are nevertheless comparatively small. An acceptable procedure is to neglect them initially and then, when the resulting simplified calculations have been carried out, to make allowances for these omitted effects. This is an approach used in most engineering-design problems. Under these conditions the diameter becomes

$$D = 2.35 \text{ ft}$$

The corresponding velocity and Reynolds number become

$$V = 27.6 \text{ ft/s}$$

$$\text{Re}_D = 5.33 \times 10^6$$

For the diameter above in the case of steel commercial pipe, we have a relative roughness 6.4×10^{-5}. The corresponding friction factor for these conditions is read from the friction diagram as 0.0139. With this new value the procedure is repeated. A second iteration results in the following quantities:

$$D = 2.32 \text{ ft}$$

$$V = 28.4 \text{ ft/s}$$

$$\text{Re}_D = 5.42 \times 10^6$$

$$f = 0.0139$$

The rapidity of the convergence is apparent in this example. Further iteration is unnecessary in light of the approximation already made.

The neglected head loss and velocity heads will now be partially accounted for in the following manner. By trial and error trying diameters slightly larger than 2.32 ft a value will be found that satisfies Eq. (d), wherein the latest friction factor has been used. In this case the value of 2.57 ft is found to be reasonably close. Since the friction-factor information as contained in the charts is not employed simultaneously with this last computation, it may be argued that the final result is not quite correct. However, the error in most cases will be of an order no greater than other inaccuracies inherent in pipe-flow analysis, as discussed earlier. Furthermore, since commercial pipe is to be used, a size nearest to the available standard inside diameter will necessarily have to be selected. This variation from theoretical requirements may very well be the largest "error."

Inherent in the preceding computation is the assumption that the minor losses are a small but not necessarily a negligible source of head loss. This assumption applies to all the problems thus far undertaken. Actually, large minor losses imply many fittings and consequently short lengths of pipe. The assumption of fully developed parallel flow throughout the pipe system becomes less tenable: hence, formulations of head loss then become less reliable for such undertakings. Thus, the restriction in the preceding example is actually in line with the basic limitations of the entire analysis set forth thus far.

No doubt students will find doing these pipe problems a bit laborious because of the iteration with the Moody diagram. If you have to solve many different pipe problems which at the same time are of the *same* general family (such as all kinds of single branch pipes from reservoir to reservoir or from a given pressure inlet to a given pressure outlet) you should consider developing your own interacting software using the friction formulas instead of the Moody diagrams. We have presented such computer projects in the computer projects of the instructors manual to do that very thing. You can then ask a nontechnical person to solve your problems after you present him/her with the appropriate disk.

9.11 HYDRAULIC AND ENERGY GRADE LINES

Civil engineers find it useful to plot various portions of the *head* terms from mechanical energy of the flow in a pipe. Thus, the locus of points at a vertical distance p/γ *above the pipe centerline* is called the *hydraulic grade* line. If one measures the height z of the pipe centerline from some convenient horizontal datum, then the ordinate $(p/\gamma + z)$ measured above this datum gives the same curve as described above and is also considered the hydraulic grade line for the aforementioned datum. This is shown in Fig. 9.28, where we use gage pressure. Note that the value of the hydraulic grade line, which we denote as H_{hyd}, starts from a height corresponding to the free surface of the reservoir where $p/\gamma =$

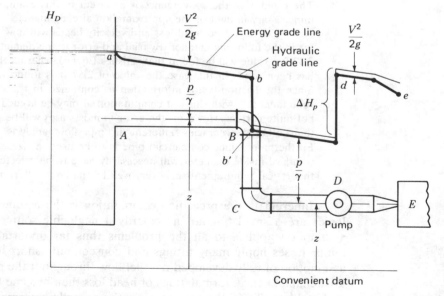

FIGURE 9.28
Energy and hydraulic grade lines.

$V^2/2g = 0$ leaving only the elevation head of the free surface. As we enter the pipe, there is a head loss, and we show a decrease in H_{hyd} because pressure is lost in the entrance. Note next that the hydraulic grade line slopes downward as we move along from A to B in the pipe. This is due to the head loss generated in the pipe AB which decreases p/γ along the pipe. At B, there is a drop in value from b to b' due to the minor losses in the two elbows plus the head loss in pipe BC. When we get to D, there is a sudden increase in hydraulic head due to the input of the pump. Let us examine the contribution of the pump more carefully. The head ΔH_D contributed by the pump clearly must be

$$\Delta H_D = \Delta \left(\frac{V^2}{2g} + \frac{p}{\gamma} + z \right) \tag{9.29}$$

where we have on the right side of the equation the changes in mechanical head terms resulting from the action of the pump on the flow. The expression in parentheses may be evaluated as follows utilizing the first law of thermodynamics:

$$\Delta \left(\frac{V^2}{2g} + \frac{p}{\gamma} + z \right) = \frac{\Delta}{g} \left(\frac{V^2}{2} + \frac{p}{\rho} + gz \right)$$

$$= -\left(\frac{1}{g} \right) \left(\frac{dW_S}{dm} \right) = -\frac{(1/g)(dW_S/dt)}{(dm/dt)} = -\frac{(1/g)(dW_S/dt)}{(\rho V A)}$$

Hence we can say

$$\Delta H_D = -\frac{(1/g)(dW_S/dt)}{(\rho V A)} \tag{9.30}$$

For the pump of this case, the term dW_s/dt is negative so that ΔH_D is positive, and we show this rise to point d. In going through the nozzle, there is a steep slope of the diagram as pressure head is converted to velocity head in the nozzle. When you get to e you may check your calculations by computing $z_E + p_E/\gamma$ for the plenum chamber E. The height $(H_{hyd})_e$ that you reach should be close to the separately computed value of $z_E + p_E/\gamma$.

If your hydraulic grade line dips below your pipe centerline, then clearly you must have negative gage pressures in those parts of the pipe system when this happens.

If one adds the velocity head $V^2/2g$ to the hydraulic grade line, we get the *energy grade line* shown in Fig. 9.28. The uppermost curve then represents the total mechanical head H_D.

To draw the hydraulic and energy grade lines, we must determine the volumetric flow q for the pipe system if q is not already known. Using the first law of thermodynamics and continuity, we can then ascertain $p/\gamma + z$ and $V^2/2g$ at key points in the system. Finally, one can connect the key points by appropriate lines. We will ask you to do this straightforward procedure in some of the homework problems.

Finally, we point out that the hydraulic and energy grade lines for a pipe system give the hydraulic engineer an overall view about energy flow. This is analogous to shear and bending moment diagrams in strength of materials, which give the structural engineer an overall view of the flow of force in a structure.

9.12 NONCIRCULAR CONDUITS

Theoretically one can solve velocity profiles and friction factors for fully developed *laminar* flow in noncircular conduits. Also one can use numerical methods such as finite or boundary elements. However, we have shown in Fig. 9.29 friction factors for laminar flow through several cross-sections. These results stem from theoretical and experimental investigations. The Reynolds numbers Re_H used in these results employ the *hydraulic diameter D_H* defined as

$$D_H = \frac{4A}{P_w} \tag{9.31}$$

where A is the cross-sectional area of the conduit and P_w is the length of the

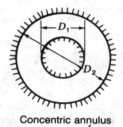

Concentric annulus

D_1/D_2	$f\,\text{Re}_H$
0.0001	71.78
0.01	80.11
0.10	89.37
0.60	95.59
1.00	96.00

Rectangle

h/b	$f\,\text{Re}_H$
0	96.00
0.10	84.68
0.25	72.93
0.50	62.19
1	56.91

Isosceles triangle

α (deg)	$f\,\text{Re}_H$
0	48.0
20	52.9
40	52.9
60	51.1
80	48.9
90	48.0

FIGURE 9.29

Friction factor for laminar flow for a few noncircular class sertions. (From R. M. Olsen and S. J. Wright, Essentials of Engineering Fluid Mechanics, 5th ed., Harper and Row, New York, 1990.)

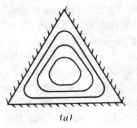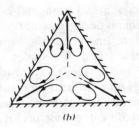

FIGURE 9.30
(*a*) Axial contour velocity lines. (*b*) Schematic of secondary flow. (*Both curves from Nikuradse.*)

wetted[20] perimeter of the conduit cross section. Let us calculate a few hydraulic diameters for some cross sections.

1. *Circular* cross section:

$$D_H = \frac{(4)(\pi D^2/4)}{\pi D} = D$$

We see that the hydraulic diameter becomes the ordinary inside diameter for the circular pipe.

2. *Circular annulus* diameters D_1 (largest) and D_2 (smallest):

$$D_H = \frac{4(\pi D_1^2/4 - \pi D_2^2/4)}{\pi D_1 + \pi D_2} = \frac{D_1^2 - D_2^2}{D_1 + D_2} = D_1 - D_2$$

3. *Rectangular* cross section of sides b and h:

$$D_H = \frac{4bh}{2b + 2h} = \frac{2bh}{b + h}$$

4. *Isosceles triangle* with vertex angle 2α (degrees)

$$D_H = \frac{(4)(\frac{1}{2})(2R \sin \alpha)(R \cos \alpha)}{2R + 2R \sin \alpha} = \frac{R \sin 2\alpha}{1 + \sin \alpha}$$

It is to be pointed out that the wall shear stress for these laminar flows is maximum near the midpoints of the sides and is zero at the corners with large variation along the walls.

For *turbulent* flows we may use the Moody diagrams with the hydraulic diameter replacing D. In these flows the wall shear stress is again zero at the corners as in laminar flow, but along the sides the wall shear stress is close to being uniform. The turbulent mean flow is more complicated than laminar flow in that there will be present in the plane of the cross section a complex flow superposed over the mean-time axial flow. This superposed flow is called *secondary flow*. The contour lines of the axial flow are shown in Fig. 9.30*a* in the

---

[20] For flows in this chapter the wetted perimeter is the entire perimeter, whereas in channel flows it would clearly exclude the free surface portion of the boundary.

case of an equilateral triangular cross section (from Nikuradse); in Fig. 9.30b is shown a schematic representation of the secondary flow in the same section. What is interesting about the latter is that the fluid flows toward the corners along the bisectrix of the angles and, on approaching the corners, then moves outward. There develops as a result high velocities at the corners. There is thus a continuous flow of linear momentum from center to corners. Another interesting phenomenon is that in narrow noncircular sections transition from laminar to turbulent flow does not occur simultaneously over the entire cross section. That is, part of the flow may be still laminar over a small portion of the section whereas the bulk of the flow may have gone over to turbulent flow.

In closing we wish to point out that the results for this cross section are better the closer the cross section is to a square.

We have presented several problems for noncircular conduits in the problems section. Note you may use the Darcy-Weisbach formula for head loss wherein

$$h_l = f \frac{L}{D_H} \frac{V^2}{2} \quad \text{(ft-lb/slug or N-m/kg)}$$

or

$$H_l = f \frac{L}{D_H} \frac{V^2}{2g} \quad \text{(ft or m)}$$

Before going to Part E of the chapter, we wish to point out that if a conduit has in it a flow of liquid that does not wet the entire perimeter, there will then be a free surface present. Such flows are considered in Chap. 14 on channel flow. The hydraulic diameter as defined in this section is still of use in Chap. 14.

PART E
HIGH REYNOLDS NUMBER TURBULENT FLOWS

9.13 APPARENT STRESS

In Sec. 9.7, we pointed out that we could at times consider turbulent flow to be thought of as laminar flow with an additional stress called *apparent* stress superimposed on the comparatively small viscous induced stresses. The apparent stress in such a situation accounts for the manifestations of turbulence. We wish now to examine this hypothesis further.

We have already pointed out in Sec. 9.2 that a velocity component V_x can be considered as the superposition of the mean-time-average velocity $\bar{V}_x$ plus a fluctuation component V_x'. The mean-time-average velocity has been defined for steady flow as

$$\bar{V}_x = \frac{1}{\Delta t} \int_0^{\Delta t} V_x \, dt$$

where Δt is a large enough time interval so that $\overline{V}_x$ is a constant. We wish to relate $\overline{V}_x$ with the apparent stress for parallel flow in the x direction.

Boussinesq first approached this problem. In the case of a steady two-dimensional, turbulent flow he hypothesized that

$$(\tau_{xy})_{\text{app}} = A\left(\frac{d\overline{V}_x}{dy}\right) \tag{9.32}$$

where A is called the *coefficient of eddy viscosity*. This is analogous to the laminar-flow case where Newton's viscosity law is applied. That is,

$$\tau_{xy} = \mu\left(\frac{dV_x}{dy}\right)$$

Furthermore, just as μ/ρ is called the kinematic viscosity ν, so it is that A/ρ is called the *kinematic eddy viscosity*, given by the symbol ϵ. Hence, the Boussinesq formulation becomes

$$(\tau_{xy})_{\text{app}} = \rho\epsilon\left(\frac{d\overline{V}_x}{dy}\right) \tag{9.33}$$

The coefficient of viscosity has been described in Chap. 1 as a property depending almost entirely on the type of fluid and the temperature. This is to be expected from the microscopic nature of its origin. For this reason computations involving Newton's viscosity law are direct and comparatively simple. However, the eddy viscosity, stemming from macroscopic actions, depends on *local conditions of flow*. Since these are unknown, it appears that despite the familiar appearance of the Boussinesq hypothesis, it is nevertheless of little help in this form.

Let us now relate the apparent shear stress $(\tau_{xy})_{\text{app}}$ with the results developed above. If one assumes that the turbulent flow velocity components satisfy the same differential continuity equation as for laminar flow and, further, satisfy the same constitutive law as do the velocity components for general three-dimensional laminar flows, then one can show (this is done later in Sec. 10.10) that the apparent stresses we have been talking about are related to the fluctuation velocity components as follows:

$$\begin{bmatrix} \tau_{xx} & \tau_{xy} & \tau_{xz} \\ \tau_{yx} & \tau_{yy} & \tau_{yz} \\ \tau_{zx} & \tau_{zy} & \tau_{zz} \end{bmatrix}_{\text{app}} = \rho \begin{bmatrix} -\overline{V_x'^2} & -\overline{V_x'V_y'} & -\overline{V_x'V_z'} \\ -\overline{V_y'V_x'} & -\overline{V_y'^2} & -\overline{V_y'V_z'} \\ -\overline{V_z'V_x'} & -\overline{V_z'V_y'} & -\overline{V_z'^2} \end{bmatrix} \tag{9.34}$$

In particular, for our parallel flow in the x direction we get

$$(\tau_{xy})_{\text{app}} = -\rho\overline{V_x'V_y'} \tag{9.35}$$

We can heuristically justify the relationship above by considering our parallel, steady, turbulent flow as shown in Fig. 9.31 where we have shown a

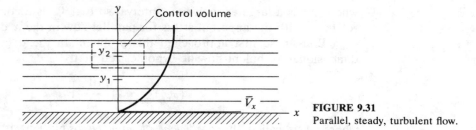

FIGURE 9.31
Parallel, steady, turbulent flow.

velocity profile for mean-time-average flow $\overline{V}_x$. We will explain the presence of $(\tau_{xy})_{\text{app}}$ by considering a small control volume about y_2 moving with a constant speed having the value $\overline{V}_x(y_2)$. When a chunk of fluid from y_1, where the mean-time-average velocity is smaller than that at y_2, migrates to y_2 as a result of the random turbulent motion, it crosses the control surface and so carries linear momentum through the control surface. The flow of the x component of linear momentum passing up through the control surface can be related to a vertical mass flow, which is proportional to V_y', times the velocity in the x direction, V_x'. This relation is that of a proportionality. Thus we can say for the average linear momentum flow in the x direction passing through the surface that

$$(\overline{\text{Momentum flow}})_x \propto \overline{V_x'V_y'}$$

Now equal to this momentum flow through the control surface is a traction force intensity on the control surface which clearly must be a stress τ_{xy}. We consider this stress to be an *apparent stress* to distinguish it from viscous shear stress from random molecular motion on the microscopic scale. We can then say

$$(\tau_{xy})_{\text{app}} = (\text{const})\overline{V_x'V_y'} \tag{9.36}$$

where the constant is the *eddy viscosity*. The apparent shear stress is much greater than viscous shear stress from molecular action and the latter is neglected.

9.14 VELOCITY PROFILES FOR HIGH REYNOLDS NUMBER TURBULENT FLOWS

We shall present two approaches to arrive at the same results. The first is longer but is based on a mechanism perhaps familiar to you from elementary considerations pertaining to the kinetic theory of gases. This will be followed by a much shorter development via dimensional analysis with no physical model. The first approach is starred so the reader may bypass it if so desired.

*Approach 1. Prandtl's Mixing Length.** We shall now present Prandtl's mixing length concept. However, we must clearly point out that the ensuing development, while physically appealing and useful, has validity only for parallel steady

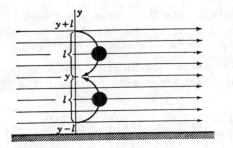

FIGURE 9.32
Mixing length at y.

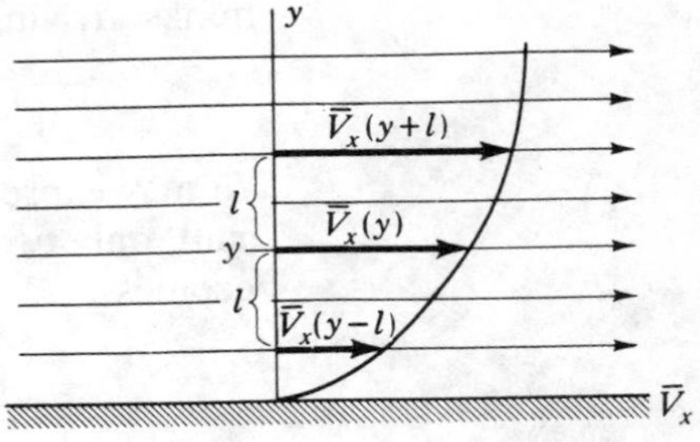

FIGURE 9.33
Mixing length and profile.

flows wherein the mean velocity varies only with one spatial coordinate. But we are primarily interested here in pipe flow away from the entrance, making this study of direct use to us. Later we shall present similar results via dimensional analysis that will justify the results we will obtain using Prandtl's mixing length approach.

Prandtl was able to restate Eq. (9.33) in a more desirable manner by replacing the eddy-viscosity term with a quantity more readily subject to direct physical interpretation which he called the *mixing length*. The advantage is that intuitive statements about the behavior of the mixing length in certain regions of flow may be more successfully made.

To accomplish this, Prandtl established a highly simplified model of linear momentum transfer for turbulent flow. He hypothesized that at any point y, such as is shown in Fig. 9.32, there arrive at random intervals chunks of fluid from positions at a distance l, the mixing length, above and below the point. These chunks are presumed to retain their original mean-time-average velocities $\overline{V}_x(y + l)$ and $\overline{V}_x(y - l)$ until arrival at point y, at which time, sudden exchanges of linear momentum are imagined to take place with the fluid already at y.[21] The result is that a random fluctuating longitudinal component of velocity is created at this position. The amplitude of this velocity fluctuation depends on the *mean-time-average distribution near y and also the choice of value of the mixing length*. In other words, the mixing length is of such a size that the model momentum exchange causes the same fluctuations in the direction of flow as in the actual flow with its more complicated actions. Thus the mixing length is a local function, as is the eddy viscosity, but with the preceding model it is more susceptible to estimation by physical argument. As a result of the aforementioned longitudinal fluctuations, continuity considerations indicate that there will be vertical fluctuations as well, with comparable amplitude.

The preceding concepts may be expressed mathematically with the aid of Fig. 9.33. The difference between the mean-time-average velocities of fluid

---

[21]This is analogous to the mean-free-path concept where molecules on the average move a distance called the mean free path before colliding with other molecules.

chunks arriving from $y + l$ and the fluid at y is given as

$$\left(\Delta \overline{V}_x\right)_1 = \overline{V}_x(y + l) - \overline{V}_x(y)$$

We may express $\overline{V}_x(y + l)$ as a Taylor series about the point y. Since l will be small, only two terms of the series will be retained. The expression above then becomes

$$\left(\Delta \overline{V}_x\right)_1 = \left[\overline{V}_x(y) + \left(\frac{d\overline{V}_x}{dy}\right)_y l\right] - \overline{V}_x(y) = l\left(\frac{d\overline{V}_x}{dy}\right)_y$$

A similar expression may be reached for the fluid arriving from position $y - l$. Thus,

$$\left(\Delta \overline{V}_x\right)_2 = l\left(\frac{d\overline{V}_x}{dy}\right)_y$$

The mean-time average of the magnitude of velocity fluctuations V'_x in the direction of flow at position y can be considered the average magnitude of the velocity differences given above. Thus,

$$\left|\overline{V'_x}\right| = \tfrac{1}{2}\left[\left|\left(\Delta \overline{V}_x\right)_1\right| + \left|\left(\Delta \overline{V}_x\right)_2\right|\right] = l\left|\frac{d\overline{V}_x}{dy}\right|_y \tag{9.37}$$

As pointed out earlier, the velocity fluctuations in the y direction are of comparable magnitude. That is,

$$\left|\overline{V'_y}\right| \approx \left|\overline{V'_x}\right|$$

This relation may be stated as

$$\left|\overline{V'_y}\right| = \text{const}\left|\overline{V'_x}\right| = \text{const } l\left|\frac{d\overline{V}_x}{dy}\right| \tag{9.38}$$

where we have used Eq. (9.37) without the now superfluous subscript y in forming the last expression. Now multiplying together Eqs. (9.37) and (9.38) gives us

$$\left|\overline{V'_y}\right|\left|\overline{V'_x}\right| = \text{const } l^2\left(\frac{d\overline{V}_x}{dy}\right)^2 \tag{9.39}$$

The magnitude sign about the derivative has been dropped in the formulation above because the squaring action makes it superfluous. The quantity $\overline{V'_x V'_y}$ must be of a magnitude which is some fraction of $|V'_x||V'_y|$. Furthermore, we can show that $\overline{V'_x V'_y}$ is negative by the following argument: If a chunk of fluid from the faster region above y in Fig. 9.33 moves into region y, then V'_x at y will be positive. This causes the average velocity in the y direction $\bar{V}_y$ to decrease from continuity considerations, so V'_y is then negative. The product of $V'_x V'_y$ is then negative, on the average, so the mean-time average is negative. On the other hand, a negative V'_x will, on the average, induce a positive V'_y, so again, for the profile shown in Fig. 9.33 the mean-time average of the product $V'_x V'_y$ is negative. We can therefore write the following relation:

$$\overline{V'_x V'_y} = -k|\overline{V'_x}||\overline{V'_y}|$$

where k represents some fraction. Using the above relation we may then give the shear stress in Eq. (9.35) as

$$(\tau_{xy})_{\text{app}} = \rho k|\overline{V'_x}||\overline{V'_y}|$$

Now employing Eq. (9.39) to replace $|\overline{V'_x}||\overline{V'_y}|$, we get

$$(\tau_{xy})_{\text{app}} = \rho(\text{const } kl^2)\left(\frac{d\bar{V}_x}{dy}\right)^2 \qquad (9.40)$$

The term $(\text{const } kl^2)$ is combined to form the square of new mixing length given as $\tilde{l}^2$. The major variations in the expression $(\text{const } kl^2)$ arise from the variations of l^2. Hence, *the new mixing length $\tilde{l}$ may be associated with the same mechanism described earlier, without serious error.* Upon using this quantity, we find that the result above becomes

$$(\tau_{xy})_{\text{app}} = \rho\tilde{l}^2\left(\frac{d\bar{V}_x}{dy}\right)^2 \qquad (9.41)$$

We have thus developed an equation equivalent to the Boussinesq formulation [Eq. [9.33], with the advantage of having greater physical meaning.

Near a boundary, there must be a decrease in linear momentum exchange, since the effects of turbulence tend to be suppressed as one gets closer to a boundary. This means that the mixing length decreases toward a boundary. Prandtl hypothesized that

$$\tilde{l} = \alpha y \qquad (9.42)$$

where y is the distance normal to the boundary and α is proportionality constant, later to be determined by experiment. Furthermore, he assumed that the *apparent* shear stress τ_0 is *constant* near a boundary. The justification of this assumption lies in the experimental correlation of the resulting formulation. Substituting the assumed $\tilde{l}$ function of Eq. (9.42) in Eq. (9.41), we get the following equation

$$\tau_0 = \rho(\alpha y)^2\left(\frac{d\bar{V}_x}{dy}\right)^2 \qquad (9.43)$$

Upon taking the root of both sides and separating variables, we may write the following differential equation:

$$d\bar{V}_x = \frac{1}{\alpha}\sqrt{\frac{\tau_0}{\rho}}\frac{dy}{y} = \frac{1}{\alpha}V_*\frac{dy}{y} \tag{9.44}$$

where the term $\sqrt{\tau_0/\rho}$ denoted as V_* has the dimensions of velocity and is called the *shear-stress velocity* or the *friction velocity*. Integrating, we get

$$\bar{V}_x = \frac{1}{\alpha}V_* \ln y + C \tag{9.45}$$

A useful form of the equation above may be established by considering the maximum mean-time-average velocity to occur at some position h. That is,

$$(\bar{V}_x)_{max} = \frac{1}{\alpha}V_* \ln h + C \tag{9.46}$$

Subtracting (9.45) from (9.46) results in a *velocity-difference* or *velocity defect* equation involving maximum conditions in place of the constant of integration:

$$\frac{(\bar{V}_x)_{max} - \bar{V}_x}{V_*} = -\frac{1}{\alpha}\ln\frac{y}{h} \tag{9.47}$$

It might appear, in the light of the development of the mixing length and the assumptions of this section, that the velocity-difference equation is applicable only to parallel two-dimensional flow in a region restricted close to the solid boundary. It must be kept in mind that many of the considerations leading to this equation are hypothetical in nature and are yet to be justified. Equally important is the fact that the range of application is also open to question. Experimental evidence indicates that we have the happy situation that this relation is valid over the entire region of two-dimensional flow except near to the wall. This region is called the *outer region*. Furthermore, this result is valid also for turbulent *pipe flow* away from the wall. This region is also denoted as the outer region. Thus, where y is considered the radial distance from the pipe wall and h is replaced by R, the pipe radius, this relation shows excellent agreement with pipe-flow data. In Fig. 9.34 data from Nikuradse have been compared with the results of the equation

$$\frac{(\bar{V}_x)_{max} - \bar{V}_x}{V_*} = -\frac{1}{0.4}\ln\frac{y}{R} \tag{9.48}$$

where the value of α has been taken as 0.4 to achieve the best correlation for pipes ($\alpha = 0.417$ for flat plates). From this diagram, it is clear that the equation above is independent of the pipe friction factor, since the experimental points,

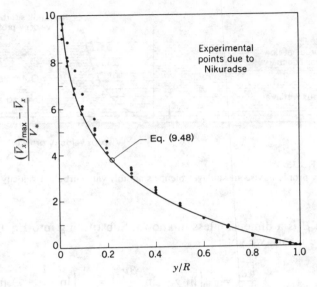

FIGURE 9.34
Velocity difference equation compared with experimental results of Nikuradse.

of which only a small number are shown, have been taken over a wide range of roughness coefficients.

Let us now return to Eq. (9.45) with a view to determining the constant of integration in order to arrive at a velocity distribution rather than the less-informative velocity-difference distribution. *It will be found that whereas the velocity-difference relation was noted to be independent of roughnesses, the velocity distribution itself is definitely affected by the pipe roughness*. For this reason, it becomes necessary to consider the thin viscous sublayer introduced in Sec. 9.7 and to express the final results separately for each of the three zones of pipe flow also described in Sec. 9.7.

Since the formulations of this section are for turbulent-flow conditions, it would not be expected that the results have any validity in the viscous sublayer. The constant of integration of Eq. (9.45) may then be adjusted to yield a zero velocity at a position y_0 *inside* the sublayer. This position is presumed chosen so that the turbulent-velocity profile blends smoothly with the viscous sublayer profile somewhere in what we shall call the *overlap layer* between the sublayer and the turbulent flow. This is illustrated in Fig. 9.35. Under these conditions, Eq. (9.45) becomes

$$\overline{V}_x = \frac{1}{\alpha} V_* (\ln y - \ln y_0) \tag{9.49}$$

It will be convenient for dimensional reasons to replace the yet unknown quantity y_0 in the following manner:

$$y_0 = \beta \frac{\nu}{V_*}$$

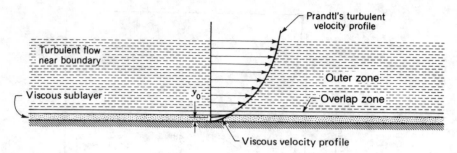

FIGURE 9.35
Velocity profile in viscous sublayer blends smoothly with turbulent-velocity profile.

where β is a dimensionless unknown. Substituting into Eq. (9.49) and rearranging the terms, we get[22]

$$\frac{\bar{V}_x}{V_*} = \frac{1}{\alpha}\left(\ln y - \ln \frac{\beta\nu}{V_*}\right) = \frac{1}{\alpha}\left(\ln \frac{yV_*}{\nu} - \ln \beta\right) \tag{9.50}$$

Letting $-\ln \beta/\alpha = B$ we then have for the above equation

$$\boxed{\frac{\bar{V}_x}{V_*} = \frac{1}{\alpha}\ln \frac{yV_*}{\nu} + B} \tag{9.51}$$

The constant α has been shown to equal 0.4 and to be independent of boundary roughness. We will later establish values of B for certain conditions.

Approach 2. Dimensional Analysis. We now proceed via a much shorter approach to reach the same results of Approach 1 via dimensional analysis followed by experiment. For the *viscous sublayer* region (see Fig. 9.35) we can assume that the velocity V_x is a function of four variables. That is,

$$V_x = f(\tau_w, \mu, \rho, y)$$

We leave it for you to show that dimensional analysis relates two π's. In particular, we have

$$\frac{V_x}{V_*} = f\left(\frac{yV_*}{\nu}\right) \tag{9.52}$$

---

[22] Note that by replacing y_0 by $\beta\nu/V_*$ we get a Reynolds number yV_*/ν in Eq. (9.51) that plays an important role in turbulent flow analysis.

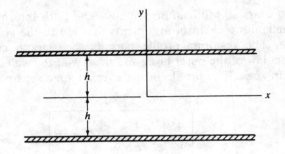

FIGURE 9.36
Turbulent flow between two parallel infinite plates.

where $V_* = \sqrt{\tau_w/\rho}$ has dimensions of velocity and as pointed out in Approach 1 for this reason is called the *friction velocity*. This equation is called the *law at the wall*. Experiment then indicates that near the wall

$$\frac{V_x}{V_*} = \left(\frac{yV_*}{\nu}\right) \tag{9.53}$$

The region directly beyond the viscous sublayer in Fig. 9.35 is the *overlap zone* where viscous and turbulent effects are significant, and, beyond this zone, is the *outer layer* where turbulent effects predominate.

For the *outer layer* we have for the *velocity defect*, $[(\bar{V}_x)_{max} - \bar{V}_x]$, the following functional relation pertaining to a flow of height $2h$ (see Fig. 9.36) where the flow over the height h away from the wall is considered:

$$\left[(\bar{V}_x)_{max} - \bar{V}_x\right] = F(\tau_0, h, \rho, y)$$

From dimensional analysis we then get

$$\frac{(\bar{V}_x)_{max} - \bar{V}_x}{V_*} = F\left(\frac{y}{h}\right) \tag{9.54}$$

This is the *velocity defect law*.[23] For a pipe the h is replaced by R, the pipe radius. Experiment then permits one to arrive at the equation

$$\frac{(\bar{V}_x)_{max} - \bar{V}_x}{V_*} = -\frac{1}{\alpha}\ln\frac{y}{R} \tag{9.55}$$

where α is a constant. Note the good agreement achievable in Fig. 9.34 when using $\alpha = 0.4$.

---

[23]For a more careful development of the law at the wall and the velocity defect equations see H. Tennekes and J. L. Lumley, *A First Course in Turbulence*, MIT Press, Cambridge, Mass., 1972.

In 1939 Clark B. Millikan presented at the Fifth International Congress of Applied Mechanics a chain of arguments that led to the velocity profile in the *overlap zone*. The resulting profile permits the smooth continuation of the velocity defect law of the outer layer and the law at the wall at their conjunction in the overlap zone. The profile in the overlap zone set forth by Milliken is as follows:

$$\frac{\bar{V}_x}{V_*} = \frac{1}{\alpha} \ln \frac{yV_*}{\nu} + B \qquad (9.56)$$

where α and B are constants. As a result of this successful formulation, the overlap zone is often called the *logarithmic-overlap layer*. This development is considered one of the major achievements in turbulence theory. Furthermore, with values of $\alpha \approx 0.4$ and $B \approx 5$ this formulation can *also* be considered valid for the *outer zone*. We have thus reproduced the key results of approach 1 in the outer layer plus the law at the wall for the viscous sublayer.

9.15 DETAILS OF VELOCITY PROFILES FOR SMOOTH AND ROUGH PIPES

Smooth Pipes

In the *smooth pipe*, experiments by Nikuradse indicate for Eq. (9.56) that B equals 5.5 and $\alpha = 0.4$. Thus, we have the so called logarithmic velocity law given as:

$$\frac{\bar{V}_x}{V_*} = 2.5 \ln \frac{yV_*}{\nu} + 5.5 \qquad \text{Smooth pipes} \qquad (9.57)$$

The equation (9.57) has been plotted in Fig. 9.37, along with experimental data taken over a wide range of Reynolds numbers. *Very good correlation is achieved as long as* $\ln(yV_*/\nu)$ *exceeds a value of* 3. In the region to the left of this position on the graph it will next be shown that viscous effects are significant, thus rendering the turbulent-flow computation invalid here.

Assume that the laminar shear stress in the sublayer is constant and equal to τ_0. Newton's viscosity law may be employed. Thus

$$\tau_0 = \mu \frac{dV_x}{dy}$$

Integrating,

$$V_x = \frac{\tau_0 y}{\mu} + \text{const}$$

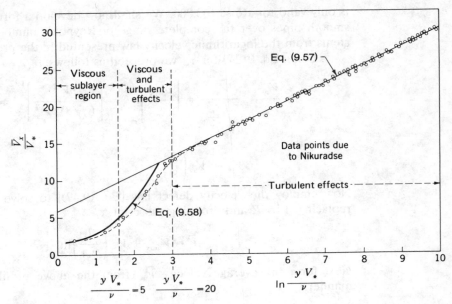

FIGURE 9.37
Velocity profile for turbulent flow in *smooth* pipes at *high* Reynolds number.

The constant integration must be zero to give a zero velocity at the pipe wall. Dividing both sides by V_* and replacing μ by $\rho\nu$ results in a relation involving the variables of the plot in Fig. 9.37. Thus we reproduce Eq. (9.53)

$$\frac{V_x}{V_*} = \frac{yV_*}{\nu} \tag{9.58}$$

This relation has been shown in the diagram as Eq. (9.58). Note that there is reasonably good agreement with experimental data in the region of the diagram identified as the viscous sublayer region. Hence, we may assume with confidence that in this region viscous effects dominate. The outer fringe of the sublayer may be considered to correspond to the position $yV_*/\nu \approx 5$, as indicated in the diagram. From this an estimation of the viscous-sublayer thickness λ for this zone of flow may be made. Thus

$$\lambda \approx \frac{5\nu}{V_*} \tag{9.59}$$

Between $yV_*/\nu = 5$ and $yV_*/\nu = 20$ (the overlap zone), we have both viscous and turbulence effects with the viscous effects decreasing as we approach $yV_*/\nu = 20$.

Earlier, for *smooth pipes* we have given a formula for f in lieu of the Moody diagram. This is the *Blasius* formula [see Eq. (9.21)]. However this result is only valid for Re < 100,000. We shall now develop a formula for f valid for smooth pipes over the complete range of Reynolds numbers. This new result stems from the logarithmic velocity law presented in the previous section.

Using Eq. (9.17) for τ_w we proceed as follows:

$$\frac{\overline{V}_{av}}{V_*} = \frac{\overline{V}_{av}}{\sqrt{\dfrac{\tau_w}{\rho}}} = \left[\frac{\rho \overline{V}_{av}^2}{\tau_w}\right]^{1/2} = \left[\frac{\rho V^2}{(f/8)\rho V^2}\right]^{1/2} = \left[\frac{8}{f}\right]^{1/2}$$

$$\therefore V_* = \overline{V}_{av}\left(\frac{f}{8}\right)^{1/2} \tag{9.60}$$

Now employ the velocity defect law [Eq. (9.47)] to solve for $\overline{V}$. We get on replacing h by R and dropping subscript x,

$$\overline{V} = \overline{V}_{max} + V_*\left(\frac{1}{0.4}\right)\ln\frac{y}{R}$$

Next form the average velocity $\overline{V}_{av}$ from the above profile in the following manner:

$$\overline{V}_{av} \equiv \frac{1}{\pi R^2}\int_0^R \overline{V}(2\pi)(R - y)\,dy = \frac{\overline{V}_{max}}{\pi R^2}\int_0^R 2\pi(r - y)\,dy$$

$$+ \frac{V_*}{(0.4\pi R^2)}\int_0^R 2\pi(R - y)\ln\frac{y}{R}\,dy$$

Carrying out the integration we get

$$\overline{V}_{av} = \overline{V}_{max} - 3.75V_* \tag{9.61}$$

As a next step we rewrite Eq. (9.56) for *smooth pipes* ($\alpha = 0.4$, $B = 5.5$) for $y = R$ to give $\overline{V}_{max}$. Thus

$$\overline{V}_{max} = \frac{V_*}{0.4}\left[\ln\frac{RV_*}{\nu} + 2.2\right] \tag{9.62}$$

Replacing $\overline{V}_{max}$ in Eq. (9.61) using the above result gives us

$$\overline{V}_{av} = \frac{V_*}{0.4}\left[\ln\frac{RV_*}{\nu} + 2.2\right] - 3.75V_*$$

$$\therefore \overline{V}_{av} = V_*\left[2.5\ln\frac{RV_*}{\nu} + 1.750\right] \tag{9.63}$$

Let us now examine the Reynolds number RV_*/ν. Using Eq. (9.60) to replace V_* we get

$$\frac{RV_*}{\nu} = \frac{R\sqrt{f}}{\nu\sqrt{8}}\overline{V}_{av} = \frac{D\sqrt{f}}{4\sqrt{2}\,\nu}\overline{V}_{av} \tag{9.64}$$

Going to Eq. (9.62) and replacing RV_*/ν with the above result and simultaneously replacing $\bar{V}_{av}$ using Eq. (9.63), we get for f, after some arithmetic and after shifting from ln (base e) to log (base 10),

$$f = \left[2.035 \log \left(\frac{\bar{V}_{av} D}{\nu} \sqrt{f} \right) - 0.91 \right]^{-2}$$

$$\therefore \frac{1}{\sqrt{f}} = 2.035 \log \left[\frac{\bar{V}_{av} D}{\nu} \sqrt{f} \right] - 0.91 \qquad (9.65)$$

This equation may be further adjusted to fit experimental data as follows:

$$\boxed{\frac{1}{\sqrt{f}} = 2.0 \log \left[\frac{\bar{V}_{av} D}{\nu} \sqrt{f} \right] - 0.8} \qquad (9.66)$$

This result is called the *Prandtl universal law of friction* for smooth pipes. We have thus described the velocity field and friction factor for smooth pipes for high Reynolds numbers. We next consider rough pipes.

Rough Pipes

For rough pipes, we must consider the three zones of flow set forth in Sec. 9.7. For the *smooth-pipe zone*[24] we may use Eq. (9.57) outside the viscous sublayer with $B = 5.5$ and $\alpha = 0.4$ as we did for smooth pipes. Thus

$$\boxed{\frac{\bar{V}_x}{V_*} = 2.5 \ln \frac{y V_*}{\nu} + 5.5} \qquad \text{Smooth pipe zone} \qquad (9.67)$$

For *completely rough* zone of flow, where the roughness has been completely exposed to the turbulent region of flow, we still use Eq. (9.56) wherein we have the following relation:

$$B = \frac{1}{\alpha} \left(3.4 - \ln \frac{e V_*}{\nu} \right) \qquad (9.68)$$

Note that eV_*/ν is a Reynolds number based on the shear-stress velocity V_* and now with the roughness measure e. The resulting velocity profile now extends over the entire cross section except in the region at the wall protur-

---

[24] Recall that in the smooth-pipe zone, the viscous sublayer covers the roughness of the pipe surface. If the Reynolds number is high enough, the roughness of the surface protrudes almost entirely outside the viscous sublayer and we have rough-pipe zone of flow.

bances and is given as follows:

$$\frac{\bar{V}_x}{V_*} = \frac{1}{\alpha}\left[\ln\frac{yV_*}{\nu} + 3.4 - \ln\frac{eV_*}{\nu}\right]$$

$$\therefore \boxed{\frac{\bar{V}_x}{V_*} = \frac{1}{\alpha}\ln\frac{y}{e} + 8.5} \quad \text{Rough zone} \qquad (9.69)$$

Note that ν no longer appears. This means that for completely rough zone of flow the profile is independent of Reynolds number. This is consistent with the Moody diagram where the friction factor is independent of the Reynolds number for the rough pipe zone.

In Prob. 9.105 we ask you to show that the average velocity $\bar{V}_{av}$ is given for the rough zone using $\alpha = 0.4$ as follows:

$$\frac{\bar{V}_{av}}{V_*} = 2.5\ln\frac{D}{e} + 3.0 \qquad (9.70)$$

We can rewrite the left side of the above equation by using Eq. (9.60) to form the following result

$$\frac{\bar{V}_{av}}{V_*} = \left[\frac{8}{f}\right]^{1/2} \qquad (9.71)$$

Hence, substituting the above result into Eq. (9.70) and solving for f we get

$$\boxed{\frac{1}{f^{1/2}} = -2.0\log\frac{e/D}{3.6}} \quad \text{Rough zone} \qquad (9.72)$$

Finally, in the *transition zone*, an approximate curve for B has been developed from the data of Nikuradse. This is shown in Fig. 9.38. These data

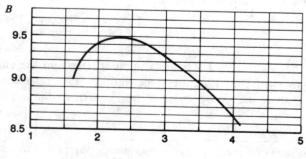

FIGURE 9.38
Plot of B for transition zone from data by Nikuradse.

are used in Eq. (9.56) outside the viscous sublayer. We can then use the formula

$$\boxed{\frac{\bar{V}_x}{V^*} = \frac{1}{\alpha} \ln \frac{yV_*}{\nu} + B} \qquad \text{Transition zone} \qquad (9.73)$$

How do we know whether we are in the smooth-pipe zone, transition-zone flow, or completely rough-pipe zone of the flow? It has been found that when the Reynolds number eV_*/ν was less than 5, the flows in the pipes were the same as if the pipes were smooth, and for $eV_*/\nu > 70$ the flows correspond to completely rough pipe flow. Thus we have

$$\frac{eV_*}{\nu} < 5 \quad \text{Smooth-pipe zone}$$

$$5 < \frac{eV_*}{\nu} < 70 \quad \text{Transition zone} \qquad (9.74)$$

$$\frac{eV_*}{\nu} > 70 \quad \text{Rough-pipe zone}$$

9.16 PROBLEMS FOR HIGH REYNOLDS NUMBER FLOW

We have formulated velocity profiles in pipe flow having high Reynolds number for both smooth and rough pipes. In all the formulations, we have the stress τ_0 present, which, for our general purposes, we can consider as the wall friction τ_w. Additionally, in the rough-pipe cases we have the roughness factor e, which we must determine along with τ_w in order to permit us to give the velocity profiles. The factor e is easily found for standard pipes by consulting the Moody chart in the same way that we did earlier. The friction factor f we then find using the Moody curves or by using the various friction factor formulas presented in Part E. The shear stress is then available from Eq. (9.17) which we again rewrite for convenience

$$\tau_w = \frac{f}{4} \frac{\rho V^2}{2}$$

where $V = q/A$.

We can now compute V_* $(= \sqrt{\tau_w/\rho})$ and then the Reynolds number eV_*/ν. Next, use the criteria for determining the zone of flow given by (9.74). The proper profile equation can then be singled out for use. We now illustrate the use of the formulations of this section.

Example 9.7. Consider a flow of gasoline through a 152-mm cast-iron pipe. What is the velocity profile for a volume flow of 170 L/s? Also, ascertain the drag induced by the flow on a unit length of the pipe. Take the kinematic viscosity of gasoline as 0.37×10^{-6} m^2/s and the mass density as 670 kg/m^3.

Our first step in this analysis will be to determine the Reynolds number:

$$\text{Re}_D = \frac{VD}{\nu} = \frac{\{0.170/[\pi(0.152^2)/4]\}(0.152)}{0.37 \times 10^{-6}} = \frac{(9.37)(0.152)}{0.37 \times 10^{-6}} = 3.85 \times 10^6$$

For this high Reynolds number we use the velocity-profile formulations of the preceding sections. Our next step is to determine what zone of pipe flow we are in. We get for e, using Table 9.1, the value of 0.26 mm and so for a 152-mm diameter pipe, $e/D = 0.0017$. We require the evaluation of τ_w, and for this we employ Eq. (9.17). Hence,

$$\tau_w = \frac{f}{4}\frac{\rho V^2}{2} = \frac{f}{4}\left[\frac{(670)(9.37^2)}{2}\right] = 7353f \qquad (a)$$

We determine f from the Moody diagram (Fig. 9.16) for an e/D of 0.0017 and a Reynolds number of 3.85×10^6, getting the value $f = 0.0223$. The shear stress τ_w is then 164.0 Pa.

To determine the zone of flow, we calculate eV_*/ν. Thus, using $e = 0.26 \times 10^{-3}$ mm, we have

$$\frac{eV_*}{\nu} = \frac{e\sqrt{\tau_w/\rho}}{\nu} = (0.26 \times 10^{-3})\frac{\sqrt{164.0/670}}{0.37 \times 10^{-6}} = 348 \qquad (b)$$

We are accordingly well into the rough-pipe zone. We can then say for the velocity profile using Eqs. (9.69) that

$$\bar{V} = V^*\left[2.5 \ln \frac{y}{e} + 8.5\right]$$

$$\bar{V} = \sqrt{\frac{164.0}{670}}\left[2.5 \ln y - 2.5 \ln(0.00026) + 8.5\right]$$

$$\therefore \bar{V} = 1.237 (\ln y + 11.66)$$

The maximum velocity occurs at $y = 0.152/2$ m. Thus,

$$(\bar{V})_{max} = 11.24 \text{ m/s}$$

Had this been laminar flow, the maximum velocity would have been 18.73 m/s. We see that the turbulent profile is then "flatter" than the paraboloidal profile of laminar flow.

The drag per unit length of pipe is now easily determined. Thus

$$\text{Drag} = (\tau_w)(\pi D) = (164.0)(\pi)(0.152) = 78.4 \text{ N/m}$$

Example 9.8. Water at 60°F is flowing in a 12-in commercial steel pipe at the rate of 70 ft^3/s. Compute $(\bar{V})_{max}$ and $(\bar{V})_{max}/V$ where $V = q/A$.

We start with Re_D. Thus

$$\text{Re}_D = \frac{VD}{\nu} = \frac{[70/(\pi 1^2/4)](1)}{1.217 \times 10^{-5}} = 7.32 \times 10^6$$

From Table 9.1, we see that $e = 0.00015$ ft. From the Moody chart for $\text{Re}_D = 7.32$

$\times 10^6$ and $e/D = 0.00015$, we have for f the value 0.013. Now we can compute τ_w to be

$$\tau_w = \frac{f}{4}\frac{\rho V^2}{2} = \frac{0.013}{4}\frac{(1.938)\left[70/\left(\pi 1^2/4\right)\right]^2}{2} = 25 \text{ lb/ft}^2$$

Next calculate V_*:

$$V_* = \sqrt{\frac{\tau_w}{\rho}} = \sqrt{\frac{25}{1.938}} = 3.59 \text{ ft/s}$$

We now can compute the Reynolds number eV^*/ν to be

$$\frac{eV_*}{\nu} = \frac{(0.00015)(3.59)}{1.217 \times 10^{-5}} = 44.3$$

Clearly, we are in the transition zone. We get for B from Fig. 9.38 the value 9. Hence from Eq. (9.73) we have

$$\begin{aligned}\overline{V} &= \frac{V_*}{\alpha}\ln\left(\frac{yV_*}{\nu}\right) + V_* B \\ &= 8.98\ln\left[2.95 \times 10^5 y\right] + 32.3\end{aligned} \qquad (a)$$

The maximum velocity is found at $y = 6$ in. Thus

$$\overline{V}_{max} = 8.98\ln\left[(2.95 \times 10^5)(0.5)\right] + 32.3 = 139.2 \text{ ft/s}$$

$$\therefore \frac{\overline{V}_{max}}{V} = \frac{132.9}{70/\left(\pi 1^2/4\right)} = 1.491$$

PART F
MULTIPLE PATH PIPE FLOW

*9.17 MULTIPLE-PATH PIPE PROBLEMS

Of significant industrial importance are multiple-path pipe systems, of which a simple example is shown in Fig. 9.39. To use the parlance of electric circuits, this network has two nodes, three branches, and two loops. We will set forth a

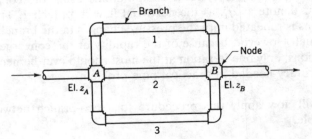

FIGURE 9.39
Multiple-path pipe system.

method of analyzing pipe networks with only two nodes but with any number of branches. However, for simplicity, in our discussion we consider the three-branch network indicated in the diagram.

We may now pose two problems for two-node-pipe networks.

1. Flow conditions at nodal point A, ahead of the pipe grid, are known, as are pipe-roughness coefficients and the entire pipe geometry. The pressure at nodal point B, downstream of the grid is desired.
2. Pressures at A and B are known, as well as pipe roughnesses and geometry. Determine the total flow q.

Case 1. A direct solution will not be used for such problems in the turbulent-flow range; instead we use Moody's chart. An effective iterative procedure may then be performed in the following manner:

1. Assume a flow q_1' through branch 1 and solve for $(p_A - p_B)/\rho$ for branch 1 using the first law of thermodynamics or modified Bernoulli (Case 1 of single-path pipe examples).
2. Using this computed pressure change, determine flows for all other branches using the first law of thermodynamics or modified Bernoulli. We denote these flows here as q_2' and q_3' (Example 9.4 of Case 2 for single-path examples). Generally $q_1' + q_2' + q_3' \neq q_{in}$, where q_{in} is the inlet flow and hence continuity will not be satisfied.
3. For this reason, assume that the actual flow divides up into flows q_1, q_2, and q_3 in the same proportions as q_1', q_2', and q_3' as required by the first law of thermodynamics, but such that $q_1 + q_2 + q_3 = q_{in}$ as required by conservation of mass.
4. With these new terms, solve for the pressure change $(p_A - p_B)/\rho$ for each branch. For a correct formulation of the problem, these should be mutually equal. A reasonably good choice of q_1' based on considerations of pipe geometries and roughnesses will yield a set of q's giving pressure changes for each branch very close to each other in value. The actual value of $(p_A - p_B)/\rho$ may then be taken as the average of the computed value for the various branches.
5. Should the pressure-change check not be satisfactory, it is necessary to repeat the whole cycle of computations using the computed q_1, which we now denote as q_1'', as the next trial flow in branch 1. This cycle of computations is repeated until the pressure changes in the branches are close to being equal. However, because of the rapidity of the convergence of these computations, only one iteration at the most should ever be necessary to achieve the accuracy needed in practical pipe problems.

We will now apply this procedure to a two-branch network in the following example.

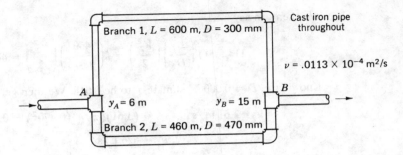

FIGURE 9.40
Two-branch pipe system.

Example 9.9. A flow of 570 L/s is proceeding through the pipe network shown in Fig. 9.40. For a pressure of 690 kPa gage at node A, what pressure may be expected at node B? Neglect minor losses. Take $\rho = 1000$ kg/m³.

Assume that a flow q_1' of 170 L/s is proceeding through branch 1. The velocity V_1 and other data are then

$$V_1 = \frac{q_1'}{A_1} = \frac{(170 \text{ L/s})(1 \text{ m}^3/1000 \text{ L})}{(\pi)(0.300^2)/4} = 2.405 \text{ m/s}$$

$$\text{Re}_D = \frac{V_1 D_1}{\nu} = \frac{(2.405)(0.300)}{0.0113 \times 10^{-4}} = 6.38 \times 10^5$$

$$\frac{e}{D_1} = 0.0009$$

$$f = 0.0198$$

$$h_l = f\frac{L_1}{D_1}\frac{V_1^2}{2} = (0.0198)\left(\frac{600}{0.300}\right)\left(\frac{2.405^2}{2}\right) = 114.5 \text{ N} \cdot \text{m/kg}$$

For the *first law of thermodynamics* (or modified Bernoulli) applied to branch 1, we get

$$\frac{(V_1^2)_A}{2} + \left(\frac{p_A}{\rho}\right)_1 + 6g = \frac{(V_1^2)_B}{\rho} + \left(\frac{p_B}{\rho}\right)_1 + 15g \mp 114.5$$

$$\therefore \left(\frac{p_A - p_B}{\rho}\right)_1 = 114.5 + (15 - 6)g = 202.8 \text{ N} \cdot \text{m/kg}$$

Using this pressure formulation, consider the *first law of thermodynamics* for branch 2. Thus,

$$202.8 + 6g + \frac{(V_2^2)_A}{2} = \frac{(V_2^2)_B}{2} + 15g + (h_l)_2$$

$$\therefore (h_l)_2 = 114.5 \text{ N} \cdot \text{m/kg}$$

We then can say

$$114.5 = f\left(\frac{L_2}{D_2}\right)\left(\frac{V_2^2}{2}\right) = f\left(\frac{460}{0.470}\right)\left(\frac{V_2^2}{2}\right) = 489 f V_2^2$$

Knowing $e/D = 0.0006$, estimate f to be 0.018. We then get for V_2 and q_2'

$$V_2 = 3.61 \text{ m/s} \qquad q_2' = (3.61)(\pi/4)(0.470^2) = 0.626 \text{ m}^3/\text{s}$$

$$\therefore q_2' = 626 \text{ L/s}$$

Now the desired actual flows q_1 and q_2 may be computed so as to have the ratio q_1/q_2 equal to q_1'/q_2' as determined by the first law of thermodynamics, and should also satisfy continuity so that $q_1 + q_2 = q_A = 570$ L/s. This may be accomplished in the following manner[25]:

$$q_1 = \frac{q_1'}{q_1' + q_2'} 570 = \frac{170}{170 + 626} 570 = 121.7 \text{ L/s} \qquad (a)$$

$$q_2 = \frac{q_2'}{q_1' + q_2'} 570 = \frac{626}{170 + 626} 570 = 448.3 \text{ L/s} \qquad (b)$$

The corresponding velocities for these flows are

$$V_1 = \frac{121.7}{(\pi)(0.300^2/4)} \frac{1}{1000} = 1.722 \text{ m/s}$$

$$V_2 = \frac{448.3}{(\pi/4)(0.470^2)} \frac{1}{1000} = 2.58 \text{ m/s}$$

Hence,
$$(\text{Re}_D)_1 = \frac{(1.722)(0.300)}{0.0113 \times 10^{-4}} = 4.57 \times 10^5$$

$$(\text{Re}_D)_2 = \frac{(2.58)(0.470)}{0.0113 \times 10^{-4}} = 1.073 \times 10^6$$

---

[25]Note that q_1 and q_2 have the same ratio as q_1' and q_2' from the given relation. Thus, dividing Eq. (a) by Eq. (b), we get

$$\frac{q_1}{q_2} = \frac{[q_1'/(q_1' + q_2')](570)}{[q_2'/(q_1' + q_2')](570)} = \frac{q_1'}{q_2'}$$

Clearly, the new q's have the desired ratio. Now, add Eqs. (a) and (b) to show that the new q's add up to 570 L/s, thus satisfying continuity.

$$q_1 + q_2 = \frac{q_1'}{q_1' + q_2'} 570 + \frac{q_2'}{q_1' + q_2'} 570$$

$$= \frac{q_1' + q_2'}{q_1' + q_2'} 570 = 570 \text{ L/s}$$

The friction factors for the two flows are then

$$f_1 = 0.0198$$

$$f_2 = 0.018$$

Now compute $(p_A - p_B)/\rho$ for each branch using the *first law of thermodynamics*.

$$\frac{(V_1^2)_A}{2} + \frac{(p_A)_1}{\rho} + 6g = \frac{(V_1^2)_B}{2} + \frac{(p_B)_1}{\rho} + 15g + (h_l)_1$$

$$\therefore \left[\frac{p_A - p_B}{\rho}\right]_1 = 9g + (0.0198)\left(\frac{600}{0.300}\right)\frac{1.722^2}{2} = 147.0 \text{ N} \cdot \text{m/kg}$$

$$\frac{(V_2^2)_A}{2} + \frac{(p_A)_2}{\rho} + 6g = \frac{(V_2^2)_B}{2} + \frac{(p_B)_2}{\rho} + 15g + (h_l)_2$$

$$\left[\frac{p_A - p_B}{\rho}\right]_2 = 9g + (0.0180)\left(\frac{460}{0.470}\right)\left(\frac{2.58^2}{2}\right) = 146.9 \text{ N} \cdot \text{m/kg}$$

In comparing $(p_A - p_B)/2$ for each branch, note that we are quite close, so no further work need be done. If there is considerable divergence in the results, we take the q_1 from Eq. (*a*) and, calling it q_1'', start all over again in the cycle. We repeat until the $(p_A - p_B)/\rho$ is close to the same value for each branch.

We can now get the desired pressure p_B using the average of the computed pressure changes throughout the branches. Thus we have

$$p_B = p_A - \rho(146.95) = 690 \times 10^3 - (1000)(146.95)$$

$$\therefore p_B = 543 \text{ kPa}$$

Case 2. This degenerates to Case 2 of the single-path pipe problem (see Example 9.4) for each path, because the end pressure of the grid may with good accuracy be considered the end pressures in the pipes and thus there is complete flow information at two points in each pipe.

For the handling of more complicated pipe networks such as the network shown in Fig. 9.41, numerical procedures have been developed by Hardy

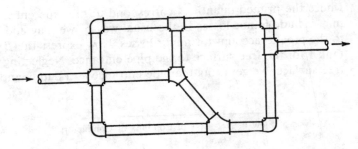

FIGURE 9.41
Complex pipe network.

Cross.[26] The reader is referred to his work for further information. Also there are computer programs for handling complex networks.

*9.18 BRANCHING PIPES

We now consider another kind of multiple-path-pipe problem as shown in Fig. 9.42. Note that three or more large tanks at various elevations and under various pressures are interconnected, forming a joint J at some unspecified elevation above the datum.[27] We wish to determine the various flows. We do not know the direction of the flows in or out of the tanks.

As a first step, we compute the total head H_D at the free surface of each tank. Neglect kinetic energy head at each free surface. The tank having the largest head of the three at the free surface can be considered a tank from which fluid must issue. Let us say that it is tank 1. At the joint, the flow q_1 goes to tanks 2 and 3 if an estimated total head at joint J, $(H_D)_J$, exceeds $(H_D)_2$ as well as $(H_D)_3$. If this is the case, we can say from continuity that

$$q_1 = q_2 + q_3 \tag{9.75}$$

We cannot estimate $(H_D)_J$ to have a value such that $(H_D)_J < (H_D)_2$ *and* $(H_D)_J < (H_D)_3$ since fluid must then issue *from* tanks 2 and 3 as well as *from* tank 1 (latter deduced earlier)—clearly an impossibility. If, on the other hand, $(H_D)_J < (H_D)_2$ only or if $(H_D)_J < (H_D)_3$ only, we can say for these respective cases:

$$q_1 + q_2 = q_3 \tag{9.76a}$$

$$q_1 + q_3 = q_2 \tag{9.76b}$$

Assume here $(H_D)_J > (H_D)_2$, and $(H_D)_J > (H_D)_3$. Hence, the flow for this premise must satisfy Eq. (9.75). To test the validity of an estimate $(H_D)_J$ satisfying these inequalities, we employ the first law of thermodynamics (or modified Bernoulli) for each pipe. Look at pipe 1 first. The first law states that

$$\frac{p_1'}{\gamma} + \frac{V_1^2}{2g} + z_1' = \left(\frac{p_J}{\gamma} + \frac{V_1^2}{2g} + z_J \right) + \frac{(h_l)_1}{g} \tag{9.77}$$

where the primed quantities correspond to the pipe entrances to the tank. We may include minor losses in the usual way or we may add an equivalent length to the pipe to account for minor losses. Now express the *Bernoulli's* equation in tank 1 for the free surface to the pipe entrance. Neglecting kinetic energy at the free surface, we get using the pipe entrance as a datum and noting that h refers

---

[26]"Analysis of Flow in Networks of Conduits or Conductors," *Univ. Illinois Eng. Expt. Sta. Bull.* 286, 1936.

[27]A special case of this would be interconnecting reservoirs at different elevations where the pressures p_i would all be the same value—p_{atm}.

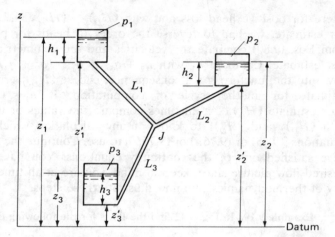

FIGURE 9.42
Branched pipes.

to the height of liquid in a tank

$$\frac{p_1}{\gamma} + h_1 = \frac{p'_1}{\gamma} + \frac{V_1^2}{2g} \tag{9.78}$$

Hence,

$$\frac{p'_1}{\gamma} = \frac{p_1}{\gamma} + h_1 - \frac{V_1^2}{2g} \tag{9.79}$$

Substituting into Eq. (9.77), we get on noting that $(p_J/\gamma + V_1^2/2g + z_J) = (H_D)_J$ on the right side of Eq. (9.77):

$$\frac{p_1}{\gamma} + (h_1 + z'_1) + \frac{V_1^2}{2g} - \frac{V_1^2}{2g} = (H_D)_J + \frac{(h_l)_1}{g}$$

Noting that $h_1 + z'_1 = z_1$ and that $p_1/\gamma + z_1 = (H_D)_1$, we have

$$(H_D)_1 - (H_D)_J = \frac{(h_l)_1}{g} \tag{9.80a}$$

Similarly,

$$(H_D)_J - (H_D)_2 = \frac{(h_l)_2}{g} \tag{9.80b}$$

$$(H_D)_J - (H_D)_3 = \frac{(h_l)_3}{g} \tag{9.80c}$$

where for positive head loss and with $(H_D)_J > (H_D)_2$ and $(H_D)_J) > (H_D)_3$ in our estimate, we had to reverse the order of heads for pipes 2 and 3. Now from Eqs. (9.80) compute the velocities and the volumetric flow rates for the first estimate. Compare q_1 with $q_2 + q_3$. If $q_1 > (q_2 + q_3)$ there is too much flow into the juncture. As a consequence, choose $(H_D)_J$ greater than your first estimate for another cycle of computation. If $q_1 < (q_2 + q_3)$, decrease your estimate $(H_D)_J$. At all times compare the values of the estimated $(H_D)_1$ with $(H_D)_2$ and $(H_D)_3$ to know at any calculation which of the continuity equations [(9.75) or (9.76a) or (9.76b)] to use. Continue the procedure until you have satisfied one of the continuity equations. You have then reached the desired flow picture since, keep in mind, you are at all times satisfying the first law of thermodynamics. We now illustrate these steps.

Example 9.10. In Fig. 9.42 find the flows for the following data:

$$L_1 = 200 \text{ m} \qquad D_1 = 300 \text{ mm} \qquad \frac{e_1}{D_1} = 0.0002 \qquad z_1 = 700 \text{ m} \qquad p_1 = 7 \text{ atm}$$

$$L_2 = 300 \text{ m} \qquad D_2 = 350 \text{ mm} \qquad \frac{e_2}{D_2} = 0.00015 \qquad z_2 = 400 \text{ m} \qquad p_2 = 2 \text{ atm}$$

$$L_3 = 400 \text{ m} \qquad D_3 = 400 \text{ mm} \qquad \frac{e_3}{D_3} = 0.0001 \qquad z_3 = 100 \text{ m} \qquad p_3 = 3 \text{ atm}$$

The fluid is water with $\nu = 0.113 \times 10^{-5} \text{ m}^2/\text{s}$.
    We first calculate $(H_D)_1$, $(H_D)_2$, and $(H_D)_3$. Thus,

$$(H_D)_1 = z_1 + \frac{(7)(p_{\text{atm}})}{\gamma} = 700 + \frac{(7)(101,325)}{9806} = 772.3 \text{ m}$$

$$(H_D)_2 = z_2 + \frac{(2)(p_{\text{atm}})}{\gamma} = 400 + \frac{(2)(101,325)}{9806} = 420.7 \text{ m}$$

$$(H_D)_3 = z_3 + \frac{(3)(p_{\text{atm}})}{\gamma} = 100 + \frac{(3)(101,325)}{9806} = 131.0 \text{ m}$$

Now estimate $(H_D)_J = 380$ m. We will hence use continuity equation (9.76a) for test purposes. We can then say in place of Eqs. (9.80) that

$$(772.3 - 380) = \frac{1}{g} f_1 \left(\frac{200}{0.300} \right) \left(\frac{V_1^2}{2} \right)$$

$$(420.7 - 380) = \frac{1}{g} f_2 \left(\frac{300}{0.350} \right) \left(\frac{V_2^2}{2} \right)$$

$$(380 - 131.0) = \frac{1}{g} f_3 \left(\frac{400}{0.400} \right) \left(\frac{V_3^2}{2} \right)$$

Estimating $f_1 = 0.014$, $f_2 = 0.013$, and $f_3 = 0.012$, we can compute velocities and the flows q_i. We get

$$V_1 = 28.72 \text{ m/s} \qquad q_1 = \left(\frac{\pi}{4}\right)(0.300^2)(28.72) = 2.03 \text{ m}^3/\text{s}$$

$$V_2 = 8.465 \text{ m/s} \qquad q_2 = \left(\frac{\pi}{4}\right)(0.350^2)(8.465) = 0.814 \text{ m}^3/\text{s}$$

$$V_3 = 20.18 \text{ m/s} \qquad q_3 = \left(\frac{\pi}{4}\right)(0.400^2)(20.18) = 2.536 \text{ m}^3/\text{s}$$

We see that $(q_1 + q_2) > q_3$. As a second estimate, we will increase $(H_D)_J$ and use more accurate friction factors. Using the preceding velocities, we now find a second set of friction factors.

$$(\text{Re}_D)_1 = \frac{(28.72)(0.300)}{0.0113 \times 10^{-4}} = 7.625 \times 10^6 \qquad f_1 = 0.014$$

$$(\text{Re}_D)_2 = \frac{(8.465)(0.350)}{0.0113 \times 10^{-4}} = 2.622 \times 10^6 \qquad f_2 = 0.0134$$

$$(\text{Re}_D)_3 = \frac{(20.18)(0.400)}{0.0113 \times 10^{-4}} = 7.143 \times 10^6 \qquad f_3 = 0.012$$

Suppose we choose $(H_D)_J$ to be 400 m. Thus, we still use Eq. (9.76a) for continuity. Omitting the details now we get the following results:

$$V_1 = 27.98 \text{ m/s} \qquad q_1 = 1.977 \text{ m}^3/\text{s}$$

$$V_2 = 5.95 \text{ m/s} \qquad q_2 = 0.5721 \text{ m}^3/\text{s}$$

$$V_3 = 20.97 \text{ m/s} \qquad q_3 = 2.635 \text{ m}^3/\text{s}$$

Note that $(q_1 + q_2) < q_3$. We now *interpolate* to get the final result. That is, $(H_D)_J = 380$ m gave us a value $(q_1 + q_2) - q_3 = 0.308$, while $(H_D)_J = 400$ m gave us a value $(q_1 + q_2) - q_3 = -0.0859$. Hence we choose a final $(H_D)_J$ to be

$$(H_D)_J = 380 + \left(\frac{0.308}{0.308 + 0.0859}\right)(400 - 380)$$

$$= 396 \text{ m}$$

For this we get

$$q_1 = 1.988 \text{ m}^3/\text{s} = 1988 \text{ L/s}$$

$$q_2 = 0.6249 \text{ m}^3/\text{s} = 624.9 \text{ L/s}$$

$$q_3 = 2.616 \text{ m}^3/\text{s} = 2616 \text{ L/s}$$

We come very close to satisfying the continuity equation [Eq. (9.76a)] so the above are the desired flows.

9.19 CLOSURE

In this chapter we have studied the very important topic of pipe flow. The importance of this topic arises from two factors. First there is the obvious practical aspect that pipe flow occurs in most devices and systems and must be understood by the engineer. The second factor is perhaps less obvious and that stems from the fact that much useful information and many concepts carry over from pipe flow to other fluid studies, the most appropriate of which is boundary layer theory to be studied in Chap. 13. In fact so similar are the various concepts between pipe flow and boundary layer flow that the reader will have to be on guard to keep importance distinctions between these flows clearly fixed in his or her mind. With this stated let us reiterate the salient features of this chapter.

After examining certain key differences between laminar and turbulent flows by reviewing Reynold's classic experiment, we then considered *laminar flow* in a pipe which occurs when $R \leq 2300$. We used for laminar flow the *Newton viscosity law*. Note that Newton's viscosity law is valid only for *parallel* flow such as is found in straight pipes. In more general flows, one must use a more general viscosity law for which the Newton viscosity law is a special case. This will be undertaken in the next chapter when we study *Stoke's viscosity law*. From this law along with Newton's law of motion, we will set forth the well-known Navier-Stokes equations; and from these equations, we can solve, among other things, laminar flow in a pipe—thus duplicating the results of Sec. 7.8. Also, we can prove rigorously the assumptions made for parallel flow concerning pressure at a section in a pipe. Thus for laminar pipe flow we were able to analytically formulate the velocity profile to be a paraboloidal surface of revolution and we were able to derive a head loss formula.

Next we considered *turbulent pipe flow*. We explained how turbulence gives rise to a so-called apparent stress just as the transport of molecules gives rise to a viscous stress. We pointed out that near a boundary the viscous stress dominates and further out from the boundary the apparent stress dominates with a region of overlap in between where both viscous effects and turbulence effects are significant. We then delineated three zones of pipe flow, namely, the *smooth pipe zone*, the *rough pipe zone*, and the *transition zone*. Hopefully you fully comprehend the mechanisms at work causing the need for these categories since they will come up again in boundary layer flow. The head loss for fully developed turbulent flow requires experimental data in the form of the Moody diagram or some formulations for the friction factor f found by curve-fitting experimental data. As for the velocity profile, we presented the so-called *one-seventh law* stemming from curve fitting also and restricted to "*low*" *Reynolds number* flow, i.e., under 3×10^6. We then undertook the solution of a large variety of pipe problems.

Finally, with the aid of the *Prandtl mixing length and/or dimensional analysis as* hypotheses, we were able with some experimental data to arrive at velocity profiles for turbulent flows at *high* Reynolds numbers (above 3×10^6) for smooth and rough pipes. In the latter we had to consider the aforementioned three zones of flow. You will recall that some of the concepts in this study applied to two-dimensional turbulent flow as well as to pipe flow. We make reference to these concepts when we study boundary-layer theory in Chap. 13.

TABLE 9.5
Pipe flow-summary sheet

I. **Laminar flow** Re < 2300

  A. $V = \dfrac{P_1 - P_2}{4\mu l}\left(\dfrac{D^2}{4} - r^2\right)$

  B. $h_l = \dfrac{128 q\mu L}{\pi D^4 \rho}$

  C. $f = \dfrac{64}{\text{Re}}$

II. **Turbulent flow**

  A. $h_l = f\dfrac{L}{D}\dfrac{V^2}{2}$

  B. $\tau_w = \dfrac{f}{8}(\rho V^2)$

    1. $\boxed{Low\ Reynolds\ number\ flow}$ Re < 3×10^6

    a. $f = \dfrac{0.3164}{\text{Re}^{1/4}}$ Re < 100,000   (Hydraulically smooth zone)

    b. $f = \dfrac{1}{[1.14 - 2.0\log(e/D)]^2}$ Re > 100,000   (Rough zone)

    c. $\dfrac{\overline{V}}{V_{\max}} = \left(\dfrac{y}{D/2}\right)^{1/n}$ Power law

    d. $\tau_w = 0.03325\rho V^2\left(\dfrac{\nu}{RV}\right)^{1/4}$ Smooth pipes

    2. $\boxed{High\ Reynolds\ number\ flow}$ Re > 3×10^6

    a. Smooth pipes

$$\frac{\overline{V}_x}{V_*} = 2.5\ln\frac{yV_*}{\nu} + 5.5$$

    b. Rough pipes

      i. Smooth zone $\dfrac{eV_*}{\nu} < 5$

$$\frac{\overline{V}_x}{V_*} = 2.5\ln\frac{yV_*}{\nu} + 5.5$$

      ii. Smooth-rough transition zone $5 < \dfrac{eV_*}{\nu} < 70$

$$\frac{\overline{V}_x}{V_*} = 2.5\ln\frac{yV_*}{\nu} + B$$

TABLE 9.5 (*Continued*)

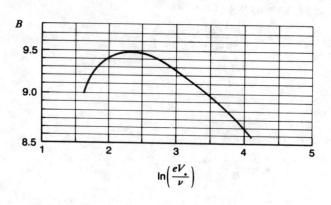

iii.   Rough zone $\dfrac{eV_*}{\nu} > 70$

$$\frac{\bar{V}_x}{V_*} = 2.5 \ln \frac{y}{e} + 8.5$$

$$\frac{1}{f^{1/2}} = -2.0 \log \frac{e/D}{3.6}$$

Prandtl universal law of friction for smooth pipes $Re > 2300$

$$\frac{1}{\sqrt{f}} = 2.0 \log[\frac{\bar{V}_{av} D}{\nu} \sqrt{f}] - 0.8$$

For rough zone and laminar-turbulent transition zone

$$f = \frac{0.25}{\{\log[e/(3.7D) + 5.74/Re^{0.9}]\}^2} \qquad \begin{pmatrix} 5 \times 10^3 \le Re \le 10^8 \\ 10^{-6} \le \left(\dfrac{e}{D}\right) \le 10^{-2} \end{pmatrix}$$

To help the reader in applying the diversity of results of this chapter a summary outline of key formulas is given in Table 9.5.

We present in the next chapter a study of incompressible flow of isotropic Newtonian fluids wherein we shall develop the well-known Navier-Stokes equations. Those readers that examine the starred section on Stokes' viscosity law will see strong unmistakable similarities between the constitutive laws of Newtonian fluids and Hookean solids, the two key continua that engineers deal with.

PROBLEMS

Problem Categories

Reynolds Number 9.1–9.6
Head Loss 9.7–9.9
Laminar Flows 9.10–9.28
Turbulent Flows (Single Branch) 9.29–9.52
Minor Loss Formulas 9.53–9.56
Pipe Flows (Single Branch) 9.57–9.64
Problems Involving Head 9.65–9.68
Pump Matching Problems 9.69–9.71
More Complex Pipe Problems 9.72–9.78
Hydraulic Grade Lines 9.79–9.81
Noncircular Ducts 9.82–9.86
High Reynolds Number Flows 9.87–9.95
General Problems (Single Branch) 9.96–9.105
Multibranch Pipe Problems 9.106–9.112

Starred Problems

9.60, 9.110

9.1. What is the Reynolds number of a flow of oil in a 6-in pipe of 20 ft^3/s, where $\mu = 200 \times 10^{-5}$ lb $\cdot$ s/ft^2 for the oil? Is the flow laminar or turbulent? Specific gravity = 0.8.

9.2. Gasoline at a temperature of 20°C flows through a flexible pipe from the gas pump to the gas tank of a car. If 3 L/s are flowing and the pipe has an inside diameter of 60 mm, what is the Reynolds number?

9.3. The Reynolds number for fluid in a pipe of 10-in diameter is 1800. What will be the Reynolds number in a 6-in pipe forming an extension of the 10-in pipe? Take the flow as incompressible.

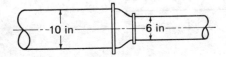

FIGURE P9.3

9.4. Water is flowing through capillary tubes A and B into tube C. If $q_A = 2 \times 10^{-3}$ L/s in tube A, what is the largest q_B allowable in tube B for laminar flow in tube C? The water is at a temperature of 40°C. With the calculated q_B, what kind of flow exists in tubes A and B?

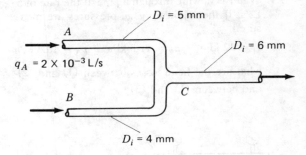

FIGURE P9.4

9.5. Do Prob. 9.4 for the case where the fluid is kerosene.

9.6. A fluid is at 50°F and is flowing through a 3-in tube at the rate of 1 ft^3/s. Determine if the flow is laminar or turbulent for the following fluids:
(*a*) saturated steam
(*b*) hydrogen
(*c*) air
(*d*) mercury

9.7. An incompressible steady flow of water is present in a tube of constant cross section. What is the head loss between positions A and B along the tube?

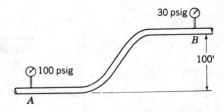

FIGURE P9.7

9.8. In Prob. 9.7, the cross-sectional area of the tube is $\frac{1}{2}$ in^2 and the average velocity is 5 ft/s. If there is an increase of internal energy of the water from A to B of 1 Btu/slug, what is the total heat transfer through the tube between these two points in 1 min? The head loss between A and B was computed to be 1976 ft^2/s^2. Do not use the first law of thermodynamics directly.

9.9. Water is flowing through a pipe at the rate of 5 L/s. If the following gage pressures are measured,

$$p_1 = 12 \text{ kPa} \qquad p_2 = 11.5 \text{ kPa} \qquad p_3 = 10.3 \text{ kPa}$$

what are the head losses between ① and ② and between ① and ③?

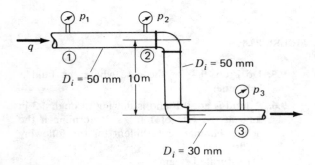

FIGURE P9.9

9.10. A large oil reservoir has a pipe of 3-in diameter and 7000-ft length connected to it. The free surface of the reservoir is 10 ft above the centerline of the pipe and can be assumed to remain at this fixed elevation. Assuming laminar flow through the pipe, compute the amount of flow issuing out of the pipe as a free jet. Compute V, and then check to see whether or not the Reynolds number is less than critical. Kinematic viscosity of the oil is 1×10^{-4} ft^2/s. Neglect entrance losses to the pipe.

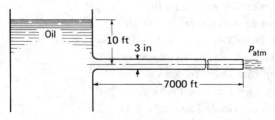

FIGURE P9.10

9.11. In Prob. 9.10, if the fluid is kerosene, do you have laminar or turbulent flow? The temperature is 50°F.

9.12. In a 10-in pipe at what radius is the velocity equal to 80% of the mean velocity for Poiseuille flow?

9.13. If 140 L/s of water flows through the pipe system, what total head loss is being developed over the length of the pipe?

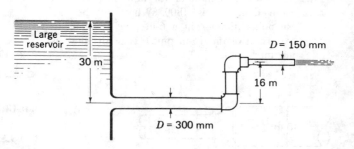

FIGURE P9.13

9.14. Water at 40°C flows from tank A to tank B (Fig P9.14). What is the volumetric flow at the configuration shown? Neglect entrance losses to the capillary tube as well as exit losses.

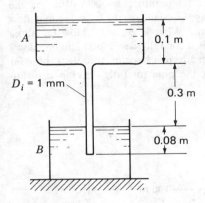

FIGURE P9.14

9.15. In Prob. 9.14, what should the internal diameter of the tube be to permit a flow of 6×10^{-4} L/s?

9.16. A hypodermic needle has an inside diameter of 0.3 mm and is 60 mm in length, (Fig. P9.16).

If the piston moves to the right at a speed of 18 mm/s and there is no leakage, what force F is needed on the piston. The medicine in the hypodermic has a viscosity μ of 0.980×10^{-3} N · s/m² and the density ρ is 800 kg/m³. Consider flows in both needle and cylinder. Neglect exit losses from the needle as well as losses at the juncture of the needle and cylinder.

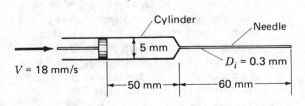

FIGURE P9.16

9.17. In Prob. 9.16 suppose that you are drawing the medicine from a container at atmospheric pressure. What is the largest flow, q, of fluid if the fluid has a vapor pressure of 4700 Pa absolute? Neglect losses in the cylinder. What is the speed of the piston for the maximum flow of medicine if there is a 10 percent leakage around the piston for the pressure of 4700 Pa absolute in the cylinder?

9.18. In Prob. 9.16 it took a force of 2.05 N to move the piston to the right at a speed of 18 mm/s. What should the inside diameter be for the needle if the force needed is only 1 N for the same piston speed? Neglect losses in cylinder.

9.19. What drag is developed by oil having a viscosity of 50×10^{-5} lb · s/ft² as it moves through a pipe of diameter 3 in and length 100 ft at an average speed of 0.2 ft/s? The specific weight of the oil is 50 lb/ft³.

9.20. Kerosene is flowing from a tank and out through two capillary tubes as shown in Fig. P9.20. Determine the heights h for the flow to just become laminar for each capillary. The temperature of the kerosene is 40°C. Neglect entrance losses to capillaries and treat problem as quasi-static.

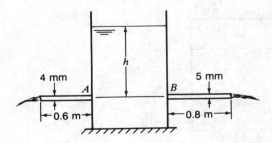

FIGURE P9.20

9.21. Shown in Fig. P9.21 is part of a robotic device. Heavy oil is pumped at A to move a piston having a constant frictional resistance of 20 N. What pressure as a function of time is needed at A to move the piston to a speed of 5 m/s in

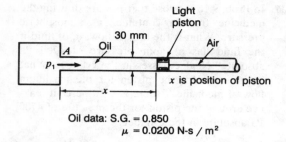

Oil data: S.G. = 0.850
μ = 0.0200 N-s / m²

FIGURE P9.21

10 s such that the velocity varies as t^2 with t in seconds? Proceed as follows:

1. At what position x_{lam} of the piston does the flow of oil cease to be laminar?
2. What is the head loss at x as a function of time for laminar flow?
3. Determine p_1 as a function of time to cause the motion of the piston during the laminar part of flow. Remember that the oil must accelerate. Do not use a control volume here. A system approach is better.
4. What is the pressure p_1 at $t = 5$ s? The piston starts at $x = 0$.

9.22. What is the pressure p_1 in Fig. P9.22 for a Reynolds number of 10 in the tube? Tank A is

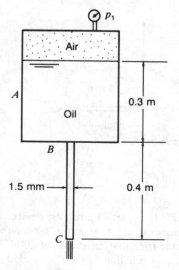

FIGURE P9.22

large. The specific gravity of the oil is 0.65 and the kinematic viscosity of the oil is 0.00018 m²/s.

9.23. Water is flowing from a large tank through a pipe having an internal diameter of 25 mm. The water temperature is 70°C. What is the highest gage pressure p_1 to have laminar flow in the pipe? What is p_1 abs.? How far along the pipe is it for fully developed laminar flow? If flow were to just become turbulent, how far is it along the pipe for fully developed turbulent flow?

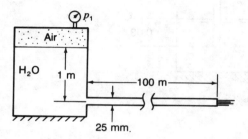

FIGURE P9.23

9.24. Do Example 9.1 for the case of mercury. Take S.G. = 13.6 and $\nu = 1.4 \times 10^{-6}$ m²/s for mercury. Use for p_A the value 458 kPa gage.

9.25. In Example 9.1, if a tiny pump were inserted in the tube, what is the power needed to cause the maximum laminar flow from tank B to tank A?

9.26. A pipe receives water from a reservoir at the rate of q L/s. The water is at 40°C. If the ratio of the distance L' for flow to become fully developed to total length, $L = 50$ m, is to be no greater than 10% for laminar flow, what is the largest q? Do the same for turbulent flow for a diameter of 0.4 m.

9.27. In a 50-mm pipe having a length of 50 m transporting crude oil at a temperature of 40°C at an average velocity of 0.02 m/s, what percentage of the pipe length is taken before fully developed viscous flow is present? What is the percentage if the average velocity is 0.30 m/s? The specific gravity of the oil is 0.86.

9.28. Consider the pipe entrance flow of water at 60°C into a pipe of diameter 100 mm.

1. What is the maximum distance in diameters for fully developed laminar flow to be established?

2. What is the minimum distance in diameters for fully developed turbulent flow to be established?

9.29. Compute the friction factors for flow having a Reynolds number and relative roughness given for the following two cases as

$$(a) \begin{cases} \text{Re} = 5 \times 10^3 \\ \dfrac{e}{D} = 0.015 \end{cases} \quad \text{Transition zone}$$

$$(b) \begin{cases} \text{Re} = 4 \times 10^6 \\ \dfrac{e}{D} = 0.0001 \end{cases} \quad \text{Rough-pipe zone}$$

Use the Colbrook formula, the Swamee-Jain formula, and the Moody diagram. Comment on the comparison of results.

9.30. What horsepower is required for a pump which will move 0.1 ft^3/s of water having a viscosity of 2.11×10^{-5} lb · s/ft^2 through a 3-in pipe of length 200 ft to discharge at the same elevation at a pressure of 20 lb/in^2 absolute? Pipe is commercial steel. The intake pressure is atmospheric.

9.31. A fire truck has its hose connected to a hydrant where the pressure is 7×10^4 Pa gage. The hose then connects up to a pump run by the engine of the fire truck. From here there extends hose to a fire fighter who, while crouching, is aiming the water at a 60° angle with the ground to enter a window on the third floor 13 m above the nozzle at the end of the hose. As the water goes through the window, it is moving parallel to the ground. The total length of hose is 65 m with 200-mm diameter. The exit diameter of the nozzle is 100 mm. Take e/D for hose as 0.0001. What power is needed from the pump on the water? Take the nozzle exit to have same elevation as the hydrant outlet. Neglect minor losses. Take $\nu = 0.113 \times 10^{-5}$ m^2/s.

9.32. If 565 L/s of flow is to be moved from A to B, what power is needed from pump to water? Take $\nu = 0.1130 \times 10^{-5}$ m^2/s.

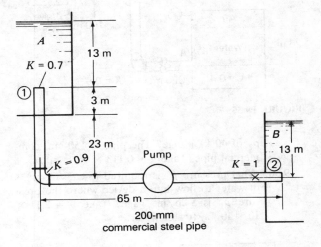

FIGURE P9.32

9.33. What gage pressure p_1 is required to cause 5 ft^3/s of water to flow through the system? Assume that the reservoir is large. Neglect minor losses. Take $\mu = 2.11 \times 10^{-5}$ ft^2/s.

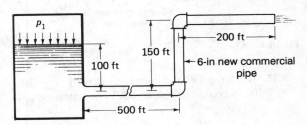

FIGURE P9.33

9.34. In Prob. 9.33 take the diameter of the pipe to be the *nominal* diameter. For the entrance fitting, $r/d = 0.06$. Calculate the pressure p_1. The elbows are screwed elbows and there is now an open globe valve in the pipe system. Include minor losses.

9.35. What pressure p_1 is needed to cause 100 L/s of water to flow into the device at a pressure

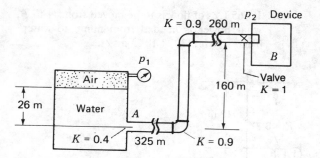

FIGURE P9.35

p_2 of 40 kPa gage? The pipe is 150-mm commercial pipe. Take $\nu = 0.113 \times 10^{-5}$ m²/s.

9.36. What pressure p_1 is required to cause 1 ft³/s of water to flow into a device where the pressure p_2 is 5 lb/in² gage? Take $\mu = 2.11 \times 10^{-5}$ lb · s/ft² for water.

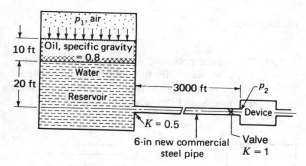

FIGURE P9.36

9.37. We have oil of kinematic viscosity 8×10^{-5} ft²/s going through an 80-ft horizontal pipe. If

the initial pressure is 5 lb/in² gage and the final pressure is 3.5 lb/in² gage, compute the mass flow if the pipe has a diameter of 3 in. At a point 10 ft from the end of the pipe a vertical tube is attached to the pipe to be flush with the inside radius of the pipe. How high will the oil rise in the tube? $\rho = 50$ lbm/ft³. Pipe is commercial steel.

9.38. What is the flow q from A to B for the system shown? Get up to the second iteration. Take $\nu = 0.113 \times 10^{-5}$ m²/s.

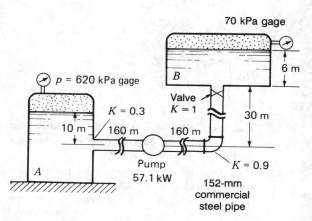

FIGURE P9.38

9.39. What thrust does the water exert in Fig. P9.39 on the pipe in the horizontal direction? Water issues out as a free jet. Take $\mu = 2.11 \times 10^{-5}$ lb · s/ft². Neglect minor losses.

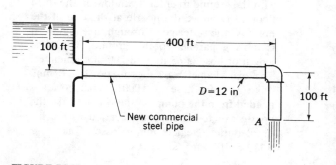

FIGURE P9.39

9.40. Do Prob. 9.39 when considering minor losses. Take the pipe diameter to be the nominal diameter and for the flanged elbow $r/d = 14$. What is the percentage error incurred in this problem by neglecting the minor losses? The r/d for the entrance fitting is 0.04. The thrust in the previous problem is 6212 lb.

9.41. How much water is flowing through the 150-mm commercial steel pipe? Take $\nu = 0.113 \times 10^{-5}$ m²/s.

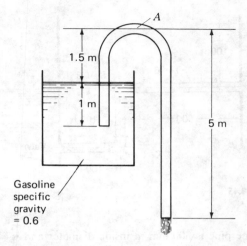

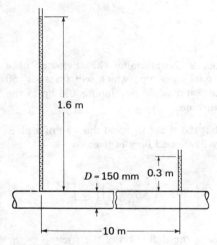

FIGURE P9.41

9.42. Gasoline at 20°C is being siphoned in Fig. P9.42 from a tank through a rubber hose having an insider diameter of 25 mm. The relative roughness for the hose is 0.0004. What is the flow of gasoline? What is the minimal pressure

FIGURE P9.42

in the hose? The total length of hose is 9 m and the length of hose to point A is 3.25 m. Neglect minor loss at hose entrance.

9.43. An *equivalent length* of pipe is one whose head loss for the same value of flow is equal to that of some other system of different geometry for which it is the equivalent. Consider a steel pipe of *nominal* diameter 10 in having in it an open globe valve and four screwed 90° elbows. The length of the pipe is 100 ft, and 5 ft³/s of water at 60°F flows through the pipe. What is the equivalent length of pipe of nominal diameter 14 in?

9.44. What is the horizontal force in Fig. P9.44 on the pipe system from the water flowing inside?

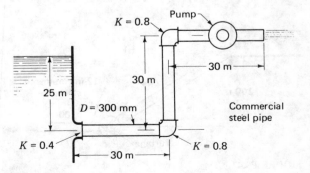

FIGURE P9.44

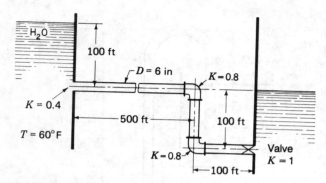

FIGURE P9.45

The pipe is 300 mm in inside diameter and is new. The pump is known to be developing 65 kW of power on the water. The temperature of the water is 5°C.

9.45. What should be the flow through the system shown in Fig. P9.45? We have commercial steel pipe, 6 in in diameter.

9.46. Put a pump on the line of Prob. 9.45. What is the new flow through the system if the pump develops 10 hp on water? Does it make any difference where the pump is placed with respect to mass flow?

9.47. How much water flows from the reservoir through the pipe system? The water drives a water turbine which develops 100 hp. Take $\mu = 2.11 \times 10^{-5}$ lb · s/ft^2.

9.48. In Prob. 9.47 determine the diameter of a commercial steel pipe which will transport 50 ft^3/s of water while developing 100 hp in the water turbine.

9.49. Show that the shear stress at the wall of a pipe for fully developed flow is given as

$$\tau_w = \left(\frac{p_1 - p_2}{L_{1\text{-}2}} \right) \left(\frac{D}{4} \right)$$

Then using the definition of the friction factor f show that

$$\tau_w = \frac{1}{8} f \rho \bar{V}^2_{\text{mean}}$$

Flow is turbulent.

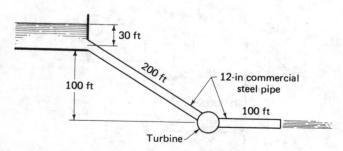

FIGURE P9.47

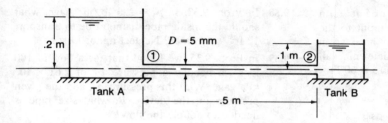

FIGURE P9.51

9.50. Show that for the power law

$$\frac{\bar{V}}{\bar{V}_{max}} = \left(\frac{y}{R}\right)^{1/n}$$

that the ratio of average mean time velocity to $\bar{V}_{max}$ over the cross section is given as

$$\frac{\bar{V}_{mean}}{\bar{V}_{max}} = \frac{2n^2}{(n+1)(2n+1)}$$

Note that $\bar{V}_{mean} = Q/A$ and that y is measured from the wall.

9.51. Oil having a specific gravity of 0.7 flows at the rate of 0.05 m³/h (see Fig. P9.51). Compute the viscosity of the oil. Do this problem two ways:

(1) Assume laminar flow.
(2) Assume turbulent flow.

Procedure: First find h_l from *modified Bernoulli* valid for either case. Now using the head loss, compute μ from the pipe head loss formula for laminar flow. Check Re to justify assumption (1). Next use *Darcy-Weisbach* to get f for turbulent flow and use Moody to get Re and then μ. Compare with μ from laminar flow.

Note we have the possibility of *two-valid solutions here.* What can you conclude about the flows close to Re = 2300?

9.52. In a turbulent flow for a Reynolds number of 5000, at what radius is the velocity no smaller than 90% of the mean velocity. Use $n = 7$ for the velocity profile.

9.53. For $D_1 = 200$ mm and $D_2 = 100$ mm what angle θ gives the largest head loss for the

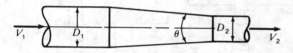

FIGURE P9.53

reducing member in Fig. P9.53? What is the head loss for $V_1 = 3$ m/s?

9.54. For the diffuser in Fig. 9.21, what is the head loss with $D_1 = 12$ in and $D_2 = 18$ in for $V_1 = 5$ ft/s? The length of the diffuser is 5 ft. Explain the opposing roles of skin friction and separation in developing this head loss. Water at 60°F is flowing.

9.55. Determine the flow Q (see P9.55). Use text for minor loss coefficients.

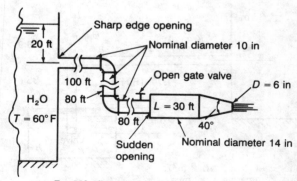

For 90° elbows $r/d = 3$
Pipes are all commercial steel

FIGURE P9.55

9.56. For a volume flow from A to B of 5 ft^3/s in Fig. P9.56, determine the power input to the flow of the pump. Note that we have given *nominal* diameters. Use text to determine all minor loss coefficients. Temperature is 60°F.

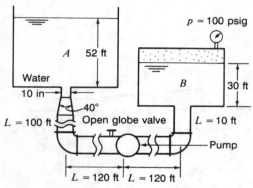

FIGURE P9.56

9.57. A flow q of 170 L/s is to go from tank A to tank B. If $\nu = 0.113 \times 10^{-5}$ m^2/s, what should the diameter be for the horizontal section of pipe?

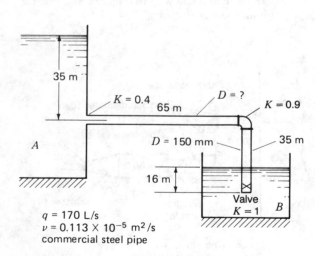

$q = 170$ L/s
$\nu = 0.113 \times 10^{-5}$ m^2/s
commercial steel pipe

FIGURE P9.57

9.58. In Prob. 9.33, if $p_1 = 200$ lb/in^2 gage, what should the inside pipe diameter be to transport 12 ft^3/s of water? Neglect minor losses.

9.59. In Prob. 9.35, we found that for a flow of 100 L/s we needed a pressure p_1 of 2.61×10^3 kPa gage. With this pressure p_1, and the given pressure on the device B, what size pipe is needed to double the flow?

*9.60. Choose the inside diameter of the pipe such that the horizontal thrust on the pipe from the water does not exceed the value of 30 kN. The water is at 5°C. Neglect minor losses.

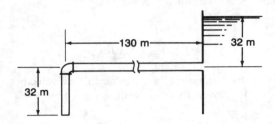

FIGURE P9.60

9.61. What should be the flow through the system shown in Fig. P9.61? We have commercial steel pipe 6 in in diameter.

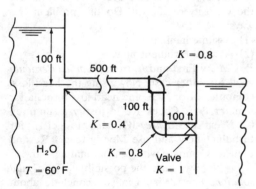

FIGURE P9.61

9.62. The pump in Fig. P9.62 develops 5×10^3 N-m/kg on the mass flow. How many liters per second flow through the commercial steel pipe

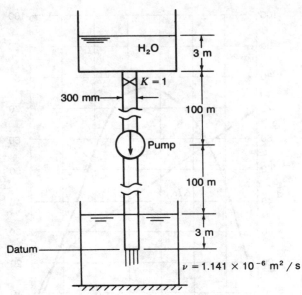

FIGURE P9.62

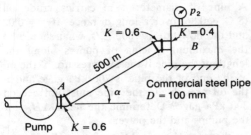

FIGURE P9.64

from the upper tank to the lower tank? Take $e = 0.046$ mm.

9.63. Find the volume flow in Fig. P9.63. The pump delivers 70-kW power on the flow.

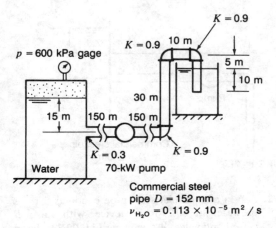

FIGURE P9.63

9.64. If the exit pressure of the pump is 250 kPa gage and the desired pressure at B is 120 kPa

gage, what is the largest angle permitted for these conditions for $V = 1$ m/s? The fluid is water at a temperature of 20°C. If the pressure going into the pump is 100 kPa gage with the same diameter pipe, what power is the pump developing?

9.65. Determine the *volume of flow Q* from A to B in Fig. P9.65 if the *pump* at E has the following *input* characteristics:

$$\Delta H_D = -\tfrac{1}{8}Q + 30 \ \text{N-m/N}$$

with Q in liters per second.

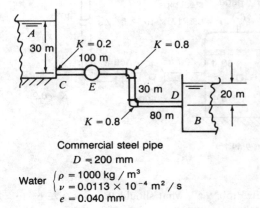

FIGURE P9.65

9.66. A pump is developing 100 kW on a flow of water of 500 L/s. What is the head ΔH_D required by the pump. Water is at 5°C.

9.67. A pipe of diameter 1 m carries crude oil (S.G. = 0.86) over a long distance. If $f = 0.02$ and the flow rate is 1000 L/s, what should be the maximum spacing of pumps along the length of the pipe if the exit pressure of the oil at the end of the pipe is 200 kPa gage and if the oil pressure in the pipe must not exceed 300 kPa gage? Determine the head ΔH_D for the pumps and the power.

9.68. In the preceding problem the pipeline has a slope of 0.2° upward from the horizontal. What are the maximum spacing of pumps, the ΔH_D needed for the pumps, and the required power?

9.69. What is the flow q for the system shown in Fig. P9.69? The pump has the characteristics shown in Fig. P9.69A. What is the power required?

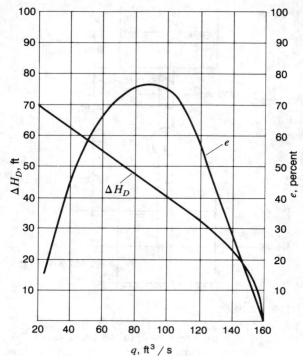

FIGURE P9.69A

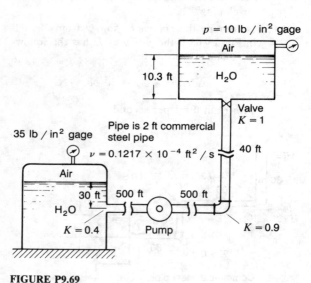

FIGURE P9.69

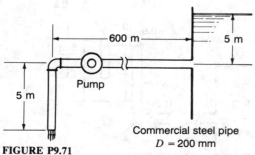

FIGURE P9.71

9.70. In Prob. 9.69, what should the pipe size be to cause a volumetric flow of 120 ft³/s?

9.71. If the pump shown in Fig. 9.71 has flow characteristics corresponding to Fig. 9.26 of Example 9.5, what is the volumetric flow and the power needed by the pump? For this problem, take the pipe diameter to be 200 mm throughout. $T = 20\,°C$. Neglect minor losses.

9.72. A 6-in commercial steel pipe conducts 5 ft³/s of water at 60°F to a device, with valve B closed and valve A open (see Fig. P9.72). In an emergency, valve A is closed and valve B is *opened* so the 5 ft³/s hits surface EG steadily to run off. For this latter case, what is the horsepower required from the pump, and what is the force on EG? Consider the exit to be a

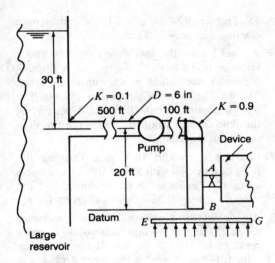

FIGURE P9.72

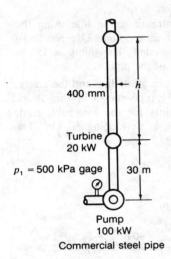

FIGURE P9.74

free jet and neglect the distance from B to EG.

9.73. Fuel oil is pumped in the winter through an exposed pipe of diameter 200 mm and of length 20 m at a temperature of 5°C. The pipe is commercial steel. The flow is at the rate of 220 L/s. What change in head ΔH_D and in power of the pump is needed to do the same job in the summer with a fuel temperature of 35°C? Use ν at 5°C = 2.323×10^{-3} m²/s and ν at 35°C = 3.252×10^{-4} m²/s. The specific gravity of fuel oil is 0.97.

9.74. A pump is developing 100 kW of power on a vertical flow in Fig. P9.74 for a skyscraper. At 30 m a turbine draws off 20 kW of power. How high can the pipe go to the next pump if we require an inlet pressure for this pump of 10,000 Pa gage? The flow q is 1 m³/s. Take $\nu = 0.01141 \times 10^{-4}$ m²/s.

9.75. Water enters a pump in a 600-mm pipe and leaves in a 400-mm pipe at an elevation of 0.3 m above the entrance. The flow is 600 L/s. If there is a static pressure rise of 50,000 Pa, what is the head ΔH_D developed by the pump? What is the power needed to run the pump if the efficiency is 65%? Water is at 30°C.

9.76. Crude oil (S.G. = 0.86) is being transported in a 500-mm steel pipe over a distance of 100 km. At a position about midway, someone has tapped in on the pipe and is drawing off some oil illegally. If the pressure drop noted by pressure gages stationed every 2 km is 3000 Pa before the suspected point and is 2800 Pa after the suspected point, how much oil is being taken away illegally? The temperature is 20°C.

9.77. In Example 9.5 suppose you had two identical pumps directly in series in the system whose performance per pump is as in Fig. 9.26. If the flow is 100 L/s, *what is the increase in head from the pumps*? The pipe diameter is 200 mm. What is the power input to the pumps?

9.78. Do the preceding problem for the pumps directly in parallel for a total flow of 160 L/s. What is the power delivered to the flow? What is the power input to the pumps?

9.79. Sketch the hydraulic and energy grade lines for the pipe in Example 9.2. Calculate the values of key points of the hydraulic grade line. Take as the datum the centerline of the lowest pipe. Use results of the calculations in Example 9.2.

9.80. Sketch the hydraulic and energy grade lines for the pipe in Example 9.3. Calculate the key points of the hydraulic grade line using the lowest pipe point as a datum. Use results of Example 9.3 as needed. The pump is 150 m from the left end of the pipe.

9.81. Sketch the hydraulic grade line and the energy grade line for the pipe shown in Fig. P9.81. Evaluate key points for the hydraulic grade line. The turbine is developing 50 kW. The water is at 5°C.

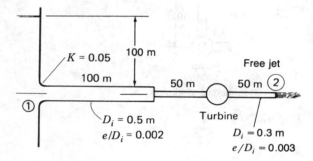

FIGURE P9.81

9.82. Consider an equiangular triangular duct which is 0.4 m on a side. What diameter circular pipe will yield the same flow characteristics? What size square pipe will do the same?

9.83. A trapezoidal duct (Fig. P9.83) is transporting 2 ft^3/s of kerosene. The roughness e is 0.0004 ft. What is the drop in pressure in 100 ft of duct? The temperature is 50°F. Use data from Figs. B.1 and B.2 for kerosene.

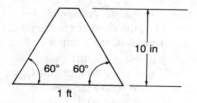

FIGURE P9.83

9.84. Oil of kinematic viscosity 2×10^{-4} ft^2/s flows through a smooth 3-in circular pipe delivering 0.02 ft^3/s flow. What dimension a should a smooth square pipe be to deliver the same volume flow and head loss?

9.85. Do Problem 9.84 for $q = 2$ ft^3/s. Iterate once starting with $a = 0.227$ in.

9.86. A fluid having the specific gravity 0.60 and a viscosity of 3.5×10^{-4} N-s/m^2 flows through a circular duct which is 300 mm in diameter. The average speed of the fluid is 15 m/s. The duct is commercial steel. Find the velocity of the fluid 30 mm from the wall. Find the drag on 5 m of the duct.

9.87. Consider a smooth 18-in pipe carrying 100 ft^3/s of crude oil with S.G. = 0.86 at a temperature of 50°F. Estimate the thickness λ of the viscous sublayer. Use Moody's diagram for f.

9.88. Air is moving in a smooth duct having a diameter of 200 mm at a temperature of 40°C and a pressure of 110,325 Pa abs. If the maximum velocity is 6 m/s, what is the shear stress at the wall using the logarithmic velocity law?

9.89. If the volume of air flowing in the previous problem is 0.5 m^3/s, find the maximum velocity
(*a*) using the one-seventh law
(*b*) using the logarithmic velocity law
The duct is made of steel.

9.90. Do Prob. 9.84 for the case where the roughness e is 0.003 in for both conduits and the flow is 2.0 ft^3/s. *Procedure:* Guess at the dimension a of the square section. Get Re$_H$ and then f from Moody. With this f, see if you get the same head loss with circular conduit flow. If not, make second guess at a, etc. until agreement is reached.

9.91. In an air-conditioning system there is a length of duct of 200 ft transporting air at 50°F at the rate of 8000 ft^3/min. The duct has a cross section of 2 ft by 1 ft and is made of galvanized iron. The pressure at the inlet of the duct is 2 lb/in^2 gage. What is the pressure drop in millimeters of mercury over the length of the duct, if we hypothesize that the temperature remains very close to 50°F and the pressure varies only slightly along the duct—i.e., we treat the flow as isothermal and incompressible? Consider that the flow is entirely turbulent.

9.92. In a heating system, there is a run of insulated duct of 50 m carrying air at a temperature of

35°C at a pressure at the inlet of 100 kPa. The duct has a rectangular cross section of 650 mm by 320 mm. If there is a pressure drop from inlet to outlet of 5 mm of mercury, what is the volumetric flow? *Hint:* For such a small pressure drop in the duct, treat the flow as incompressible. Take $R = 287$ N · m/(kg)(K). The duct is galvanized iron. Consider that the flow is entirely turbulent.

9.93. A fluid having specific gravity 0.60 and a viscosity of 3.5×10^{-4} N-s/m² flows through a circular duct which is 300 mm in diameter. The average speed of the fluid is 15 m/s. The duct is commercial steel. Find the velocity of the fluid 30 mm from the wall. Find the drag on 5 m of the duct.

9.94. Using Eq. (9.50), develop a formula for the average velocity Q/A in a turbulent pipe flow. Now relate this average velocity V_{av} with the maximum velocity V_{max} in this flow. Show that

$$\overline{V}_{av} = \frac{V_*}{\alpha}\left[\ln\frac{V_* R}{\nu} - \ln\beta - \frac{3}{2}\right]$$

9.95. Using the results of the preceding problem show that

$$\frac{\overline{V}_{av}}{\overline{V}_{max}} = \left\{1 - \frac{3/2}{\ln(V_* R/\nu) - \ln\beta}\right\}$$

Get the ratio of average velocity to maximum velocity for smooth pipe flow of crude oil at 60°C. Take S.G. of oil as 0.86. Get this ratio for $\overline{V}_{av} = 3$ m/s and $D = 20$ mm. Note $\ln\beta = -2.2$ for smooth pipe flow. *Hint:* Can you use Eq. (9.21)? Compare the result with 0.5 for laminar flow. What can you conclude about the shapes of profile for laminar and turbulent flow?

9.96. Consider flow of water at 60°C in a smooth 100-mm-diameter pipe. Examine the situations when Re = 100,000 and when Re = 50,000. Compute the friction factor f using Blasius' formula and then compute f using the Prandtl universal law of friction. Finally, look at Moody curves. Compare results.

9.97. For Re ≤ 100,000, show that for smooth pipes we can give the pressure drop due to friction as follows:

$$\Delta p = 0.2414 L \mu^{1/4} D^{-4.75} Q^{1.75} \rho^{3/4}$$

Find the pressure drop in a smooth 100-mm pipe for a flow of 0.5 m³/s over a distance of 50 m. The fluid is water at 30°C.

9.98. A pump has performance given by the plot in Fig. 9.26. If it is pumping 120 L/s of water at 10°C, what is the term dW_s/dm needed for the energy equation using energy per unit mass. What is the power input in kilowatts?

9.99. Water at a temperature of 20°C flows through an inclined pipe of length 80 m and emerges as a free jet. What is the roughness of the pipe? Neglect losses at entrance to the pipe. The fluid velocity is 2 m/s.

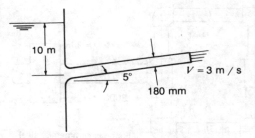

FIGURE P9.99

9.100. In Example 9.5 determine the pipe size needed for a flow of 100 L/s. Use a trial-and-error procedure for different diameters to get the proper ΔH_D required by the pump. Only try two diameters.

9.101. Flowing through a pipe of 12-in diameter is 10 ft³/s of water at 60°F. The pipe is very smooth. Estimate the shear stress τ_w at the wall and the thickness λ of the viscous sublayer. *Hint:* Use Eq. (9.23) for τ_w.

9.102. Do Prob. 9.101 using Eq. (9.17) for τ_w. Compare your estimation of λ with that of the preceding problem where we got $\lambda = 0.001983$ in.

9.103. Flowing through commercial steel pipe of diameter 200 mm are 1600 L/s of water at 10°C. What is the wall friction τ_w? Express the velocity profile as a function of y, the radial distance in from the wall. What is the maximum velocity $\overline{V}$? Use the Moody diagram to get τ_w.

9.104. Gasoline having a specific gravity of 0.68 and a viscosity of 3×10^{-4} N $\cdot$ s/m² flows through a 250-mm pipe at an average rate of 10 m/s. What drag is developed per meter of pipe by the gasoline? What is the velocity of the gasoline 25 mm radially in from the pipe wall? The pipe is new cast iron.

9.105. Derive Eq. (9.70) using Eq. (9.69). Take $\alpha = 0.4$.

9.106. If 1 ft³/s of water flows into the system at A at a pressure of 100 lb/in² gage, what is the pressure at B if one neglects minor losses? The pipe is commercial steel. Take $\mu = 2.11 \times 10^{-5}$ lb $\cdot$ s/ft².

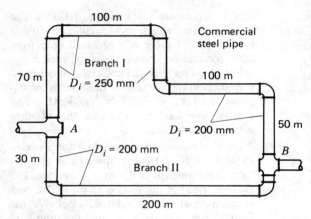

FIGURE P9.108

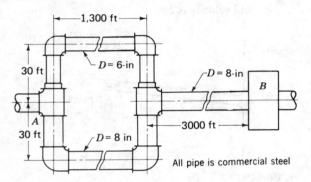

FIGURE P9.106

9.107. Do Prob. 9.106 for the case where we have nominal pipe sizes and we have an open globe valve in the 8-in pipe just before reaching the apparatus B. The fittings are all screwed fittings. Use K for a 6-in standard tee.

9.108. A two-branch pipe system in Fig. P9.108 is to deliver 400 L/s of water at 5°C. The pressure at B is 20 kPa gage. What is the pressure at A? Note the different diameters of the pipes. Neglect minor losses in this problem.

9.109. A flow q of 800 L/s goes through the pipe system in Fig. P9.109. What is the pressure drop between A and B if the elevation of A is 100 m and of B is 200 m? Neglect minor losses here. The water is at 5°C.

*9.110. Going into the system of pipes at A, we have 200 L/s. In the 150-mm pipe, there is a turbine, as shown in Fig. P9.110. The turbine has

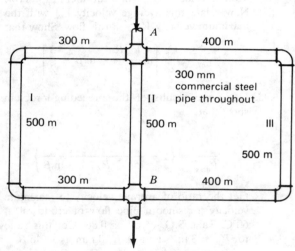

FIGURE P9.109

performance characteristics as shown in Fig. 9.26. (The head ΔH_D is now the *decrease* in head rather than the *increase* in head that would be the case for a pump.) The water is at 5°C. What is the power developed by the turbine? *Hint:* Choose a flow q in the lower branch. Read off ΔH_D for this flow from the performance chart. Now compute $(p_A - p_B)/\rho$ for this branch. Plot a curve of $(p_A - p_B)/\rho$

versus q using about five values of q. Now compute for the upper branch $(p_A - p_B)/\rho$ for flows $(0.200 - q)$ for the same set of q's as for the lower branch and again plot $(p_A - p_B)/\rho$ versus q. The intersection of the two curves is the operating point. This is another example of *matching systems* described in Example 9.5. Neglect minor losses.

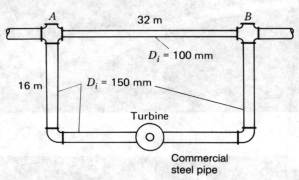

FIGURE P9.110

9.111. Consider three interconnected reservoirs in Fig. P9.111. The following additional data apply.

$L_1 = 2000$ m $\qquad L_2 = 2300$ m $\qquad L_3 = 2500$ m

$D_1 = 1$ m $\qquad D_2 = 0.60$ m $\qquad D_3 = 1.20$ m

$\dfrac{e_1}{D_1} = 0.00015 \qquad \dfrac{e_2}{D_2} = 0.001 \qquad \dfrac{e_3}{D_3} = 0.002$

The water is at 5°C. What is the volumetric flow through the pipes. Neglect minor losses.

9.112. Reservoirs 1 and 2 in Fig. P9.112 are connected to a tank which has on the free surface a pressure p of 50 lb/in² gage. The following data apply:

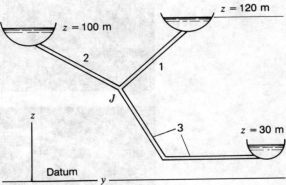

FIGURE P9.111

$z_1 = 650$ ft $\qquad L_1 = 2000$ ft $\qquad D_1 = 3$ ft $\qquad \left(\dfrac{e}{D}\right)_1 = 0.001$

$z_2 = 600$ ft $\qquad L_2 = 2500$ ft $\qquad D_2 = 3.5$ ft $\qquad \left(\dfrac{e}{D}\right)_2 = 0.002$

$z_3 = 50$ ft $\qquad L_3 = 2200$ ft $\qquad D_3 = 4$ ft $\qquad \left(\dfrac{e}{D}\right)_3 = 0.002$

What are the flows in the pipes? Water is at 60°F. Neglect minor losses.

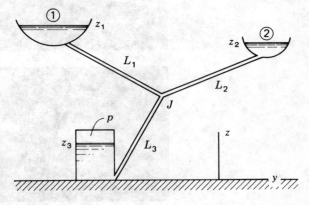

FIGURE P9.112

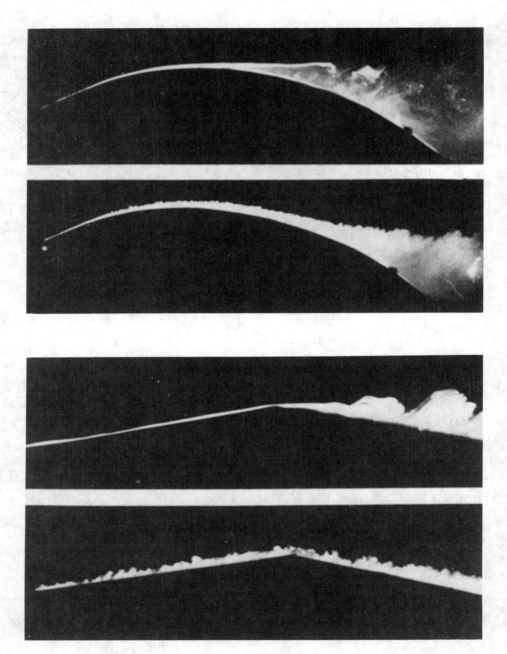

Separation process involving laminar and turbulent boundary layers. (*From M. R. Head, Flow Visualization in Cambridge University Engineering Department.* Wolfgang Merzkirch, Flow Visualization II, Hemisphere Publishing Corporation, N.Y., 1980, p. 403.)

The four photos show certain key characteristics of laminar and turbulent boundary layers. The upper one shows a laminar boundary layer separating at the crest of a convex surface, whereas the turbulent boundary layer in the second photo remains attached. In the next two figures we see the same action for a sharp corner. In these demonstrations titanium chloride is painted on the forepart of the model.

434

GENERAL INCOMPRESSIBLE VISCOUS FLOW: THE NAVIER-STOKES EQUATIONS

10.1 INTRODUCTION

In the previous chapter, we considered the flow in pipes. The use of a simple viscosity law, namely, Newton's viscosity law sufficed for such flows. In this chapter, we present the general Stokes' viscosity law starting from a very basic beginning wherein we have 36 viscosity coefficients. The starting point is similar to the beginning of the development of the *generalized Hooke's law* of solid mechanics. Newton's viscosity law is a special case of the Stokes' viscosity law. From the aforementioned starting point, by a series of steps, we limit the formulation to that of an *isotropic fluid* in exactly the same way we instituted isotropy in the generalized Hooke's law.[1] All but two of the 36 viscosity

[1] See I. H. Shames, *Introduction to Solid Mechanics*, 2d ed., Prentice-Hall, Inc., Englewood Cliffs, N.J., Chap. 6.

coefficients are shown to be zero. These nonzero coefficients consist of the familiar coefficient of viscosity μ called now the *first coefficient of viscosity* and the so-called *second coefficient of viscosity* μ'. They correspond to the two mechanical constants of elastic solids, namely, the *Lamé constants*. Now by using the result presented in Chap. 2 wherein for many fluids $p = -\frac{1}{3}(\tau_{xx} + \tau_{yy} + \tau_{zz})$ we can show that $\mu' = -(2/3)\mu$ so that we end up with only one viscosity coefficient, μ. Next, we restrict ourselves to *incompressible* flows. Thus, arriving at Eqs. (10.10) at the end of our development, we have limited ourselves to isotropy and incompressibility. The subsequent developments of this chapter will be limited to these conditions and so we will only use this restricted form of the Stokes' viscosity law. Those readers not interested in the more general forms of Stokes' viscosity law or who wish to skip the development of the isotropic, incompressible form of Stokes' viscosity law may proceed to Sec. 10.3 and accept Eq. (10.10) simply as an extrapolation of Newton's viscosity law.

We then formulate the incompressible Navier-Stokes equations, undoubtedly the most important equations of fluid mechanics. From these equations, we prove the validity of some key assumptions made in the last chapter concerning parallel flow. Furthermore, we solve several interesting flow problems using the Navier-Stokes equations.

We also consider turbulent flow in a more detailed manner than heretofore with the aid of the Navier-Stokes equations. We show that turbulent flows can be viewed as mean-time-average flows with the addition of stresses called *apparent* stresses, which reflect the effects of turbulence.

**PART A
LAMINAR FLOW**

---

*10.2 STOKES' VISCOSITY LAW

For parallel flow we have indicated that the shear stress τ on an interface parallel to the streamlines for Newtonian fluids is given as

$$\tau = \mu \frac{\partial V}{\partial n} \tag{10.1}$$

where n is the coordinate direction normal to the interface and μ is the coefficient of viscosity. This is the well-known Newton's viscosity law. In more general flows, there are more general relations between the stress field and the velocity field. Any such relation is called a *constitutive law*; the one we will consider here is *Stokes' viscosity law*. We start by assuming that each stress is *linearly* related through a set of constants to each of the six *strain rates* discussed in Chap. 4. In addition, each normal stress is directly related to the

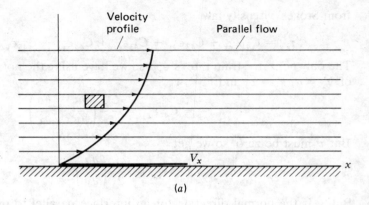

FIGURE 10.1
Deforming element in a parallel flow.

pressure p (or bulk stress $\bar{\sigma}$).[2] Thus:

$$\tau_{xx} = -p + C_{11}\dot{\epsilon}_{xx} + C_{12}\dot{\epsilon}_{yy} + C_{13}\dot{\epsilon}_{zz} + C_{14}\dot{\epsilon}_{xy} + C_{15}\dot{\epsilon}_{yz} + C_{16}\dot{\epsilon}_{xz}$$

$$\tau_{yy} = -p + C_{21}\dot{\epsilon}_{xx} + C_{22}\dot{\epsilon}_{yy} + C_{23}\dot{\epsilon}_{zz} + C_{24}\dot{\epsilon}_{xy} + C_{25}\dot{\epsilon}_{yz} + C_{26}\dot{\epsilon}_{xz}$$

$$\tau_{zz} = -p + C_{31}\dot{\epsilon}_{xx} + C_{32}\dot{\epsilon}_{yy} + C_{33}\dot{\epsilon}_{zz} + C_{34}\dot{\epsilon}_{xy} + C_{35}\dot{\epsilon}_{yz} + C_{36}\dot{\epsilon}_{xz}$$

$$\tau_{xy} = \quad\quad C_{41}\dot{\epsilon}_{xx} + C_{42}\dot{\epsilon}_{yy} + C_{43}\dot{\epsilon}_{zz} + C_{44}\dot{\epsilon}_{xy} + C_{45}\dot{\epsilon}_{yz} + C_{46}\dot{\epsilon}_{xz}$$

$$\tau_{yz} = \quad\quad C_{51}\dot{\epsilon}_{xx} + C_{52}\dot{\epsilon}_{yy} + C_{53}\dot{\epsilon}_{zz} + C_{54}\dot{\epsilon}_{xy} + C_{55}\dot{\epsilon}_{yz} + C_{56}\dot{\epsilon}_{xz}$$

$$\tau_{xz} = \quad\quad C_{61}\dot{\epsilon}_{xx} + C_{62}\dot{\epsilon}_{yy} + C_{63}\dot{\epsilon}_{zz} + C_{64}\dot{\epsilon}_{xy} + C_{65}\dot{\epsilon}_{yz} + C_{66}\dot{\epsilon}_{xz}$$

$$(10.2)$$

The constants C_{ij} are called *viscosity coefficients*. Fluids behaving according to these relations are again called *Newtonian fluids*.

We now point out that Stokes' viscosity law degenerates to Newton's viscosity law for the special case of parallel flow. We have shown such a flow in Fig. 10.1. We have shown a rectangular parallelepiped of fluid at time t in Fig. 10.1a. This element for the velocity profile shown deforms as shown in Fig. 10.1b. We can expect a shear stress τ_{xy} to cause this deformation, so we have

---

[2] In this way, for static equilibrium Stokes' law satisfies Pascal's law for hydrostatics.

from Stokes' viscosity law

$$\tau_{xy} = C_{41}\dot{\epsilon}_{xx} + C_{42}\dot{\epsilon}_{yy} + C_{43}\dot{\epsilon}_{zz} + C_{44}\dot{\epsilon}_{xy} + C_{45}\dot{\epsilon}_{yz} + C_{46}\dot{\epsilon}_{xz}$$

The only nonzero strain rate is $\dot{\epsilon}_{xy}$, so we have using the notation is this chapter of $V_x = u$, $V_y = v$, and $V_z = w$.

$$\tau_{xy} = C_{44}\dot{\epsilon}_{xy} = \frac{C_{44}}{2}\left(\frac{\partial u}{\partial y} + \frac{\partial v}{\partial x}\right)$$

But v must be zero, so we get

$$\tau_{xy} = \frac{C_{44}}{2}\left(\frac{\partial u}{\partial y}\right)$$

But y is the normal direction for an interface parallel to the xz plane, so the result above is the same as Newton's viscosity law for this case with $C_{44}/2$ becoming the familiar coefficient of viscosity.

It would seem that the constitutive law presented with its 36 constants is frightfully complicated. However, most Newtonian fluids that we deal with have flow properties which are *not dependent* in any way on *direction*. That is to say, Stokes' viscosity law for another reference $x'y'z'$ rotated relative to xyz would retain the same constants C_{ij} as when the law is given for xyz. Thus for $\tau_{x'x'}$ we have

$$\tau_{x'x'} = -p + C_{11}\dot{\epsilon}_{x'x'} + C_{12}\dot{\epsilon}_{y'y'} + C_{13}\dot{\epsilon}_{z'z'} + C_{14}\dot{\epsilon}_{x'y'} + C_{15}\dot{\epsilon}_{y'z'} + C_{16}\dot{\epsilon}_{x'z'}$$

$$(10.3)$$

where the C's are the same as those in Eq. (10.2) for the same fluid. When a fluid behaves in the same way for all directions at a point, we say that the fluid is *isotropic*. We will now show that for isotropic fluids the 36 viscosity constants degenerate to just 2, rendering great simplication for the resulting constitutive law.

In order to do this, we impose a certain sequence of rotation of axes xyz, requiring each time that the stress-strain rate relations for the new orientation have the same form and constants as for the original set of axes. First consider a 180° rotation about the z axis as shown in Fig. 10.2. The direction cosines between the various axes is given in the following tabular form:

	x	y	z
x'	-1	0	0
y'	0	-1	0
z'	0	0	1

$$(10.4)$$

Thus the direction cosine between the x' and the x axes (that is, $a_{x'x}$) is -1. Using the transformation equations given in Chap. 2 for stress under a rotation of axes we have for $\tau_{x'x}$

$$\begin{aligned}
\tau_{x'x'} = {} &\tau_{xx}a_{x'x}^2 & + \tau_{xy}a_{x'x}a_{x'y} &+ \tau_{xz}a_{x'x}a_{x'z} + \\
&\tau_{yx}a_{x'y}a_{x'x} + \tau_{yy}a_{x'y}^2 & &+ \tau_{yz}a_{x'y}a_{x'z} + \\
&\tau_{zx}a_{x'z}a_{x'x} + \tau_{zy}a_{x'z}a_{x'y} &+ \tau_{zz}a_{x'z}^2
\end{aligned}$$

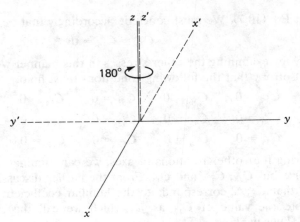

FIGURE 10.2
First rotation of axes (180°).

Inserting numerical values for the direction cosines from Eq. (10.4), we drop all terms on the right side of the equation except the first term to get

$$\tau_{x'x'} = \tau_{xx}$$

Similarly, considering the transformation formulas for the other stresses we find for all stresses that

$$
\begin{aligned}
\tau_{x'x'} &= \tau_{xx} & \tau_{x'y'} &= \tau_{xy} \\
\tau_{y'y'} &= \tau_{yy} & \tau_{y'z'} &= -\tau_{yz} \\
\tau_{z'z'} &= \tau_{zz} & \tau_{x'z'} &= -\tau_{xz}
\end{aligned}
\tag{10.5}
$$

Now we have pointed out in Chap. 4 that strain rates transform exactly as the stress terms, so we can say for $\dot{\epsilon}_{x'x'}$ that

$$
\begin{aligned}
\dot{\epsilon}_{x'x'} = \dot{\epsilon}_{xx}a_{x'x}^2 &+ \dot{\epsilon}_{xy}a_{x'x}a_{x'y} + \dot{\epsilon}_{xz}a_{x'x}a_{x'z} + \\
\dot{\epsilon}_{yx}a_{x'y}a_{x'x} &+ \dot{\epsilon}_{yy}a_{x'y}^2 + \dot{\epsilon}_{yz}a_{x'y}a_{x'z} + \\
\dot{\epsilon}_{zx}a_{x'z}a_{x'x} &+ \dot{\epsilon}_{zy}a_{x'z}a_{x'y} + \dot{\epsilon}_{zz}a_{x'z}^2
\end{aligned}
$$

Hence, for the rotation of axes at hand,

$$\dot{\epsilon}_{x'x'} = \dot{\epsilon}_{xx}$$

Similary we get for all the strain rates

$$
\begin{aligned}
\dot{\epsilon}_{x'x'} &= \dot{\epsilon}_{xx} & \dot{\epsilon}_{x'y'} &= \dot{\epsilon}_{xy} \\
\dot{\epsilon}_{y'y'} &= \dot{\epsilon}_{yy} & \dot{\epsilon}_{y'z'} &= -\dot{\epsilon}_{yz} \\
\dot{\epsilon}_{z'z'} &= \dot{\epsilon}_{zz} & \dot{\epsilon}_{x'z'} &= -\dot{\epsilon}_{xz}
\end{aligned}
\tag{10.6}
$$

We next replace the primed quantities in Eq. (10.3) using the results of Eqs. (10.5) and (10.6). We thus get

$$\tau_{xx} = -p + C_{11}\dot{\epsilon}_{xx} + C_{12}\dot{\epsilon}_{yy} + C_{13}\dot{\epsilon}_{zz} + C_{14}\dot{\epsilon}_{xy} - C_{15}\dot{\epsilon}_{yz} - C_{16}\dot{\epsilon}_{xz} \tag{10.7}$$

Compare Eq. (10.7) with the corresponding equation of Eqs. (10.2). For isotropy, these equations must be identical, since all we did was rotate axes at a point to arrive at Eq. (10.7). We must conclude accordingly that

$$C_{15} = C_{16} = 0$$

Similarly, by examining the other stresses in this manner we conclude additionally for isotropy that the following conditions must hold:

$$C_{25} = 0 \qquad C_{36} = 0 \qquad C_{51} = 0 \qquad C_{54} = 0 \qquad C_{63} = 0$$
$$C_{26} = 0 \qquad C_{45} = 0 \qquad C_{52} = 0 \qquad C_{61} = 0 \qquad C_{64} = 0$$
$$C_{35} = 0 \qquad C_{46} = 0 \qquad C_{53} = 0 \qquad C_{62} = 0$$

Considering four other rotations of axes, we can eliminate as zero all viscosity coefficients but C_{44}, C_{12}, and C_{11}. From the earlier discussion on parallel flow, we note that $C_{44}/2$ corresponds to the familiar coefficient of viscosity μ. And we denote the value of C_{12} as μ', which we call the *second coefficient of viscosity*. Thus we have

$$\frac{C_{44}}{2} = \mu$$
$$C_{12} = \mu'$$

As for C_{11}, we find, in addition, that $C_{11} = C_{44} + C_{12}$, so that we can say

$$C_{11} = 2\mu + \mu'$$

We thus have reduced the number of constants from 36 to 2; additional rotations of axes coupled with isotropy requirements yield no more additional information. We then have for Stokes' viscosity law in the case of isotropy

$$\tau_{xx} = -p + 2\mu\dot{\epsilon}_{xx} + \mu'\left(\dot{\epsilon}_{xx} + \dot{\epsilon}_{yy} + \dot{\epsilon}_{zz}\right) \qquad \tau_{xy} = 2\mu\dot{\epsilon}_{xy}$$
$$\tau_{yy} = -p + 2\mu\dot{\epsilon}_{yy} + \mu'\left(\dot{\epsilon}_{xx} + \dot{\epsilon}_{yy} + \dot{\epsilon}_{zz}\right) \qquad \tau_{xz} = 2\mu\dot{\epsilon}_{xz} \qquad (10.8)$$
$$\tau_{zz} = -p + 2\mu\dot{\epsilon}_{zz} + \mu'\left(\dot{\epsilon}_{xx} + \dot{\epsilon}_{yy} + \dot{\epsilon}_{zz}\right) \qquad \tau_{yz} = 2\mu\dot{\epsilon}_{yz}$$

Recall from Chap. 2 that for many fluids the thermodynamic pressure p equals $-\bar{\sigma} = (-1/3)(\tau_{xx} + \tau_{yy} + \tau_{zz})$. For *such fluids we can eliminate the second coefficient of viscosity in terms of the first coefficient of viscosity.* To do this, add the three equations above for τ_{xx}, τ_{yy}, and τ_{zz}. We then get

$$\left(\tau_{xx} + \tau_{yy} + \tau_{zz}\right) = -3p + 2\mu\left(\dot{\epsilon}_{xx} + \dot{\epsilon}_{yy} + \dot{\epsilon}_{zz}\right) + 3\mu'\left(\dot{\epsilon}_{xx} + \dot{\epsilon}_{yy} + \dot{\epsilon}_{zz}\right)$$

Noting that $\tau_x + \tau_{yy} + \tau_{zz} = 3\bar{\sigma} = -3p$, we can conclude that

$$0 = 2\mu\left(\dot{\epsilon}_{xx} + \dot{\epsilon}_{yy} + \dot{\epsilon}_{zz}\right) + 3\mu'\left(\dot{\epsilon}_{xx} + \dot{\epsilon}_{yy} + \dot{\epsilon}_{zz}\right)$$

$$\therefore \mu' = -\tfrac{2}{3}\mu$$

Returning to Eq. (10.8), we get

$$\tau_{xx} = -p + \mu\left[2\dot{\epsilon}_{xx} - \tfrac{2}{3}\left(\dot{\epsilon}_{xx} + \dot{\epsilon}_{yy} + \dot{\epsilon}_{zz}\right)\right] \qquad \tau_{xy} = 2\mu\dot{\epsilon}_{xy}$$
$$\tau_{yy} = -p + \mu\left[2\dot{\epsilon}_{yy} - \tfrac{2}{3}\left(\dot{\epsilon}_{xx} + \dot{\epsilon}_{yy} + \dot{\epsilon}_{zz}\right)\right] \qquad \tau_{xz} = 2\mu\dot{\epsilon}_{xz}$$
$$\tau_{zz} = -p + \mu\left[2\dot{\epsilon}_{zz} - \tfrac{2}{3}\left(\dot{\epsilon}_{xx} + \dot{\epsilon}_{yy} + \dot{\epsilon}_{zz}\right)\right] \qquad \tau_{yz} = 2\mu\dot{\epsilon}_{yz}$$

Noting that

$$\dot{\epsilon}_{xx} = \frac{\partial u}{\partial x} \quad \dot{\epsilon}_{yy} = \frac{\partial v}{\partial y} \quad \text{and} \quad \dot{\epsilon}_{zz} = \frac{\partial w}{\partial z}$$

we can say using vector notation

$$\boxed{
\begin{aligned}
\tau_{xx} &= \mu\left(2\frac{\partial u}{\partial x} - \frac{2}{3}\operatorname{div}\mathbf{V}\right) - p & \tau_{xy} &= \mu\left(\frac{\partial u}{\partial y} + \frac{\partial v}{\partial x}\right) \\
\tau_{yy} &= \mu\left(2\frac{\partial v}{\partial y} - \frac{2}{3}\operatorname{div}\mathbf{V}\right) - p & \tau_{xz} &= \mu\left(\frac{\partial u}{\partial z} + \frac{\partial w}{\partial x}\right) \\
\tau_{zz} &= \mu\left(2\frac{\partial w}{\partial z} - \frac{2}{3}\operatorname{div}\mathbf{V}\right) - p & \tau_{yz} &= \mu\left(\frac{\partial v}{\partial z} + \frac{\partial w}{\partial y}\right)
\end{aligned}}
\tag{10.9}$$

This is a common form of *Stokes' viscosity law*.

If the flow is *incompressible*, one notes that the expression div $\mathbf{V}$ is zero from continuity considerations, so that Eqs. (10.9) become

$$\boxed{
\begin{aligned}
\tau_{xx} &= 2\mu\frac{\partial u}{\partial x} - p & \tau_{xy} &= \mu\left(\frac{\partial u}{\partial y} + \frac{\partial v}{\partial x}\right) \\
\tau_{yy} &= 2\mu\frac{\partial v}{\partial y} - p & \tau_{xz} &= \mu\left(\frac{\partial u}{\partial z} + \frac{\partial w}{\partial x}\right) \\
\tau_{zz} &= 2\mu\frac{\partial w}{\partial z} - p & \tau_{yz} &= \mu\left(\frac{\partial v}{\partial z} + \frac{\partial w}{\partial y}\right)
\end{aligned}}
\tag{10.10}$$

The corresponding equations for cylindrical coordinates are given as follows:

$$\begin{aligned}
\tau_{rr} &= -p + 2\mu\frac{\partial v_r}{\partial r} & \tau_{r\theta} &= \mu\left[r\frac{\partial}{\partial r}\left(\frac{v_\theta}{r}\right) + \frac{1}{r}\frac{\partial v_r}{\partial\theta}\right] \\
\tau_{\theta\theta} &= -p + 2\mu\left(\frac{1}{r}\frac{\partial v_\theta}{\partial\theta} + \frac{v_r}{r}\right) & \tau_{\theta z} &= \mu\left[\frac{\partial v_\theta}{\partial z} + \frac{1}{r}\frac{\partial v_z}{\partial\theta}\right] \\
\tau_{zz} &= -p + 2\mu\frac{\partial v_z}{\partial z} & \tau_{rz} &= \mu\left[\frac{\partial v_r}{\partial z} + \frac{\partial v_z}{\partial r}\right]
\end{aligned}
\tag{10.11}$$

10.3 NAVIER-STOKES EQUATIONS FOR LAMINAR INCOMPRESSIBLE FLOW

Restricting ourselves to laminar incompressible flow of isotropic fluids, we now study the motion of an infinitesimal system of fluid which at time t is a rectangular parallelepiped, as is shown in Fig. 10.3. To avoid cluttering the

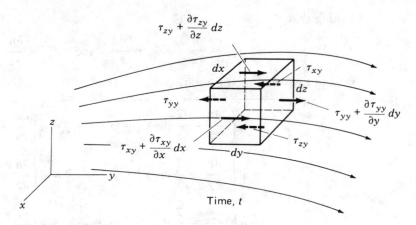

FIGURE 10.3
Element showing stresses in y direction.

diagram, only stresses contributing a force in the y direction have been indicated. Body force per unit mass will be taken as $B_x\mathbf{i} + B_y\mathbf{j} + B_z\mathbf{k}$, but this is not shown in Fig. 10.3. Note that first-order stress variations have now been included, since they are significant in the ensuing dynamic formulations.

Let us now express Newton's law for this system. The general vector equation using the velocity field has been worked out in Sec. 7.3 as

$$\mathbf{df} + \mathbf{B}\,dm = dm\left(u\frac{\partial \mathbf{V}}{\partial x} + v\frac{\partial \mathbf{V}}{\partial y} + w\frac{\partial \mathbf{V}}{\partial z} + \frac{\partial \mathbf{V}}{\partial t}\right)$$

where $\mathbf{df}$ is the resultant surface force and $\mathbf{B}$ is the body force per unit mass. In the y direction, this equation becomes

$$df_y + B_y\,dm = dm\left(u\frac{\partial v}{\partial x} + v\frac{\partial v}{\partial y} + w\frac{\partial v}{\partial z} + \frac{\partial v}{\partial t}\right) \qquad (10.12)$$

The surface force df_y is easily determined with the help of Fig. 10.3. Thus,

$$df_y = \frac{\partial \tau_{yy}}{\partial y}\,dy\,dx\,dz + \frac{\partial \tau_{zy}}{\partial z}\,dz\,dx\,dy + \frac{\partial \tau_{xy}}{\partial x}\,dx\,dy\,dz$$

Now replace the shear and normal stresses in this equation by employing the general Stokes'-viscosity-law relations [Eq. (10.9)]. The equation then becomes, after collecting and arranging the terms,[3]

$$df_y = dx\,dy\,dz\left[-\frac{\partial p}{\partial y} + \mu\nabla^2 v + \frac{\mu}{3}\left(\frac{\partial^2 u}{\partial x\,\partial y} + \frac{\partial^2 v}{\partial y^2} + \frac{\partial^2 w}{\partial z\,\partial y}\right)\right]$$

---

[3]Note that $\nabla \cdot \nabla \equiv \nabla^2$ is the so-called Laplacian operator which for Cartesian coordinates is $(\partial^2/\partial x^2 + \partial^2/\partial y^2 + \partial^2/\partial z^2)$.

It can readily be shown that the expression $\partial^2 u/(\partial x\,\partial y) + \partial^2 v/\partial y^2 + \partial^2 w/(\partial z\,\partial y)$ is zero in the preceding equation for *incompressible flow*. Thus, extracting the operator $\partial/\partial y$ from this expression, we get $(\partial/\partial y)(\partial u/\partial x + \partial v/\partial y + \partial w/\partial z)$. *Continuity* then imposes the condition $\partial u/\partial x + \partial v/\partial y + \partial w/\partial z = 0$, so the expression is zero. Hence, the preceding equation becomes

$$df_y = \left(-\frac{\partial p}{\partial y} + \mu \nabla^2 v \right) dx\,dy\,dz$$

Upon substituting this result in Eq. (10.12), then replacing dm by $\rho\,dx\,dy\,dz$ and, finally, using the notation D/Dt as the substantial derivative, the following equation may be found:

$$\left(-\frac{\partial p}{\partial y} + \mu \nabla^2 v \right) dx\,dy\,dz + \rho B_y\, dx\,dy\,dz = \rho\,dx\,dy\,dz\,\frac{Dv}{Dt}$$

Canceling and rearranging terms, we get

$$\frac{Dv}{Dt} = B_y + \frac{1}{\rho}\left(-\frac{\partial p}{\partial y} + \mu \nabla^2 v \right)$$

This is the y component of the Navier-Stokes equation for incompressible flow, and the other components may be similarly ascertained. Thus we have

$$\rho\frac{Du}{Dt} = \rho B_x + \left(-\frac{\partial p}{\partial x} + \mu \nabla^2 u \right) \qquad (10.13a)$$

$$\rho\frac{Dv}{Dt} = \rho B_y + \left(-\frac{\partial p}{\partial y} + \mu \nabla^2 v \right) \qquad (10.13b)$$

$$\rho\frac{Dw}{Dt} = \rho B_z + \left(-\frac{\partial p}{\partial z} + \mu \nabla^2 w \right) \qquad (10.13c)$$

The general case of Navier-Stokes equations for compressible laminar flow is considerably more complicated and will not be undertaken in this text.

Note that the Navier-Stokes equations reduce to Euler's equations for the case of nonviscous flow. Also, these equations are given as follows using vector and tensor notation, respectively:

$$\rho\frac{D\mathbf{V}}{Dt} = \rho\mathbf{B} + \left(-\nabla p + \mu \nabla^2 \mathbf{V} \right) \qquad (10.14)$$

$$\rho\frac{DV_i}{Dt} = \rho B_i + \left(-p,_i + \mu V_i,_{jj} \right) = \rho B_i + \left(-\frac{\partial p}{\partial x_i} + \mu \frac{\partial^2 V_i}{\partial x_j\,\partial x_j} \right) \qquad (10.15)$$

Equations (10.13) and the continuity equation form four simultaneous differential equations from which the four unknowns u, v, w, and p could conceivably be solved for many problems were it not for the nonlinear nature and general complexity of the equations. [The nonlinearity is present in the substantial derivative, wherein one finds terms such as $u(\partial u/\partial x)$, etc.] In cylindrical coordinates the Navier-Stokes equations are given as follows:

$$\rho\left(\frac{\partial v_r}{\partial t} + v_r\frac{\partial v_r}{\partial r} + \frac{v_\theta}{r}\frac{\partial v_r}{\partial \theta} - \frac{v_\theta^2}{r} + v_z\frac{\partial v_r}{\partial z}\right)$$

$$= \rho B_r - \frac{\partial p}{\partial r} + \mu\left[\frac{1}{r}\frac{\partial}{\partial r}\left(r\frac{\partial v_r}{\partial r}\right) - \frac{v_r}{r^2} + \frac{1}{r^2}\frac{\partial^2 v_r}{\partial \theta^2} - \frac{2}{r^2}\frac{\partial v_\theta}{\partial \theta} + \frac{\partial^2 v_r}{\partial z^2}\right] \quad (10.16a)$$

$$\rho\left(\frac{\partial v_\theta}{\partial t} + v_r\frac{\partial v_\theta}{\partial r} + \frac{v_\theta}{r}\frac{\partial v_\theta}{\partial \theta} - \frac{v_r v_\theta}{r} + v_z\frac{\partial v_\theta}{\partial z}\right)$$

$$= \rho B_\theta - \frac{1}{r}\frac{\partial p}{\partial \theta} + \mu\left[\frac{1}{r}\frac{\partial}{\partial r}\left(r\frac{\partial v_\theta}{\partial r}\right) - \frac{v_\theta}{r^2} + \frac{1}{r^2}\frac{\partial^2 v_\theta}{\partial \theta^2} + \frac{2}{r^2}\frac{\partial v_r}{\partial \theta} + \frac{\partial^2 v_\theta}{\partial z^2}\right]$$

$$(10.16b)$$

$$\rho\left(\frac{\partial v_z}{\partial t} + v_r\frac{\partial v_z}{\partial r} + \frac{v_\theta}{r}\frac{\partial v_z}{\partial \theta} + v_z\frac{\partial v_z}{\partial z}\right)$$

$$= \rho B_z - \frac{\partial p}{\partial z} + \mu\left[\frac{1}{r}\frac{\partial}{\partial r}\left(r\frac{\partial v_z}{\partial r}\right) + \frac{1}{r^2}\frac{\partial^2 v_z}{\partial \theta^2} + \frac{\partial^2 v_z}{\partial z^2}\right] \quad (10.16c)$$

In this chapter we will analyze several laminar flows using the Navier-Stokes equations.

10.4   PARALLEL FLOW: GENERAL CONSIDERATIONS

We will now examine the special incompressible flow wherein all streamlines are straight and parallel (Fig. 10.4a). The direction of flow is taken along the z axis, so that $u = v = v_r = v_\theta = 0$. It will be shown that certain general conclusions may be made, with the aid of the Navier-Stokes equations, concerning the pressure distribution for such flows.

Examine the Navier-Stokes equation [Eq. (10.13b)] in the y direction, which will correspond to minus the direction of gravity for this analysis. The

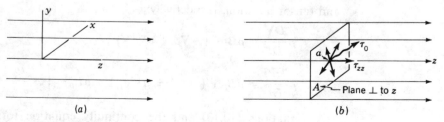

FIGURE 10.4
Normal stress at a point in parallel flow.

body force per unit mass is that of gravity, so that B_y may be replaced by $-g$. The resulting equation is then

$$-\frac{\partial p}{\partial y} = \rho g \qquad (10.17)$$

It will be convenient to consider the pressure as the superposition of two separate pressures. One is a hydrostatic pressure distribution p_g called the *geometric pressure* due *only* to gravitational influence on a *static* fluid having the same geometry as the flow under consideration. The remaining portion is a quantity which when superposed on p_g yields the proper pressure distribution. This quantity can be considered to represent dynamic effects as well as some pressure from the boundaries; we denote it as $\bar{p}$. Thus,

$$p = \bar{p} + p_g \qquad (10.18)$$

Substituting Eq. (10.18) into Eq. (10.17) and replacing ρg by γ results in the following equation:

$$\frac{\partial \bar{p}}{\partial y} + \frac{\partial p_g}{\partial y} = -\gamma \qquad (10.19)$$

But in the study of hydrostatics in Chap. 3, it was shown that in the direction of gravity $\partial p_g / \partial y = -\gamma$, so that Eq. (10.19) simplifies to

$$\frac{\partial \bar{p}}{\partial y} = 0 \qquad (10.20)$$

Next, considering the Navier-Stokes equation [Eq. (10.13a)] for the x direction, normal to the direction of flow so that $u = 0$, we get

$$\frac{\partial p}{\partial x} = 0 \qquad (10.21)$$

Using the pressure $\bar{p}$ and geometric pressure p_g in Eq. (10.21), we get

$$\frac{\partial \bar{p}}{\partial x} + \frac{\partial p_g}{\partial x} = 0 \qquad (10.22)$$

The term $\partial p_g / \partial x$ must be zero, since it was shown that hydrostatic pressure can vary only in the direction of gravity which here is the y direction. Thus the nonhydrostatic pressure $\bar{p}$ behaves according to the relation

$$\frac{\partial \bar{p}}{\partial x} = 0 \qquad (10.23)$$

Equations (10.20) and (10.23) lead us to conclude that the nonhydrostatic pressure $\bar{p}$ can vary *only* in the direction of flow z. That is, at any section normal to this direction, $\bar{p}$ is of uniform magnitude over the entire section. Hence, the total pressure p at any section normal to the flow can depart from the aforementioned uniform distribution $\bar{p}$ only to the extent of an additional hydrostatic pressure variation p_g in the y direction.

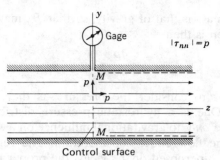

FIGURE 10.5
Control surface MM normal to the flow.

With this conclusion in mind, let us now examine *Stokes' viscosity* law for normal stresses [Eq. (10.10)] as applied to parallel flow in the z direction. Since u, v, v_r, v_θ, and $\partial v_z/\partial z$ equal zero everywhere,[4] the following relations are found to apply to any point:

$$\tau_{xx} = -p$$
$$\tau_{yy} = -p \qquad\qquad (10.24)$$
$$\tau_{zz} = -p$$

Since v_r and v_θ are zero, it would have been entirely possible to rotate the xy direction about the z axis to any position without altering the above relation. It may then be concluded that the normal stress *in the direction of flow z* equals all normal stresses *at right angles to the direction of flow z* at a given point. This is illustrated in Fig. 10.4*b*, where all normal stresses shown straight for point *a* in plane A equal the stress τ_{zz}. Other stresses, such as τ_0 *inclined* both to plane A and the z direction, however, need not equal τ_{zz}.

The conclusions of the preceding paragraphs become very significant for cases where the *hydrostatic* pressure p_g is a *small* part of the pressure p. This occurs for the flow of air through ducts or liquids at appreciable head through other than very large pipes. It is then permissible to consider the pressure p *equal* to the nonhydrostatic pressure $\bar{p}$. Under these conditions, one may consider the pressure p to be *uniform* over the entirety of any cross section normal to the direction of flow. Furthermore, according to the previous paragraph, this means that the stress τ_{nn}, normal to this section, must *also* be uniformly distributed over the entire section and equal to minus the pressure p. Finally, note in Fig. 10.5 that a pressure gage mounted at right angles to the flow will record a value which equals the thermodynamic pressure p and thus minus the stress τ_{nn} across the control surface.

---

[4]It was shown in the previous chapter that since $\nabla \cdot \mathbf{V} = 0$ for incompressible, parallel flow, then $\partial v_z/\partial z = 0$ where z is the direction of flow. Hence, the *velocity profile* does not change in the direction of flow.

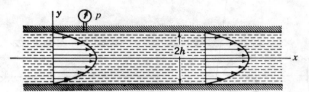

FIGURE 10.6
Flow between parallel plates.

10.5 PARALLEL LAMINAR FLOW PROBLEMS

We now examine a series of parallel flows. The first two will be the parallel plate flow and the circular duct flow that were considered in Chap. 7.

Case 1. Steady, laminar, two-dimensional incompressible flow between infinite parallel plates. We will now apply the material of the previous sections to the case of a two-dimensional flow between infinite parallel plates (Fig. 10.6) wherein you will note that x has been taken as the direction of flow. The restrictions of incompressible, steady, laminar flow will be applied to this analysis. To be investigated are the velocity field and the head loss. The former information will be complete if the velocity profile for a section normal to the flow is ascertained, since the profile is invariant for all such sections in the case of parallel flow. As for the head loss, it will be necessary to evaluate only the variation of the pressure along the direction of flow. The Navier-Stokes equation [Eq. (10.13a)] for parallel flow involves the aforementioned quantities. With the further restriction of two-dimensional steady flow, this equation simplifies to

$$-\frac{\partial p}{\partial x} = -\mu \frac{\partial^2 u}{\partial y^2} \tag{10.25}$$

Now p may be replaced by $\bar{p} + p_g$. However, since p_g can vary only in the y direction, the expression $\partial p/\partial x$ becomes equal to $\partial \bar{p}/\partial x$. It was shown in parallel flow that $\bar{p}$ is uniform over any section normal to the direction of flow. Consequently, $\partial \bar{p}/\partial x = \partial p/\partial x$ must be constant for any section and can vary only in the direction of flow.

We thus arrive at the same differential equation for this flow as in Case 1 of Chap. 7. We can now go back to Case 1 in Chap. 7 for the details of the solution.[5]

---

[5]For turbulent flow we use the hydraulic diameter which here becomes for width b and height $2h$ and with $b \to \infty$,

$$D_H = \lim_{b \to \infty} \frac{4(2h)(b)}{2b + 2h} = 4h$$

The critical Reynolds number remains 2300 using the hydraulic diameter. We use e/D_H for the roughness ratio. For even better results, use for D_H the value $0.667 D_H$ in the calculations involving the Moody diagram.

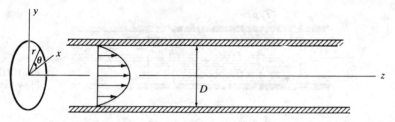

FIGURE 10.7
Fully developed laminar flow in a circular pipe.

Case 2. Incompressible, laminar, steady flow in a pipe. As a second and particularly vital example of parallel flow we now investigate the velocity profile and head loss for the case of steady, incompressible, laminar flow through a straight pipe of constant cross section. The flow may be considered axially symmetric with respect to the pipe centerline. Neglecting hydrostatic variations, it is then a flow wherein all parameters are independent of θ, as shown in Fig. 10.7. Note that for this case we have returned to z for the direction of flow so as to be consistent with polar coordinates involving θ.

The component of Navier-Stokes equations in the direction of flow embodies the desired information. Rewriting Eq. (10.13c), we have

$$\frac{Dv_z}{Dt} = \frac{1}{\rho}\left(-\frac{\partial p}{\partial z} + \mu \nabla^2 v_z\right) \tag{10.26}$$

We may readily transform this equation to cylindrical coordinates by using the form of the Laplacian operator corresponding to cylindrical coordinates or use Eq. (10.16c). Since v_z does not depend on z, we then have for axially symmetric flow

$$\frac{Dv_z}{Dt} = \frac{1}{\rho}\left[-\frac{\partial p}{\partial z} + \mu\left(\frac{\partial^2 v_z}{\partial r^2} + \frac{1}{r}\frac{\partial v_z}{\partial r}\right)\right] \tag{10.27}$$

For steady parallel flow in the z direction the substantial derivative is zero, and Eq. (10.27) simplifies to

$$\frac{\partial p}{\partial z} = \mu\left(\frac{\partial^2 v_z}{\partial r^2} + \frac{1}{r}\frac{\partial v_z}{\partial r}\right) \tag{10.28}$$

We thus arrive at the same differential equation formed in Case 2 of Chap. 7. Again we refer now to this problem in Chap. 7 for the details of the solution of this equation.

Case 3. Laminar incompressible flow in an annulus. We next consider laminar incompressible flow in an annulus (see Fig. 10.8). As in the pipe flow, we have axial symmetry so that Eq. (10.28) is valid here. Furthermore, all the steps taken in Case 2 of Sec. 7.8 from Eq. (7.59) to (7.64) are also valid. Our starting point

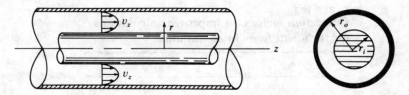

FIGURE 10.8
Laminar flow through an annulus.

for the problem is then

$$v_z = -\frac{\beta}{4\mu}r^2 + C_1 \ln r + C_2 \qquad (10.29)$$

The boundary condition for determining C_1 and C_2 are then

$$v_z = 0 \quad \text{when} \quad r = r_i$$
$$v_z = 0 \quad \text{when} \quad r = r_o$$

We accordingly must satisfy the following equations:

$$0 = -\frac{\beta}{4\mu}r_i^2 + C_1 \ln r_i + C_2$$

$$0 = -\frac{\beta}{4\mu}r_o^2 + C_1 \ln r_o + C_2$$

Solving for the constants, we get for v_z

$$v_z = \frac{\beta}{4\mu}\left\{-r^2 + r_o^2 + \frac{r_o^2 - r_i^2}{\ln(r_i/r_o)}\ln\frac{r}{r_o}\right\} \qquad (10.30)$$

We now wish to get the separation constant, β. Accordingly we compute q:

$$q = \int_{r_i}^{r_o} 2\pi r\left\{\frac{\beta}{4\mu}\left[-r^2 + r_o^2 + \frac{r_o^2 - r_i^2}{\ln(r_i/r_o)}\ln\left(\frac{r}{r_o}\right)\right]\right\} dr$$

We leave it for you to show that

$$q = \frac{\beta\pi}{8\mu}\left[r_o^4 - r_i^4 - \frac{(r_o^2 - r_i^2)^2}{\ln(r_o/r_i)}\right] \qquad (10.31)$$

Hence

$$\beta = \frac{8\mu q}{\pi}\left/\left[r_o^4 - r_i^4 - \frac{(r_o^2 - r_i^2)^2}{\ln(r_o/r_i)}\right]\right. \qquad (10.32)$$

We thus have the profile in terms of q and the geometry.

TABLE 10.1
Correction factors for improved calculations
for turbulent flows in an annulus*

r_o / r_i	κ
0.0001	0.892
0.01	0.799
0.05	0.742
0.1	0.716
0.2	0.693
0.4	0.676
0.6	0.670
0.8	0.667

*See O. C. Jones, Jr. and J. C. Leung, "An Improvement in the Calculation of Turbulent Friction in Smooth Concentric Annuli," *Journal of Fluid Mechanics Eng.*, Dec. 1981.

We have assumed that the flow is laminar. We require a Reynolds number under 2300. However in this regard we must use the *hydraulic diameter D_H* for this case as discussed in Sec. 9.12. Thus we have for D_H

$$D_H = \frac{(4)\left(\pi r_o^2 - \pi r_i^2\right)}{2\pi r_o + 2\pi r_i} = 2(r_o - r_i) \qquad (10.33)$$

In a homework problem we ask you to show that the maximum velocity occurs at a radius $\tilde{r}$

$$\tilde{r} = \left[\frac{r_o^2 - r_i^2}{2 \ln(r_o/r_i)}\right]^{1/2} \qquad (10.34)$$

This value of $\tilde{r}$ does not occur at the center of the annulus flow section. Finally, we note that the pressure change along the annulus over span L is

$$\Delta p = (-\beta)(L) = -\frac{8\mu qL}{\pi} \Bigg/ \left[r_o^4 - r_i^4 - \frac{r_o^2 - r_i^2}{\ln(r_o/r_i)}\right] \qquad (10.35)$$

For turbulent flow use the hydraulic diameter D_H, which is $2(r_o - r_i)$, along with the Moody diagram. For better results modify the hydraulic diameter by multiplying by the factor κ given in Table 10.1. Note that as $r_o/r_i \to 1$ we get $\kappa \to 0.667$ of the infinite parallel plate case.

Case 4. Couette flow. We now consider flow between two infinite parallel plates, this time with the upper plate having a constant velocity U in the x direction (see Fig. 10.9). We consider the flow only after steady flow has been achieved.

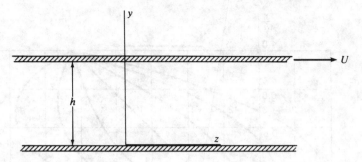

FIGURE 10.9
Viscous flow between an infinite moving and an infinite stationary plate.

Navier-Stokes equations simplified for the x direction is as follows for this case:

$$\frac{\partial p}{\partial x} = \mu \frac{\partial^2 u}{\partial y^2} \qquad (10.36)$$

Furthermore all steps taken in Case 1 of Sec. 7.8 from Eq. (7.45) to Eq. (7.48) are valid here. Hence we have for u,

$$u = -\frac{\beta}{\mu}\left(\frac{y^2}{2} + C_1 y + C_2\right) \qquad (10.37)$$

The boundary conditions that we must now impose are

$$\text{when} \quad y = 0 \quad u = 0$$
$$\text{when} \quad y = h \quad u = U$$

Thus we have

$$0 = -\frac{\beta}{\mu}(C_2)$$

$$U = -\frac{\beta}{\mu}\left(\frac{h^2}{2} + C_1 h + C_2\right)$$

Clearly $C_2 = 0$. Solving for C_1 we have

$$C_1 = -\left(\frac{U\mu}{\beta h} + \frac{h}{2}\right)$$

The velocity field is then, on replacing β by $-\partial p/\partial x$,

$$u = U\frac{y}{h} + \frac{1}{2\mu}\left(\frac{\partial p}{\partial x}\right)(y^2 - hy) \qquad (10.38)$$

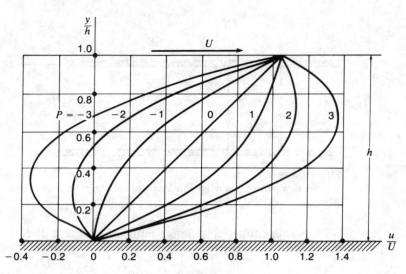

FIGURE 10.10
Couette flow between parallel plates. $P < 0$, adverse pressure gradient; $P > 0$, decreasing pressure in direction of flow; $P = 0$ zero pressure gradient. (*From H. Schlichting, Boundary Layer Theory, 7th ed.*, McGraw-Hill, N.Y., 1979, *with permission.*)

In dimensionless form, we have

$$\frac{u}{U} = \frac{y}{h} - \left[\frac{h^2}{2\mu U} \left(\frac{\partial p}{\partial x} \right) \right] \left(\frac{y}{h} \right) \left(1 - \frac{y}{h} \right) \qquad (10.39)$$

Note that the shape of the velocity profile is determined by the dimensionless expression in square brackets which we shall denote as P. Thus

$$P = -\frac{h^2}{2\mu U} \left(\frac{\partial p}{\partial x} \right)$$

is a dimensionless pressure gradient.

If P is negative, it means we have an increasing pressure in the direction of flow (from outside agents at the end of the flow) opposing the movement of fluid induced by the drag of the moving plate. In Fig. 10.10 we have shown velocity profiles u/U for various values of P including the aforementioned negative values (from H. Schlichting). Calling this increasing pressure an adverse pressure gradient we see that for $P = -1$ the velocity u is less than U below the moving plate but it is still positive. For $P = -2$ there is a back-flow over half the section. Additionally, for a larger adverse pressure gradient such that $P = -3$, the back-flow extends over two-thirds of the section. With $P = 0$ the plate is the sole mover of the fluid and we have a linear profile.

Back in Chap. 1 we discussed the torsional drag of a shaft rotating concentrically in a journal separated by a thin film of oil. The fluid undergoes a motion very much like the Couette flow here. That is, in the case of the rotating shaft it is the shaft surface dragging the fluid with a zero pressure gradient present. Furthermore if the film of fluid is very thin compared to the radius of the shaft, the effects of curvature are second order. Hence we then have Couette flow with $P = 1$, thus justifying the linear profile we used in Chap. 1.

10.6   A NOTE

In the next (starred) section we shall present an interesting detailed order-of-magnitude study to develop a simplified form of the Navier-Stokes equations for a very thin layer of flow such as a boundary layer (see Fig. 10.11). We start with the two-dimensional steady flow form of Navier-Stokes with no body force. This is shown in the next section as Eqs. (10.41). The continuity equation is then Eq. (10.42). We next examine the extreme values of the various terms in these equations in order to estimate the order of magnitude of the various derivatives. Computing these orders of magnitude, certain derivatives can be deleted as negligible. We thus learn from this study that:

(a) Changes in pressure p in the y direction normal to the film are very small compared to changes of p in the x direction of flow.

(b) Additionally, as a consequence, pressure p in the boundary layer will equal p just outside the boundary layer.

(c) Furthermore, the velocity v is small compared to the velocity u so that we have close to parallel flow.

(d) Finally, we can delete $\partial^2 u / \partial x^2$ as negligible.

The results of the study are presented below:

$$
\boxed{
\begin{aligned}
u\frac{\partial u}{\partial x} + v\frac{\partial u}{\partial y} &= -\frac{1}{\rho}\frac{dp}{dx} + \nu\frac{\partial^2 u}{\partial y^2} \quad (a) \\
\frac{\partial u}{\partial x} + \frac{\partial v}{\partial y} &= 0 \quad (b)
\end{aligned}
}
\qquad (10.40)
$$

In the next chapter, we will use these equations to present the work of Blasius for an exact formulation of the boundary layer thickness of a laminar boundary layer. Again we will offer the reader a choice of a thumbnail sketch and a detailed starred development.

*10.7   SIMPLIFIED NAVIER-STOKES EQUATIONS FOR A VERY THIN LAYER OF FLOW

We will now consider the two-dimensional, steady, incompressible, laminar flow forming a very thin layer. Such a flow might represent the flow in a boundary

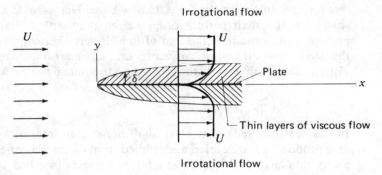

FIGURE 10.11
Boundary-layer flow about a flat plate.

layer over a flat plate (see Fig. 10.11).[6] Note that U is the free-stream velocity just outside the layer. Also, it is the velocity reached by fluid in the layer at its largest vertical distance δ from the plate.

We start by expressing the *Navier-Stokes* equation for two-dimensional flow with no body forces.

$$\rho\left(u\frac{\partial u}{\partial x} + v\frac{\partial u}{\partial y}\right) = -\frac{\partial p}{\partial x} + \mu\left(\frac{\partial^2 u}{\partial x^2} + \frac{\partial^2 u}{\partial y^2}\right) \qquad (10.41a)$$

$$\rho\left(u\frac{\partial v}{\partial x} + v\frac{\partial v}{\partial y}\right) = -\frac{\partial p}{\partial y} + \mu\left(\frac{\partial^2 v}{\partial x^2} + \frac{\partial^2 v}{\partial y^2}\right) \qquad (10.41b)$$

The continuity equation meanwhile is

$$\frac{\partial u}{\partial x} + \frac{\partial v}{\partial y} = 0 \qquad (10.42)$$

We wish to simplify these equations further by utilizing the fact that the thickness δ of the layer is such that the ratio δ/x is very small (x is a distance downstream away from the beginning of the layer). We do this by making a very rough order-of-magnitude study of the terms in the differential equations [Eqs. (10.41) and (10.42)] with the purpose of dropping those terms which we feel are very small in the computations as compared with other terms in the equations.

---

[6] We study boundary-layer theory in Chapter 13.

We use the concept of the order of magnitude to enable us to compare terms quickly and efficiently in the ensuing discussion.

To simplify the discussion, let us choose a scale of measure of time such that U is of the same order of magnitude as x, some position away from the front edge, where we will make an order-of-magnitude study. And, in addition, we choose a scale for the measure of distance such that the magnitude of x is of the order unity. We then consider that the velocity U and x both have the order of magnitude of unity, which we designate as $O(1)$. The magnitudes of the vertical distances with in the layer are much less than x according to the opening statement, so we call the order of magnitude of these distances $O(\delta)$, which is considerably smaller than $O(1)$. We next examine the *extreme values* of the various terms of Eqs. (10.41) and (10.42) for purposes of comparison. For instance, the maximum value of u at x is U, so that

$$u = O(1) \tag{10.43}$$

Next, the extreme value of the difference of the velocity u in the y direction is also of the order unity, since this velocity component goes from zero to U. This is similarly true for the extreme of the second difference of u in the y direction. Finally, since the maximum of the difference of y in the layer is of order $O(\delta)$, we can then say that

$$\frac{\partial u}{\partial y} = O\left(\frac{1}{\delta}\right) \tag{10.44a}$$

$$\frac{\partial^2 u}{\partial y^2} = O\left(\frac{1}{\delta^2}\right) \tag{10.44b}$$

Furthermore, u may change from close to U as it enters the thin stream at $x = 0$ and is close to zero at x when one is near the plate surface as shown in Fig. 10.11. We can then say that the extreme order of magnitude of $\partial u/\partial x$ is unity, so we have

$$\frac{\partial u}{\partial x} = O(1) \tag{10.45}$$

and by a similar argument we can conclude that

$$\frac{\partial^2 u}{\partial x^2} = O(1) \tag{10.46}$$

Looking at the continuity equation [Eq. (10.42)], we can see, since $\partial u/\partial x = O(1)$, that $\partial v/\partial y$ is also of order $O(1)$ to satisfy the equation. From this result, we can conclude that since changes in y in the boundary layer are of the order $O(\delta)$, it is necessary that changes in v in the y direction be also of the order $O(\delta)$. And since $v = 0$ for $y = 0$, it is then clear that for v itself $v = O(\delta)$. We may then make the following statements:

$$v = O(\delta) \tag{10.47a}$$

$$\frac{\partial v}{\partial y} = O(1) \tag{10.47b}$$

$$\frac{\partial^2 v}{\partial y^2} = O\left(\frac{1}{\delta}\right) \tag{10.47c}$$

$$\frac{\partial v}{\partial x} = O(\delta) \tag{10.47d}$$

$$\frac{\partial^2 v}{\partial x^2} = O(\delta) \tag{10.47e}$$

Let us now go back to the differential equation [Eq. (10.41a)]. Dividing through by ρ, we rewrite the equation, and directly below the equation, we put in the order of magnitudes:

$$u\frac{\partial u}{\partial x} + v\frac{\partial u}{\partial y} = -\frac{1}{\rho}\frac{\partial p}{\partial x} + \nu\left(\frac{\partial^2 u}{\partial x^2} + \frac{\partial^2 u}{\partial y^2}\right) \tag{10.48a}$$

$$[O(1)][O(1)] + [O(\delta)]\left[O\left(\frac{1}{\delta}\right)\right] = -\frac{1}{\rho}\frac{\partial p}{\partial x} + \nu\left[O(1) + O\left(\frac{1}{\delta^2}\right)\right] \tag{10.48b}$$

In the last bracket of Eq. (10.48b), the term having an order of magnitude $O(1)$ is much smaller than the term having an order of magnitude $O(1/\delta^2)$, so we can drop the former. This means that we can neglect $\partial^2 u/\partial x^2$ in the differential equation. The last bracketed expression in Eq. (10.48b) stems from frictional effects, and since frictional effects are significant, this expression must have an order of magnitude comparable with the other expressions in the equation, which is unity. Thus

$$\nu\left[O\left(\frac{1}{\delta^2}\right)\right] = O(1)$$

From this we conclude that the coefficient of kinematic viscosity ν has an order of magnitude $O(\delta^2)$ for the units we have imposed on the discussion.

Now consider Eq. (10.41b). We divide through by ρ and then express the order of magnitude of the corresponding terms directly below the equation.

$$u\frac{\partial v}{\partial x} + v\frac{\partial v}{\partial y} = -\frac{1}{\rho}\frac{\partial p}{\partial y} + \nu\left(\frac{\partial^2 v}{\partial x^2} + \frac{\partial^2 v}{\partial y^2}\right) \tag{10.49a}$$

$$[O(1)][O(\delta)] + [O(\delta)][O(1)] = -\frac{1}{\rho}\frac{\partial p}{\partial y} + [O(\delta^2)]\left[O(\delta) + O\left(\frac{1}{\delta}\right)\right] \tag{10.49b}$$

It is again clear that the term $\partial^2 v / \partial x^2$ may be neglected in the last bracket. Assuming that the normal stress distribution is significant, the term $-(1/\rho)(\partial p/\partial y)$ must be of the order $O(\delta)$, like the other expressions in the equations. Now the mass density ρ can always be taken as being of the order of magnitude $O(1)$ or less, and, consequently, the order of magnitude of the product of the quantity ρ times changes in the value of y inside the very thin layer is of the order $O(\delta)$ or less. In considering the expression $(1/\rho)(\partial p/\partial y)$ to have an order of magnitude $O(\delta)$ as the other significant terms, we see that changes of p must then be of the order $O(\delta^2)$ or less than $O(\delta^2)$ in the vertical direction, inside the very thin layer. By a similar argument, you can show in Eq. (10.48a) that changes in p in the direction of flow x are of the order $O(1)$. Thus *changes of p in the y direction are very small indeed compared with changes in the x direction and may be neglected.* Hence, we have a situation approaching the condition discussed in reference to parallel flow. Clearly, if changes of p are to be neglected in the y direction, then *p will be equal in magnitude to the pressure p at position x, just outside the layer.* For the boundary layer, this pressure, as has been pointed out earlier, may sometimes be developed from irrotational-incompressible-flow theory wherein the boundary layer is completely neglected.

Furthermore, as a result of this study we consider the velocity v, having order of magnitude $O(\delta)$, as a negligible quantity compared with u, having order of magnitude $O(1)$, so we do not trouble with Eq. (10.49a), which has v as the principal dependent variable. Instead, we consider Eq. (10.48a) only, with the term $\partial^2 u / \partial x^2$ deleted, and we assume that the main-stream flow imposes its pressure distribution on the thin layer. We can then write the simplified thin-layer equation and the continuity equation as

$$u\frac{\partial u}{\partial x} + v\frac{\partial u}{\partial y} = -\frac{1}{\rho}\frac{dp}{dx} + \nu\frac{\partial^2 u}{\partial y^2} \qquad (10.50a)$$

$$\frac{\partial u}{\partial x} + \frac{\partial v}{\partial y} = 0 \qquad (10.50b)$$

*10.8 DYNAMIC SIMILARITY LAW FROM THE NAVIER-STOKES EQUATIONS

We will now develop the dynamic similarity law, by considering the Navier-Stokes equation for a steady, incompressible, laminar flow with no body force. For simplicity, we work with the x component of the equation:

$$\rho\frac{Du}{Dt} = -\frac{\partial p}{\partial x} + \mu \nabla^2 u \qquad (10.51)$$

Expanding out the expressions in the equation we have

$$\rho\left(u\frac{\partial u}{\partial x} + v\frac{\partial u}{\partial y} + w\frac{\partial u}{\partial z}\right) = -\frac{\partial p}{\partial x} + \mu\left(\frac{\partial^2 u}{\partial x^2} + \frac{\partial^2 u}{\partial y^2} + \frac{\partial^2 u}{\partial z^2}\right) \quad (10.52)$$

We wish to introduce into this equation dimensionless variables which we define as follows:

$$u^* = \frac{u}{U} \qquad x^* = \frac{x}{L} \qquad p^* = \frac{p}{\rho U^2}$$

$$v^* = \frac{v}{U} \qquad y^* = \frac{y}{L} \qquad\qquad (10.53)$$

$$w^* = \frac{w}{U} \qquad z^* = \frac{z}{L}$$

where U is the free-stream velocity and L is some typical distance characterizing the geometry of the flow pattern. Replacing u by Uu^*, x by x^*L, and so forth, we have for Eq. (10.52)

$$\rho\frac{U^2}{L}\left(u^*\frac{\partial u^*}{\partial x^*} + v^*\frac{\partial u^*}{\partial y^*} + w^*\frac{\partial u^*}{\partial z^*}\right)$$

$$= -\frac{\rho U^2}{L}\frac{\partial p^*}{\partial x^*} + \frac{\mu U}{L^2}\left(\frac{\partial^2 u^*}{\partial x^{*2}} + \frac{\partial^2 u^*}{\partial y^{*2}} + \frac{\partial^2 u^*}{\partial z^{*2}}\right)$$

Multiplying through by $L/(\rho U^2)$, we may express this equation as

$$\frac{Du^*}{Dt^*} = -\frac{\partial p^*}{\partial x^*} + \frac{\mu}{\rho UL}\nabla^{*2}u^*$$

Noting that the coefficient of the last expression is the reciprocal of the Reynolds number, we have for the full set of Navier-Stokes equation

$$\frac{Du^*}{Dt^*} = -\frac{\partial p^*}{\partial x^*} + \frac{1}{\text{Re}}\nabla^{*2}u^*$$

$$\frac{Dv^*}{Dt^*} = -\frac{\partial p^*}{\partial y^*} + \frac{1}{\text{Re}}\nabla^{*2}v^* \qquad (10.54)$$

$$\frac{Dw^*}{Dt^*} = -\frac{\partial p^*}{\partial z} + \frac{1}{\text{Re}}\nabla^{*2}w^*$$

Let us now consider flows about geometrically similar boundaries as shown in Fig. 10.12 where we have prototype and model flows around spheres. The typical length L that we use is the diameter D of the spheres. *We will stipulate that the Reynolds number is duplicated between the flows.* Consider the x

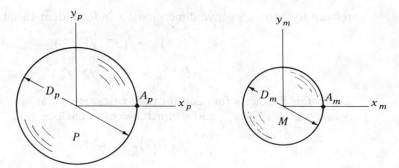

FIGURE 10.12
Geometrically similar boundaries.

component of Navier-Stokes equation for prototype and model flows.

For prototype:
$$\frac{D(u^*)_p}{Dt_p^*} = -\frac{\partial p_p^*}{\partial x_p^*} + \frac{1}{\text{Re}}\nabla^{*2}(u^*)_p \tag{10.55}$$

For model:
$$\frac{D(u^*)_m}{Dt_m^*} = -\frac{\partial p_m^*}{\partial x_m^*} + \frac{1}{\text{Re}}\nabla^{*2}(u^*)_m \tag{10.56}$$

Note that the starred dependent variables for the prototype flow $[(u^*)_p, p_p^*)]$ satisfy the *same differential equation* in terms of the starred independent variables (x_p^*, y_p^*, z_p^*) as do the dependent variables of the model flow $[(u^*)_m, p_m^*]$ in terms of the independent variables (x_m^*, y_m^*, z_m^*).

Furthermore, consider points A_p and A_m in the prototype and model flows, respectively, in Fig. 10.12. They represent two corresponding points in the flows. Now evaluate x_p^* and x_m^* for these points. Thus,

$$(x_p^*)_{A_p} = \left[\left(\frac{x}{D}\right)_p\right]_{A_p} = \left(\frac{D_p/2}{D_p}\right) = \frac{1}{2}$$

$$(x_m^*)_{A_m} = \left[\left(\frac{x}{D}\right)_m\right]_{A_m} = \left(\frac{D_m/2}{D_m}\right) = \frac{1}{2}$$

We can conclude from this that the *dimensionless coordinates at corresponding points have identical values*. Then the boundaries of the spheres are characterized by the same values of dimensionless coordinates between the flows. Hence, the model flow and the prototype flow have identically the *same boundary conditions*.

With the same differential equation governing the flows subject to the same boundary conditions, we can conclude that the solution $(u^*)_p$ for the prototype flow and the solution $(u^*)_m$ for the model flow should be identically related to their respective dimensionless independent variables. That is,

$$(u^*)_p = f(x_p^*, y_p^*, z_p^*)$$

$$(u^*)_m = f(x_m^*, y_m^*, z_m^*)$$

where f is the *same* function for both cases. Hence at any set of corresponding points where $x_p^* = x_m^*$, and so forth, we can conclude that

$$(u^*)_p = (u^*)_m$$

That is, in coordinates without stars,

$$\left(\frac{u}{U}\right)_p = \left(\frac{u}{U}\right)_m$$

Hence,

$$\frac{(u)_m}{(u)_p} = \frac{(U)_m}{(U)_p}$$

Similarly,

$$\frac{(v)_m}{(v)_p} = \frac{(U)_m}{(U)_p}$$

$$\frac{(w)_m}{(w)_p} = \frac{(U)_m}{(U)_p}$$

Thus between any set of corresponding points,

$$\frac{(u)_m}{(u)_p} = \frac{(v)_m}{(v)_p} = \frac{(w)_m}{(w)_p}$$

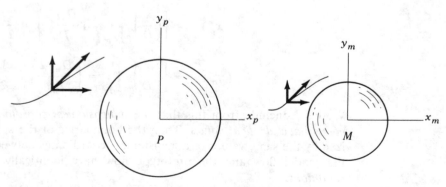

FIGURE 10.13
Velocity components at corresponding points.

From this we can conclude that with the components of **V** between the flows having the *same* ratios at corresponding points, then the velocities at these corresponding points must be parallel (see Fig. 10.13). This can happen only if the flows are *kinematically similar*.

Next consider *Stokes'* viscosity law for τ_{xx} [Eq. (10.10)]:

$$\tau_{xx} = 2\mu \frac{\partial u}{\partial x} - p$$

Expressing the right side in terms of starred variables, we can say that

$$\tau_{xx} = 2\mu \frac{U}{L} \frac{\partial u^*}{\partial x^*} - p^* \rho U^2 \tag{10.57}$$

Hence,

$$\frac{\tau_{xx}}{\rho U^2} = \frac{2}{\text{Re}} \frac{\partial u^*}{\partial x^*} - p^*$$

Similarly,

$$\frac{\tau_{xy}}{\rho U^2} = \frac{1}{\text{Re}} \left(\frac{\partial v^*}{\partial x^*} + \frac{\partial u^*}{\partial y^*} \right)$$

Now we have shown that for the same Reynolds numbers between model and prototype flows, the dependent dimensionless variables relate to dimensionless independent variables in exactly the same manner. Hence, at corresponding points between prototype and model flows $(\partial u^*/\partial x^*)_p = (\partial u^*/\partial x^*)_m$, etc., so we can say that

$$\left(\frac{\tau_{xx}}{\rho U^2} \right)_m = \left(\frac{\tau_{xx}}{\rho U} \right)_p$$

$$\left(\frac{\tau_{xy}}{\rho U^2} \right)_m = \left(\frac{\tau_{xy}}{\rho U^2} \right)_p$$

We can conclude that at corresponding points between the flows

$$\frac{(\tau_{xx})_m}{(\tau_{xx})_p} = \frac{(\tau_{xy})_m}{(\tau_{xy})_p} = \frac{(\rho U^2)}{(\rho U^2)_p} = \text{const}$$

The ratio of each of the components of the stress tensor τ_{ij} at corresponding points between the flows clearly equals a constant. And this constant $(\rho U^2)_m/(\rho U^2)_p$ is the same value for any and all sets of corresponding points. Corresponding forces (shear, normal, etc.) thus have the same ratio between the flows at all corresponding points. We have *dynamic similarity* between the flows. In this case, as pointed out earlier in Chap. 8, integrated forces such as lift and drag have the same ratio $(\rho U^2)_m/(\rho U^2)_p$, making it possible to give prototype data for lift and drag from model data for these quantities.

In retrospect, the dimensionless form of the differential equation yields us the dimensionless group that has to be duplicated for similarity. For the differential equation we were also able to deduce the *nature of the similarity law* —in our case, dynamic similarity.

Suppose now that there is a free surface in the region of interest of the previous flow. Then the Navier-Stokes equation for the y direction becomes

$$\rho \frac{Dv}{Dt} = -\frac{\partial p}{\partial y} - \rho g + \mu \nabla^2 u \qquad (10.58)$$

In terms of dimensionless variables, we have

$$\rho \frac{U^2}{L} \frac{Dv^*}{Dt^*} = -\frac{\rho U^2}{L} \frac{\partial p^*}{\partial x^*} - \rho g + \mu \frac{U}{L^2} \nabla^{*2} u^*$$

Dividing through by $\rho U^2/L$, we get

$$\frac{Dv^*}{Dt^*} = -\frac{\partial p^*}{\partial x^*} - \frac{1}{\text{Fr}} + \frac{1}{\text{Re}} \nabla^{*2} u^* \qquad (10.59)$$

We see now that for dynamic similarity we must duplicate the Reynolds number and the Froude number between the model and prototype flow exactly as indicated in Chap. 8.

*PART B
TURBULENT FLOW

10.9 A COMMENT

It has been pointed out that a complete theoretical approach to turbulent flow paralleling that of laminar flow is impossible because of the complexity and apparently random nature of the velocity fluctuations in turbulent flow. Nevertheless, semitheoretical analysis aided by limited experimental data was presented in Chap. 9 and extensions will be made in this portion of the chapter. It will be important, to understand the concepts and methods of approach in this difficult phase of fluid mechanics. Since much research is under way in this area, such a background is desirable for evaluating the current literature.

10.10 MEAN-TIME AVERAGES FOR STEADY TURBULENT FLOW

As it is futile to deal with actual velocities and other fluctuating quantities in turbulent flow, it becomes necessary to employ some sort of statistical approach. We have been using the mean-time average of these quantities, which we define for steady flows in the following manner for quantity B:

$$\bar{B} = \frac{1}{\Delta t} \int_t^{t+\Delta t} B \, dt \tag{10.60}$$

where Δt is the interval of time over which the average is carried out and is large enough to render the quantity $\bar{B}$ independent of time. Thus, the mean-time average represents the well-ordered part of the flow described earlier in Chap. 9. It is such quantities that an observer measures with standard instrumentation. The fluctuating part of the flow is denoted by primed quantities. Thus, the velocity field may then be represented as

$$u = \bar{u} + u'$$
$$v = \bar{v} + v' \tag{10.61}$$
$$w = \bar{w} + w'$$

It may be clear from the preceding definition that the mean-time average of primed quantities must be zero. To show this, replace B in Eq. (10.60) by $\bar{B} + B'$. Thus

$$\bar{B} = \frac{1}{\Delta t} \int_t^{t+\Delta t} (\bar{B} + B') \, dt = \frac{1}{\Delta t} \int_t^{t+\Delta t} \bar{B} \, dt + \frac{1}{\Delta t} \int_t^{t+\Delta t} B' \, dt$$

Since $\bar{B}$ is constant, it may be extracted from the integral. Carrying out the integration for this expression permits the following statement:

$$\bar{B} = \bar{B} + \frac{1}{\Delta t} \int_t^{t+\Delta t} B' \, dt$$

The last term of this equation is the mean-time average of B', making it evident that

$$\bar{B'} = 0$$

The following operations involving mean-time average may be shown to be valid by considering the definition of mean-time averages along with elementary rules of the calculus. If C and D are turbulent-flow quantities and n represents a spatial coordinate, we may say

$$\overline{C + D} = \bar{C} + \bar{D} \tag{10.62a}$$

$$\overline{\text{const } C} = \text{const } \bar{C} \tag{10.62b}$$

$$\overline{\frac{\partial C}{\partial n}} = \frac{\partial \bar{C}}{\partial n} \tag{10.62c}$$

Use of these relations will be made shortly.

It is important to note a certain restriction on the mean-time average of the product CD. If $\bar{C}$ is zero and $\bar{D}$ is zero, it is not necessarily true that $\overline{CD}$ is zero. For example, consider the case where both quantities are sinusoids of the same wavelength and in phase, as is illustrated in Fig. 10.14. Clearly, the mean-time average of each quantity is zero but the mean-time average of the product, shown dashed, must be a positive number.

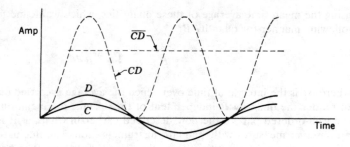

FIGURE 10.14
Mean-time averages of C, D and CD.

10.11 NAVIER-STOKES EQUATIONS FOR MEAN-TIME-AVERAGE QUANTITIES: APPARENT STRESS

Stokes' viscosity law and the resulting Navier-Stokes equations are considered valid for turbulent flow if actual velocities, etc., are used. However, it will be more significant from a physical standpoint to reach relations for mean-time averages from these equations, since it is these averages which are readily observed and more meaningful in practical computations. For convenience, the Navier-Stokes equations will be rewritten at this time. Since the actual turbulent-flow quantities are functions of time, it is necessary to express the complete substantial derivative. Thus

$$u\frac{\partial u}{\partial x} + v\frac{\partial u}{\partial y} + w\frac{\partial u}{\partial z} + \frac{\partial u}{\partial t} = B_x + \frac{1}{\rho}\left(-\frac{\partial p}{\partial x} + \mu\nabla^2 u\right) \qquad (10.63a)$$

$$u\frac{\partial v}{\partial x} + v\frac{\partial v}{\partial y} + w\frac{\partial v}{\partial z} + \frac{\partial v}{\partial t} = B_y + \frac{1}{\rho}\left(-\frac{\partial p}{\partial y} + \mu\nabla^2 v\right) \qquad (10.63b)$$

$$u\frac{\partial w}{\partial x} + v\frac{\partial w}{\partial y} + w\frac{\partial w}{\partial z} + \frac{\partial w}{\partial t} = B + \frac{1}{\rho}\left(-\frac{\partial p}{\partial y} + \mu\nabla^2 w\right) \qquad (10.63c)$$

As the first step, replace the dependent variables in these equations by the mean-time average plus fluctuation components. For instance, the expression $u(\partial u/\partial x)$ in the first equation is replaced by $(\bar{u} + u')[\partial(\bar{u} + u')/\partial x]$, which expands out to the group of terms

$$\bar{u}\frac{\partial \bar{u}}{\partial x} + \bar{u}\frac{\partial u'}{\partial x} + u'\frac{\partial \bar{u}}{\partial x} + u'\frac{\partial u'}{\partial x}$$

Next, making use of rules (10.62), take the mean-time average of each term of the expanded equations. For instance, in the above quantity the first term $\bar{u}(\partial\bar{u}/\partial x)$ is not changed by the averaging process, since the entire expression is constant with time. The next expression is that of a constant times a quantity whose mean-time average is zero, thus making the mean-time average of the product zero. Similarly, the next term, $u'(\partial\bar{u}/\partial x)$ becomes zero upon averaging. Finally the last term, being a product of two fluctuating quantities, is expressed as $\overline{u'(\partial u'/\partial x)}$ and is not necessarily zero. By carrying out such operations on all the remaining terms, the following equations may be formed:

$$\rho\left(\bar{u}\frac{\partial\bar{u}}{\partial x} + \bar{v}\frac{\partial\bar{u}}{\partial y} + \bar{w}\frac{\partial\bar{u}}{\partial z}\right)$$

$$= \rho\bar{B}_x + \left(-\frac{\overline{\partial p}}{\partial x} + \mu\nabla^2\bar{u}\right) - \rho\left(\overline{u'\frac{\partial u'}{\partial x}} + \overline{v'\frac{\partial u'}{\partial y}} + \overline{w'\frac{\partial u'}{\partial z}}\right) \qquad (10.64a)$$

$$\rho\left(\bar{u}\frac{\partial\bar{v}}{\partial x} + \bar{v}\frac{\partial\bar{v}}{\partial y} + \bar{w}\frac{\partial\bar{v}}{\partial z}\right)$$

$$= \rho\bar{B}_y + \left(-\frac{\overline{\partial p}}{\partial y} + \mu\nabla^2\bar{v}\right) - \rho\left(\overline{u'\frac{\partial v'}{\partial x}} + \overline{v'\frac{\partial v'}{\partial y}} + \overline{w'\frac{\partial v'}{\partial z}}\right) \qquad (10.64b)$$

$$\rho\left(\bar{u}\frac{\partial\bar{w}}{\partial x} + \bar{v}\frac{\partial\bar{w}}{\partial y} + \bar{w}\frac{\partial\bar{w}}{\partial z}\right)$$

$$= \rho\bar{B}_z + \left(-\frac{\overline{\partial p}}{\partial z} + \mu\nabla^2\bar{w}\right) - \rho\left(\overline{u'\frac{\partial w'}{\partial x}} + \overline{v'\frac{\partial w'}{\partial y}} + \overline{w'\frac{\partial w'}{\partial z}}\right) \qquad (10.64c)$$

To help reformulate these equations, let us consider the equation of *continuity*, using mean-time average plus fluctuating components to represent the velocity components. Upon convenient arrangement, this equation may be expressed as

$$\left(\frac{\overline{\partial u}}{\partial x} + \frac{\overline{\partial v}}{\partial y} + \frac{\overline{\partial w}}{\partial z}\right) + \left(\frac{\partial u'}{\partial x} + \frac{\partial v'}{\partial y} + \frac{\partial w'}{\partial z}\right) = 0 \qquad (10.65)$$

Taking the mean-time average leaves the first parenthetical expression unaffected while rendering the second parenthetical expression equal to zero. Thus it is evident that

$$\frac{\overline{\partial u}}{\partial x} + \frac{\overline{\partial v}}{\partial y} + \frac{\overline{\partial w}}{\partial z} = 0 \qquad (10.66)$$

This means that the mean-time-average velocity field satisfies the same continuity equation as the actual velocity field. Furthermore, upon reexamining Eq. (10.65) in the light of Eq. (10.66), it becomes clear that

$$\frac{\partial u'}{\partial x} + \frac{\partial v'}{\partial y} + \frac{\partial w'}{\partial z} = 0 \qquad (10.67)$$

Thus, the fluctuating velocity components also satisfy the same continuity equation as should now be expected.

To return to the Navier-Stokes development, it is now possible to replace the last group in parentheses in each of Eqs. (10.64) by a more convenient form. For example, in Eq. (10.64a) the expression

$$\overline{u'\frac{\partial u'}{\partial x}} + \overline{v'\frac{\partial u'}{\partial y}} + \overline{w'\frac{\partial u'}{\partial z}}$$

may be replaced by

$$\frac{\overline{\partial(u')^2}}{\partial x} + \frac{\overline{\partial(u'v')}}{\partial y} + \frac{\overline{\partial(u'w')}}{\partial z}$$

That this is true may be directly verified by the reader by carrying out the differentiation in the latter expression and employing the continuity equation [Eq. (10.67)] in conjunction with Eqs. (10.62). Incorporating these changes into Eq. (10.64a) and employing the notation $\overline{D}/\overline{D}t$ for the differentiation operations on the left side of the equations,[7] we may write the Navier-Stokes equations in terms of time-averaged quantities in the following manner[8]:

$$\rho\frac{\overline{D}\bar{u}}{\overline{D}t} = \rho\bar{B}_x + \left(-\frac{\partial p}{\partial x} + \mu\nabla^2\bar{u}\right) - \rho\left(\frac{\overline{\partial(u')^2}}{\partial x} + \frac{\overline{\partial(u'v')}}{\partial y} + \frac{\overline{\partial(u'w')}}{\partial z}\right) \quad (10.68a)$$

$$\rho\frac{\overline{D}\bar{v}}{\overline{D}t} = \rho\bar{B}_y + \left(-\frac{\partial p}{\partial y} + \mu\nabla^2\bar{v}\right) - \rho\left(\frac{\overline{\partial(v'u')}}{\partial x} + \frac{\overline{\partial(v')^2}}{\partial y} + \frac{\overline{\partial(v'w')}}{\partial z}\right) \quad (10.68b)$$

$$\rho\frac{\overline{D}\bar{w}}{\overline{D}t} = \rho\bar{B}_z + \left(-\frac{\partial p}{\partial z} + \mu\nabla^2\bar{w}\right) - \rho\left(\frac{\overline{\partial(w'u')}}{\partial x} + \frac{\overline{\partial(w'v')}}{\partial y} + \frac{\overline{\partial(w')^2}}{\partial z}\right) \quad (10.68c)$$

It is quite clear that the equations for mean-time-average quantities are formidably difficult, involving nonlinearities as well as little known mean-time averages of fluctuating velocity products. Note, however, that these equations are *identical* in form to the case of laminar flow [Eq. (10.13)] except for the presence of the aforementioned fluctuating velocity products. The unit of these expressions is force per unit volume, so a convenient viewpoint is to consider that the mean-time-average flow behaves in the same way as a laminar flow except that additional hypothetical body-force[9] distributions are present which account for the effects of turbulence. Acting on volume element dv, these forces which we call *apparent forces*, have the values

$$(df_x)_{\text{app}} = -\rho\left(\frac{\overline{\partial(u')^2}}{\partial x} + \frac{\overline{\partial(u'v')}}{\partial y} + \frac{\overline{\partial(u'w')}}{\partial z}\right)dv$$

$$(df_y)_{\text{app}} = -\rho\left(\frac{\overline{\partial(v'u')}}{\partial x} + \frac{\overline{\partial(v')^2}}{\partial y} + \frac{\overline{\partial(v'w')}}{\partial z}\right)dv$$

$$(df)_{\text{app}} = -\rho\left(\frac{\overline{\partial(w'u')}}{\partial x} + \frac{\overline{\partial(w'v')}}{\partial y} + \frac{\overline{\partial(w')^2}}{\partial z}\right)dv \quad (10.69)$$

---

[7]We cannot properly use the familiar substantial derivative (D/Dt) here, since our definition of the substantial derivative requires the use of the actual velocity components as coefficients of the derivatives, and not the mean-time-average velocity components as we have here.

[8]This equation is often called the Reynolds equation.

[9]You will recall in mechanics that we did a similar thing when dealing with noninertial references, where we formulated hypothetical forces (inertia forces) in order to think in terms of the familiar form of Newton's law.

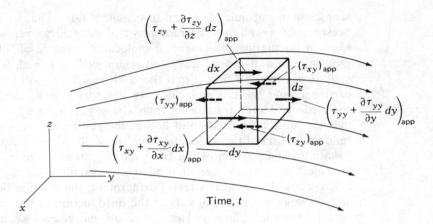

FIGURE 10.15
Apparent stress in the y direction. (Other stresses omitted for clarity.)

The stresses which may be associated with these forces are called *apparent*, or *Reynolds*, *stresses*. Considering an infinitesimal cube of fluid such as is shown in Fig. 10.15, the following relation may be made between the apparent stresses and the above force components:

$$(df_x)_{\text{app}} = \left(\frac{\partial (\tau_{xx})_{\text{app}}}{\partial x} + \frac{\partial (\tau_{xy})_{\text{app}}}{\partial y} + \frac{\partial (\tau_{xz})_{\text{app}}}{\partial z} \right) dv$$

$$(df_y)_{\text{app}} = \left(\frac{\partial (\tau_{yx})_{\text{app}}}{\partial x} + \frac{\partial (\tau_{yy})_{\text{app}}}{\partial y} + \frac{\partial (\tau_{yz})_{\text{app}}}{\partial z} \right) dv \qquad (10.70)$$

$$(df_z)_{\text{app}} = \left(\frac{\partial (\tau_{zx})_{\text{app}}}{\partial x} + \frac{\partial (\tau_{zy})_{\text{app}}}{\partial y} + \frac{\partial (\tau_{zz})_{\text{app}}}{\partial z} \right) dv$$

Comparing Eqs. (10.69) and Eqs. (10.70), one sees that the apparent stress tensor is related to the velocity fluctuations in the following manner:

$$\begin{bmatrix} \tau_{xx} & \tau_{xy} & \tau_x \\ \tau_{yx} & \tau_{yy} & \tau_y \\ \tau_x & \tau_y & \tau \end{bmatrix}_{\text{app}} = \begin{bmatrix} -\rho\overline{(u')^2} & -\rho\overline{(u'v')} & -\rho\overline{(u'w')} \\ -\rho\overline{(v'u')} & -\rho\overline{(v')^2} & -\rho\overline{(v'w')} \\ -\rho\overline{(w'u')} & -\rho\overline{(w'v')} & -\rho\overline{(w')^2} \end{bmatrix} \qquad (10.71)$$

10.12 MANIFESTATION OF APPARENT STRESS: EDDY VISCOSITY

We can now see that the idea of apparent stress is not so artificial and arbitrary as what might at first appear to be the case. Let us first consider what causes the familiar and accepted shear stresses occurring in laminar flow. Here there is a motion which we observe macroscopically as a well-ordered flow. The ever present random molecular motion is similar to the random motion of macroscopic chunks of fluid, as found in turbulent flow. That is, we can consider the presence of a well-ordered movement of molecules over which is superposed random fluctuating movement of molecules. As a result of these fluctuations molecules in a macroscopically fast region of flow may drift into a macroscopically slow region of flow. From the collisions, there will then be a momentum exchange, and the fast molecules are slowed up, while the slow molecules are speeded up. Molecules moving from a slow region to a fast region in a similar manner cause a retardation of flow in the fast region. And so you see that this random motion of the molecules from the well-ordered pattern causes an effort toward an equalization of the gross velocity profile. This action and the effects of molecular cohesion are then microscopic actions which manifest themselves macroscopically as shear stress. Furthermore, the relation between shear stress and the velocity field brings about the introduction of the property of viscosity.

In turbulent flow we have *macroscopic chunks of fluid* which have a random fluctuating velocity superposed on *a well-ordered mean-time-average flow*. Thus there is an analogy between molecular action described in the previous paragraph and the turbulent action of fluid elements in turbulent flow. The former is a *microscopic* action of molecules that gives rise to *macroscopic* stresses and viscosity. We can then consider that the latter (namely, turbulence), being a macroscopic action, manifests itself as an effect which is *one order more gross* than macroscopic effect, and that is the *mean-time-average* effect which we call *apparent stress*. Thus, the concept of apparent stress, when considered in terms of this extrapolation, is not an unnatural one. Furthermore one might next try to relate apparent stress with the mean-time average velocity field and in so doing set forth a *mean-time-average* property paralleling the *macroscopic* property of viscosity. This mean-time-average property is called the *eddy viscosity*. Clearly, if apparent stresses can be related to the mean-time-average velocity field through the eddy viscosity, we see from Eqs. (10.68) that the number of unknowns will be decreased in the Navier-Stokes equations. At this point, investigators have in some instances been able to promulgate hypothetical relations between these quantities which in simple, limited cases lead to meaningful results. The mixing length discussion of the previous chapter is one such example.

10.13 CLOSURE

In Sec. 14.3, the advanced reader will be able to use the thin-film Navier-Stokes equations presented in this chapter for purposes of developing the celebrated Blasius equation for laminar boundary-layer flow. And in Chap. 16 (starred), the advanced reader will have the opportunity of solving the Blasius equation numerically.

We finish Part II by investigating one-dimensional compressible flows, particularly that of a perfect gas. The reader will be introduced to the various peculiar effects of supersonic flow that will often be contrary to intuition.

PROBLEMS

Problem Categories

Stokes Viscosity Law 10.1–10.2
Navier-Stokes Equations 10.3–10.6
Lubrication Theory 10.7–0.10
Problems Using Navier-Stokes 10.11–10.13

10.1. Show that $C_{25} = C_{26} = 0$ in the development of Sec. 10.1.

10.2. Consider flow of a fluid having the following velocity field:

$$\mathbf{V} = (3y^2x\mathbf{i} + 10yz^2\mathbf{j} + 5\mathbf{k}) \quad \text{m/s}$$

What are the normal stresses at $(2, 4, 3)$ m? The stress τ_{xx} at this point is known to be -10 lb/ft^2 gage. Take $\mu = 10^{-2}$ lb · s/ft^2.

10.3. Using Navier-Stokes equations, determine the thickness of a film of fluid moving at constant speed and thickness down an infinite inclined wall in terms of volume flow q per unit width. *Hint*: Use the fact that the shear stress is zero at the free surface.

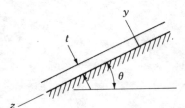

FIGURE P10.3

10.4. The Navier-Stokes equation for *cylindrical coordinates* are presented for the case of incompressible flow with constant viscosity μ as follows:

$$\rho\left(\frac{Dv_r}{Dt} - \frac{v_\theta^2}{r}\right) = \rho B_r - \frac{\partial p}{\partial r}$$

$$+ \mu\left(\nabla^2 v_r - \frac{v_r}{r^2} - \frac{2}{r^2}\frac{\partial v_\theta}{\partial \theta}\right)$$

$$\rho\left(\frac{Dv_\theta}{Dt} + \frac{v_r v_\theta}{r}\right) = \rho B_\theta - \frac{\partial p}{r\,\partial \theta} \qquad (a)$$

$$+ \mu\left(\nabla^2 v_\theta + \frac{2}{r^2}\frac{\partial v_r}{\partial \theta} - \frac{v_\theta}{r^2}\right)$$

$$\rho\frac{Dv_z}{Dt} = \rho B_z - \frac{\partial p}{\partial z} + \mu \nabla^2 v_z$$

Simplify these equations to apply to the rotational flow (only) between two infinite concentric cylinders where the smaller cylinder has angular velocity ω_1 and the outside cylinder has angular velocity ω_2. Use the fact that in cylindrical coordinates

$$\nabla^2 \equiv \frac{1}{r}\frac{\partial r(\partial/\partial r)}{\partial r} - \frac{1}{r^2}\frac{\partial^2}{\partial \theta^2} + \frac{\partial^2}{\partial z^2} \qquad (b)$$

Neglect body force of gravity. Show that we get

$$\frac{\rho v_\theta^2}{r} = \frac{dp}{dr}$$
$$\qquad (c)$$
$$\frac{d^2 v_\theta}{dr^2} + \frac{d}{dr}\left(\frac{v_\theta}{r}\right) = 0$$

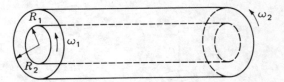

FIGURE P10.4

10.5. In Prob. 10.4 find the velocity field for the conditions stated.

10.6. As an introduction to *lubrication theory*, consider a slipper block moving with speed U over a bearing surface. With $h \ll L$, the film of oil between the slipper block and bearing surface may be considered *very thin*, so that Eqs. (10.50) apply. Clearly the velocity component v must be much less than u and with $\partial u/\partial x$ of the same order of magnitude as $\partial v/\partial y$, we can drop the second expression of Eq. (10.50a). Next, consider the expression

$$\frac{\rho u(\partial u/\partial x)}{\mu(\partial^2 u/\partial y^2)}$$

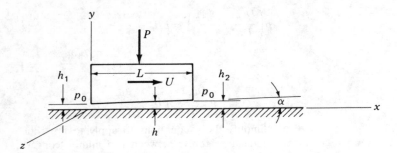

FIGURE P10.6

in dimensionless form so that $u^* = u/U$, $x^* = x/L$, and $y^* = y/h$. If UL/ν in bearing problems is seldom greater than 2.5×10^4, indicate why $u(\partial u/\partial x)$ can be deleted in Eq. (10.50a) compared with $\nu \, \partial^2 u/\partial y^2$ for $(h/L) < 0.001$. We are then left with the equation

$$\frac{dp}{dx} = \mu \frac{\partial^2 u}{\partial y^2} \qquad (a)$$

for lubrication problems. *Hint*: What are the orders of magnitude of u^*, y^*, and x^*?

10.7. In Prob. 10.6 take reference xyz as fixed to the slipper block and show that the velocity profile is given as

$$u = \frac{1}{2\mu} \frac{dp}{dx} (y^2 - hy) + U\left(\frac{y}{h} - 1\right) \qquad (a)$$

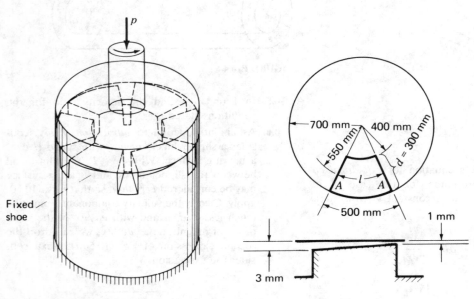

FIGURE P10.9

Next, show that the volumetric flow q per unit width of the slipper block is

$$q = -\frac{1}{12\mu}\frac{dp}{dx}h^3 - \frac{Uh}{2} \qquad (b)$$

Hence, show that

$$\frac{dp}{dx} = -\frac{12\mu}{h^3}\left(\frac{Uh}{2} + q\right) \qquad (c)$$

From this, show on integrating from 0 to L that

$$q = -\frac{U}{2}\frac{\int_0^L (dx/h^2)}{\int_0^L (dx/h^3)} = -\left(\frac{h_1 h_2}{h_1 + h_2}\right)U$$

Finally, determine the pressure as a function of x given as

$$p = p_0 + \frac{6\mu UL}{h^2(h_1^2 - h_2^2)}(h - h_2)(h - h_1) \quad (d)$$

10.8. In Probs. 10.6 and 10.7 determine the lifting force P on the slipper block per unit width of the slipper block. Get the following result.

$$P = p_0 L$$
$$+ \frac{6\mu UL^2}{(h_2 - h_1)^2}\left(\ln\frac{h_2}{h_1} - 2\frac{h_2 - h_1}{h_2 + h_1}\right) \quad (a)$$

10.9. A thrust bearing with six fixed shoes is shown in Fig. P10.9. If the shaft rotates at a speed of 6000 r/min, approximately what load can safely be supported by the bearing? Neglect side leakage. The dimension of the shoe and the film of oil is shown. The μ for the oil is 0.0958 N · s/m². *Hint*: See Prob. 10.8.

10.10. Do Prob. 7.30 using the Navier-Stokes equations. Note for a thin film we can neglect the term $v_\theta/\bar{r}^2$ in the Navier-Stokes equations.

10.11. In Prob. 7.33 using Navier-Stokes equations, find the velocity profile of the oil, for the case where the shaft attains terminal velocity V_T.

10.12. Do Prob. 7.29 using the Navier-Stokes equations.

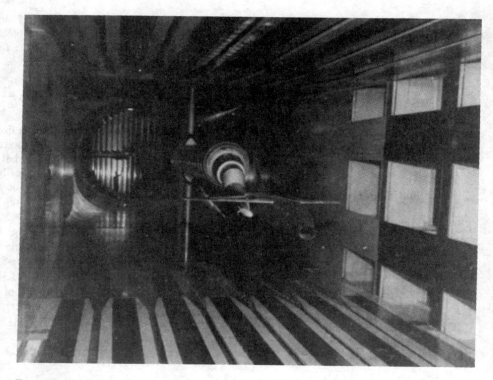

Transonic wind tunnel with an F-8 Crusader aircraft under test. (*Courtesy Naval Ship and Research Facility, Carderock, Md.*)

The wind tunnel depicted above is a **transonic** wind tunnel with a test section 7 feet high, 10 feet wide, and 15 feet long. To achieve dynamic similarity, both the Mach number and the Reynolds numbers must be duplicated. There is no difficulty in getting the desired Mach number. To achieve also a correct Reynolds number, one may increase the stagnation pressure, but this introduces structural problems in design of model and support. Also, one can reduce the operating temperature to a very low value. The latter is more feasible but has limited useful application. Usually only the Mach number is duplicated and theoretical corrections are made for the Reynolds number discrepancy. Notice the slots in the top and bottom of the test section. In operation air is drawn through these slots from the test section since this is necessary to avoid **blockage** during the transonic flow range.

CHAPTER 11

ONE-DIMENSIONAL
COMPRESSIBLE FLOW

11.1 INTRODUCTION

For the first time, we will include the effects of compressibility in dynamic problems. Since the variation of density is usually accompanied by temperature changes as well as heat transfer, we will require far greater use of thermodynamics than heretofore, particularly with respect to the second law of thermodynamics.

The study of compressible flow is broken down into the usual categories which have already been utilized in incompressible analysis. These are:

1. One-dimensional, two-dimensional, and three-dimensional flows
2. Steady and unsteady flows
3. Rotational and irrotational flows

In addition to these familiar categories it is useful to form the following additional classifications:

4. *Subsonic compressible flow*. The Mach number somewhere in the flow region should exceed the approximate value of 0.4 and not exceed 1 anywhere in the flow region.

5. *Transonic flow.* This flow includes Mach numbers slightly less than and slightly greater than unity.

6. *Supersonic flow.* Flows with Mach numbers exceeding unity but less than about 3.

7. *Hypersonic flow.* The Mach number exceeds the approximate value of 3.

When the Mach number exceeds unity, one finds a startling change in fluid behavior as compared with fluid flow below Mach 1. The direct experience of most people is limited to fluid phenomena in the subsonic range; they may build up, as a result of this, an intuitive feeling which may betray them in the supersonic range, since in this range radically different fluid behavior occurs from what is "normally" expected. We will have occasion to note these differences in this chapter.

PART A
BASIC PRELIMINARIES

11.2 THERMODYNAMIC RELATIONS FOR A PERFECT GAS

There is ample opportunity in high-speed flow to use the concept of the perfect gas. It will be recalled from Chap. 1 that such fluids have a simple relation between properties which is embodied in the familiar equation of state for a perfect gas, a relation which has been used many times in the preceding chapters. Other important relations will be briefly presented at this time.

Experiments with gases approaching perfect-gas behavior as well as thermodynamic considerations, which we will delete, show that the internal energy u of a perfect gas behaves according to the relation

$$\left(\frac{\partial u}{\partial v}\right)_T = 0 \tag{11.1}$$

where the subscript T indicates an isothermal process. This relation means, essentially, that the internal energy of a perfect gas depends only on the temperature. That is,

$$u = u(T) \tag{11.2}$$

A similar relation may be made for the enthalpy for a perfect gas. Expressing the definition of enthalpy and using the equation of state, we have

$$h = u + pv = u + RT$$

The right side is clearly a function of temperature so that $h = h(T)$. Hence you may recall from your thermodynamics course we can express the specific heats

c_v and c_p as

$$c_v = \frac{du}{dT} \tag{11.3a}$$

$$c_p = \frac{dh}{dT} \tag{11.3b}$$

Furthermore, a simple relation between c_p and c_v for a perfect gas may be found be writing h in the form $u + RT$, as was done earlier, and substituting it into Eq. (11.3b). Thus

$$c_p = \frac{d(u + RT)}{dT} = c_v + R$$

Therefore

$$\boxed{c_p - c_v = R} \tag{11.4}$$

The ratio of specific heats c_p and c_v is a useful dimensionless parameter given as

$$\frac{c_p}{c_v} = k \tag{11.5}$$

The value of k is a constant whose value depends on the fluid. It is a simple algebraic manipulation, using the above definition along with Eq. (11.4), to form the following relations applicable to perfect gases:

$$c_p = \frac{k}{k-1} R \tag{11.6a}$$

$$c_v = \frac{R}{k-1} \tag{11.6b}$$

The perfect gas need not have constant specific heat. The kinetic theory of gases indicates and experience verifies the fact that specific heats of perfect gases depend mainly on the temperature, as pointed out earlier.[1] However, if the temperature range is not extreme, considerable simplification may be found by considering the specific heats to be constant.

We will have occasion to consider processes with zero heat transfer to the surroundings, called *adiabatic processes*. A useful relation between p and v for a perfect gas with constant specific heat may be evaluated for an *adiabatic reversible*, or *isentropic*, *process* in the following manner: Express first-law equations for an adiabatic process with dh and du put in terms of specific heats

---

[1] However, R is constant, and consequently the difference in specific heats is constant, according to Eq. (11.4).

and temperatures according to Eqs. (11.3a) and (11.3b). Thus

$$d\cancel{Q}^0 = du + p\,dv$$

$$\therefore c_v\,dT = -p\,dv$$

$$d\cancel{Q}^0 = dh - v\,dp$$

$$\therefore c_p\,dT = v\,dp$$

Now, dividing the fourth of the above equations by the second and again rearranging the terms, we get

$$\frac{dp}{p} = -\frac{c_p}{c_v}\frac{dv}{v} = -k\frac{dv}{v}$$

With a constant value of k, we may now integrate to form the equation

$$\ln p = \ln v^{-k} + \ln \text{const}$$

Hence

$$pv^k = \text{const} \tag{11.7}$$

Another way of expressing this relation between states 1 and 2 is as follows:

$$\frac{p_1}{p_2} = \left(\frac{v_2}{v_1}\right)^k \tag{11.8}$$

By using the equation of state, other forms of this result in terms of temperature and density may be evolved. Thus

$$\frac{T_1}{T_2} = \left(\frac{\rho_1}{\rho_2}\right)^{k-1} \tag{11.9a}$$

$$\frac{T_1}{T_2} = \left(\frac{p_1}{p_2}\right)^{(k-1)/k} \tag{11.9b}$$

Since the equation of state is a relationship between fluid properties while at thermodynamic equilibrium, and since we employ quasi-static restrictions on many of the formulations of this chapter, one might suppose that such relations as given above would not be valid under nonequilibrium conditions. Experience indicates that it is permissible to use these relations for nonequilibrium conditions in most practical situations. However, under extreme conditions such as in explosions or shock waves the departure from equilibrium is so great as to preclude use of the thermodynamic relations herein presented. In other words, an expression such as $pv^k = $ const could be applied with little error to a compressor (if there is little turbulence and heat transfer) to relate properties during the compression at nonequilibrium conditions as well as at the end conditions, where there may be equilibrium.

11.3 PROPAGATION OF AN ELASTIC WAVE

An immediate consequence of allowing for variation in density is that a system of fluid elements can now occupy varying volumes in space, and this possibility means that a group of fluid elements can spread out into a larger region of space without requiring a simultaneous shift to be made of *all* fluid elements in the flow. However, you will remember from your earlier studies of physics that a small shift of fluid elements in compressible media will induce in due course similar small movements in adjacent elements and that in this way a distur- bance, called an *acoustic wave*, propagates at a relatively high speed corre- sponding to the speed of sound through the medium. In the incompressible flows we have studied up to now these propagations have infinite velocity; that is, adjustments take place instantaneously throughout the entire flow in the manner explained above; so in the usual sense, there are no acoustic or elastic waves to be considered. With the admission of compressibility, we thus permit the possibility of elastic waves having a celerity which is finite. The value of this celerity is of prime importance in compressible-flow theory.

We will first set forth formulations for the determination of the celerity of an infinitesimal elastic wave in a fluid and then observe some vital consequences of having such waves. Accordingly, in Fig. 11.1, we show a constant-area duct containing fluid initially at rest. Assume, now that an infinitesimal pressure increase dp is established and maintained in some manner on the left side of position *A-A*. Two things will now happen. Due to *molecular action*, the pressure will increase to the right of *A-A* and this increase in pressure will move downstream in the channel at a high speed c, called the acoustic celerity. We thus have a *pressure wave* of speed c moving to the right due to microscopic action. We have shown this wave in Fig. 11.2*a*. The second effect is on the *macroscopic* level. According to Newton's law, the fluid just to the right of the

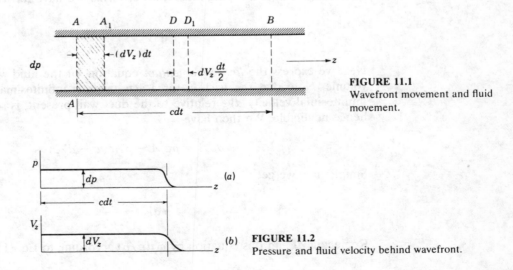

FIGURE 11.1
Wavefront movement and fluid movement.

FIGURE 11.2
Pressure and fluid velocity behind wavefront.

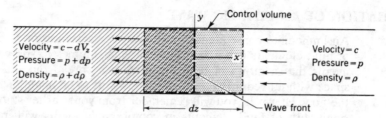

FIGURE 11.3
Stationary wavefront with control volume.

wavefront described above must accelerate as a result of the pressure difference dp to a velocity dV_z. Once the pressure rise dp has been established in the fluid, there is no further change in velocity, so it remains at dV_z. Behind the wavefront, the fluid is thus moving to the right at speed dV_z (see Fig. 11.2b).

Consider what takes place during a time interval dt in the apparatus. The wave has progressed a distance $c\,dt$ and is shown at position B in Fig. 11.1. Meanwhile, fluid particles at A move a distance $dV_z\,dt$ to position A_1. At an intermediate position, such as halfway between A and B, shown in the diagram as D, the fluid velocity dV_z has persisted for a time interval $dt/2$. Consequently, fluid initially at D has moved a distance $(dV_z\,dt)/2$ to position D_1.

In order to make a steady-flow analysis, we place a reference on the wavefront and a moving infinitesimal control volume enclosing the front is established as indicated in Fig. 11.3. The equation of *continuity* for this control volume is

$$c
\rho A = (\rho + d\rho)(c - dV_z)A$$

Canceling A and dropping second-order terms, we have for dV_z

$$dV_z = c\frac{d\rho}{\rho} \tag{11.10}$$

Next we express the *linear momentum* equation for the fluid within the control volume. Note that the friction force acts over an infinitesimal area and, with infinitesimal velocity dV_z relative to the duct wall present, is second order and hence negligible. We then have

$$dp\,A = \rho c^2 A - \rho Ac(c - dV_z)$$

Simplifying, we get

$$\rho\,dV_z = \frac{dp}{c} \tag{11.11}$$

Replacing dV_z in this equation by $c(d\rho/\rho)$ according to Eq. (11.10) and solving for the velocity of propagation, we get

$$c^2 = \frac{dp}{d\rho} = \left(\frac{\partial p}{\partial \rho}\right)_s \qquad (11.12)$$

The reason for changing from a total derivative to a partial derivative in Eq. (11.12) is explained as follows: The variations in density, pressure, and temperature incurred during the wave propagation are of infinitesimal proportions. Furthermore, the friction developed by the infinitesimal velocity of flow relative to the wall caused by the wave propagation is of second order and negligible. Finally, the rapidity of the wave propagation and the infinitesimal temperature variations preclude the possibility of other than a second-order contribution from heat transfer as a result of this action. All these considerations then mean that an infinitesimal acoustic wave is very close to being an isentropic process for any flow in which the action may be generated; so instead of describing the circumstances under which we get $dp/d\rho$, we simply use the partial-derivative notation, which efficiently gives the equivalent information. In *any fluid flow* this would be true for an infinitesimal *acoustic-wave propagation* irrespective of the *nonisentropic effects* existing in the flow itself.

In the case of a *perfect gas*, we may employ Eq. (11.7) to relate pressure with the specific volume or the density for an isentropic process. Using the latter, we can say that

$$p\left(\frac{1}{\rho}\right)^k = \text{const} \qquad (11.13)$$

We take the logarithm first and then take the differential[2] of the equation.

$$\ln p - k \ln \rho = \ln \text{const}$$

$$\therefore \frac{dp}{p} - k\frac{d\rho}{\rho} = 0 \qquad (11.14)$$

Rearranging the terms and using a partial derivative to indicate the isentropic restriction inherent in this relation, we thus have

$$\left(\frac{\partial p}{\partial \rho}\right)_s = \frac{kp}{\rho} \qquad (11.15)$$

Another useful form of this equation may be found by replacing p by ρRT as a

---

[2] This operation of taking the logarithm first and then the differential, you will note, separates the variables in the equation. We shall use this sequence of operations in succeeding chapters, and we shall term the sequence the *logarithmic differential*.

result of the *equation of state*. Thus

$$\left(\frac{\partial p}{\partial \rho}\right)_s = kRT \tag{11.16}$$

Substituting the last two relations back into Eq. (11.12) gives us two formulations for the celerity of a pressure wave of infinitesimal strength in terms of fluid properties for the case of a perfect gas. Thus

$$c = \sqrt{\frac{kp}{\rho}} \tag{11.17a}$$

$$c = \sqrt{kRT} \tag{11.17b}$$

Although the action herein employed was a rather artificial contrivance for the purpose of simplifying the computations, it is nevertheless true that the results are valid for any *small* disturbance of pressure fluctuation. Thus, weak spherical and cylindrical waves move with speed which may be computed for a perfect gas by employing Eqs. (11.17a) and (11.17b) and for other fluids by employing Eq. (11.12). It should also be noted that the celerity c is measured *relative to the fluid in which the front is propagating*. Finally note that our proof assumed a constant value of c; that is, we used an inertial control volume with steady flow relative to this control volume. We have then really considered an isothermal fluid. We can show that for *any nonisothermal case*, the results we developed are applicable, since noninertial effects and unsteady contributions to our control-volume analysis would be of second order, permitting us to reach the same results given in this section.

The waves we have been considering involve infinitesimal (or very small) pressure variations. In a later section, we study waves where comparatively large pressure variation occurs over a very narrow front; we call such waves *shock waves*. These waves are not isentropic, and they move relative to the fluid at speeds in excess of the acoustic speeds we have been discussing. We may think of our acoustic waves as limiting cases of shock waves where the change in pressure across the waves becomes infinitesimal.

In our study of a free-surface flow in Chap. 14, we will see that similar actions can occur in such flows. Thus the *gravity wave* of small amplitude corresponds to the *acoustic wave* herein presented, while the celebrated *hydraulic jump* corresponds to the *shock wave*.

11.4 THE MACH CONE

We are now ready to observe some interesting and important consequences of acoustic waves which will help us to understand supersonic flow.

First, consider that at some point P in a stationary fluid an instantaneous, small, spherically symmetric disturbance is emitted. The front will proceed out spherically with the speed of sound, as evaluated in the preceding section. This

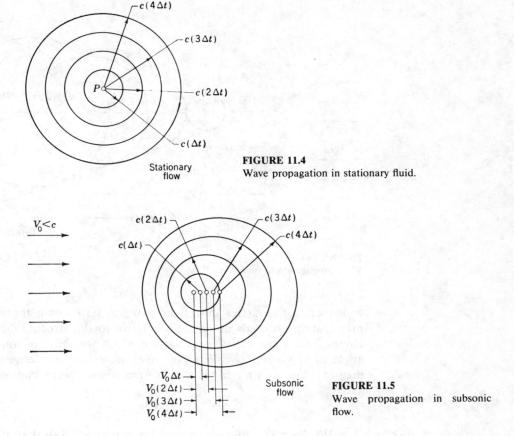

FIGURE 11.4
Wave propagation in stationary fluid.

FIGURE 11.5
Wave propagation in subsonic flow.

is shown in Fig. 11.4, where the front has been indicated at succeeding time intervals, thus forming a set of concentric circles in our diagram. Next we consider that the disturbance is emitted in a fluid moving from left to right at a uniform velocity V_0 which is *less* than c. Observing the wavefront from a stationary position at succeeding time intervals (Fig. 11.5), we no longer have concentricity of the circles for this case since the propagation moves out spherically *relative* to the fluid and therefore moves downstream[3] with velocity V_0. This means that the center for drawing the circles is moving with velocity V_0 from left to right. Clearly, if $V_0 < c$, the circles can never intersect. This represents a simple action in a *subsonic flow*. Finally, let us consider the case where the fluid is moving with a speed V_0 *exceeding* the value of c. This will represent a simple action in a *supersonic flow*. The wavefront of a disturbance is again observed at successive intervals, as shown in Fig. 11.6. The center for

---

[3]If we moved with the fluid, we should again see concentric circles.

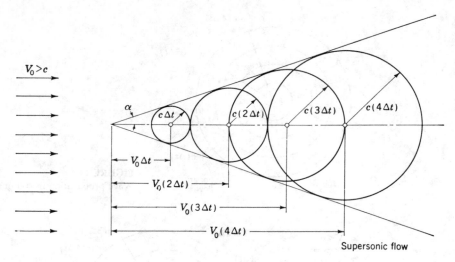

FIGURE 11.6
Wave propagation in supersonic flow.

constructing the circles moves downstream faster than the rate at which the propagation proceeds out radially relative to the stream, and we see that the circles form a conical tangent surface which we call the *Mach cone*. The half angle of this cone is called the *Mach angle* and is designated as α in the diagram. You may readily ascertain from trigonometric considerations that

$$\sin \alpha = \frac{c}{V} = \frac{1}{\text{M}} \qquad (11.18)$$

We have considered a single instantaneous disturbance up to this time. Let us now imagine that the source is continually emitting sound. In supersonic flow an observer outside the Mach cone does not hear this signal, so this region is called a *zone of silence*. Inside this cone, on the other hand, there will be ample evidence of the disturbance, so this region is called *the zone of action*. In the case of the subsonic flow the disturbance will spread out in all directions, although in an unsymmetrical manner, so that there cannot be such a thing as a zone of silence. Clearly, this is a basic difference between subsonic and supersonic flow. In the stationary fluid there is a symmetrical distribution of sound waves resulting from the continuous disturbance. An *incompressible* flow will also yield a symmetrical response to the signal, but this time the signal is "heard" everywhere at the instant when it is produced as a result of the infinite speed of propagation.[4] If this seems "unreal," keep in mind that the model is also unreal and such departures from "natural action" are then to be expected.

---

[4] It should also be clear from Eq. (11.12) that for an incompressible flow $d\rho = 0$ and so $\partial p/\partial \rho = \infty$ even if there are pressure changes. This in turn means that the velocity $c = \infty$.

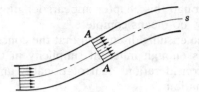

FIGURE 11.7
One-dimensional flow.

It should now be evident from this discussion why an airplane moving at supersonic speed is not heard by a stationary observer until the plane has passed.

11.5 A NOTE ON ONE-DIMENSIONAL COMPRESSIBLE FLOW

The most elemental of compressible flows is one-dimensional compressible flow, which, in Sec. 4.9, we define as a flow describable in terms of one spatial coordinate and time. You must not confuse one-dimensional flow with the parallel-flow model employed in earlier studies. For the latter type of flow, there is no restriction on the number of spatial coordinates necessary to describe the flow. For instance, certain pipe-flow considerations of Chap. 9 required two spatial coordinates. On the other hand, unlike the parallel-flow case, the streamlines in a one-dimensional flow need not be straight. Thus, in the duct shown in Fig. 11.7 the flow may be considered as one-dimensional if flow parameters are reasonably close to being constant over each section A-A at any time and change only with position s and time t. Even if the aforementioned uniformity is not met, an idealization of one-dimensional flow will yield average values of these parameters at sections along s. Such information is often of prime interest, thus making the one-dimensional model with its inherent simplification of analysis a valuable tool. Analysis of nozzles, diffusers, high-speed ducts, and wind tunnels are but a few examples where one-dimensional analysis is of value.

In this chapter we study a number of types of one-dimensional flow, each involving separate effects such as area change, friction, and heating. Such effects must be mild enough so as not to create too radical a departure from the one-dimensional model stated at the outset of the section. For instance, friction will be confined to the boundary layer and to regions inside shock waves.[5] Otherwise, it might be necessary to consider a viscous compressible three-dimensional flow of a type analogous to that undertaken in Chap. 9 for the

---

[5]Because the friction is confined to a small region of the flow, namely, the boundary layer, leaving the major portion of the flow inviscid, we replace σ_{nn} by $-p$ at all times even though the flows in this chapter may not be parallel flows.

incompressible case. Although this chapter appears lengthy, it must be pointed out that the coverage is by no means complete.[6]

It will be noted in the ensuing discussion that the concept of irrotationality is not examined. There is enough inherent simplicity in the one-dimensional assumption to make the considerations of irrotationality unnecessary for ascertaining the desired information.

PART B
ISENTROPIC FLOW WITH SIMPLE AREA CHANGE

11.6 BASIC AND SUBSIDIARY LAWS FOR ISENTROPIC FLOW

First we develop the general equations governing the *isentropic* one-dimensional steady flow of *any* compressible fluid, and then in greater detail we consider the case of the perfect gas.

Let us then consider a duct with a straight centerline in which there is isentropic steady flow, as shown in Fig. 11.8. At the enlarged region at the left we may consider the velocity to be negligibly small so that conditions here are termed *stagnation conditions*, denoted with a subscript zero. This might correspond to a storage tank in a practical situation. Since the stagnation conditions are usually known, many of the results in this section will be expressed in terms of these conditions. Two control volumes have been indicated, for which the basic laws will now be applied.

First law of thermodynamics. Using both control volumes we may arrive at the following equations:

$$h_0 = h_1 + \frac{V_1^2}{2} = h_2 + \frac{V_2^2}{2} \tag{11.19}$$

Note that these equations would not change form with the inclusion of friction. Hence, if the fluid is slowed up *adiabatically* at a later time, the first law of thermodynamics indicates that *the original stagnation enthalpy would again be approach* even with friction present.

Second law of thermodynamics. There is no restriction for these processes beyond our own restrictions that there is no friction and zero heat transfer. As the name implies, there is then a constant entropy along the direction of flow.

---

[6]For a more complete discussion see A. Shapiro, *The Dynamics and Thermodynamics of Compressible Fluid Flow*, Ronald, New York, 1953, vol. I, pt. II.

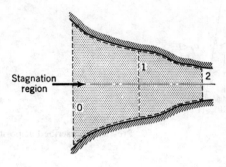

FIGURE 11.8
Control volumes for simple area change.

That is,

$$s_0 = s \tag{11.20}$$

Continuity equation. The mass flow through all sections must be equal. Thus

$$\rho V A = \text{const} = w \tag{11.21}$$

A useful formulation is the flow per unit area, w/A, denoted as G. Hence

$$\rho V = G = \frac{w}{A} \tag{11.22}$$

Linear momentum equation. For a duct with a straight centerline, the linear momentum equation for a control volume with end sections 1 and 2 as shown in Fig. 11.8 becomes

$$p_1 A_1 - p_2 A_2 + R = \rho_2 V_2^2 A_2 - \rho_1 V_1^2 A_1 \tag{11.23}$$

where R is the force from the duct wall on the fluid. The reaction to this force is the thrust on the wall from fluid flow between the sections chosen.

Equation of state. For pure gaseous substances, it is known that two properties will determine the state uniquely; that is, these substances have two degrees of freedom. Knowing v and T, for example, one can determine p from a p-v-T relation. It is similarly possible to relate any property in terms of any other two properties by relations which, in general, we call equations of state. For instance, we may consider

$$h = h(s, p) \tag{11.24}$$

or

$$\rho = \rho(s, p) \tag{11.25}$$

as equations of state in this discussion. For other than perfect gases, these relations may be quite involved or, even worse, may be expressible only in the form of experimentally derived curves and charts.

It will now be shown that for isentropic flow with a given set of stagnation conditions, an area may be associated with each pressure less than the stagnation pressure for a stipulated mass flow. Thus, it will be assumed that h_0, s_0,

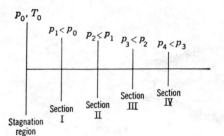

FIGURE 11.9
Pressures prescribed at positions along flow.

etc., are known, as well as the mass flow w. Then, choosing a pressure p_1 less then p_0 and knowing $s_1 = s_0$, we may employ the equations of state [Eqs. (11.24) and (11.25)] to evaluate h_1 and ρ_1. Now the velocity V_1 may readily be ascertained from Eq. (11.19), the first law of thermodynamics. Finally the correct cross-sectional area of flow may be determined from the continuity equation [Eq. (11.21)].

We can thus formulate theoretically possible isentropic flows by first stipulating pressure decrements along a centerline such that the pressure decreases from stagnation to the ambient (outside) pressure (Fig. 11.9) and next computing, in the aforementioned manner, the corresponding areas for an isentropic expansion. The question then is: How close to isentropic will an actual flow be for a passage computed in this manner when it is subjected to the same stagnation and ambient conditions used in the computations? This important consideration will be examined carefully in a later section of the chapter. Meanwhile, by using superheated steam, an illustration of the preceding formulations which we have outlined will be carried out in the following example.

Example 11.1. Superheated steam at a pressure 100 lb/in^2 absolute and a temperature of 815°F is to be expanded isentropically to a pressure of 100 lb/in^2 absolute at a rate of 1 lbm/s. Map out a theoretically possible passage for such a flow.

We choose a set of pressure decrements of 25 lb/in^2. At each pressure it is a simple matter to find the corresponding enthalpy and temperature by using the well-known *Mollier* chart,[7] which is an equation of state in the form of an elaborate set of curves relating, among other things, enthalpy, entropy, temperature, and pressure. Since the enthalpy is the ordinate and entropy is the abscissa, this is called an *h-s* diagram. We can locate points in the diagram corresponding to the state at each section by móving along the vertical line corresponding to the known entropy s_0 until the proper pressure is reached. The enthalpy can then be read from the ordinate, and the temperature can be determined from the isothermal

[7]See chart in back cover of J. H. Keenan and F. G. Keyes, *Thermodynamic Properties of Steam*, Wiley, New York, 1936.

line going through the point. The following chart gives the results of these evaluations in columns 2 and 3:

	(1)	(2)	(3)	(4)	(5)	(6)
					Specific	
	Pressure,	Temperature,	Enthalpy,	Velocity,	volume,	Area,
Section	lb/in²	°R	Btu/lbm	ft/s	ft³/lbm	in²
1	300	815	1428	0	2.47	
2	275	790	1417	735	2.65	0.520
3	250	764	1404	1092	2.85	0.376
4	225	735	1391	1355	3.07	0.326
5	200	704	1375	1623	3.39	0.301
6	175	670	1359	1850	3.77	0.294
7	150	630	1341	2080	4.24	0.294
8	125	584	1319	2330	4.84	0.300
9	100	530	1295	2570	5.78	0.324

The velocities in column 4 may be computed from the *first law of thermodynamics* in the following manner:

$$V_i = \sqrt{2(h_0 - h_i)}$$

Thus, at section 3, where the pressure is 250 lb/in² absolute, we have

$$V_3 = \sqrt{2(1428 - 1404)g_0(778)} = 1092 \text{ ft/s}$$

To determine the specific volume listed in column 5, we may employ the well-known *steam tables*. Finally, the areas for these sections may be evaluated for the stipulated mass flow of 1 lbm/s by using the equation of *continuity*. Hence at section 3 the computation becomes

$$w = 1 = \frac{1}{v_3}V_3 A_3 = \frac{1}{2.85}(1092)(A_3)$$

Therefore $\qquad\qquad A_3 = 0.00261 \text{ ft}^2 = 0.376 \text{ in}^2$

If the nozzle is rectangular in cross section with a uniform width, the longitudinal section would then appear as shown in Fig. 11.10.

We will now state and later prove (Sec. 11.8) several important features of isentropic flow, using Fig. 11.10. First the expansion proceeds from stagnation

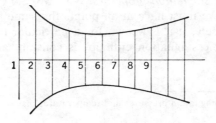

FIGURE 11.10
Computed heights for rectangular nozzle.

conditions as subsonic flow with a decreasing cross-sectional area until a minimum area has been reached, at which time the Mach number is unity. This section is called the *throat*, or *sonic* section, and the flow properties here are termed *critical properties*. We use an asterisk to denote these quantities (p^*, T^*, etc.). After the sonic section, the area increases, and *supersonic* flow conditions are found. This portion of the expansion is called the *diverging* portion, in contrast to the region ahead of the throat, which is termed the *converging* portion. Nozzles designed to conduct a fluid in an isentropic expansion to an ambient (outside) pressure exceeding the critical pressure will have a shape similar to the convergent outline shown and are called *convergent nozzles*. Those which are designed to conduct isentropic expansions to ambient pressure less then critical will also have a diverging section like the one in Fig. 11.10 and are called *convergent-divergent*, or *De Laval*, nozzles. You may deduce from the discussion that the function of a nozzle is to convert the enthalpy of a fluid into kinetic energy in an efficient, effective manner. Opposite to this is the *diffuser*, which converts kinetic energy into enthalpy.

A very significant consideration is the fact that any number of shapes could have evolved in the previous example by choosing different spacings and pressure decrements at the outset. However, for a set of conditions of the type given, all the geometries will have the same *throat* area and the same *exit* area. This will be more evident in the analytical formulations to be carried for the case of a perfect gas.

11.7 LOCAL ISENTROPIC STAGNATION PROPERTIES

In our discussion of isentropic flow in the preceding section, we termed the properties of a fluid having zero velocity as stagnation properties. And we pointed out that in an adiabatic, one-dimensional flow the same stagnation enthalpy is approached whenever the flow approaches zero velocity. It should then be simple to suppose that if *at any point* in this adiabatic flow the flow were *imagined* locally to slow up to a zero velocity *isentropically*, this same stagnation enthalpy would be reached. On the other hand, if the actual flow were not adiabatic and not one-dimensional, we should in all probability get a different value of stagnation enthalpy at each point when stopping the flow locally at these points in an isentropic manner. Hence, we could in all cases assign by this means a stagnation enthalpy or for that matter any other isentropic stagnation property at each point in any given flow. Such values are termed *local isentropic stagnation properties*. Now we know that in an adiabatic one-dimensional flow we must have at all points the same isentropic stagnation enthalpy,[8] and conversely, if it is known for a particular one-dimensional flow that the local isentropic stagnation enthalpy is constant at all points, we may

---

[8] However, local isentropic stagnation pressures and temperatures may vary from point to point.

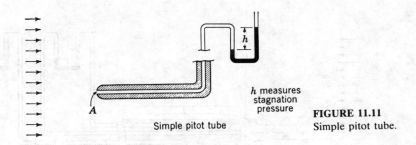

h measures stagnation pressure

Simple pitot tube

FIGURE 11.11
Simple pitot tube.

conclude that the flow is adiabatic. In general the knowledge of the variation of local isentropic stagnation properties may shed much light on the nature of the flow, as we will see as we continue into this chapter. We may then state formally: *The local isentropic stagnation properties are those properties which would be reached at a point in any given flow by a hypothetical isentropic retardation process ending at zero velocity and having as the initial condition that corresponding to the actual flow at the point in question.*

The notation for local stagnation properties will be that of the zero subscript, as employed in the previous section, where stagnation conditions were presumed actually to be present in an isentropic flow.

The local isentropic stagnation pressure may be measured with a simple pitot tube, shown in Fig. 11.11. When it is oriented so that the axis is parallel to the flow, there will be very close to an isentropic retardation of flow to a stagnation condition at A. This pressure is transmitted through the tube and is measured with the aid of a manometer.

At a position in the flow where the *actual undisturbed fluid velocity* is zero, the stagnation pressure corresponds to the *undisturbed pressure* p. At other positions, however, we may now distinguish between two pressures p and p_0. It is often the practice to call the undisturbed pressure at such points the *static pressure* and to call the stagnation pressure the *total pressure*. In measuring the undisturbed pressure, it is necessary that the measuring device not appreciably alter the flow conditions at the position of measurement. This is in direct contrast to the stagnation-pressure measurement, wherein a deliberate, carefully controlled disturbance is created at the position of interest, as was described in the previous paragraph concerning the pitot tube. A third pressure may be associated with the flow conditions at a point by taking the difference between the stagnation pressure and the undisturbed pressure; we call this the *dynamic pressure*. All three pressures may be measured in a region of flow if holes are added to the sides of the simple pitot tube (which then becomes a *pitot-static tube*), as is shown in Fig. 11.12. Pressure measurement at B gives the undisturbed pressure, and the difference between p_A and p_B determines the dynamic pressure. The position of measurement for this device is clearly not a point but a region small enough to be considered as a point in many calculations.

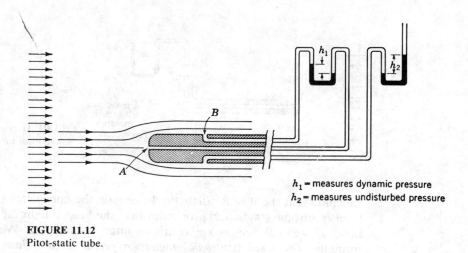

h_1 = measures dynamic pressure
h_2 = measures undisturbed pressure

FIGURE 11.12
Pitot-static tube.

We have used in Sec. 10.4 yet a fourth pressure in our basic considerations of parallel flow, which we called the *geometric pressure*. You will recall that this pressure is due only to the gravitational action on a static fluid having the same geometry as the actual flow. In Chap. 14 on free-surface flow we again have occasion to use the geometric pressure.

11.8   AN IMPORTANT DIFFERENCE BETWEEN ONE-DIMENSIONAL SUBSONIC AND SUPERSONIC FLOW

Let us now ascertain how a variation in area, consistent with isentropic conditions of flow, affects an existing subsonic or supersonic flow as to velocity and pressure. To do this, we apply the basic laws for steady flow to a stationary control volume of infinitesimal thickness, as shown in Fig. 11.13, where z is the direction of flow. The result will be an expression of these laws in differential form. It will be assumed in the development of the basic equations that all dependent quantities increase with increasing z. This means that a negative sign for a differential quantity resulting from these formulations indicates that the quantity itself is decreasing in the direction of flow. Upon observing the

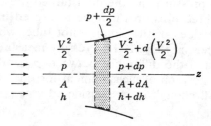

FIGURE 11.13
Infinitesimal control volume.

diagram, we can give the basic laws as:

First law of thermodynamics.

$$h + \frac{V^2}{2} = (h + dh) + \frac{V^2}{2} + d\left(\frac{V^2}{2}\right)$$

(11.26)

$$\therefore dh = -d\left(\frac{V^2}{2}\right)$$

Second law of thermodynamics.

$$ds = 0$$

(11.27)

Continuity equation. A desirable form of this law may be found by taking the logarithmic differential of Eq. (11.21). We get

$$\ln \rho + \ln V + \ln A = \ln \text{const}$$

$$\therefore \frac{d\rho}{\rho} + \frac{dV}{V} + \frac{dA}{A} = 0$$

(11.28)

Linear momentum equation. By considering the average pressure on the infinitesimal circumferential area of the control volume to be $p + dp/2$, the linear momentum equation becomes

$$pA + \left(p + \frac{dp}{2}\right)dA - (p + dp)(A + dA) = (\rho V A)(V + dV) - \rho V^2 A$$

Upon canceling terms and dropping second-order expressions, the following formulation may then be written:

$$dp = -\rho V \, dV$$

(11.29)

Now, in Eq. (11.28), replace dV, using Eq. (11.29). Solving for dA/A and arranging the terms for convenience, we get

$$\frac{dA}{A} = \frac{dp}{\rho V^2} - \frac{d\rho}{\rho} = \frac{dp}{\rho V^2}\left(1 - V^2 \frac{d\rho}{dp}\right)$$

(11.30)

For an isentropic process, it has been shown earlier that $dp/d\rho$ equals the square of the local velocity of sound. Hence, inserting this quantity and using the Mach number for V/c, we get

$$\frac{dA}{A} = \frac{dp}{\rho V^2}(1 - \text{M}^2)$$

(11.31)

For a *nozzle*, dp is negative, since it is a device which permits a fluid to expand from high to low pressure, and from Eq. (11.29) we see that dV must then be positive, indicating an increasing velocity in the direction of flow. Finally, for subsonic flow $\text{M}^2 < 1$ and with dp for a nozzle negative, we see in

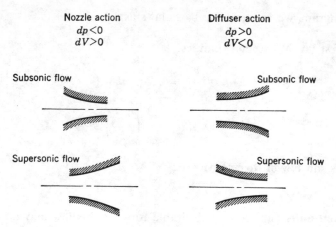

FIGURE 11.14
Area changes for nozzles and diffusers.

Eq. (11.31) that dA must be negative. Thus, in a nozzle, the area should be decreasing in the direction of flow for subsonic flow conditions. Furthermore, when M = 1, that is at the *sonic condition*, dA is zero from the above equation, and we thus have reached the *minimum area*. For M exceeding unity, dA is positive, so we have an increasing area during a supersonic expansion in a nozzle. We thus prove the statements made in Sec. 11.6 in our discussion of the nozzle evaluated in Example 11.1.

For a *diffuser*, dp is positive, since it is a device to increase pressure, indicating from Eq. (11.29) that the flow is slowing up. For subsonic flow this means, according to Eq. (11.31), that there is an increasing area and for supersonic flow that there is a decreasing area. Most persons will find the latter "strange" in that a narrowing passage is associated with a velocity retardation.[9] This is but one instance of the radical difference between subsonic and supersonic flow. These results are shown in Fig. 11.14.

11.9 ISENTROPIC FLOW OF A PERFECT GAS

Using the familiar equation of state for a perfect gas, we can now formulate very useful formulas giving isentropic-flow characteristics in terms of stagnation conditions and the local Mach number. To do this, we reconsider certain of the equations of Sec. 11.6, using perfect-gas relations. Assuming constant specific

---

[9]Wouldn't you expect that the narrowing passage would cause the fluid to "squirt" through faster?

heats, we may then rewrite the *first law of thermodynamics* in the following manner:

$$c_p T_0 = c_p T + \frac{V^2}{2}$$

From this we conclude that for a perfect gas *the stagnation temperature is constant for an adiabatic flow*. We compute now the ratios T/T_0, p/p_0, and ρ/ρ_0. We begin by dividing the equation above by $c_p T$. Thus

$$\frac{T_0}{T} = 1 + \frac{V^2}{2c_p T}$$

Replacing c_p by $[k/(k-1)]R$ according to Eq. (11.6a), we get

$$\frac{T_0}{T} = 1 + \frac{k-1}{2} \frac{V^2}{kRT}$$

We have learned in Sec. 11.3 that $c^2 = kRT$ for a perfect gas so that the equation above may be written to give the stagnation-temperature ratio in terms of the Mach number in the following way:

$$\boxed{\frac{T}{T_0} = \frac{1}{1 + [(k-1)/2]\mathrm{M}^2}} \tag{11.32}$$

The temperature ratio in the preceding formula may be replaced by a pressure ratio in accordance with isentropic property relations for a perfect gas as given by Eq. (11.9b). That is,

$$\frac{T}{T_0} = \left(\frac{p}{p_0}\right)^{(k-1)/k} \tag{11.33}$$

Substituting into Eq. (11.32) and solving for p/p_0, the stagnation-pressure ratio, we get

$$\boxed{\frac{p}{p_0} = \frac{1}{\{1 + [(k-1)/2]\mathrm{M}^2\}^{k/(k-1)}}} \tag{11.34}$$

The temperature ratio in Eq. (11.32) is next replaced by using the following relation coming from Eq. (11.9a):

$$\frac{T}{T_0} = \left(\frac{\rho}{\rho_0}\right)^{k-1} \tag{11.35}$$

Substituting and solving for ρ/ρ_0,

$$\boxed{\frac{\rho}{\rho_0} = \frac{1}{\left\{1 + [(k-1)/2]M^2\right\}^{1/(k-1)}}} \tag{11.36}$$

Note that the boxed-in formulations are also useful in the case of a nonisentropic flow of a perfect gas where *local* stagnation pressures and temperatures may be computed by employing *actual local* Mach numbers, pressures, and temperatures.

As a final computation, the ratio of areas A/A^*, where A^* is the throat area, will be ascertained. Starting with the continuity equation, we may say that

$$G = \rho V = \frac{p}{RT}V$$

Now G may be put in terms of Mach number and stagnation conditions by first performing algebraic manipulations on the equation above in the following way:

$$G = \frac{p}{\sqrt{RT}\sqrt{RT}}V\frac{\sqrt{k}}{\sqrt{k}} = p\frac{V}{\sqrt{kRT}}\frac{\sqrt{k}}{\sqrt{R}}\frac{1}{\sqrt{T}} = pM\sqrt{\frac{k}{RT}}$$

Next, multiply the right side of this equation by $\sqrt{T_0/T_0}$, and replace the resulting term $\sqrt{T_0/T}$ with the aid of Eq. (11.32). Similarly, multiply by p_0/p_0, and replace p/p_0, using Eq. (11.34). The resulting expression for G may then be given as

$$G = \sqrt{\frac{k}{R}}\frac{p_0}{\sqrt{T_0}}\frac{M}{\left\{1 + [(k+1)/2]M^2\right\}^{(k+1)/[2(k-1)]}} \tag{11.37}$$

The value of G at the throat is found by setting $M = 1$. Thus

$$G^* = \sqrt{\frac{k}{R}}\frac{p_0}{\sqrt{T_0}}\left(\frac{2}{k+1}\right)^{(k+1)/[2(k-1)]} \tag{11.38}$$

The ratio of G^*/G is next formed from the preceding equations. Noting that $G = w/A$, we may write

$$\boxed{\frac{G^*}{G} = \frac{A}{A^*} = \frac{1}{M}\left[\frac{2}{k+1}\left(1 + \frac{k-1}{2}M^2\right)\right]^{(k+1)/[2(k-1)]}} \tag{11.39}$$

Relations (11.32), (11.34), (11.36), and (11.39) have been tabulated in terms of M for $k = 1.4$ in Table B.5 in Appendix B under the heading "One-Dimensional Isentropic Relations." We find these tables very useful in solving problems, so you are urged to familiarize yourself with them.

It should be remembered that in isentropic flow all stagnation properties are conserved, and we can attribute these conditions to the fact that the flow is *adiabatic* and *reversible*. We will later see that in some other flows we can relate the change in local stagnation temperature with heat transfer and change in local stagnation pressure with friction and other such dissipative actions. These quantities then are extremely important in one-dimensional compressible-flow analysis.

It was pointed out earlier than an infinite number of isentropic-flow geometries could be formed analytically with given stagnation conditions, final pressure, and mass flow but that for *all these there would be a common throat area and a common exit area*. Thus, examination of Eq. (11.38) indicates that the throat area for a given mass flow is fixed by the stagnation conditions and the nature of the fluid (as given by the quantities k and R). Furthermore, Eq. (11.34) indicates that specifying the exit pressure fixes the exit Mach number so that according to Eq. (11.37) a definite area will be required for a given mass flow, as we have pointed out in an earlier discussion.

Example 11.2. A nozzle for an ideal rocket is to operate at a 15,250-m altitude in a standard atmosphere where the pressure is 11.60 kPa and is to give a 6.67-kN thrust when the chamber pressure is 1345 kPa and the chamber temperature is 2760°C. What are the throat and exit areas and the exit velocity and temperature? Take $k = 1.4$ and $R = 355$ N · m/(kg)(K) for this calculation. Take the exit pressure to be at ambient pressure.

We have p/p_0 for the exit,

$$\frac{p}{p_0} = \frac{11.60}{1345} = 0.008625$$

From the *isentropic tables*, we see that $M_{exit} = 3.8$, and we have an area ratio $A_{exit}/A_{throat} = 8.95$. Finally, we get for the exit temperature T_e

$$\frac{T_e}{T_0} = 0.257$$

Therefore, $T_e = (2760 + 273)(0.257) = 779$ K

And so $T_e = 779 - 273 = 506°C$

We can also determine the exit velocity easily now. Thus

$$V_e = M_e c = M_e \sqrt{kRT_e}$$

Therefore $V_e = 3.8\sqrt{(1.4)(355)(779)} = 2364$ m/s

To ascertain the *throat* and *exit* areas, we must consider the thrust. Using a control volume comprising the interior of the combustion chamber and nozzle and considering this control volume to be inertial,[10] we have from *linear momentum* considerations

$$6670 = (\rho_e V_e A_e)V_e = \rho_e A_e (2364^2) \tag{a}$$

---

[10]There will be only a small error here, since the mass of gas inside the control volume will be small.

We can get ρ_e by using the *equation of state* at the exit conditions. Thus

$$p_e = \rho_e R T_e$$

Therefore,

$$\rho_e = \frac{11,600}{(355)(779)} = 0.04195 \text{ kg/m}^3$$

Hence, going back to Eq. (*a*), we can say

$$A_e = \frac{6670}{(2364^2)(0.04195)} = 0.02845 \text{ m}^2$$

And using the area ratio of 8.95 as determined earlier from the tables, we have for the throat area

$$A^* = \frac{0.02845}{8.95} = 0.00318 \text{ m}^2$$

Note that a lower value of k will usually be needed for many of the fuels used in rockets, so we would have to use isentropic tables for other values of k.[11] Also the value of R can generally be expected to be higher for the rocket-fuel combustion products.

11.10 REAL NOZZLE FLOW AT DESIGN CONDITIONS

We now have available equations for determining the throat area of a nozzle which is to pass a specified mass flow from given stagnation conditions to an ambient pressure in an isentropic manner. However, in real operation there is always a certain amount of friction present in the boundary layer which would prevent the nozzle from operating in the prescribed manner even if the exact design conditions are imposed on the device. Fortunately, it is in most cases a deviation which is small enough to require only a mild correction of the isentropic analysis. This correction, as will soon be indicated, stems from experimental evidence developed for various types of nozzles.

To illustrate the action of friction in a nozzle operating under otherwise ideal conditions, examine the first law of thermodynamics for a control volume bounded by sections 1 and 2.

$$h_1 + \frac{V_1^2}{2} = h_2 + \frac{V_2^2}{2}$$

In the absence of heat transfer, friction will tend to increase the temperature of the fluid above that which corresponds to isentropic flow, and hence the enthalpy is similarly increased. Thus with a larger h_2 it is necessary according to the equation above that V_2 become smaller. Since the purpose of a nozzle is to develop high kinetic energy at the expense of enthalpy, it is clear that friction

[11]As given in J. H. Keenan and J. Kaye, *Gas Tables*, Wiley, New York, 1945.

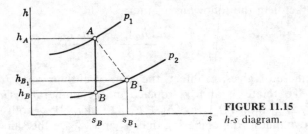

FIGURE 11.15
h-s diagram.

decreases the effectiveness of a nozzle. This may also be seen by examining a typical enthalpy-entropy diagram of a gas, as is shown in Fig. 11.15. An isentropic expansion from some initial pressure p_1 to p_2 is shown as a vertical line. In the case of an irreversible adiabatic process, the second law dictates that the entropy increase. Hence, the final state must lie on constant-pressure line p_2 *to the right* of B as shown by point B_1. Note that a higher enthalpy is established. Consequently, a smaller conversion to kinetic energy is achieved.

As a measure of frictional effects in a nozzle one ordinarily uses the *nozzle efficiency*, defined as the ratio of actual kinetic energy leaving the nozzle per unit of mass flow to the theoretical kinetic energy per unit of mass flow that could be achieved by an isentropic expansion for the same inlet conditions and exit pressure. Thus, by using the first law of thermodynamics above to replace $(V_2^2/2)_{\text{isen}}$, this efficiency can be given as

$$\eta = \frac{\left(V_2^2/2\right)_{\text{act}}}{\left(V_2^2/2\right)_{\text{isen}}} \frac{\left(V_2^2/2\right)_{\text{act}}}{\left[V_1^2/2 + (h_1 - h_2)\right]_{\text{isen}}} \tag{11.40}$$

where subscript 2 refers to exit conditions and subscript 1 refers to the conditions on entering the nozzle.

It is often the case that the kinetic energy per unit mass entering the nozzle, $V_1^2/2$, is negligibly small compared with $h_1 - h_2$, so it is dropped from the formulation. Thus, by omitting the "approach velocity" term, the nozzle efficiency becomes

$$\eta = \frac{\left(V_2^2/2\right)_{\text{act}}}{(h_1 - h_2)_{\text{isen}}} \tag{11.41}$$

In the case of a perfect gas this becomes

$$\eta = \frac{\left(V_2^2/2\right)_{\text{act}}}{c_p(T_1 - T_2)_{\text{isen}}} \tag{11.42}$$

In employing vapors, we occasionally use another formulation, called the *reheat factor y*, which is simply related to the definition above. We define the reheat factor in the following manner:

$$y = \frac{(h_2)_{act} - (h_2)_{isen}}{(h_1 - h_2)_{isen}} \tag{11.43}$$

The numerator gives, in effect, the gain of enthalpy as a result of irreversible effects. To relate η and y, replace in the numerator $(h_2)_{act}$ by $(h_1 + V_1^2/2 - V_2^2/2)_{act}$ and $(h_2)_{isen}$ by $(h_1 + V_1^2/2 - V_2^2/2)_{isen}$ both from the first law of thermodynamics. Remembering that h_1 and V_1 are the same for the actual flow and isentropic flow, we have

$$y = \frac{\left[h_1 + V_1^2/2 - (V_2^2/2)_{act}\right] - \left[h_1 + V_1^2/2 - (V_2^2/2)_{isen}\right]}{(h_1 - h_2)_{isen}}$$

$$= \frac{(V_2^2/2)_{isen} - (V_2^2/2)_{act}}{(h_1 - h_2)_{isen}} \tag{11.44}$$

You can readily show that, if you neglect the approach velocity, the right side of Eq. (11.44) equals $1 - \eta$. Thus, we have

$$y = 1 - \eta \tag{11.45}$$

Since the decreasing pressure in a nozzle is favorable in creating a thinner boundary layer, it is found to be the case that for well-designed nozzles operating at design conditions, nozzle efficiencies of 90 to 95 percent may be achieved.

Thus, when nozzle efficiencies and reheat factors are available, it is possible to make adjustments in isentropic considerations to take into account frictional effects. The frictional effects are confined principally to the *divergent* portion of the nozzle, so we will use the formulations above to correct the *exit area*. The remaining geometry (other than the throat) is usually a matter of experience. The convergent part is fairly arbitrary, while the divergent portion has a shape which is a compromise between two effects. A short length for the divergent section means that the flow will have an appreciable velocity component in the direction normal to the centerline. This means a loss in thrust and is consequently undesirable. The name for this effect is *divergence* (see Prob. 5.65). In the case of a long divergent section, there is less divergence, but there is the disadvantage that a greater amount of wall friction will be present.

The preceding consideration of nozzles results from simplified one-dimensional analysis. For more accurate computations, particularly in the supersonic range, the flow must be considered as two- or three-dimensional, depending on the shape of the cross section.

It must be pointed out that a nozzle will operate according to previous computations *only if the nozzle is subject to conditions reasonably close to that for which it has been designed*. Subjecting a nozzle to *off-design* conditions means that predictions from isentropic-flow analysis modified by friction corrections may not be valid. Usually there appear shock patterns in the supersonic part of the expansion which must be considered. Hence, we defer discussions of off-design performance to Part D after the normal and oblique shocks have been studied.

Example 11.3. If the anticipated nozzle efficiency of the nozzle computed in Example 11.1 is 90 percent, what should the exit area be to pass 1 lbm/s of superheated steam under the same initial stagnation conditions and exit pressure?

The losses may be assumed to take place essentially between the throat and exit, where there are greater wall distance and supersonic flow. Therefore, we will consider an adjustment of only the exit area. Using the definition of *nozzle efficiency*, we may calculate the actual exit velocity in the following manner on using results given in the table of Example 11.1 in Eq. (11.41):

$$\eta = 0.90 = \frac{(V_e^2/2)_{\text{act}}}{(h_0 - h_e)_{\text{isen}}} \tag{a}$$

Note that

$$(V_2)_{\text{act}} = \sqrt{(2)(0.90)(1428 - 1295)(778)g_0}$$

$$= 2449 \text{ ft/s}$$

We next compute the actual exit *enthalpy*. The *first law of thermodynamics* stipulates that $h_0 = (h_e)_{\text{act}} + (V^2/2)_{\text{act}}$. Hence $(V^2/2)_{\text{act}} = h_0 - (h_e)_{\text{act}}$. Substituting the result for $(V^2/2)_{\text{act}}$ in Eq. (a), we get

$$0.90 = \frac{h_0 - (h_e)_{\text{act}}}{h_0 - (h_e)_{\text{isen}}}$$

Hence

$$(h_e)_{\text{act}} = h_0 - 0.90[h_0 - (h_e)_{\text{isen}}] = 1428 - 0.90(1428 - 1295) = 1308 \text{ Btu/lbm}$$

We can now get the actual exit specific volume $(v_e)_{\text{act}}$ from the steam tables, using the pressure $p_c = 100 \text{ lb/in}^2$ absolute and the enthalpy above. We get $(v_e)_{\text{act}} = 5.95$ ft^3/lbm. Now, employing the continuity equation, we can compute the corrected exit area. Thus

$$w = 1 \text{ lbm/s} = \left(\frac{V_e A_e}{v_e}\right)_{\text{act}} = \frac{2449(A_e)_{\text{act}}}{5.95}$$

Therefore

$$(A_e)_{\text{act}} = 0.350 \text{ in}^2$$

This is an increase in area of 7.4 percent over that computed entirely by isentropic considerations.[12]

---

[12] If the fluid were a perfect gas and not steam, we would use $c_p T$ instead of h and we would use the equation of state $pv = RT$ instead of tables to get $v \equiv 1/\rho$.

PART C
THE NORMAL SHOCK

11.11 INTRODUCTION

We mentioned in Part A that a shock wave is similar to an acoustic wave except that it has finite strength. And we pointed out that variations in flow properties in the wave occur over a very small distance. In fact, so small is the wave thickness that we can consider in our computations that discontinuous changes in flow properties take place across the wavefront. We also pointed out in Part A that the shock wave moves relative to the fluid at a speed greater than the speed of an acoustic wave. We now continue our discussion of shock waves, but we are not concerned in this text with actions in the shock itself, i.e., the so-called shock structure. This is quite a difficult area of study, requiring, among other things, the use of nonequilibrium thermodynamics. What we will do instead is to ascertain how the flows before and after a shock are related in certain cases by employing the familiar basic and subsidiary laws. Unfortunately, the prediction of the position and shape of a shock can be determined in a reasonable straightforward manner only for very simple situations.

Shocks may be expected in almost all actual supersonic flows. This will be seen when we discuss nozzles under off-design operation.

11.12 FANNO AND RAYLEIGH LINES

As an aid in arriving at shock relations, we will now study two important processes with the aid of an enthalpy-entropy diagram. Let us first consider a steady compressible *adiabatic* flow through a passage of constant cross section A wherein the nonisentropic effect of boundary-layer friction is permitted. A control volume has been designated with end sections 1 and 2 as shown in Fig. 11.16. We express three of the basic laws for this control volume (we delete the second law of thermodynamics now) and a general equation of state. Thus for adiabatic flow we have

First law of thermodynamics.

$$h_1 + \frac{V_1^2}{2} = h_2 + \frac{V_2^2}{2} \tag{11.46}$$

FIGURE 11.16
Finite-control volume.

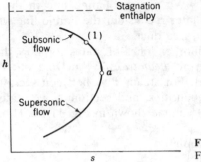

FIGURE 11.17
Fanno line.

Continuity.
$$\rho_1 V_1 = \rho_2 V_2 = G = \text{const} \tag{11.47}$$

Linear momentum equation.
$$(p_1 - p_2) + \frac{R}{A} = G(V_2 - V_1) \tag{11.48}$$

Equations of state.
$$h = h(s, \rho) \tag{11.49a}$$
$$\rho = \rho(s, p) \tag{11.49b}$$

Let us suppose that flow conditions at section 1 are known so that we can identify this state as point 1 on the h-s diagram in Fig. 11.17. Our purpose now is to locate on this h-s diagram other states which in accordance with the equations above can be reached by having the fluid start at section 1, with the aforesaid given flow conditions, and undergo varying amounts of boundary-layer friction as the fluid flows to section 2. This means we want to ascertain flow conditions at section 2 in the equations above for fixed conditions at section 1 and for varying values of R, which is the negative of the drag. We may choose some velocity V_2 and from Eq. (11.47) determine the corresponding density ρ_2. Also in Eq. (11.46), we can determine h_2. Now, going to the equations of state, we can get the entropy s_2 and the pressure p_2. Finally, we can find the drag force R from Eq. (11.48), so we have a point on the h-s diagram for a particular frictional effect. The locus of points denoted in Fig. 11.1 as the *Fanno line* then represents the locus of states starting from 1 that may thus be reached by simply changing the amount of boundary-layer friction in an adiabatic flow. The shape of the curve shown in the diagram is typical of most fluids. Clearly, at any point in the h-s diagram one could establish such a curve in the manner outlined.

As an exercise (Prob. 11.42) you will be asked to demonstrate that the extremum point a in the Fanno line, i.e., the point of maximum entropy, corresponds to sonic conditions, $M = 1$. The branch of the curve above a approaches the stagnation enthalpy, as shown in the diagram, and this part of the curve represents subsonic conditions, while the lower branch represents supersonic conditions.[13] We use the Fanno line in the present discussion on normal shocks and will later reconsider the Fanno line when we study adiabatic flow through a constant-area duct.

We now set up another locus of states for flow through a constant-area duct. This time we permit *heat transfer* but rule out boundary-layer friction from our considerations. For steady flow between section 1, where flow conditions are given fixed quantities, and section 2, we state the three basic laws again, using the control volume shown in Fig. 11.16. Thus

First law of thermodynamics.

$$h_1 + \frac{V_1^2}{2} + \frac{dQ}{dm} = h_2 + \frac{V_2^2}{2} \tag{11.50}$$

Continuity.
$$\rho_1 V_1 = \rho_2 V_2 = G = \text{const} \tag{11.51}$$

Linear momentum equation.

$$p_1 - p_2 = G(V_2 - V_1) \tag{11.52}$$

Equation of state.

$$h = h(s, \rho) \tag{11.53a}$$

$$\rho = \rho(s, p) \tag{11.53b}$$

Again, we assume the initial conditions fixed thus providing a starting point (1) on the *h-s* diagram (Fig. 11.18) and inquire about the possible states reached at section 2 by variations in heating. The procedure may again begin by choosing a velocity V_2 for which a density ρ_2 is immediately apparent from Eq. (11.51). From the linear momentum equation [Eq. (11.52)], we may next evaluate pressure p_2. By using the equation of state, the entropy and enthalpy at 2 may then by computed. Finally, the first law of thermodynamics yields the heat-transfer rate dQ/dm for this process. In this way a locus of points on the *h-s* diagram may be plotted called the *Rayleigh line* as shown in Fig. 11.18. You will also be asked to show (Prob. 11.43) that point *b*, corresponding to the maximum entropy, is the sonic point for the curve. The portion of the curve extending from above point *b*, corresponds to subsonic conditions because of higher enthalpy, while the remaining portion represents supersonic conditions as indicated in the diagram. Clearly, a Rayleigh line may be drawn starting at any point in the diagram by the methods above.

The Rayleigh line will be of use in the next section for shock evaluations and will also be of use later when simple heating in constant-area ducts is undertaken.

---

[13]Higher enthalpy means smaller velocity and a smaller Mach number.

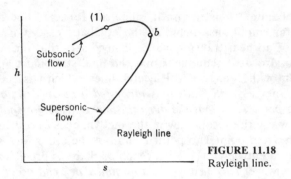

FIGURE 11.18
Rayleigh line.

11.13 NORMAL-SHOCK RELATIONS

The normal shock can be pictured as a plane surface of discontinuity of flow characteristics oriented normal to the direction of flow as shown diagrammatically in Fig. 11.19. The flow shown in this diagram is steady and the normal shock is stationary relative to the boundary. We therefore consider a fixed control volume of infinitesimal length dl which encloses the shock wave. Although this control volume is of infinitesimal size, the basic laws pertaining to it will not be differential equations, as was the case in Sec. 11.8 because there exist, in this case, *finite* changes in the flow properties across the infinitesimal length of the control volume. This means that differential quantities in the equations resulting from such things as area change are negligible in the resulting equations. Thus the flow can be considered as constant area through the control volume, and, in addition, boundary-layer friction and heat transfer can be disregarded for the chosen control volume. Let us now assume that conditions at section 1 are known. What are the conditions at section 2 consistent with one-dimensional fluid-flow theory? A very informative procedure to answer this is to make use of the h-s diagram in Fig. 11.20 containing a Fanno and a Rayleigh line going through point 1 corresponding to the fluid state just ahead of the shock. We have just concluded that the final state at point 2 may be considered a result of constant-area flow with no friction and no heat transfer. Now the condition of no friction corresponds to the Rayleigh line,

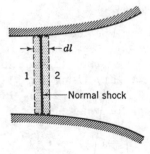

FIGURE 11.19
Infinitesimal control volume around normal shock.

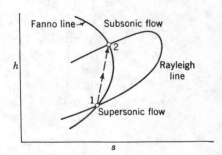

FIGURE 11.20
Fanno and Rayleigh lines.

so the final state 2 must lie on it, since the Rayleigh line includes all possible heat-transfer conditions, including the adiabatic case. On the other hand, the condition of no heat transfer makes it necessary that the final state be on the Fanno line also. It is then clear that the final state must appear at the second intersection of the Fanno and Rayleigh lines, which has been shown as point 2 in this diagram. For *all fluids thus far tested it has been found that point 2 is to the right of point 1, so that the diagram may be used for making generalizations.* We can now use the second law of thermodynamics to conclude that the process can go *only* from point 1 to point 2 and not the other way, since there must be an increase in entropy for any irreversible adiabatic process such as is occurring in the shock wave. *This means that a normal shock can occur only in a supersonic flow, with the result that after the normal shock there must be subsonic flow.*

Under each set of supersonic initial conditions there will then be a pair of Fanno and Rayleigh lines which give information concerning the changes of properties through the shock. If, however, an analytical equation is available for the equation of state of a fluid, one may compute these changes without the use of *h-s* charts. Accordingly, we will study the flow of a perfect gas through a normal shock. With the aid of the resulting equations for the perfect gas, further generalizations will be set forth for all fluids concerning changes of property and flow characteristics through the normal shock.

11.14   NORMAL-SHOCK RELATIONS FOR A PERFECT GAS

The *first law of thermodynamics* for the control volume of the previous section (Fig. 11.19) may be written as follows for the perfect gas:

$$c_p T_1 + \frac{V_1^2}{2} = c_p T_2 + \frac{V_2^2}{2} \tag{11.54}$$

The isentropic stagnation temperature for conditions on each side of the shock are given as

$$c_p (T_0)_1 = c_p T_1 + \frac{V_1^2}{2}$$

$$c_p (T_0)_2 = c_p T_2 + \frac{V_2^2}{2} \tag{11.55}$$

Examining Eqs. (11.54) and (11.55), one sees that $(T_0)_2$ and $(T_0)_1$ must be equal, so *there is no change in stagnation temperature across a shock for a perfect gas.* (The reader will recall that we reached this conclusion earlier for any steady adiabatic flow of a perfect gas.) Using Eq. (11.32) to express the stagnation temperature in terms of M and T, we may then equate stagnation temperatures across the shock in the following manner:

$$(T_0)_1 = T_1 \left(1 + \frac{k-1}{2} M_1^2 \right) = (T_0)_2 = T_2 \left(1 + \frac{k-1}{2} M_2^2 \right) \tag{11.56}$$

Rearranging the terms, we get

$$\frac{T_2}{T_1} = \frac{1 + [(k-1)/2]M_1^2}{1 + [(k-1)/2]M_2^2} \tag{11.57}$$

It will be desirable to form ratios of flow characteristics across the shock in terms of only the *initial* Mach number. By using Eq. (11.57) and also the continuity and momentum equations, we can relate M_2 directly to M_1, and this will permit us to determine T_2/T_1 as a function only of M_1 and k. Thus, considering the *continuity* equation first, we replace ρ by p/RT, and V by cM from the definition of the Mach number. That is,

$$\rho_1 V_1 = \left(\frac{p_1}{RT_1}\right)(c_1 M_1) = \rho_2 V_2 = \left(\frac{p_2}{RT_2}\right)(c_2 M_2)$$

Next we replace c by $\sqrt{kRT}$. Canceling terms, we can then form the relation

$$\frac{p_1}{p_2}\sqrt{\frac{T_2}{T_1}} = \frac{M_2}{M_1} \tag{11.58}$$

Now using Eq. (11.57) to replace $\sqrt{T_2/T_1}$, and solving for p_2/p_1, we get

$$\frac{p_2}{p_1} = \frac{M_1}{M_2}\frac{\{1 + [(k-1)/2]M_1^2\}^{1/2}}{\{1 + [(k-1)/2]M_2^2\}^{1/2}} \tag{11.59}$$

Next, examine the *linear momentum* equation for constant-area adiabatic flow with no friction. Equation (11.52) is applicable, and we rewrite it in the following way:

$$p_1 - p_2 = G(V_2 - V_1) = \left(\rho_2 V_2^2 - \rho_1 V_1^2\right)$$

We replace V by $M\sqrt{kRT}$, as before, and ρ by p/RT. Collecting terms,

$$p_1\left(1 + k\,M_1^2\right) = p_2\left(1 + k\,M_2^2\right)$$

Solving for p_2/p_1, we get

$$\frac{p_2}{p_1} = \frac{1 + k\,M_1^2}{1 + k\,M_2^2} \tag{11.60}$$

Thus, along with Eq. (11.59) there is now a second independent expression for p_2/p_1. Equating the right-hand sides of Eqs. (11.59) and (11.60) gives the desired relation between the Mach numbers.

$$\frac{1 + k\,M_1^2}{1 + k\,M_2^2} = \frac{M_1}{M_2}\left\{\frac{1 + [(k-1)/2]M_1^2}{1 + [(k-1)/2]M_2^2}\right\}^{1/2} \tag{11.61}$$

We may solve for M_2 in terms of M_1 by algebraic manipulations to form the relation

$$M_2^2 = \frac{[2/(1-k)] - M_1^2}{1 + [2k/(1-k)]M_1^2} \tag{11.62}$$

With the availability of this equation we now may determine T_2/T_1 and p_2/p_1 in terms of the initial Mach number M_1 and the constant k only. Thus we get from Eqs. (11.57) and (11.60), respectively along with Eq. (11.62),

$$\boxed{\frac{T_2}{T_1} = \frac{\{1 + [(k-1)/2]M_1^2\}\{[2k/(k-1)]M_1^2 - 1\}}{\left[(k+1)^2/2(k-1)\right]M_1^2}} \tag{11.63}$$

$$\boxed{\frac{p_2}{p_1} = \frac{2k}{k+1}M_1^2 - \frac{k-1}{k+1}} \tag{11.64}$$

Next we express ρ_2/ρ_1 using the equation of state as follows:

$$\frac{\rho_2}{\rho_1} = \frac{p_2/RT_2}{p_1/RT_1} = \frac{p_2}{p_1}\frac{T_1}{T_2}$$

Using Eqs. (11.63) and (11.64) we get

$$\frac{\rho_2}{\rho_1} = \frac{\left\{\dfrac{2k}{k+1}M_1^2 - \dfrac{k-1}{k+1}\right\}\left\{\dfrac{(k+1)^2}{2(k-1)}M_1^2\right\}}{\left\{1 + \dfrac{k-1}{2}M_1^2\right\}\left\{\dfrac{2k}{k-1}M_1^2 - 1\right\}}$$

Extracting $(k-1)/(k+1)$ from the first bracketed expression in the numerator permits us to cancel $\{[2k/(k-1)]M_1^2 - 1\}$ from numerator and denominator. We may accordingly reach the following desired value of ρ_2/ρ_1:

$$\boxed{\frac{\rho_2}{\rho_1} = \frac{k+1}{2}\frac{M_1^2}{1 + [(k-1)/2]M_1^2}} \tag{11.65}$$

While it is true that stagnation temperature is unchanged across a normal shock, this is definitely *not* the case for *stagnation pressure*, which may undergo an appreciable change, and this will be true for all adiabatic processes. In fact, the loss of stagnation pressure in such flows is a good indication of the frictional effects. To compute the ratio of stagnation pressures across a shock, we perform the following manipulative step:

$$\frac{(p_0)_2}{(p_0)_1} = \frac{(p_0)_2}{p_2}\frac{p_2}{p_1}\frac{p_1}{(p_0)_1} \tag{11.66}$$

Now each of the ratios on the right side has been evaluated in terms of either

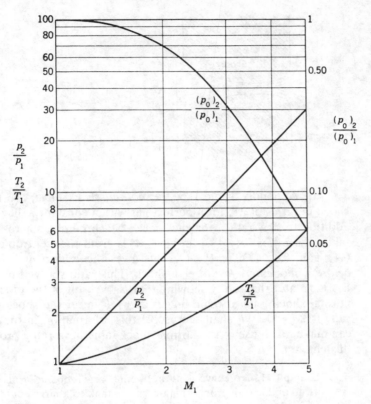

FIGURE 11.21
Curves for normal shock. $k = 1.4$.

M_1 or M_2. For instance $(p_0)_2/p_2$ and $p_1/(p_0)_1$ may be replaced with the use of Eq. (11.34), and p_2/p_1 may be replaced with the use of Eq. (11.64). Finally, using Eq. (11.62) to replace M_2 by M_1, we get

$$\frac{(p_0)_2}{(p_0)_1} = \frac{\left\{ \dfrac{[(k+1)/2]M_1^2}{1 + [(k-1)/2]M_1^2} \right\}^{k/(k-1)}}{\left\{ [2k/(k+1)]M_1^2 - (k-1)/(k+1) \right\}^{1/(k-1)}} \tag{11.67}$$

Figure 11.21 is a plot of the previously developed ratios as functions of the initial Mach number M_1 for air having $k = 1.4$. It can be seen from this plot that the higher the initial Mach number, the more extreme the change of properties and flow characteristics across the shock. From these curves one sees that there is a higher temperature and a higher undisturbed pressure after the shock and a lower stagnation pressure after the shock.

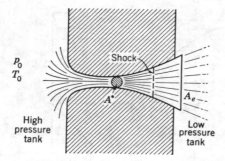

p_0
T_0

Shock

A^*

A_e

High
pressure
tank

Low
pressure
tank

FIGURE 11.22
Convergent-divergent nozzle with normal shock.

In Appendix Table B.6, the *normal shock table*, are flow ratios across a shock in terms of M_1 for a perfect gas with a constant k having the value 1.4. In addition to flow ratios, for each set of Mach numbers across the shock we have taken the ratios $(A/A^*)_1$ and $(A/A^*)_2$ from the isentropic tables and formed $(A/A^*)_1/(A/A^*)_2 = A_2^*/A_1^*$. This gives the *critical area ratios* for flows *just before* a shock and *just after* a shock. This ratio you will notice from the Table B.6 is greater than unity showing that, as a result of the dissipative action in the normal shock, one needs a bigger area A_2^* after the shock to return isentropically to $M = 1$. We shall see use of this information in the next example. You can make much use of the normal shock table in working problems at the end of this chapter.

Example 11.4. A convergent-divergent nozzle is operating at off-design conditions[14] as it conducts air from a high-pressure tank to a large container as shown in Fig. 11.22. A normal shock is present in the divergent part of the nozzle. We are to find the exit pressure as well as the loss in stagnation pressure during the flow between the tanks for the following data:

$$p_0 = 207 \text{ kPa absolute}$$

$$T_0 = 38°C$$

$$A^* = 1290 \text{ mm}^2$$

$$A_e = 2580 \text{ mm}^2$$

The shock is at a position where the cross-sectional area is 1935 mm^2.

We will assume isentropic flow everywhere except through the normal shock and consider the fluid as a perfect gas having constant specific heat.

First we ascertain the Mach number of the flow *ahead* of the shock. The value of A/A^* at this section is $1935/1290 = 1.5$. From Appendix Table B.5 on *isentropic flow* we get $M_1 = 1.853$ for the approach Mach number. Also from this table we get for p_1/p_0 the value 0.1607 so that the pressure p_1 just ahead of the shock is $(0.1607)(207) = 33.3$ kPa. Note that p_0 does not change in isentropic flow.

---

[14]This type of operation will be discussed in detail in Part D of this chapter.

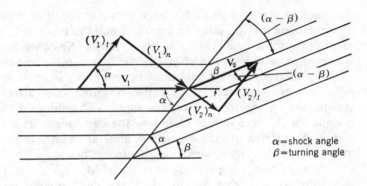

FIGURE 11.23
2-D oblique shock.

Next we get conditions *behind* the shock from Appendix Table B.6, the *normal-shock table*. Thus for $M_1 = 1.853$ we find that $M_2 = 0.605$, $p_2/p_1 = 3.84$, and $(p_0)_2/(p_0)_1 = 0.789$. Thus, we get for p_2 the value 127.9 kPa, and the stagnation pressure $(p_0)_2$ after the shock is 163.3 kPa.

We now turn to the *isentropic* flow relations to relate conditions *after* the shock with those at the *exit*. First we must find A^* for this flow. It is *not* the 1290 mm^2 of the nozzle because the flow has undergone the nonisentropic effects of the normal shock.

From the normal shock tables we compute $A_2^* = 1635$ mm^2. At the exit, A_e/A^* is then 1.578, and in the subsonic part of isentropic table B.5 we get $M = 0.404$ and p/p_0 of 0.894. The exit pressure is then $(0.894)(163.3) = 146.0$ kPa.

The loss in stagnation pressure occurs only across the shock, so $\Delta p_0 = 207 - 163.3 = 43.7$ kPa. Note that since the flow is entirely adiabatic there is no loss in stagnation temperature.

11.15 A NOTE ON OBLIQUE SHOCKS

Before discussing the operation of nozzles, it will be helpful to consider briefly the oblique shock, which is a plane shock whose normal is inclined at an angle to the direction of flow. Consider Fig. 11.23, where a two-dimensional supersonic flow is shown undergoing an oblique shock oriented at an angle α with the free stream. The velocity of a fluid particle prior to the shock is shown as V_1. It has been decomposed into a component normal to the shock $(V_1)_n$ and a component parallel to the shock $(V_1)_t$. In passing through the shock, the component $(V_1)_t$ is not affected, while the component $(V_1)_n$ undergoes a change prescribed by the normal-shock relations, as described in the previous section.[15]

---

[15]It should be clear that $(V_1)_n > c_1$, since a normal shock can occur only in a supersonic flow.

The velocity after the shock has been shown in the diagram with the subscript 2. The component $(V_2)_n$ must be subsonic, and the result is that the direction and magnitude of the final velocity $\mathbf{V}_2$ must change. *Note that the change of direction is toward the shock, as will always be the case in two-dimensional oblique shocks.*

Furthermore, since only the component V_n is affected by the shock, it is possible, for weak shocks, that the resulting flow after the shock is still supersonic. Thus, either subsonic or supersonic conditions are possible after the oblique shock. Note in Fig. 11.23 how the streamlines have turned in a manner completely foreign to subsonic action, illustrating once more the vast difference in behavior between subsonic and supersonic flows.

We will now compute $\mathbf{M}_2$ in terms of $\mathbf{M}_1$, k, α, and β (see Fig. 11.23 for these angles). As a first step we shall consider the components of $\mathbf{M}_1$ and $\mathbf{M}_2$ normal to the oblique shock. Thus considering Fig. 11.23 we have

$$(V_1)_n = V_1 \sin \alpha$$

$$(V_2)_n = V_2 \sin(\alpha - \beta)$$

Hence
$$(\mathbf{M}_1)_n = \frac{V_1 \sin \alpha}{c_1} = \mathbf{M}_1 \sin \alpha \tag{11.68a}$$

$$(\mathbf{M}_2)_n = \frac{V_2 \sin(\alpha - \beta)}{c_2} = \mathbf{M}_2 \sin(\alpha - \beta) \tag{11.68b}$$

To have an oblique shock, $(\mathbf{M}_1)_n$ must be > 1. Thus we require

$$(\mathbf{M}_1)_n = \mathbf{M}_1 \sin \alpha > 1$$

Hence, for a given $\mathbf{M}_1$ the *minimum* value of α clearly is

$$\alpha_{\min} = \sin^{-1}\left(\frac{1}{\mathbf{M}_1}\right) \tag{11.69}$$

The *maximum* value would correspond to a normal shock wherein $(\mathbf{M}_1)_n = \mathbf{M}_1$ so that $\sin \alpha = 1$ with $\alpha = 90°$. Now go to Eq. (11.62) relating $\mathbf{M}_2$ to $\mathbf{M}_1$ for a normal shock. Using normal components of $\mathbf{M}_1$ and $\mathbf{M}_2$ in this equation and then substituting for these quantities using Eqs. (11.68a) and (11.68b), we get

$$\mathbf{M}_2^2 \sin^2(\alpha - \beta) = \frac{[2/(1-k)] - \mathbf{M}_1^2 \sin^2 \alpha}{1 + [2k/(1-k)](\mathbf{M}_1 \sin^2 \alpha)} \tag{11.70}$$

$$\therefore \mathbf{M}_2^2 = \frac{[2/(1-k)] - \mathbf{M}_1^2 \sin^2 \alpha}{\sin^2(\alpha - \beta)\{1 + [2k/(1-k)]\mathbf{M}_1^2 \sin^2 \alpha\}}$$

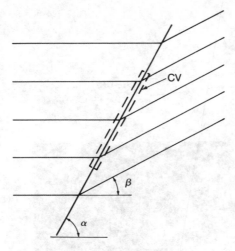

FIGURE 11.24
An infinitesimal control volume across the oblique shock.

Thus we relate M_2 with M_1 for a given turning angle α and a given shock angle β.

Next we wish to be able to compute the shock angle α for an oblique shock knowing M_1 and the turning angle β of the streamlines. From Fig. 11.23 we have

$$\frac{(V_1)_n}{(V_1)_t} = \tan \alpha \qquad \frac{(V_2)_n}{(V_2)_t} = \tan(\alpha - \beta)$$

Hence, noting that $(V_1)_t = (V_2)_t$, we can say

$$\frac{\tan(\alpha - \beta)}{\tan \alpha} = \frac{(V_2)_n/(V_2)_t}{(V_1)_n/(V_1)_t} = \frac{(V_2)_n}{(V_1)_n} \tag{11.71}$$

We now consider a control volume in the shape of a plate of infinitesimal thickness and transverse area A as shown in Fig. 11.24 enclosing part of the oblique shock. *Continuity* for this control volume requires that

$$\rho_1(V_1)_n A = \rho_2(V_2)_n A$$

$$\therefore \frac{(V_2)_n}{(V_1)_n} = \frac{\rho_1}{\rho_2}$$

Applying this result to Eq. (11.71) and using Eq. (11.65), we get

$$\frac{\tan(\alpha - \beta)}{\tan \alpha} = \frac{\rho_1}{\rho_2} = \left(\frac{2}{k + 1}\right)\left(\frac{1 + [(k - 1)/2]M_1^2 \sin^2 \alpha}{M_1^2 \sin^2 \alpha}\right) \tag{11.72}$$

$M_1 > c$ $M_2 > c$

FIGURE 11.25
Expansion wave.

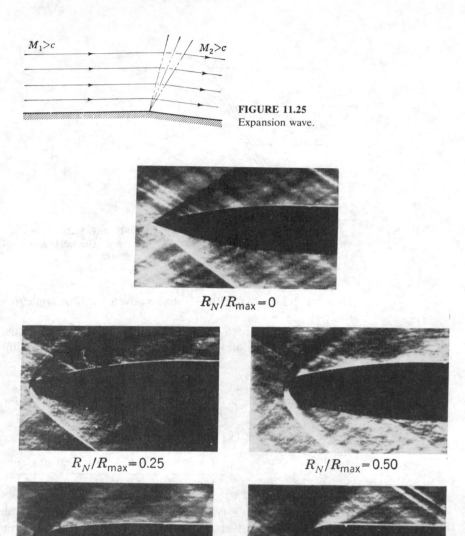

$R_N/R_{max} = 0$

$R_N/R_{max} = 0.25$ $R_N/R_{max} = 0.50$

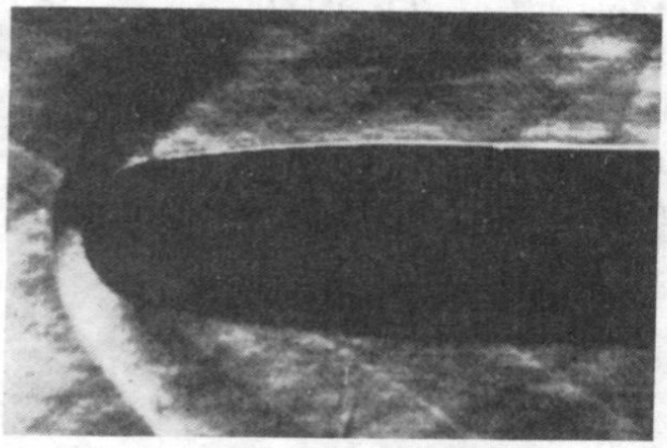

$R_N/R_{max} = 0.75$ 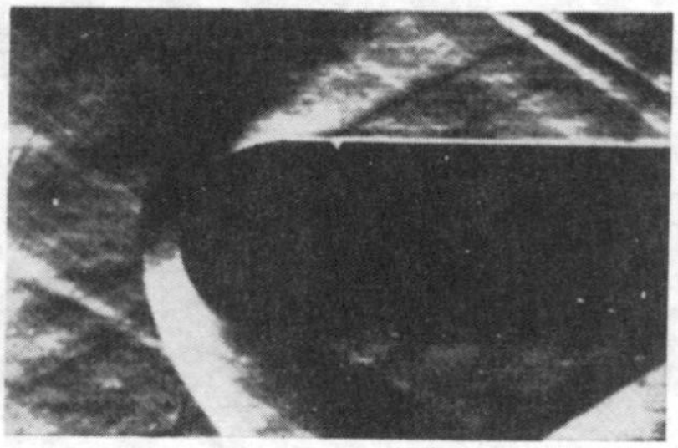 $R_N/R_{max} = 1.00$

FIGURE 11.26
Two-dimensional shock waves. The captions above show the effect of nose radius R_N on the bow wave for a supersonic flow of Mach 1.86. The sharp-edge profile on top develops an oblique shock, while the remaining profiles develop curved shocks with varying shapes and position relative to the body. (*Courtesy David Taylor Research Facility, Carderock, Md.*)

FIGURE 11.27
Conical shock wave. A cone-cylin-
der model at a Mach number of
2.0 and a Reynolds number per
inch of 3×10^5. (*Courtesy David
Taylor Research Facility, Carderock, Md.*)

Thus we relate α in terms of β, M_1, and k. In homework problems we will use
the above formula in evaluating oblique shock geometry for supersonic flow
around wedges and along a wall with a sharp inside corner. Also there is an
interesting computer project in the instructor's manual which will show that for
a given M_1 and β there may be two possible values of α.

In the case of symmetric three-dimensional supersonic flows one encounters shock surfaces in the form of cones rather than the oblique plane-shock
surfaces of two-dimensional supersonic flow. These are usually called *conical
shocks*. We merely point out at this time that for such shocks, streamlines
undergo a change in direction similar to what was described in the previous
paragraph. Also, there may be supersonic or subsonic flow behind the conical
shock, depending on its strength.

It follows from the preceding discussion on normal shocks that there
cannot be an oblique expansion shock. However, one finds rapid expansions in
supersonic flows which take place over a thin fan-shaped region, as is shown in
Fig. 11.25. These are called oblique expansion waves, or *Prandtl-Meyer* expansions.

Finally, note that *curved shocks* are to be found in supersonic flows. These
various types of shocks have been shown in the Schlieren photographs in Figs.
11.26 to 11.28.

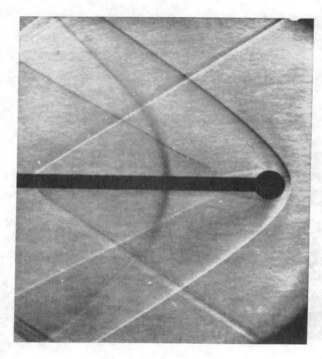

FIGURE 11.28
Three-dimensional curved wave. A $\frac{3}{8}$-in-diameter sphere at a Mach number of 2.0 and a Reynolds number per inch of 3×10^3. (*Courtesy David Taylor Research Facility, Carderock, Md.*)

PART D
OPERATION OF NOZZLES

11.16   A NOTE ON FREE JETS

A free jet will be considered in this text to be a flow of fluid issuing from a duct into a comparatively large region containing fluid which is stationary or which has a velocity relative to the jet that is parallel to the direction of flow in the jet. Before discussing the operation of nozzles and diffusers we must consider certain elementary characteristics of the free jet.

First consider the case of fluid moving out of a nozzle with subsonic flow into the atmosphere. We will show *that the exit pressure must be that of the surrounding atmosphere for such flows.* Let us consider for the moment the alternatives to such a condition. If the pressure of the atmosphere were to be less than that of the jet, there would take place a lateral expansion of the jet. This action would decrease the velocity in the jet according to isentropic-flow theory, and consequently the pressure in the jet would necessarily increase, thus aggravating the situation further. A continuation of this action would then clearly be catastrophic. On the other hand, consider the hypothesis that the pressure of the atmosphere exceeds that of the jet. Then there would have to be a contraction of the jet according to isentropic-flow theory and an increase in

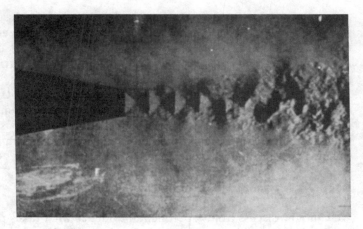

FIGURE 11.29
Convergent nozzle. $M_e = 1.00$, $p_j/p_e = 1.41$. (*Courtesy National Aeronautics and Space Administration, NASA TR R-6.*)

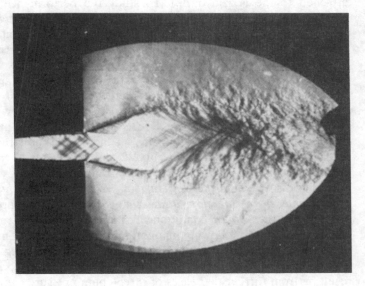

FIGURE 11.30
Two-dimensional convergent-divergent nozzle operating underexpanded at an exit Mach number 2.33. (*Courtesy S. I. Pai, University of Maryland.*)

velocity. This would result in a further decrease in jet pressure, again aggravating the situation. It is obvious that either supposition leads us to expect an *instability* in the jet flow. Since it is known that the free subsonic jet is stable, we may conclude that the jet pressure must equal the surrounding (ambient) pressure.

If the jet emerges supersonically, however, the exit pressure need *not* equal the surrounding pressure. An adjustment may be made to the outside pressure through a succession of oblique expansions and shocks for the two-dimensional case or corresponding conical waves in the symmetric three-dimensional case (see Figs. 11.29 and 11.30).

11.17   OPERATION OF NOZZLES

We can now discuss from experimental evidence what takes place when a nozzle is subject to conditions other than those for which it was designed. Consider first the convergent nozzle arranged as shown in Fig. 11.31 where the design calls for an exit pressure equal to the critical pressure and hence an exit Mach number of unity. Note that a large chamber, called the *plenum chamber*, is attached to the nozzle outlet. To study the performance of this nozzle, the pressure in this chamber, p_B, called the *plenum*, or *back*, *pressure*, will be varied while stipulated stagnation conditions will be maintained at the nozzle entrance. We start with a plenum pressure slightly less than the stagnation pressure. This results in flow which is entirely subsonic, as shown by curve 1 in Fig. 11.32, which is a plot of the ratio of nozzle to stagnation pressure at positions along the flow. The fluid emerges at the ambient pressure $(p_B)_1$ as a free subsonic jet. Nonisentropic effects for this type of flow may be considered very small. As the plenum pressure is lowered, there is an increase in Mach number throughout the flow. Finally, a sonic condition is reached at the nozzle throat. This has been shown as curve 2 and represents operation of the nozzle at design conditions where the plenum pressure now corresponds to the critical pressure. *A further lowering of pressure in the plenum chamber has no effect whatever on the flow inside the nozzle*, and the nozzle is said to be operating in the *choked* condition. A simple physical explanation of this action can be given in the following manner: When sonic conditions are established at the throat, this means that the fluid in this region is moving downstream as fast as a pressure propagation can move upstream. Hence, pressure variations resulting from further decreases of the plenum pressure cannot "communicate"

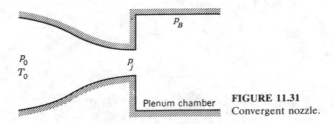

FIGURE 11.31
Convergent nozzle.

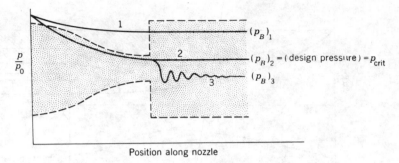

FIGURE 11.32
Operation of convergent nozzle.

up ahead of the throat, which is thus acting as a barrier. Thus, no changes can occur ahead of the throat under these conditions. As the plenum pressure is lowered further, the jet pressure continues to remain at the critical pressure on emergence into the plenum chamber. There is now a pressure difference between the jet and surroundings, a condition possible in a free jet only when the flow has a Mach number equal to or greater than unity. As indicated earlier, an adjustment to ambient pressure then takes place in the jet through a series of oblique expansion waves and oblique shocks. Such a flow is shown in the graph as plot 3, where we employ a wavy line to represent these shock effects outside the nozzle. Further decreases in pressure serve only to amplify the strength of the shock pattern.

Thus, we see from this discussion that the converging nozzle can act as a *limiting valve*, allowing only a certain maximum mass flow for a given set of stagnation conditions.

Let us turn next to the convergent-divergent nozzle (Fig. 11.33), which we will consider in a manner paralleling the previous case. Again, the stagnation conditions will be held fixed, while the plenum pressure is to be varied. A high plenum pressure permits a subsonic flow throughout the nozzle, and the flow emerges as a free jet with a pressure equal to the surrounding pressure. This flow is shown as curve 1 in Fig. 11.34, where the ratio p/p_0 is plotted along the flow. By further decreasing the plenum pressure, a condition is reached with sonic flow at the throat and a return to subsonic flow in the diverging section of the nozzle. This is shown as curve 2, which is the limiting curve for entirely

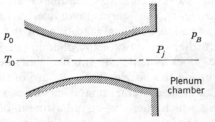

FIGURE 11.33
Convergent-divergent nozzle.

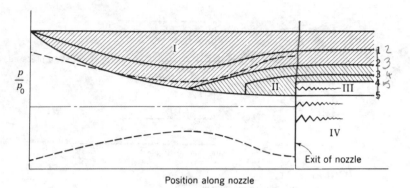

FIGURE 11.34
Operation of convergent-divergent nozzle.

subsonic flow. Note that the region in the plot representing subsonic flow throughout has been identified as region I. A further decrease in plenum pressure does not affect the flow in the converging portion of the nozzle as in the previous case of the convergent nozzle. The mass flow consequently cannot be increased after region I has been passed, and the nozzle is now considered to be operating in the *choked* condition. Beyond the throat, however, there is now an isentropic supersonic expansion which, you will notice for curve 3, is suddenly interrupted by a plane shock, as indicated by a discontinuous pressure rise in the curve. After the shock there is a subsonic expansion out to the plenum pressure. This subsonic portion of the flow may be considered isentropic if no undue boundary-layer growth has taken place as a result of the unfavorable pressure gradient of the shock. The position of the shock may be determined by the following procedure: Starting from known conditions at the throat and at the exit, consider isentropic-flow relations inward from both ends of the divergent section of the nozzle. Somewhere along this portion of the nozzle there will be a position where the subsonic flow computed from the exit conditions and the supersonic flow computed from the throat conditions will have relations corresponding to those across a normal shock as evaluated in Sec. 11.14. It is at this position that the shock may be expected. A computer project has been presented for this in the instructor's manual.

When we lower the back pressure further, the shock will move downstream, becoming stronger as the shock takes place at higher Mach number. Finally, it will appear just at the nozzle exit, as indicated by curve 4 in the diagram. This curve and curve 2 form the limiting expansions where shocks will be found inside the nozzle, and the corresponding enclosed area in our diagram has been designated as region II. Flow conditions in the *entire nozzle* are now fixed with respect to further decrease of the plenum pressure. Such a decrease in p_B now brings the shock outside the nozzle, with the result that we have supersonic flow just outside the nozzle. The pressure of the jet is now less than the ambient pressure, and the aforementioned shock becomes part of a complex oblique pattern during which there is an adjustment of the jet pressure to ambient conditions. As the plenum pressure is decreased further, the shocks diminish in intensity, until a pressure is reached whereby no appreciable shocks appear. This is shown as curve 5 in the plot and corresponds to conditions for which the nozzle was designed. Thus, another region is formed shown as region III, where shock patterns are found outside the nozzle, with a pressure adjustment in the jet taking place from a lower to a higher value. The nozzle is said to be operating *overexpanded* for this region. Decreasing the back pressure below design conditions then results in the necessity of an adjustment from a higher jet pressure to a lower ambient pressure through a series of expansion waves and oblique shocks which increases in intensity with decreasing plenum pressure. Thus, we may denote another region in the plot, region IV, where the nozzle is said to be operating *underexpanded* (see Fig. 11.34).

You may ask what is the advantage of operation at the design condition of the nozzle. In the case of a steam-turbine application, it is clear that the shocks at nondesign operation cause a decrease in the kinetic energy available for developing power. Also, a very interesting example of the deleterious nature of off-design nozzle operation may be shown for the jet-engine nozzle of a high-speed plane, where the purpose of the nozzle in such an application is twofold. Operated in its choked condition, it limits the flow to a value which is properly matched to the requirements of other components of the jet-engine system. The size of the throat section is the controlling factor for this phase of nozzle function. The second purpose is to provide for a flow which yields the greatest thrust consistent with external drag and structural considerations. Considering only internal flow, we can readily demonstrate that operating the nozzle at design conditions gives the best thrust from this single viewpoint. Let us consider first the case of overexpanding operation of the nozzle, for which the pressure variation over a portion of the nozzle has been shown in Fig. 11.35. Note between sections A and B that the pressure inside the nozzle is less than ambient, contributing in effect a negative thrust in the direction of flight. Cutting this section from the nozzle would then increase the thrust to its maximum. However, this would mean that the fluid leaves the nozzle at ambient pressure, so we are at design conditions of the new nozzle geometry. In the case of the underexpanded operation the exit pressure exceeds the ambient pressure, as is shown in Fig. 11.36. Now, if the nozzle were to be extended so that the

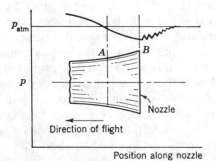

FIGURE 11.35
Nozzle too long.

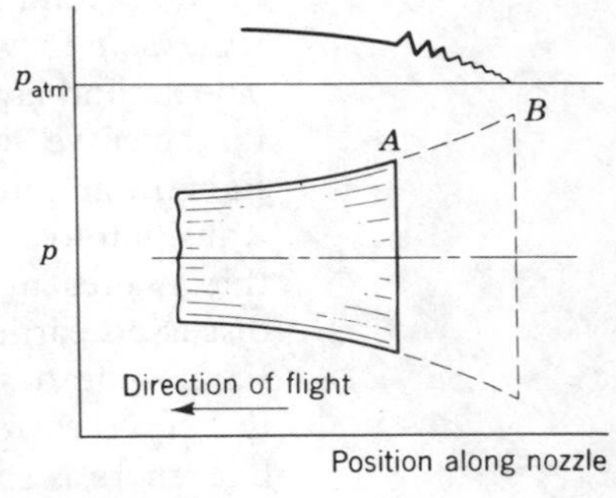

FIGURE 11.36
Nozzle too short.

expansion proceeds to ambient pressure, there would clearly be an added thrust coming from pressure exceeding ambient pressure acting on the nozzle wall. Again this takes us back to design operation.

Since a rocket or jet plane operates over a range of conditions, it may be necessary to use variable geometry to achieve desirable performance. This will usually include the variations of throat and exit areas. The proper control of this geometry in flight, as well as the control and matching of other components of a propulsion system, poses a challenging problem for engineers resulting in the need for on-board sophisticated computers.

*PART E
FLOW THROUGH A CONSTANT-AREA
DUCT WITH FRICTION

11.18 INTRODUCTION

We now examine one-dimensional flow where frictional effects in the boundary layer are significant enough to require consideration. The flow will be considered steady and adiabatic. While the cross section is constant in magnitude, there is no restriction as to its shape.

Presented earlier was a locus of states called the Fanno line for adiabatic, constant-area, steady flow such as is here to be undertaken. Thus, for a given state at a section in the duct and a given flow rate, the fluid must change state continuously along the duct to follow the Fanno line, except in the case of a shock. Hence, by observing a Fanno line (Fig. 11.37) we may learn much concerning the flow, without resorting to computations. Let us start with a *subsonic* flow indicated as state 1 on the Fanno curve corresponding to flow rate G_1. It is clear that any change of state that may occur must be such as to increase the entropy, so you see that the flow must go toward the sonic point at a. Thus, strangely enough, friction results in an increase in Mach number. There cannot be a continuation of change of state beyond the sonic point into the supersonic regime, since this would entail a decrease in entropy and a violation of the second law of thermodynamics. If, on the other hand, the initial flow is *supersonic*, it will mean that the fluid is at some position 2 on the Fanno line. The second law now demands that all changes of state be such that we proceed up the curve again toward a, the sonic point. The flow cannot proceed along the Fanno line into the subsonic flow without a violation of the second law. For a transition to subsonic flow in the duct a normal shock is necessary. Suppose that this is present. Then the state will change, according to the shock relations discussed earlier, into the subsonic regime and then again may move along the Fanno line toward the sonic condition. This possibility has been illustrated in the curve corresponding to G_2 in Fig. 11.37, where the help of the Rayleigh line, there is shown the locus of states traversed by fluid. Note that at all times the entropy is increasing. Now let us consider actions in an actual duct.

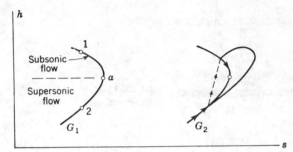

FIGURE 11.37
Fanno and Rayleigh lines.

For a subsonic flow at the duct entrance, there will be an increasing Mach number as the fluid moves downstream. If the flow has not reached sonic conditions at exit, it will proceed from the duct subsonically as a free jet at ambient pressure. Let us suppose now that the duct is long enough to have sonic conditions at the exit. Next assume that the duct is further increased in length. According to conclusions stemming from Fanno considerations the flow cannot simply continue to increase the Mach number into the supersonic regime in the new duct section. Instead, there is a *readjustment* in flow such that there is a decrease in mass flow and the sonic condition once again is established at the exit of the duct. Also, subsonic flow in the duct is *choked* when sonic flow is at the exit. In terms of the *h-s* diagram, the action described above means a shift to another Fanno line corresponding to a different value of G.

If the flow is initially supersonic, there is a decrease in Mach number along the duct and at the limiting condition there will be sonic conditions at the duct exit. The flow is *choked*. Now, further increase in length will be accompanied by a *rearrangement* in flow to maintain the sonic condition at the exit. This is accomplished by a plane shock forming near the exit, after which the subsonic flow then accelerates to sonic conditions at the exit. There will ultimately be a change in flow rate for sufficiently large increases in duct length.

The conclusions reached may be considered valid for all gases. We next turn to considerations of the perfect gas, wherein analytic expressions may be formulated.

11.19 EQUATIONS OF ADIABATIC CONSTANT-AREA FLOW FOR A PERFECT GAS

Let us develop equations which will give the Mach numbers as well as the variations in pressure to be expected along a duct for a given flow of a perfect gas with *constant specific heat*. Consider an infinitesimal control volume for duct flow, as shown in Fig. 11.38. The basic laws and equation of state may then be expressed in the following differential forms:

First law of thermodynamics.

$$c_p \, dT + d\left(\frac{V^2}{2}\right) = 0 \tag{11.73}$$

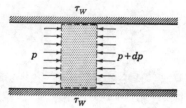

FIGURE 11.38
Infinitesimal control volume.

Continuity.

$$\rho V = \text{const}$$

Taking the logarithmic differential, we get

$$\frac{d\rho}{\rho} + \frac{dV}{V} = 0 \tag{11.74}$$

Linear momentum equation.

$$-A\,dp + dR = (\rho V A)\,dV$$

where dR represents frictional forces on the control surface. Dividing by A, we get

$$-dp + \frac{dR}{A} = \rho V\,dV \tag{11.75}$$

If Θ is the perimeter of the duct cross section, dR may be given as

$$dR = -\Theta\,dz\,\tau_w \tag{11.76}$$

Now consider for a moment Fig. 11.39 showing a chunk of fluid of length L. For steady flow, Newton's law requires that for a circular duct

$$\Delta p\left(\frac{\pi D^2}{4}\right) = \tau_w(\pi D L)$$

Next, we divide and multiply the left side of the equation above by ρ to get

$$\rho\left(\frac{\Delta p}{\rho}\right)\left(\frac{\pi D^2}{4}\right) = \tau_w(\pi D L) \tag{11.77}$$

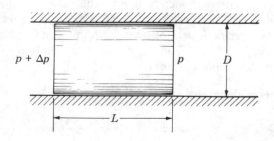

FIGURE 11.39
A chunk of fluid in a circular duct.

Just as in the case of pipe flow, we can consider $\Delta p/\rho$ to be the *head loss* in the duct and expressible, as before, in terms of a friction factor f. That is,

$$\frac{\Delta p}{\rho} = (f)\left(\frac{L}{D}\right)\left(\frac{V^2}{2}\right) \tag{11.78}$$

Substituting the result above in Eq. (11.77), we get on solving for τ_w

$$\tau_w = \left(\frac{f}{4}\right)\left(\frac{\rho V^2}{2}\right) \tag{11.79}$$

This relation is identical to the result for τ_w and f as developed for incompressible viscous pipe flow. We use this formulation for noncircular as well as circular ducts. Experience indicates that the *friction factors f for subsonic flows can be evaluated from the friction charts of Nikuradse and Moody.*[16] For noncircular ducts, we use the *hydraulic diameter* D_H which has been defined in Chap. 9 as $4A/\Theta$. Recall that for a circular cross section the hydraulic diameter D_H becomes identical to the ordinary diameter D. The friction factor f for noncircular ducts is then given in terms of Re with D_H as the length parameter for various values of roughness e/D_H. Replacing dR in Eq. (11.75), using the preceding relations, Eqs. (11.76) and (11.79), and $D_H = 4A/\Theta$, we may form the following relation:

$$-dp = \frac{f}{D_H}\frac{\rho V^2}{2}\,dz + \rho V\,dV \tag{11.80}$$

Equation of state.

$$p = \rho RT$$

Taking the logarithmic differential now gives us

$$\frac{dp}{p} = \frac{d\rho}{\rho} + \frac{dT}{T} \tag{11.81}$$

Equations (11.73), (11.74), (11.80), and (11.81) constitute four independent equations involving T, V, ρ, p, z, and f. They may be combined into one differential equation involving the variables M, f, and z by means of algebraic manipulations. The resulting equation is

$$\frac{1-M^2}{1+[(k-1)/2]M^2}\frac{d\mathrm{M}^2}{\mathrm{M}^4} = \frac{fk}{D_H}\,dz \tag{11.82}$$

---

[16]A dimensional analysis which includes effects of compressibility yields the relation $f = f(\mathrm{Re}, \mathrm{M}, e/D_H)$. The data given by the Moody chart ignores the M dependence. Therefore the data taken from the chart are good only for relatively low Mach number subsonic flow.

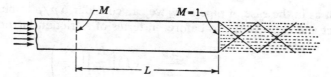

FIGURE 11.40
Choked duct.

This equation may be integrated using an integration formula to be found in math handbooks if one assumes that we can employ average values of f which may be considered constant over the interval for purposes of computations.[17] Let us suppose that a *sonic condition exists at one end of the duct*. That is, the duct is in a *choked condition*. The integration will then be carried out from any position along the duct, where the Mach number is given as M, to the end of the pipe, where M = 1. The results of the integration then are given by the equation

$$\left(\frac{1}{M^2} - 1\right) + \frac{k+1}{2} \ln \frac{(k+1)M^2}{2\{1 + [(k-1)/2]M^2\}} = \frac{fk}{D_H} L \qquad (11.83)$$

where L is the distance from the position corresponding to Mach number M to the end of the duct, as is shown in Fig. 11.40. Thus, by stipulating the position from the end of a duct in a choked condition, the Mach number may reasonably be estimated in terms of an average friction factor at any position L. Values of M (subsonic only) for a range of expression $(fk/D_H)L$ is presented in Table 11.1.

Along with this result, it will be of interest to know the pressure variation along the duct for subsonic flow. It will often be the case that the stagnation temperature (constant for the entire length of duct) and mass flow are known. We may then say at any position along the duct that

$$\frac{p}{p_0} = \frac{\rho T}{\rho_0 T_0} = \frac{\rho}{\rho_0} \frac{1}{1 + [(k-1)/2]M^2} \qquad (11.84)$$

wherein T/T_0 has been replaced by the use of Eq. (11.32). Using the definition of G, we replace ρ by G/V, and replacing V by $M\sqrt{kRT}$, we then get

$$\frac{p}{p_0} = \frac{G}{M\sqrt{kRT}} \frac{1}{\rho_0} \frac{1}{1 + [(k-1)/2]M^2} \qquad (11.85)$$

---

[17]In subsonic flow the Reynolds number will not change appreciably along the duct, so that for such flows this assumption is an entirely reasonable one.

TABLE 11.1
Friction data for Eq. (11.83) with $k = 1.4$

$\cdot\dfrac{fk}{D_H}L$	M	$\dfrac{fk}{D_H}L$	M
0.02	0.901	0.56	0.625
0.04	0.865	0.58	0.621
0.06	0.839	0.60	0.617
0.08	0.818	0.62	0.613
0.10	0.800	0.64	0.609
0.12	0.786	0.66	0.605
0.14	0.772	0.68	0.601
0.16	0.760	0.70	0.598
0.18	0.749	1.00	0.462
0.20	0.739	2.00	0.409
0.22	0.729	3.00	0.373
0.24	0.720	4.00	0.346
0.26	0.712	5.00	0.324
0.28	0.704	6.00	0.307
0.30	0.697	7.00	0.292
0.32	0.690	8.00	0.279
0.34	0.683	9.00	0.268
0.36	0.677	10.00	0.258
0.38	0.671	11.00	0.249
0.40	0.665	12.00	0.241
0.42	0.659	13.00	0.234
0.44	0.654	14.00	0.227
0.46	0.649	15.00	0.221
0.48	0.644	16.00	0.216
0.50	0.639	17.00	0.211
0.52	0.634	18.00	0.206
0.54	0.630	19.00	0.201

Now we replace $T^{1/2}$ by $(T_0/\{1 + [(k - 1)/2]M^2\})^{1/2}$ and ρ_0 by p_0/RT_0. This gives us

$$p = \frac{G}{M}\left(\frac{RT_0}{k\{1 + [(k - 1)/2]M^2\}}\right)^{1/2} \tag{11.86}$$

Having the Mach number at *any position* in *any* constant-area, adiabatic, one-dimensional flow, one could thus determine the pressure at this position in the flow in terms of the constants T_0 and G for the flow.

Note that in ascertaining M and p in a *choked flow* using Eqs. (11.83) and (11.86) we would have to estimate a friction factor for a given roughness ratio e/D_H, as was the case when we worked with incompressible viscous flows through pipes. We can set up an iterative computational process in this analysis similar to what we did in Chap. 9 to achieve greater accuracy in our results.

Thus, after calculating M based on an estimated value of f,

1. We determine the *temperature* at the location in question by employing the computed Mach number, the stagnation temperature T_0 (recall T_0 is constant for adiabatic flow), and the isentropic tables.
2. With this temperature, we look up the viscosity μ in the appendix.
3. Knowing G, we now compute Reynolds number at the location of interest. That is, $\text{Re}_H = GD_H/\mu$.
4. We now look up a new friction factor from the Moody chart.

The new friction factor should be employed in a new set of computations if it is far out of line with the initial estimate. Although it may be argued that the friction factor thus ascertained is a result of considerations only at a specific location, it is nevertheless true that for subsonic flow the Reynolds number is close to being constant for a given flow and hence the friction factor determined at a specific location may be used for the entire length of the duct, with a resulting accuracy consistent with the other steps in the analysis.

Now let us consider a *subsonic flow* which is *not* in a choked condition. For this case, *we know that the exit pressure must be that of the surroundings, and consequently a known quantity*. Equation (11.83), which was developed for the choked case, may still be of use by allowing for the presence of a *hypothetical extension* of the duct to a critical section. This equation will then give the length out to this hypothetical exit if one knows the Mach number at some position in the duct and can make a good estimate of the friction factor. On the other hand, if one knows G and T_0, *the Mach number at the actual exit may then be computed from Eq. (11.86), since p is known* (it is the ambient pressure) and since this equation is in no way restricted to choked flow. The hypothetical extension L_{hyp} (Fig. 11.41) would then be a direct computation using Eq. (11.83). This brings us back to the choked-flow problem that we have discussed and the procedures that were developed apply again.

As an aid in computations, we have given the tabulation of the Fanno line in Appendix B.7 where for $k = 1.4$, we have the ratios T/T^*, p/p^*, p_0/p_0^*, and V/V^* in terms of M. The quantities with asterisks correspond as usual to sonic conditions.

In the case of supersonic flow, we do not have comparable data for friction factors as we have for subsonic flow. It is to be pointed out that large stagnation pressure losses will be encountered in supersonic flow through ducts, and consequently long runs of supersonic flow are to be avoided. It is better to diffuse the flow to subsonic conditions during transport and when supersonic flow is required to expand through a convergent-divergent nozzle.

Example 11.5. A duct having a square cross section 0.300 m on a side has 25 kg of air per second flowing in it. The air, originally in a chamber where the temperature is 90°C, has been insulated by the duct walls against heat transfer to the outside.

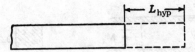

FIGURE 11.41
Hypothetical extension of duct to M = 1.

The duct is operating in a choked condition. If the duct has a relative roughness of 0.002, determine the Mach number at a position 6 m from the exit of the duct.

We may solve for M at this position by employing Eq. (11.83). To do this, we estimate f to be 0.024, from the Moody diagram. The hydraulic diameter D_H is

$$D_H = \frac{4A}{\Theta} = \frac{(4)(0.3^2)}{(4)(0.3)} = 0.3 \text{ m}$$

Using $k = 1.4$, we then have

$$\left(\frac{1}{M^2} - 1\right) + 1.2 \ln \frac{1.2M^2}{1 + 0.2M^2} = \frac{(0.024)(1.4)}{0.3}(6) = 0.672$$

From Table 11.1 or solving by trial and error, we get M = 0.60.

To check our friction factor, we must compute other conditions at this section of the pipe. The temperature T is determined from isentropic Appendix Table B-5 for M = 0.60. Thus

$$\frac{T}{T_0} = 0.933$$

Therefore

$$T = 339 \text{ K} = 65.7°C$$

We may determine the viscosity of air at that temperature to be 2.15×10^{-5} N · s/m^2 by making use of Appendix Fig. B.1, since viscosity does not depend greatly on the pressure. Noting that

$$G = \rho V = \frac{w}{A} = \frac{25}{0.300^2} = 277.8 \text{ kg/(m}^2)(s)$$

we get for Re$_H$

$$\text{Re}_H = \frac{GD_H}{\mu} = \frac{(277.8)(0.300)}{2.15 \times 10^{-5}}$$

$$= 3.88 \times 10^6$$

Returning to the Moody diagram, we see that our choice of f is close enough not to require further computation and the desired Mach number is 0.60.

Example 11.6. If, in Example 11.5, the flow leaves the duct subsonically into an ambient pressure of 101.4 kPa, what are the flow condition 6 m from the end of the duct?

We first compute the exit Mach number of the flow, using Eq. (11.86), since we know the exit pressure of the jet. Thus

$$101,400 = \frac{25/0.3^2}{M}\left[\frac{(278)(363)}{(1.4)(1 + 0.2M^2)}\right]^{1/2} \qquad (a)$$

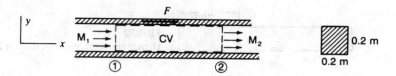

FIGURE 11.42
Adiabatic flow over air through a constitute area duct.

which becomes

$$1.336 = \frac{1}{M}\left(\frac{1}{1 + 0.2M^2}\right)^{1/2} \tag{b}$$

Solving for M, we get M = 0.713. Now using the value M = 0.713 at the actual exit and knowing T_0, we can find T at the actual exit from isentropic tables. With known T, we can find the viscosity μ from the charts in the appendix. Next, using the known value G, we get the Reynolds number GD_H/μ. Finally, we can look up f from the Moody diagram. We assume as in previous discussions that this f is to be taken as the average value over the hypothetical extension of the duct. We now use Eq. (11.83) to first find L_{hyp} and then M at any section of the *actual* duct. In doing this, we must remember to attach the hypothetical duct length onto the actual duct to ensure M = 1 at the hypothetical exit as is required by Eq. (11.83). This means that we measure L from the choked end of the hypothetical duct. We leave the actual computation to the reader. The computation follows that of Example 11.5 from this point onward.

In the final example for this section we will illustrate the use of the Fanno line table in Appendix B.3.

Example 11.7. Air is flowing in a highly insulated constant-area duct (see Fig. 11.42) where at section ① the following data apply:

$$M_1 = 0.3$$
$$T_1 = 60°C$$
$$p_1 = 1.8 \times 10^5 \text{ Pa abs.}$$

If at section ② the Mach number is $M_2 = 0.7$,
(*a*) What is the mass flow rate?
(*b*) What is the drag on the duct between sections ① and ②?
   For part (*a*) we will use the formula

$$w = \rho_1 V_1 A_1$$

and for part (*b*) we will use linear momentum and continuity equations for the control volume in Fig. 11.42. Thus

$$(p_1 - p_2)A + F = \rho_1 V_1 A_1(V_2 - V_1)$$

We shall now proceed to find the quantities needed in the above equations for the mass flow w and the drag $D = -F$.

We start by going to the *isentropic tables*. For $M_1 = 0.3$, we get from these tables

$$\frac{T_1}{(T_0)_1} = 0.982$$

$$\therefore (T_0)_1 = \frac{(273 + 60)}{0.982} = 339.1 \text{ K}$$

This stagnation temperature is constant in the flow. Thus

$$(T_0)_1 = (T_0)_2 = 339.1 \text{ K}$$

For $M_2 = 0.7$,

$$\frac{T_2}{(T_0)_2} = 0.911$$

$$\therefore T_2 = (339.1)(0.911) = 308.9 \text{ K}$$

Before going to the Fanno line tables we shall use the *equation of state* to get ρ_1. Thus

$$\rho_1 = \frac{p_1}{RT_1} = \frac{1.8 \times 10^5}{(287)(333)} = 1.883 \text{ kg/m}^3$$

Also the velocity of sound c_1 is

$$c_1 = \sqrt{kRT_1} = \sqrt{(1.4)(287)(333)} = 365.8 \text{ m/s}$$

Finally, for the velocity V_1 we have

$$M_1 = \frac{V_1}{c_1}$$

$$\therefore V_1 = (0.3)(365.8) = 109.7 \text{ m/s}$$

We are now ready for the *Fanno* data.

For $M = 0.3$.

$$\frac{T_1}{T^*} = 1.179$$

$$\therefore T^* = \frac{333}{1.179} = 282.4 \text{ K}$$

$$\frac{p_1}{p^*} = 3.62$$

$$\therefore p^* = \frac{1.8 \times 10^5}{3.62} = 4.972 \times 10^4 \text{ Pa abs.}$$

$$\frac{V_1}{V^*} = 0.326$$

$$\therefore V^* = \frac{109.7}{0.326} = 336.5 \text{ m/s}$$

For $M = 0.7$. We may use the critical values found above. Thus

$$\frac{T_2}{T^*} = 1.093$$

$$\therefore T_2 = (282.4)(1.093) = 308.7 \text{ K}$$

$$\frac{p_2}{p^*} = 1.49$$

$$\therefore p_2 = (4.972 \times 10^4)(1.49) = 7.408 \times 10^4 \text{ Pa abs.}$$

$$\frac{V_2}{V^*} = 0.732$$

$$\therefore V_2 = (336.5)(0.732) = 246.3 \text{ m/s}$$

We can now find desired results.

(a) $$w = \rho_1 V_1 A_1 = (1.883)(109.7)(0.04) = 8.263 \text{ kg/s}$$

(b) $$(p_1 - p_2)A + F = \rho_1 V_1 A_1 (V_2 - V_1)$$

$$(1.8 \times 10^5 - 7.408 \times 10^4)(0.04) + F = 8.263(246.3 - 109.7)$$

$$\therefore F = -3108 \text{ N}$$

Hence

$$\text{Drag} = 3108 \text{ N}$$

*PART F
STEADY FLOW THROUGH A CONSTANT-AREA
DUCT WITH HEAT TRANSFER

11.20 INTRODUCTION

We will now study the case of steady flow through a constant-area duct, where we consider heat transfer with the surroundings, but where we neglect friction.

Remember that the Rayleigh line includes all possible states for a fluid having given flow conditions at a section wherein the changes of state are developed by heat transfer. As in the previous analysis involving the Fanno line much can be learned by considering a typical Rayleigh line (Fig. 11.43), where it will be remembered that point b, the point of maximum entropy, corresponds to sonic conditions. In connection with the Rayleigh line, let us consider for simplicity that the heating is reversible.[18] From the *second law of thermodynamics* considerations we know that

$$dS = \frac{dQ}{T} \tag{11.87}$$

---

[18]Reversible heating means that only an infinitesimal temperature difference between reservoir and the fluid is permitted.

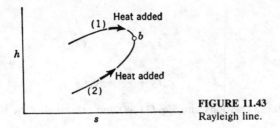

FIGURE 11.43
Rayleigh line.

and consequently adding heat means an increase in entropy, while extracting heat means a decrease in entropy. If the flow is subsonic at a section, it will have a state represented at some position 1 on the Rayleigh diagram and adding heat then means that we move toward point b, since the entropy must increase. However, we clearly cannot go beyond this point into the supersonic regime by the process of adding heat. In an actual flow in a duct, this indicates that sonic conditions have been reached at the end of the duct, and the flow is said to be *choked*, since further addition of heat will result in a reduction of flow. Such a change means that the flow conditions shift to another Rayleigh line, with sonic conditions still present at the exit. Next, starting in the supersonic regime, the opposite effect takes place when heat is added; that is, a steady decrease in Mach number is achieved. The limiting condition, for which there is no major change in the flow pattern, occurs when Mach number unity is reached at the duct exit. Thus the flow may be choked in the supersonic regime.

11.21 RELATIONS FOR A PERFECT GAS

Let us now consider the case of a perfect gas, with a constant value of k, undergoing steady flow through a constant-area duct as shown in Fig. 11.44. Considering the finite control volume shown in the diagram, we can form useful algebraic equations from the basic laws so as to relate the pertinent variables. Thus:

First law of thermodynamics.

$$c_p T_1 + \frac{V_1^2}{2} + \frac{dQ}{dm} = c_p T_2 + \frac{V_2^2}{2} \tag{11.88}$$

Let us now imagine that the flow is slowed isentropically to stagnation conditions at sections 1 and 2. The first law of thermodynamics for these actions then gives us the

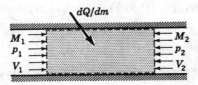

FIGURE 11.44
Finite control volume.

relations

$$c_p T_1 + \frac{V_1^2}{2} = c_p (T_0)_1 \qquad (11.89a)$$

$$c_p T_2 + \frac{V_2^2}{2} = c_p (T_0)_2 \qquad (11.89b)$$

Substituting these relations into Eq. (11.88) and rearranging the terms, we then get

$$\frac{dQ}{dm} = c_p (T_0)_2 - c_p (T_0)_1 = c_p [(T_0)_2 - (T_0)_1] \qquad (11.90)$$

Thus we see that *the change in the local stagnation temperature reflects, in this type of flow, the amount of heat added per unit mass of the flow.*

Second law of thermodynamics (reversible heating).

$$\begin{aligned} \Delta S > 0 \text{ heat added} \\ \Delta S < 0 \text{ heat extracted} \end{aligned} \qquad (11.91)$$

Linear momentum equation.

$$p_1 - p_2 = G(V_2 - V_1) = \rho_2 V_2^2 - \rho_1 V_1^2 \qquad (11.92)$$

Continuity equation.

$$\rho_1 V_1 = \rho_2 V_2 \qquad (11.93)$$

Let us now determine the ratios of flow parameters at any two sections of the flow in terms of the Mach numbers at these sections. This may readily be done for the pressure ratio by considering Eq. (11.92). We arrange this equation as follows:

$$p_1 + \rho_1 V_1^2 = p_2 + \rho_2 V_2^2$$

Now replace the ρ's using the equation of state so that $\rho = p/RT$. Next replace the V's so that $V^2 = M^2(kRT)$. We get

$$p_1 + \frac{p_1}{RT_1} M^2 (kRT_1) = p_2 + \frac{p_2}{RT_2} M_2^2 (kRT_2)$$

Canceling terms, we have

$$p_1(1 + kM_1^2) = p_2(1 + kM_2^2)$$

We may now form the desired ratio of pressures.

$$\frac{p_1}{p_2} = \frac{1 + kM_2^2}{1 + kM_1^2} \qquad (11.94)$$

The temperature ratio may next be formed by replacing p in the equation above by ρRT. Using the continuity equation [Eq. (11.93)] to eliminate the density ratio gives us

$$\frac{T_1}{T_2} \frac{V_2}{V_1} = \frac{1 + kM_2^2}{1 + kM_1^2} \qquad (11.95)$$

Now replacing V by $M\sqrt{kRT}$ once again permits the elimination of the velocity ratio in the equation above and the solution of the temperature ratio in the form

$$\frac{T_1}{T_2} = \left(\frac{M_1}{M_2} \frac{1 + kM_2^2}{1 + kM_1^2} \right)^2 \tag{11.96}$$

To solve for V_1/V_2, replace the temperatures in Eq. (11.95) by using the Eq. (11.96). The following equation may then be written:

$$\frac{V_1}{V_2} = \left(\frac{M_1}{M_2} \right)^2 \frac{1 + kM_2^2}{1 + kM_1^2}$$

Finally, the ratio of the stagnation pressures is ascertained from Eqs. (11.34) and (11.94). Thus

$$\frac{(p_0)_1}{(p_0)_2} = \frac{p_1}{p_2} \left\{ \frac{1 + [(k-1)/2]M_1^2}{1 + [(k-1)/2]M_2^2} \right\}^{k/(k-1)}$$

$$= \frac{1 + kM_2^2}{1 + kM_1^2} \left\{ \frac{1 + [(k-1)/2]M_1^2}{1 + [(k-1)/2]M_2^2} \right\}^{k/(k-1)} \tag{11.97}$$

To ease the work of computation, it is convenient to determine the ratio of flow parameters between any section of the duct and the critical section. This may be done by setting $M_2 = 1$ at section 2 in the preceding equations and deleting the subscript 1. Thus

$$\frac{p}{p^*} = \frac{1 + k}{1 + kM^2} \tag{11.98a}$$

$$\frac{T}{T^*} = \left[\frac{M(1 + k)}{1 + kM^2} \right]^2 \tag{11.98b}$$

$$\frac{V}{V^*} = \frac{M^2(1 + k)}{1 + kM^2} \tag{11.98c}$$

$$\frac{p_0}{p_0^*} = \frac{1 + k}{1 + kM^2} \left(\frac{2\{1 + [(k-1)/2]M^2\}}{1 + k} \right)^{k/(k-1)} \tag{11.98d}$$

As a final step, use Eq. (11.32) for T and T^*. That is,

$$T = \frac{T_0}{1 + [(k-1)/2]M^2}$$

$$T^* = \frac{T_0^*}{1 + (k-1)/2}$$

Taking the ratio of left and right sides of the equations above, we get

$$\frac{T}{T^*} = \frac{T_0}{T_0^*} \frac{1 + (k-1)/2}{1 + [(k-1)/2]M^2}$$

Now we replace T/T^* using Eq. [11.98b)] and solving for T_0/T_0^*, we get the following equation, which completes our list:

$$\frac{T_0}{T_0^*} = \frac{2(k+1)M^2\{1 + [(k-1)/2]M^2\}}{(1+kM^2)^2} \tag{11.99}$$

These ratios have been tabulated as the Rayleigh table in Appendix Table B.8, for $k = 1.4$.

It should be understood that Appendix Table B.8 may be used for flows which are not choked. The critical conditions p^*, T^*, etc., then apply to some hypothetical extension of the duct to Mach 1 in a manner similar to the hypothetical extension to the critical section discussed in the adiabatic-flow cases of Part E.

An interesting aspect of frictionless constant-area flow with heat transfer has to do with the temperature variation along the flow. We start by taking the differential of Eq. (11.98b). Thus for a given T^* we have

$$dT = T^*2\left[\frac{M(1+k)}{1+kM^2}\right]\left\{\frac{1+k}{1+kM^2} - \frac{M(1+k)(2kM)}{(1+kM^2)^2}\right\}dM \tag{11.100}$$

For subsonic flow with heat added, clearly dM is positive. If we examine the second bracketed expression, we see that it becomes zero when

$$M = \frac{1}{\sqrt{k}} \tag{11.101}$$

Furthermore, if $M < 1/\sqrt{k}$ this bracketed expression is positive whereas when $M > 1/\sqrt{k}$ it is negative. We can conclude that as heat is added and the Mach number increases from a value less than $1/\sqrt{k}$, the temperature T increases until it reaches a maximum at $M = 1/\sqrt{k}$ after which it decreases as M exceeds $1/\sqrt{k}$. Thus we reach the peculiar result where for part of the Rayleigh line *heat addition* will *cool* the gas. This is consistent with the fact (see Fig. 11.43) that on part of the subsonic portion of the Rayleigh line *adding heat decreases* the *enthalpy h* signifying for a perfect gas a decrease in temperature. Conversely in this region *removing heat* and thus moving away from the sonic point *b* means an *increase* in temperature.

Example 11.8. Air is moving as a steady flow through a duct (see Fig. 11.45) having a constant rectangular cross section measuring 0.600 m by 0.300 m. At a position 6 m from the end, the pressure is 12 kPa gage, and the temperature is 260°C. The

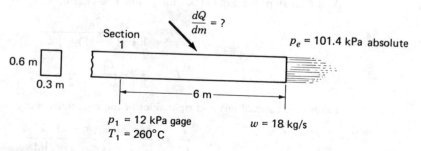

FIGURE 11.45
Constant-area duct with subsonic flow and heat transfer.

fluid leaves the duct subsonically at a pressure of 101.4 kPa. If there is 18 kg of fluid flow per second, what is the heat transfer per kilogram of fluid between the aforementioned section and the exit? Assume a constant specific heat c_p of 1.0865 kJ/(kg)(K) and neglect friction.

What we must do is to determine T_1, M_1 at section 1 and T_e, M_e at the exit. Then we can get the stagnation temperatures and thus dQ/dm from Eq. (11.90).

Our first step will be to ascertain the Mach number at section 1 inside the duct where we have known data. We need c_1 and V_1. Thus

$$c_1 = \sqrt{kRT_1} = \sqrt{(1.4)(287)(260 + 273)} = 463 \text{ m/s}$$

Using the *continuity* equation, we have

$$\rho_1 V_1 A_1 = 18 \text{ kg/s}$$

Therefore,

$$V_1 = \frac{18}{(\rho_1)(0.300)(0.600)} \text{ m/s}$$

Since $\rho_1 = p_1/RT_1$, we have for V_1

$$V_1 = \frac{(287)(260 + 273)(18)}{(113,400)(0.300)(0.600)} = 134.9 \text{ m/s}$$

Thus,

$$M_1 = \frac{V_1}{c_1} = \frac{134.9}{463} = 0.921$$

We can now make good use of the *Rayleigh* line data, Appendix Table B.8. Thus we have for the above Mach number

$$\frac{p_1}{p^*} = 2.14 \qquad \frac{T_1}{T^*} = 0.3901$$

Therefore,

$$p^* = 52.9 \text{ kPa} \qquad T^* = 1366 \text{ K} \tag{a}$$

We now know the pressure and temperature that would be reached by extending the duct hypothetically to sonic conditions, as shown in Fig. 11.46. Knowing p_e and p^*, we can get M_e, the Mach number, at the actual exit from the Rayleigh line

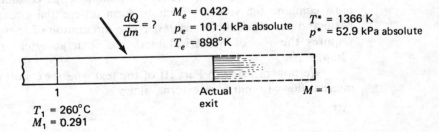

FIGURE 11.46
Duct showing hypothetical extension out to sonic conditions.

tables. Thus, for $p_e/p^* = 101.4/52.9 = 1.917$, we get $M_e = 0.422$. Also, we note that $T_e/T^* = 0.657$ from the same table for the actual exit. Using T^* from Eq. (a), we have for T_e

$$T_e = (0.657)(1366) = 898 \text{ K}$$

We thus have the Mach number and temperature at both section 1 and the actual exit, so from the *isentropic* tables, we can get the desired stagnation temperatures at these locations. Thus we have

$$(T_0)_1 = \frac{260 + 273}{0.983} = 542 \text{ K}$$

$$(T_0)_e = \frac{898}{0.966} = 929.6 \text{ K}$$

(b)

We then have for our heat transfer dQ/dm:

$$\frac{dQ}{dm} = c_p[(T_0)_e - (T_0)_1]$$

$$= (1.0865 \times 10^3)(1414 - 542)$$

$$= 948 \text{ kJ/kg}$$

(c)

11.22 CLOSURE

In this chapter we restricted our discussion, on the whole, to one-dimensional steady flow under conditions of simple area change, friction, and simple heating, each of which we considered separately. It was feasible to proceed in this way since in many practical problems one of these effects will dominate all others, and our formulations are then of great value. Sometimes we can modify results based primarily on one effect in a simple way to take into account a secondary but nonnegligible effect. The means we presented in Sec. 11.10 of modifying the isentropic results in a nozzle to account for friction is an example of such a procedure. However, there will be times when several effects of comparable significance will be present, requiring simultaneous consideration.[19] We then must combine the several effects to form differential equations satisfying the basic and pertinent subsidiary laws. The integration of these equations usually requires the use of numerical methods. Such analyses are to be found in Chap. 16.

We now proceed to Part III of the text where we shift our attention from internal flows primarily to external flows.

---

[19]This will be especially true when we have heat transfer, since friction, having a similar mechanism as heat transfer, will often be of equal significance.

PROBLEMS

Problem Categories

Perfect Gas Problems 11.1–11.8
Mach Cone and Velocity of Sound Problems 11.9–11.13
Simple Area Change and Stagnation Pressures
    11.14–11.41
Fanno and Rayleigh Lines 11.42–11.43
Normal Shocks and Nozzle Operation 11.44–11.66
Oblique Shocks 11.67–11.71
Adiabatic Constant Area Flows 11.72–11.84
Constant Area Ducts with Heat Flow 11.85–11.92

Starred Problems

11.66, 11.71, 11.80, 11.81

Derivations or Verifications

11.42, 11.43, 11.51, 11.52, 11.57

PROBLEMS

11.1. The internal energy of a hypothetical perfect gas is given as

$$u = \tfrac{1}{50}T^{1/2} + 100$$

Determine c_v and c_p. Take $R = 50$ ft · lb/(lbm)(°F).

11.2. Show the validity of Eqs. (11.6a) and (11.6b).

11.3. Show the validity of Eqs. (11.9a) and (11.9b) using Eq. (11.8) and the equation of state.

11.4. Air at 15°C and 101,325 Pa is compressed to a pressure of 345,000 Pa absolute. If the compression is adiabatic and reversible, what is the final specific volume? How much work is done per kilogram of the gas?

11.5. Do Prob. 11.4 for an isothermal compression.

11.6. An airplane is capable of attaining a flight Mach number of 0.8. When it is flying at an altitude of 1000 ft in standard atmosphere, what is the ground speed if the air is not moving relative to the ground? What is the ground speed if the plane is at an altitude of 35,000 ft in standard atmosphere?

11.7. Do the first part of Prob. 11.6 if the air is moving at 60 mi/h directly opposite to the direction of flight.

11.8. What is the value of k for standard atmosphere at an altitude of 30,000 ft?

11.9. Suppose that a plane is moving horizontally relative to the ground at a speed of twice the velocity of sound and that the air is moving in the opposite direction at a speed of one-half the velocity of sound relative to the ground. What is the Mach angle?

11.10. Suppose that a cruise missile under test is moving horizontally at $M = 2$ in the atmosphere at an elevation of 305 m above the earth's surface. How long does it take for an observer on the ground to hear the disturbance from the instant when it is directly overhead? Assume standard atmosphere.

11.11. Suppose in Prob. 11.10 that an observer in a plane is moving in the same direction as the missile at a speed of one-half the speed of sound at an elevation of 305 m above the missile. What is the time elapsed between the instant when the missile is directly below and the instant when the observer hears the sound? Neglect changes of c from 305- to 610-m elevation.

11.12. What is the velocity of sound for each of the gases in Table B.3 at 60°F in feet per second?

11.13. What is the velocity of sound for each of the gases in Table B.3 at 15°C in meters per second?

11.14. The inlet velocity of an isentropic diffuser is 305 m/s and the undisturbed pressure and temperature are 34,500 Pa absolute and 235°C, respectively. If the pressure is increased by 30 percent at the exit of the diffuser, determine the exit velocity and temperature. Use tables.

11.15. A rocket has an area ratio A_{exit}/A^* of 3.5 for the nozzle, and the stagnation pressure is 50×10^5 Pa absolute. Fuel burns at the rate of 45 kg/s and the stagnation temperature is 2870°C. What should the throat area and exit area be? Take $R = 355$ N · m/(kg)(K) and $k = 1.4$.

11.16. A missile is moving at Mach number 3 at an altitude of 11,280 m in standard atmosphere. What temperature is the nose of the rocket exposed to if one assumes no detached shocks?

11.17. Determine the throat and the exit areas of an ideal rocket motor to give a static thrust of 6670 N at 6100-m altitude standard atmosphere if the chamber pressure is 1.035×10^6 Pa absolute and chamber temperature is 3315°C. Find the velocity at the throat. Take $k = 1.4$ and $R = 355$ N · m/(kg)(K). Assume that exit pressure is that of surroundings.

11.18. Determine the throat and exit area of an *ideal* rocket motor for flight conditions at 30,000 ft. The chamber pressure is 160 lb/in² absolute and the temperature is 6000°F. If the mass flow in the engine is 50 lbm/s, what is the thrust of the rocket? Take $k = 1.4$ and $R = 66$ ft · lb/(lbm)(°R).

11.19. What static thrust can be expected ideally from a rocket motor burning 25 kg of fuel per second in a test on the ground? The chamber pressure is 1.379×10^6 Pa absolute, and the chamber temperature is 2760°C. Take $k = 1.4$ and $R = 355$ N · m/(kg)(K) for the products of combustion. What are the exit velocity and the exit temperature of the gases, and what should the exit area be? The exit pressure of the jet is atmospheric pressure?

11.20. An airplane is diving at a speed of 225 m/s and is at an elevation of 3050 m in standard atmosphere. The air-speed indicator which converts the dynamic pressure to a velocity is calibrated for incompressible flow corresponding to conditions at that altitude. What is the percentage of error in the reading given by the indicator?

11.21. Air is kept in a tank at a pressure of 6.89×10^5 Pa absolute and a temperature of 15°C. If one allows the air to issue out in a one-dimensional isentropic flow, what is the greatest possible flow per unit area? What is the flow per unit area at the exit of the nozzle where $p = 101,325$ Pa?

11.22. A nozzle expands air from a pressure $p_0 = 200$ lb/in² absolute and temperature $T_0 = 100°F$ to a pressure of 20 lb/in² absolute. If the mass flow w is 50 lbm/s, what is the throat area and the exit area? Take $k = 1.4$ and $R = 53.3$ ft · lb/(lbm)(°R).

11.23. Determine the exit area in Prob. 11.17 for a nozzle efficiency of 85 percent. Take $c_p = 1214$ N · m/(kg)(K).

11.24. A supersonic diffuser, operating ideally, diffuses air from a Mach number of 3 to a Mach number of 1.5. The pressure of the incoming air is 10 lb/in² absolute and the temperature is 20°F. If 280 lbm/s of air flows through the diffuser, determine the inlet and exit areas for the diffuser and the static pressure for the outlet air.

11.25. A De Laval nozzle is connected to a tank containing air at a pressure of 550,000 Pa absolute at a temperature of 25°C. The throat area of the nozzle is 0.0015 m² and the exit area is 0.0021 m². What should the *ambient pressure* be if the nozzle is to operate at *design conditions*? What is the *mass flow* if friction is completely neglected? What is the *critical pressure*?

11.26. Using *ideal* theory, what is the thrust of the rocket engine in Fig. P11.26. What is the exit Mach number? Use $R = 287$ N-m/(kg)(K).

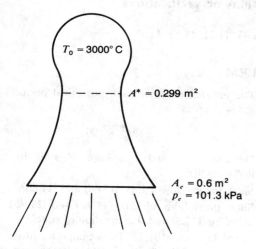

$T_0 = 3000°$ C

$A^* = 0.299$ m²

$A_c = 0.6$ m²
$p_c = 101.3$ kPa

FIGURE P11.26

11.27. A convergent nozzle is operating in the choked condition. The exit area is 0.05 ft². The reservoir temperature is 150°F. The reservoir pressure is varied slowly so that

$$p_0 = 101,325 + (1000)t^{1/2} \text{ Pa}$$

with t in hours. What is the mass flow of air as a function of time? What is the mass flow at $t = 6$ h?

11.28. Air is drawn from a tank of air having a temperature of 60°C and a pressure of 101.3 kPa absolute through a convergent-divergent nozzle. At one point in the nozzle the static pressure is 40.0 kPa absolute and the cross-sectional area is 0.02 m². What is the mass flow through the nozzle?

11.29. A blow-down supersonic wind tunnel has a reservoir tank at the entrance having air at $T = 60°F$ and $p = 60$ lb/in² absolute. At the exit the plenum pressure is 5 lb/in² absolute. If in the region between entrance and exit there is a nozzle and the test section is just before the exit and has an area of 0.5 m², what must be the area of the throat for ideal isentropic flow? What is the Mach number at the test section?

11.30. A convergent-divergent nozzle is attached to a reservoir where the temperature is 70°C and the pressure is 5×10^5 Pa absolute. If the exhaust pressure is at ambient pressure of 3×10^5 Pa absolute, what is the exhaust temperature?

11.31. From Eq. (11.62) show that as $M_1 \to \infty$ we get a limiting minimum Mach number M_2. What is this Mach number for air?

11.32. Water is flowing in a channel at a uniform speed of 6 m/s. At a depth 3 m from the free surface determine (a) stagnation pressure, (b) undisturbed pressure, (c) dynamic pressure, and (d) geometric pressure. Take the flow as incompressible and frictionless.

11.33. Determine the exit area and the exit velocity for isentropic flow of a perfect gas with $k = 1.4$ in the nozzle shown. How small can we make the exit area and still have isentropic flow with the given conditions entering the nozzle? Use tables.

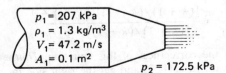

$p_1 = 207$ kPa
$\rho_1 = 1.3$ kg/m³
$V_1 = 47.2$ m/s
$A_1 = 0.1$ m²
$p_2 = 172.5$ kPa

FIGURE P11.33

11.34. Sketch a passage which will (a) increase the pressure in a subsonic flow isentropically, (b) increase the pressure in a supersonic flow isentropically, (c) increase the Mach number in a supersonic flow isentropically, (d) decrease the Mach number in a subsonic flow isentropically.

11.35. Air is moving at high subsonic speed in a duct (Fig. 11.35). If h is 200 mm of mercury, what is the speed of the flow at point A if there is no pitot tube present? The static pressure is 800 mm of mercury. The temperature is 80°C in the undistributed flow.

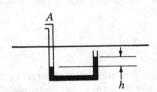

FIGURE P11.35

11.36. Determine the exit area in Prob. 11.33 for a nozzle efficiency of 90 percent. Take $c_p = 0.24$ Btu/(lbm)(°F), $k = 1.4$, and $R = 287$ N · m/(kg)(K). Give result in square meters.

11.37. For a plane moving at M = 2 at 30,000 ft in the standard atmosphere, what is the dynamic pressure its pitot tube will record? What is the stagnation temperature?

11.38. In an isentropic flow of air the temperature of some point in the flow is 10°C and the velocity is 130 m/s. Determine the velocity of sound at stagnation conditions and at M = 1 for the flow.

11.39. Air is to expand isentropically in a nozzle from a stagnation temperature of 30°C. What is the Mach number when the speed of flow reaches 200 m/s? What is the velocity at the throat?

11.40. The pitot tube on a plane gives the following data:

$$p_0 = 5 \text{ lb/in}^2 \text{ gage}$$

$$p = 13.8 \text{ lb/in}^2 \text{ absolute}$$

The ambient temperature is 35°F. What are the Mach number and speed of the plane?

11.41. A convergent nozzle is to discharge air at sonic conditions at a pressure of the atmosphere

101,325 Pa. What is the reservoir pressure required? If a nozzle is to discharge at sonic conditions at a pressure of 150,000 Pa, what is the reservoir pressure?

11.42. Prove that the point of maximum entropy on the Fanno line corresponds to sonic conditions. *Hint:*

(a) Express the first law of thermodynamics in differential form at point a, using enthalpy (steady-flow equation).

(b) Express the continuity equation in differential form.

(c) Now express the first law of thermodynamics for a system in differential form, and reach the relation

$$dh = \frac{dp}{\rho}$$

(d) Noting that $ds = 0$ at the point of interest, indicating that the flow there is isentropic, we can reach the following result by combining the preceding equations:

$$V = \sqrt{\left(\frac{\partial p}{\partial \rho}\right)_s}$$

The flow is thus sonic at the point of maximum entropy.

11.43. Prove that the point of maximum entropy on the Rayleigh line corresponds to sonic conditions. Study the hint in Prob. 11.42.

11.44. Explain why, despite the use of an infinitesimal control volume in Fig. 11.19, we get algebraic relations rather than differential relations for the basic laws as applied to the normal shock.

11.45. Air is moving at Mach number 3 in a duct and undergoes a normal shock. If the undisturbed pressure ahead of the shock is 69,000 Pa absolute, what is the increase in pressure after the shock? What is the loss in stagnation pressure across the shock?

11.46. An airplane having a diffuser designed for subsonic flight has a normal shock attached to the edge of the diffuser when the plane is flying at a certain Mach number. If at the exit of the diffuser the Mach number is 0.3, what must the flight Mach number be for the plane, assuming isentropic diffusion behind the shock? The inlet area is 0.25 m² and the exit area is 0.4 m².

11.47. A normal shock forms ahead of the diffuser of a turbojet plane flying at Mach number 1.2. If the plane is flying at 35,000-ft altitude in standard atmosphere, what is the entering Mach number for the diffuser and the stagnation pressure?

11.48. Consider a supersonic flow through a stationary duct wherein a stationary shock is present. The Mach number ahead of the shock is 2 and the pressure and temperature are 103,500 Pa absolute and 40°C, respectively. What is the velocity of propagation of the shock relative to the fluid ahead of the shock? The fluid is air.

11.49. A jet plane is diving at supersonic speed at close to constant speed. There is a curved shock wave ahead of it. A static pressure gage near the nose of the plane measures 30.5 kPa absolute. The ambient pressure and temperature of the atmosphere are 10 kPa and 254K, respectively. What are the flight Mach number for the plane and its speed if one assumes that in front of the static pressure gage the shock wave is plane?

11.50. Which of the following quantities depend on the spatial reference of observation, and which do not?

(a) Pressure (undistributed)
(b) Temperature (undistributed)
(c) Stagnation pressure
(d) Enthalpy
(e) Stored energy
(f) Entropy
(g) Work

Explain your decisions.

11.51. An important equation which relates the pressure ratio across a shock with the density ratio across the shock for a perfect gas with constant specific heat is called the *Rankine-Hugoniot relation* and is usually given as

$$\frac{p_2}{p_1} = \frac{[(k+1)/(k-1)](\rho_2/\rho_1) - 1}{[(k+1)/(k-1)] - \rho_2/\rho_1}$$

Verify the correctness of this equation. *Hint:* Start with Eq. (11.57) and get the right side in terms of p_2/p_1 by first replacing M_2, using Eq. (11.62), and then using Eq. (11.64). Use the equation of state to replace T_2/T_1 on the left side of the equation. This gives a relation between p_2/p_1 and ρ_2/ρ_1, which by algebraic manipulation can be brought into the above form. (If this proves too laborious, substitute arbitrary values for k, ρ_2, and ρ_1 into your equation and check to see if you get the same value for p_2/p_1 predicted by the equation above.)

11.52. Show in the Rankine-Hugoniot equation that a shock of infinite strength, that is, $p_2/p_1 \rightarrow \infty$, implies, for $k = 1.4$, that $\rho_2/\rho_1 \rightarrow 6$. This gives the maximum increase in density through a normal shock possible for a perfect gas. (See Prob. 11.51.)

11.53. In a convergent-divergent nozzle a normal shock occurs at M = 2.5. If the pressure just after the shock is 500 kPa absolute, what must the reservoir pressure be?

11.54. What is the static thrust for a rocket engine burning 25 kg/s of fuel on a test stand at the earth's surface where $p = 101{,}325$ Pa. The chamber pressure is 1.38×10^6 Pa absolute and the chamber temperature is 2760°C. Take $k = 1.4$ and $R = 355$ N-m/kg(K) for products of combustion. The throat area is 0.02 m², the exit area is 0.05 m², and there is a normal shock at $A_s = 0.04$ m².

11.55. Air flows in a convergent-divergent nozzle and a normal shock occurs where the pressure is 90 kPa absolute. The reservoir conditions are $T = 60°C$ and $p = 201.3$ kPa absolute. What is the loss in stagnation pressure across the shock? In what sense is this a "loss"?

11.56. A duct in Fig. P11.56 has "swallowed" a stationary normal shock as shown. Find p_e, V_e, T_e, and c_e for the plenum chamber.

11.57. Start with the combined first and second law ($T\,ds = du + p\,dv$). Use the definition of the enthalpy to replace du. Using the equation of state, eliminate v and use Eq. (11.3b) to eliminate h. Now integrate to get the entropy up to a constant of integration in the form

$$s = c_p \ln T - R \ln p + C$$

Hence the change in entropy across a shock is

$$\Delta s = c_p \ln \frac{T_2}{T_1} - R \ln\left(\frac{p_2}{p_1}\right)$$

11.58. Figure P11.58 is a convergent-divergent nozzle attached to a chamber (tank 1) where the pressure is 100 lb/in² absolute and the temperature is 200°F. The area of the throat is 3 in² and A_1, where we happen to have a normal shock, is 4 in². Finally A_e is 6 in². What is the Mach number right after the shock wave? What is the Mach number at exit? Compute the stagnation pressure and actual pressure for the jet in tank 2. What is the stagnation temperature at exit? The fluid is air.

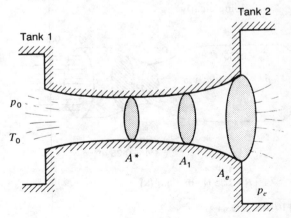

FIGURE P11.58

11.59. Explain why a free subsonic jet should have the same pressure as the surroundings.

11.60. A convergent nozzle has an exit area of 1.3×10^{-3} m². It permits flow of air to proceed from a large tank in which the pressure of the air is 138,000 Pa absolute and the temperature is

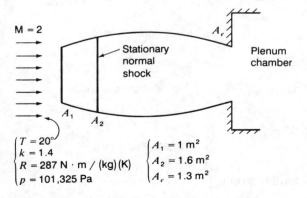

$$\begin{cases} T = 20° \\ k = 1.4 \\ R = 287 \text{ N} \cdot \text{m} / \text{(kg)(K)} \\ p = 101{,}325 \text{ Pa} \end{cases}$$

$$\begin{cases} A_1 = 1 \text{ m}^2 \\ A_2 = 1.6 \text{ m}^2 \\ A_e = 1.3 \text{ m}^2 \end{cases}$$

FIGURE P11.56

20°C. If the ambient pressure outside the tank is 101,325 Pa, what are the velocity of the flow on leaving the nozzle and the mass flow? Neglect friction.

11.61. In Prob. 11.60 suppose that you are changing the ambient pressure. What is the largest pressure that will permit the maximum flow through the nozzle? What are the maximum mass flow and temperature of the air leaving the nozzle? Neglect friction.

11.62. A convergent-divergent nozzle with a throat area of 0.0013 m^2 and an exit area of 0.0019 m^2 is connected to a tank wherein air is kept at a pressure of 552,000 Pa absolute and a temperature of 15°C. If the nozzle is operating at design conditions, what should be the ambient pressure outside and the mass flow? What is the critical pressure? Neglect friction.

11.63. In Prob. 11.62 what is the ambient pressure at which a shock will first appear just inside the nozzle? What is the ambient pressure for the completely subsonic flow of maximum mass flow? Neglect friction.

11.64. A rocket is operating with a normal shock positioned as shown in Fig. P11.64. What is the thrust of the rocket? Explain what you are doing clearly.

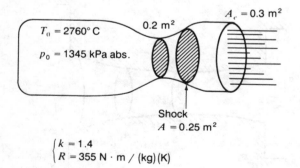

$T_0 = 2760°$ C

$p_0 = 1345$ kPa abs.

0.2 m^2

$A_e = 0.3$ m^2

Shock
$A = 0.25$ m^2

$\begin{cases} k = 1.4 \\ R = 355 \text{ N} \cdot \text{m} / \text{(kg)(K)} \end{cases}$

FIGURE P11.64

11.65. Suppose that you are given the data shown for the convergent-divergent nozzle. A shock is present in the nozzle, as shown. Set up formulations by which one could proceed to ascertain the approximate position and strength of the shock.

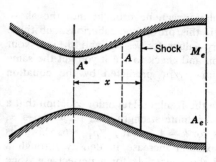

A^* A ← Shock M_e

x

A_e

FIGURE P11.65

*11.66. In Prob. 11.65, $M_{exit} = 0.5$, $A^* = 2$ in^2, and $A_e = 3$ in^2, with the divergent part of the nozzle having a conical shape between these two areas. This cone has a half angle of 20°. Take p_{amb} as 70 lb/in^2. What is the position of the normal shock? Neglect friction outside the normal shock. Give the result as a distance from the exit of the nozzle.

11.67. For a flow shown in Fig. P11.67:
   (a) Find the Mach number of the flow directly after the oblique shocks.

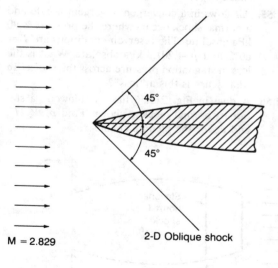

45°

45°

M = 2.829

2-D Oblique shock

$p = 101,325$ Pa $\Big\}$ Static
$T = 20°$ C $\Big\}$ values

FIGURE P11.67

(b) Find the direction of streamlines directly after the shocks. Give angle between streamline and shock.

11.68. A supersonic flow of air at Mach number 2.5 approaches a double-wedge airfoil. What is the possible smaller shock angle α? What is M_2? Does an oblique shock occur as a result of corner B? Of corner C?

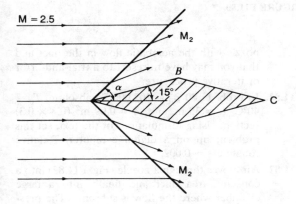

11.69. A two-dimensional supersonic flow of air at $M = 3$ forms an oblique shock as a result of the corner A at the boundary. Verify that one possible angle α is 52°. Compute M_2 for this α. Verify that another possible angle is $\alpha = 75.2°$. Compute M_2 for this α. Note there are two solutions possible: one leading to subsonic flow and another leading to supersonic flow.

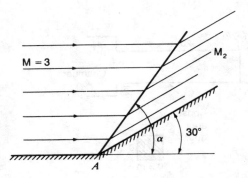

FIGURE P11.69

11.70. An oblique shock is oriented at an angle α of 70° to the flow (see Fig. 11.23). The approach Mach number is 3. What is the final Mach number after the shock? The temperature ahead of the shock is 5°C and the fluid is air.

*11.71. Supersonic flow at Mach number 4 undergoes an oblique shock and then has a Mach number of 1.815. What is the inclination of the shock, α, with the initial direction of the stream lines? The initial temperature is 10°F and the fluid is air. (Solve by trial and error.)

11.72. A constant-area duct having a circular cross-sectional area of 0.19 m^2 is operating in a choked condition. It is highly insulated against heat transfer, and the inside surface has a relative roughness of 0.002. At a distance 9 m from the end of the duct, what is the Mach number of 35 kg of air flow per second? The stagnation temperature is 95°C. Perform one iteration. Fluid is air.

11.73. In Prob. 11.72, determine where the Mach number is 0.5, and ascertain the pressure and temperature of the flow there.

11.74. A constant-area duct is operating in the choked condition. The cross section is rectangular, having sides 6 ft by 4 ft, and the surface has a relative roughness of 0.0001. At 20 ft from the end of the duct, the pressure is 18 lb/in^2 absolute. If there is no heat transferred through the walls, determine the Mach number and Reynolds number at this section for air flow. The exit pressure is that of ambient pressure of the surroundings, which is 14.7 lb/in^2.

11.75. A circular duct having a diameter of 0.6 m conducts air adiabatically so that it emerges subsonically into the atmosphere at a pressure of 101,325 Pa. At a position 12 m from the end of the duct, the pressure and temperature are 117,200 Pa absolute and 38°C, respectively. If the duct has a relative roughness of 0.002, what are equations for the Mach numbers at the aforementioned section and at the exit? *Hint:* Find two simultaneous equations involving the Mach numbers at both sections.

11.76. Air is flowing in a well-insulated duct (see Fig. P11.76) into a large plenum chamber A. The mass flow is 15 kg/s. What is the Mach number at exit? How far into the chamber can the

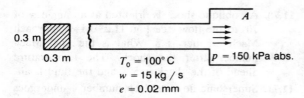

FIGURE P11.76

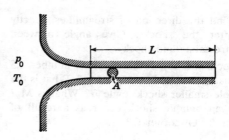

FIGURE P11.80

duct length be extended before the flow is choked? The roughness is 0.02 mm.

11.77. What is the exit Mach number in Fig. P11.77 and the exit pressure? Air flows adiabatically with $k = 1.4$ and $R = 287$ N-m/(kg)(K).

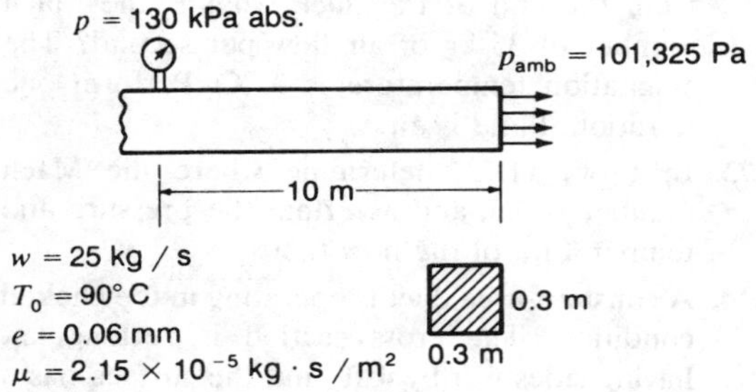

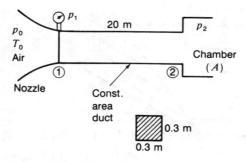

FIGURE P11.77

11.78. Air flows through a constant area duct from an initial Mach number of 0.3 at a temperature of 50°C to a Mach number of 0.7. Using only Tables B.5 and B.7, find the temperature and stagnation temperature at $M = 0.7$. There is no heat transfer.

11.79. Air is flowing through a heavily insulated constant area duct from a Mach number $M = 0.3$ at a pressure of 1.8×10^5 Pa absolute to a Mach number of 0.5. What is the pressure at $M = 0.5$ and what is the loss in stagnation pressure?

*11.80. Consider a high-pressure chamber connected to a short convergent nozzle, which in turn is attached to an insulated constant-area duct, as shown. Outline a method for ascertaining the mass flow out of the tank assuming the geometry and stagnation conditions are known. Note that you must match the isentropic flow of the

nozzle with the adiabatic flow in the duct and that you may have to resort to a trial-and-error or iterative procedure.

*11.81. In Prob. 11.80 $p_0 = 207,000$ Pa absolute, $T_0 = 40°C$, $L = 30$ m, and $A = 0.09$ m² (0.3×0.3) section. Using formulations of the text, set this problem up on a digital computer. Relative roughness = 0.002.

11.82. Air moves through a nozzle (Fig. P11.82) into a constant area duct and finally into a large chamber where the flow is subsonic. The pressure in the chamber p_2 is *unspecified*. But pressure p_1 at entrance to the duct is given. What is the *Mach number* of the air entering chamber A? There is no heat transfer in constant area duct. The following data apply.

$$p_0 = 146,727 \text{ Pa absolute}$$

$$T_0 = 20°C$$

$$p_1 = 130,000 \text{ Pa absolute}$$

$$e = 0.0006 \text{ m for duct}$$

FIGURE P11.82

Proceed as follows:

1. Consider isentropic flow in the nozzle. Calculate at the *entrance of the constant area duct*:
   (*a*) $(Re)_1$ (for the duct)
   (*b*) f (for the duct)
   (*c*) G_1
2. Now find the *additional length* of duct needed to add on to the 20 m so as to cause a *choked* condition at exit into chamber A.
3. Finally, we get M_2 for the *actual duct*.

11.83. A diffuser slows fluid to go through a constant area long duct in Fig. P11.83 so as to reduce losses in stagnation pressure. The duct has a roughness e of 0.4 mm. To go to a supersonic test section the flow is expanded through a nozzle. The geometries are given in the diagram. If there is no normal shock:
   (*a*) What is the pressure at the test section?
   (*b*) What is the Mach number entering the duct from the diffuser?

11.84. If in Prob. 11.83 the Mach number at B is 0.485, what are the Mach number and pressure at the diffuser inlet at D? What is the critical area for the diffuser flow?

11.85. How much heat per unit mass of flow of air is needed to increase the initial Mach number from 0.4 to a Mach number of 0.7 in a constant area duct, if we neglect friction? The tempera-ture at $M = 0.3$ is 70°C. Take $c_p = 1.0865$ kJ/(kg)(K) as a constant specific heat.

11.86. In the preceding problem the initial pressure is known to be $p_1 = 2.00 \times 10^5$ Pa absolute. The initial temperature is not known. What is the change in stagnation pressure in going from the initial Mach number = 0.3 to final Mach number = 0.7?

11.87. A choked flow of air in a constant area duct has an initial Mach number of 0.3 with a temperature of 60°C.
   (*a*) What is the heat added per unit mass?
   (*b*) What is the exit temperature?
   (*c*) What is the maximum temperature in the flow?
   Take $c_p = 1.0865$ kJ/(kg)(K) and neglect friction.

11.88. Air flows through a heavily insulated constant area duct having a 0.3-m × 0.3-m cross section. Initially the following data apply:

$$M_1 = 0.4$$

$$p_1 = 2.5 \times 10^5 \text{ Pa absolute}$$

$$T_1 = 80°C$$

At a distance of 4 m downstream, the Mach number is 0.6. What is the drag per unit length developed on the duct by the flow?

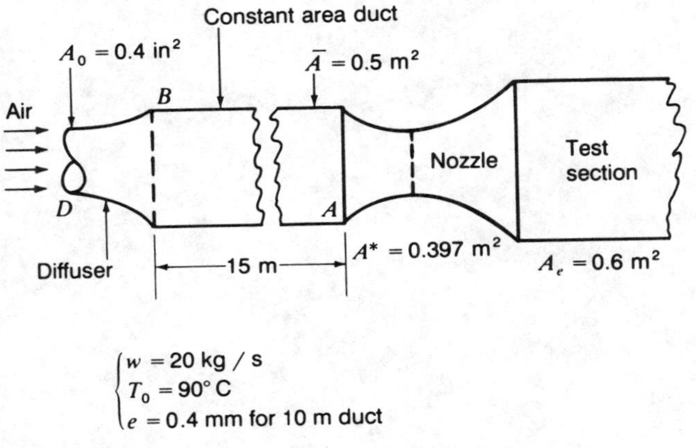

Constant area duct

$A_0 = 0.4$ in^2

$\bar{A} = 0.5$ m^2

Air

B

Nozzle

Test section

D

A

Diffuser

15 m

$A^* = 0.397$ m^2

$A_e = 0.6$ m^2

$\begin{cases} w = 20 \text{ kg / s} \\ T_0 = 90° \text{ C} \\ e = 0.4 \text{ mm for 10 m duct} \end{cases}$

FIGURE P11.83

11.89. In Example 11.8, determine the heat transfer per unit mass between section 1 and the end that will cause the flow to choke.

11.90. In Example 11.8, suppose that for the same condition of temperature at section 1 (p not specified here) we change the mass flow to 22.7 kg/s and note that the temperature on leaving the duct is 193°C at ambient pressure of 101,325 Pa. What is then the heat flow per unit mass between section 1 and the end of the duct? Use a constant specific heat c_p of 1046 N · m/(kg)(K).

11.91. Very cold air, to be used in a air-conditioning system of a test chamber, passes through a rectangular duct of cross-sectional area A square feet and of length L feet. It enters the duct at a temperature of T_1 degrees Fahrenheit at a pressure of p_1 pounds per square inch absolute. It is estimated that Q Btu per unit length per pound-mass will be transferred from the surroundings into the flow of air in the duct. If the exit temperature is to be T_2 degrees Fahrenheit at ambient pressure p_2, how much flow should there be? Set up equations only. Explain how you might go about solving the equations.

11.92. A constant-area duct conducts a gaseous mixture of air and fuel. At the entrance to the duct the mixture has a velocity of 30 m/s, a temperature of 40°C, and a pressure of 124,100 Pa absolute. If the heat of reaction for the mixture is 465,180 J/kg of the mixture, what are the exit Mach number, the temperature, and the velocity of the flow? We assume that the properties of the reactants and products of combustion are the same as air as far as the specific heat and the gas constant R are concerned. Take $c_p = 1005$ J/(kg)(K) for the calculations. Pipe is insulated.

ANALYSIS OF IMPORTANT EXTERNAL FLOWS

Now that we have studied internal flows under a number of useful circumstances, we turn next to external flows. We start out by considering inviscid two-dimensional flow and also axisymmetric inviscid flow. The flows will be incompressible and the mathematics that attends this material is often called potential theory. Students will see similarities from both these considerations with what the student may have studied in electromagnetic theory. Next we go into boundary layer theory whereby we incur viscous effects and consider incompressible steady flow over plates. Despite the limited geometry, the reader will see that with some ingenuity, one can nevertheless use these theories to solve some interesting problems. Furthermore, the flow around bodies giving rise to concepts of drag and lift is then undertaken to round out the chapter. We are then ready to go into channel flow or what is often called free surface flows. This is generally an external type flow wherein part of the boundary has a specified pressure acting on it rather than fixed geometry, as for example a river or a canal. The reader is urged to look for the connections between free surface flow and the compressible flow studied in Part II. There are some very interesting useful connections.

We close Part III with a chapter on turbomachinery that deals with external and internal flows. The final chapter is on numerical techniques for solving both internal and external flow.

The V-22 Osprey experimental aircraft. (*COURSTEY BELL-BOEING CORP.*)

The V-22 Osprey developed by Bell-Boeing is part helicopter and part airplane. It is powered by large rotor blades at the wing tips and can hover or fly at subsonic speeds. Although designed for military uses, the Osprey could be used as an intercity shuttle at speeds up to 350 mph which is faster than any helicopter. One of the most severe technical hurdles in the development of this plane was that of aeroelastic instabilities. This problem was overcome by the use of composite materials which can be tailored to have required stiffness properties in certain directions. That this design has potential is evidenced by the fact that a European consortium is designing such a plane as is a Japanese company.

CHAPTER
12

POTENTIAL FLOW

12.1 INTRODUCTION

All engineering science is a compromise between reality and the necessary simplifications needed for mathematical computations. We have already employed many simplifying idealizations. At this time a highly idealized flow will be studied which is amenable to mathematical treatment and which at the same time is useful in the comprehension of certain flows. The following are the key assumptions in the ensuing treatment:

Incompressibility. The density and specific weight are to be taken as constant.

Irrotationality. This implies a nonviscous fluid whose particles are initially moving without rotation. Therefore **curl V = 0**.

Steady flow. This means that all properties and flow parameters are independent of time.

The chapter will be divided into six parts in undertaking this analysis. They are:

Part A. Mathematical considerations.

Part B. The two-dimensional stream function and important relations. Here important concepts are established which are useful for elementary studies and necessary for advanced analysis.

Part C. Basic analysis of two-dimensional, steady, irrotational, incompressible flows. The four basic laws are separately and individually examined for this flow, and key equations are developed in terms of the quantities introduced in Part A. Boundary conditions are introduced.

Part D. Simple two-dimensional flows. Employing the equations of Part C, we introduce simple flows and note their various characteristics. These flows are important in that they are useful in forming more-complicated flows by superposition.

Part E. Superposition of two-dimensional simple flows. Two analyses are undertaken which serve to illustrate the superposition method. As a by-product, these examples demonstrate one of the most important equations in aerodynamics—the basic equation of lift.

Part F. Axisymmetric three-dimensional flows.

PART A
MATHEMATICAL CONSIDERATIONS

12.2   CIRCULATION: CONNECTIVITY OF REGIONS

Circulation is defined as the line integral about a closed path at time t of the tangential velocity component along the path. Calling the circulation Γ, we have

$$\Gamma = \oint_c \mathbf{V} \cdot \mathbf{ds} \tag{12.1}$$

where c is the closed path. Figure 12.1 illustrates the terms involved in the integration.

In discussions involving circulation, it is useful to classify regions of flow into simply connected regions and multiply connected regions. A *simply connected region* is one wherein every closed path forms the edge of a family of hypothetical surfaces called *capping* surfaces which do not cut through the physical boundaries of the flow. Figure 12.2 illustrates such a region of flow. Paths a and b satisfy the above requirement, as do all other conceivable paths in the flow. A region not having this property is called a *multiply connected region*. For example, a region including the two-dimensional airfoil (Fig 12.3) is

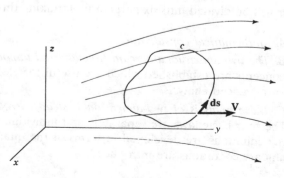

FIGURE 12.1
Closed path c for determining circulation.

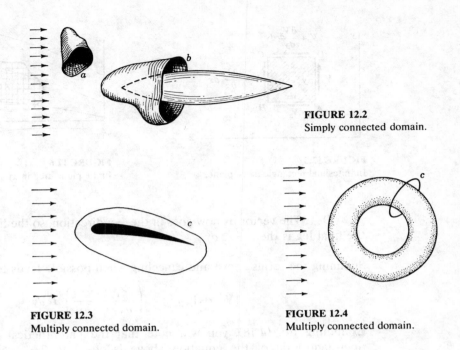

FIGURE 12.2
Simply connected domain.

FIGURE 12.3
Multiply connected domain.

FIGURE 12.4
Multiply connected domain.

multiply connected, since path c cannot be associated with a surface which does not cut "the body of the airfoil." The same holds true for path c in Fig. 12.4, which surrounds the three-dimensional torus, making this region multiply connected also.

12.3 STOKES' THEOREM

To develop Stokes' theorem, examine the circulation about an infinitesimal rectangular path whose plane is parallel to the xy plane as shown in Fig. 12.5. We carry out the line integration in four steps:

A-B. ds corresponds to dx, and because of the infinitesimal length of path no integration is necessary. Hence for this section we have $V_x\,dx$ as the contribution to circulation.

B-C. Similarly ds is now dy, and the tangential component of V is in the y direction, having a value which can be given as a Taylor expansion of conditions at corner A (Fig. 12.5). Thus we get $(V_y)_{BC} = V_y + (\partial V_y/\partial x)\,dx$, so the contribution to the circulation from side BC is then $[V_y + (\partial V_y/\partial x)\,dx]\,dy$.

C-D. The vector $\mathbf{ds}$ is now in the negative x direction. In a manner paralleling the previous discussion in the tangential velocity becomes $V_x + (\partial V_x/\partial y)\,dy$, making the line integration equal to $-[V_x + (\partial V_x/\partial y)\,dy]\,dx$ for CD.

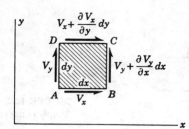

FIGURE 12.5
Infinitesimal rectangle in xy plane.

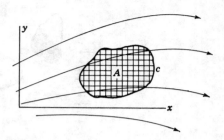

FIGURE 12.6
Finite plane area in xy plane.

D-A. The vector **ds** now goes in the $-y$ direction, so the line integration of the final leg is then $-V_y\, dy$.

Summing the terms above and canceling when possible leads to the result

$$[\mathbf{V} \cdot \mathbf{ds}]_{ABCD} = \left(\frac{\partial V_y}{\partial x} - \frac{\partial V_x}{\partial y} \right) dx\, dy \tag{12.2}$$

Examining Eq. (4.18), you will note that the parenthetical quantity on the right-hand side of the equation above is equal to $2\omega_z$, that is, twice the component of the angular-velocity vector normal to the plane of the area element. Remembering that $2\boldsymbol{\omega} = \mathbf{curl\ V}$, we can then say

$$d\Gamma = [\mathbf{V} \cdot \mathbf{ds}]_{ABCD} = (\mathbf{curl\ V})_z\, dA \tag{12.3}$$

We used a rectangular shape for our area element in the development only for convenience; the relation above holds true for any infinitesimal area element.

Next consider a hypothetical finite plane area A in a flow. We break up this area into infinitesimal-area elements, as shown in Fig. 12.6 and apply Eq. (12.2) to each of the elements. If we move in the same sense, i.e., clockwise or counterclockwise, about each element, there will be a cancelation of the line integration everywhere except for the outer boundary c when we integrate the terms in Eq. (12.2) to cover the entire region A. This should be clear, since we have, for each inner boundary, integration in opposite directions with the net internal result of zero for the total circulation. We can then say

$$\Gamma = \oint_c \mathbf{V} \cdot \mathbf{ds} = \iint_A (\mathbf{curl\ V})_z\, dA \tag{12.4}$$

This is a two-dimensional Stokes' theorem. It equates the circulation of a vector field about a plane path with the curl component of the field *normal* to the plane surface enclosed by the path.

To arrive at the general Stokes' theorem, examine a hypothetical curved surface in a flow bounded by the noncoplanar curve c. This surface is also subdivided into infinitesimal-area elements, as shown in Fig. 12.7. In consider-

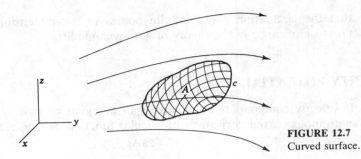

FIGURE 12.7
Curved surface.

ing Eq. (12.3) for one of these area elements we can say

$$d\Gamma = (\text{curl V})_n \, dA = \text{curl V} \cdot \text{dA} \tag{12.5}$$

where the component of the curl vector required clearly is normal to the area element. In integrating Eq. (12.5) for all area elements on the surface we get a cancelation of the line integration over internal boundaries if we maintain the same sense of line integration for each area element, as explained earlier. The result is then the well-known Stokes' theorem in three dimensions. Thus

$$\oint_c \text{V} \cdot \text{ds} = \iint_A \text{curl V} \cdot \text{dA} \tag{12.6}$$

We thus relate a line integral, which is the circulation of the velocity field for any curve c, with a surface integral, which is the integration of the normal component of the **curl** of the velocity field over any capping surface for which c is an edge. It should be kept in mind that although we derived Stoke's theorem using a velocity field, it is properly stated for any continuous vector field **B**. Thus

$$\oint_c \text{B} \cdot \text{ds} = \iint_A \text{curl B} \cdot \text{dA} \tag{12.7}$$

12.4 CIRCULATION IN IRROTATIONAL FLOWS

With the aid of Stokes' theorem and the definition of irrotationality, we see that the *circulation in any simply connected region of irrotational flow must be zero for all paths*, since one may always associate a capping surface with such a path on which $(\text{curl V})_n$ is zero by the condition of irrotationality.

For multiply connected regions, circulation about certain paths, such as those shown in Figs. 12.3 and 12.4 cannot be ascertained by making use of Stokes' theorem. However, paths which do not enclose the infinite body in Fig. 12.3 or the torus in Fig. 12.4 give zero circulation for irrotational flow. We find that the circulation about infinite bodies in two-dimensional flows will have great significance in the theory of aerodynamic lift.

12.5 THE VELOCITY POTENTIAL

If velocity components at all points in a region of flow can be expressed as continuous partial derivatives of a scalar function $\phi(x, y, z, t)$ thusly

$$V_x = \frac{\partial \phi(x, y, z, t)}{\partial x}$$

$$V_y = \frac{\partial \phi(x, y, z, t)}{\partial y} \tag{12.8}$$

$$V_z = \frac{\partial \phi(x, y, z, t)}{\partial z}$$

then the flow must be irrotational. The scalar function is termed the *velocity potential*. To show that such a flow is irrotational, substitute the above formulations of velocity into the equation of irrotationality [Eq. (4.21)]. Thus

$$\frac{\partial^2 \phi}{\partial y\, \partial z} - \frac{\partial^2 \phi}{\partial z\, \partial y} = 0$$

$$\frac{\partial^2 \phi}{\partial z\, \partial x} - \frac{\partial^2 \phi}{\partial x\, \partial z} = 0 \tag{12.9}$$

$$\frac{\partial^2 \phi}{\partial x\, \partial y} - \frac{\partial^2 \phi}{\partial y\, \partial x} = 0$$

It is known that the order of partial differentiation makes no difference provided that each partial derivative is a continuous function. Since this will be the case, it is then true that the Eqs. (12.9) are satisfied.

Vectorially, we may state Eq. (12.8) as

$$\mathbf{V} = \mathbf{grad}\ \phi = \nabla \phi \tag{12.10}$$

This form no longer binds us to thinking in terms of a particular coordinate system and is to be preferred.[1]

---

[1] By carrying out the differentiation and vector algebraic operations using Cartesian components you may readily show that **curl** (**grad** ϕ) = **0** for any function ϕ having continuous first and second derivatives. This means that *any field* which is expressible as the gradient of a scalar in this manner must be an irrotational field. Thus the conservative force fields studied in mechanics would have to be irrotational fields.

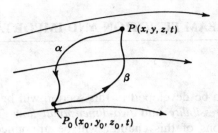

FIGURE 12.8
Closed path formed by α and β.

It may also be shown that the converse to this theorem is true. That is, *any irrotational flow may be expressed as the gradient of a scalar function ϕ.* Consider, then, a fixed point $P_0(x_0, y_0, z_0)$ and a movable point $P(x, y, z)$ in an irrotational flow at some time t, as shown in Fig. 12.8. Now connect these points by arbitrary paths α and β, thus forming a closed loop. Since the flow is irrotational, the circulation for this loop is zero and we can say

$$\oint \mathbf{V} \cdot \mathbf{ds} = \int_{P_0 \atop \beta}^{P} \mathbf{V} \cdot \mathbf{ds} + \int_{P \atop \alpha}^{P_0} \mathbf{V} \cdot \mathbf{ds} = 0 \tag{12.11}$$

Reversing the limits on the last integral, we may state

$$\int_{P_0 \atop \beta}^{P} \mathbf{V} \cdot \mathbf{ds} = \int_{P_0 \atop \alpha}^{P} \mathbf{V} \cdot \mathbf{ds} \tag{12.12}$$

We can conclude from this relation that the integral $\int_{P_0}^{P} \mathbf{V} \cdot \mathbf{ds}$ is independent of the path and depends only on the coordinates of the end points P and P_0 and the time. Thus we can introduce a function ϕ such that

$$\int_{P_0}^{P} \mathbf{V} \cdot \mathbf{ds} = \left[\phi(x, y, z, t) - \phi(x_0, y_0, z_0, t) \right] \tag{12.13}$$

For any infinitesimal displacement $\mathbf{ds}$ this equation becomes:

$$\mathbf{V} \cdot \mathbf{ds} = d\phi = \nabla\phi \cdot \mathbf{ds} \tag{12.14}$$

Thus
$$\mathbf{V} = \nabla\phi \tag{12.15}$$

We have thus shown that the velocity field of an irrotational flow is always expressible in terms of the gradient of some scalar function ϕ. That this is a distinct advantage can be appreciated by the fact that a vector field $\mathbf{V}(x, y, z, t)$ may now be analyzed in terms of a scalar field ϕ. Great use of the velocity potential will be made in succeeding sections.

12.6   THE STREAM FUNCTION

The stream function to be developed in this section will be subject only to the restrictions of *incompressibility* and *two-dimensional flow*. The introduction of the remaining conditions of this chapter will be announced as they become necessary to the discussion. Accordingly, a two-dimensional continuous flow subject only to the additional condition of incompressibility is shown in Fig. 12.9 at some time t. A point (x_0, y_0) is chosen arbitrarily as a reference, or "anchor point." From any point (x, y) two arbitrary paths a and b are now drawn to the anchor point (x_0, y_0). Each path, in this two-dimensional study, may be imagined as the profile of a prismatic surface extending indefinitely in the z direction. Hence, the area bounded by the two paths may be interpreted as the cross section of a prismatic volume extending unchanged in the z direction. It will be convenient to consider a unit slice of the prismatic volume as a control volume.

Let us examine the *continuity equation* for this control volume. Since the flow is incompressible, there can be no change in the quantity of fluid inside the control volume at any time. Furthermore, the restriction of two-dimensionality means that each end surface of the unit slice (i.e., cross-sectional surfaces parallel to the xy plane) is subject to the same flow pattern. Consequently, the influx through such a sectional surface at one end of the slice must equal the efflux through the sectional surface at the other end, thus canceling these flow terms in the continuity equations. The general continuity equation [Eq. (5.1)] at time t then becomes

$$\iint_{\alpha} \rho \mathbf{V} \cdot d\mathbf{A} + \iint_{\beta} \rho \mathbf{V} \cdot d\mathbf{A} = 0 \tag{12.16}$$

where the subscripts α and β refer to the lateral surfaces of the slice associated with paths a and b proceeding from x_0, y_0 to x, y of Fig. 12.9. Upon canceling

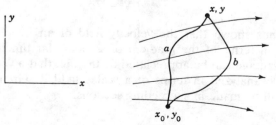

FIGURE 12.9
Two-dimensional incompressible flow.

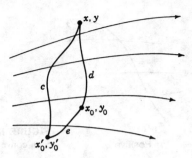

FIGURE 12.10
New anchor point.

out the density ρ (it is a constant), and introducing the notation $q_\alpha = -\iint_\alpha \mathbf{V} \cdot \mathbf{dA}$ and $q_\beta = \iint_\beta \mathbf{V} \cdot \mathbf{dA}$, Eq. (12.16) becomes

$$q_\alpha = q_\beta \tag{12.17}$$

wherein q in this discussion represents the volume of flow per unit time *through* a surface extending a unit distance in the z direction. Thus, the flows through *all* unit strips terminating at these points must be equal. This flow q for a given reference will depend on the position of the roving point x, y and the time t. Functionally, this may be expressed as

$$q_{x_0 y_0} = \psi_{x_0 y_0}(x, y, t) \tag{12.18}$$

where the subscripts identify the anchor point. The function ψ is the *stream function*.

Next, we investigate the effect of changing the anchor point from x_0, y_0 to another point x_0', y_0'. This is shown in Fig. 12.10. The flow associated with any path c between anchor point x_0', y_0' and roving point x, y is expressed as the function $\psi_{x_0' y_0'}'(x, y, t)$. However, another path which traverses the point x_0, y_0 can easily be established between points $x_0' y_0'$ and xy. This is indicated by paths e and d in the diagram. According to the preceding discussion, the flows associated with the paths c and $e + d$ can be equated in the following manner:

$$q_c = q_e + q_d$$
$$\therefore \psi_{x_0' y_0'}'(x, y, t) = \psi_{x_0' y_0'}'(x_0, y_0, t) + \psi_{x_0 y_0}(x, y, t) \tag{12.19}$$

where it will be noted that $\psi_{x_0 y_0}(x, y, t)$ is the stream function using the old anchor point. The expression $\psi_{x_0' y_0'}'(x_0, y_0, t)$ is only a function of time which we denote as $g(t)$, since end points x_0', y_0' and x_0, y_0 are fixed. Hence the expression above may be written as

$$\psi_{x_0' y_0'}'(x, y, t) = \psi_{x_0 y_0}(x, y, t) + g(t) \tag{12.20}$$

Thus it is clear that the stream function changes only by a function of time when the anchor point is altered. However, in most computations involving the stream function there will be partial differentiation of this function with respect to the

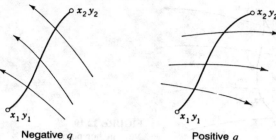

FIGURE 12.11
Sign convention for flow.

spatial variables x and y. The function $g(t)$ will, under these operations, differentiate out to zero and, hence, will usually be of little significance. Accordingly, it is often the procedure to delete it, implying that all stream functions depicting a given flow are equivalent to each other for aforementioned computations. Hence, in most cases any known stream function for a given flow may be employed for all anchor points. The notation is then simply $\psi(x, y, t)$ for the stream function.

For steady flow, the expression $\psi'_{x'_0 y'_0}(x_0, y_0, t)$ becomes a constant, and hence Eq. (12.20) becomes

$$\psi'_{x'_0 y'_0}(x, y) = \psi_{x_0 y_0}(x, y) + \text{const} \tag{12.21}$$

showing that the stream function changes by a constant when the anchor point is changed. As in the general case, this constant is usually of no practical significance in computations, so it is dropped. The steady-flow stream function is then usually taken as $\psi(x, y)$ without regard to the anchor point.

Also, it will be convenient to establish a sign convention for volume flow q traversing a strip between two points. It will be the practice in this text to consider as positive flow that which passes from an observer's left to his/her right as he/she looks out from one point to the other. Figure 12.11 gives examples of positive and negative flows established by this sign convention.

Finally, in stating the volume flow between any two points 1 and 2 in the flow we can use the following simplified notation,

$$q_{1,2} = \psi_2 - \psi_1 \tag{12.22}$$

where it will be understood that the same anchor point is used for the evaluation of ψ at points 1 and 2 but where the identity of the anchor point is of no consequence. The sequence of subscripts of q indicates that we are looking out from point 1 to point 2; in other words, we are integrating along the path from point 1 to point 2.

12.7 RELATIONSHIP BETWEEN THE STREAM FUNCTION AND THE VELOCITY FIELD

It will now be demonstrated that a relatively simple relation exists between the stream function $\psi(x, y, t)$ and the velocity field $\mathbf{V}(x, y, t)$. Figure 12.12 shows a

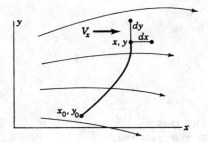

FIGURE 12.12
dx and dy extensions to path.

two-dimensional incompressible flow at time t. A fixed point x_0, y_0 has been shown, as well as some "roving" point x, y to which have been added increments dx and dy. The flow q associated with the path from x_0, y_0 to the extremity of dy may be expressed in two ways. These are equated as

$$\psi_{x_0 y_0}(x, y, t) + \frac{\partial \psi(x, y, t)}{\partial y}\, dy = \psi_{x_0 y_0}(x, y, t) + V_x\, dy$$

Note that the subscript $x_0 y_0$ is not employed in the partial derivative of ψ according to the previous discussion. Also note that the aforementioned sign convention has been employed in the last term of this equation, $V_x\, dy$, which is the volume flow through the strip associated with the segment dy. After cancelation of equal terms on opposite sides of the equation, we get one of the desired relations between the stream function and the velocity field,

$$\boxed{V_x = \frac{\partial \psi}{\partial y}} \qquad (12.23)$$

Performing the same computation for the extreme point of the dx segment leads to the result

$$\boxed{V_y = -\frac{\partial \psi}{\partial x}} \qquad (12.24)$$

As in the previous section, these relations have been developed independently of conditions of rotationality.

12.8 RELATION BETWEEN THE STREAM FUNCTION AND STREAMLINES

It will now be shown that a locus of points corresponding to a constant value of the stream function at time t (i.e., a contour line of ψ) is nothing more than a streamline. Such a locus of points has been indicated in Fig. 12.13. A series of

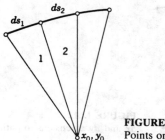

FIGURE 12.13
Points on ψ = constant.

adjacent control volumes of unit thickness have been formed employing the fixed point x_0, y_0 and infinitesimally separated points on the locus as shown. Since the value of ψ is the same for all locus points, it is clear that the flows across all the radial lines shown emanating from x_0, y_0 are equal. Hence, by continuity there can be no flow through sections ds_1 and ds_2, and so forth. Thus, the velocity vectors along the contour line ψ = const must be tangent to it. Returning to the definition of the streamline, one sees that the locus of points ψ = const must be a streamline.

In steady flow, lines of constant ψ will form a fixed array of lines. For unsteady flow the array will change continuously with time, as pointed out in an earlier chapter.

12.9 RELATION BETWEEN THE STREAM FUNCTION AND VELOCITY POTENTIAL FOR FLOWS WHICH ARE IRROTATIONAL AS WELL AS TWO-DIMENSIONAL AND INCOMPRESSIBLE

We will now add the restriction of *irrotationality* to the previous restrictions. This means the existence of a velocity potential ϕ. For two-dimensional flow, ϕ must be a function of x, y, and t. By equating the corresponding expressions of velocity involving the stream function and velocity potential, the following relations may be established[2]:

$$\frac{\partial \psi}{\partial y} = \frac{\partial \phi}{\partial x} \qquad (12.25a)$$

$$\frac{\partial \psi}{\partial x} = -\frac{\partial \phi}{\partial y} \qquad (12.25b)$$

---

[2]These are the well-known Cauchy-Riemann equations of complex variable theory.

If either ϕ or ψ is known, it is possible from these equations to solve for the other function. We will now demonstrate one method which is usually quick and effective. Assuming that ϕ is known, integrate both sides of Eq. (12.25a) with respect to the variable y (i.e., consider x as constant during integration). We get

$$\psi = \int \left(\frac{\partial \phi}{\partial x} \right) dy + f(x) \qquad (12.26)$$

where $f(x)$ is an arbitrary function analogous to the constant of integration for ordinary integration. Now in Eq. (12.25b), carry out a partial integration in terms of the variable x. We get

$$\psi = -\int \left(\frac{\partial \phi}{\partial y} \right) dx + g(y) \qquad (12.27)$$

where $g(y)$ is the arbitrary integration function of the variable y. By comparing Eqs. (12.26) and (12.27), it is usually not difficult to ascertain the integration functions by inspection. This will be illustrated in the following example.

Example 12.1. Assume that $\phi = \ln(x^2 + y^2)^{1/2}$ is the velocity potential of a two-dimensional, irrotational, incompressible flow defined everywhere but at the origin. Determine the stream function for this flow.
Employing Eqs. (12.26) and (12.27) gives us

$$\psi = \int \frac{x}{x^2 + y^2} \, dy + f(x) \qquad (a)$$

$$\psi = -\int \frac{y}{x^2 + y^2} \, dx + g(y) \qquad (b)$$

Carrying out the partial integrations,

$$\psi = \tan^{-1} \frac{y}{x} + f(x) \qquad (c)$$

$$\psi = -\tan^{-1} \frac{x}{y} + g(y) \qquad (d)$$

From simple trigonometric relations we may say that

$$\tan^{-1} \frac{x}{y} = \frac{\pi}{2} - \tan^{-1} \frac{y}{x} \qquad (e)$$

Substituting the expression (e) into (d) and equating the right-hand sides of (c) and (d) gives us

$$\tan^{-1} \frac{y}{x} + f(x) = -\frac{\pi}{2} + \tan^{-1} \frac{y}{x} + g(y) \qquad (f)$$

From this it is clear that $f(x) - g(y) = -\pi/2$. Since $f(x)$ is a function of only x and $g(y)$ is a function of only y, it is necessary that each be constant in order that the expression $f(x) - g(y)$ be constant over the entire range of the independent variables x and y. As described in Sec. 12.6 it is usually permissible to drop the additive constant in the stream function. Hence, the stream function is seen to be

$$\psi = \tan^{-1} \frac{y}{x} \qquad (g)$$

If the function ψ is known, the velocity potential may be ascertained by a similar procedure. Briefly, one integrates Eqs. (12.25a) and (12.25b) with respect to the variables x and y, respectively. Following this, an inspection of the resulting equations for ϕ will usually reveal the correct formulation for the velocity potential.

12.10 RELATIONSHIP BETWEEN STREAMLINES AND LINES OF CONSTANT POTENTIAL

In a previous section, it was learned that lines of constant ψ formed a set of streamlines. It will now be shown that lines of constant ϕ (i.e., contour lines of ϕ) or *potential lines*, form a family of curves which intersect the streamlines in such a manner as to have the tangents of the respective contour lines always at right angles at the points of intersection. The two sets of curves hence form an *orthogonal grid system*, or *flow net*.

Examine a system of potential lines and streamlines in Fig. 12.14. To verify the orthogonality relation, it is necessary to prove that the slopes of the streamlines and potential lines are negative reciprocals of each other at any intersection A. On the line $\phi = K_1$ it is clear that $d\phi = 0$. Employing rules of calculus, we may then say at any time t

$$\frac{\partial \phi}{\partial x} \, dx + \frac{\partial \phi}{\partial y} \, dy = 0$$

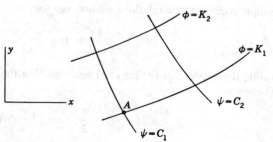

FIGURE 12.14
Intersecting streamlines and potential lines.

Solving for dy/dx gives the slope of the line $\phi = K_1$. Thus

$$\left(\frac{dy}{dx}\right)_{\phi=K_1} = -\frac{\partial\phi/\partial x}{\partial\phi/\partial y} \qquad (12.28)$$

Similarly, the slope of the line $\psi = C_1$ at time t is

$$\left(\frac{dy}{dx}\right)_{\psi=C_1} = -\frac{\partial\psi/\partial x}{\partial\psi/\partial y} \qquad (12.29)$$

In Eq. (12.28) we may replace derivatives of the velocity potential by derivatives involving the stream function with the aid of Eqs. (12.25a) and (12.25b). Thus

$$\left(\frac{dy}{dx}\right)_{\phi=K_1} = \frac{\partial\psi/\partial y}{\partial\psi/\partial x} \qquad (12.30)$$

Now at the point of intersection, A, the coordinates of the points on the two contour lines $\psi = C_1$ and $\phi = K_1$ become identical, and the right-hand sides of Eqs. (12.29) and (12.30) then become related as negative reciprocals. This means that at point A

$$\left(\frac{dy}{dx}\right)_{\phi=K_1} = -\left(\frac{dy}{dx}\right)_{\psi=C_1}^{-1}$$

so the streamline and potential line meet at right angles since slopes are negative reciprocals of each other. Since point A was any point of intersection, we may conclude that the streamlines and potential lines form an orthogonal grid system.

PART C
BASIC ANALYSIS OF TWO-DIMENSIONAL INCOMPRESSIBLE IRROTATIONAL FLOW

12.11 A DISCUSSION OF THE FOUR BASIC LAWS

In the previous sections, the stream function was established, as were pertinent relations linking this function with the velocity field and the velocity potential. The four basic laws will now be discussed in terms of these concepts for irrotational, incompressible, two-dimensional flows.

1. Conservation of mass. Examine the differential continuity equation in Cartesian coordinates. In Sec. 7.2 it was shown that for incompressible flow this equation becomes

$$\frac{\partial V_x}{\partial x} + \frac{\partial V_y}{\partial y} + \frac{\partial V_z}{\partial z} = 0 \qquad (12.31)$$

The stream function, which stems from continuity considerations, automatically satisfies this equation when proper derivatives of it are substituted for the velocity components. This has been done in the following equation. Note that the term $\partial V_z / \partial z$ must be zero because of the two-dimensionality of the analysis. Thus

$$\frac{\partial^2 \psi}{\partial x \, \partial y} - \frac{\partial^2 \psi}{\partial y \, \partial x} = 0 \tag{12.32}$$

Clearly, if ψ has continuous first partial derivatives, one may change the order of partial differentiation, so Eq. (12.31) has been satisfied.

However, employing derivatives of the velocity potential for the velocity components in the continuity equation [Eq. (12.31)] (and, hence, restricting the flow to irrotational flow) leads us to the following partial differential equation:

$$\boxed{\frac{\partial^2 \phi}{\partial x^2} + \frac{\partial^2 \phi}{\partial y^2} = 0} \tag{12.33}$$

This is called the *two-dimensional Laplace equation*. Solutions for the Laplace equation are known as *harmonic functions*. Another means of denoting this equation is $\nabla^2 \phi = 0$, where ∇^2 is the *Laplace operator* or *Laplacian*. In the general three-dimensional case for Cartesian coordinates

$$\nabla^2 \equiv \frac{\partial^2}{\partial x^2} + \frac{\partial^2}{\partial y^2} + \frac{\partial^2}{\partial z^2} \tag{12.34}$$

(We develop ∇^2 for polar coordinates in Sec. 12.13.) *It is thus seen that continuity considerations impose the necessary condition that the velocity potential be a harmonic function.*[3]

Furthermore, we can also show that for *irrotational* flow the stream function must also satisfy Laplace's equation. For this purpose, we express again Eqs. (12.25a) and (12.25b). Thus

$$\frac{\partial \psi}{\partial y} = \frac{\partial \phi}{\partial x} \tag{12.35a}$$

$$\frac{\partial \psi}{\partial x} = -\frac{\partial \phi}{\partial y} \tag{12.35b}$$

---

[3] In electrostatics, the electric field **E** is given as the gradient of the scalar function V, called the *electric potential*, just as in our studies of the velocity field for irrotational incompressible flow. The electric potential is harmonic in regions where there is no electric charge. We shall point out certain analogies in the footnotes in succeeding pages between the present study of fluid flow and electrostatics.

Now, we differentiate Eq. (12.35*a*) with respect to *y* and Eq. (12.35*b*) with respect to *x*; and, finally, add the equation. We get

$$\frac{\partial^2 \psi}{\partial y^2} + \frac{\partial^2 \psi}{\partial x^2} = \frac{\partial^2 \phi}{\partial x\, \partial y} - \frac{\partial^2 \phi}{\partial y\, \partial x} \tag{12.36}$$

Since the right side is zero, we see that the stream function ψ is harmonic. That is,

$$\boxed{\nabla^2 \psi = 0} \tag{12.37}$$

2. Newton's law. The preceding considerations have not been subject to the condition of steady flow. However, in the interests of simplicity, this restriction will be imposed during considerations of Newton's law. In Chap. 7, it was learned that for steady, irrotational, incompressible flow, Newton's law for an infinitesimal system (Euler's equation) could be integrated to form a very useful equation applicable throughout the entire flow. This was called the steady incompressible *Bernoulli equation*. It being noted that *y* is the elevation coordinate in the notation of this chapter, this equation becomes

$$\frac{p}{\gamma} + y + \frac{V^2}{2g} = \text{const} \tag{12.38}$$

By using this equation, it is a simple matter to evaluate pressures, once the velocity distribution has been established. The manner of use of the Bernoulli equation, in conjunction with the velocity potential and stream function, will presently be explained.

3. First law of thermodynamics. It was learned in Chap. 6 that the application of the first law of thermodynamics to a streamtube control volume for this flow leads to an equation identical to Bernoulli's equation. Thus, the first law and Newton's law under the flow conditions of this chapter lead essentially to the same result. Hence, if Newton's law is satisfied, one may assume that the first law of thermodynamics is satisfied under these circumstances.

4. Second law of thermodynamics. In the absence of heat transfer and friction there are no restrictions imposed by the second law.

To summarize the salient points of the preceding paragraphs, it may be said that for irrotational, incompressible, two-dimensional flows the stream

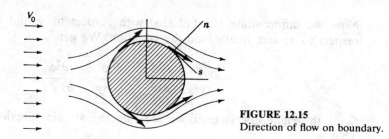

FIGURE 12.15
Direction of flow on boundary.

function and velocity potential must be harmonic functions. Once these functions are known, one may, for steady flows, easily determine pressure distributions with the aid of Bernoulli's equation.

12.12 BOUNDARY CONDITIONS FOR NONVISCOUS FLOWS

In addition to being harmonic, the stream function and velocity potential of a particular flow must not violate the boundary conditions. For instance, in the case of the cylinder shown in Fig. 12.15 the velocities at the boundary surface of the cylinder must be such as to have a zero component for the direction normal to the surface at all times. Since friction is entirely deleted from present considerations, there is no restriction on the tangential velocity components at the boundary surface, i.e., no "sticking" to the surface. In addition to these considerations at a solid boundary, there may be other conditions imposed on ϕ and ψ at infinity. For instance, in the flow illustrated, the velocity must become uniform and equal to V_0 at large distances from the cylinder.

Mathematically, the local boundary conditions (denoted by subscript b) and distant conditions are given as

$$\left(\frac{\partial \phi}{\partial n}\right)_b = 0 \qquad \left(\frac{\partial \psi}{\partial s}\right)_b = 0 \qquad (12.39a)$$

$$\left(\frac{\partial \phi}{\partial x}\right)_{\substack{x \to \infty \\ y \to \infty}} = V_0 \qquad \left(\frac{\partial \psi}{\partial y}\right)_{\substack{x \to \infty \\ y \to \infty}} = V_0 \qquad (12.39b)$$

It is this consideration of boundary conditions that makes for the major difficulty in analysis of flow, a difficulty arising whenever partial differential equations are solved to satisfy certain boundary conditions.

12.13 POLAR COORDINATES

In order to decrease difficulties arising from boundary considerations, other coordinate systems are employed whose coordinate lines "fit" some of the boundaries. For two-dimensional flows the polar-coordinate system is useful in

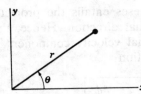

FIGURE 12.16
Polar coordinates.

many problems involving circular boundaries. This is shown in Fig. 12.16, where the modulus r (always positive) and the angle θ constitute polar coordinates.[4] We will now develop the essential relations of the preceding sections of this chapter in terms of polar coordinates. Of general use will be the transformation equations

$$x = r \cos \theta \tag{12.40a}$$

$$y = r \sin \theta \tag{12.40b}$$

$$\theta = \tan^{-1} \frac{y}{x} \tag{12.40c}$$

$$r = \sqrt{x^2 + y^2} \tag{12.40d}$$

The radial and transverse components of *velocity* may readily be determined in terms of polar coordinates. Using the relations above, we may properly carry out the following differentiation:

$$\frac{\partial \phi}{\partial r} = \left(\frac{\partial \phi}{\partial x}\right)\left(\frac{\partial x}{\partial r}\right) + \left(\frac{\partial \phi}{\partial y}\right)\left(\frac{\partial y}{\partial r}\right) \tag{12.41}$$

Employing (12.40a) and (12.40b) to eliminate the terms $\partial x/\partial r$ and $\partial y/\partial r$ from the equation above we get

$$\frac{\partial \phi}{\partial r} = \frac{\partial \phi}{\partial x} \cos \theta + \frac{\partial \phi}{\partial y} \sin \theta \tag{12.42}$$

Substituting velocity components for the derivatives of the velocity potential, we get

$$\frac{\partial \phi}{\partial r} = (V_x \cos \theta + V_y \sin \theta) \tag{12.43}$$

---

[4] In using cylindrical coordinates we have employed $\bar{r}$, θ, and z as our coordinates (see Fig. 7.2). Polar coordinates are the same as cylindrical coordinates when z is identically zero. Since there is here no longer a need to distinguish the coordinate $\bar{r}$ from the magnitude of the position vector $\mathbf{r}$, we do not trouble to put a bar over the r.

Note that the quantity in parentheses entails the projections of the velocity components V_x and V_y in the radial direction. Hence, the right side of this equation may be replaced by radial velocity component V_r, so we have on interchanging the sides of the equation

$$V_r = \frac{\partial \phi}{\partial r} \qquad (12.44)$$

In a similar manner, it may readily be established that[5]

$$V_\theta = \frac{\partial \phi}{r \, \partial \theta} \qquad (12.45)$$

Furthermore, such computations may be made for the stream function, leading to the relation

$$V_r = \frac{\partial \psi}{r \, \partial \theta} \qquad (12.46a)$$

$$V_\theta = -\frac{\partial \psi}{\partial r} \qquad (12.46b)$$

The polar relations between the stream function and the velocity potential are now available from the preceding equations. Thus

$$\frac{\partial \phi}{\partial r} = \frac{\partial \psi}{r \, \partial \theta} \qquad (12.47a)$$

$$\frac{\partial \phi}{r \, \partial \theta} = -\frac{\partial \psi}{\partial r} \qquad (12.47b)$$

The technique of integration for the determination of either function when the other is known as proposed in the work with Cartesian coordinates is equally valid for Eqs. (12.47a) and (12.47b).

---

[5]As a by-product of our work here, we see that since $\mathbf{V} = \nabla \phi$ the gradient operator for polar coordinates is

$$\nabla \equiv \frac{\partial}{\partial r} \boldsymbol{\epsilon}_r + \frac{\partial}{r \, \partial \theta} \boldsymbol{\epsilon}_\theta$$

and, for cylindrical coordinates, we get

$$\nabla \equiv \frac{\partial}{\partial \bar{r}} \boldsymbol{\epsilon}_{\bar{r}} + \frac{\partial}{\bar{r} \, \partial \theta} \boldsymbol{\epsilon}_\theta + \frac{\partial}{\partial z} \boldsymbol{\epsilon}_z$$

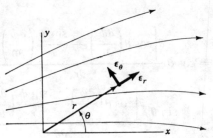

FIGURE 12.17
Polar unit vectors.

The condition of *irrotationality* in polar coordinates will now be discussed. We could proceed as before using the transformation formulas given by Eq. (12.40) to put Eqs. (4.21) in terms of polar coordinates. However, a more simple approach is to employ vector methods. First, we note that the curl operator can be given as $\nabla \times$. Next, using the gradient operator given in footnote 5, we may express the equation **curl V = 0** in the following way:

$$\left(\epsilon_r \frac{\partial}{\partial r} + \epsilon_\theta \frac{\partial}{r \, \partial \theta} \right) \times (V_r \epsilon_r + V_\theta \epsilon_\theta) = \mathbf{0}$$

Carrying out the cross product in the equation above while noting that the unit vectors vary only with θ and, furthermore, that $\partial \epsilon_r / \partial \theta = \epsilon_\theta$ and $\partial \epsilon_\theta / \partial \theta = -\epsilon_r$ (see Fig. 12.17), we have on dropping ϵ_z

$$\frac{\partial V_\theta}{\partial r} + \frac{V_\theta}{r} - \frac{\partial V_r}{r \, \partial \theta} = 0$$

Multiplying through by r, we may rewrite this equation in the following more compact form:

$$\frac{\partial (r V_\theta)}{\partial r} - \frac{\partial V_r}{\partial \theta} = 0 \tag{12.48}$$

Let us now develop Laplace's equation in terms of polar coordinates. Note that both stream function and velocity potential may be expressed in terms of polar coordinates by employing the transformation equations [Eq. (12.40)]. Hence, using the chain rule, we may carry out the partial differentiation of ϕ (or ψ) with respect to x in the following way:

$$\frac{\partial \phi}{\partial x} = \left(\frac{\partial \phi}{\partial r} \right) \left(\frac{\partial r}{\partial x} \right) + \left(\frac{\partial \phi}{\partial \theta} \right) \left(\frac{\partial \theta}{\partial x} \right) = \cos \theta \frac{\partial \phi}{\partial r} - \frac{\sin \theta}{r} \frac{\partial \phi}{\partial \theta}$$

Differentiating the right side of this equation with respect to x again in a similar manner, we get

$$\frac{\partial^2 \phi}{\partial x^2} = \cos\theta \frac{\partial\left[\cos\theta\left(\dfrac{\partial\phi}{\partial r}\right) - \dfrac{\sin\phi}{r}\left(\dfrac{\partial\phi}{\partial\theta}\right)\right]}{\partial r}$$

$$- \frac{\sin\theta}{r}\frac{\partial\left[\cos\theta\left(\dfrac{\partial\phi}{\partial r}\right) - \dfrac{\sin\phi}{r}\left(\dfrac{\partial\phi}{\partial\theta}\right)\right]}{\partial\theta} \qquad (12.49)$$

A similar computation may be made for ascertaining $\partial^2\phi/\partial y^2$. Thus

$$\frac{\partial^2 \phi}{\partial y^2} = \sin\theta \frac{\partial\left[\sin\theta\left(\dfrac{\partial\phi}{\partial r}\right) + \dfrac{\cos\theta}{r}\left(\dfrac{\partial\phi}{\partial\theta}\right)\right]}{\partial r}$$

$$+ \frac{\cos\theta}{r}\frac{\partial\left[\sin\theta\left(\dfrac{\partial\phi}{\partial r}\right) + \dfrac{\cos\theta}{r}\left(\dfrac{\partial\phi}{\partial\theta}\right)\right]}{\partial\theta} \qquad (12.50)$$

Carrying out the differentiation in both Eqs. (12.49) and (12.50) and adding and collecting terms leads us to the Laplace equation in polar coordinates[6]:

$$\frac{\partial^2 \phi}{\partial x^2} + \frac{\partial^2 \phi}{\partial y^2} = \nabla^2\phi = \frac{1}{r}\frac{\partial\left(r\dfrac{\partial\phi}{\partial r}\right)}{\partial r} + \frac{1}{r^2}\frac{\partial^2 \phi}{\partial\theta^2} = 0 \qquad (12.51)$$

**PART D
SIMPLE FLOWS**

12.14 NATURE OF SIMPLE FLOWS TO BE STUDIED

In this section, several important flows will be presented. The reader is urged to think of these as mathematically contrived flows whose physical meaning at the

---

[6]Thus the Laplacian operator ∇^2 in polar coordinates is

$$\nabla^2 \equiv \frac{1}{r}\frac{\partial\left(r\dfrac{\partial}{\partial r}\right)}{\partial r} + \frac{1}{r^2}\frac{\partial^2}{\partial\theta^2}$$

and for cylindrical coordinates we have

$$\nabla^2 = \frac{1}{\bar{r}}\frac{\partial\left(\bar{r}\dfrac{\partial}{\partial\bar{r}}\right)}{\partial\bar{r}} + \frac{1}{\bar{r}^2}\frac{\partial^2}{\partial\theta^2} + \frac{\partial^2}{\partial z^2}$$

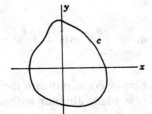

FIGURE 12.18
Path surrounding origin.

present time is of secondary importance. However, these flows will be super-posed in Part E of this chapter to yield flow patterns of physical significance.

One of the difficulties in the ensuing flow patterns is the presence of points at which the velocity is infinite. For instance, in Example 12.1 (Sec. 12.9) the velocity potential of the flow in polar coordinates is $\phi = \ln r$. The velocity components are then $V_t = 1/r$ and $V_\theta = 0$. Clearly at the origin, V_r becomes infinite. Such points are called *singular points*, which, of course, have no physical counterpart. Singular points are not considered part of an irrotational flow region, thus making a region containing such points multiply connected regions as described in Sec. 12.2. Therefore, in the example just cited, one cannot employ Stokes' theorem for a path surrounding the origin as shown in Fig. 12.18.

However, it will now be demonstrated that if the origin is the *only* singular point (with irrotational flow present at all other points) then the circulations for *all* paths about the origin are of *equal* value. Fig. 12.19 shows two arbitrary paths about the origin for such a flow. Two parallel lines *AB* and *ED* are drawn between the inner and outer curves to form a narrow strip. In this way a simply connected region may be formed (shown crosshatched in Fig. 12.19) which entails most of the boundaries of C_1 and C_2 except for the strip extremities *DB* and *EA*. Since the flow is entirely irrotational and without singularities in this region, we may employ Stokes' theorem to give a zero circulation for the boundary. Hence, by starting at *A* and moving clockwise on boundary C_1, etc.,

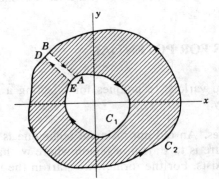

FIGURE 12.19
Formation of simply connected region.

it may be said that

$$\int_A^E \mathbf{V} \cdot \mathbf{ds} + \int_E^D \mathbf{V} \cdot \mathbf{ds} + \int_D^B \mathbf{V} \cdot \mathbf{ds} + \int_B^A \mathbf{V} \cdot \mathbf{ds} = 0 \qquad (12.52)$$
$$C_1 \qquad\qquad\qquad\qquad C_2$$

Arrows in the diagram have been employed to indicate the direction of the line integrations above. If the strip is now made thinner without limit, the integrations along *ED* and *BA* (the second and forth of the integrations above) become equal in magnitude but opposite in sign. Hence, in the limit these integrations must cancel out. Meanwhile, the integral $\int_A^E \mathbf{V} \cdot \mathbf{ds}$ becomes effec-
$$C_1$$
tively a closed line integral in the clockwise direction, while the integral $\int_D^B \mathbf{V} \cdot \mathbf{ds}$ becomes effectively a closed line integral in the counterclockwise
$$C_2$$
direction. In the limit, Eq. (12.52) then becomes

$$\oint_{C_1} \mathbf{V} \cdot \mathbf{ds} + \oint_{C_2} \mathbf{V} \cdot \mathbf{ds} = 0$$

Reversing the direction of integration on C_2 changes the sign of the closed integral so that

$$\oint_{C_1} \mathbf{V} \cdot \mathbf{ds} = \oint_{C_2} \mathbf{V} \cdot \mathbf{ds} \qquad (12.53)$$

Since these C_1 and C_2 are arbitrary paths about the origin, it may be concluded that the circulation for all paths surrounding the origin are equal if taken in the same sense.

The preceding is true only if the paths surround the origin as the lone singular point. However, should there be a distant additional singular point and should the path be large enough to include both singularities, then circulation for this path may be different from that of the family of curves including only the singular point at the origin. *As a way of generalizing, we may state that the circulation for a given path in an irrotational flow containing a finite number of singular points is constant as long as the path is altered in such a way as to retain the same singular points.*

12.15   SOLUTION METHODOLOGIES FOR POTENTIAL FLOW

We shall now examine various techniques for effecting a solution. The techniques vary in accuracy.

1. Graphical techniques. An old approximate technique is to draw a *flow net* "by eye." The procedure is to start in a region where we have uniform flow or where a known flow exists. For the former, we start in the uniform flow region

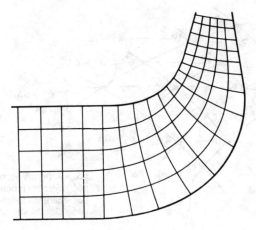

FIGURE 12.20
Flow net—two-dimensional flow.

with a set of equally spaced streamlines. For the latter where a known flow is present, space the streamlines so that equal volumes of flow (Δq) are present between the streamlines. In this regard, note that for such flows $\Delta q = (\Delta_N)(V)$ where Δ_N is the distance between streamlines and V is the mean velocity between the streamlines. The streamlines are then continued so as to conform to the boundaries. Note that the boundaries are streamlines and form part of the pattern. The remaining pattern is adjusted so as to "look right." Equipotential lines are drawn and, with streamlines, adjustments are made so that the grid will have a system of *curvilinear squares* of varying size. The finer the flow net, the closer will the curvilinear squares approach actual squares. We have shown a flow net in Fig. 12.20. When the flow net has been adjusted to seem reasonable, we can compute the velocity crossing the potential line segment in each curvilinear square by using simple continuity considerations, knowing the Δq for the pair of associated streamlines and measuring Δ_N of the potential line segment between these streamlines at the position of interest. Also, we can compute pressures using Bernoulli. We leave these simple considerations to the student. Finally, if we draw diagonals to the curvilinear squares we have a means of checking the accuracy of the flow net by gaging how close the diagonals form squares—the closer to the latter, the more valid is the flow net.

2. Analytical techniques. In the sections to follow, we shall simply present harmonic functions for either a potential function or a stream function. Once such a function is presents, say a stream function, we can get the associated potential function or vice versa using the Cauchy-Riemann equations. Each such function then generates a valid theoretical flow.

Next we can *superpose* certain of the simple flows to form other useful flows. This is possible because of the linearity of Laplace's equation. We shall do this in Part E of this chapter. For such flows, we wish next to point out a useful technique for sketching the streamlines of the *combined* flow. First draw streamlines for the simple constituent flows using contour lines of the stream

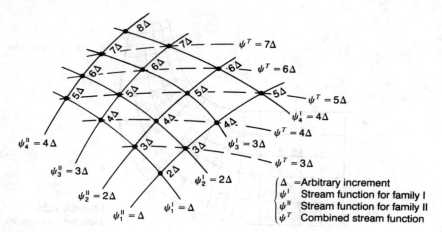

FIGURE 12.21
Construction of a combined streamline pattern from constituent streamline patterns.

functions from each constituent family having *identical* sets of constants, the successive difference of which are the *same value* (see Fig. 12.21). Where the streamlines intersect, the combined contour line passing through will have the value of the *sum* of the ψ's for the constituents at that point. By simply passing curves through the points having the *same total* ψ at the corners of the curvilinear parallelpipeds formed by the intersecting constituent streamlines (see dashed lines in Fig. 12.21) we can readily sketch the streamlines for the combined flow.

Another analytic approach is to use *analytic complex variable functions* $f(z) = \phi + i\psi$. For such functions the real part ϕ and the imaginary part ψ are harmonic functions which are related to each other by the Cauchy-Riemann equations. Hence each analytic function can thus represent some two-dimensional potential flow. By a technique called *conformal mapping*, ϕ and ψ can be changed or "mapped" to form another different set of harmonic functions still related by the Cauchy-Riemann equations and thus representing yet another two-dimensional potential flow. By such successive mappings, one can sometimes find useful flows. However, the prospects of success for any given geometry are usually very dim. Now because of the computer we have effective numerical methods that can readily be used for specific problems, making the complex variable approach of very limited value as a direct tool.

Finally there is the approach of direct integration using the theory of the partial differential equation.

3. Numerical techniques. The most important direct approaches via numerical methods are:

Finite difference. This is an old method made more useful with the advent of the high-speed computer. In Chap. 16 we have presented an introduction to this approach with some examples and projects.

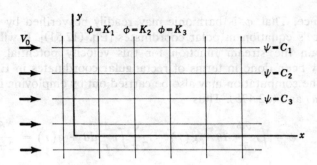

FIGURE 12.22
Flow net for uniform flow.

Finite elements. This is an approach developed in the 1950s in the aircraft industry for structural design. It requires the use of the computer. It is now widely used in many fields, including fluid flow and heat transfer.[7]

Boundary elements. This is a newer development that is growing in popularity. Again, it relies on the computer and can be very effective when the boundaries are small compared to the size of the domain.

We now present certain simple flows.

12.16 UNIFORM FLOW

The most elementary flow is given by the velocity potential $\phi = V_0 x$. Clearly this represents a uniform flow in the positive x direction of magnitude V_0. The corresponding stream function by the methods of Sec. 12.9 is easily evaluated as $V_0 y$. Streamlines are formed by loci of points corresponding to constant values of $V_0 y$. These must be horizontal lines as shown in Fig. 12.22. The lines of constant potential $V_0 x = \text{const}$ are vertical lines. There are no singular points in this flow; so Stokes' theorem indicates that there is a zero circulation for all paths in the flow.

12.17 TWO-DIMENSIONAL SOURCES AND SINKS

Let us now explore the flow represent by the velocity potential

$$\phi = \frac{\Lambda}{2\pi} \ln r \tag{12.54}$$

where Λ is a positive constant and r is the modulus from the origin of a

---

[7]See I. H. Shames and C. L. Dym, *Energy and Finite Element Methods in Structural Mechanics*, Hemisphere Publishing Corp., 1985. See Chap. 17, Part D for two-dimensional potential flow.

reference. That ϕ is harmonic may readily be verified by substituting it into Laplace's equation in polar coordinates [Eq. (12.51)]. It will be very useful to ascertain the stream function for this velocity potential. Actually this has already been done in terms of rectangular coordinates in Example 12.1. However, the computation may also be carried out by employing integrations of Eqs. (12.47a) and (12.47b). Thus

$$\psi = \int r \frac{\partial \phi}{\partial r} \, d\theta + g(r) = \frac{\Lambda}{2\pi} \int r \frac{1}{r} \, d\theta + g(r) = \frac{\Lambda\theta}{2\pi} + g(r)$$

$$\psi = -\int \frac{1}{r} \frac{\partial \phi}{\partial \theta} \, dr + h(\theta) = 0 + h(\theta)$$

Comparing both ψ's one sees that $h(\theta)$ must equal $\Lambda\theta/2\pi$ and $g(r)$ must be a constant, which is of no consequence in the ensuing computations. Hence

$$\psi = \frac{\Lambda\theta}{2\pi} \tag{12.55}$$

where θ is $\tan^{-1}(y/x)$.

Let us first establish the flow net. Streamlines will be a family of lines of the equation

$$\frac{\Lambda\theta}{2\pi} = \text{const} \tag{12.56}$$

By choosing different constants, it is seen that the resulting loci of points will form a family of radial lines emanating from the origin. These are shown in Fig. 12.23. It is immediately inferred from the streamline pattern and the symmetry of the stream function that the fluid is either emanating at the origin and then spreading radially out to infinity or possibly the reverse. The former type of flow

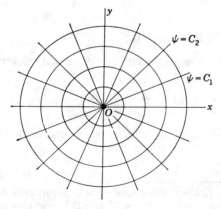

FIGURE 12.23
Flow net for source.

is called a *source*, while the latter is called a *sink*.[8] The lines of constant potential are given as

$$\frac{\Lambda}{2\pi} \ln r = \text{const}$$

Therefore

$$r = \text{const} \tag{12.57}$$

We see that a family of concentric circles is formed by the potential lines. Note the orthogonality of the resulting grid system in Fig. 12.23 as stipulated in Sec. 12.10.

The velocity components in the radial and transverse direction may be evaluated simply.

$$V_r = \frac{\partial \phi}{\partial r} = \frac{\Lambda}{2\pi r}$$

$$V_\theta = \frac{\partial \phi}{r\partial \theta} = 0 \tag{12.58}$$

Note that in this case the radial velocity is positive and hence is directly away from the origin O, making the flow that of source.[9] Also, this velocity is zero at infinity and increases towards the origin, reaching an infinite value at the origin. As pointed out in Sec. 12.14, the origin is thus a singular point. In addition to the presence of infinite velocity, which does not exist in nature, the question about what happens to the fluid "going into point O" leads us to conclude that point O of the flow plane is entirely fictitious and devoid of physical counterpart.

The significance of the constant Λ will now be investigated. Compute the rate of flow q through a control surface associated with any circular boundary about the origin and extending a unit length in the z direction, as shown in Fig. 12.24. The evaluation of q becomes

$$q = \oint\!\!\!\oint \mathbf{V} \cdot d\mathbf{A} = \int_0^{2\pi} V_r\, r\, d\theta = \int_0^{2\pi} \frac{\Lambda}{2\pi r} r\, d\theta = \Lambda \tag{12.59}$$

Since r could have been any value in the computation, it is clear that volume

---

[8] The two-dimensional source is analogous to the positive two-dimensional *line charge* of electrostatics, while the two-dimensional sink is analogous to the two-dimensional negative line charge of electrostatics. If λ is the strength of the line charge in units of charge per unit length, we have for the potential of the electric field

$$V = -\frac{\lambda}{2\pi\epsilon_0} \ln r$$

[9] For a sink the velocity potential is $(-\Lambda/2\pi)\ln r$, and the stream function is $\psi = -\Lambda\theta/2\pi$.

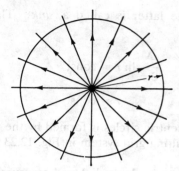

FIGURE 12.24
Control surface around source.

efflux per unit time through any circular strip about the origin equals the constant Λ. In the case of the sink, Λ is similarly shown to equal influx rate for all unit circular strips. Furthermore, the reader may ascertain from continuity that Λ is the volume flow through *any* noncircular unit strip about the origin. The constant Λ is usually called the *strength* of the source or sink.

Since the origin represents the sole singularity of this flow, we may establish the circulation about the origin by using *any* surrounding path. Therefore, employing a circle of radius r, we have

$$\Gamma = \oint \mathbf{V} \cdot \mathbf{ds} = \int_0^{2\pi} V_\theta r\, d\theta = 0 \tag{12.60}$$

since V_θ is everywhere zero along the path. Source and sink flows thus have zero circulation for all possible paths.

12.18 THE SIMPLE VORTEX

The simple vortex is a useful flow which may easily be established by selecting the stream function of the source to be the velocity potential of the vortex. Thus

$$\phi = \frac{\Lambda}{2\pi}\theta \tag{12.61}$$

and considering Eq. (12.47*b*), we see that the stream function of the vortex is

$$\psi = -\frac{\Lambda}{2\pi}\ln r \tag{12.62}$$

The flow net will be the same form as in the source-sink analysis except that the concentric circles will now be the streamlines and the family of radial lines, the

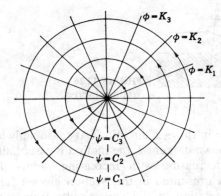

FIGURE 12.25
Flow net for vortex.

potential lines as shown in Fig. 12.25. The streamline pattern indicates that the fluid is moving in circular paths about the origin. Computing the velocities from the preceding function, we get

$$V_\theta = \frac{\Lambda}{2\pi r} \qquad (12.63a)$$

$$V_r = 0 \qquad (12.63b)$$

Here a singularity again appears at the origin, where you will note that the tangential velocity approaches infinity. Also note that the choice of sign of Eq. (12.61) has led to a counterclockwise flow about the origin, i.e., a counterclockwise vortex. Obviously, a change in the sign of Eq. (12.61) will yield a clockwise vortex.

As in the previous section, the origin is a singular point and is thus excluded from the region of irrotational flow, making this a multiply connected region for our purposes. Hence, to ascertain the circulation for any path about the origin, we must carry out the line integration. Using a circular path of radius r we then get

$$\Gamma = \oint \mathbf{V} \cdot \mathbf{ds} = \int_0^{2\pi} V_\theta r\, d\theta = \int_0^{2\pi} \frac{\Lambda}{2\pi r} r\, d\theta = \Lambda \qquad (12.64)$$

Since there are no other singularities, this must be the circulation for all paths about the origin. Consequently, Λ in the case of the vortex is the measure of circulation about the origin and is also referred to as the *strength* of the vortex.

As the name implies, a portion of this flow resembles part of the common whirlpool found while paddling a boat. An approximate surface profile of a whirlpool is indicated in Fig. 12.26. If z is taken as vertical, there is an outer region shown crosshatched where the flow is approximately two-dimensional, as

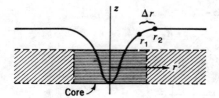

FIGURE 12.26
Whirlpool vortex.

we observe from above in the $-z$ direction. At reasonably large distances from the "core," which is shown darkened, the motion is essentially along the circular paths with velocity decreasing with increasing radius. In this region the "mathematical" vortex presented in this section gives a fair representation of flow. The lack of correlation in the core results from the increased viscous action due to the high velocity gradients here, which are proportional to $1/r^3$. Recall that it is essentially this mechanism which accounts for lack of correlation between nonviscous and real flows of gases having even small viscosity.

Now consider the free surface of the whirlpool (see Fig. 12.26). Look at points at r_1 and r_2 separated by a short distance Δr near the core of the whirlpool. Bernoulli's equation for those flow particles then becomes, on noting that the pressure at the free surface is p_{atm},

$$\frac{p_2 - p_1}{\gamma} = 0 = \frac{\Lambda^2}{8g\pi^2}\left(\frac{1}{r_1^2} - \frac{1}{r_2^2}\right) + (z_1 - z_2) \qquad (12.65)$$

Clearly the expression $(1/r_1^2 - 1/r_2^2)$ with $r_1 < r_2$ becomes larger for equal increments of Δr as one approaches the core. From Eq. (12.65) we see that to satisfy the equation z_1 must be increasingly smaller than z_2 as we get to the core. This indentation of the free surface is shown in Fig. 12.26.

12.19 THE DOUBLET

The doublet will be the final simple flow of Part D of this chapter. It is formed by a limiting procedure in a manner which may seem very artificial to you. Nevertheless, the doublet is an extremely important flow used in analyses of practical flows, as you will soon see.

To develop the doublet, imagine a source and sink of equal strength Λ at equal distances a from the origin along the x axis as shown in Fig. 12.27. From

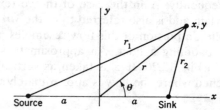

FIGURE 12.27
Doublet development.

any point (x, y) lines indicated as r_1 and r_2, respectively, are drawn to the source and the sink. Also the polar coordinates (r, θ) of this point are shown. From the law of cosines, the following relations may be established relating r_1 and r_2 with the polar coordinates of the point (x, y):

$$r_1 = \left(r^2 + a^2 + 2ra \cos \theta \right)^{1/2}$$
$$r_2 = \left(r^2 + a^2 - 2ra \cos \theta \right)^{1/2} \tag{12.66}$$

Now the potential function of the two flows may be superposed to give us for the combined flow

$$\phi = \frac{\Lambda}{2\pi} (\ln r_1 - \ln r_2) \tag{12.67}$$

Or, employing the relations of Eq. (12.66), we have

$$\phi = \frac{\Lambda}{2\pi} \left[\frac{1}{2} \ln \left(r^2 + a^2 + 2ra \cos \theta \right) \right.$$
$$\left. - \frac{1}{2} \ln \left(r^2 + a^2 - 2ra \cos \theta \right) \right] \tag{12.68}$$

In each logarithmic expression, we divide and multiply by $r^2 + a^2$. This gives us

$$\phi = \frac{\Lambda}{4\pi} \left\{ \ln \left[(r^2 + a^2) \left(1 + \frac{2ra \cos \theta}{r^2 + a^2} \right) \right] - \ln \left[(r^2 + a^2) \left(1 - \frac{2ra \cos \theta}{r^2 + a^2} \right) \right] \right\}$$

Since the logarithm of a product of two quantities equals the sum of logarithms, the following may be said:

$$\phi = \frac{\Lambda}{4\pi} \left[\ln(r^2 + a^2) + \ln \left(1 + \frac{2ra \cos \theta}{r^2 + a^2} \right) \right.$$
$$\left. - \ln(r^2 + a^2) - \ln \left(1 - \frac{2ra \cos \theta}{r^2 + a^2} \right) \right] \tag{12.69}$$

Note that the first and third terms cancel inside the brackets. The term $2ra \cos \theta / (r^2 + a^2)$ for point P is less than unity everywhere except[10] when P is at the singularity points—i.e., at $x = \pm a$, $y = 0$. Hence avoiding the positions for P at the singularities, each logarithmic expression may be expanded

---

[10]To show this explicitly, divide numerator and denominator of the expression by $2ra$, and consider the magnitude of the resulting expression. Thus

$$\left| \frac{2ra \cos \theta}{r^2 + a^2} \right| = \frac{|\cos \theta|}{|\frac{1}{2}(r/a + a/r)|}$$

The resulting numerator cannot exceed unity. But the denominator will exceed unity if $r \neq a$ as you may easily verify yourself. We can conclude further that the magnitude of the quantity is less than unity everywhere except at $\theta = 0$ or π and $r = a$.

into a power series.

$$\phi = \frac{\Lambda}{4\pi} \left\{ \left[\frac{2ra\cos\theta}{r^2 + a^2} - \frac{1}{2}\left(\frac{2ra\cos\theta}{r^2 + a^2}\right)^2 + \frac{1}{3}\left(\frac{2ra\cos\theta}{r^2 + a^2}\right)^3 + \cdots \right] \right.$$

$$\left. - \left[-\frac{2ra\cos\theta}{r^2 + a^2} - \frac{1}{2}\left(\frac{2ra\cos\theta}{r^2 + a^2}\right)^2 - \frac{1}{3}\left(\frac{2ra\cos\theta}{r^2 + a^2}\right)^3 + \cdots \right] \right\} \quad (12.70)$$

Collecting terms, we get

$$\phi = \frac{\Lambda}{4\pi} \left[\frac{4ra\cos\theta}{r^2 + a^2} + \frac{2}{3}\left(\frac{2ra\cos\theta}{r^2 + a^2}\right)^3 + \cdots \right] \quad (12.71)$$

Now bring the source and sink together,[11] i.e., let $a \to 0$, and at the same time increase the strength Λ to infinite value such that in the limit the product of $a\Lambda/\pi$ is some finite number χ. That is,

$$a \to 0$$
$$\Lambda \to \infty \quad (12.72)$$

such that $a\Lambda/\pi \to \chi$. Under such circumstances the potential becomes, in the limit,

$$\boxed{\phi = \frac{\chi\cos\theta}{r}} \quad (12.73)$$

Since the higher-order terms all have a's which are not incorporated with Λ, these terms have vanished in the limiting process.

By partial integration of the polar-coordinate relations involving the stream function and velocity potential [Eq. (12.47a)], we may readily establish the stream function as

$$\boxed{\psi = -\frac{\chi\sin\theta}{r}} \quad (12.74)$$

The streamlines associated with the doublet[12] are then

$$\frac{\chi\sin\theta}{r} = C \quad (12.75)$$

Replacing $\sin\theta$ by y/r, we get

$$\chi\frac{y}{r^2} = C \quad (12.76)$$

---

[11]Since $a \to 0$, the previously made restriction of $r \neq a$ merely means that we must exclude the origin from our results.

[12]The two-dimensional doublet is analogous to the *two-dimensional electric dipole*, where positive and negative line charges of equal strength are merged in the manner described in this section.

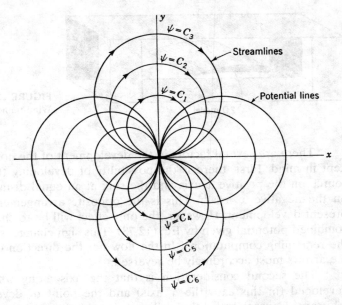

FIGURE 12.28
Flow net for doublet.

In terms of Cartesian coordinates only, this equation may be rearranged to form

$$x^2 + y^2 - \frac{\chi}{C}y = 0 \qquad (12.77)$$

From analytic geometry this is recognized as a family of circles. For $x = 0$ there are two values of y, one of which is zero; the circles clearly have their centers along the y axis. At the position on the circle where $y = 0$, note that x is zero for all values of the constant. Therefore, the family of circles formed by choosing various values of C must be tangent to the x axis at the origin. The family of streamlines are shown in Fig. 12.28 in accordance with the previous conclusions. From the positions of the source and sink in the development of the doublet, it should be obvious that flow must proceed out from the origin in the negative x direction, as is shown in the diagram.

In a similar manner, lines of constant potential are given by the equation

$$\frac{\chi \cos \theta}{r} = K \qquad (12.78)$$

This equation becomes, in Cartesian coordinates,

$$x^2 + y^2 - \frac{\chi}{K}x = 0 \qquad (12.79)$$

Again we have a family of circles, this time with centers along the x axis. As shown in Fig. 12.28 these circles are tangent to the y axis at the origin. Note that we have an orthogonal grid system.

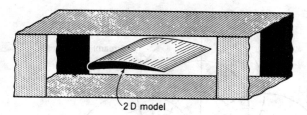

FIGURE 12.29
Wind-tunnel test section.

2 D model

There are several factors in the development of the doublet that should be kept in mind. First, there is the possibility of developing the doublet with the source on the positive x axis and the sink at an equal distance from the origin on the negative x axis.[13] This is the opposite arrangement from that in the present development. However, the only effect will be to change the sign of the combined potential given by Eq. (12.73). This sign change is carried throughout the remaining computations. In the flow net the direction of flow indicated by the arrows must accordingly be reversed.

The second consideration is that the axis along which the doublet is developed (in this case the x axis) and the point of development are quite arbitrary. Consequently, the collinear diameters of the potential lines must coincide with the axis of development, and the intersection of such streamline and potential line diameters corresponds to the doublet origin.

Upon observing the stream function or velocity potential, it is seen that radial velocities must become infinite as one approaches the center of the doublet, in this case the origin of the reference. Hence, doublet flow has a singularity which makes it for our purposes, a flow in a multiply connected region. However, since the circulation about the singular point of a source or sink is zero for any strength, it is then seen that the circulation about the singular point in a doublet flow must be zero, so for all paths in a doublet flow $\Gamma = 0$.

It is known that at large distances from a doublet, the flow approximates the disturbances of a two-dimensional airfoil of infinite aspect ratio (Fig. 12.29). This is useful in wind-tunnel work, where the effects of an airfoil as felt at the distant walls may be approximated mathematically by substituting a doublet of proper strength in place of the wing. In this manner, design of wind tunnels may be made to minimize "wall interference" or "blockage" to render results which more clearly duplicate free-flight conditions.

Thus, although "artificial," the doublet has immediate use in aerodynamics. Furthermore, use will be made in the following section in the development

---

[13]The direction of the doublet is taken from the sink to the source. Thus we have developed in the preceding discussion the negative doublet, i.e., a doublet pointing in the negative coordinate direction. Our reason for doing this is that it will be the negative doublet that will be used in the remainder of the chapter.

TABLE 12.1

	ϕ	ψ	Γ	V	
Uniform flow (toward $+x$)	$V_0 x$	$V_0 y$	0 everywhere	$V_x = V_0$	$V_y = 0$
Source	$\dfrac{\Lambda}{2\pi} \ln r$	$\dfrac{\Lambda}{2\pi} \theta$	0 everywhere	$V_r = \dfrac{\Lambda}{2\pi r}$	$V_\theta = 0$
Counter-clock-wise vortex	$\dfrac{\Lambda}{2\pi} \theta$	$-\dfrac{\Lambda}{2\pi} \ln r$	Λ around singular point	$V_r = 0$	$V_\theta = \dfrac{\Lambda}{2\pi r}$
Doublet issuing in $-x$ direction	$\dfrac{\chi \cos \theta}{r}$	$-\dfrac{\chi \sin \theta}{r}$	0 everywhere	$V_r = -\dfrac{\chi \cos \theta}{r^2}$	$V_\theta = -\dfrac{\chi \sin \theta}{r^2}$

of flows of superposition. As an aid in the ensuing computations we have tabulated the results of Part D in Table 12.1.

PART E
SUPERPOSITION OF 2-D SIMPLE FLOWS

12.20   INTRODUCTORY NOTE ON THE SUPERPOSITION METHOD

The course of action to be followed in this section will be to examine certain combinations of simple flows with the purpose of interpreting physically significant flow patterns from each combination. This will seem to the reader as a procedure of "going backward," i.e., a solution appearing first (the choice of the simple flows to be superposed) followed by efforts to ascertain what problem has been solved (the interpretation of the combined flow pattern). The following considerations may help to justify this procedure.

1. Actually to establish a two-dimensional, irrotational, incompressible flow for any given problem is usually extremely difficult if not impossible by straight analytical methods. However, numerical methods such as finite difference, finite elements, and boundary elements may be used. Also, *approximate solutions* may be made for given problems by an extension of the superposition technique.

2. Despite the ineffectiveness of this study as a direct tool for providing solutions to given problems, it is nevertheless necessary as background for exploiting the use of more sophisticated mathematical and numerical techniques. Furthermore, experimental investigations can be more effectively initiated and results more readily interpreted with the aid of this fundamental background.

3. Finally, we will presently see the demonstration of an important law of aerodynamics by this superposition study.

12.21 SINK PLUS A VORTEX

We now consider the flow resulting from the superposition of a *sink plus a vortex*. We will see that the resulting flow will simulate the flow exiting into a drain from a tank and also possibly that of a tornado. Accordingly, we now superpose these flows to get the following stream function and velocity potential using Γ to be the strength of the vortex:

$$\phi = -\frac{\Lambda}{2\pi} \ln r + \frac{\Gamma}{2\pi} \theta \qquad (12.80a)$$

$$\psi = -\frac{\Lambda}{2\pi} \theta - \frac{\Gamma}{2\pi} \ln r \qquad (12.80b)$$

In Fig. 12.30 we have drawn streamlines for both the sink and the vortex and then combined these in the manner described earlier (see Fig. 12.21) to show a graphical representation of the combined streamline flow. The similarity of the combined flows with what one would expect to see looking down on the flow in an emptying tank or a tornado is quite clear.

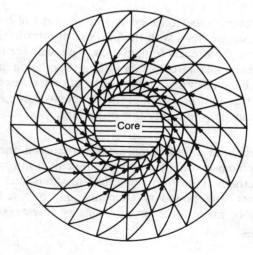

FIGURE 12.30
Streamlines for combined sink plus vortex flow.

The velocity components for this flow are

$$V_r = -\frac{\Lambda}{2\pi r}$$

$$V_\theta = \frac{\Gamma}{2\pi r}$$

(12.81)

If we wish to determine the pressure, we may use Bernoulli's equation. Thus, the possible change $\Delta p/r$ from position 1 to position 2 is

$$\frac{\Delta p}{\gamma} = \left(\frac{V_2^2}{2g} + z_2\right) - \left(\frac{V_1^2}{2g} + z_1\right)$$

$$= \frac{\Lambda^2 + \Gamma^2}{8g\pi^2 r_2^2} - \frac{\Lambda^2 + \Gamma^2}{8g\pi^2 r_1^2} + z_2 - z_1$$

$$= \frac{\Lambda^2 + \Gamma^2}{8g\pi^2}\left[\frac{1}{r_2^2} - \frac{1}{r_1^2}\right] + (z_2 - z_1)$$

(12.82)

By choosing the correct strengths Λ and Γ of the sink and the vortex using actual known data for two points, the above equations will give a good representation of the velocity and pressure variation of the combined flow except near the center where r becomes small. This is easily understood since the velocity gradients for V_r and V_θ become very large and viscous effects now become significant, thus invalidating the inviscid theory we are using. As in the case of the vortex flow describing a whirlpool this viscous region is called the core.

12.22   FLOW ABOUT A CYLINDER WITHOUT CIRCULATION

Let us now examine the combination of a uniform flow and a doublet. The latter has its axis of development parallel to the direction of the uniform flow and is so oriented that the direction of the efflux opposes the uniform flow. This is shown in Fig. 12.31, where portions of the streamline patterns of both simple flows have been indicated. The combined potential for the combination is then given as

$$\phi = V_0 x + \frac{\chi \cos \theta}{r}$$

(12.83)

From previous work, the stream function of the combined flow becomes

$$\psi = V_0 y - \frac{\chi \sin \theta}{r}$$

(12.84)

In this type of analysis the streamline is of particular importance. In two-dimensional flow the streamline has already been interpreted as the edge of

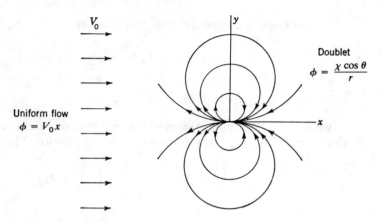

FIGURE 12.31
Superposition of uniform flow and doublet.

a surface extending without limit in the z direction. The velocity must always be tangent to this surface, so there can be no flow penetration through it. Is this not identically the characteristic of a solid impervious boundary? Hence, one may consider a streamline as the contour of an impervious, two-dimensional body. For instance, Fig. 12.32 shows a set of streamlines. Any streamline such as *A-B* may be considered as the edge of a two-dimensional body shown crosshatched in the diagram. The remaining streamlines then form the flow pattern about this boundary. The essential procedure in the superposition technique may now be stated. A streamline is to be found which encloses an area whose shape is of practical significance in fluid flow. This streamline will then be considered as the boundary of a two-dimensional solid object, thus establishing the problem. The streamline pattern outside this region then constitutes the flow about this body.

Accordingly, let us examine the streamline whose constant is zero. Thus

$$V_0 y - \frac{\chi \sin \theta}{r} = 0 \qquad (12.85)$$

Replacing y by $r \sin \theta$ and factoring out $\sin \theta$ gives us

$$\sin \theta \left(V_0 r - \frac{\chi}{r} \right) = 0 \qquad (12.86)$$

What curve in the xy plane satisfies this equation? Clearly, if $\theta = 0$ or $\theta = \pi$, the equation is satisfied, indicating that the x axis is part of the streamline $\psi = 0$. Also, when the quantity in parentheses is zero, the equation is satisfied. That is, when

$$r = \left(\frac{\chi}{V_0} \right)^{1/2} \qquad (12.87)$$

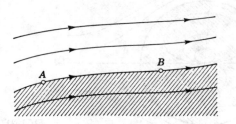

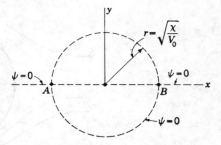

FIGURE 12.32
Streamline becomes boundary.

FIGURE 12.33
Streamline $\psi = 0$.

the equation is satisfied. Hence, there is a circle of radius $\sqrt{\chi/V_0}$ which may also be considered part of the streamline $\psi = 0$. This is shown in Fig. 12.33. It will be of interest to examine points A and B, the points of intersection of the circle and the x axis. The polar coordinates for these points are $r = (\chi/V_0)^{1/2}$, $\theta = \pi$ for A and $r = (\chi/V_0)^{1/2}$, $\theta = 0$ for B. The velocity at these points may be found by first taking partial derivatives of the velocity potential or the stream function in two orthogonal directions and combining the velocity components. Thus

$$V_r = \frac{\partial \phi}{\partial r} = V_0 \cos \theta - \frac{\chi \cos \theta}{r^2} \qquad (12.88a)$$

$$V_\theta = \frac{\partial \phi}{r \, \partial \theta} = -V_0 \sin \theta - \frac{\chi \sin \theta}{r^2} \qquad (12.88b)$$

At points A and B it is seen from these formulations that substituting the proper values of the coordinates, we get zero values for the velocity components, so $\mathbf{V}_A = \mathbf{V}_B = \mathbf{0}$. Such points clearly are *stagnation points*. As in the present case, stagnation points will usually be located at positions through which streamlines "open up" and thus form a region of possible physical interest. Use will be made of this fact in Sec. 12.23.

According to the earlier discussion of this section the circular region enclosed by part of the streamline $\psi = 0$ could be taken as a solid cylinder in a frictionless flow which at a large distance from the cylinder is moving uniformly in a direction transverse to the cylinder axis.

Figure 12.34 shows other streamlines of the flow. Those outside the circle form the flow pattern of the aforementioned flow, while the streamlines inside may be disregarded. However, it is entirely possible to imagine the region *outside* the circle as solid material with a fluid flow taking place inside the circular boundary. But this flow has a point of infinite velocity at the center of the circle and hence is of little interest in physical situations. Furthermore,

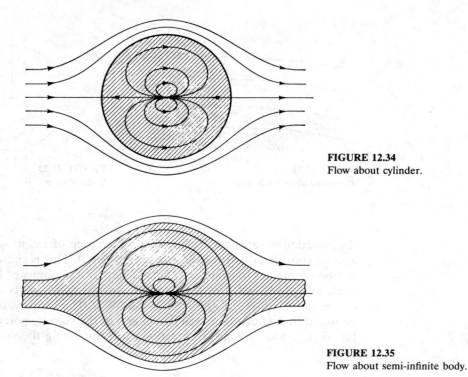

FIGURE 12.34
Flow about cylinder.

FIGURE 12.35
Flow about semi-infinite body.

there are other regions which may be chosen to represent solid bodies in a flow, as in the illustration given in Fig. 12.35. However, this body and the remaining other possibilities similar to it extend infinitely in the x direction and for this reason are of little interest for physical problems.

To return to the flow around a cylinder, it should be noted that each of the basic flows used has zero circulation everywhere and, therefore, the circulation for a path about the cylinder must also be zero.

12.23 LIFT AND DRAG FOR A CYLINDER WITHOUT CIRCULATION

Lift and drag have been defined as forces per unit length on the cylinder in directions normal and parallel, respectively, to the uniform flow.

The pressure for the combined flows becomes uniform at large distances from the cylinder where the effects of the doublet becomes diminishingly small. Knowing this pressure p_0, as well as the uniform velocity V_0, we may profitably employ *Bernoulli*'s equation between infinity and points on the boundary of the cylinder in order to ascertain pressures on the boundary. Thus disregarding

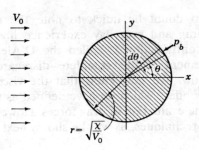

$$V_0$$

$$r = \sqrt{\frac{\chi}{V_0}}$$

FIGURE 12.36
Compute drag on cylinder.

potential energy changes as negligible

$$\frac{V_0^2}{2g} + \frac{p_0}{\gamma} = \frac{V_b^2}{2g} + \frac{p_b}{\gamma} \tag{12.89}$$

where the subscript b refers to the cylindrical boundary. Since fluid cannot penetrate into the boundary, the velocity V_b must be in the transverse direction and hence equal to its component V_θ. Thus at $r = (\chi/V_0)^{1/2}$

$$V_b = \frac{\partial \phi}{r \, \partial \theta}\bigg|_{r=(\chi/V_0)^{1/2}} = -2V_0 \sin \theta \tag{12.90}$$

Substituting into Eq. (12.89) and solving for p_b gives us

$$p_b = \gamma \left[\frac{V_0^2}{2g} + \frac{p_0}{\gamma} - \frac{(2V_0 \sin \theta)^2}{2g} \right] \tag{12.91}$$

The computation of the drag is carried out by integrating the force components in the x direction stemming from pressure on the boundary. Using Fig. 12.36, we then get for the drag D

$$D = -\int_0^{2\pi} p_b \cos \theta \left(\frac{\chi}{V_0} \right)^{1/2} d\theta$$

$$= -\int_0^{2\pi} \gamma \left(\frac{\chi}{V_0} \right)^{1/2} \left[\frac{V_0^2}{2g} + \frac{p_0}{\gamma} - \frac{(2V_0 \sin \theta)^2}{2g} \right] \cos \theta \, d\theta$$

Carrying out this integration leads to a drag D of zero magnitude.[14] The lift L may similarly be computed to be of zero magnitude.

---

[14] You will recall that we worked out this problem in Sec. 6.6 as an illustration of the use of Bernoulli's equation.

The reader will no doubt be quick to note the discrepancy between theoretical results for drag and everyday experience involving fluids moving about all bodies. This contradiction is called the D'Alembert paradox. The reason for the disagreement is the complete disregard of viscous effects throughout the entire flow. We now know that the viscous action in a thin region adjacent to the boundary which we have defined as the boundary layer is of prime importance in the evaluation of drag forces. However, lift may often be predicted by the present techniques, as will be shown next.

12.24 CASE OF THE ROTATING CYLINDER

A very interesting experiment is to set a light cardboard cylinder into motion in such a manner that:

1. The axis of the cylinder is perpendicular to the direction of motion.

2. The cylinder is set spinning rapidly about its own axis as the axis moves.

Fig. 12.37 illustrates a procedure of instituting this motion. Also a portion of the resulting trajectory of the cylinder is indicated. It is clear that there is a lift present which is associated with the rotation of the cylinder, since moving the cylinder in the above manner without inducing rotation results in the cylinder moving in the usual trajectory of bodies lacking a lifting force. This was known many years ago and has resulted in early attempts to employ rotating cylinders for airfoils in aircraft and for sails in ships.

The preceding experiment involves a real fluid with viscous action in regions of high velocity gradients, namely, the boundary layer. The rotation of the cylinder, as a result of this action, has developed a certain amount of rotary motion of the air about the cylinder. Hence, it is this aspect of the fluid motion that is responsible for the lift effect.

This rotary motion may be simulated in irrotational analysis by superposing a vortex onto the doublet of the analysis in the preceding section. The presence of lift stemming from this motion will then be demonstrated. Accord-

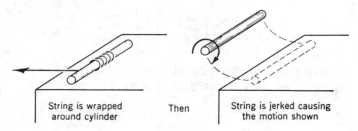

FIGURE 12.37
Experiment showing lift for rotating cylinder.

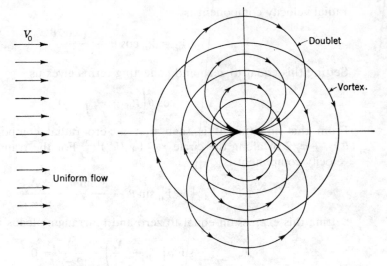

FIGURE 12.38
Superposition of uniform flow, vortex, and doublet.

ingly, the stream function and velocity potential for the combination of doublet, vortex, and uniform flow are

$$\phi = V_0 r \cos \theta + \frac{\chi \cos \theta}{r} - \frac{\Lambda}{2\pi} \theta \qquad (12.92)$$

$$\psi = V_0 r \sin \theta - \frac{\chi \sin \theta}{r} + \frac{\Lambda}{2\pi} \ln r \qquad (12.93)$$

where the last terms of these equations correspond to a clockwise vortex. The streamline patterns of the respective flow components are shown in Fig. 12.38.

The procedure in studying this flow will be essentially that of the previous section, i.e., finding a streamline enclosing a region of physical interest. It has already been implied that this will be a circle, but let us proceed as if this were not known. As suggested in the last section, it will be useful to ascertain the positions of the stagnation points and then to determine the streamlines proceeding through these points.[15] The velocity components in two orthogonal directions will be set equal to zero in computing the location of the stagnation points. By employing either the stream function or the velocity potential, the

---

[15] $\psi = 0$ will not be the streamline of interest to us here.

radial-velocity component is

$$V_r = V_0 \cos \theta - \frac{\chi \cos \theta}{r^2} \tag{12.94}$$

Setting this equal to zero and collecting terms gives us

$$\cos \theta \left(V_0 - \frac{\chi}{r^2} \right) = 0 \tag{12.95}$$

From this equation it is seen that a zero radial component may occur at $\theta = \pm \pi/2$ or along the circle $r = (\chi/V_0)^{1/2}$. For the transverse direction, the velocity component is

$$V_\theta = -V_0 \sin \theta - \frac{\chi \sin \theta}{r^2} - \frac{\Lambda}{2\pi r} \tag{12.96}$$

Setting this expression equal to zero and rearranging leads to

$$-\sin \theta \left(V_0 + \frac{\chi}{r^2} \right) - \frac{\Lambda}{2\pi r} = 0 \tag{12.97}$$

From this equation it is seen that for a zero transverse velocity

$$\theta = \sin^{-1} \frac{-\Lambda/2\pi r}{V_0 + \chi/r^2} \tag{12.98}$$

For a stagnation point, both radial and transverse components must be zero so that the location of the stagnation points occurs at

$$r = \left(\frac{\chi}{V_0} \right)^{1/2}$$

$$\theta = \sin^{-1} \frac{\dfrac{-\Lambda}{2\pi(\chi/V_0)^{1/2}}}{V_0 + [\chi/(\chi/V_0)]} = \sin^{-1} \frac{-\Lambda}{4\pi(\chi V_0)^{1/2}} \tag{12.99}$$

There will generally be two stagnation points since there are two angles for a given sine except for $\sin^{-1}(\pm 1)$.

Now the streamline proceeding through these points may be found by evaluating ψ at these points. Substituting the coordinates above into the function ψ [Eq. (12.93)] and collecting terms, we get

$$\psi_{\text{stag}} = \frac{\Lambda}{2\pi} \ln \left(\frac{\chi}{V_0} \right)^{1/2} \tag{12.100}$$

By setting the stream function [Eq. (12.93)] equal to this constant we obtain the desired streamlines.

$$V_0 r \sin \theta - \frac{\chi \sin \theta}{r} + \frac{\Lambda}{2\pi} \ln r = \frac{\Lambda}{2\pi} \ln \left(\frac{\chi}{V_0} \right)^{1/2} \tag{12.101}$$

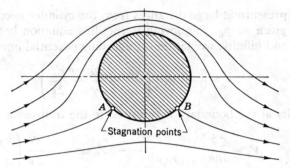

FIGURE 12.39
Flow around cylinder with circulation.

Rearranging to a more convenient form, we get

$$\sin\theta\left(V_0 r - \frac{\chi}{r}\right) + \frac{\Lambda}{2\pi}\left[\ln r - \ln\left(\frac{\chi}{V_0}\right)^{1/2}\right] = 0 \qquad (12.102)$$

All points along the circle $r = (\chi/V_0)^{1/2}$ satisfy Eq. (12.102) since for this value of r each bracketed quantity in the equation is zero. Although there are other curves satisfying the equation, we will not consider them, since it is sufficient for our purposes to have found a streamline enclosing an area of physical interest. Upon choosing the interior of the circle to represent a solid cylinder, the outer streamline pattern is then shown in Fig. 12.39. As in the preceding case, the effects of the vortex and doublet become negligibly small as one goes to large distances from the cylinder. The flow then becomes uniform at infinity.

The only flow component capable of having circulation is that of the vortex, where, it will be remembered, the circulation about the singular point equals Λ, the strength of the vortex. Hence, the circulation about the cylinder must also equal Λ. In the following section it will be learned that the circulation is important in the evaluation of lift.

12.25 LIFT AND DRAG FOR A CYLINDER WITH CALCULATION

It can readily be deduced that the presence of the vortex will result in a definite lift for the cylinder. Clearly, *above* the cylinder the vortex motion adds to the velocity of flow stemming from the uniform flow and the doublet, while *below* the cylinder the vortex takes away from the velocity stemming from the uniform flow and the doublet. From Bernoulli's equation, we would then expect the pressure on the lower half of the cylinder to be larger than the pressure on the upper half—thus giving us the lift upward.

Note also that changing the strength Λ of the vortex changes the flow patterns, particularly the position of the stagnation points, but *does not* change the radius of the cylinder. Thus, we can have an infinity of flows about a cylinder each different from the other by the strength of circulation present.

The pressure at large distances from the cylinder becomes uniform and at infinity is given as p_0. Employing Bernoulli's equation between points on the boundary and infinity, we get on disregarding potential energy change

$$p_b = \gamma \left(\frac{V_0^2}{2g} + \frac{p_0}{\gamma} - \frac{V_b^2}{2g} \right) \tag{12.103}$$

The velocity at the boundary V_b must be in the transverse direction. Thus

$$V_b = \frac{\partial \phi}{r \partial \theta}\bigg|_{r=(\chi/V_0)^{1/2}} = -2V_0 \sin \theta - \frac{\Lambda}{2\pi} \left(\frac{V_0}{\chi} \right)^{1/2}$$

Substituting this velocity into Bernoulli's equation establishes the values of pressure on the cylindrical boundary. The lift may be formulated with the help of Fig. 12.36 in the following manner:

$$L = -\int_0^{2\pi} p_b \sin \theta \left(\frac{\chi}{V_0} \right)^{1/2} d\theta$$

Substituting for p_b gives us the following expression:

$$l = -\int_0^{2\pi} \gamma \left\{ \frac{V_0^2}{2g} + \frac{p_0}{\gamma} - \frac{\left[-2V_0 \sin \theta - (\Lambda/2\pi)(V_0/\chi)^{1/2} \right]^2}{2g} \right\} \left(\frac{\chi}{V_0} \right)^{1/2} \sin \theta \, d\theta$$

Although at first glance the integration may appear a bit formidable, one may proceed nevertheless with ease, since most of the terms integrate out to zero. The result is a very simple expression

$$L = \rho V_0 \Lambda \quad \text{lift per unit length} \tag{12.104}$$

Thus, the theoretical model also shows a lift, as in the case of the earlier demonstration. Furthermore, the lift becomes a simple formula involving only the density of the fluid, the free-stream velocity, and the circulation. In addition, it can be shown that in two-dimensional, incompressible, steady flow about a boundary of *any* shape the lift always equals the product of these quantities. That is,

$$\boxed{L = \rho V_0 \Gamma} \tag{12.105}$$

This consideration is of prime importance in aerodynamics and is called the *Magnus effect*.

With the aid of conformal mapping as pointed out earlier, it is possible to establish a two-dimensional, irrotational, incompressible flow about bodies resembling subsonic airfoils. As in the case of the cylinder, one has in this analysis a family of flows for a given boundary and a given free-stream velocity. These flows differ primarily in the amount of circulation. However, all but one of these flows have an infinite velocity at the trailing edge of the airfoil. This point of the airfoil section is shown in Fig. 12.40. The Kutta-Joukowski hypothe-

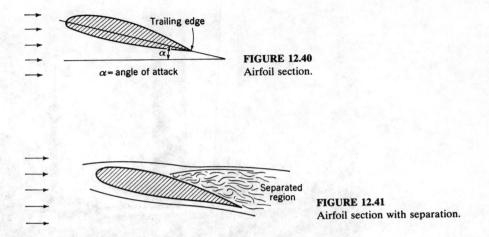

FIGURE 12.40
Airfoil section.

Trailing edge

α = angle of attack

Separated region

FIGURE 12.41
Airfoil section with separation.

sis states that the correct flow pattern of any family of flows is the one flow with a finite velocity at the trailing edge. Experiment has indicated that the theory restricted in this way gives good results. That is, the lift computed theoretically for the theoretical airfoil checks with reasonable accuracy the actual lift of a two-dimensional airfoil of the same shape as the mathematically contrived airfoil. However, the correspondence between theory and experiment breaks down when the angle of attack is made large enough to induce a condition of separation in the actual flow. This is illustrated in Fig. 12.41. Two theoretical airfoil sections developed with the aid of conformal mapping are shown in Fig. 12.42. The theory provides for the possibility of a continuous variation of the angle of attack, but this is unfortunately accompanied by a change in the shape of the airfoil section.

Example 12.2. In 1927, a man named Flettner had a ship built with two rotating cylinders to act as sails (see Fig. 12.43a). If the height of the cylinders is 15 m and the diameter $2\frac{3}{4}$ m, find the maximum possible propulsive thrust on the ship from the cylinders. The wind speed is 30 km/h, as shown in Fig. 12.43b, and the speed of the ship is 4 km/h. The cylinders rotate at a speed of 750 r/min by the action of a steam engine below deck.

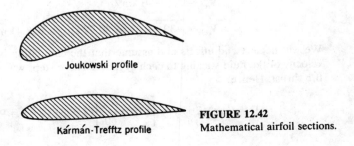

Joukowski profile

Kármán-Trefftz profile

FIGURE 12.42
Mathematical airfoil sections.

(a)

(b)

FIGURE 12.43
Flettner's ship. (*From Palmer Coslett, Power from the Wind, New York, Putnam, Van Nostrand Co.*)

The velocity V_{rel} relative to the ship is easily determined as

$$\mathbf{V}_{rel} = 30\mathbf{j} - 4\mathbf{i} \text{ km/h}$$

$$\therefore \mathbf{V}_{rel} = \sqrt{30^2 + 4^2} = 30.27 \text{ km/h} = 8.41 \text{ m/s}$$

We will neglect end effects and assume that the circulation is computed using the velocity of the fluid sticking to each cylinder. These steps will give an upper limit to the thrust. Hence,

$$\Gamma = (\omega r)(2\pi r) = \left[(750)\frac{2\pi}{60}\right](1.375)^2(2\pi) = 933 \text{ m}^2/\text{s}$$

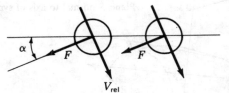

FIGURE 12.44
Lift on rotors of Flettner's ship.

The force on each cylinder per unit length is

$$F = \rho V_{\text{rel}} \Gamma = (1.229 \text{ kg/m}^3)(8.41 \text{ m/s})(933 \text{ m}^2)$$

$$= 9643 \text{ kg/s}^2 = 9643 \text{ N/m}$$

The total thrust is then:

$$F_T = (2)(9643)(15) = 289 \text{ kN}$$

This force is at right angles to the direction of the velocity relative to the ship (see Fig. 12.44). Hence, the propulsive force in the direction of motion is

$$(F_T)_{\text{prop}} = F_T \cos \alpha = 289 \frac{30}{(30^2 + 4^2)^{1/2}} = 287 \text{ kN}$$

In the actual case, we would not have a Γ as large as we have used but somewhat less than half this value. Also, in the real case, there would be some drag in the direction of V_{rel} in Fig. 12.44. If α were small, this drag would be a side thrust that is counteracted by the keel of a boat as in a sailboat. The expected thrust would then be expected to be about 40 percent of the theoretical upper limit which we have calculated.

Lest anyone scoff at this unique ship, it should be pointed out that Flettner sailed successfully across the Atlantic with his ship.

*PART F
AXISYMMETRIC THREE-DIMENSIONAL FLOWS

12.26 INTRODUCTION

In this section, we undertake the study of three-dimensional, axially symmetric, ideal flow in a manner paralleling the two-dimensional analysis of preceding parts. The axis of symmetry will be chosen as the z axis, as shown in Fig. 12.45.

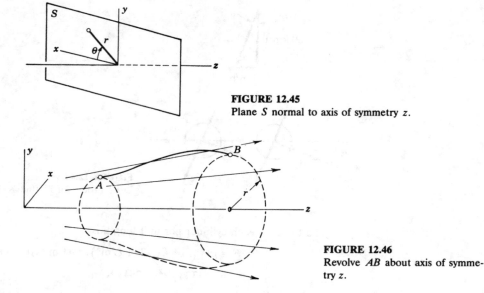

FIGURE 12.45
Plane S normal to axis of symmetry z.

FIGURE 12.46
Revolve AB about axis of symmetry z.

On planes normal to this axis, such as plane S, the flow characteristics depend only on the radial distance r and time and are independent of the angle θ. Examples of practical problems where the material of this section is of value include flow around blimps, missiles, and fully submerged submarines.

12.27 STOKES' STREAM FUNCTION

It will be useful to develop a stream function for this flow. To do this, choose an "anchor point" A and a roving point B, as indicated in Fig. 12.46. Now join both points with any curve A-B. The surface that may be profitably associated with this curve is not so obvious, perhaps, as in the case of two-dimensional flow. In this case, it is a *surface of revolution* generated by revolving the curve about the axis of symmetry. It will now be shown that the flow through this surface is independent of the generating curve for a given set of terminal points. Imagine two arbitrary surfaces formed in the above manner between points A and B. These surfaces may be considered to form a control volume in the shape of a ring about the z axis. This has been illustrated in Fig. 12.47. Conservation of mass in the case of incompressible flow demands that the flow through each surface of revolution be equal. Since the generating paths are arbitrary, one may conclude that flows associated, in the manner described, with all paths between A and B are equal. Furthermore, other points B', with the same coordinates (z,r) as B, may be used to generate surfaces of revolution with A which by the preceding argument have the same flow at any given time as in the case of surfaces generated by using point B. Thus, for a given anchor point, all these flows can be considered as a function of spatial coordinates (z, r) of the roving point and the time. Hence, it is now possible to establish a stream function. For

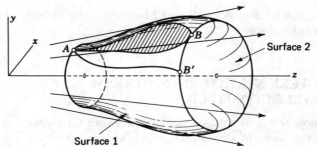

FIGURE 12.47
Control volume about axis of symmetry.

convenience in ensuing computations, this function is related to the flow in the following manner:

$$q = 2\pi\psi_A(z, r, t) \tag{12.106}$$

where ψ is *Stokes' stream function*.

As in the case of two-dimensional flow, the use of the stream function almost invariably involves the operation of partial differentation with respect to a spatial coordinate. Since changes in anchor point result at most in the stream function changing by a pure function of time, the location of the anchor point is of little consequence for such operations. Hence, the subscript A in the presentation above is usually deleted.

The preceding discussion has been undertaken by using cylindrical coordinates. However, spherical coordinates are useful in this work. These are illustrated in Fig. 12.48. Because we are considering axially symmetric flow, only the radial distance from the origin noted as R and the angle β between R and the axis of symmetry are needed. The stream function may then be expressed as

$$q = 2\pi\psi(R, \beta, t) \tag{12.107}$$

Finally, in stating the volume flow through a surface of revolution with edges going through any two points 1 and 2 in the flow we can use the following simplified notation:

$$q_{1,2} = 2\pi(\psi_2 - \psi_1) \tag{12.108}$$

where it will be understood that the same anchor point is used for the

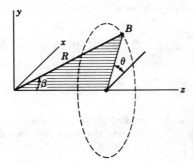

FIGURE 12.48
Spherical coordinates R, β, and θ.

evaluation of ψ at points 1 and 2 but where the identity of the anchor point is of no consequence.

12.28 RELATION BETWEEN STREAMLINES, STREAM FUNCTION, AND THE VELOCITY FIELD

We now relate Stokes' stream function with the streamlines of a flow. Consider, first, the locus of streamlines which at some position z have the same coordinate r. Owing to the symmetry of the flow about the z axis these streamlines will form a surface of revolution which we call a *stream surface*. Clearly there can be no flow through the stream surface, so it acts like an impervious container. From Eq. (12.108), it must then be true that all points along the surface have the same value of stream function ψ, leading us to conclude that *stream surfaces are the same as contour surfaces of Stokes' stream function*. But the stream surface could also have been generated by revolving a streamline about the z axis. Thus we see that each streamline is part of a contour surface of ψ. To represent a flow graphically, we draw the traces of the yz plane and the contour surfaces of ψ.[16]

Up to this point note that no restriction has been made in the development for the irrotationality of the flow. Hence, Stokes' stream function exists and can be used for rotational as well as irrotational flows. However, if the flow is irrotational, there must exist a velocity potential, as discussed earlier. Owing to the axial symmetry of the flow, it is necessary that surfaces of constant velocity potential also be surfaces of revolution. The graphical representation of these surfaces is given by their intersections with the yz plane. The resulting lines are called potential lines. Streamlines and potential lines drawn in the preceding manner then form a grid system similar to those discussed in the previous chapter. *The condition of orthogonality, however, is no longer present, as will be seen in the forthcoming examples.*

The sign convention adopted in Part E may be extended conveniently to the present discussion. That is, looking out from one position to another, positive flow is considered to pass from the observer's left to his/her right.

It will be most useful to formulate the relations between ψ, $\mathbf{V}$, and ϕ in terms of spherical coordinates, as shown in Fig. 12.49. The flow between A and B plus the segment $R\,d\beta$ may be written in equivalent ways, these being equated to yield the following statement:

$$2\pi\left[\psi_A(R,\beta,t) + \frac{\partial\psi}{\partial\beta}\,d\beta\right] = 2\pi\psi_A(R,\beta,t) + 2\pi(R\sin\beta)V_R R\,d\beta$$

---

[16]If $V_\theta = 0$, that is, there is no rotation of the flow about the z axis, these traces will correspond to actual streamlines in the yz plane. If $V_\theta \neq 0$, these traces are still generally called streamlines.

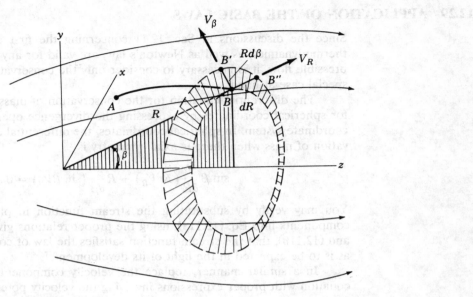

FIGURE 12.49
Axisymmetric flow with spherical coordinates.

where $2\pi(R \sin \beta)$ is the circumference of the strip $R\,d\beta$. This simplifies to

$$V_R = \frac{1}{R^2 \sin \beta} \frac{\partial \psi}{\partial \beta} \qquad (12.109)$$

Similarly, using the segment dR, we get

$$V_\beta = -\frac{1}{R \sin \beta} \frac{\partial \psi}{\partial R} \qquad (12.110)$$

Finally, for irrotational flows the following relations may be reached between the stream function and the velocity potential by equating corresponding expressions for the velocities V_β and V_R:

$$\frac{1}{\sin \beta} \frac{\partial \psi}{\partial R} = -\frac{\partial \phi}{\partial \beta} \qquad (12.111)$$

$$\frac{1}{R^2 \sin} \frac{\partial \psi}{\partial \beta} = \frac{\partial \phi}{\partial R} \qquad (12.112)$$

It is now possible to solve for either function, once the other is known, by integration, as discussed in Part B.

12.29 APPLICATION OF THE BASIC LAWS

Since the discussions in Sec. 12.11 concerning the first and second laws of thermodynamics, as well as Newton's law, are valid for any irrotational, incompressible flow, it is necessary to consider only the conservation of mass, for the special case at hand.

The differential equation for the conservation of mass can be ascertained for spherical coordinates by expressing the divergence operator in the desired coordinate system. In spherical coordinates, the differential equation for conservation of mass when there is axial symmetry is

$$\sin \beta \frac{\partial}{\partial R} \left(R^2 V_R \right) + R \frac{\partial}{\partial \beta} \left(\sin \beta V_\beta \right) = 0 \tag{12.113}$$

You may verify by substituting the stream function in place of the velocity components into Eq. (12.113), using the proper relations given in Eqs. (12.109) and (12.110), that the stream function satisfies the law of conservation of mass, as is to be expected in the light of its development.

In a similar manner, replace the velocity components of the preceding equation with proper expressions involving the velocity potential. The result is

$$\frac{\partial}{\partial R} \left(R^2 \frac{\partial \phi}{\partial R} \right) + \frac{1}{\sin \beta} \frac{\partial}{\partial \beta} \left(\sin \beta \frac{\partial \phi}{\partial \beta} \right) = 0 \tag{12.114}$$

This equation is Laplace's equation in spherical coordinates, and solutions of this equation are called *spherical harmonics*.

Since the *stream function in this case does not satisfy Laplace's equation, as was the case in the two-dimensional analysis*, we usually first solve for a velocity potential which is harmonic and which satisfies the boundary conditions of a given problem. One useful procedure is the superposition of simple known solutions in such a manner as to approximate the action of a given boundary. We will explore this approach.

12.30 UNIFORM FLOW

Consider a uniform flow in the positive z direction with a velocity V_0. The spherical components of this velocity may then be expressed as

$$V_R = V_0 \cos \beta \tag{12.115}$$

$$V_\beta = -V_0 \sin \beta \tag{12.116}$$

Replacing the quantities V_R and V_β by using Eqs. (12.109) and (12.110), we get

$$\frac{1}{R^2 \sin \beta} \frac{\partial \psi}{\partial \beta} = V_0 \cos \beta \tag{12.117}$$

$$\frac{1}{R \sin \beta} \frac{\partial \psi}{\partial R} = V_0 \sin \beta \tag{12.118}$$

Integrating Eq. (12.117) with respect to β and Eq. (12.118) with respect to R, we get

$$\psi = \frac{R^2 \sin^2 \beta}{2} V_0 + g(R)$$

$$\psi = \frac{R^2 \sin^2 \beta}{2} V_0 + h(\beta)$$

Comparing these above equations, we conclude that the arbitrary functions of integration, $g(R)$ and $h(\beta)$, must equal the same constant. Omitting this constant, we then have as the desired stream function

$$\boxed{\psi = \tfrac{1}{2} V_0 R^2 \sin^2 \beta} \tag{12.119}$$

The corresponding velocity potential may readily be found by integrating Eqs. (12.111) and (12.112) for the above stream function. Thus

$$\boxed{\phi = R V_0 \cos \beta} \tag{12.120}$$

12.31 THREE-DIMENSIONAL SOURCES AND SINKS

Let us examine the following spherical harmonic function, in which m is a constant:

$$\boxed{\phi = -\frac{m}{4\pi} \frac{1}{R}} \tag{12.121}$$

Note that this function is symmetric about the origin and hence is axially symmetric about any axis proceeding through this point. The only nonzero spherical velocity component is V_R, having the value

$$V_R = \frac{m}{4\pi} \frac{1}{R^2} \tag{12.122}$$

Consideration of this velocity field leads to the following conclusions:

1. The flow proceeds radially out from the origin, and the flow is hence that of a three-dimensional *source*. Using $(m/4\pi)(1/R)$ as a velocity potential gives us a radially inward flow and, hence, the flow of a *sink*.
2. At the source point the velocity becomes infinite. It is then necessary to exclude this point from regions of physically possible flows. Far from the source point, the velocity becomes diminishingly small.

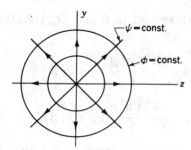

FIGURE 12.50
Three-dimensional source.

3. Employing a spherical control surface of radius R_0 about the source point to compute the efflux of fluid from the source leads one to the statement

$$q = \oint V_R \, dA = \int_0^\pi \left(\frac{m}{4\pi} \frac{1}{R_0^2} \right) (2\pi R_0 \sin \beta)(R_0 \, d\beta) = m \quad (12.123)$$

As in the case of two-dimensional sources and sinks, m is termed the *strength* of the source or sink.

The corresponding stream function may be ascertained by integrating Eqs. (12.111) and (12.112). Thus

$$\psi = -\frac{m}{4\pi} \cos \beta \quad (12.124)$$

Surfaces of constant velocity potential form a family of concentric spherical surfaces about the source point. Stream surfaces, on the other hand, form a family of concentric cones about the z axis, with the apex at the source point. A trace of these surfaces at the yz plane is shown in Fig. 12.50.

12.32 THREE-DIMENSIONAL DOUBLET[17]

As in the previous two-dimensional case the doublet is developed by a limiting process. A three-dimensional source and a sink of equal strength are positioned on the z axis a distance $2a$ apart, as shown in Fig. 12.51. To form the doublet, bring the source and sink together (that is, $2a \to 0$), while simultaneously increasing the strength without limit (that is, $m \to \infty$), in such a way that the product $2am$ in the limit has the finite value $4\pi\mu$. Mathematically, this

---

[17]The three-dimensional doublet is analogous to the electric dipole in electrostatics where positive and negative point charges are merged in a manner described in this section.

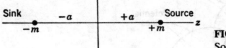

FIGURE 12.51
Source and sink along z.

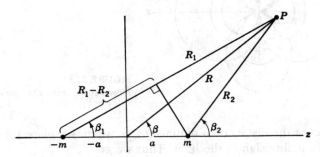

FIGURE 12.52
Coordinates for development of three-dimensional doublet.

operation is indicated as

$$\lim_{\substack{2a \to 0 \\ m \to \infty}} (2am) = 4\pi\mu \tag{12.125}$$

The constant μ is known as *the strength of the doublet*.[18]

The velocity potential for this doublet may be established by first computing the combined potential of the sink-source arrangement before the limiting action. In Fig. 12.52 convenient notation has been indicated for this formulation, in terms of which the velocity potential for the combination is

$$\phi = \frac{m}{4\pi}\left(\frac{1}{R_1} - \frac{1}{R_2}\right) = \frac{m}{4\pi}\frac{R_2 - R_1}{R_1 R_2} \tag{12.126}$$

If the distance a is chosen small compared with R, the following approximation may be made (see Fig. 12.52):

$$R_1 - R_2 \approx 2a \cos \beta_1 \tag{12.127}$$

Substituting this approximation into Eq. (12.126), we get

$$\phi \approx -\frac{m}{4\pi}\frac{2a \cos \beta_1}{R_1 R_2} \tag{12.128}$$

Now let the source and sink come together at the origin in the manner described earlier. Note that $2am$ becomes $4\pi\mu$, that $R_1 R_2$ becomes R^2, and finally that β_1 becomes β. Furthermore, the statement above becomes more

---

[18]Since there cannot be a coefficient of viscosity in frictionless flow, we may use the letter μ without fear of misinterpretation.

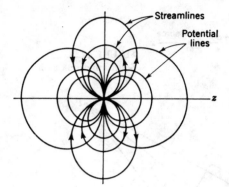

FIGURE 12.53
Flow net for three-dimensional doublet.

correct as the distance $2a$ diminishes, so we may replace the approximation sign by an equality sign in the limit. Thus we get[19]

$$\phi = -\frac{\mu \cos \beta}{R^2} \qquad (12.129)$$

It is clear from the way the doublet was formed that it is a spherically harmonic function. The doublet source must be excluded from physical flow regions because you can readily demonstrate that the velocity goes to infinity there. The corresponding stream function may be evaluated to be

$$\psi = \frac{\mu \sin^2 \beta}{R} \qquad (12.130)$$

The traces of the stream surfaces and constant-potential surfaces are shown in Fig. 12.53. Note that there is a system of closed curves tangent to the z axis resembling the circles for the two-dimensional doublet.

12.33 STEADY FLOW ABOUT A SPHERE

It will now be demonstrated that the superposition of a uniform flow and a three-dimensional doublet results in the solution of irrotational incompressible flow about a sphere. The strength μ of the doublet is taken as $V_0 b^3/2$, wherein

---

[19]It is clear that the doublet has a directional property. If μ is considered a vector directed from sink to source, it will be left for you to show with the aid of Fig. 12.52 that $\phi = (\boldsymbol{\mu} \cdot \mathbf{R})/R^3$. In electrostatics the *dipole moment* $\mathbf{p}$ is analogous to the doublet strength vector $\boldsymbol{\mu}$, and we have, for the potential of an electric field due to a dipole,

$$\phi = \frac{1}{4\pi\epsilon_0} \frac{\mathbf{p} \cdot \mathbf{R}}{R^3}$$

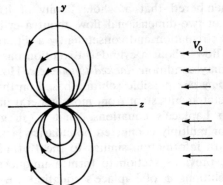

FIGURE 12.54
Superpose 3-D doublet and uniform flow.

V_0 is the velocity of the uniform flow and b is a constant which will later be shown to be the radius of the sphere. Taking the axis of development of the doublet along the z axis, we show the trace of the streamlines for each flow in Fig. 12.54. The stream function of the combination is

$$\psi = -\frac{V_0 R^2 \sin^2 \beta}{2} + \frac{V_0 b^3}{2R} \sin^2 \beta \qquad (12.131)$$

Let us examine the stream surface corresponding to the constant $\psi = 0$. Collecting terms, we have

$$\frac{V_0 \sin^2 \beta}{2}\left(-R^2 + \frac{b^3}{R}\right) = 0 \qquad (12.132)$$

Note that for all values of R, the conditions $\beta = \pi$ and $\beta = 0$ satisfy the equation. Hence, the axis of symmetry may be considered as part of the stream surface. Furthermore, the condition $R = b$ satisfies the equation for all values of β. Part of the stream surface $\psi = 0$ is shown in Fig. 12.55. The streamlines outside the spherical surface may then be considered as the desired flow.

Computations of lift and drag on the sphere may now be carried out by using Bernoulli's equation and known flow conditions at infinity. As an exercise you will be asked to show that both lift and drag are zero in value.

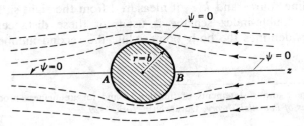

FIGURE 12.55
$\psi = 0$ forms a sphere.

It will be remembered that a whole family of irrotational flows was possible in the case of two-dimensional flow about a cylinder. Each of these flows had a different circulation and consequently a different lift. In short, for the two-dimensional flows about a cylinder a solution based only on the basic laws and given boundary conditions *lacked uniqueness*. However, in the case of the sphere, there is *only one* possible solution based on these considerations.

The preceding statements stem from more-general uniqueness considerations of solutions to Laplace's equations. That is, in general, solutions of Laplace's equation for multiply connected boundaries lack uniqueness. Invariably one requires extra information similar to the Kutta-Joukowski condition that was used in the previous section to form a unique solution. For simply connected regions, solutions ϕ of Laplace's equation are unique when ϕ has been specified on the boundaries (Dirichlet's problem). Also when $\partial\phi/\partial n$ has been specified on the boundary (Neumann's problem), as in our case, solutions are unique up to an additive constant.[20]

12.34 FLOW AROUND BODIES OF REVOLUTION

Using an arrangement of line sources and sinks in conjunction with a uniform flow, it is possible to develop approximate solutions for flows about bodies of revolution. In Fig. 12.56 is shown the boundary in the yz plane of such a body. This body is exposed to a flow which far away from the body is uniform, with a direction parallel to the axis of symmetry of the body. The technique is to establish a series of line sources and sinks along the axis of symmetry, which we take as the z axis. The strengths of the line sources and sinks are so adjusted that when combined with the uniform flow one can form a stream surface with a shape approximating the given body. The number of line sources and sinks that are employed depends on the desired degree of accuracy. For convenience, these elements are usually chosen of equal lengths, each with a strength η_i. The sign of η_i, as pointed out earlier, indicates whether the segment is a source or sink type of flow. We shall henceforth refer to them simply as line sources. The proposed arrangement then is shown in Fig. 12.57. At the center of each line source the radial distance to the desired boundary has been indicated as r. Notice that points at the radial extremities have been numbered. We shall refer to these points as *body points*. It should be clear that there are as many body points as there are line sources. Finally, distances from the ends of each line source m to a body point n are indicated as $^{\lceil}k_{nm}$, if measured from the left side of the line source and $k_{nm}^{\rceil}$, if measured from the right side of the line source. Thus a double-index scheme is used for these distances whereby the first number identifies the body point, while the second number indicates the line source.

---

[20] For more rigorous statements of uniqueness and for uniqueness proofs see O. D. Kellogg, *Foundations of Potential Theory*, Dover, New York, 1929.

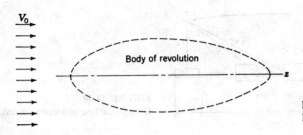

FIGURE 12.56
Body of revolution.

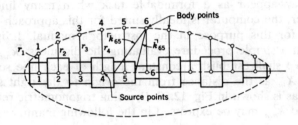

FIGURE 12.57
Line sources and sinks for approximate solution.

The stream function at any body point n resulting from line source m can be expressed in the following manner by employing the result of Problem 12.82

$$\Delta\psi_{nm} = -\frac{\eta_m}{4\pi}(\ulcorner k_{nm} - k_{nm}\urcorner) \tag{12.133}$$

The combined stream function at body point n from all f line sources and the uniform flow can be formed by the following summation:

$$\psi_n = -\sum_{m=1}^{m=f} \frac{\eta_m}{4\pi}(\ulcorner k_{nm} - k_{nm}\urcorner) + \frac{V_0}{2}r_n^2 \tag{12.134}$$

The next step is to adjust the strengths of the f line sources $\eta_1, \eta_2, \ldots, \eta_f$ so that the value of the stream function at body point n is some constant. Using zero as the constant, we have

$$-\sum_{m=1}^{m=f} \frac{\eta_m}{4\pi}(\ulcorner k_{nm} - k_{nm}\urcorner) + \frac{V_0}{2}r_n^2 = 0 \tag{12.135}$$

If this is done for all f body points, there will be a set of f simultaneous equations for the f unknown strengths. Thus

$$\frac{\ulcorner k_{11} - k_{11}\urcorner}{4\pi}\eta_1 + \frac{\ulcorner k_{12} - k_{12}\urcorner}{4\pi}\eta_2 + \cdots + \frac{\ulcorner k_{1f} - k_{1f}\urcorner}{4\pi}\eta_f = \frac{V_0}{2}r_1^2$$

$$\vdots \qquad\qquad\qquad\qquad\qquad\qquad\qquad \vdots \tag{12.136}$$

$$\frac{\ulcorner k_{f1} - k_{f1}\urcorner}{4\pi}\eta_1 + \frac{\ulcorner k_{f2} - k_{f2}\urcorner}{4\pi}\eta_2 + \cdots + \frac{\ulcorner k_{ff} - k_{ff}\urcorner}{4\pi}\eta_f = \frac{V_0}{2}r_f^2$$

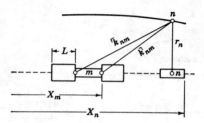

FIGURE 12.58
Details for line sources and sinks.

This may appear as a formidable task when many line sources are used. However, the computer is ideally suited for this approach and we have listed a project for this purpose in the instructor's manual. It is an easy matter to establish a simple procedure evaluating the distances $^\ulcorner k_{nm}$ in the above equations. We shall establish a constant length of each line source as L and shall denote X_m as the distance from the origin to the right side of the mth line source, as is shown in Fig. 12.58. By using trigonometric relations, the distances $^\ulcorner k_{nm}$ and $k_{nm}^\urcorner$ may be expressed in the following manner:

$$^\ulcorner k_{nm} = \left\{ r_n^2 + \left[\left(X_n - \frac{L}{2} \right) - (X_m - L) \right]^2 \right\}^{1/2}$$

$$= \left[r_n^2 + \left(X_n - X_m + \frac{L}{2} \right)^2 \right]^{1/2} \tag{12.137a}$$

$$k_{nm}^\urcorner = \left\{ r_n^2 + \left[\left(X_n - \frac{L}{2} \right) - X_m \right]^2 \right\}^{1/2}$$

$$= \left[r_n^2 + \left(X_n - X_m - \frac{L}{2} \right)^2 \right]^{1/2} \tag{12.137b}$$

By placing the origin at the beginning of the first line source, we may say, for the ith source, $X_i = iL$. Consequently, the preceding equations simplify further to

$$^\ulcorner k_{nm} = \left\{ r_n^2 + \left[L(n - m) + \frac{L}{2} \right]^2 \right\}^{1/2} \tag{12.138a}$$

$$k_{nm}^\urcorner = \left\{ r_n^2 + \left[L(n - m) - \frac{L}{2} \right]^2 \right\}^{1/2} \tag{12.138b}$$

It is clear from the diagrams, and from Eqs. (12.138) that $^\ulcorner k_{nm}$ represents the same line segment as $k_{n,m-1}^\urcorner$ so that it would appear that only $f + 1$ distances must be computed in the above manner for each body point. Calling the

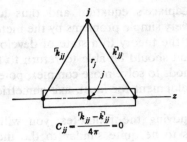

$$C_{jj} - \frac{\ulcorner k_{jj} - k_{jj}^{\urcorner}}{4\pi} = 0$$

FIGURE 12.59
Source point j and body point j.

coefficients of Eq. (12.136) C_{nm}, we may write

$$C_{11}\eta_1 + C_{12}\eta_2 + \cdots + C_{1f}\eta_f = \frac{V_0}{2}r_1^2 \tag{12.139a}$$

$$\vdots$$

$$C_{f1}\eta_1 + C_{f2}\eta_2 + \cdots + C_{ff}\eta_f = \frac{V_0}{2}r_f^2 \tag{12.139b}$$

Finally, it should be pointed out that coefficients with repeated indices such as C_{jj} are zero. That is, the principal diagonal of the matrix of coefficients of η is a set of zeros.[21] This is seen by examining Fig. 12.59.

Once the line-source strengths have been evaluated, the approximately correct stream function for the entire region outside the body of revolution is available. Thus, the value of the stream function at any position P in space is available from Eq. (12.134) by using the computed values of η_m and by using the distances k and r to the position P instead of to the body point n. We then have

$$\psi_P = -\sum_{m=1}^{m=f} \frac{\eta_m}{4\pi}\left(\ulcorner k_{pm} - k_{pm}^{\urcorner}\right) + \frac{V_0}{2}r_p^2 \tag{12.140}$$

However, in most instances, information will be desired only along the boundary. Pressure distributions may be ascertained by employing Bernoulli's equation in the same manner as in the two-dimensional problems undertaken in the previous chapter. The instructor's manual will have two computer projects using the material of this section.

12.35 CLOSURE

In this chapter we have considered the two-dimensional potential-flow case, wherein we have set forth the concept of the stream function and related it to the velocity potential. We found, in going back to the basic laws, that both

---

[21]We now see that only $f - 1$ distances need be computed for each body point because of the fact that $C_{jj} = 0$. Thus it is unnecessary to calculate $\ulcorner k_{jj}$ (or $k_{jj}^{\urcorner}$).

functions had to satisfy Laplace's equation and thus had to be harmonic functions. We solved some very simple problems by the method of superposition and have generally brought the theory to a state of development so that when you study finite elements you should be able to return to this material and be able to apply powerful methods to solve more complex, potential-flow problems. We then had a similar development for ideal, axisymmetric, three-dimensional flow.

If you have been studying the footnotes, you will have noticed that potential-flow theory seems to be quite similar to the theory of electrostatic fields. Indeed there should be this similarity, since each field of study essentially involves the solution of Laplace's equation for certain boundary conditions. This differential equation also characterizes the steady flow of heat in homogeneous media and the flow of steady electric currents and is of significance in the theory of elasticity. And so, in proper perspective, we have been studying part of a general body of knowledge involving Laplace's equations, called *potential theory*, as it applies specifically to fluid mechanics. The names of certain corresponding items and their physical meaning differ from one area of study to the other, but you should not allow this to obscure the fundamental relations of potential theory pervading all these studies.

This brings to a conclusion our formal studies of potential or ideal flow. In Chap. 13 on boundary layers, we have explained how potential-flow theory and boundary-layer theory may be coupled to reach useful results for certain problems.

Having studied viscous flow in a conduit in Chap. 9, we now proceed to consider viscous flow in the boundary layer. The understanding of such flows in the boundary layer is vital for predicting the performance of vehicles and turbomachines. Also interesting and important phenomena occurring when there is flow about a body may be understood in terms of the boundary-layer flow.

PROBLEMS

Problem Categories

Starred Problems

Derivations and Justifications

12.21, 12.33, 12.35, 12.70, 12.76, 12.72

12.1. Indicate whether the following regions are simply or multiply connected:
  (a) The region between a solid sphere and an enveloping spherical shell [Fig. P12.1(a)].
  (b) The region between two tubes [Fig. P12.1(b)].
  (c) The region between two tubes with end A closed [Fig. P12.1(c)].
  (d) The region inside the big tube and inside the little tube with end A open [Fig. P12.1(c)].

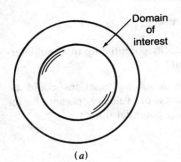

(a)

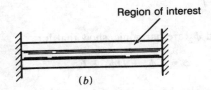

(b)

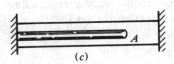

(c)

FIGURE P12.1

12.2. Given the following velocity field

$$\mathbf{V} = 2x^2y\mathbf{i} + 2y^2x\mathbf{j} + 10\mathbf{k} \ \text{m/s}$$

what is the circulation Γ around a square path in the xy plane about the origin at the center of the square? The square is 2 m per side. What can you say about **curl V**?

12.3. A stream function ψ is given as $\psi = -(x^2 + 2xy + 4t^2y)$. When $t = 2$, what is the flow rate across the semicircular path shown in Fig. P12.3? What is the flow across the x axis from A to $x = 10$?

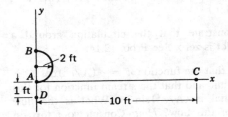

FIGURE P12.3

12.4. In Prob. 12.3 point A, the origin of xy, has been used as the anchor point. Suppose that we use point D at $y = -1$ as the anchor point. What will be the difference between the flows through paths to points B and C from the new anchor point as compared with using the old anchor point at any time t?

12.5. Show that $V_y = -\partial\psi/\partial x$.

12.6. If the stream function of a flow is $\psi = -2xy$, what is the velocity at position $x = 2$, $y = 5$?

12.7. In Prob. 12.6 sketch the streamline pattern. What is the significance of the spacing between your streamlines?

12.8. If the flow in Prob. 12.6 is irrotational, what should the proper velocity potential be within a constant?

12.9. Sketch the flow net for the flow given in Prob. 12.6.

12.10. Draw the flow net for the flow given in Example 12.1.

12.11. Show that $\phi = (\Lambda/2\pi)\ln r$ is a harmonic function.

12.12. Consider the potential function

$$\phi = -A(x^2 - y^2).$$

Show that it is a harmonic function. What is the stream function ψ? Express this stream function in polar coordinates. Draw the flow net. Is there a stagnation point? If so, where? Is there a singular point? If so, where?

12.13. Show that the stream function for $\phi = (\Lambda\theta/2\pi)$ is $\psi = -(\Lambda/2\pi)\ln r$. (This is the simple vortex.)

12.14. Show that the stream function for $\phi = (\chi \cos \theta)/r$ is $\psi = -(\chi \sin \theta)/r$. (This is a doublet.)

12.15. Demonstrate that the circulation around a doublet is zero. See Prob. 12.14.

12.16. Show that the function $\phi = -(\Lambda/r^2)\cos 2\theta$ is harmonic and that the stream function for this potential is $(\Lambda/r^2)\sin 2\theta$. What is the flow net for this flow? *Hint:* Consult your text on analytic geometry for its discussion on lemniscates.

12.17. In Example 4.1 we presented the following two-dimensional velocity field:

$$V_x = -Ax$$
$$V_y = Ay$$

where A is a constant. Is the flow irrotational? Does it satisfy continuity? Determine ϕ and ψ if the answers to these questions are "yes."

12.18. A two-dimensional flow has the following velocity field:

$$\mathbf{V} = x^2\mathbf{i} + (-2xy + 4x)\mathbf{j} \text{ m/s}$$

Is this an irrotational flow? Does it satisfy continuity? If so, what is ψ up to an arbitrary constant?

12.19. Consider the potential function

$$\phi = A(x^2 - y^2)$$

Show that it is a harmonic function. What is the stream function ψ? Express this stream function in polar coordinates. Draw the flow net. Is there a stagnation point? If so, where?

12.20. A velocity potential ϕ is given as

$$\phi = 5x^2y + 10yz^3 + 3zt^2 \text{ m}^2/\text{s}$$

Formulate the acceleration field.

12.21. Show that the curl of any potential flow must be zero by examining **curl**$(\nabla\phi)$.

12.22. If the velocity field is given as

$$\mathbf{V} = (2x^2 + 3y)\mathbf{i} + 10yz^2\mathbf{j} + 10z^2t\mathbf{k}$$

is the flow irrotational? If not, what is the **curl V**?

12.23. If we have an irrotational steady flow with the following velocity field:

$$\mathbf{V} = 2x\mathbf{i} + 2\mathbf{j} \text{ m/s}$$

what is the velocity potential up to an undetermined constant?

12.24. Which of the following functions could be stream functions or velocity potentials for two-dimensional potential flows?
(a) $x^2 - y^2$
(b) $\sin(x + y)$
(c) $\ln(x - y)$
(d) $K \ln \bar{r}$
(e) $D\theta$

12.25. If ϕ_1 and ϕ_2 are harmonic, show that for

$$\phi = C\phi_1 + D\phi_2 + E$$

then ϕ is also harmonic if C, D, and E are constants.

12.26. For any two-dimensional incompressible flow (not necessarily irrotational) show that the vorticity vector ω has the magnitude equal to $\frac{1}{2}\nabla^2\psi$.

12.27. In Sec. 7.8, Case 1, we examined the laminar flow between two parallel infinite plates (see Fig. 7.12) and found that

$$u = \frac{\beta}{2\mu}(h^2 - y^2)$$

What is the stream function ψ for this flow?

12.28. In Sec. 10.5, Case 4, we examined Couette flow. Recall that here the flow between two infinite parallel plates is instituted by moving the upper plate. For viscous fluids we found that

$$\frac{u}{U} = \frac{y}{h} - \left[\frac{h^2}{2\mu U} \frac{\partial p}{\partial x} \right] \frac{y}{h}\left(1 - \frac{y}{h}\right)$$

What is the stream function?

12.29. A two-dimensional potential flow is shown flowing radially in toward a small opening. What is the velocity potential for the flow if 15 m³/s is flowing into the opening per unit length of the opening?

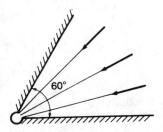

FIGURE P12.29

12.30. If $\psi = -\Lambda r^{\pi/\alpha} \sin(\pi\theta/\alpha)$, show that it satisfies Laplace's equation and that $\phi = -\Lambda r^{\pi/\alpha} \cos(\pi\theta/\alpha)$. Show that this may represent a flow over a straight boundary as shown in Fig. P12.30.

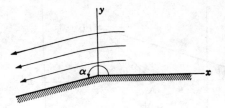

FIGURE P12.30

12.31. (a) Show that the flow given in Prob. 12.30 reduces to the uniform flow of Sec. 12.16 when $\alpha = \pi$.

(b) When $\alpha = \pi/2$, draw an approximate flow net, making use of the fact that the potential lines and streamlines are orthogonal. Show that this can represent a flow in a corner. If $\Lambda = 1$, what is the velocity of flow along the horizontal wall at 10 ft from the corner?

12.32. A source and sink of equal strength of 20 ft³/s is shown in Fig. P12.32. What is the velocity at position $(15, 15)$?

12.33. Consider a source and a sink of equal strength Λ positioned at $x = \pm a$ along the x axis as shown in Fig. P12.33. Show that the combined stream function is given as follows:

$$\psi = \frac{\Lambda}{2\pi}\left(\tan^{-1}\frac{y}{x+a} - \tan^{-1}\frac{y}{x-a} \right) \quad (a)$$

Using the identity,

$$\tan^{-1} A - \tan^{-1} B = \tan^{-1}\frac{A-B}{1+AB} \quad (b)$$

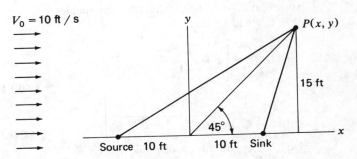

FIGURE P12.32

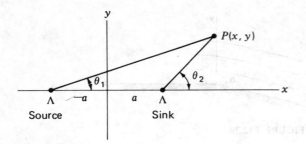

FIGURE P12.33

show that the streamlines are give by the equation

$$x^2 + y^2 - a^2 = \left[\frac{2ay}{\tan\left(\dfrac{2\pi\psi}{\Lambda} \right)} \right] \quad (c)$$

Show that the streamlines are circles with centers on the y axis by putting Eq. (c) in the form $x^2 + (y - C_1)^2 = C_2^2$, where C_1 and C_2 are constants for a streamline. Show that the radii of the circles are given as

$$R = a \csc \frac{2\pi\psi}{\Lambda}$$

Also, show that the circles all pass through the source and sink points.

12.34. In Prob. 12.30, what is the velocity at the corner for $0 < \alpha < \pi$ and for $\pi < \alpha < 2\pi$?

12.35. Examine the flow of a pair of two-dimensional opposite vortices placed at $(a, 0)$ and $(-a, 0)$, respectively, of strength Λ. Take the counterclockwise vortex at $x = -a$ and the clockwise vortex at $x = a$. Show that the streamlines $\psi = C$ are given as

$$y^2 + \left[x + a\left(\frac{1 + e^{4\pi C/\Lambda}}{1 - e^{4\pi C/\Lambda}} \right)^2 \right]$$

$$= \left[\frac{2a}{e^{-2\pi C/\Lambda} - e^{2\pi C/\Lambda}} \right]^2$$

Hence they are circles of radii $2a/(e^{-2\pi C/\Lambda} - e^{2\pi C/\Lambda})$.

12.36. What is the circulation about the singular point in the flow given in Prob. 12.16?

12.37. We wish to represent the potential flow about a cylinder of radius 2 ft without circulation, where the free-stream velocity at infinity is 10 ft/s. What should the stream function be? What is the drop in pressure at the top of the cylinder from the free-stream pressure at infinity? What is the increase in pressure above the free-stream pressure at the stagnation point? Take the fluid as air having $\rho = 0.002378$ slug/ft^3.

12.38. Show that the lift of a cylinder without circulation is zero.

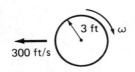

FIGURE P12.38

12.39. A whirlpool has a velocity of 2 m/s at a distance $r = 1$ m from the center of the core. What is the decrease in elevation of the free surface from $r = 3$ m to $r = 1$ m?

12.40. Draw a flow net for a two-dimensional potential flow for the geometry shown in Fig. P12.40.

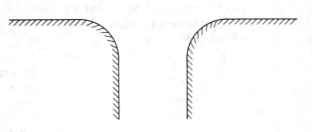

FIGURE P12.40

12.41. Show that the superposition of equal sources at a distance $2a$ apart gives the flow of a single source with an infinite wall normal to the line of connection of the sources and at a position halfway between the sources. (This method of creating boundaries mathematically is called the *method of images*.)

12.42. Using the method of images as presented in Prob. 12.41, what is the speed of flow at position $(3, 5)$ m from two sources near an infinite wall as shown. The strengths of the sources are $\Lambda_1 = 10$ m^2/s and $\Lambda_2 = 5$ m^2/s.

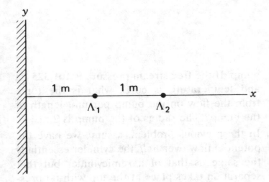

FIGURE P12.42

12.43. What is the velocity of flow at position $(5, 5)$ from a source of strength $\Lambda = 10$ m^2/s involving two infinite walls at right angles to each other? See Prob. 12.41

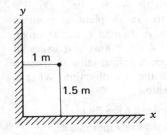

FIGURE P12.43

12.44. In the previous problems, we have used point sources or sinks in conjunction with the method

of images. Explain why this would not work if we were to consider the flow about something like a cylinder near a wall as shown in Fig. P12.44.

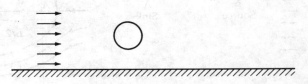

FIGURE P12.44

12.45. Consider the superposition of a uniform flow in the positive x direction with velocity of 2 m/s and a source at the origin with a strength $\Lambda = 3$ m^2/s. Where is the stagnation point? What is the shape of the boundary about which this might represent a flow pattern? Sketch the boundary.

12.46. In the preceding problem, if the free stream pressure is 101,325 Pa, what is the gage pressure on the boundary at $r = 4$ m? The boundary of the body is found from the preceding problem to be

$$V_0 r \sin \theta + \frac{\Lambda}{2\pi} \theta = \frac{\Lambda}{2}$$

The fluid is water at 30°C.

12.47. In Probs. 12.45 and 12.46 we considered the flow about a half body formed by a source and a uniform flow. If the strength of the source is 3 m^3/(s)(m) and the uniform velocity is 5 m/s, where is the stagnation point and what is the maximum total width of the half body?

12.48. A source at the origin and normal to the xy plane of strength 10 m^3/(s)(m) and a uniform flow are superposed to form a half body. What should the uniform velocity be to have a maximum width of 0.8 m? Use results of Probs. 12.45 and 12.46.

12.49. Consider a uniform flow in the x direction at speed V_0 superposed on a source of strength Λ and at $x = -a$ and a sink of strength Λ at $x = +a$. The streamline along the x axis will

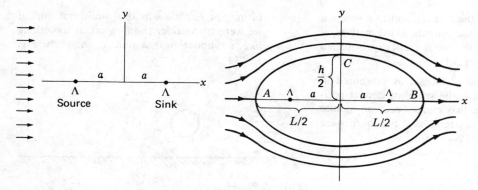

split and form an oval-shaped region which is called a *Rankine oval*. First, show that the streamlines that open at points A and B must be $\psi = 0$. Then, show that the width of the oval h satisfies the equation

$$\frac{h}{a} = 2 \tan\left(\frac{\pi}{2} - \frac{\pi V_0 h}{2\Lambda}\right)$$

12.50. In Prob. 12.49 determine the length L as a function of V_0, Λ, and a for the Rankine oval. *Hint:* Superpose Eq. (12.68) for ϕ of the source and sink combination with a uniform flow of velocity V_0 using cylindrical coordinates for the latter. Show that

$$\frac{L}{a} = 2\left(1 + \frac{\Lambda}{\pi V_0 a}\right)^{1/2}$$

12.51. A Rankine oval is formed by a uniform flow V_0 of 5 m/s and a source and sink with strengths Λ equal to 8 m³/(s)(m). If the source and sink are at distance $a = 0.2$ m, respectively, what is the maximum width of the oval and what is the maximum velocity of the flow? (See Probs. 12.49 and 12.50.)

12.52. We wish to have a Rankine oval of length $L = 4$ m (see Fig. P12.49). If the free stream velocity is 8 m/s and the strengths of source and sink are 6 m²/s, what will be width of the oval h? (See Probs. 12.49 and 12.50.)

12.53. A potential flow with a free stream uniform velocity of 5 m/s flows over a long semicircular

bump. If the free stream pressure is 101,325 Pa and temperature is 60°C, what is the force from the flow on the bump per unit length of the bump? The radius of the bump is 2 m.

12.54. In the previous problem, assume we have the potential flow over *half* the cylinder essentially the same as that of a semicylinder but that separation takes place at the top with no pressure recovery. What would the drag be on the half cylinder per unit length?

12.55. A doublet distributed along the x axis is of strength $\chi = 6$ m³/(s)(m). It is developed along and issues in the direction

$$\boldsymbol{\nu} = 0.6\mathbf{j} + 0.8\mathbf{k}$$

What is the velocity at position

$$\mathbf{r} = 3\mathbf{i} + 2\mathbf{j} + 4\mathbf{k} \text{ m}$$

12.56. A source of strength $\Lambda_1 = 3$ m³/(s)(m) is oriented normal to the xy plane at position $x = 3$ m, $y = 5$ m. A doublet is at the origin with strength $\chi = 4$ m³/(s)(m) and issues in the plus x direction. A uniform flow of speed 6 m/s is directed in the $+y$ direction. What is the velocity at position

$$\mathbf{r} = -4\mathbf{i} + 6\mathbf{j} + 3\mathbf{k} \text{ m}$$

when these two flows are superposed?

12.57. A thin-walled tank in Fig. P12.57 formed from two semicylinders having an outside diameter of 3 m and a length of 10 m sits on end outdoors exposed to a wind of speed 10 km/h. If the inside pressure is 200 Pa gage and

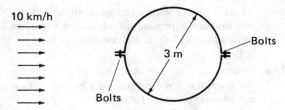

FIGURE P12.57

if there are 10 bolts each side, what is tensile stress per bolt as a result of the inside and outside pressures? The cross-sectional area of each bolt is 12 mm². Take $\rho_{air} = 1.225$ kg/m³ and note that $\int \sin^3 \theta \, d\theta = -\frac{1}{3}\cos \theta(\sin^2 \theta + 2)$.

12.58. A horizontal circular cylinder of diameter 6 ft is rotating at a rate of speed ω of 400 r/min and is moving through the air at a speed of 300 ft/s. What is the lift per unit length of the cylinder if the circulation is 40 percent of the maximum possible circulation? $\rho = 0.002378$ slug/ft³.

12.59. There were early attempts in the development of the airplane to use two rotating cylinders as airfoils. Consider such cylinders each having a diameter of 3 ft and length of 30 ft. If each cylinder is rotated at 800 r/min while the plane moves at a speed of 60 mi/h through the air at 2000 ft standard atmosphere, estimate the lift that could be developed on the plane disregarding end effects. Assume the circulation for the cylinder is 35 percent of the theoretical maximum.

12.60. In Example 12.2, suppose that the wind is oriented at an angle of 30° as shown in Fig.

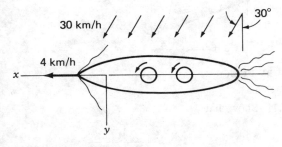

FIGURE P12.60

P12.60 and that 45 percent of the maximum circulation is developed by the rotors. If the drag coefficient is one-third the lift coefficient for the rotors, what is the thrust in the x direction from the rotors? All other data are unchanged.

12.61. In Prob. 12.60, locate the stagnation points for the condition of circulation that was employed.

12.62. A windmill in Fig. P12.62 is composed of rotating cylinders and operates according to the Magnus effect. The windmill rotates with angular speed $\omega_1 = 40$ r/min relative to the ground. The cylinders rotate with angular speed $\omega_2 = 750$ r/min relative to the windmill. A wind of velocity 50 km/h goes directly toward the windmill. If the circulation around the cylinder is 60 percent of the theoretical maximum, what is the total torque on the windmill?

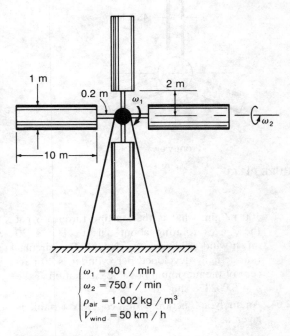

$$\begin{cases} \omega_1 = 40 \text{ r / min} \\ \omega_2 = 750 \text{ r / min} \\ \rho_{air} = 1.002 \text{ kg / m}^3 \\ V_{wind} = 50 \text{ km / h} \end{cases}$$

FIGURE P12.62

12.63. A wind turbine with three rotating cylinders in Fig. P12.63 has been designed and built by Thomas Hansen under a federal grant. If each cylinder is rotating about each of the axes at

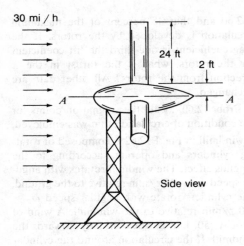

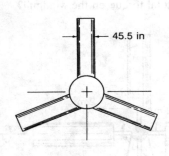

Front view

FIGURE P12.63

100 r/min, what is the starting torque to get these axes rotating about axis A-A if a 30 mi/h wind is blowing as shown? The circulation actually developed per cylinder is 40 percent of the maximum possible circulation. Take $\rho = 0.002378$ slug/ft^3.

12.64. An analytic function w of a complex variable is given as

$$w = A/z = A/(x + iy)$$

Show that it can represent a doublet by duplicating Eqs. (12.77) and (12.79).

12.65. Consider the complex function

$$w = -\Lambda \ln(z - z_0)$$

where z_0 is a complex constant. Show that this function can represent a vortex. Λ is a positive real constant. Use polar coordinates for $(z - z_0)$.

12.66. Determine the velocity at $r = 3\mathbf{i} + 2\mathbf{j}$ m for the potential flows represented by the following analytic complex functions:
(a) $w = iz^2 + 2z$
(b) $w = 1/z + \ln z$

12.67. Describe the families of streamlines and potential lines generated by the following analytic complex function:

$$w = 2z^2 + 4z$$

12.68. Show that $\phi = (m/4\pi)(1/R)$ is harmonic and then determine the corresponding stream function.

12.69. Given $\phi = (\mu \cos \beta)/R^2$, show that $\psi = (\mu \sin^2 \beta)/R$.

12.70. If the origin of our reference is an "anchor point," what is the flow through a truncated conical surface of half angle 20° shown if the stream function for the flow is

$$\psi = \frac{\mu \sin^2 \beta}{R}$$

where μ is the constant?

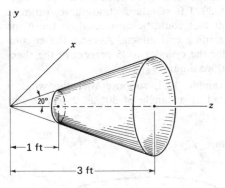

FIGURE P12.70

12.71. Show that

$$V_\beta = -\frac{1}{R \sin \beta} \frac{\partial \psi}{\partial R}$$

12.72. Given the stream function $\psi = -\frac{1}{2}V_0 R^2 \sin^2 \beta$, show that $\phi = -RV_0 \cos \beta$.

12.73. What are the maximum and minimum pressures on a sphere of radius 2 ft in a potential flow of air where the free-stream pressure is 5 lb/in^2 absolute and the free-stream velocity is 100 ft/s? Take $\rho = 0.002378$ slug/ft^3

12.74. Consider the superposition of a source of strength m at the origin of xyz and a uniform flow in the $+z$ direction with a velocity V_0. Locate the stagnation point. Show that the value of the stream function going through this point is $m/(4\pi)$. Show that the simple flows combine to give the flow around a semi-infinite body of revolution having the equation

$$\cos \beta - \frac{2\pi V_0}{m}R^2 \sin^2 \beta = 1$$

12.75. Show that the drag on a sphere in a potential flow is zero.

12.76. A three-dimensional source and sink are superposed on a uniform flow as shown. The stream function $\psi = 0$ will form a body of revolution called a *Rankine body of revolution*. Give the length of the body for the case where $m = 5$ m^3/s, $V_0 = 10$ m/s, and $a = 0.3$ m. *Suggestions:* First show that

$$\beta_A = \tan^{-1}\frac{R \sin \beta}{R \cos \beta + \alpha}$$

$$\beta_B = \tan^{-1}\frac{R \sin \beta}{R \cos \beta - \alpha}$$

Get the stream function in terms of R and β as well as the constants a, V_0, and m. Now compute V_R, set it equal to zero, and set $\beta = \pi$

or 0 to get a stagnation point. The desired result for R to the ends of the body is

$$\frac{R}{(R^2 - a^2)^2} = \frac{\pi V_0}{am}$$

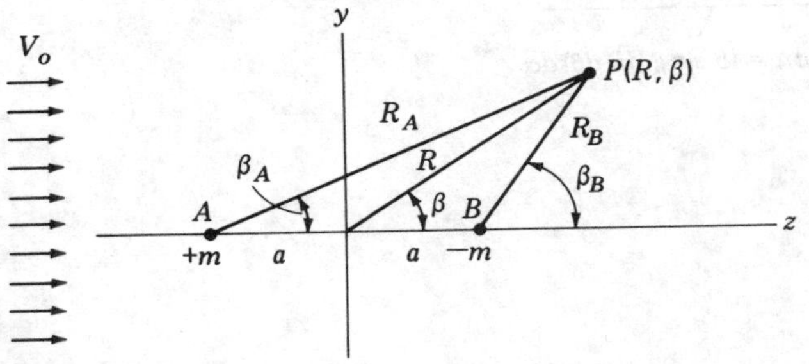

12.77. Show that ψ, Stokes' stream function, satisfies the differential form of conservation of mass in spherical coordinates.

12.78. In Prob. 12.72, transform the stream function and velocity potential to cylindrical coordinates.

12.79. In Prob. 12.76 the stream function was found to be

$$\psi = -\frac{m}{4\pi}\frac{R\cos \beta + a}{(R^2 + 2aR\cos \beta + a^2)^{1/2}}$$
$$+\frac{m}{4\pi}\frac{R\cos \beta - a}{(R^2 - 2aR\cos \beta + a^2)^{1/2}}$$
$$+\frac{1}{2}V_0 R^2 \sin^2 \beta$$

What is the diameter D at the largest circular cross section normal to the axis of symmetry of the Rankine body of revolution?

12.80. Shown is a sphere of radius b in the stream of fluid having a uniform velocity V_0 far from the sphere. If separation occurs along the peripheral line shown as AB at the indicated angle α, give an approximation of the form drag (drag due to normal stresses only) encountered by the sphere. *Hint:* Assume irrotational-flow data for the unseparated region and zero pressure recovery in the separated region.

FIGURE P12.76

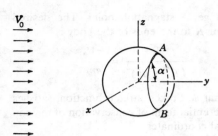

FIGURE P12.80

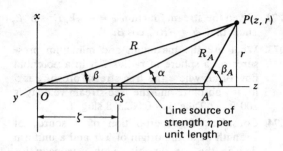

FIGURE P12.82

12.81. A spherical dome (Fig. P12.81) enclosing a radar installation has an outside radius b. What is the lift on the dome resulting from a uniform flow of air at speed V_0? *Hint:* Use one quadrant and multiply by 4. Note that β goes from zero to $\pi/2$ for such an integration.

12.82. A *line* source or sink in Fig. 12.82 is a *continuous* distribution of finite length of sources or sinks of strength η units cubed per unit length, so that the element $d\zeta$ is a source of strength $\eta\,d\zeta$. State the stream function for the uniform line source shown expressed as an integral. Relate ζ in terms of r and α. Take the differential for the case of a fixed point P to get $d\zeta$. Substitute this into the integral for ψ and integrate to get for a *constant* line strength η the result $\psi = (\eta/4\pi)(R - R_A)$. *Hint:*

$$\int \frac{\cos\beta}{\sin^2\beta}\,d\beta = -\frac{1}{\sin\beta} \qquad (a)$$

12.83. A line source issuing $1\ \text{ft}^3/\text{s}\cdot\text{ft}$ has a length of 6 ft and lies along the z axis. What is the flow crossing the conical surface which is 10 ft in length and has a half angle of 20°? See Prob. 12.82.

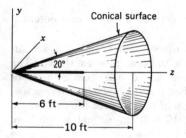

FIGURE P12.83

12.84. In Prob. 12.83, what is the flow through the conical surface if a line sink having strength of

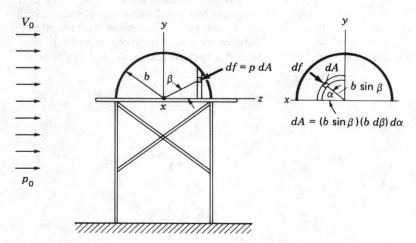

FIGURE P12.81

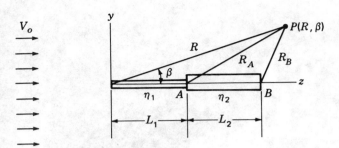

FIGURE P12.85

$\frac{1}{2}$ ft^3/s per unit length is added between $z = 6$ and $z = 10$? From Prob. 12.83, $q = 5.61$ ft^3/s.

12.85. Superpose a uniform flow V_0 with a line source or sink strength η_1 and a line source or sink of strength η_2. Depending on η_1, η_2, L_1, L_2, one can give the flow about a body of revolution of finite or infinite length. (If $\eta_1 L_1 = -\eta_2 L_2$, you will get a finite body of revolution.) Express R_A and R_B in terms of R, β, L_1, and L_2. Express ψ in terms of R, β, L_1, L_2, and the strengths η_1, η_2. Compute V_R and set it equal to zero. The end points of the body of revolution are at positions on the z axis. Show that the positions at the ends are at a distance R along the z axis given by the following equations:

$$\pm V_0 R - \frac{\eta_1 - \eta_2}{4\pi} \frac{L_1}{(R \pm L_1)}$$

$$- \frac{\eta_2}{4\pi} \frac{L_1 + L_2}{[R \pm (L_1 + L_2)]} = 0$$

See Prob. 12.82. By using many short line sources and sinks, one can get the approximate flow about any body of revolution.

12.86. In Prob. 12.85, for which

$$\psi = \frac{1}{2} V_0 R^2 \sin^2 \beta - \frac{\eta_1}{4\pi} R$$

$$+ \frac{\eta_1 - \eta_2}{4\pi} \left(R^2 - 2RL_1 \cos \beta + L_1^2 \right)^{1/2}$$

$$+ \frac{\eta_2}{4\pi} \Big[R^2 - 2R(L_1 + L_2)$$

$$\times \cos \beta + (L_1 + L_2)^2 \Big]^{1/2}$$

take η_1 as a line source of strength 1 m^3/(s)(m) and η_2 as a line sink of $\frac{1}{2}$ m^3/(s)(m). Also take $L_1 = 0.2$ m and $L_2 = 0.4$ m. The speed V_0 is 10 m/s. How long is the body of revolution? What is its radius at $\beta = 45°$?

*12.87. For a short project, find the radius of cross section of the body of revolution of the preceding problem at points separated by 0.05 m along the x axis.

Dirigible *Akron* Emerging from hanger. (*Courtesy U.S. Navy.*)

In 1924 the Goodyear Tire and Rubber Company formed the Goodyear-Zeppalin Company as a subsidiary enterprise in Akron, Ohio. This company built two identical airships—the *Macon* and the *Akron*—having each a volume of 6,500,000 cubic feet. Thrust was developed from eight 560-hp diesel engines fitted with exhaust condensers to retain water from the exhaust and thus to minimize weight loss due to fuel consumption. The *Akron* crashed in 1933 after 1200 hours of service, while the *Macon* was lost at sea in 1935. In Examples 13.4 and 13.7 you will find interesting problems, as well as more information about these famous dirigibles.

CHAPTER
13

BOUNDARY-LAYER
THEORY

13.1 INTRODUCTORY REMARKS

You will recall that in Chap. 4 we considered cases where we have irrotational, incompressible flow. Viscous action was then completely ignored. It was pointed out that for fluids with small viscosity such as air and water we could, with a high degree of accuracy at times, consider frictionless flow to exist over virtually the entire flow, except for thin regions around the bodies themselves. Here, because of the high velocity gradients, we could not properly neglect friction, so we considered these regions apart from the main flow, terming them boundary layers. It is often the case for streamlined bodies that these layers are extremely thin so that we can neglect them entirely in computing the irrotational main flow. Once the irrotational flow has been established, we can then compute the boundary-layer thickness and velocity profile in the boundary layer, etc., by first finding the pressure distributions evaluated from irrotational flow theory that was introduced in Chap. 12 and then using the results of these solutions to evaluate the boundary-layer flow.

In this way, we can at times still employ inviscid flow formulations and yet account for the drag, which we know must always be present, by separate considerations.

You will see the boundary-layer flow is even more complex than the flows studied heretofore. For this reason, we have to limit ourselves to very simple

627

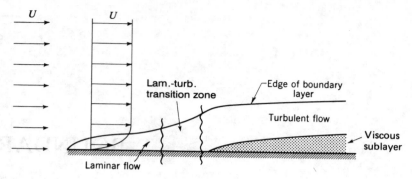

FIGURE 13.1
Details of boundary layer.

situations in our introductory study. The difficulty in boundary-layer theory is readily understandable when one realizes that in the irrotational, incompressible flows we neglect friction completely and account for only inertial effects of the fluid, so to speak. In Chap. 9, on pipe flow, we accounted for frictional effects, but when we recall that for the parallel flows studied we had constant velocity profiles, it is clear that inertial effects were not significant. In the present case, as in general viscous flows, we have both frictional *and* inertial effects of significance, and we have a more difficult situation. Thus, there may be laminar or turbulent flow in the layer, *and* the thickness and profile will change along the direction of flow. In general, we will focus much of our discussion on *steady, incompressible* flow over a flat plate at zero angle of attack.[1] Extrapolations to other conditions must be done in an appropriate manner on account of the inherent complexity of the flow.

Examining Fig. 13.1, we will now consider qualitatively the boundary-layer flow over a flat plate. Note that a laminar region begins at the leading edge and grows in thickness, as shown in the diagram. A laminar-to-turbulent transition region is reached where the flow changes from laminar to turbulent, with a consequent thickening of the boundary layer. We consider later in the chapter the question of when the transition occurs. We will see that transition depends partly on the Reynolds number, Ux/ν, where x is the distance downstream from the leading edge. Transition occurs in the range $\text{Re}_x = 3 \times 10^5$ to $\text{Re}_x = 10^6$. In the turbulent region we find, as in the turbulent pipe flow of the previous chapter, that as we get near the boundary the turbulence becomes suppressed to such a degree that viscous effects predominate, leading us to formulate the concept of a *viscous sublayer*. This very thin region is shown darkened in the diagram. You should not get the impression that these various

---

[1] For further study, you are referred to H. Schlichting, *Boundary Layer Theory*, McGraw-Hill, New York, 1979.

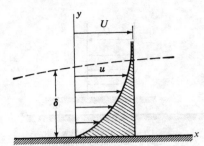

FIGURE 13.2
Boundary-layer thickness.

regions shown in our diagram are sharp demarcations of different flows. There is actually a smooth variation from regions where certain effects predominate to other regions where other effects predominate. It is merely easier to think of the action in terms of distinct regions separated by sharp boundaries.

Although the boundary layer is thin, it plays a vital role in fluid dynamics. The drag on ships and missiles, the efficiency of compressors and turbines in jet engines, the effectiveness of air intakes for ram- and turbojets—these vital considerations depend on the behavior of the boundary layer and its effects on the main flow.

13.2 BOUNDARY-LAYER THICKNESSES

We have talked about boundary-layer thickness in a qualitative manner as the elevation above the boundary which covers a region of flow where there is a large velocity gradient and consequently nonnegligible viscous effects. As pointed out, the velocity profile merges smoothly into the main-stream profile as shown in Fig. 13.2, so there is no obvious demarcation for permitting the measurement of a boundary-layer thickness in a simple manner. However, there are several definitions of boundary-layer thickness that are quite useful. One such measure is to consider that the thickness is the distance δ from the wall out to where the fluid velocity is 99 percent of the main-stream velocity.[2]

Another measure is the *displacement thickness* δ^*, defined as the distance by which the boundary would have to be displaced if the *entire* flow were imagined to be frictionless and the *same mass flow* maintained at any section. Thus, considering a unit width along z across an infinite flat plate at zero angle of attack. (Fig. 13.3), we have for incompressible flow

$$\int_0^\infty u\,dy = q = \int_{\delta^*}^\infty U\,dy$$

---

[2] Note that the boundary-layer outline as shown in Fig. 13.1 and other succeeding figures *does not* correspond to a streamline.

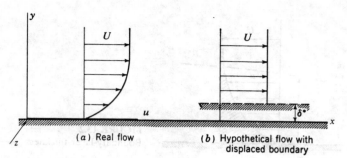

FIGURE 13.3
Displacement thickness.

Hence changing the lower limit on the second integral, we have

$$\int_0^\infty u\, dy = \int_0^\infty U\, dy - U\delta^*$$

Solving for δ^*, we get

$$\delta^* = \int_0^\infty \left(1 - \frac{u}{U}\right) dy \qquad (13.1)$$

These results are shown in Fig. 13.3. The motivation for ascertaining the displacement thickness is to permit the use of a "displaced" body in place of the actual body, such that the frictionless mass flow around the displaced body is the same as the actual mass flow around the real body. Use is made of the displacement thickness in the design of wind tunnels, air intakes for airplane jet engines, etc. We shall present several homework problems involving displacement thickness.

Yet another thickness is the *momentum thickness* θ. It is defined as the distance θ from the actual boundary such that the linear momentum flow rate for uniform velocity U through a section of height θ (see Fig. 13.4a) equals a momentum flow rate over the entire section wherein we use the *actual* profile $u(y)$ for the mass flow but use the *velocity deficit* $[U - u]$ for the velocity in computing this momentum flow (see Fig. 13.4b). That is,

$$\rho U^2 \theta = \int_0^\infty (U - u)(\rho u\, dy)$$

Canceling ρ, we get for θ

$$\theta = \int_0^\infty \frac{u}{U}\left(1 - \frac{u}{U}\right) dy \qquad (13.2)$$

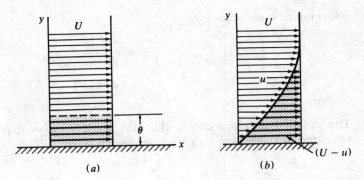

FIGURE 13.4
Momentum thickness θ.

In Prob. 13.12 we show that the momentum thickness θ is useful in evaluating the drag $D(x)$ along a distance x of a plate as well as the shear stress at the surface of the plate. The following formulas are developed in this problem for these quantities:

$$D(x) = \rho b U^2 \theta \tag{13.3}$$

$$\tau_w = \rho U^2 \left(\frac{d\theta}{dx} \right) \tag{13.4}$$

In the next section (starred) we present in some detail the work of Blasius for computing the thickness δ of the laminar boundary layer. For readers not wishing this detailed development at this time, we will now give a cursory outline of this development.

We start with the Navier-Stokes equations, simplified in Sec. 10.7 for thin films, and the continuity equations. Thus we have, for $\partial p / \partial x = 0$, the following results:

$$u \frac{\partial u}{\partial x} + v \frac{\partial u}{\partial y} = \nu \frac{\partial^2 u}{\partial y^2} \qquad \text{boundary layer equation} \tag{13.5}$$

$$\frac{\partial u}{\partial x} + \frac{\partial v}{\partial y} = 0 \qquad \text{continuity} \tag{13.6}$$

The procedure is to reduce the above partial differential equations to a *single ordinary* differential equation. This is done by first introducing the *stream function* ψ defined as you will recall by the relations

$$u = \frac{\partial \psi}{\partial y} \qquad v = -\frac{\partial \psi}{\partial x} \tag{13.7}$$

Recall that the continuity equation (13.6) is identically satisfied when giving the velocity field in terms of ψ as we have done above. Then we introduce the following new variables

$$\eta = y\sqrt{\frac{U}{\nu x}} \qquad (13.8)$$

$$f(\eta) = \frac{\psi(x, y)}{\sqrt{\nu x U}} \qquad (13.9)$$

The detailed development in the next section shows that Eq. (13.5) can be rewritten as the ordinary differential equation for f

$$2f''' + ff'' = 0 \qquad (13.10)$$

where η is the independent variable. This equation is called *Blasius' equation*. The proper boundary conditions are

$$f(0) = 0 \qquad f'(0) = 0 \qquad f'(\infty) = 1.$$

Blasius solved this equation analytically. We have not pursued the details of the actual solution in the next section since these details are beyond the level of this text. However, in Chap. 16 we solve this equation numerically. From this analytical solution for f, Blasius was able to give the following exact results for laminar boundary layers using as the thickness δ the height for which $u = 0.99U$.

$$\frac{\delta}{x} = 4.96\left(\frac{Ux}{\nu}\right)^{-1/2} = 4.96(\mathrm{Re}_x)^{-1/2} \qquad (13.11a)$$

$$\frac{\delta^*}{x} = 1.73\left(\frac{Ux}{\nu}\right)^{-1/2} = 1.73(\mathrm{Re}_x)^{-1/2} \qquad (13.11b)$$

In Sec. 13.4, the reader will be presented with approximate methodology for getting quite close to the above results for δ/x and δ^*/x for laminar flow. Most importantly, this method gives good results for turbulent boundary layers where we do not have an exact solution for the boundary layer thickness.

*13.3 SIMPLIFIED BOUNDARY-LAYER EQUATIONS FOR LAMINAR FLOW; BLASIUS' EQUATION[3]

In Sec. 10.7, we set forth from the Navier-Stokes equations a simplified set of equations for a thin layer of flow over a flat surface that we could apply to lubrication films as well as to boundary layers. We thus set forth the following equations:

$$u\frac{\partial u}{\partial x} + v\frac{\partial u}{\partial y} = -\frac{1}{\rho}\frac{dp}{dx} + \nu\frac{\partial^2 u}{\partial y^2} \qquad (13.12a)$$

$$\frac{\partial u}{\partial x} + \frac{\partial v}{\partial y} = 0 \qquad (13.12b)$$

---

[3]This section assumes that the reader has studied Sec. 10.7. If not, it should be bypassed.

Note that for a slightly curved boundary these equations are valid at a point for boundary-layer flow if one uses curvilinear coordinates one fitting the boundary and the other normal to the boundary.

It is furthermore apparent from the discussion in Sec. 10.7 that a Reynolds number denoted as Re_x of the form Ux/ν, where the length dimension is the distance from the plate leading edge to any point along the boundary, is of the order of magnitude $O(1/\delta^2)$. This is so because U and x were of order unity in the discussion of Sec. 10.7 and ν turned out to be of order of magnitude $O(\delta^2)$. We can accordingly state, on noting that $x = O(1)$, that the boundary-layer thickness is related to this Reynolds number as

$$\frac{\delta}{x} = O\left(\frac{1}{\delta^2}\right)^{-1/2} = O(\text{Re}_x)^{-1/2} \tag{13.13a}$$

$$\therefore \delta = O\left(\frac{x^2}{\text{Re}_x}\right)^{1/2} = O\left(\sqrt{\frac{\nu x}{U}}\right) \tag{13.13b}$$

This result is *independent* of the units involved. We will find such a Reynolds number significant in determining the position along the plate where transition is likely to occur from laminar to turbulent flow. It is clear from above that at any fixed position x, the larger the Reynolds number, i.e., the greater the U or the smaller the ν, the thinner the boundary layer.

We will consider the boundary layer forming on a thin plate in the presence of a uniform inviscid flow of velocity U approaching the plate (see Fig. 13.5). In this case, the pressure in the uniform flow is constant, so there will be no pressure gradient dp/dx in the boundary layer. Accordingly, we have to work with the following set of equations.

$$u\frac{\partial u}{\partial x} + v\frac{\partial u}{\partial y} = \nu\frac{\partial^2 u}{\partial y^2} \tag{13.14a}$$

$$\frac{\partial u}{\partial x} + \frac{\partial v}{\partial y} = 0 \tag{13.14b}$$

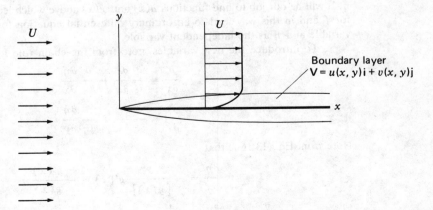

FIGURE 13.5
Uniform flow past a plate.

Consider now the velocity profile $u(y)$ in the boundary layer. Let us assume we can find a pair of *scale factors* to change the ordinate y and the abscissa x, respectively, so that at any position x_0 in the flow, the scaled profile is congruent (identical) with the scaled profile at any other position x. Such a flow is called a *self-similar*, or simply, *similar flow*. The velocity field in the scaled variable will be a function of only the scaled ordinate; the abscissa does not appear. Hence, rather than have a partial differential equation [Eq. (13.14a)] with two independent variables, we will have for the scaled velocity an *ordinary* differential equation with the scaled ordinate as the independent variable. By working on Eq. (13.14a) toward replacing it with an ordinary differential equation, we will find the appropriate scale factors to render the flow self-similar. We would expect at this early stage that the scaled velocity field should be (u/U) and the scaled ordinate should be y/δ. From Eq. (13.13b) we would further expect, the scaled ordinate to be $[y/(\sqrt{\nu x/U}\,)]$. We will now proceed toward this end by concentrating on getting to the aforementioned ordinary differential equation, which is the well-known *Blasius* equation. As a first step, we will reduce Eqs. (13.14) to a single equation. We do this by introducing the *stream function* ψ, such that

$$u = \frac{\partial \psi}{\partial y} \qquad v = -\frac{\partial \psi}{\partial x} \qquad (13.15)$$

By using ψ, the continuity equation is automatically satisfied as you may verify, so we will work solely with Eq. (13.14a).

We wish by a similarity transformation of variables to form an ordinary differential equation to replace the partial differential equation. To do this, we transform the independent variables x and y and the dependent variable ψ to the following forms:

$$\xi = \xi(x) \qquad (13.16a)$$

$$\eta = \frac{y}{g(x)} \qquad (13.16b)$$

$$f = \frac{\psi(x,y)}{h(x)} \qquad (13.16c)$$

In the ensuing process, we will consider f to be *independent* of ξ and *only* a function of η. It will be our job to find functions $h(x)$ and $g(x)$ above which lead to such a condition for f and in this way lead to an ordinary differential equation for f as the dependent variable and η as the independent variable.

To introduce the new variables, note from the chain rule for differentiation that

$$\frac{\partial}{\partial x} = \frac{\partial}{\partial \xi}\frac{\partial \xi}{\partial x} + \frac{\partial}{\partial \eta}\frac{\partial \eta}{\partial x}$$

$$\frac{\partial}{\partial y} = \frac{\partial}{\partial \xi}\frac{\partial \xi}{\partial y} + \frac{\partial}{\partial \eta}\frac{\partial \eta}{\partial y} \qquad (13.17)$$

Note from Eq. (13.16b) that

$$\frac{\partial \eta}{\partial x} = -\frac{y}{[g(x)]^2}g'(x) = -\frac{g'(x)}{g(x)}\eta \qquad (13.18a)$$

$$\frac{\partial \eta}{\partial y} = \frac{1}{g(x)} \qquad (13.18b)$$

Additionally, we *impose* the condition $\partial(\ )/\partial\xi = 0$ in the expectation of ending up with an ordinary differential equation. Equations (13.17) then become, with the aid of Eqs. (13.18),

$$\frac{\partial}{\partial x} = -\frac{g'}{g}\eta\frac{d}{d\eta} \qquad (13.19a)$$

$$\frac{\partial}{\partial y} = \frac{1}{g}\frac{d}{d\eta} \qquad (13.19b)$$

We now express u and v as follows, on employing Eqs. (13.15), (13.16), and (13.19),

$$u = \frac{\partial\psi}{\partial y} = \frac{\partial}{\partial y}[h(x)f] = h\frac{\partial f}{\partial y} = \frac{h}{g}\frac{df}{d\eta} \qquad (13.20a)$$

$$v = -\frac{\partial\psi}{\partial x} = -\frac{\partial}{\partial x}[h(x)f] = -h'f - h\frac{\partial f}{\partial x} = -h'f + \frac{hg'}{g}\eta\frac{df}{d\eta} \qquad (13.20b)$$

We can now learn something about the functions f and g. First we can readily see from Eq. (13.16b) that when

$$y \to 0 \qquad \text{then } \eta \to 0 \qquad (13.21a)$$

$$y \to \infty \qquad \text{then } \eta \to \infty \qquad (13.21b)$$

Note next that for $u(x,y)$ the boundary conditions are

$$u(x,0) = 0 \qquad (13.22a)$$

because of the "sticking" condition on the plate and

$$u(x,\infty) = U \qquad (13.22b)$$

where far from the plate we have the undistributed velocity. Finally, it is clear for no penetration into the plate heat

$$v(x,0) = 0 \qquad (13.22c)$$

It then follows from Eqs. (13.22a), (13.20a), and (13.21a) that

$$u(x,0) = 0 = \frac{h}{g}\left(\frac{df}{d\eta}\right)_{\eta=0}$$

We then conclude from above that[4]

$$f'(0) = 0 \qquad (13.23)$$

Similarly, from Eqs. (13.22b), (13.20a), and (13.21b),

$$u(x,\infty) = U = \frac{h}{g}\left(\frac{df}{d\eta}\right)_{\eta=\infty} \qquad (13.24)$$

Since h and g are functions of x, the only way to satisfy Eq. (13.24) is to require that

---

[4]Note that we are considering f as only a function of η in the expectation of finding $h(x)$ and $g(x)$ for this condition. For this reason, $df/d\eta$ will be noted as f', etc.

h/g be a *constant*. We choose U to be the constant at this time so that

$$h(x) = Ug(x) \tag{13.25}$$

It then follows from Eq. (13.24) that

$$f'(\infty) = 1 \tag{13.26}$$

Finally, noting that $v(x,0) = 0$ from Eq. (13.22c), we have from Eqs. (13.20b) and (13.21a),

$$v(x,0) = 0 = -h'f(0) + \frac{hg'}{g}\left[\eta\left(\frac{df}{d\eta}\right)\right]_{\eta=0}$$

Using the result from Eq. (13.23) which sets $(df/d\eta)_{\eta=0} = 0$, we see from the above that

$$f(0) = 0 \tag{13.27}$$

Summarizing the conditions on f, which we reiterate is set as a function of η alone, we have

$$
\begin{array}{ll}
f(0) = 0 & \text{(13.28}a\text{)}\\
f'(0) = 0 & \text{(13.28}b\text{)}\\
f'(\infty) = 1 & \text{(13.28}c\text{)}
\end{array}
$$

while for u and v we have from Eqs. (13.20) on replacing h in terms of g from Eq. (13.25),

$$u = U\frac{df}{d\eta} = Uf' \tag{13.29a}$$

$$v = U(g')(\eta f' - f) \tag{13.29b}$$

We will now see whether we indeed arrive at the desired ordinary differential equation involving f as the dependent variable and η as the independent variable. Note, using Eqs. (13.29) and (13.19b), that

$$\frac{\partial u}{\partial x} = \frac{\partial}{\partial x}(Uf') = -\frac{g'}{g}\eta Uf'' \tag{13.30a}$$

$$\frac{\partial u}{\partial y} = \frac{\partial}{\partial y}(Uf') = \frac{1}{g}Uf'' \tag{13.30b}$$

$$\frac{\partial^2 u}{\partial y^2} = \frac{\partial}{\partial y}\left(\frac{1}{g}Uf''\right) = \frac{1}{g}\left(\frac{1}{g}Uf'''\right) = \frac{U}{g^2}f''' \tag{13.30c}$$

Substituting from Eqs. (13.29) and (13.30) into Eq. (13.14a), we get

$$(Uf')\left(-\frac{g'}{g}\eta Uf''\right) + Ug'(\eta f' - f)\left(\frac{1}{g}Uf''\right) = \nu\frac{U}{g^2}f'''$$

Collecting terms,

$$f''' + \left(\frac{gg'U}{\nu}\right)ff'' = 0$$

Now, since f is to be a function of η, the expression $gg'U/\nu$ *cannot* be a function of x, so it must be a constant. We take the constant as $\frac{1}{2}$, so that

$$\frac{gg'U}{\nu} = \frac{1}{2} \tag{13.31}$$

We can separate variables for this first-order differential equation, integrate, and solve for g. Setting the constant of integration equal to zero, we get

$$g = \sqrt{\frac{\nu x}{U}} \tag{13.32}$$

Hence, from Eq. (13.25) we have, on using the above result,

$$h = U\sqrt{\frac{\nu x}{U}} \tag{13.33}$$

We have thus been able to find functions $g(x)$ and $h(x)$ to permit us to carry out our scheme set forth at the outset! It should be noted that with the above functions, f and η are dimensionless. The function f must satisfy the sought-for ordinary differential equation:

$$\boxed{2f''' + ff'' = 0} \tag{13.34}$$

The three boundary conditions for this equation are given by Eqs. (13.28). This equation is the *Blasius* equation. In Eq. (13.20a), note that $u/U = f'$. Note further that f (and hence f') is a function of only η, which from Eq. (13.16b) equals $y/g(x)$. Thus we can say

$$\left(\frac{u}{U}\right) = f' = \phi(\eta) = \phi\left(\frac{y}{g}\right)$$

where ϕ is some function. Replace g with Eq. (13.32). We get

$$\left(\frac{u}{U}\right) = \phi\left(\frac{y}{\sqrt{\nu x/U}}\right) = \phi\left(\frac{y}{\delta}\right)$$

where we have replaced $\nu x/U$ on the basis of Eq. (13.13b). We have thus arrived at the scale factors proposed heuristically at the outset of this section.

Returning to the Blasius equation, we note that it is nonlinear. Blasius solved this equation using a series approximation as part of his doctoral dissertation at Göttingen. The methodology of the solution is quite complex and beyond the level of this text.[5] We will now simply give the pertinent results reached by Blasius for the boundary-layer thickness and the displacement thickness

$$\frac{\delta}{x} = 4.96 \mathrm{Re}_x^{-1/2} \tag{13.35a}$$

$$\frac{\delta^*}{x} = 1.73 \mathrm{Re}_x^{-1/2} \tag{13.35b}$$

---

[5]S. Goldstein, *Modern Developments in Fluid Dynamics*, Oxford, New York, 1938.

13.4   VON KÁRMÁN INTEGRAL MOMENTUM EQUATION AND SKIN FRICTION

As discussed in the previous (starred) section, Blasius solved for the thickness δ of laminar boundary layers for $dp/dx = 0$ arriving at Eqs. (13.35), which we will find useful for comparison purposes to check approximate methods of finding the thickness δ as some function of x. Specifically, we consider now the von Kármán integral momentum equation which will give us very good results for δ not only in the laminar flow range but also in the turbulent range as well.

Consider a control volume of unit thickness of length dx with a height corresponding to thickness of the boundary layer, as is shown in Fig. 13.6. We consider the linear momentum equation in the x direction for this control volume in the case of a steady flow. The forces on the control surface in the x direction are shown in Fig. 13.7. Because the flow is almost *parallel flow*, we can assume, as in pipe flow, that there is a uniform pressure at a section if we neglect hydrostatic pressure. Furthermore, because the boundary layer is thin, this pressure at x equals the pressure in the main-stream flow at position x just outside the boundary layer.[6]

The *force* in the x direction can be written as

$$df_x = p\delta - \left(p + \frac{dp}{dx}\,dx\right)(\delta + d\delta) + \left(p + \frac{1}{2}\frac{dp}{dx}\,dx\right)d\delta - \tau_w\,dx$$

where τ_w is the shear stress at the wall. Canceling terms and dropping second-order expressions, we get

$$df_x = -\left(\delta\frac{dp}{dx} + \tau_w\right)dx \tag{13.36}$$

Next, we consider the *linear momentum efflux* through the control volume in the x direction. At the vertical side of the control volume at x we have

$$-\int_0^\delta \rho u^2\,dy$$

and at the other vertical section at $(x + dx)$ we can give the linear momentum efflux as a Taylor series with two terms[7]

$$\int_0^\delta \rho u^2\,dy + \frac{d}{dx}\left(\int_0^\delta \rho u^2\,dy\right)dx$$

This means that we are considering the linear momentum flow to vary continuously in the x direction. On the top surface of the control volume using u_m and v_m as the main stream velocity components, there is a mass efflux rate $\rho\mathbf{V}\cdot\mathbf{dA}$ given here as $\rho(u_m\mathbf{i} + v_m\mathbf{j})\cdot(-d\delta\,\mathbf{i} + dx\,\mathbf{j})$, so the x component of momentum

---

[6]The use of pressure p instead of τ_{nn} has been justified in Sec. 10.4. The uniformity of the pressure when neglecting hydrostatic pressure has also been shown to be valid in Sec. 10.4.

[7]Note $\int_0^\delta \rho u^2\,dy$ is a function of its upper limit δ and is hence a function of x only.

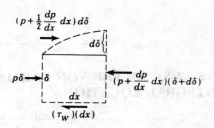

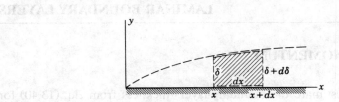

FIGURE 13.6
Control volume in the boundary layer.

FIGURE 13.7
Traction forces on control volume.

leaving the control volume along the upper surface is $\rho u_m v_m \, dx - \rho u_m^2 \, d\delta$. The *linear momentum* equation can now be given as

$$-\left(\delta \frac{dp}{dx} + \tau_w\right) dx = \frac{d}{dx}\left(\int_0^\delta \rho u^2 \, dy\right) dx + u_m(\rho_m v_m \, dx - \rho_m u_m \, d\delta) \quad (13.37)$$

wherein u_m and v_m are to be considered as *local* main-stream velocity components.

Let us next consider the *continuity* equation for the chosen control volume. We have

$$-\int_0^\delta \rho u \, dy + \rho_m(u_m \mathbf{i} + v_m \mathbf{j}) \cdot (-d\delta \, \mathbf{i} + dx \, \mathbf{j})$$

$$+\left[\int_0^\delta \rho u \, dy + \frac{d}{dx}\left(\int_0^\delta \rho u \, dy\right) dx\right] = 0 \quad (13.38)$$

Canceling terms where possible and rearranging, we get

$$(\rho_m v_m \, dx - \rho_m u_m \, d\delta) = -\frac{d}{dx}\left(\int_0^\delta \rho u \, dy\right) dx \quad (13.39)$$

Now substitute this result back into Eq. (13.37) to replace the last parenthetical expression. We get, after canceling dx,

$$-\delta \frac{dp}{dx} - \tau_w = \frac{d}{dx}\left(\int_0^\delta \rho u^2 \, dy\right) - u_m \frac{d}{dx}\left(\int_0^\delta \rho u \, dy\right) \quad (13.40)$$

This is a general form of the von Kármán integral-momentum equation. If $dp/dx = 0$, then u_m is essentially a constant which we will now denote as U and we may combine the integrations on the right side of Eq. (13.40) to give us the following more restricted form of the von Kármán integral momentum equation:

$$-\tau_w = \frac{d}{dx}\left[\int_0^\delta \rho(u^2 - Uu) \, dy\right] \quad (13.41)$$

PART A
LAMINAR BOUNDARY LAYERS

13.5 USE OF THE VON KÁRMÁN MOMENTUM INTEGRAL EQUATION

In general, we may estimate the boundary-layer thickness from Eq. (13.40) for incompressible, steady, *laminar* flow over a flat plate by

1. Using for dp/dx the gradient of the pressure in the x direction determined from irrotational-flow analysis of the flow outside the boundary layer.
2. Assuming some reasonable velocity-profile shape of the flow inside the boundary layer applicable to *any* section x so as to have similar profiles. That is, we assume u as a reasonable function of (y/δ).
3. Using Newton's viscosity law to replace τ_w by $\mu(\partial u/\partial y)_w$.

Taking ρ as constant then permits the formulation of a differential equation where δ, the boundary-layer thickness, is the dependent variable to be solved for.

To illustrate the method, we will consider the special case where $dp/dx = 0$, that is, the case for which the analytical solution from Blasius exists. Thus we will now formulate results which can be compared with the results of Blasius.

For the approximate analysis, we first assume that the profile is a second-degree curve of the form

$$u = \alpha y + \beta y^2 \tag{13.42}$$

where α and β must be determined from imposed boundary conditions. These conditions are:

When $y = 0$, $\qquad\qquad\qquad\qquad\qquad u = 0$ $\qquad\qquad\qquad$ (13.43*a*)

When $y = \delta$, $\qquad\qquad\qquad\qquad\qquad u = U$ $\qquad\qquad\qquad$ (13.43*b*)

When $y = \delta$, $\qquad\qquad\qquad\qquad\qquad \dfrac{\partial u}{\partial y} = 0$ $\qquad\qquad\qquad$ (13.43*c*)

This last condition has the edge of the boundary layer corresponding to zero slope of u in the vertical direction. This gives zero shear stress from the main stream on the boundary-layer flow. Condition (13.43*a*), you will note, is already satisfied by Eq. (13.42). Conditions (13.43*b*) and (13.43*c*), when imposed on Eq. (13.42), give us

$$U = \alpha\delta + \beta\delta^2$$

$$0 = \alpha + 2\beta\delta$$

We can then determine the constants.

$$\beta = -\frac{U}{\delta^2}$$

$$\alpha = 2\frac{U}{\delta}$$

The equation for the profile shape then becomes[8]

$$u = 2U\left(\frac{y}{\delta}\right) - U\left(\frac{y}{\delta}\right)^2 \qquad (13.44)$$

We thus have u as a function of y/δ. This function reasonably satisfies the conditions to be expected in the profile.

Going back to Eq. (13.41), we divide through by ρ; replace τ_w by $\mu(\partial u/\partial y)_w$; and, using the profile given above, we then get

$$-\nu\left\{\frac{\partial\left[2U(y/\delta) - U(y/\delta)^2\right]}{\partial y}\right\}_{y=0} = \frac{d}{dx}\int_0^\delta\left\{\left[2U\left(\frac{y}{\delta}\right) - U\left(\frac{y}{\delta}\right)^2\right]^2\right.$$
$$\left. - U\left[2U\left(\frac{y}{\delta}\right) - U\left(\frac{y}{\delta}\right)^2\right]\right\}dy \quad (13.45)$$

Carrying out the differentiation and setting $y = 0$ on the left side of Eq. (13.45) and then expanding the integral on the right side, we get

$$-\nu\frac{2U}{\delta} = \frac{d}{dx}\int_0^\delta U^2\left[4\left(\frac{y}{\delta}\right)^2 - 4\left(\frac{y}{\delta}\right)^3 + \left(\frac{y}{\delta}\right)^4 - 2\left(\frac{y}{\delta}\right) + \left(\frac{y}{\delta}\right)^2\right]dy$$

Integrating and putting in limits, we get

$$-2\frac{\nu U}{\delta} = U^2\frac{d}{dx}\left[\left(\frac{4}{3}\right)\delta - \delta + \frac{\delta}{5} - \delta + \frac{\delta}{3}\right]$$

Collecting and rearranging the terms gives us

$$\frac{2\nu}{\delta} = \left(\frac{2}{15}\right)U\frac{d\delta}{dx}$$

where, since δ is a function only of x, we have used ordinary derivatives. We have, then, a first-order differential equation in δ which is easily put in separated form. Thus

$$\delta\,d\delta = \frac{15\nu}{U}\,dx \qquad (13.46)$$

---

[8]Note that by having imposed conditions on u only in the y direction and in terms of δ, the resulting profile is then valid for the profile at any section x in the laminar-flow range. That is, we have imposed similar velocity profiles.

Integrating, we get

$$\frac{\delta^2}{2} = \frac{15\nu}{U}x + C_1 \tag{13.47}$$

If we elect to have the origin of our reference at the leading edge, we have $\delta = 0$ when $x = 0$ so that $C_1 = 0$. Solving for δ, we get

$$\delta = \sqrt{30\left(\frac{\nu x}{U}\right)} \tag{13.48}$$

Dividing both sides by x to get into dimensionless form,

$$\frac{\delta}{x} = 5.48\sqrt{\frac{\nu}{Ux}} = 5.48\mathrm{Re}_x^{-1/2} \tag{13.49}$$

We note that the boundary-layer thickness increases with the square root of the distance from the leading edge. In comparing the result above with that of Blasius [Eq. (13.35a)], we see that we are about 10 percent too high.

In Problem 13.13 we have asked you to show that the above result could also be reached using Eq. (13.4) using the momentum thickness θ with the same parabolic profile (13.44).

A better comparison to make involves the displacement thickness. To evaluate this, we return to Eq. (13.1), which becomes, on dividing by δ and changing the limits for our approximate profile,

$$\frac{\delta^*}{\delta} = \int_0^1 \left(1 - \frac{u}{U}\right)d\left(\frac{y}{\delta}\right)$$

We substitute for u/U in the above integral, using Eq. (13.44) and we get

$$\frac{\delta^*}{\delta} = \int_0^1 \left[1 - 2\left(\frac{y}{\delta}\right) + \left(\frac{y}{\delta}\right)^2\right]d\left(\frac{y}{\delta}\right)$$

Integrating and putting in the limits, we get

$$\frac{\delta^*}{\delta} = \frac{1}{3}$$

It is then apparent, upon using the result above in Eq. (13.49) to replace δ that

$$\frac{\delta^*}{x} = 1.835\mathrm{Re}_x^{-1/2} \tag{13.50}$$

In comparing the result above with Eq. (13.35b), we see that we are only about 6 percent off on the displacement thickness from the result of Blasius.

One of the problems at the end of this chapter asks for the evaluation of the boundary-layer thickness and the displacement thickness for a cubic profile of the form $\alpha y + \beta y^3$. The results from this calculation check even more closely with the results of Blasius. The results for this profile are

$$u = \frac{3}{2}U\left(\frac{y}{\delta}\right) - \frac{U}{2}\left(\frac{y}{\delta}\right)^3 \tag{13.51a}$$

$$\frac{\delta}{x} = 4.64\mathrm{Re}_x^{-1/2} \tag{13.51b}$$

$$\frac{\delta^*}{x} = 1.740\mathrm{Re}_x^{-1/2} \tag{13.51c}$$

We have thus checked our approximate method against the exact solution for the special case of a zero pressure gradient with good success. We may now, with confidence, use this approximation procedure for situations where we do not have a zero pressure gradient. And by using curvilinear coordinates, it is possible to extend this method of procedure to boundaries of mild curvature. The various velocity profiles discussed in this section are shown in Fig. 13.8.

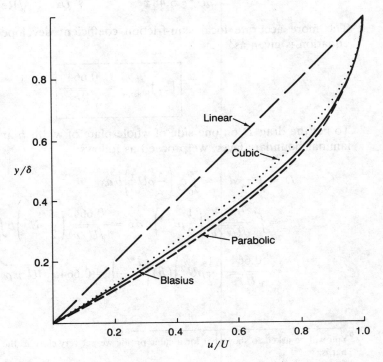

FIGURE 13.8
Laminar boundary-layer velocity profiles.

13.6 SKIN FRICTION FOR LAMINAR BOUNDARY-LAYER FLOW

We now introduce the local *skin-friction coefficient* c_f defined as

$$\boxed{c_f = \frac{\tau_w}{\frac{1}{2}\rho U^2}} \tag{13.52}$$

Note that c_f is dimensionless and is kind of an Euler number. Using the approximate profile that we used in the von Kármán momentum integral approximation, we have for c_f using Newton's viscosity law

$$c_f = \frac{\mu\left(\dfrac{\partial u}{\partial y}\right)_{y=0}}{\frac{1}{2}\rho U^2} = \frac{\mu\left\{\dfrac{\partial}{\partial y}\left[2U\dfrac{y}{\delta} - U\left(\dfrac{y}{\delta}\right)^2\right]\right\}_{y=0}}{\frac{1}{2}\rho U^2} = \frac{\mu 2U(1/\delta)}{\frac{1}{2}\rho U^2}$$

We now use Eq. (13.49) for δ to get

$$c_f = \frac{4\mu U}{\rho U^2}\frac{\sqrt{Ux/\nu}}{5.48x} = 0.730\sqrt{\frac{\nu}{Ux}} = \frac{0.730}{\sqrt{\mathrm{Re}_x}}$$

The more accurate local skin-friction coefficient developed from the Blasius equation is given as[9]

$$\boxed{[c_f]_{\text{Blasius}} = \frac{0.664}{\sqrt{\mathrm{Re}_x}}} \tag{13.53}$$

To get the drag D on one side of whole plate of width b and length l having a laminar boundary layer, we proceed as follows:

$$D = \int_A \tau_w \, dA = \int_0^l c_f\left(\frac{1}{2}\rho U^2\right)b\, dx$$

$$= \int_0^l \frac{0.664}{\sqrt{Ux/\nu}}\left(\frac{1}{2}\rho U^2\right)b\, dx = \frac{0.664}{\sqrt{U/\nu}}\left(\frac{1}{2}\rho U^2\right)b\int_0^l x^{-1/2}\, dx$$

$$= \frac{0.664}{\sqrt{U/\nu}}\left(\frac{1}{2}\rho U^2\right)(b)\left(\frac{l^{1/2}}{\frac{1}{2}}\right) = 0.664b\sqrt{lU^3\mu\rho} \tag{13.55}$$

---

[9]You will be asked to show that for a cubic profile we get very close to the Blasius solution for c_f. That is,

$$c_f = \frac{0.647}{\sqrt{\mathrm{Re}_x}} \tag{13.54}$$

FIGURE 13.9. Flow around a French turbotrain model with floor moving with free stream speed U. The stream lines are made visual by introduction of fine aluminum particles in the flow. (*Courtesy Dr. Henri Werlé, ONERA, France.*)

Note that the drag depends on $l^{1/2}$. This means that the drag toward the end of the plate contributes less proportionately than the drag near the leading edge of the plate. This is due to the fact that the boundary layer is thicker at the end, giving a smaller slope of the velocity profile at the plate boundary. This in turn results in a smaller shear stress.

The coefficient of skin friction for the *whole plate* is called the *plate skin friction coefficient* and is denoted as C_f. This useful coefficient is defined as

$$C_f = \frac{D}{\frac{1}{2}\rho U^2 A} \quad\quad D \cdot 0.5^x$$

(13.56)

Using the results of Eq. (13.55) for drag, we get for one side of the plate

$$C_f = \frac{(0.664)b\sqrt{lU^3\mu\rho}}{(\frac{1}{2}\rho U^2)(bl)} = 1.328\left[\frac{\nu}{Ul}\right]^{1/2}$$

Hence we get

$$C_f = \frac{1.328}{\sqrt{\mathrm{Re}_L}}$$

(13.57)

where Re_L is the *plate Reynolds number*. We now consider examples.

Example 13.1. In Fig. 13.9, we show flow around a model of high-speed French turbotrain.[10] Note the boundary-layer growth on the top surface of the vehicle. In this test, the "floor" of the wind tunnel moves with the speed corresponding to that of the main flow. This prevents a boundary layer from building up on the floor as the fluid reaches the turbotrain and more closely resembles the actual flow relative to a moving train.

---

[10] This train is the fastest in the world with a capability up to 270 mi/h.

Suppose that the floor were *not* moving. There is a distance of 2.5 m from the leading edge of the floor up to the front of the train. The velocity of the free stream is 6 m/s. What is the boundary-layer thickness as one reaches the train? The Reynolds number Re_x for transition to turbulent boundary-layer flow is 10^6. The air is at 20°C. Comment on the usefulness of the moving floor.

We first calculate Re_x just as the flow reaches the train. For this purpose, we will need ν, the kinematic viscosity of the air. We get this from Appendix Fig. B.2 to be 1.55×10^{-5} m²/s. Hence we have for Re_x

$$Re_x = \frac{(6)(2.5)}{1.55 \times 10^{-5}} = 9.68 \times 10^5$$

We thus still have a laminar boundary layer when the flow approaches the model. We can then give δ as follows, using Eq. (13.35a),

$$\delta = \frac{(x)(4.96)}{Re_x^{1/2}} = \frac{(2.5)(4.96)}{(9.68 \times 10^5)^{1/2}} = 1.260 \times 10^{-2} \text{ m}$$

$$= 12.60 \text{ mm}$$

If the train is approximately 12 mm or less above the floor, then the flow on the bottom portion of the turbotrain will be distorted from that which would actually occur around a moving train. This would affect the boundary-layer growth on the bottom of the train, the flow around the wheels, and the wakes that develop underneath (one of which can readily be seen in the figure). These effects will yield error in estimating the drag.

Example 15.2. The fixed keel of your author's Columbia 22 sailboat is about 38 in long (see Fig. 13.10). Moving in Lake Ontario at a speed of 3 knots, what is the skin drag from the keel? The water is at 40°F.

Do this problem two ways. First, use a rectangular plate of length 38 in and width 24.5 in, which is the average width of the keel. Then solve for the drag using the actual dimensions of the keel as shown in Fig. 13.10. Compare answers and comment on the results. Transition takes place at a Reynolds number Re_x of 10^6.

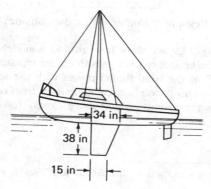

FIGURE 13.10
Sailboat with fixed keel.

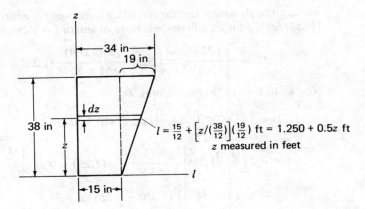

FIGURE 13.11
Detailed analysis of keel

Calculation 1. Compute the plate Reynolds number for the rectangular model of the keel.

$$\text{Re}_L = \frac{Ul}{\nu} = \frac{[(3)(1.689)](24.5/12)}{1.664 \times 10^{-5}} = 6.22 \times 10^5 \qquad (a)$$

We thus have a laminar boundary layer. We get the plate coefficient of drag using Eq. (13.57).

$$C_f = \frac{1.328}{\sqrt{6.22 \times 10^5}} = 1.684 \times 10^{-3} \qquad (b)$$

Next from Eq. (13.56) we get the skin drag, realizing that there are two sides to be considered. Thus,

$$
\begin{aligned}
D &= 2\left[(C_f)\left(\frac{1}{2}\rho U^2\right)(A)\right] \\
&= 2\left\{(1.684 \times 10^{-3})\left(\frac{1}{2}\right)(1.940)[(3)(1.689)]^2\left[\frac{(38)(24.5)}{144}\right]\right\} \\
&= 0.542 \text{ lb}
\end{aligned}
\qquad (c)
$$

Calculation 2. We will now perform another calculation of the drag. For this purpose, consult Fig. 13.11 in which a keel with an infinitesimal slice dz is depicted. The length $l(z)$ of the slice is:

$$l = \frac{15}{12} + \left(\frac{z}{38}\right)(19) = 1.250 + 0.5z \quad \text{ft} \qquad (d)$$

with z in feet. Now let us see if we have transition anywhere on the keel. Looking at the uppermost portion, we have for $(\text{Re}_x)_{\max}$

$$(\text{Re}_x)_{\max} = \frac{Ul_{\max}}{\nu} = \frac{[(3)(1.689)](\frac{34}{12})}{1.664 \times 10^{-5}} = 8.63 \times 10^5 \qquad (e)$$

We accordingly have a laminar boundary layer over the entire surface. We use Eq. (13.57) for C_f for an infinitesimal plate of length $l = 1.250 + 0.5z$ ft. Thus,

$$C_f = \frac{1.328}{(Ul/\nu)^{1/2}} = 1.328\left[\frac{(3)(1.689)}{1.664 \times 10^{-5}}(1.250 + 0.5z)\right]^{-1/2}$$

Now go to Eq. (13.56) for the drag D.

$$D = 2\int_0^{38/12} C_f\left(\frac{1}{2}\rho U^2\right)(1.250 + 0.5z)(dz)$$

$$= 2\int_0^{38/12} 1.328\left[\frac{(3)(1.689)}{1.664 \times 10^{-5}}(1.250 + 0.5z)\right]^{-1/2}\left(\frac{1}{2}\right)(1.940)$$

$$\times [(3)(1.689)]^2(1.250 + 0.5z)\,dz$$

$$= 0.1199\int_0^{3.17}(1.250 + 0.5z)^{1/2}\,dz$$

Let

$$1.250 + 0.5z = \eta$$
$$\therefore 0.5\,dz = d\eta$$
$$dz = 2\,d\eta$$

Hence,

$$D = \frac{0.1199}{0.5}\int_{1.250}^{2.833}\eta^{1/2}\,d\eta$$

$$= \frac{0.1199}{0.5}\eta^{3/2}\left(\frac{2}{3}\right)\Big|_{1.250}^{2.833} = 0.539 \text{ lb}$$

The averaging process in calculation 1 has given a very good result when compared with the result above.

In this section, we have considered only laminar boundary layers. We have indicated that there can be a transition from a laminar boundary layer to a turbulent boundary layer dependent in some way on the Reynolds number Re_x. We will look into this process in the next section and then we will use the von Kármán momentum integral method to investigate turbulent boundary-layer flow. However, we will not have an exact solution to afford us a check as was the case for laminar boundary-layer flow.

13.7 TRANSITION FOR FLAT-PLATE FLOW

The transition from laminar to turbulent flow in the boundary layer of a flat plate depends on many things. The more important factors are listed as:

1. Reynolds number, Ux/ν
2. Free-stream turbulence
3. Roughness of the plate
4. Heat transfer to or from the plate

Furthermore, the process of transition is intermittent, consisting of bursts of turbulence in small regions in the boundary layer. These travel with the boundary-layer flow appearing and disappearing in an irregular manner in increasing numbers until the flow is fully turbulent, having the microscopic random fluctuations described in Chap. 9 on pipe flow.

Figure 13.12 is a plot of $\delta/\sqrt{\nu x/U}$ (note that the denominator is not the Reynolds number) versus the Reynolds number from the early experimental data of Hansen for a flow over a smooth plate. Note that the ordinate is constant until a point corresponding to a Reynolds number of about 3.2×10^5

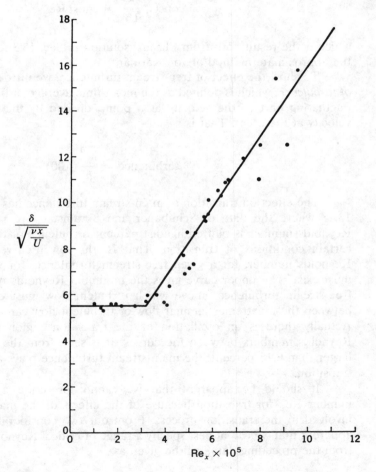

FIGURE 13.12
Variation of boundary-layer thickness with local Reynolds number for flow over a flat plate. (*From Hansen, NACA TM 585, 1930.*)

is reached, at which time there is a sudden change in that the ordinate goes up rapidly beyond the point. We can take this point as the transition point along the plate, after which turbulent flow with its thickened boundary layer is to be found.

The fact that $\delta/\sqrt{\nu x/U}$ is constant in the laminar region checks with the theory presented thus far. Thus we have from the experiments

$$\frac{\delta}{\sqrt{\nu x/U}} = \text{const}$$

so

$$\frac{\delta}{x} = \text{const}\sqrt{\frac{\nu}{Ux}} = \text{const}\,\text{Re}_x^{-1/2} \qquad (13.58)$$

which is the result both from Blasius' solution in Sec. 13.3 for laminar flow and the approximate method of von Kármán.

To show the effect of free-stream turbulence, we introduce the *percentage of turbulence*, which is defined as the mean-time average of the *magnitude* of the fluctuating part of the velocity at a point, divided by the mean-time-average velocity at the point. That is,

$$\% \text{ turbulence} = \frac{\overline{|u'|}}{\bar{u}}(100) \qquad (13.59)$$

The effect on transition of main-stream turbulence has been shown in Fig. 13.13 where the data of Schubauer and Skramstad are reported. The local Reynolds number is plotted against percent turbulence of the main stream for certain conditions of transition. That is, the lower curve gives the highest Reynolds number, for a given free-stream turbulence, for which laminar flow must exist. The upper curve gives the minimum Reynolds number, for a given free-stream turbulence, above which turbulent flow must exist. In the region between these extremes laminar flow or turbulent flow can exist at any instant. Actually there is an oscillation of the transition region over the range of Reynolds numbers between the curves. It is seen from the diagram that once larger than 0.15 percent, the main-stream turbulence plays a significant role in transition.

It should be apparent that we cannot prescribe a specific Reynolds number Re_x for transition because of the effects of the many factors that are involved in the transition process. From our two considerations thus far, it is apparent that we can at best specify a range of critical Reynolds numbers which, from the preceding data, can be given as

$$\text{Re}_{cr} = \left(\frac{Ux}{\nu}\right) = 3.2 \times 10^5 \text{ to } 10^6 \qquad (13.60)$$

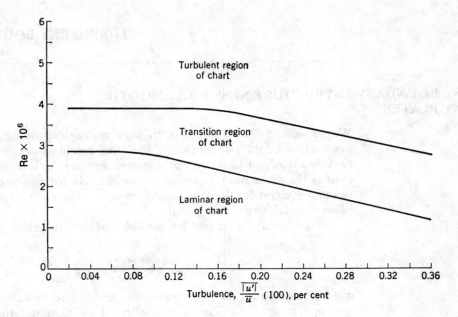

FIGURE 13.13
Effect of free-stream turbulence on transition for flow over a flat plate. (*Schubauer and Skramstad.*)

The high value of 10^6 can be reached by having very small free-stream turbulence, plate smoothness, etc. In our work we generally use as the critical Reynolds number Re_{cr}, the value 500,000 unless known local circumstances indicate some other number in the above-stated range.

Finally, note that roughness of the plate surface will bring on an earlier transition, as one might expect. Also heating the plate in the laminar region will hasten a transition to turbulent flow.

Transition from laminar to turbulent flow is even more complex for two-dimensional flow with curvature or three-dimensional flow. There is information in the literature giving experimental results relating to transition for simple shapes such as cylinders, spheres, and ellipsoids. One interesting and important feature of these flows is that the pressure gradient exerts a significant influence on the position of the point of transition. A decreasing pressure will tend to delay transition from laminar to turbulent flow. Since the laminar boundary layer offers less skin friction than the turbulent boundary, bodies so shaped as to have a pressure gradient which delays transition will then have less skin drag. Airfoils so designed are called *laminar airfoils*. We will have more to say about these important factors for curved boundaries later in this chapter.

13.8   BOUNDARY-LAYER THICKNESS FOR SMOOTH FLAT PLATES

We showed in Chaps. 9 and 10 that the mean-time averages of turbulent-flow parameters could be computed like the actual quantities in laminar flow provided that you included the *apparent* stresses. In this section, we therefore employ the momentum equation of von Kármán for mean-time-average quantities and account for the apparent stress by employing an experimentally derived value of shear stress at the wall.

Blasius has found that for *smooth* surfaces the result

$$\tau_w = 0.0225\rho U^2\left(\frac{\nu}{U\delta}\right)^{1/4} \tag{13.61}$$

can be used for turbulent flow considerations. This was determined by Blasius for pipe flow[11] and was later confirmed by Schultz-Grunow for Reynolds numbers Re_x *between 5×10^5 and 10^7 for flat plates*. As before, we assume that the main-stream pressure just outside the boundary layer prevails as the mean-time-average uniform pressure over a section inside the turbulent boundary layer.

We need an approximate velocity profile for use in the von Kármán momentum equation, so we use the one-seventh law, which we found acceptable for pipe flow of comparatively low Reynolds numbers. Thus,

$$\bar{u} = U\left(\frac{y}{\delta}\right)^{1/7} \tag{13.62}$$

For the case of a zero pressure gradient in the main flow, we can then say for incompressible flow on using Eq. (13.41) that

$$-\frac{1}{\rho}\tau_w = \frac{d}{dx}\int_0^\delta (\bar{u}^2 - U\bar{u})\,dy \tag{13.63}$$

Substituting from Eqs. (13.61) and (13.62) into Eq. (13.63), we then have

$$-0.0225U^2\left(\frac{\nu}{U\delta}\right)^{1/4} = \frac{d}{dx}\int_0^\delta U^2\left[\left(\frac{y}{\delta}\right)^{2/7} - \left(\frac{y}{\delta}\right)^{1/7}\right]dy \tag{13.64}$$

---

[11]This was presented as Eq. (9.23), where R, the radius of the pipe, was used in place of δ and where a different constant is used.

Canceling U^2 and integrating, we get

$$-0.0225\left(\frac{\nu}{U\delta}\right)^{1/4} = \frac{d}{dx}\left(\frac{7}{9}\delta - \frac{7}{8}\delta\right) = -\frac{7}{72}\frac{d\delta}{dx}$$

Separating variables, we get

$$\left(\frac{\nu}{U}\right)^{1/4} dx = 4.32\,\delta^{1/4}\,d\delta \tag{13.65}$$

Integrating,

$$\left(\frac{\nu}{U}\right)^{1/4} x = (4.32)\left(\frac{4}{5}\right)\delta^{5/4} + C_1 = 3.46\delta^{5/4} + C_1 \tag{13.66}$$

We must decide what to do with the constant of integration. You will recall that the turbulent boundary layer starts at the transition location and has some thickness at this position. The location of this transition is a difficult problem. Furthermore, we do not know in any simple way the initial thickness of the turbulent boundary layer. It would appear that we have no known initial condition by which to ascertain the constant of integration. To escape this dilemma, Prandtl has shown that by imagining the turbulent boundary layer to start at the edge of the plate at $x = 0$, one gets reasonably good results at locations beyond the transition location where the turbulent boundary layer actually exists. Newer measurements show that this is not the best assumption, but in the interests of simplicity we will follow Prandtl and set $C_1 = 0$ accordingly. Solving for δ in Eq. (13.66), we then have

$$\delta = \left(\frac{\nu}{U}\right)^{1/5}(x^{4/5})(0.37)$$

Rearranging, we get

$$\boxed{\frac{\delta}{x} = \frac{0.37}{(Ux/\nu)^{1/5}} = 0.37\mathrm{Re}_x^{-1/5}} \tag{13.67}$$

You are asked to show (Prob. 13.34), using the one-seventh law of turbulent flow, that the displacement thickness δ^* is given as

$$\boxed{\frac{\delta^*}{x} = 0.0463\mathrm{Re}_x^{-1/5}} \tag{13.68}$$

From Eqs. (13.67) and (13.35a), you can see that if you compute the boundary-layer thickness at the same location for laminar and turbulent flow, respectively, the turbulent layer will be thicker.

Example 13.3. In Example 13.1, consider that the free-stream turbulence is such that transition takes place at a Reynolds number predicted by Hansen, namely, at 3.2×10^5. Compute the boundary-layer thickness at transition for laminar boundary layer and compare it to the boundary-layer thickness computed from turbulent flow at the same position. Next, find the boundary-layer thickness at the leading position of the turbotrain.

First we find the poisson x_T for transition.

$$\text{Re}_{cr} = 3.2 \times 10^5 = \left[\frac{(6)(x_T)}{1.55 \times 10^{-5}} \right] \qquad (a)$$

$$\therefore x_T = 0.827 \text{ m}$$

We next compute δ at x_T for a laminar boundary layer using Eq. (13.35a):

$$\delta_T = \frac{x_T(4.96)}{\sqrt{\text{Re}_{cr}}}$$

$$= \frac{(0.827)(4.96)}{\sqrt{3.2 \times 10^5}} \qquad (b)$$

$$= 7.25 \text{ mm}$$

As for the thickness δ for the turbulent boundary layer at transition, we have from Eq. (13.67),

$$\delta_T = \frac{x_T(0.37)}{(\text{Re}_{cr})^{1/5}}$$

$$= \frac{(0.827)(0.37)}{(3.2 \times 10^5)^{1/5}} \qquad (c)$$

$$= 24.25 \text{ mm}$$

We see the growth of boundary-layer thickness as we move through the transition region. Now let us go to the front of the turbotrain model. We get for δ

$$\delta = \frac{x(0.37)}{(\text{Re}_x)^{1/5}}$$

$$= \frac{(2.5)(0.37)}{\{[(6)(2.5)]/(1.55 \times 10^{-5})\}^{1/5}}$$

$$= 58.75 \text{ mm}$$

It should now be evident why the moving floor in the test section of the wind tunnel is so desirable.

13.9 SKIN-FRICTION DRAG FOR SMOOTH PLATES

Low Reynolds Number Flow $< 10^7$

Let us now consider the *local* skin-friction coefficient c_f and the drag D due to turbulent flow. We can say for c_f

$$c_f = \frac{\tau_w}{\frac{1}{2}\rho U^2} = \frac{0.0225\rho U^2(\nu/U\delta)^{1/4}}{\frac{1}{2}\rho U^2} \qquad (13.69)$$

where we have used Eq. (13.61) for τ_w. Now solving for δ in Eq. (13.67) and substituting into Eq. (13.69), we get on canceling ρU^2,

$$c_f = (2)(0.0225)\left[\frac{\nu}{(U)(x)(0.37)[\nu/Ux]^{1/5}}\right]^{1/4}$$

$$= 0.0577\left[\frac{\nu^{4/5}}{U^{4/5}x^{4/5}}\right]^{1/4}$$

$$= \frac{0.0577}{[\mathrm{Re}_x]^{1/5}} \qquad \text{for } 5 \times 10^5 < \mathrm{Re}_x < 10^7 \qquad (13.70)$$

where because of the limitations on the shear-stress formula used in Eq. (13.69), the result in Eq. (13.70) is valid only in the range of Reynolds numbers from 5×10^5 to 10^7, as has been indicated. The drag D is next calculated to be for a plate of width b and length l. Start with Eq. (13.69).

$$D = \int_A \tau_w \, dA = \int_0^l (c_f)\left(\frac{1}{2}\rho U^2\right) b \, dx$$

$$= \frac{0.0577}{2}\rho U^2 b \int_0^l \left[\frac{\nu}{Ux}\right]^{1/5} dx$$

$$= \frac{0.0577}{2}\rho U^2 b \left[\frac{\nu}{U}\right]^{1/5} x^{4/5}\left(\frac{5}{4}\right)\Bigg|_0^l$$

$$= 0.0361\rho U^2 bl\left(\frac{\nu}{Ul}\right)^{1/5}$$

$$= 0.0361\rho U^2 bl[\mathrm{Re}_L]^{-1/5} \qquad (13.71)$$

where we note again that Re_L is the *plate* Reynold's number, Ul/ν.

The coefficient of skin friction for the *plate* is defined as in the previous section as

$$C_f = \frac{D}{\frac{1}{2}\rho U^2(bl)} \qquad (13.72)$$

so that

$$\boxed{D = C_f\left(\tfrac{1}{2}\rho U^2\right)(bl)} \qquad (13.73)$$

Inserting the result from Eq. (13.71) into Eq. (13.72), we get

$$C_f = \frac{0.0361\rho U^2(bl)[\mathrm{Re}_L]^{-1/5}}{\frac{1}{2}\rho U^2(bl)}$$

$$= \frac{0.072}{[\mathrm{Re}_L]^{1/5}} \qquad (13.74)$$

TABLE 13.1

Re_L	300,000	500,000	10^6	3×10^6
A	1050	1700	3300	8700

For a better fit with experimental data, Eq. (13.74) is slightly changed to read

$$C_f = \frac{0.074}{[Re_L]^{1/5}} \quad \text{for } 5 \times 10^5 < Re_L < 10^7 \qquad (13.75)$$

where the range of the plate Reynolds number has once more been indicated.

The formulations above can be used assuming that there is turbulent boundary layer over the *entire* plate. We do know however that there will generally be a region of laminar boundary-layer flow near the leading edge having *less* skin drag than were the boundary layer turbulent over that region. Following Prandtl, the coefficient of skin friction can be adjusted to take this laminar boundary layer into account. Of course, the formula has to involve the value of the critical Reynolds number Re_{cr} for transition. Thus, we have for C_f, taking the laminar region into account, with A determined from Table 13.1

$$C_f = \frac{0.074}{[Re_L]^{1/5}} - \frac{A}{[Re_L]} \quad \text{for } 5 \times 10^5 < Re_L < 10^7 \qquad (13.76)$$

High Reynolds Number Flow $> 10^7$

The results above for C_f and c_f we repeat, are valid for Reynolds numbers Re_L from 5×10^5 to 10^7 because of the limitations on the friction formula given by Eq. (13.61). However, for modern aircraft and fast ships, the Reynolds numbers can far exceed the above limits, so we need for such cases more accurate coefficients in the range of Reynolds numbers exceeding 10^7. In pipe flow, we discussed turbulent velocity profiles for very high Reynolds numbers and we showed that the *universal logarithmic velocity profile* could be extrapolated to arbitrarily large Reynolds numbers. We now use this profile for the boundary-layer flow, since the earlier discussion in Chap. 9 showed it first to be valid for two-dimensional turbulent flows and then later extended it to pipe flow.

From Eq. (9.51) we have[12]

$$\frac{\bar{u}}{V_*} = B_1 \log\left(\frac{yV_*}{\nu}\right) + B_2 \qquad (13.77)$$

---

[12] Note that we are now using logarithm to the base 10 (log) rather than the logarithm to the base e (ln). For the base e, as was used in pipe flow, $B_1 = 1/0.4 = 2.5$.

where $V_* = \sqrt{\tau_w/\rho}$. For pipe flow, $B_1 = 5.76$ and $B_2 = 5.5$. For the flat plate, extensive experimentation has indicated that $B_1 = 5.85$ and $B_2 = 5.56$. From this profile, we could conceivably formulate the total skin-friction coefficient C_f. But this results in a formulation which is very inconvenient to use. Instead, we have an empirical equation due to H. Schlichting that relates C_f and Re_L to fit experimental data.

$$C_f = \frac{0.455}{(\log \text{Re}_L)^{2.58}} \qquad \text{for Re}_L > 10^7 \qquad (13.78)$$

This equation is plotted as curve 3 in Fig. 13.14. This figure shows C_f as a function of Re_L over the laminar boundary-layer flow and turbulent boundary-layer flow with a transition region. The curve 3 is valid to the right of the transition zone where $\text{Re}_L > 10^7$. Curve ① is the plot of Eq. (13.57) valid for laminar boundary-layer flow.

If we wish to include in our formulation of C_f the laminar region near the leading edge in the boundary-layer flow as well as the turbulent region, we present the following empirical formula, called the *Prandtl-Schlichting* skin-friction formula:

$$C_f = \frac{0.455}{(\log \text{Re}_L)^{2.58}} - \frac{A}{\text{Re}_L} \qquad (13.79)$$

where A depends on the transition position and is given in Table 13.1. The formula for C_f given by Eq. (13.79), with the proper A, is valid over a range of Reynolds numbers from initial transition up to $\text{Re}_L = 10^9$. We now illustrate the use of the preceding results.

Example 13.4. The United States at one time in the thirties had three large dirigibles—the *Los Angeles*, the *Macon*, and the *Akron*.[13] Two of them were destroyed by accidents. The largest were the *Akron* and the *Macon* each having a length of 785 ft and a maximum diameter of 132 ft. The maximum speed was 84 mi/h. The useful lift was 182,000 lb.

Moving at top speed, estimate the power needed to overcome skin friction, which is a significant part of the drag. Disregard effects of protrusion from engine cowlings, cabin region, etc. Assume that the surface is smooth. Take the critical Reynolds number to be 500,000. Consider the Akron at 10,000 ft standard atmosphere.

---

[13]The Akron maintained five fighter planes that could come aboard or leave, using a hook mechanism. Also, there was a trapeze arrangement which permitted the lowering of an observer well below the airship to guide the airship while the airship remained hidden in cloud cover. This was pretty heady stuff in the author's day.

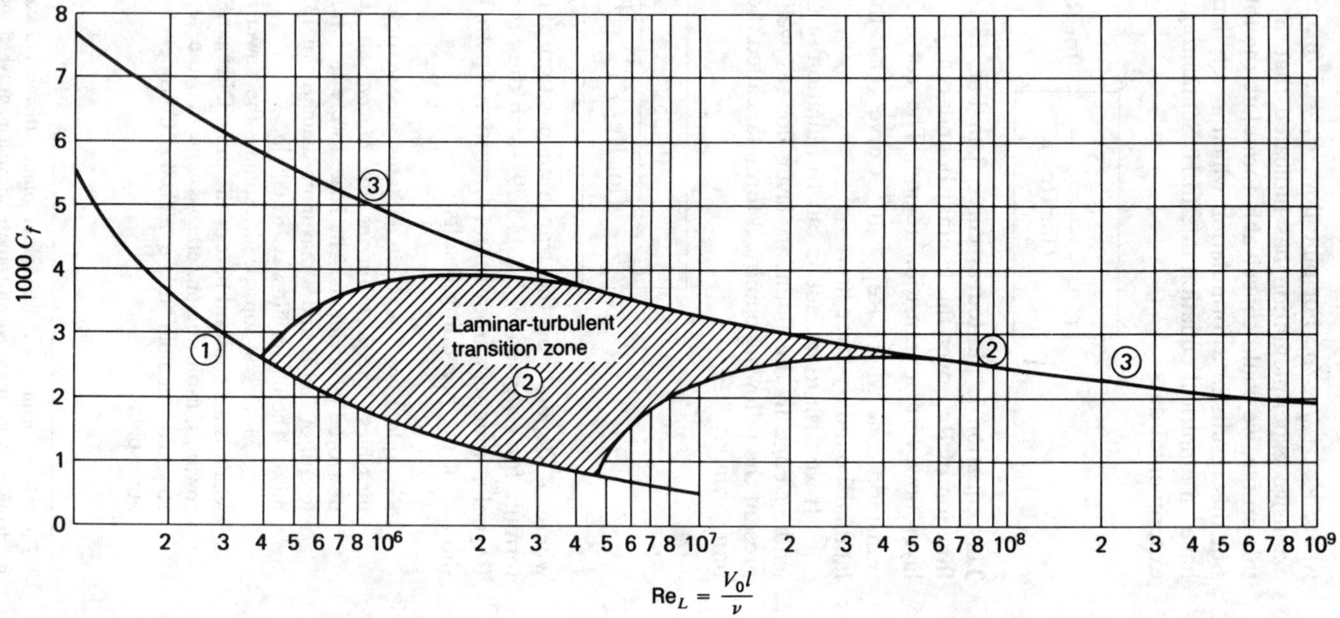

FIGURE 13.14

Skin-friction coefficient for smooth plates. Curve ①, laminar Eq. (13.57). Curves in ②, transition to turbulent flow. Curve ③, turbulent flow, Eq. (13.78).

We can make a reasonable estimate of the skin drag by "unwrapping" the outer surface of the *Akron* to form a flat plate. As a first step, we wish to calculate the plate Reynolds number. For this we find from the standard atmosphere table that

$$\rho = (0.7385)(0.002378) = 0.001756 \text{ slugs/ft}^3$$

$$T = 23.3°F$$

From the viscosity curves in Appendix B, we then find that

$$\mu = 3.7 \times 10^{-7} \text{ lb} \cdot \text{s/ft}^2$$

We can now compute Re_L

$$\text{Re}_L = \frac{(0.001756)[(84)(5280/3600)](785)}{3.7 \times 10^{-7}}$$

$$= 4.59 \times 10^8 \qquad\qquad (a)$$

We will use the Prandtl-Schlichting skin-friction formula [Eq. (13.79)] with $A = 1700$ in accordance with Table 13.1 for $\text{Re}_{cr} = 500,000$. We get for C_f

$$C_f = \frac{0.455}{\left[\log(4.59 \times 10^8)\right]^{2.58}} - \frac{1700}{4.59 \times 10^8}$$

$$= 1.730 \times 10^{-3} \qquad\qquad (b)$$

Now going to Eq. (13.73), we have for the drag D owing to skin friction using the maximum diameter of 132 ft:

$$D = (1.730 \times 10^{-3})(\tfrac{1}{2})(0.001756)[84(5280/3600)]^2(785)[(\pi)(132)]$$

$$= 7505 \text{ lb} \qquad\qquad (c)$$

The power needed then is

$$\text{Power} = \frac{(7505)(84)(5280/3600)}{550} = 1681 \text{ hp} \qquad\qquad (d)$$

This should be a lower limit, because we have not included surface roughness, and pressure drag as well as the effects of outside protuberances. In a later example, we return to this case, using experimentally formulated drags for ellipsoidal bodies of revolution so that we can estimate the pressure drag.

Example 13.5. In Fig. 13.15 is shown an aircraft that made its appearance in the 1930's and was a forerunner to the helicopter.[14] It was called the *autogyro*. In the autogyro, the lift is developed by *freely rotating vanes*. The rotation is caused by the aerodynamic forces on the vanes themselves. When there is engine failure, the autogyro if it has sufficient forward velocity can simply "parachute" downward under the support of the rotating vanes (no crashes). Using flat-plate theory, what

---

[14]The autogyro gave way to the helicopter because the autogyro could not hover. However much of the complex rotor technology of the helicopter was developed on the autogyro by a Spanish engineer named Juan de la Cierva.

is the aerodynamic torque needed to overcome skin friction for an angular speed of the vanes of 80 r/min. Take each vane to be a flat plate of dimension 4.5 m by 0.3 m. The air is at a temperature of 10°C. Transition in the boundary layer takes place at $\text{Re}_{cr} = 5 \times 10^5$. Take $\nu = 1.55 \times 10^{-5}$ m²/s.

The angular velocity in radians per second is

$$\omega = 80\left(\frac{2\pi}{60}\right) = 8.378 \text{ rad/s}$$

The maximum plate Reynolds number is

$$(\text{Re}_L)_{max} = \frac{(8.378)(4.5)(.3)}{1.55 \times 10^{-5}} = 7.297 \times 10^5$$

Hence for part of the vane, that is, beyond a certain radius $\bar{R}$ we have a turbulent boundary layer. Clearly then some portion of the vane up to this radius $\bar{R}$ will have plate Reynolds numbers less than 5×10^5 and hence will be subject to a laminar boundary layer. We next determine $\bar{R}$. Hence

$$\text{Re}_{cr} = 5 \times 10^5 = \frac{(\bar{R})(8.378)(.3)}{1.55 \times 10^{-5}} \tag{a}$$

$$\therefore \bar{R} = 3.083 \text{ m}$$

We will now compute the torque due to skin friction in two stages. We start with the laminar boundary-layer region, $0 \le x \le 3.083$. Noting that we must account for six flat surfaces we have using Eq. (13.57),

$$T_{\text{lam}} = 6\int_0^{3.083} (C_f)\left(\frac{1}{2}\right)(\rho u^2)\underbrace{(0.3)(dR)}_{dA}(R)$$

$$\underbrace{\hspace{4cm}}_{d\,(\text{Drag})}$$

$$\underbrace{\hspace{6cm}}_{d\,(\text{Torque})}$$

$$= 6\int_0^{3.083} \frac{1.328}{\left[\dfrac{(R)(8.378)(0.3)}{1.55 \times 10^{-5}}\right]^{1/2}}\left(\frac{1}{2}\right)(\rho)[(R)(8.378)]^2(0.3)R\,dR$$

$$= 0.2083\rho\int_0^{3.083} R^{5/2}\,dR$$

$$= 0.2083\rho(R^{7/2})\left(\frac{2}{7}\right)\Bigg|_0^{3.083} = 3.063\rho \tag{b}$$

To get the density ρ we use the *equation of state* for air. Thus

$$\rho = \frac{p}{RT} = \frac{101,325}{(287)(283)} = 1.248 \text{ kg/m}^3 \tag{c}$$

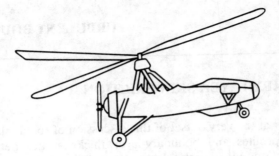

FIGURE 13.15
An autogyro of the type used during the
1930's.

Hence, from Eq. (b) we get

$$T_{\text{lam}} = 3.82 \text{ N-m}$$

Next we compute the skin friction torque for the portion of the vane along $3.083 \le R \le 4.5$ where we have turbulent boundary-layer flow. Thus using Eq. (13.76) we have

$$T_{\text{turb}} = 6\int_{3.083}^{4.5} (C_f)\left(\frac{1}{2}\right)(\rho U^2)(0.3)(dR)R$$

$$= 6\int_{3.083}^{4.5} \left\{ \frac{0.074}{\left[\dfrac{(R)(8.378)(0.3)}{1.55 \times 10^{-5}}\right]^{1/5}} - \frac{1700}{\left[\dfrac{(R)(8.378)(0.3)}{1.55 \times 10^{-5}}\right]} \right\}$$

$$\times \left(\frac{1}{2}\right)(1.248)[(R)(8.378)]^2(0.3)(R) \, dR$$

$$= 78.84 \left\{ \int_{3.083}^{4.5} [6.718 \times 10^{-3}R^{14/5} - 1.048 \times 10^{-2}R^2] \, dR \right.$$

$$= 78.84 \left[6.718 \times 10^{-3}R^{19/5}\left(\frac{5}{19}\right) - 1.048 \times 10^{-2}\frac{R^3}{3} \right]\Bigg|_{3.083}^{4.5}$$

$$= 78.84[0.4091 - 0.2160] = 15.23 \text{ N-m}$$

The total skin friction torque is then

$$T_{\text{total}} = 3.82 + 15.23 = \underline{19.05 \text{ N-m}}$$

TURBULENT BOUNDARY LAYERS:
ROUGH PLATES

13.10   TURBULENT BOUNDARY-LAYER SKIN-FRICTION DRAG FOR ROUGH PLATES

We point out at the very outset of this discussion of rough plates that such items as velocity profiles and boundary-layer thickness are determined as in the previous section on smooth plates. We will center our discussion here on skin-friction drag which does depend on roughness of the plate. You may recall this was the case with pipe flow.

Let us now proceed to consider the roughness of the plate. Recall in pipe flow that we used the relative roughness parameter e/D, where e was the average height of the sand particles for artificially roughened pipes, or the average height of protuberances in naturally roughened pipes, while D is the inside pipe diameter. In plate flow, we use for the relative roughness the ratio l/e, where l is the length of the plate and e is the mean height of the protuberances. Also, we use for the roughness the coefficient C_f instead of f.

Recall furthermore that we had three zones of pipe flow, namely, the *smooth-pipe zone*, where the viscous sublayer completely covered the protuberances; the *transition zone*,[15] where the protuberances were partially outside of the viscous sublayer; and finally the *rough-pipe zone*, where the protuberances were for the most part exposed to the main flow. We have the same zones in the plate-flow boundary layer. Thus, we have the *hydraulically smooth zone* where, like the pipe, the skin-friction factor does not depend on the relative roughness because of the submergence in the viscous sublayer of the protuberances. Additionally, we have a *transition zone*. Finally, we have the *rough zone*, where analogous to pipe flow the skin-friction factor does not depend on the plate Reynolds number Re_L.

In Fig. 13.16, we show C_f versus Re_L for different values of relative roughness expressed here as l/e. The three zones of flow discussed above are shown as distinct regions in the diagram. Note that in the rough zone C_f is constant with respect to Re_L as indicated earlier. The greater the roughness, the sooner does the curve C_f versus Re_L detach from the smooth zone to enter the transition zone on its way to the rough zone.

There is an important difference to be pointed out here from that of pipe flow. Assume that the boundary layer is turbulent near the leading edge of the plate. The sublayer is very thin, and we would start out near the leading edge of

---

[15]Do not confuse transition, when it applies to a change from laminar flow to turbulent flow, with the transition zone in Fig. 13.16, which represents turbulent boundary-layer flow wherein the protuberances are partially exposed to the main flow.

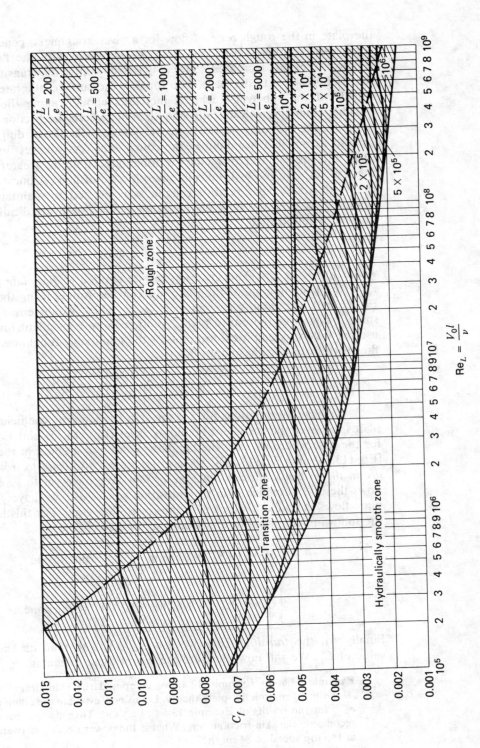

FIGURE 13.16
Three zones of flow for a rough plate.

the plate in the rough zone of flow for a given roughness, generating locally a comparatively high skin-friction drag. As we move with the flow, the viscous sublayer thickness increases, so we may possibly enter the transition zone where the friction starts to decrease. The rougher the surface, the later this happens. Finally, if the sublayer thickness increases enough along the flow, we may possibly enter the hydraulically smooth zone, where the friction is still less. To minimize the skin drag for the plate, we should make the leading edge region as smooth as possible to minimize the rough zone. As you get further along the plate and the sublayer thickens, the roughness could be greater and still allow the flow to be in the hydraulically smooth zone of flow. A question can now be put forth about what relative roughness is needed to maintain hydraulically smooth flow over the bulk of the *entire* plate. For hydraulically smooth *pipe* flow we required that

$$\frac{V_* e}{\nu} \leq 5 \tag{13.80}$$

where $V_* = \sqrt{\tau_w / \rho}$. This criterion can be applied to *flat-plate flow*. Thus, for τ_w at the leading edge of the plate, we can determine, using the equal sign in Eq. (13.80), the *admissible roughness*, e_{adm}, which would ensure hydraulically smooth flow over the entire plate. Another more convenient formulation is to use the following empirical formula for the admissible roughness e_{adm} for plate flow with hydraulically smooth flow:

$$e_{\text{adm}} \leq l \left[\frac{100}{\text{Re}_L} \right] \tag{13.81}$$

We have formulations for C_f, the skin-friction drag coefficient, for *smooth* plates to cover the turbulent boundary layer [Eq. (13.75) and Eq. (13.78)] and for the case where we take the laminar region into account at the leading edge [Eq. (13.76) and Eq. (13.79)]. *These formulas are also valid for rough plates that are in the hydraulically smooth zone of flow.* We have yet to give C_f for plate flows the bulk of which is in the transition zone. Also, we have yet to consider plate flows the bulk of which is in the rough-flow zone. For this latter, we have the following empirical formulations due to Schlichting:

$$c_f = \left(2.87 + 1.58 \log \frac{x}{e} \right)^{-2.5} \quad \text{rough zone} \tag{13.82a}$$

$$C_f = \left(1.89 + 1.62 \log \frac{l}{e} \right)^{-2.5} \quad \text{rough zone} \tag{13.82b}$$

Finally, for the *transition zone* we use Fig. 13.16 to read off the appropriate value of C_f. We will now illustrate the use of these formulas.

Example 13.6. In Example 13.4, we computed the skin drag for the dirigible *Akron* using smooth flat-plate theory. Let us now evaluate the admissible roughness e_{adm} for the results of Example 13.4 to be valid. Then using a roughness 0.05 in, recompute the skin-friction drag. What is the power needed to overcome this drag at the top speed of 84 mi/h?

Using Eq. (13.81), we can immediately solve for the admissible roughness and hence the largest roughness that will still give hydraulically smooth flow. Using the plate Reynolds number from Example 13.4, we have

$$e_{adm} = l \left[\frac{100}{Re_L} \right] = (785) \left[\frac{100}{4.59 \times 10^8} \right]$$

$$= 1.710 \times 10^{-4} \text{ ft} = 0.00205 \text{ in} \tag{a}$$

For the dirigible, the actual roughness coefficient is 0.05 in. Now l/e is $785/(0.05/12) = 1.884 \times 10^5$, so that on consulting Fig. 13.16 we are clearly in the rough zone. We accordingly use Eq. (13.82b) to find C_f. Thus,

$$C_f = \left[1.89 + 1.62 \log \frac{785}{0.05/12} \right]^{-2.5}$$

$$= 0.002843 \tag{b}$$

For the skin drag, we have, using 0.001756 slug/ft^3 for ρ,

$$D = (0.002843)(\tfrac{1}{2})(0.001756)[(84)(5280/3600)]^2(785)(\pi)(132)$$

$$= 12,331 \text{ lb}$$

The power needed to overcome skin friction is

$$\text{Power} = \frac{(12,331)(84)(5280/3600)}{550} = 2762 \text{ hp}$$

Example 13.7. A ground-effects vehicle in Fig. 13.17 is moving over water at a speed of 100 km/h. While over water, a pair of retractable rudders are inserted into the water. The width of the rudder is a constant equal to 0.75 m and a length of 1 m extends into the water. What is the skin drag on the rudders if transition occurs at $Re_{cr} = 5 \times 10^5$? The water is fresh water at a temperature of 15°C. The rudders have a roughness five times the admissible roughness.

We start by computing the plate Reynolds number. Thus

$$Re_L = \frac{\left(100 \dfrac{1000}{3600} \right)(0.75)}{1.141 \times 10^{-6}} = 1.826 \times 10^7$$

The admissible roughness then is

$$e_{adm} = (0.75) \left(\frac{100}{1.826 \times 10^7} \right) = 4.107 \times 10^{-6} \text{ m}$$

The actual roughness accordingly is

$$e = (5)(4.107 \times 10^{-6}) = 2.054 \times 10^{-5} \text{ m}$$

To find what zone we are in we go to Fig. 13.16. Using

$$\frac{l}{e} = \frac{0.75}{2.054 \times 10^{-5}} = 3.652 \times 10^4$$

FIGURE 13.17
A ground effects machine moving with rudders inserted.

we see that we are in the *transition zone*. Here we must read off C_f from the diagram. We have for this case

$$C_f = 0.0034$$

We can now get the total skin friction drag for the two rudders:

$$D = 4\left[(0.0034)\left(\frac{1}{2}\right)(\rho)\left(100\frac{1000}{3600}\right)^2(0.75)(1)\right]$$

Using $\rho = 999.1 \text{ kg/m}^3$ for water at 15°C (see Table B.2 in Appendix B) we have the desired result.

$$D = 3932 \text{ N}$$

PART C
FLOW OVER IMMERSED CURVED BODIES

13.11 FLOW OVER CURVED BOUNDARIES; SEPARATION

We have restricted our attention up to this point to flow over a flat plate with a zero pressure gradient in order to present some of the basic formulations in a simple way and to be able to check approximate methods with theory. We will now consider an important effect associated with flow over boundaries other than that of the flat plate oriented parallel to the main flow: this is the phenomenon of *separation*, which we described roughly on earlier occasions as the breaking away of the main flow from a boundary.

We will try first to establish a better physical picture of the *onset* of separation by considering incompressible flow around an airfoil at a high angle of attack, as shown in Fig. 13.18. We will focus our attention on the top surface. Theory and experiment indicate that between point A and a point B ahead of the position of maximum thickness C, there is a continual increase of the main-stream velocity just outside the boundary layer.[16] Beyond point B, there is a continual decrease in the main-stream velocity just outside the boundary layer, so that the maximum velocity just outside the boundary layer occurs at the point B. According to Bernoulli's equation, which is applicable for the main-stream flow, the pressure must decrease from close to stagnation pressure at A to a minimum pressure at B and must then increase again beyond point B. *Thus the boundary layer beyond B feels a pressure which is increasing in the direction of flow, and such a pressure variation we call an adverse pressure gradient.* The fluid moving in the boundary layer in this region is subject to this increasing pressure, so this fluid also shows up. But since the fluid in the boundary layer has small kinetic energy, it may possibly reach a condition of stopping and reversing its direction and thus cause the boundary layer to deflect away from the boundary. This is the *onset of separation*. There may then take place a considerable readjustment of the flow, where separation having started downstream of D

---

[16]See H. Schlichting, *Boundary Layer Theory*, McGraw-Hill, New York, 1979.

FIGURE 13.18
Flow around an airfoil.

FIGURE 13.19
Separation is present for airfoil.

(see Fig. 13.19), in the manner described, results in a thick region of highly irregular, milling flow, as shown by the darkened region. We can also see in this diagram how, effectively, a new boundary is formed for the regular main-stream flow. From this discussion, it should be clear that *separation may occur when one has an adverse pressure gradient*; *we later show that it can occur only under this condition*.

What are the disadvantages of separation? First, there is an increase in drag as a result of separation, since the front edge of the separated region will tend toward *B*, where there is a low pressure. This low pressure persists in the separated region, since fluid entering it from the main stream has a very low recovery of pressure from kinetic energy, much as was the case in the analysis of the minor loss developed in a sudden opening in Sec. 9.9.[17] Thus, as a result of this lowered pressure in the rear of the airfoil over what would be the case had there been no separation, there is an increase in drag which overshadows, in its bad effects, any increase in lift that one might associate with the action of having increased the angle of attack to a point where separation has been instituted. We will consider the performance of airfoils later in the chapter and will discuss further the effects of separation for this case.

It is also clear that we cannot use irrotational-flow theory once serious separation has taken place, since the effective boundary of the irrotational flow is no longer the body but some unknown shape encompassing part of the body and the separated region as well.

Let us next go back to the model process for describing the onset of separation to set forth certain mathematical criteria that may be of use for analytically determining the onset of separation. First, it is to be pointed out that for curved boundaries with no large variations in curvature we can use the simplified boundary-layer equations [Eqs. (13.12)], provided that we consider x as a curvilinear coordinate having the shape of the boundary and y as the normal distance away from the boundary.[18] With this in mind, consider the flow shown in Fig. 13.20 which depicts the onset of separation. As was pointed out earlier, separation results from a reversal of flow in the boundary layer and takes place as a result of an adverse pressure gradient imposed on the boundary

---

[17]This is also an example of separation, where the increased drag is in terms of an increased head loss.

[18]See H. Schlichting, *Boundary Layer Theory*, McGraw-Hill, New York, 1979.

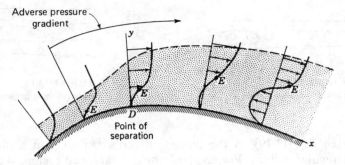

FIGURE 13.20
Onset of separation.

layer by the main flow. The point on the boundary where this action starts is shown as point D, and at this point, the velocity and its variation in the y direction are clearly both zero. Thus, using the indicated curvilinear coordinates, we can define the point of separation as that point along the boundary where

$$\left(\frac{\partial u}{\partial y}\right)_{y=0} = 0 \tag{13.83}$$

In general, it is extremely difficult, without experimental data, to determine the position along the boundary where this condition can take place. Ideally, we could first solve the steady irrotational flow around the body, and using pressures computed from the analysis, we could then analyze the boundary layer with a view to ascertaining the possibility of reaching the condition given by Eq. (13.83). Should this condition be reached, we could conclude that separation with in all likelihood take place. You should understand that beyond affording us this conclusion and giving us the approximate position of the point of separation, the aforementioned solutions of irrotational flow and boundary-layer flow will have little relation to the actual flow once separation has taken place, for reasons discussed earlier. The calculations outlined are generally very difficult, requiring numerical calculations and are beyond the scope of this text.[19]

We will show that separation occurs *only* when there is an *adverse pressure* gradient from the main flow. To do this, consider the simplified boundary-layer equation [Eq. (13.12*a*) using curvilinear coordinates and applied at the boundary, i.e., at $y = 0$. For such conditions $u = v = 0$, and we then get the result

$$\frac{1}{\rho}\frac{dp}{dx} = \nu\left(\frac{\partial^2 u}{\partial y^2}\right)_{y=0} \tag{13.84}$$

---

[19]The remaining portion of this section should be considered only if Sec. 13.3 has been covered.

Consider first the case of a favorable gradient where $dp/dx < 0$. We can then infer from Eq. (13.84) that $\partial^2 u/\partial y^2 < 0$ near the boundary. As we move toward the free stream, the velocity approaches the local free stream asymptotically, so $\partial u/\partial y$ decreases at a continuously lesser rate as you go out to the free stream. This means that $\partial^2 u/\partial y^2 < 0$ near the edge of the boundary layer. Thus we can conclude for the condition of a decreasing pressure that the curvature of the velocity profile (that is, $\partial^2 u/\partial y^2$) is always negative. Now consider the case of the adverse pressure gradient. Here we infer from Eq. (13.84) that the curvature of the profile must be positive (since $dp/dx > 0$) near the boundary. However, near the free stream our previous argument still applies, and the curvature is still negative. Thus we see that *for an adverse pressure gradient there must be present an inflection point in the profiles*, as shown in Fig. 13.20 where E indicates the inflection points.

Now let us consider the condition for the onset of separation. At the boundary $\partial u/\partial y = 0$ for this point, and since the velocity increases from zero at the boundary to a positive value directly next to the boundary in a continuous manner, we can see by carrying out a Taylor expansion of the velocity from the boundary to the region directly next to the boundary that $\partial^2 u/\partial y^2 > 0$. Since $\partial^2 u/\partial y^2 < 0$ at the edge of the boundary layer, we see that if there is a point of separation there must be an inflection point in the profile. *And since an inflection point occurs only when there is an adverse pressure gradient, we can conclude that the condition for the onset of separation can occur in the region of an adverse pressure gradient.* It should be clearly understood, however, that the presence of an adverse pressure gradient is a *necessary* but *not a sufficient* condition for separation. In other words, we can have the adverse pressure gradient and no separation; but without the adverse pressure gradient there can be no separation.

13.12 DRAG ON IMMERSED BODIES

We have already defined drag as the force component exerted on a body from a moving fluid in the direction of the free stream of the fluid far from the body. The drag on a body in a moving fluid is a difficult quantity to determine, since it will be seen soon to depend on such things as the location of the transition from laminar to turbulent flow in the boundary layer as well as the location of separation, to name just two difficulties. We are therefore very often forced to use experimental data. For this purpose, we usually express the drag D in the following form:

$$D = C_D A \frac{\rho U^2}{2} \tag{13.85}$$

where C_D is the *coefficient of drag*, A is usually the projected area in the direction of the free stream,[20] and U is the free-stream velocity. However for plates oriented parallel to the flow, the area A is bl—that is, the actual plate area. The coefficient C_D is dimensionless as you may yourself readily verify. To

---

[20] For airfoils, the area A is the *planform* area of the airfoil (i.e., the area as seen *normal* to the chord line of the airfoil). For ships it is the projection of the *wetted* area in the direction of motion of the ship.

understand why we use the form above, consider the evaluation of C_D in the form

$$C_D = 2\left(\frac{\text{Drag}/A}{\rho U^2}\right) \tag{13.86}$$

Note that in the parentheses we have an Euler number, as discussed in Chap. 8 on dimensional analysis. The only other significant dimensionless group for the low-speed flows over an immersed body is the Reynolds number[21] ($\rho U L/\mu$), where L is some convenient measurement of the immersed body. Hence for a given Reynolds number, the coefficient C_D will have the *same value* for all *dynamically similar* flows.

Thus far we have considered the drag on a flat plate oriented parallel to the flow of fluid. The drag was due entirely to shear stress on the surface and we used the plate coefficient of skin friction C_f to measure this. We rewrite Eq. (13.72) for C_f.

$$C_f = \frac{D/bl}{\frac{1}{2}\rho U^2} \tag{13.87}$$

Clearly here C_f is identical to C_D. It is merely the drag due to shear stress (skin friction). There may also be drag on a plate due to normal stress. It is called *pressure drag* and is best illustrated by considering a plate oriented normal to the flow as shown in Fig. 13.21. The drag on this plate is entirely due to normal stress and is thus entirely a pressure drag. Note that the flow actually separates at the edge since it cannot negotiate the sharp corner. This separation will always take place at such a sharp corner. Thus, we have shown two extreme positions of a flat plate where we have on the one extreme only skin-friction drag and on the other extreme only pressure drag. At any other inclination α the plate will have both kinds of drag present in varying amounts depending on α. The coefficient of drag C_D accounts for both skin and pressure drag; we have shown a plot of C_D versus α for the flat plate in Fig. 13.22.

Consider next the flow around a cylinder. The plot for C_D versus Re = $\rho U D/\mu$ is plotted in Fig. 13.23. We have both skin drag and pressure drag present. When the Reynolds number is less than 10, we have *creeping flow*, where viscous effects dominate the entire flow and the drag is overwhelmingly skin-friction drag. Note that C_D is very high. As the Reynolds number $U D/\nu$ increases, the viscous effects become restricted more and more to the boundary layer and there is a decreasing coefficient of friction C_D. However, at about Re = 5×10^3, C_D starts to increase. This is due to the separation process moving toward the top and bottom of the cylinder. As this happens, the pressure in the wake gets lower and lower. The reason for this is that the speed of flow of the main stream gets higher as you approach the top and bottom of

---

[21] For higher speeds in modern aircraft, we must also consider the Mach number.

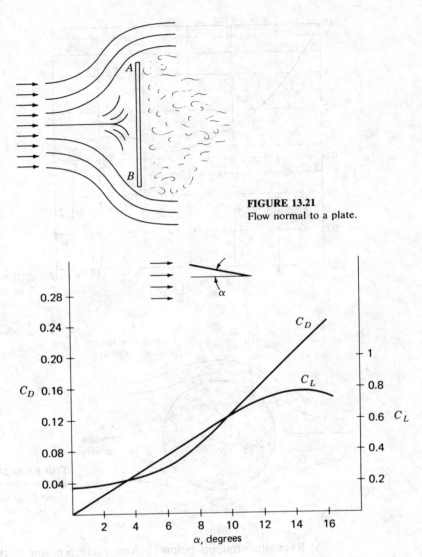

FIGURE 13.21
Flow normal to a plate.

FIGURE 13.22
Coefficients of lift and drag for a flat plate at varying inclination α.

the cylinder, so in accordance with Bernoulli's equation, the main-stream pressure gets less and less. Thus, when separation occurs closer to the top and bottom of the cylinder with little pressure recovery in the wake, there is a smaller pressure in back of the cylinder (see Fig. 13.24), thus resulting in greater drag. As we proceed further by increasing Re, we come across a sudden dip in the curve between $Re = 10^5$ and $Re = 10^6$, which we can explain qualitatively.

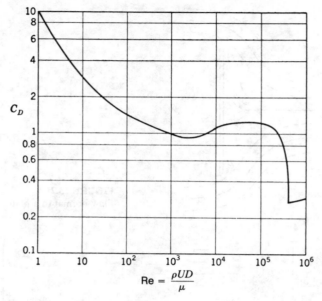

FIGURE 13.23
Drag coefficient for a cylinder.

$$\text{Re} = \frac{\rho U D}{\mu}$$

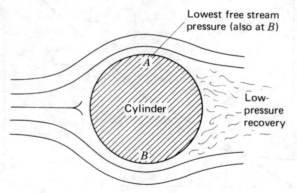

FIGURE 13.24
As the separation occurs closer and closer to A (and B), the pressure in the wake gets less and less.

At Reynolds numbers below the one corresponding to the dip, the separation taking place at the downstream side of the cylinder is one from a *laminar* boundary layer. However, as the Reynolds number of the main flow is increased, the Reynolds number in the boundary layer at any given value of x also increases. The laminar boundary layer then undergoes a transition to a turbulent boundary layer *before the separation*. Now one of the important differences between laminar boundary-layer flow and turbulent boundary-layer flow is that the profile of the latter is much steeper, as can be seen in Fig. 13.25. This means that for the same free-stream velocity and the same thickness δ, the flow in a turbulent boundary layer has appreciably more momentum than the flow of a laminar boundary layer. And for this reason the flow in a turbulent boundary

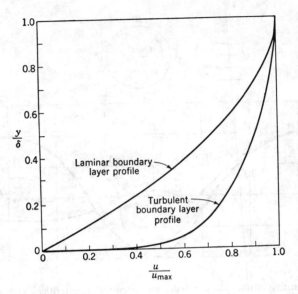

FIGURE 13.25
Laminar and turbulent boundary-layer profiles.

layer can *penetrate* an adverse pressure gradient *further* than can a laminar boundary layer before separation is likely to take place, so that when transition occurs in the boundary layer of the cylinder, the point of separation is suddenly set back to a position *farther on the downstream* surface of the cylinder, with the result that the drag and drag coefficient decrease appreciably. Thus, it is the transition in the boundary layer which accounts for the aforementioned sudden drop in the drag-coefficient curve.

The curve we have been discussing is typical, in general shape, to other such curves for three-dimensional bodies of revolution. The *main-stream Reynolds number* where the transition takes place accompanied by the drop in drag is called the *critical Reynolds number* for such flows. From our discussion of transition in the boundary layer, it should be obvious that this critical Reynolds number will depend on the main-stream turbulence, the roughness of the surface, etc., so one cannot give a precise value of this number without making a number of other specifications. As the Reynolds number UD/ν is increased further beyond the critical Reynolds number, the separation now moves again toward A and B (see Fig. 13.24) and there is further increase in the coefficient of drag C_D as explained earlier.

In Fig. 13.26 we have shown three pressure distributions around a cylinder. One is for inviscid flow, which is theoretical and which was developed in Chap. 12. The curve called subcritical has separation taking place from a laminar boundary layer and the curve called supercritical has the separation taking place from a turbulent boundary layer. The Reynolds numbers for the main flow are $(UD/\nu)_{\text{supercr}} = 6.7 \times 10^5$, which is just above the critical Reynolds number as seen in Fig. 13.23 and $(UD/\nu)_{\text{subcr}} = 1.9 \times 10^5$, which is just below the critical Reynolds number. Note that separation takes place *later* for super-

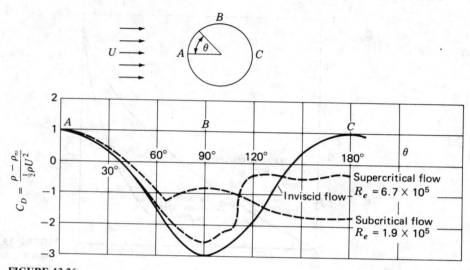

FIGURE 13.26

Pressure distributions around a cylinder for subcritical, supercritical, and completely inviscid flows.

critical flow and that a larger back pressure is developed for this flow. The latter fact explains the *decrease* in drag one gets after passing the critical Reynolds number for the main flow.

Another factor should be noticed from Fig. 13.26—the rather close correlation of pressure of the theoretical inviscid flow and the experimental data for the two real flows up to the instant of separation. We will use this fact in homework assignments (Probs 13.88 to 13.90).

How can we reduce drag in bodies? In bodies where pressure drag is very high due to separation, as in the cylinder described above, one can add material to the downwind side of the body in an effort to delay separation as long as possible. This is called *streamlining*. (We have already discussed internal streamlining briefly, you will recall, in Sec. 9.9 concerning the sudden opening in pipes.) Such a step will decrease the pressure drag, but by adding more surface, we will be increasing the skin-friction drag. The idea is to get the optimal shape in order to decrease the total drag coefficient C_D. This can be illustrated by considering Fig. 13.27 wherein we have a cross section of a strut of length L and maximum thickness t. A flow of free stream velocity U passes around this body, producing a drag. Suppose that we consider a family of such bodies each with the same maximum thickness but with varying lengths L. We have shown a plot of the pressure drag coefficient and the skin-friction drag coefficient as well as the total drag coefficient for varying length L. When L is small enough so that t/L is 0.4, we have a bluff body as shown in the diagram. The early separation yields a high-pressure drag because of the low pressure on the downstream portion of the body. The skin-friction drag on the front face is small, so we get a

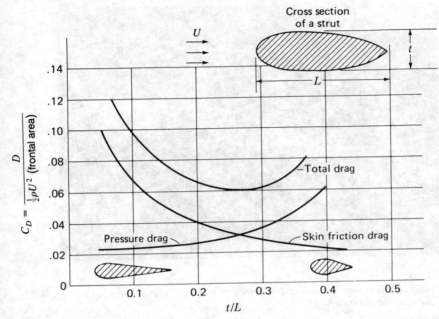

FIGURE 13.27
Drag coefficients for a family of struts. (*S. Goldstein, Modern Developments in Fluid Dynamics, Dover Publications, New York, 1965.*)

very small skin-friction drag. At the other extreme, where L is comparatively large so that $t/L \simeq 0.1$, we have successfully moved the separation toward the end of the body. The pressure recovery on the downstream portion of the body is good, and there is very little pressure drag. However, the long length of the "tail" of this body has now developed a large skin-friction drag. The optimal design for this case is where the skin-friction drag coefficient and pressure drag coefficient intersect to give a minimal total drag coefficient.

In the 1930's automobile manufacturers here and abroad became concerned about reducing drag in production model cars and thus in introducing streamlining. Most of this work was intuitive on the part of the designer; it is only since 1945 that wind tunnel testing has been extensively used in design of cars. It is pointed out that reducing drag while increasing downward force on the front of a vehicle for purposes of stability is a complicated process. Actually intuitive designs that looked streamlined often had serious airflow problems.

The first two mass-produced aerodynamic cars were American, namely the *Chrysler Airflow* (1934) (see Fig. 13.28) with a $C_D = 0.50$ and the *Lincoln Zephyr* (1936) (see Fig. 13.29) with a $C_D = 0.45$. These drag coefficients were considerably lower than other mass produced cars but the cars had very limited commercial success. In 1948, 51 of the ill-fated *Tucker Torpedo* ($C_D = 0.39$) were produced amid much fanfare. At present the *Pontiac Grand Prix* coupe

FIGURE 13.28
First streamlined production car designed with wind tunnel data, *Chrysler Airflow*, C-1; $C_D = 0.50$. (*Courtesy Chrysler Corporation.*)

has a $C_D = 0.30$. The lowest C_D for a mass-produced car is 0.29 for a *Subaru XT* coupe. The most aerodynamic sedans now have C_D's from 0.33 to 0.35.

In closing this section, we present various drag coefficients for two-dimensional bodies in Table 13.2 and three-dimensional bodies in Table 13.3.

Example 13.9. We will consider the dirigible *Akron* once again for drag (see Example 13.4). This time we will use an ellipsoidal body of revolution from Table 13.3 to represent the dirigible. Estimate the pressure drag.

The coefficient of drag should correspond to an ellipsoid whose $l/D_{max} = 785/132 = 5.95$. For turbulent flow, we estimate C_D using simple interpolation from the 4:1 to the 8:1 ellipsoids:

$$C_D = 0.06 + \frac{5.95 - 4}{4}(0.13 - 0.06)$$
$$= 0.094 \qquad (a)$$

Now the total drag is next computed using Eq. (13.86):

$$D = C_D A \frac{\rho U^2}{2} = (0.094)\frac{(\pi)(132^2)}{4}\left(\frac{1}{2}\right)(0.001756)\left(84\frac{5280}{3600}\right)^2$$
$$= 17{,}143 \text{ lb} \qquad (b)$$

We can now estimate the pressure drag on the *Akron* using the drag from Example 13.4. Thus considering hydraulically smooth flow,

$$D_{press} = 17{,}143 - 7505 = 9638 \text{ lb} \qquad (c)$$

FIGURE 13.29
Second streamlined production car designed with wind tunnel data, *Lincoln Zephyr*; $C_D = 0.45$.
(*Courtesy Ford-Lincoln Co.*)

TABLE 13.2
Drag coefficients for two-dimensional bodies at Re $\simeq 10^5$

Shape		C_D	Shape	C_D Laminar flow	Turbulent flow
Plate	→ ▯	2.0	Ellipses:		
			2:1 ⬭	0.6	0.20
Half cylinder	→ ◖	1.2	4:1 ⬬	0.35	0.10
	→ ◗	1.7	8:1 ⬬	0.28	0.10
Half tube	→ ◜	1.2	Cylinder: → ◯	1.2	0.30
	→ ◝	2.3			
Square cylinder	→ ▢	2.1	Rectangular plate → ⊢: $\left(\dfrac{b}{h}\right)$	1 → 1.18	
	→ ◇	1.6		5 → 1.2	
Equilateral triangle	→ ◁	1.6		10 → 1.3	
	→ ▷	2.0	$h\,\substack{b \\ \square}$	20 → 1.5	

The total power needed to move the Akron should be

$$\text{Power} = \frac{(17{,}143)[(84)(5280/3600)]}{550} = 3840 \text{ hp}$$

The Akron actually had eight 560-hp diesel engines.

TABLE 13.3
Drag coefficients for three-dimensional bodies

Shape	C_D	Shape	C_D Laminar flow	Turbulent flow
Disc →❙	1.17	Sphere: →○	0.47	0.27
60° cone →◁	0.49	Ellipsoidal body of revolution:		
Cube {→☐	1.05			
{→◇	0.80	2:1 → ⬬	0.27	0.06
Hollow cup {→◖	0.38	4:1 → ⬭	0.20	0.06
{→◗	1.42	8:1 → ⬭	0.25	0.13
Solid hemisphere {→◖	0.38			
{→◗	1.17			

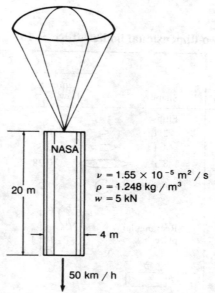

$\nu = 1.55 \times 10^{-5} \text{ m}^2/\text{s}$
$\rho = 1.248 \text{ kg/m}^3$
$w = 5 \text{ kN}$

NASA

20 m

4 m

50 km/h

FIGURE 13.30
Hollow cylinder is descending at constant speed under the restraint of the parachute.

Example 13.10. A hollow cylinder (Fig. 13.30), which is part of a spent rocket system is to fall with a speed of 50 km/h as it nears sea level. What projected area should the drag parachute have if C_D for the parachute is 1.2? The inside and outside surfaces of the cylinder are smooth. Take $\nu = 1.55 \times 10^{-5}$ m²/s for the air. Take $\text{Re}_{cr} = 10^6$ for the boundary layer along the inside and the outside surfaces of the hollow cylinder. The weight of the cylinder is 5 kN. Take $\rho = 1.248$ kg/m³ for air. There is no wind present.

Two drags are to be considered here. First there is the drag of the parachute and second there is skin drag from air flow on both the inside and outside surfaces of the cylinder. We consider the cylinder first. The plate Reynolds number is

$$\text{Re}_L = \frac{UL}{\nu} = \frac{(50)\left(\dfrac{1000}{3600}\right)(20)}{1.55 \times 10^{-5}} = 1.792 \times 10^7$$

We are accordingly in the *high Reynolds number regime* for boundary layer flow. We use Eq. (13.79) for C_f. Thus

$$C_f = \frac{0.455}{(\log \text{Re}_L)^{2.58}} - \frac{3300}{(\text{Re}_L)}$$

$$= \frac{0.455}{\left[\log(1.792 \times 10^7)\right]^{2.58}} - \frac{3300}{(1.792 \times 10^7)} = 0.002556$$

The skin drag D_{cyl} is then

$$D_{cyl} = 2\left[C_f \tfrac{1}{2}\rho V^2 A\right] = 2\left[(0.002556)(\tfrac{1}{2})(1.248)(50/3.6)^2(\pi)(4)(20)\right] = 154.7 \text{ N}$$

From Newton's law we next have for steady velocity using D_{par} as the parachute drag

$$D_{par} + 154.7 = W = 5000$$

$$\therefore D_{par} = 4845 \text{ N}$$

Hence

$$4845 = C_D\left(\tfrac{1}{2}\right)\left(\rho U^2\right) A_{par} = (1.2)\left(\tfrac{1}{2}\right)(1.248)(50/3.6)^2 A_{par}$$

$$\therefore A_{par} = 33.54 \text{ m}^2$$

Note this is the vertical *projected* area of the parachute.

Example 13.11. A nuclear submarine is moving underwater at a speed of 20 kn. It is 800 ft in length and has a maximum diameter of 80 ft. The nuclear power plant has a thermal efficiency of 40% and is operating at a steady state. The power plant is cooled by seawater with a volumetric flow rate q of 13 ft³/s. Take the specific heat c_p of the sea water to be 1.01 Btu/(lbm)(°F). How much heat is discharged to the ocean per second and what is the increase in temperature of the coolant discharged to the ocean?

For a simple first estimate assume the submarine is an ellipsoidal body of revolution and neglect the drag of appendages such as control surfaces and the coning tower. Take $Re_{cr} = 5 \times 10^5$. Also take $\rho = 62.5$ lbm/ft^3 and $\nu = 1.663 \times 10^{-5}$ ft^2/s for the sea water.

We will consider for steady state operation that the nuclear power plant generates exactly the power needed to overcome drag and discharges unusable heat through heat exchangers to the ocean which in the parlance of thermodynamics then acts as a *sink* for the power plant.

Let us next compute the plate Reynolds number for the present conditions. Thus

$$Re_L = \frac{[(20)(1.688)](800)}{1.663 \times 10^{-5}} = 1.624 \times 10^9$$

Clearly we have turbulent boundary layer flow. The ratio of length to maximum diameter is

$$\frac{L}{D_{max}} = \frac{800}{80} = 10$$

Using handbook data for ellipsoidal bodies of revolution, we estimate $C_D \approx 0.148$. We can now compute the power needed for propulsion.

$$Power = (Drag)(velocity)$$

$$= \left[(C_D)(\tfrac{1}{2})(\rho V^2)(A) \right](V)$$

$$= (0.148)(\tfrac{1}{2})(62.5/32.2)[(20)(1.688)]^3(\pi/4)(80^2)$$

$$= 2.78 \times 10^7 \text{ ft-lb/s}$$

With a thermal efficiency of 40% the heat generated is

$$Heat = \left(\frac{2.78 \times 10^7}{0.40} \right)\left(\frac{1}{778} \right) = 8.93 \times 10^4 \text{ Btu/s}$$

The rejected heat is then

$$Heat_{rej} = 8.93 \times 10^4 - \frac{2.78 \times 10^7}{778} = 5.35 \times 10^4 \text{ Btu/s}$$

The increase in temperature ΔT of the coolant can now be estimated using conservation of energy. Thus

$$(Heat)_{rej.} = q c_p \rho \, \Delta T$$

$$\therefore 5.35 \times 10^4 \text{ Btu/s} = (13 \text{ ft}^3/\text{s})(1.01 \text{ Btu}/(\text{bm})(°\text{F}))(62.5 \text{ lbm/ft}^3) \, \Delta T(°\text{F})$$

$$\Delta T = 65.2°\text{F}$$

This rejected heat is said by the submariners to form a thermal "scar" in the ocean. Research is underway to detect from earth satellites the thermal gradients created by this discharge heat so as to signal the presence of a submarine.[22]

---

[22]This example is adapted from an interesting paper entitled "Problem: Nuclear Submarine Exhaust Characteristics," by J. R. Shanebrook, *Am. J. Physics*, vol. 56, no. 9, September 1988.

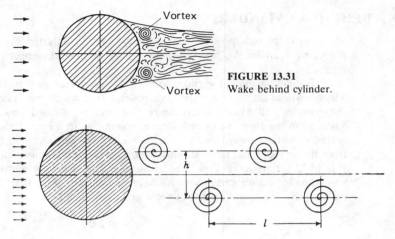

FIGURE 13.31
Wake behind cylinder.

FIGURE 13.32
Diagrammatic representation of von Kármán vortex street.

FIGURE 13.33
Vortices shedding from a cylinder. (*Courtesy Dr. Henri Werlé, ONERA, France.*)

˙13.13 WAKE BEHIND A CYLINDER

We have discussed the wake behind a cylinder in only the most simple terms in the preceding section. We now examine a very interesting aspect of the wake behind a cylinder.

Initially, a pair of vortices form in the separated region, as shown in Fig. 13.31. These vertices act like "aerodynamic rollers" over which the main stream flows. One of the vortices will always break away first and be washed downstream in the wake, whereupon another vortex will begin growing in its place. Later, the second mature vortex will break away, so a process is started whereby vortices break away alternately from the cylinder and move downstream, as shown diagrammatically in Fig. 13.32 and in Fig. 13.33 for an actual flow. The arrangement of these vortices in the wake is called a von Kármán *vortex street*. Von Kármán was able to show that that for stability of the pattern, the configuration must have a geometry such that

$$\frac{h}{l} = 0.281 \tag{13.88}$$

where h and l are dimensions shown in Fig. 13.32.

The vortex street moves downstream with a velocity u_s which is smaller than the main-stream velocity. Using perfect-fluid theory, von Kármán was able to show, furthermore, that the average value of the drag per unit length of the cylinder is

$$D = \rho U^2 h \left[2.83 \frac{u_s}{U} - 1.12 \left(\frac{u_s}{U} \right)^2 \right] \tag{13.89}$$

wherein h and u_s/U have to be known from experimental data.

It is clear that such alternate developments of vortices cause a force on the cylinder having the above average D but which repeats cyclically with a frequency dependent on the flow conditions. If the natural frequency of the cylinder in the transverse direction is close to the frequency of the variation of the force from the vortices breaking off, a lateral vibration will be induced. This accounts for "singing" telephone and transmission lines at certain wind velocities.

˙13.14 AIRFOILS; GENERAL COMMENTS

Of particular importance is the performance of airfoils, which we will now discuss briefly. We will be interested in the drag D of the airfoil and also the lift L, which is the force normal to the free stream.[23] We have shown these forces in Fig. 13.34 for a two-dimensional flow over an airfoil having theoretically infinite length. The line connecting the leading and trailing edges is the *chord line* of length C. The angle α formed between the chord line and the direction of flow U is the *angle of attack*. The complexities for determining drag described in the previous section apply to the drag of airfoils and also to the lift.

---

[23]The drag presented here represents the *total* of the drag due to *skin drag*, and the drag due to *pressure*, or *form*, *drag*. This total drag is sometimes termed the *profile drag*. In dealing with lifting surfaces such as finite airfoils, we also talk of *induced drag*, which is the part of the form drag associated with vortex motion about the lifting surface (to be considered later in this section). In supersonic flow, the drag due to normal stress is usually called *wave drag*. Discussions of these various drags can be found in more specialized texts.

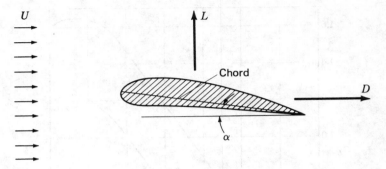

FIGURE 13.34
Airfoil with infinite aspect ratio showing lift and drag.

FIGURE 13.35
P-51D Mustang fighter plane of World War II. A modified version of this plane holds the world record for speed of propeller-driven planes (499 mi/h).

Accordingly, we must rely on wind-tunnel experimental data and use, as in the previous section, coefficients of drag and lift defined as

$$C_D = \frac{D/A}{\frac{1}{2}\rho U^2} \qquad (13.90a)$$

$$C_L = \frac{L/A}{\frac{1}{2}\rho U^2} \qquad (13.90b)$$

where for A we use the *planform* area of the wing. (For a wing of finite length where l is the length and C is the chord, the planform area is simply lC.) It is the usual practice to plot C_L and C_D versus angle of attack. We have considered a subsonic airfoil for this purpose, using data for the celebrated P-51D Mustang fighter plane of World War II (Fig. 13.35).[24] A modified version of this

---

[24] The P51 Mustang was designed at a desperate time during World War II when fighter plane protection was sorely needed by bomber squadrons on long range missions. The Mustang met this need and turned out to be an engineering triumph.

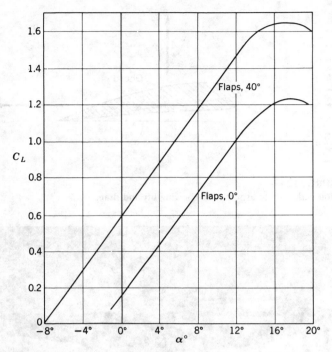

FIGURE 13.36
Wind-tunnel data for North American P-51D Mustang fighter plane.

plane is still used in racing competition around pylons. Note in Fig. 13.36 that at about 16° the lift for both flap configurations[25] drops off rapidly. Here we have serious separation taking place with resulting loss of lift and increase in drag. The condition is called *stall*. Just landing the pilot may cause stall to take place so as to drop the plane to the runway, but in normal flight this configuration can be dangerous. Note further that the lift is markedly increased with the use of flaps. In Fig. 13.37, we have a plot of C_L versus C_D called a *polar plot* whereby it is clear that with the flaps present a larger drag is developed.

The P-51D Mustang airfoil is an example[26] of a laminar airfoil, wherein by maintaining a favorable pressure gradient the transition from laminar to turbulent flow in the boundary layer has been delayed to a position farther downstream than is usual. This gives the airfoil smaller skin drag. The favorable gradient is obtained essentially by moving the point of maximum thickness of the airfoil section farther downstream than is usual. As is to be expected, the drawback to this type of airfoil is that the low-angle-of-attack drag reduction is achieved at the expense of rapid separation tendencies at high angles of attack.

---

[25]A flap is a movable portion of a wing which is extended during takeoffs and landings to give greater effective wing area so that a greater lift is possible at these slow speeds.
[26]The Lockheed Electra is another example.

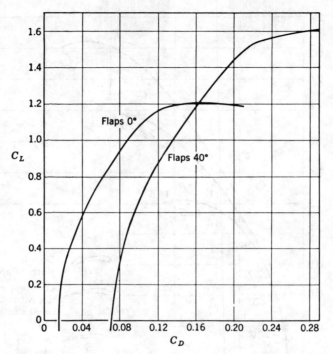

FIGURE 13.37
Polar plot for airfoil of P-51D fighter plane.

The fact that airfoils have a maximum coefficient of lift means that there is a minimum speed, called the *stall speed*, for an aircraft when the plane is at the maximum coefficient of lift and is simply supporting its own dead weight W. This mean that

$$L = W = (C_L)_{max}\left(\tfrac{1}{2}\rho V_{stall}^2\right)(A)$$

Hence

$$V_{stall} = \sqrt{\frac{2W}{\rho(C_L)_{max}A}} \tag{13.91}$$

This speed is the minimum landing speed for an aircraft. We see here that by increasing $(C_L)_{max}$ using the flaps, the pilot can effectively reduce his safe landing speed.

There are other ways to increase the coefficient of lift for airfoils other than the flap which we have mentioned for the Mustang. Fig 13.38 shows the polar plot for an ordinary airfoil A as well as one with a *leading-edge slat* identified as B. Note the increase in C_L possible. The way this works is as follows: On the main airfoil the boundary layer is given high momentum as a result of the rapid passage of flow through the space between slat and wing. This decreases the tendency of the flow to separate, so that higher coefficients of lift are possible. The third case C is that of the airfoil with a *single slotted flap* in addition to a leading-edge slat. Higher-pressure air from below rushes through the flap and is directed *along* the upper-airfoil surface. This adds

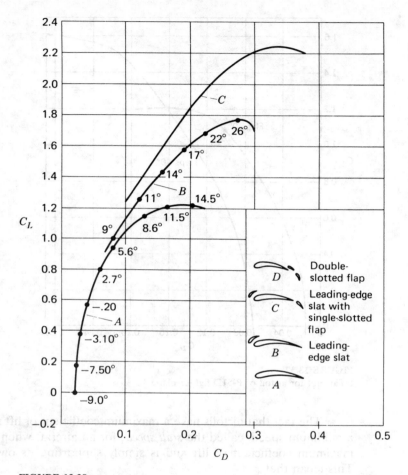

FIGURE 13.38
Performance of an airfoil with leading edge slat and flap. (*Adapted from H. Schlichting, Boundary Layer Theory, 7th ed., McGraw-Hill, New York, 1978.*)

momentum to the boundary-layer flow, again decreasing the tendency to separate. Note that there is considerable gain in the coefficient of lift possible. There can be further improvement by using a *double-slotted* airfoil flap, shown as D in Fig. 13.38.

In the years following World War II, there has been much research and development work on other means of controlling boundary-layer separation in order to develop better short-takeoff aircraft, including vertical-takeoff aircraft. Two procedures will simply be mentioned here. One is to *blow* fluid through carefully designed slots in the airfoil. This gives momentum to the boundary-layer flow and thus delays the separation tendency (see Fig. 13.39). Another method is *suction*, whereby the decelerated boundary-layer flow is removed through a slot before separation is about to take place. Just downstream of the slot a new boundary layer forms with greater momentum and can then proceed more effectively against the adverse pressure gradient. Also, suction has been used in recent years by having many small suction slits along the upper surface of

(a)

(b)

FIGURE 13.39
(*a*) Severe separation. (*b*) Same situation as (*a*) with the added blowing along the upper surface. (*Courtesy Dr. Henri Werlé, ONERA, France.*)

the airfoil, which decreases the thickness of the boundary layer. This has the tendency to delay the onset of transition from laminar to turbulent flow in the boundary layer and thus decreases skin drag, as in the earlier-discussed laminar airfoil.

*13.15 ADDITIONAL AIRFOIL TOPICS; INDUCED DRAG AND TRANSONIC FLOW

The *aspect ratio AR* of a wing is defined as

$$AR = \frac{A_p}{C^2} \tag{13.92}$$

where A_p is the planform area. For a two-dimensional wing, the aspect ratio is clearly infinite. For finite aspect ratio, there will be a reduction of maximum lift because of the *end effects* on the flow. What happens is that the higher pressure at the wing tips on the *bottom* face of the airfoil causes air to move around the wing tips to the lower pressure *above* the airfoil, thus causing a wing-tip vortex to form. This is shown in Fig. 13.40. The resulting trailing vortex then induces a downward flow behind the airfoil, as shown in Fig. 13.40. We call this flow *induced downwash*. This induced downwash decreases the effective angle of attack α-which can readily be seen by consulting Fig. 13.41. Figure 13.41*a* is an infinite aspect ratio wing and Fig. 13.41*b* is a finite aspect ratio wing for the same airfoil shape. Note that the velocity of the air relative to the airfoil in (*b*) is no

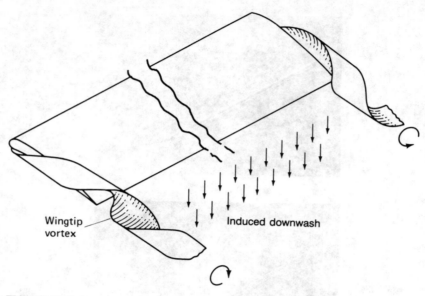

FIGURE 13.40
Wingtip-induced vortices and induced downwash velocity.

longer simply V_0 as in (*a*) but involves now the induced downwash velocity V_{ind} as well as V_0. The *effective* angle of attack α_{eff}, which is the angle between the chord line and the relative velocity V_{rel}, has decreased by the angle β, so we will have more drag and less lift. The additional drag due to the finite-aspect-ratio effect is called *induced drag*.

As a final note, we will discuss airfoils in the range of flow just above and below the speed of sound (transonic range). In Fig. 13.42*a*, we show an airfoil moving first at what we call *subcritical flow*, where there are no shock waves. In Fig. 13.42*b* we are in the *supercritical* flow even though the plane is moving with a speed less than the speed of sound because a shock wave has formed shown by the solid line. There is a region bounded by the shock wave and the sonic line (M = 1) inside of which M > 1-that is, we have local supersonic flow. In Fig. 13.42*c* as the speed of the plane increases further, the

(*a*) Infinite aspect ratio

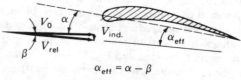

$$\alpha_{eff} = \alpha - \beta$$
(*b*) Finite aspect ratio

FIGURE 13.41
Diagrams show decrease in effective angle of attack for finite-aspect ratio.

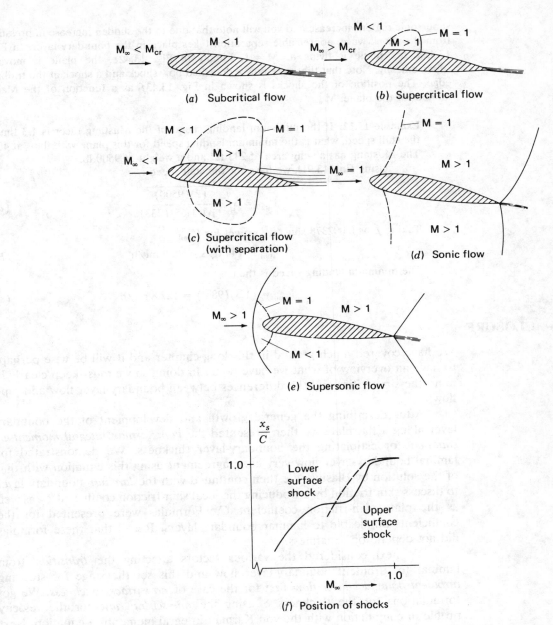

$M_\infty < M_{cr}$ → $M < 1$

(a) Subcritical flow

$M_\infty > M_{cr}$ → $M < 1$ $M > 1$ $M = 1$

(b) Supercritical flow

$M_\infty < 1$ → $M < 1$ $M = 1$ $M > 1$ $M > 1$

(c) Supercritical flow
(with separation)

$M_\infty = 1$ → $M = 1$ $M > 1$ $M > 1$

(d) Sonic flow

$M_\infty > 1$ → $M = 1$ $M > 1$ $M < 1$

(e) Supersonic flow

$\dfrac{x_s}{C}$

1.0

Lower surface shock

Upper surface shock

M_∞ ⟶ 1.0

(f) Position of shocks

FIGURE 13.42
Shock formation for flow around an airfoil as it goes through the so-called sonic barrier. (*From G. Y. Nieuwland and B. M. Spee, "Transonic Airfoils: Recent Developments in Theory, Experiment, and Design," Annual Review of Fluid Mechanics, vol. 5, 1973.*)

supersonic regions increase and you will note that due to the sudden increase in pressure from the shock wave, appreciable separation takes place in the boundary layer. In Fig. 13.42d the plane is flying at M = 1. Finally in Fig. 13.42e, the plane is moving supersonically. Note that there are now a detached bow shock and a shock at the trailing edge. The position of the shocks is shown in Fig. 13.43f as a function of the Mach number of the plane, M_∞.

Example 13.12. If the minimum landing speed of the Mustang racer is 1.3 times the stall speed, what is the minimum landing speed for this plane with flaps at 40°. The Mustang has a wing area of 233 ft^2 and a weight of 9500 lb.

Using Eq. (13.91), we have for the stall speed

$$V_{\text{stall}} = \sqrt{\frac{(2)(9500)}{(\rho)(1.65)(233)}} \qquad (a)$$

Taking ρ as 0.002378 slug/ft^3 we get for V_{stall}

$$V_{\text{stall}} = 144 \text{ ft/s} = 98.3 \text{ mi/h} \qquad (b)$$

The minimum landing speed is then

$$V_{\text{min}} = (1.3)(98.3) = 127.8 \text{ mi/h} \qquad (c)$$

13.16 CLOSURE

We have covered much ground in this long chapter and it will be wise perhaps to have an overview of what we have done. In doing so we must keep clearly in mind the similarities and the differences between boundary-layer flow and pipe flow.

After describing the general growth and development of the boundary layer along a flat plate we then presented the *von Kármán integral momentum equation* for estimating the boundary-layer thickness. We demonstrated for laminal boundary-layer flow very good agreement using this equation with that of the solution by Blasius. We then continued with the *laminar* boundary layer to discuss skin friction by introducing the local skin-friction coefficient c_f as well as the plate skin-friction coefficient C_f. Formulas were presented for the coefficients applicable to laminar boundary layers. Recall that these formulas did not depend on roughness.

We next considered the various factors affecting the *transition* from laminar to turbulent boundary-layer flow and this set the stage for studying *turbulent boundary-layer flow* first for the case of *no surface roughness*. We got formulations for the thickness δ using the *one-seventh law* for the velocity profile in conjunction with the von Kármán integral momentum equation. Next we set forth empirical formulations for the skin-friction coefficient for smooth plates. This was done first for plate Reynolds numbers *less* then 10^7. For plate Reynolds numbers *exceeding* 10^7 we presented other empirical formulations for C_f. It is hoped through the numerous examples presented here that we demonstrated how we could extend the flat-plate boundary-layer theory so as to approximately solve skin-friction drags for a variety of practical problems. The next step was to include *roughness* in the plate. This led to consideration of the

TABLE 13.4
Summary sheet for boundary layers

I. Boundary-layer thickness formulas
A. Ordinary thickness δ where $u = 0.99U$
B. Displacement thickness δ^*

$$\delta^* = \int_0^\infty (1 - \frac{u}{U})\, dy$$

C. Momentum thickness θ

$$\theta = \int_0^\infty \frac{u}{U}(1 - \frac{u}{U})\, dy$$

II. Skin friction definitions

$$c_f = \frac{\tau_w}{\frac{1}{2}\rho U^2} \quad \text{Local skin friction coefficient}$$

$$C_f = \frac{D}{\frac{1}{2}\rho U^2 Lb} \quad \text{Plate skin friction coefficient}$$

III. Laminar boundary layers

$$\frac{\delta}{x} = \frac{4.96}{\sqrt{\text{Re}_x}} \qquad c_f = \frac{0.664}{\sqrt{\text{Re}_x}}$$

$$\frac{\delta^*}{x} = \frac{1.73}{\sqrt{\text{Re}_x}} \qquad C_f = \frac{1.328}{\sqrt{\text{Re}_L}}$$

IV. Turbulent boundary layers

A. | Smooth plates: low Reynolds number flow Re $< 10^7$

$$\tau_w = 0.0225\rho U^2 (\frac{\nu}{U\delta})^{1/4} \quad 5 \times 10^5 < \text{Re} < 10^7 \quad \frac{\delta}{x} = 0.37(\text{Re}_x)^{-1/5}$$

$$\bar{u} = U\left(\frac{y}{\delta}\right)^{1/7} \qquad\qquad \frac{\delta^*}{x} = 0.0463(\text{Re}_x)^{-1/5}$$

$$c_f = \frac{0.0577}{(\text{Re}_x)^{1/5}}$$

$$C_f = \frac{0.074}{(\text{Re}_L)^{1/5}}$$

$$C_f = \frac{0.074}{(\text{Re}_L)^{1/5}} - \frac{A}{\text{Re}_L}$$

Re_{cr}	300,000	500,000	10^6	3×10^6
A	1050	1700	3300	8700

TABLE 13.4 *Continued*

B. | Smooth plates: High Reynolds number flow Re $> 10^7$ |

$$\bar{u} = 5.85 \log\left(\frac{yV_*}{\nu}\right) + 5.56$$

$$C_f = \frac{0.455}{(\log \text{Re}_L)^{2.58}}$$

$$C_f = \frac{0.455}{(\log \text{Re}_L)^{2.58}} - \frac{A}{\text{Re}_L}$$

C. | Rough plates |

1. Hydraulically smooth zone; use smooth plate results
2. Smooth-Rough Transition zone; use Fig. 13.16 for C_f
3. Rough zone

$$c_f = \left(2.87 + 1.58 \log \frac{x}{e}\right)^{-2.5}$$

$$C_f = \left(1.89 + 1.62 \log \frac{L}{e}\right)^{-2.5}$$

V. Drag on immersed bodies

$$C_D = \frac{D/A}{\frac{1}{2}\rho V^2}$$

If D from shear stress only $C_D \equiv C_f$.

viscous sublayer much as for pipe flow with the same three zones of flow, namely, the *hydraulically smooth zone*, the *transition zone*, and the *rough zone*. The particular zone of flow for any particular plate flow had to be determined by using a plot of C_f versus Re_L for different relative roughness L/e (Fig. 13.16). For the hydraulically smooth zone we could use the results developed earlier for smooth plates. For the rough zone, we presented an empirical formula for C_f. Finally, for the transition zone we used results from the plot directly (this could also have been done for the other two zones).

At this point we turned to flow over *curved surfaces* in our study of flow over *immersed curved bodies*. We took pains to discuss in some detail the crucial phenomenon of *separation*, its causes and its effects.[27] Next we formulated the drag coefficient C_D and looked at several interesting situations involving C_D. Of particular interest was the case of flow around a cylinder with emphasis on

---

[27]We admonish you at this time not to confuse *transition* and *separation*. They are two different and distinct actions. Separation can take place from a laminar or from a turbulent boundary layer if part of the boundary layer is subject to an adverse pressure gradient.

explaining the so-called critical Reynolds number signalling a sudden drop in the value of C_D. This led to discussions of the wake behind cylinders as well as an introductory discussion of airfoils. Next time you are flying in a transport plane be sure to observe changes that the pilot institutes in his wings during takeoff or landing. Can you explain now why the pilot makes these changes?

We thus come to the end of a most significant chapter. In retrospect we hope that you will now more fully appreciate the importance of the boundary layer which despite its thin size relative to the main flow, nevertheless has profound effects on the performance of many devices.

As in the closure of Chap. 9, we have prepared a review outline of key salient features for boundary-layer flow that is presented in Table 13.4.

Up to this time, we have considered flow problems with a *free surface* present only in Chap. 5 when we used *finite* control volumes for steady flows and consequently did not have to know details of the flow. In practice, flows with a free surface such as flows in canals, rivers, oceans, and partially filled pipes are of great importance. In the next chapter, we consider such flows in detail.

PROBLEMS

Problem Categories

Boundary Layer Thicknesses and Wall Shear Stress for Laminar Boundary-Layer Flow 13.1–13.18

von Kármán Momentum Integral Equation 13.19–13.20

Skin Friction Coefficients and Skin Drag for Laminar Boundary-Layer Flows 13.21–13.22, 13.24–13.27

Transition, Boundary Layer Thickness, and Drag for Turbulent Boundary Layers on Smooth Plates 13.28–13.48

Rough Plate Problems 13.49–13.64

Drag on Bodies 13.65–13.74, 13.76–13.95

Starred Problems

13.23, 13.77,

Derivations or Justifications

13.12, 13.14, 13.19, 13.75, 13.87, 13.88

For this chapter's problems, use for air $\nu = 1.50 \times 10^{-5}$ m^2/s *at* $T = 5°C$; $\nu = 1.55 \times 10^{-5}$ m^2/s *at* $T = 10°C$, *and* $\nu = 1.70 \times 10^{-5}$ m^2/s *at* $T = 20°C$.

13.1. The flow over a flat plate is shown in Fig. P13.1. The velocity outside the boundary layer is uniform and equal to U, while inside the boundary layer is a parabolic profile. What is the displacement thickness δ^* in terms of δ?

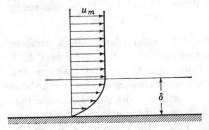

FIGURE P13.1

13.2. Using a cubic profile $\alpha y + \beta y^3$ for the laminar boundary layer, show that $\delta/x = 4.64\,\mathrm{Re}^{-1/2}$ by using the von Kármán momentum integral equation for a flat plate with a zero pressure gradient.

13.3. For the cubic profile, where $u = \frac{3}{2}U(y/\delta) - \frac{1}{2}U(y/\delta)^3$, find δ^*/x using Eq. (13.51b) for δ. Check your result against Eq. (13.51c).

13.4. Give an expression for the velocity profile in a laminar boundary layer wherein this profile is sinusoidal and fits the boundary conditions at

$y = 0$ and $y = \delta$. Determine δ^* as a function of δ.

13.5. With the profile found in Prob. 13.4 [$u = U\sin(\pi y/2\delta)$], determine the ratio δ/x, using the von Kármán momentum-integral equation for a zero pressure gradient. What is the percentage error of your result as compared with the exact solution of Blasius? Using the result $\delta^* = 0.363\ \delta$ from the preceding problem, compute δ^*/x.

13.6. Work out the expression for shear stress τ_w at the wall for flow over a flat plate wherein a laminar boundary layer is present for the case of a zero pressure gradient. Use the parabolic profile as discussed in the text. Results should be put in the form $\tau_w = 0.365\ \rho U^2\ \mathrm{Re}^{-1/2}$.

13.7. Perform the same computations as were asked for in Prob. 13.6, this time for the case of the cubic profile. Use the results of Prob. 13.2. Give result in the form $\tau_w = 0.323\ \rho U^2\ \mathrm{Re}^{-1/2}$.

13.8. Perform the same computations as were asked for in Prob. 13.6, this time for the case of the sinusoidal profile [$u = U\sin(\pi y/2\delta)$]. Use the results of Probs. 13.4 and 13.5. Put the result in the form $\tau_w = 0.327\ \rho U^2\ \mathrm{Re}^{-1/2}$.

13.9. Air moves over a flat plate with a uniform free-stream velocity of 30 ft/s. At a position 6 from the front edge of the plate, what is the boundary-layer thickness and what is the shear stress at the surface of the plate? Assume that the boundary layer is laminar. The air temperature is 100°F, and the pressure is 14.7 lb/in² absolute. Do this problem using the
(a) Parabolic profile in the boundary layer as examined in the text and Prob. 13.6.
(b) Cubic profile in the boundary layer examined in Probs. 13.2 and 13.7.
(c) Result from Blasius' analytical solution.

13.10. Water approaches a device (Fig. P3.10) for diverting a portion of the flow. The water is moving at a speed U of 3 m/s and is at a temperature of 5°C. At what distance from A along the horizontal part of the diverter will the laminar boundary layer be a thickness of 1.2 mm? Use Blasius' solution.

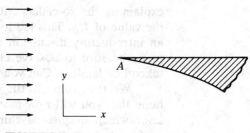

FIGURE P13.10

13.11. If the shear stress in a laminar boundary layer varies linearly from zero at $y = \delta$ to τ_w at $y = 0$, what is the momentum thickness?

13.12. Using the momentum thickness θ show that for a flow over a flat plate of width b that the drag $D(x)$ over length x of the plate is

$$D(x) = \rho b U^2 \theta \qquad (a)$$

Noting that

$$D(x) = \int_0^x b\tau_w\ dx \qquad (b)$$

differentiate (a) and (b) with respect to x and show that

$$\tau_w = \rho U^2 \frac{d\theta}{dx} \qquad (c)$$

Hint: Use the control volume shown in Fig. P13.12 with the mass flow in and out given as $\int_0^\delta \rho ub\ dy$. Thus we can see that the momentum thickness θ is a measure of drag on a plate and the gradient of θ is a measure of shear stress at the wall.

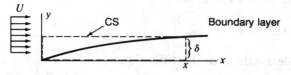

FIGURE P13.12

13.13. Using Eq. (c) from Prob. 13.12 and a parabolic profile for velocity in the boundary layer [see Eq. (13.44)] as well as Newton's viscosity law for τ_w, show that

$$\delta \, d\delta = 15 \frac{\nu}{U} \, dx$$

so that on integrating and letting $\delta = 0$ when $x = 0$ we get

$$\frac{\delta}{x} = \frac{5.48}{\sqrt{\text{Re}_x}} \qquad (a)$$

the same result reached in the text via the von Kármán integral momentum theorem.

13.14. Using Eq. (13.44) show that we get the following formula for τ_w for a laminar boundary layer:

$$\tau_w = 0.332 \frac{\rho^{1/2}\mu^{1/2}U^{1.5}}{x^{1/2}}$$

Now get the drag $D(x)$ for one surface of the plate of width b, and from this calculate C_f. Note that

$$C_f = 2c_f(L)$$

That is, the plate friction coefficient C_f equals twice the skin-friction coefficient c_f taken at the end of the plate.

13.15. Air at 5°C and atmospheric pressure is flowing into the region between two horizontal parallel plates. How close to each other must the plates be if when transition in the boundary layer is about to take place the flow is entirely viscous laminar flow over the entire height between the plates? Transition takes place at $\text{Re}_x = 10^6$. The speed of air U is 20 m/s. Assume zero pressure gradient outside the boundary layer.

13.16. Water at 20°C enters a pipe as initially frictionless irrotational flow. A boundary layer immediately forms on the inside periphery of the pipe. The inside diameter of the pipe is 20 mm and transition in the boundary layer takes place at $\text{Re}_x = 10^6$. Using a flat-plate model, find the distance x_0 from the entrance of the pipe to where there is a laminar flow over the entire cross section of the pipe and where also transition is just beginning to occur. What speed should the flow have on entering the pipe for this to be possible?

13.17. Air at a temperature of 20°C and pressure of 150 kPa absolute enters a smooth circular duct of diameter 0.5 m. If at the entrance to the duct the profile is that of a one-dimensional flow of velocity 0.3 m/s with zero boundary-layer thickness, compute the boundary-layer thickness and the displacement thickness at 0.1 m from the entrance. What is the velocity here of the flow outside the boundary layer assuming a uniform profile outside and using a cubic velocity profile inside the boundary layer? Take $\text{Re}_{cr} = 10^5$. Next, compute this uniform velocity using the displacement thickness. Compare results. Use Eq. (13.51a) for the velocity profile.

13.18. Do Prob. 13.17 for $x = 1$ m and $U = 5$ m/s.

13.19. Consider for incompressible flow that the free-stream pressure outside the boundary layer is a function of x, i.e., $p(x)$. Replace dp/dx in Eq. (13.40) using Bernoulli's equation outside the boundary layer and reach the following form of the von Kármán momentum equation with u_m as free stream velocity:

$$\tau_w = \rho \frac{d}{dx}\int_0^\delta (u_m - u)u \, dy$$

$$+ \rho \left(\frac{du_m}{dx}\right)\int_0^\delta (u_m - u) \, dy$$

Hint: Use the relation

$$u_m\left[\frac{d}{dx}\int_0^\delta \rho u \, dy\right]$$

$$= \frac{d}{dx}\left[u_m\int_0^\delta \rho u \, dy\right] - \left(\frac{du_m}{dx}\right)\int_0^\delta \rho u \, dy$$

13.20. Suppose the free-stream velocity outside the boundary layer for a flat plate is known to have the form

$$u_m = C_1 + C_2 x$$

Using the form of the von Kármán momentum equation given in Prob. 13.19, formulate the differential equation for the boundary layer thickness equation. Get the following result using the approximate velocity profile given by

Eq. (13.44):

$$\left[(C_1 + C_2 x)\frac{2u_m}{3} - \frac{\delta u_m^2}{15}\right]\frac{d\delta}{dx}$$

$$+ C_2(C_1 + C_2 x)\delta - \frac{2u_m \nu}{\delta} = 0$$

13.21. Water at 20°C moves over one side of a flat plate. The free-stream speed is 3 m/s. The plate is 0.5 m wide. What maximum length should the plate be to have only a laminar boundary layer if the transition takes place at $\text{Re}_x = 500{,}000$? For this length, what is the plate drag coefficient C_f and the drag D? At what position is the local coefficient of skin drag 1.5 that of the plate coefficient?

13.22. Show that for a cubic profile, the local coefficient of skin drag is $c_f = 0.647/\sqrt{\text{Re}_x}$.

*13.23. Explain what you mean by a two-dimensional self-similar boundary-layer flow. Draw the second profile for a self-similar flow shown. Express both velocity profiles in a single equation. Verify that using the stream function ψ, as stated in Eq. (13.7), we automatically satisfy the continuity equation. Assume zero pressure gradient outside the boundary layer.

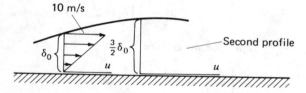

FIGURE P13.23

13.24. The rectangular rudder on the author's Columbia 22 sailboat extends 2 ft into the water and is 10 in in width. When the boat is moving at 6 knots, what is the skin drag from the rudder? The water is at 60°F. Transition takes place at $\text{Re}_x = 8 \times 10^5$.

13.25. A large, two-bladed wind turbine for generating power is stationary and has feathered its blades in a storm so as to be essentially parallel to the wind, which has a speed U of 50 km/h. What is the bending moment from skin drag at the base A of each blade if we simplify the blade to be a plate of length 30 m and width 0.3 m, as shown in the Fig. P13.25?

Consider transition to occur at $\text{Re}_x = 10^6$. The air is at 5°C.

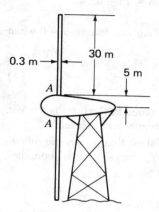

FIGURE P13.25

13.26. A helicopter during a test has its four main blades turning at 100 r/min with the blades oriented parallel to the plane of rotation. Each blade is 3.5 m long. The average width is 200 mm. The transition in the boundary layer is at a Reynolds number of 10^6 and the air is at 20°C. What power is needed to maintain this rotation of the four blades? Consider only skin drag.

13.27. The large wind turbine of Prob. 13.25 (3-mW capacity) is not self-starting.[28] If the blades are feathered so as to be parallel to the plane of motion of the blades to minimize drag, at what speed ω_c does the turbine reach constant angular velocity? The torque is 800 N · m. Note that the wind normal to the blade surface has only small effect and is neglected. Transition occurs at $\text{Re}_{cr} = 10^6$.

13.28. Air at 60°F and 14.7 lb/in² absolute moves over a flat plate at a speed of 50 ft/s. What is the boundary-layer thickness and the shear stress 2 ft from the front edge of the plate for a transition value of 3×10^5 for Re_{cr}?

13.29. Using the data of Hansen for transition, determine in Prob. 13.28 the position of transition

---

[28]Large wind turbines of this design will not start up by the action of the wind but must be brought up to a certain speed before the wind can take over.

and the ratio of the turbulent-boundary-layer thickness to the laminar-boundary-layer thickness at this position.

13.30. One of the problems in controlling a large dirigible is that the boundary layer is quite thick by the time the airflow reaches the tail control surfaces. The slow flow detracts from the ability of the tail controls to develop large forces. To get a "ball-park" estimation, replace the side of the greatest of dirigibles, namely the ill-fated *von Hindenberg*, by a flat plate of length 300 m. If it is flying at 100 km/h, what is the thickness of the boundary layer at the end of the 300 m. The air is at a temperature of 10°C and at atmospheric pressure near the earth's surface.

13.31. Air at 60°F and 14.7 lb/in^2 absolute with zero pressure gradient moves over a flat 30-ft-long plate. The main stream has a 0.2 percent turbulence. What are the minimum and maximum distances from the front edge of the plate along which one can expect laminar flow in the boundary layer? $U = 100$ ft/s.

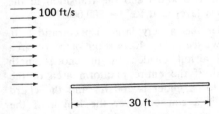

100 ft/s

|← 30 ft →|

FIGURE P13.31

13.32. In Prob. 13.31, determine the minimum and maximum possible drags on the top surface of the plate. Use the data of the cubic profile as determined in Prob. 13.7. The plate is 10 ft wide. The minimum total distance for a laminar boundary is 3.91 ft and the maximum total distance for a laminar boundary layer is 6.205 ft. Take $\rho = 0.00238$ slug/ft^3.

13.33. A smooth plate of dimensions 5 m long by 1.5 m wide is held in water at 20°C, which has a main-stream velocity of 1.5 m/s with zero pressure gradient. Using the data of Hansen, determine the drag on the upper surface of the plate. Do not use the Prandtl-Schlichting formula. *Hint*: Do you really have to consider the

laminar part of the boundary layer in this case?

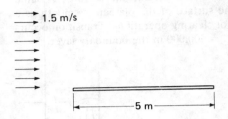

1.5 m/s

|← 5 m →|

FIGURE P13.33

13.34. Compute the ratio δ^*/x for turbulent flow using the one-seventh power velocity profile [Eq. (13.62)]. The result is $\delta^*/x = 0.0463$ Re$_x^{-1/5}$.

13.35. A water tunnel has a square test-section cross section of 3 ft by 3 ft at the entrance. The test section of the tunnel is 8 ft long. The velocity profile is uniform at the entrance of the test section at a speed U of 3 ft/s. To maintain this profile over a 3 ft by 3 ft region throughout the test section (with no model in place), we must continuously widen the test section as we move downstream of the entrance to minimize the effect of the boundary layer on this 3 ft × 3 ft stream. What should the exit dimension be of the test section? The water is at 60°F. Assume that boundary-layer thickness is zero at the entrance to the test section.

13.36. Water is moving between two plates 3 in apart. If the velocity of the air approaching the plates is 20 ft/s and the plates are smooth, how far from the leading edges of the plates does viscous action completely exists over the entire height of the plate? The air is at 60°F. Re$_{cr} = 500,000$.

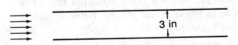

|‹ 3 in ›|

FIGURE P13.36

13.37. A wind turbine is placed on a plateau. The wind approaching the plateau has a speed U of 30 km/h on the average. Each blade of the turbine is 30 m in length. How high should you place the centerline of the turbine if you want the blades not to come within 3 m of the boundary layer? The distance d is 1000 m and the temperature of the air is 10°C. Assume that the surface of the plateau is smooth as a result of clearing operations. Transition occurs at $Re_{cr} = 500,000$ in the boundary layer.

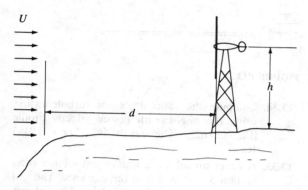

U

d

h

FIGURE P13.37

13.38. A barge having dimensions of 30 m by 12 m moves at a speed of 1 m/s in fresh water at 15°C, as shown in Fig. P13.38. Given an estimate of the skin friction drag D. Transition is at $Re_{cr} = 3 \times 10^5$.

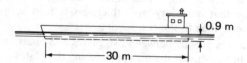

0.9 m

30 m

FIGURE P13.38

13.39. Do Prob. 13.38 for $U = 1.030$ m/s and a transition at $Re_{cr} = 3.2 \times 10^5$. Do not use the drag coefficient formulas but work from first principles with shear stress on the laminar and turbulent boundary layers.

13.40. A torpedo having a length of 4 m and an outer diameter of 0.49 m along most of its length is moving at a speed of 40 knots in seawater at 10°C. What power is needed to overcome the skin-friction drag? Transition occurs at a Reynolds number of 10^6. Take ν to be 1.361×10^{-6} m²/s and ρ to be 1025 kg/m³. Neglect friction of appendages.

13.41. Ocean liners have in the past been equipped with retractable hydrofoils for purposes of maintaining stability in heavy weather. If the ship is moving at 40 knots, what is the skin-friction drag on the hydrofoil if each is 2 m long and 2 m wide? For the seawater which is at 10°C, the coefficient of viscosity μ is 1.395×10^{-3} N · s/m² and the density ρ is 1026 kg/m³. Transition takes place at $Re_{cr} = 10^6$. Compute the skin drag of the hydrofoils taking the turbulent boundary layer over the entire length. Then calculate skin drag, taking into account the laminar portion of the boundary layer.

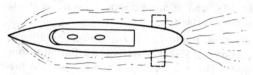

FIGURE P13.41

13.42. Do Prob. 13.41 for the case where the speed of the ship is 5 knots and the transition in the boundary layer is at $Re_{cr} = 500,000$.

13.43. A glider has a very large aspect ratio (40), which we see in Sec. 13.14 is the planform area of the airfoil divided by its chord length squared. If the entire planform area of the airfoil for a glider is 200 ft² and the chord length is a constant over the length of the airfoil, what is the skin-friction drag on the airfoil for a speed of the plane of 50 mi/h? The air temperature is 60°F. Transition takes place at $Re_{cr} = 10^6$.

13.44. A ground-effects vehicle is moving over water at a speed of 100 km/h. While over water, a pair of retractable rudders are inserted into the water. The width of the rudder is a constant equal to 0.75 m and a length of 1 m extends into the water. What is the skin drag

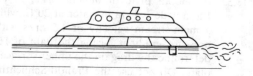

FIGURE P13.44

on the rudders if transition occurs at $Re_{cr} = 5 \times 10^5$. The water is fresh water at a temperature of 15°C.

13.45. In Prob. 13.44, the rudder has a wetted area having the shape shown in Fig. P13.45. Find the drag if the vehicle has a speed of 60 km/h. Other data are the same as in the previous problem.

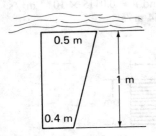

FIGURE P13.45

13.46. In the autogyro, the lift is developed by freely rotating vanes. The rotation is caused by the aerodynamic forces on the vanes themselves. Using flat-plate theory, what is the aerodynamic torque needed to overcome skin friction for an angular speed of the vanes of 50 r/min? Take each vane to be a plate of dimension 4.5 m by 0.3 m. The air is at a temperature of 10°C. Transition takes place at $Re_{crit} = 3.2 \times 10^5$. Consider as an approximation Eq. (13.76) to be valid for $Re_L < 5 \times 10^5$ for turbulent boundary layer. Take $p = 101.404$ Pa.

13.47. A barge having dimensions of 30 m by 12 m (Fig. P13.47) moves at a speed of 1 m/s in fresh water at 15°C. Estimate the skin-friction drag D. Transition is at $Re_{cr} = 3 \times 10^5$. Also give an equation with only U as the unknown for the speed U to double the skin-friction drag. Take $\nu = 1.141 \times 10^{-6}$ m²/s.

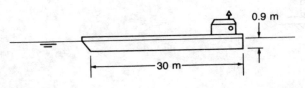

FIGURE P13.47

13.48. A mixing device in Fig. P13.48 with two blades is rotating with speed $\omega = 20$ r/min with the blades in a horizontal orientation. What is the torque needed for *smooth* blades if $Re_{cr} = 320{,}000$? The blades have a width of 0.3 m.
(a) Give torque from the purely laminar boundary-layer part of flow.
(b) Give torque from turbulent boundary-layer part of the flow.

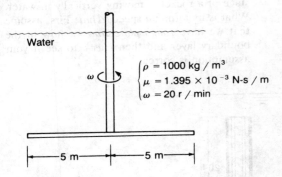

FIGURE P13.48

13.49. A spline shape in Fig. P13.49 is turning at a speed ω of 80 rad/s in air whose temperature is 20°C. Find the resisting torque from skin friction on three surfaces AB. The length of the spline is 2 m. The roughness e is 0.09 mm.

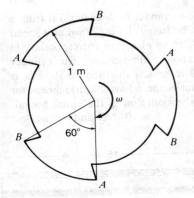

FIGURE P13.49

13.50. A wind sock in Fig. P13.51 has a length of 3 ft and a mean diameter of 1 ft. A 30 mi/h wind is blowing at a temperature of 50°F. What is your estimate of the shear drag? Take $e = 0.072$ in.

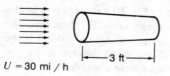

$U = 30$ mi / h

FIGURE P13.50

13.51. A hollow cylinder coming from a discarded stage of a rocket is moving vertically in water. What is its terminal speed? *Hint*: First assume that we have *rough plate zone* of flow in the boundary layer and then *check* to see if your assumption is correct.

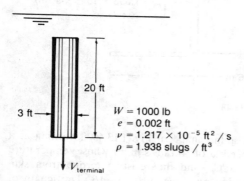

20 ft

3 ft

$W = 1000$ lb
$e = 0.002$ ft
$\nu = 1.217 \times 10^{-5}$ ft^2 / s
$\rho = 1.938$ slugs / ft^3

$V_{terminal}$

FIGURE P13.51

13.52. A 40-ft *hollow* cylinder of diameter 3 ft from a spent rocket booster is falling down at a speed of 500 mi/h at an elevation corresponding to 25,000 ft standard atmosphere. The cylinder weighs 2000 lb and is oriented so its axis is vertical. What is the admissible roughness for hydraulically smooth flow in the entire boundary layer. If e is 8×10^{-4} ft and we neglect

wave drag, what is the acceleration of the cylinder? Take $\nu = 6 \times 10^{-5}$ ft^2/s.

13.53. For an underwater "village" for research, an American flag is in place as shown in Fig. P13.53. It is of plastic material and can rotate so as to be parallel to the flow of water. If the critical Re is 500,000 and the roughness e is 0.06 mm, what is the bending moment at the base from a flow rate of water of 25 kn? Take $\rho = 1000$ kg/m^3 and $\nu = 0.0115 \times 10^{-4}$ m^2/s. Note 1 kn = 0.5144 m/s.

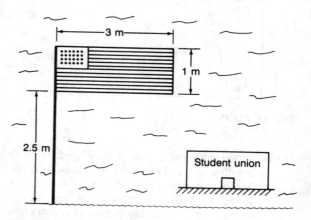

3 m

1 m

2.5 m

Student union

FIGURE P13.53

13.54. Three steel open cylinders on a flatcar moving at a speed of 50 m/s in Fig. P13.54 undergo skin drag. The 10 m/s wind is opposite to the velocity of the flatcar. If the roughness $e = 0.15$ mm, determine the skin drag. The diameter of each cylinder is 1 m. Neglect the effects on the flow about and through the cylinders arising from the flatcar and from each other. Take the

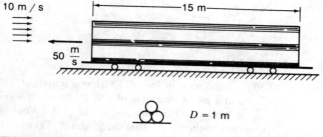

10 m / s

15 m

50 $\frac{m}{s}$

$D = 1$ m

FIGURE P13.54

transition Reynolds number to be 10^6. Take $\nu = 0.180 \times 10^{-4}$ m^2/s. The temperature is 20°C. How many kilowatts of power are needed to overcome this drag?

13.55. A thin plate moves at constant speed $V = 3$ m/s vertically into water which is at 20°C. What is the skin drag as a function of x. The plate is smooth. $\mathrm{Re}_{cr} = 500{,}000$. Do a quasi-static analysis.

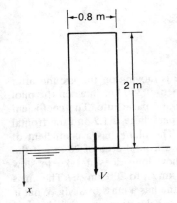

FIGURE P13.55

13.56. In Prob. 13.55 the plate is rough with $e = 0.4$ mm. Determine the skin drag as functions of x. Do not give ranges.
   (a) The hydraulically smooth zone. Note Eq. (13.76) is not valid for $\mathrm{Re}_L < 5 \times 10^5$.
   (b) The transition zone using an average value of 0.0045 for C_f from the chart.
   (c) The rough zone.
The length of the plate for this problem is 4 m.

13.57. A ship has a length of 250 m and is moving at a speed of 30 knots. The wetted area is 14,000 m^2. What is the admissible roughness? What is the minimum possible skin drag for this case? If $e = 0.1875$ mm, what is the skin drag and the power needed to overcome this drag? The kinematic viscosity of the seawater is 1×10^{-6} m^2/s and $\rho = 1010$ kg/m^3. Transition for the boundary layer is at $\mathrm{Re}_{cr} = 500{,}000$. What is the percent increase in power needed as a result of plate roughness?

13.58. In Prob. 13.31, what is the admissible roughness e_{adm} so as to have hydraulically smooth plate flow over the entire plate? Find τ_0.

13.59. In Prob. 13.38, what is the maximum roughness e that can be accepted to still have hydraulically smooth flow over the entire wetted boundary? If the drag for smooth barge is 526 N, what must it be for $e = 0.050$ mm?

13.60. Do Prob. 13.41 for the case of a rough surface where $e = 0.030$ mm.

13.61. In Prob. 13.40, what is the percentage increase in power needed for the case of a rough surface having $e = 0.050$ mm? The power needed for a smooth surface was 61.3 kW.

13.62. In Prob. 13.44, we found the drag to be 3052 N. If after a summer's use the drag was measured to be 6500 N, what would be the roughness e?

13.63. The rudders in Prob. 13.45 are rough, having a roughness coefficient e of 0.0400 mm. What is the drag on the vehicle at 60 km/h?

13.64. A delta-winged fighter plane is flying subsonically at a speed of 600 km/h. What is the skin drag from the wing, which has a roughness coefficient due to camouflage paint and flush riveting of 0.0050 mm? Transition occurs at $\mathrm{Re}_{cr} = 500{,}000$. Air is at 10°C.

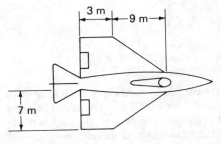

FIGURE P13.64

13.65. In your own words, describe the model we have established, describing the onset of separation.

13.66. Explain the difference between transition and separation.

13.67. An object has a projected area of 10 ft^2 in the direction of its motion. It has a drag coefficient of 0.4 for a Reynolds number of 10^7 using a characteristic length of 5 ft. For this Reynolds number, what is the drag on the object when moving through water at 60°F? What is the drag when moving through air at 60°F and 14.7 lb/in^2 absolute?

13.68. A body travels through air at 60°F at a speed of 100 ft/s, and 8 hp is required to accomplish this. If the projected area is 10 ft^2 in the direction of motion, determine the coefficient of drag.

13.69. A streamline strut (see Figs. 13.27) has a width of 3 in. What should the length be for minimal total drag from pressure and skin friction? If the strut were not streamlined, what diameter rod would give the same drag per unit length for turbulent flow? What does this tell you about the advantage of streamlining?

13.70. An anemometer is held stationary in a 50-knot wind. What torque T is required to maintain the instrument stationary? Cups A and B are oppositely oriented and each has a diameter of 75 mm. The wind is oriented perpendicular to the arm connecting the cups. Air is at a temperature of 20°C.

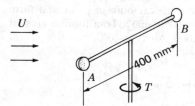

FIGURE P13.70

13.71. A sports car has a coefficient of drag of 0.40 and a frontal area of 25 ft^2. The vehicle weighs 2600 lb. If a total constant torque of 800 ft · lb is maintained on the rear wheels, how long will it take for the vehicle to reach 60 mi/h if we neglect the rotational inertia of the wheels? The air is at 60°F. Neglect rolling resistance of the tires. The tires have a 24-in diameter. *Hint*:

$$\int \frac{dx}{c^2 - x^2} = \frac{1}{2c} \ln\left(\frac{c + x}{c - x}\right)$$

13.72. A wind of 50 km/h impinges normal to the sign. What is the bending moment on each leg of the support at the ground? The temperature is 5°C. Neglect the drag on the supports.

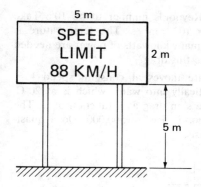

FIGURE P13.72

13.73. A fighter plane is moving on the ground after landing at a speed of 350 km/h when the pilot deploys his braking parachute. The coefficient of drag for the parachute is 1.2 and the frontal area is 30 m^2. The plane has a coefficient of drag of 0.4 and a frontal area of 20 m^2. If the engine is off, how long does it take to slow down from 350 km/h to 200 km/h? The air is at 10°C. The plane has a mass of 8 Mg. What is the maximum deceleration in g's? Neglect rolling resistance of tires.

FIGURE P13.73

13.74. In Prob. 13.73, what is the largest frontal area of the braking parachute if the maximum deceleration of the plane is to be 5 g's when at a speed of 350 km/h the parachute is first deployed?

*__**13.75.** In Sect. 7.8 we examined laminar flow between parallel infinite plates separated by distance h. For turbulent flow we can assume a logarithmic velocity profile

$$\frac{\bar{V}_x}{V_*} = \frac{1}{\alpha} \ln \frac{\eta V_*}{\nu} + B \qquad 0 < \eta < \frac{h}{2}$$

where η is measured from the bottom plate

upward to the midplane. Form the equation for the average velocity $\bar{V}_{mean}$ between the plates. To introduce f show that using $\alpha = 0.41$ and $B = 5$,

$$\frac{1}{\sqrt{f}} = 2.0\log\left(\mathrm{Re}_{D_H}\sqrt{f}\right) - 1.19$$

which is the analog of Prandtl's universal pipe friction formula. *Hint*: Use the result

$$\frac{\bar{V}_{mean}}{V_*} = \left(\frac{8}{f}\right)^{1/2}$$

13.76. A mixing rotor in Fig. P13.76 consists of cylindrical arms *BC* and blades *AB*. The blades are oriented parallel to the free surface and have a width in this direction of 0.2 m. What is your estimate of the rotational drag of this rotor at $\omega = 20$ rad/s. Take C_D for cylinder to be 0.30 and the blades to be thin plates for which $\mathrm{Re}_{cr} = 500,000$. The water is at 20°C. Take e for blades to be 0.4 mm.

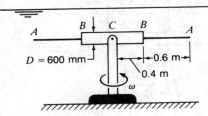

FIGURE P13.76

13.77. In the preceding problem the water is evacuated leaving air at 20°C. How long will it take for a torque of 10 N-m to bring the system up to speed if it has a radius of gyration about the vertical axis of 1.3 m and a mass of 50 kg. Set up in the form of a quadrature

$$t = \int_0^{20} \frac{d\omega}{f(\omega)}$$

where $f(\omega)$ is a function of ω.

13.78. An automobile has a coefficient of drag of 0.36 and a frontal area A. A remodeling of the sheet metal reduces C_D to 0.30 and the frontal area of $0.9A$. If the older model gets 26 mi/gal at 55 mi/h, what should the newer model get at the same speed? The drive system has not been changed.

13.79. An aquaplaning board is 2 ft long and 2 ft wide. If it is at an angle of 8° with the horizontal and has a speed of 10 mi/h, what load can it carry as a passenger? What is the drag force?

13.80. If the critical Reynolds number for flow around a smooth cylinder is 6.7×10^5, what is the velocity for this condition of air at 40°C around a smooth cylinder having a diameter of 100 mm? How about water? Both are at atmospheric pressure. If the pressure is doubled for air, what is this velocity?

13.81. A torpedo in Fig. P13.81 is moving at a speed of 50 km/h. We want to estimate the power required in kilowatts to propel the torpedo.

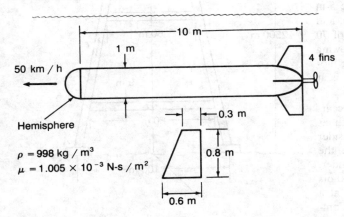

FIGURE P13.81

We do it in three steps. Proceed as follows:
1. Get skin drag over the cylindrical part of the torpedo (length 10 m). Take $e = 2 \times 10^{-3}$ m for roughness.
2. Get the skin drag of fins using an average rectangular plate for each fin. Take $e = 4.5 \times 10^{-5}$ m.
3. Get the wave drag (pressure drag) of the half sphere by considering a solid half sphere for which $C_D = 0.38$.

Now get desired power. $\text{Re}_{cr} = 500,000$.

13.82. A trailer truck is shown moving at a speed of 100 km/h. The truck weighs 53.5 kN. If it has a coefficient of drag of 0.60, how far will it have to go before the speed is reduced to 50 km/h. The motor is not engaged during the action. The rolling resistance of the tires is 2 kN. Air is at 15°C. *Hint*: Recall from sophomore mechanics that

$$\frac{d}{dt} \equiv \frac{d}{dx}\frac{dx}{dt} = V\frac{d}{dx}$$

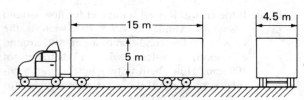

FIGURE P13.82

13.83. A flagpole is 15 m high. The lowest 5 m has a uniform diameter of 125 mm, the middle 5-m section has a uniform diameter of 90 mm, and the top section has a uniform diameter of 70 mm. If a strong wind of 50 km/ h is blowing and there is no flag, what is the bending moment at the base of the flagpole? The air is at 10°C.

13.84. In recent times, we have had disasters occurring on offshore drilling operations, such as one in the North Sea "hotel." Let us consider a hotel as shown in Fig. P13.84, where the above-water housing is idealized as a cylinder. In a storm, the air at 5°C is moving at 50 knots in the y direction and the sea is moving at a speed of 4 knots also in the y direction. Compute the total shear force at the base of the

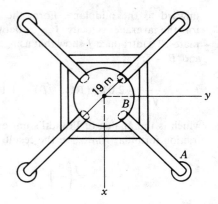

Top view showing only one side-bracing members

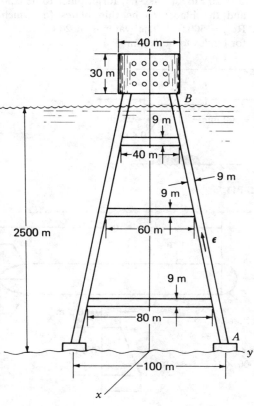

FIGURE P13.84

structure. Take ρ of the seawater to be 1025 kg/m^3. *Hint*: Because of the great height of the columns, we can assume that they are vertical without significant loss in accuracy.

13.85. In Prob. 13.84, suppose that the height is only 300 m instead of 2500 m. The approximation made in the previous problem can no longer be made without some loss in accuracy. Solve for shear force. *Hint*: Get a unit vector ϵ for a column, say AB. The desired velocity component $V_\perp$ normal to the column is then $V\mathbf{j} \cdot (\epsilon \times \mathbf{i})$. The drag component in the y direction is then the computed drag normal to the column times $(\epsilon \times \mathbf{i}) \cdot \mathbf{j}$.

13.86. A Cottrell precipitator consists of a series of horizontal, parallel rods forming a grid. The rods are held by a horizontal circular hoop. These rods are kept at a very high voltage. Flue gases from a power plant pass through the precipitator in the chimney. The fly ash particles are given a charge and are later precipitated out by being attracted to properly charged plates. If each rod has a $\frac{1}{4}$-in diameter and are spaced 2 in apart between centerlines, what is the force on the precipitator from combustion products having a density of 0.003 slug/ft^3 (this density takes flue particles into account) and a speed of 30 ft/s. The flow is turbulent. There are 35 rods in the grid and the hoop has a 6-ft diameter. The center rod goes through the center of the circular hoop. (here is a good opportunity to use a programmable calculator, but it is not necessary.)

13.87. From irrotational-incompressible-flow, theory, the velocity V_B along the surface of a sphere is

$$V_B = \frac{3}{2}U \sin \beta \qquad (a)$$

where β is the angle between the radius R and the z axis. Show that the drag on this sphere is zero by first formulating the following equation for drag.

$$D = 2\pi R^2 \left[p_\infty \frac{1}{2}(\sin^2 \beta) + \frac{\rho U^2}{2} \frac{\sin^2 \beta}{2} \right.$$
$$\left. - \frac{9}{8}\rho U^2 \left(\frac{\sin^4 \beta}{4} \right) \right] \Big|_0^\pi \qquad (b)$$

and then putting in the limits $\pi, 0$ f(Use a strip as shown in the diagram $(R\,d\beta)$ and circumference $(2\pi R)\sin \beta$. Note

$$\int \sin \beta \cos \beta \, d\beta = \frac{1}{2}\sin^2 \beta$$

and $\quad \int \sin^m \beta \cos \beta \, d\beta = \frac{\sin^{m+1} \beta}{m + 1}$

The resulting zero drag stems from no skin friction going with zero viscosity and no pressure drag because of no separation. This was called the d'Alembert paradox in early days of fluid mechanics because everyone knew there had to be drag in real flows.

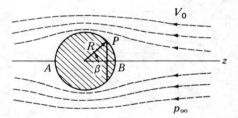

FIGURE P13.87

13.88. In Table 13.3 we listed the coefficient of drag for a solid hemisphere as 0.38. Let us assume that the flow to the right of $A - A$ (see Fig. P13.88) is essentially given by the irrotational-flow theory as presented in the previous problem *outside* the boundary layer. Compute the pressure drag on the *spherical* part of the boundary using Eq. (b) of Prob. 13.87 with the *proper* limits on β for this case of $\pi/2$ and 0. At the periphery of the hemisphere we showed in Prob. 13.87 that the pressure p for this hypothetical ideal flow is

$$p = p_\infty + \frac{\rho U^2}{2} - \left(\frac{9}{4} \right) \frac{\rho U^2}{2} \sin^2\beta \quad (a)$$

At the edge B the pressure is computed by setting $\beta = \pi/2$. Now in the region behind the hemisphere, the pressure is the same pressure as in Eq. (*a*) with $\beta = \pi/2$ with a fraction coefficient η multiplying the last expression. This means that only the fraction $(1 - \eta)$ of the kinetic energy at edge B is recovered into pressure in the wake behind the hemisphere. We call η a *recovery factor*. Show that the total drag D is given as

$$D = (\pi R^2)\left(\frac{\rho U^2}{2}\right)\left(\frac{9}{8}\right)(2\eta - 1) \quad (b)$$

Compute recovery factor η for $C_D = 0.38$. Comment on assumptions made that weaken your answer.

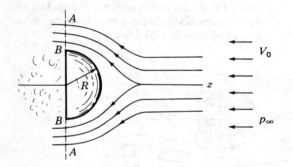

FIGURE P13.88

13.89. A bullet of mass 2 g has been shot into the air and is descending at its terminal speed. Estimate this speed by first computing the skin-friction drag on the cylindrical surface. Estimate the pressure drag by considering the front tip of the bullet to be a hemisphere and then use the results of Prob. 13.88 adapted to this problem with a recovery factor η of 0.669 in the wake. The bullet is near the earth's surface. The temperature of the air is 20°C. Transition in the boundary layer is at $\text{Re}_{cr} = 10^5$.

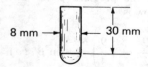

FIGURE P13.89

13.90. Recently a jet-driven vehicle broke the sound barrier, achieving a world record for land vehicles. An idealization of the general shape is shown in Fig. P13.90. If the vehicle is coasting at 400 mi/h, what is the estimated drag? *Hint:* See Prob. 13.88 and consider the drag due to the skin friction on the cylindrical surface of the vehicle plus the pressure drag of a hemisphere as developed in Prob. 13.88. Consider the surface to be smooth. The air is at 90° F on the Utah Salt Flats. Take the recovery factor $\eta = 0.669$. Transition occurs at $\text{Re}_{cr} = 500,000$.

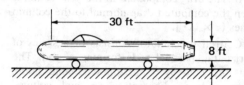

FIGURE P13.90

13.91. If the Mustang fighter plane weighs 42.7 kN, at what angle of attack should it be flown at a speed of 250 km/h? The planform area is 25 m^2. What power is required in horsepower to overcome wing drag? The temperature is 20°C. The flaps are at zero degrees.

13.92. If the takeoff speed is about 1.3 times the stall speed for the Mustang, which weighs 42.7 kN, what is the takeoff distance for a constant thrust of 9 kN and a rolling resistance of 0.5 kN? The planform area is 25 m^2. The flaps are at 40°. The air is at 20° C. The overall coefficient of drag for the plane is 0.20 and the frontal area is 15 m^2.

13.93. What weight can the Mustang plane have if it has a planform wing area of 233 ft^2 and is flying at an angle of attack of 3° at a speed of 210 mi/h? Air is at 60° F. What power is needed for overcoming wing drag at this speed? The flaps are at zero degrees.

13.94. A boat is fitted with hydrofoils having a total planform area of 1 m^2. The coefficient of lift is 1.5 when the boat is moving at 10 knots, which is the slowest speed for the hydrofoils to support the boat. The coefficient drag is 0.6 for this case. What is the maximum weight of the boat to fulfill the minimal speed for hydrofoil support? What power is needed for this speed? Water is fresh water at 5° C.

13.95. A fan used at the turn of the century and in some ice cream parlors and homes today consists of flat wooden slats rotated at a small inclination of about 15°. At what speed ω would you have to rotate the fan (with four blades) to result in zero vertical force on the bearings supporting the wooden blades each of which weighs 1.5 lb? What torque is needed for this motion? *Hint:* Use Fig. 13.22. Air is at 60° F.

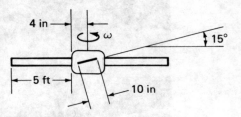

FIGURE P13.95

Two views of a hydraulic jump. (Courtesy Dr. J. Atkinson, State University of New York at Buffalo.)

A hydraulic jump is a sudden change in elevation of the free surface in channel flow. In many ways, it is analogous to a normal shock in one-dimensional compressible flow. Thus, the hydraulic jump occurs only when the flow velocity exceeds the speed of a surface wave (shooting flow), while a normal shock occurs only when there is supersonic flow with a flow velocity exceeding the speed of an acoustic wave. The slower flow after the hydraulic jump is called tranquil flow with a speed less than a surface wave, whereas the flow after the normal shock is subsonic with a flow slower than an acoustic wave. As can be seen in the photos, there is great turbulence and irregularity in the actual jump, and for this reason there is much dissipation of mechanical energy. The hydraulic jump is usually present in the runoff from a dam where its dissipative capacity is very useful.

CHAPTER
14

FREE-SURFACE FLOW

14.1 INTRODUCTION

Free-surface flow usually refers to the flow of liquid wherein a portion of the boundary of the flow called the free surface is subject to only certain prescribed conditions of pressure. The movement of oceans and rivers as well as the flow of liquids in partially filled pipes are free-surface flows where atmospheric pressures are imposed on part of the boundary surface. In analyzing free-surface flow, the geometry of the free surface is not known a priori. This shape is part of the solution, which means that we have a very difficult boundary condition to cope with. For this reason, general analyses are extremely involved and beyond the scope of this text. We will therefore restrict ourselves in this chapter to the flow of liquids under certain simplifying conditions that will later be set forth.

Although much of the material to be considered might appear at first to be of interest only to hydraulicians and civil engineers, you will later see that the water waves and the celebrated hydraulic jump are analogous to the pressure wave and shock wave, respectively, studied in compressible flow.

14.2 CONSIDERATION OF VELOCITY PROFILE

For flow in open channels, we have in addition to the difficulties at the free surface the added difficulty in long channels that friction must be accounted for

709

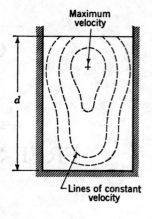

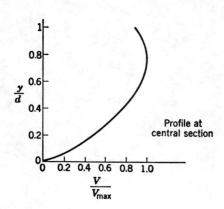

FIGURE 14.1
Velocity for narrow channel

as a result of the proximity of the wetted boundaries to the main flow. Furthermore, for such channels we have usually to consider fully developed *turbulent flow* to be present.

What can be said about velocity profiles in channel flow? There are semitheoretical approximate formulations developed for flow in channels which have a width that is large compared with the depth. Such studies may be found in more specialized texts.[1] Even more difficult is the case of narrow channels, for which we have shown a typical profile in Fig. 14.1. Note that the maximum velocity does not occur at the free surface but somewhere below, as shown in the diagram. Nor will the free surface be level at a section of the flow as we have shown it in the approximate diagram. Rather, there will be a slight elevation near the center; we call this region the *thalweg*.

In this chapter we do not delve into the empirical studies of velocity profiles but instead consider a one-dimensional flow model wherein friction and turbulence are accounted for in long channels by a particular shear stress at the wall of the channel.

14.3 NORMAL FLOW

We now consider channels that are straight and which maintain constant cross sections along the length. We call these channels *prismatic* channels. We consider flow of a liquid whose free surface maintains a constant depth y_N above the bed of the channel (see Fig. 14.2). The slope of the channel bed must have a certain value to maintain this kind of flow for a given volume flow Q.

---

[1]See H. Rouse, *Fluid Mechanics for Hydraulic Engineers*, McGraw-Hill, New York, 1938.

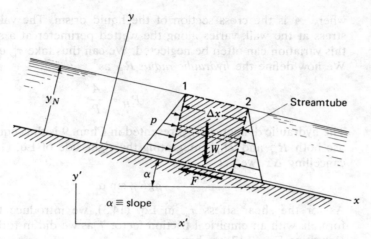

FIGURE 14.2
Normal flow in a prismatic channel.

Such a flow is called *normal*, or *uniform*, flow and we can readily deduce that there exists a balance between gravitational forces urging the flow along and frictional forces at the wetted perimeter retarding the flow. For liquids like water, the slope of the channel bed must be small.[2] We now state that the slope of the channel bed under this assumption will be taken equal to the angle of inclination α in radians (see Fig. 14.2) and will be denoted as S_0. The depth y_N is called the *normal* depth.

To analyze the flow, we use a one-dimensional flow model acted on by friction forces at the wetted boundary. The streamlines are parallel and we consider the pressure to be hydrostatic in a direction normal to the bed. A small system of fluid of length Δx has been shown in Fig. 14.2. Newton's law for this system then stipulates in the x direction

$$W \sin \alpha - F = 0$$

where your will note that the hydrostatic forces cancel. Using P as the length of the wetted perimeter of a cross section of the channel, we have for the equation above

$$\rho g A \, \Delta x \sin \alpha = \Delta x \int_0^P \tau_w \, dP \tag{14.1}$$

---

[2]The slopes generally are in the vicinity of 0.0001 rad for many rivers and canals.

where A is the cross section of the liquid prism. The value of τ_w, the shear stress at the wall, varies along the wetted perimeter at a section; nevertheless this variation can often be neglected. We can thus take τ_w constant at a section. We now define the *hydraulic radius* R_H as[3]

$$R_H = \frac{A}{P} \tag{14.2}$$

The hydraulic diameter D_H presented in Chap. 9 is thus equal to $4R_H$. We shall use both R_H and D_H. Integrating the right side of Eq. (14.1), using R_H, and cancelling Δx, we get

$$\rho g R_H \sin \alpha = \tau_w \tag{14.3}$$

As for the shear stress τ_w in Eq. (14.3), we introduce the Darcy-Weisbach formula with an empirical friction factor f as we did in turbulent flow in pipes. Rewriting Eq. (9.17) we have

$$\tau_w = \frac{f}{4}\rho\frac{V^2}{2} \tag{14.4}$$

Substituting τ_w from above into Eq. (14.3), we have on solving for V

$$V = \left(\frac{8g}{f}\right)^{1/2}(R_H \sin \alpha)^{1/2} \tag{14.5}$$

Now let us consider smooth channels. There can be fully developed laminar or fully developed turbulent flow. In Fig. 14.3 we have shown a plot of the friction factor f versus Re_H relating experimental results with certain theoretical results. Notice in the laminar range we have the theoretical result $f = 96/\text{Re}_H$ for wide channels and $f = 56/\text{Re}_H$ for 90° triangular channels. Inside the band formed by these two curves would fall the friction factor $64/\text{Re}_H$ for laminar flow in pipes [see Eq. (9.14)]. In the turbulent zone we have used Eq. (9.21) for $\text{Re} < 100,000$ and Eq. (9.66), the Prandtl universal law of friction, for pipe flow. The data points are for smooth channel flow. This plot shows the similarity between smooth pipe flow and smooth channel flow.[4] Thus, just as in boundary layers we shall be able to extend much information from our work in pipe flow. Now the friction factor generally depends on the Reynolds number Re_H of the flow (using the hydraulic diameter for the length parameter),

---

[3]The hydraulic radius, you will note, becomes *half* the ordinary radius for a circular wetted perimeter. For the case of rectangular channel flow of height y and width b we have

$$R_H = \frac{by}{2y + b}$$

[4]This also confirms the approach of using $4R_H$ for D_H in noncircular duct flows.

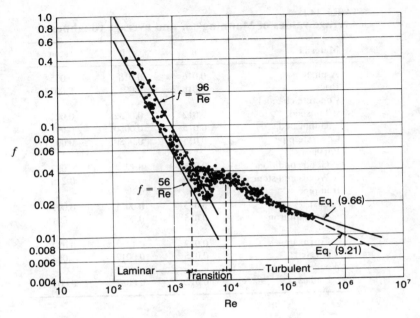

FIGURE 14.3
The friction factor versus Reynolds number from experiments in smooth open channels. (*Source: By permission from Open-Channel Hydraulics by V. T. Chow. Copyright 1959. McGraw-Hill Book Company, Inc.*)

the roughness of the channel bed, and the shape and size of the channel section. You may remember when we discussed flow through artificially roughened pipes from the work of Nikuradse, however, that for *large Reynolds numbers* and *large roughness factors* the friction factor f was *independent* of the Reynolds number and depended *only* on the roughness factor (recall this is the rough-zone flow). This is the case for many channel flows that one usually encounters. We can then say

$$\left(\frac{8g}{f}\right)^{1/2} \equiv C = \text{function of roughness factor} \qquad (14.6)$$

The term C is called the *Chézy coefficient*. Replacing $\sin \alpha$ by S_0 in Eq. (14.5) and using Eq. (14.6) to introduce C, we can write Eq. (14.5) as follows:

$$\boxed{V = C\sqrt{R_H S_0}} \qquad (14.7)$$

This is the well-known Chézy formula. From experiment, the Chézy coefficient

TABLE 14.1
Average values of Manning's *n* and average roughness *e*

Material	n	e, ft	e, m
Asphalt	0.016	0.018	0.0054
Brick	0.016	0.0012	0.0037
Concrete channel			
Finished	0.012	0.0032	0.001
Unfinished	0.015	0.0080	0.0024
Concrete pipe	0.015	0.0080	0.0024
Earth			
Good condition	0.025	0.12	0.037
Weeds and stones	0.035	0.8	0.240
Iron pipe			
Cast	0.015	0.0051	0.0016
Wrought iron	0.015	0.0051	0.0016
Steel			
Corrugated	0.022	0.12	0.037
Riveted	0.015	0.0012	0.0037
Wood			
Planed	0.012	0.0032	0.001

may be expressed as follows:

$$C = \frac{1.486}{n}(R_H)^{1/6} \quad \text{for USCS units}$$

$$C = \frac{1.000}{n}(R_H)^{1/6} \quad \text{for SI units} \tag{14.8}$$

$$\therefore C = \frac{\kappa}{n}(R_H)^{1/6}$$

where n, called *Manning's n*, is dependent[5] primarily on the relative roughness and where κ is either 1.486 or 1.000 depending on the system of units. Typical values of n are given in Table 14.1. The dependence of V on shape and size of the channel is embodied in the hydraulic radius. We then have for Eq. (14.7) on using the formulations above for C

$$\boxed{V = \left(\frac{\kappa}{n}\right)R_H^{2/3}\sqrt{S_0}} \tag{4.9}$$

---

[5]Manning's n is also dependent on the size and shape of the channel; the results given for n are primarily from canal data. Also, it is to be noted that Eq. (14.8) was wrongly attributed to Manning. Furthermore, these results climaxed much work done during the period 1869 to 1911 and for rough-zone flow such as is found in rivers, canals, etc., is still quite accurate.

Next, multiplying Eq. (14.9) by A, we get the volume flow, Q.

$$Q = \left(\frac{\kappa}{n}\right) R_H^{2/3} \sqrt{S_0}\, A \tag{14.10}$$

Finally, solving for S_0 in Eq. (14.9) we get

$$S_0 = \left(\frac{n}{\kappa}\right)^2 \frac{1}{R_H^{4/3}} V^2 = \left(\frac{n}{\kappa}\right)^2 \frac{1}{R_H^{4/3}} \frac{Q^2}{A^2} \tag{14.11}$$

We see here that for a given uniform flow Q in a given prismatic channel where there is a given cross section of flow there is one and only one slope S_0 for normal flow.

For a *wide rectangular* channel, we can replace R_H by y_N; A by $y_N b$ (where b is the width); and Q/b by q, the flow per unit width. We get from Eq. (14.10), on solving for y_N,

$$y_N = \left[\frac{nq}{\kappa \sqrt{S_0}}\right]^{3/5} \tag{14.12}$$

Solving for S_0, we have for normal flow

$$S_0 = \left(\frac{n}{\kappa}\right)^2 \frac{q^2}{y_N^{10/3}} \tag{14.13}$$

Here we relate the slope of the channel with the depth for normal flow. Keep in mind that the slope of the free surface is parallel to the slope of the bed S_0.

We now consider approximate solutions for flows which are neither canals nor rivers.

Example 14.1. Water at 60° F is flowing in a finished-concrete, semicircular channel shown in Fig. 14.4 inclined at a slope S_0 of 0.0016. What is the volume flow Q if the flow is normal?

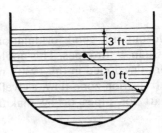

FIGURE 14.4
Uniform flow in channel.

We first compute the hydraulic radius for the flow. Thus

$$R_H = \frac{A}{P} = \frac{\frac{1}{2}\pi(10^2) + (3)(20)}{\frac{1}{2}\pi(20) + (2)(3)} = 5.80 \text{ ft}$$

Using Eq. (14.9) for a value of n equal to 0.012, we get for the average velocity V:

$$V = \frac{1.486}{n} R_H^{2/3} \sqrt{S_0} = \frac{1.486}{0.012}(5.80)^{2/3} \sqrt{0.0016} = 15.99 \text{ ft/s}$$

The volume of flow Q is then:

$$Q = VA = (15.99)\left[\frac{1}{2}\pi(10^2) + (3)(20)\right] = 3471 \text{ ft}^3/\text{s}$$

Example 14.2. In a planed-wood rectangular channel of width 4 m, water is flowing at the rate Q of 20 m³/s. If the slope of the channel is 0.0012, what is the depth d corresponding to normal flow?

The hydraulic radius for this case is given in terms of d as follows:

$$R_H = \frac{4d}{4 + 2d} \tag{a}$$

Also, we can express Q using Eq. (14.10) as follows:

$$Q = \frac{1.00}{n} R_H^{2/3} \sqrt{S_0}\, A$$

$$\therefore 20 = \frac{1.00}{0.012}\left(\frac{4d}{4 + 2d}\right)^{2/3}(\sqrt{0.0012})(4d) \tag{b}$$

Rearranging the equation, we have

$$0.687 = \left(\frac{d}{4 + 2d}\right)^{2/3} d \tag{c}$$

We may now solve by trial and error to get

$$d = 1.796 \text{ m}$$

14.4 NORMAL FLOW: NEWER METHODS

We have thus far considered old but still good formulations that apply to the rough-flow zone as discussed in Chap. 9 for pipes. This applies to most river and canal problems. More recent work developed in the 1930's may be used to cover hydraulically smooth zone flow, transition-zone flow, as well as rough-zone flow by making use of Moody's chart for pipes or by making use of empirical formulas for the friction factor f. These formulas are analogous to those presented in Chap. 9 for pipe flow.

To clarify this, consider steady open-channel flow again in Fig. 14.2, this time considering the cross-hatched region to be a stationary control volume. A streamtube is shown in this control volume. From our discussion in Sec. 9.3, we can use the *first law of thermodynamics* for this streamtube. The result is Eq. (9.5), which we now rewrite using reference $x'y'$:

$$\frac{\Delta p}{\rho g} = (y_2' - y_1') + (H_l)$$

where H_l is the head loss per unit weight of fluid. But Δp is zero in this case, so we get

$$(y_1' - y_2') = H_l \tag{14.14}$$

But $y_1' - y_2'$ is the same as $S_0 \Delta x$ and we have

$$\boxed{S_0 \, \Delta x = H_l = \frac{h_l}{g}} \tag{14.15}$$

This holds for the entire control volume, since each streamtube gives the same result. We now can replace h_l by the Darcy-Weisbach formula as was done in Eq. (9.13), thereby introducing the friction factor f. However, instead of D, the inside diameter of the pipe, we insert $4R_H$ for the channel to get

$$S_0 \Delta x = \frac{V^2}{2} \left(\frac{\Delta x}{4R_H} \right) \frac{f}{g}$$

$$\therefore \boxed{V = \left(\frac{8gS_0 R_H}{f} \right)^{1/2}} \tag{14.16}$$

The procedure is to first estimate f. Then after determining V from Eq. (14.16), compute the Reynolds number for the flow using $4R_H$ as the length parameter. With this Reynolds number and with a relative roughness ratio, $e/4R_H$ (see Table (9.1), find f in the Moody chart (Fig. 9.16). If the f from the Moody diagram does not agree with the original estimate, then go through a second cycle of the steps above using the computed f. Proceed in this way until good agreement is reached between the inserted f and the resulting computed f. You will recall that we did this in pipe-flow problems in Chap. 9.

If we wish to use formulas for f, then we must know what flow zone we are in. From the work on pipe flow in Sec. 9.16, we give the following criteria as applied to channel flow:

$$\frac{V_* e}{\nu} < 4 \qquad \text{Hydraulically smooth zone flow}$$

$$4 \leq \frac{V_* e}{\nu} \leq 100 \qquad \text{Transition-flow zone} \tag{14.17}$$

$$\frac{V_* e}{\nu} > 100 \qquad \text{Rough-flow zone}$$

where V_*, you will recall, is the *shear velocity* given as $\sqrt{\tau_w/\rho}$. From Eq. (14.3), we see that V_* can be given as

$$V_* = \sqrt{\frac{\tau_w}{\rho}} = \sqrt{\frac{\rho g R_H \sin \alpha}{\rho}} = \sqrt{g R_H S_0} \tag{14.18}$$

For *hydraulically smooth* flow, we have the Blasius formula [see Eq. (9.21)] which for $\text{Re}_H < 10^5$ is

$$f = \frac{0.316}{(\text{Re}_H)^{1/4}} \tag{14.19}$$

From Eq. (14.6) we can then give the Chézy coefficient for this flow as

$$C = 28.6(\text{Re}_H)^{1/8} \quad \text{USCS units}$$
$$C = 15.76(\text{Re}_H)^{1/8} \quad \text{SI units} \tag{14.20}$$

For $\text{Re}_H > 10^5$ the following formulas are recommended for *hydraulically smooth* channel flow.

$$\frac{1}{\sqrt{f}} = 2.0 \log\left(\frac{\text{Re}_H \sqrt{f}}{2.51}\right) \tag{14.21a}$$

$$C = 4\sqrt{2g} \log\left(\frac{\text{Re}_H \sqrt{8g}}{2.51C}\right) \tag{14.21b}$$

Next, for the *transition-zone* flow, a modification of the Colebrooke formula [Eq. (9.18)] may be used.

$$\frac{1}{\sqrt{f}} = 2.16 - 2\log\left[\frac{e}{R_H} + \frac{30}{\text{Re}_H \sqrt{f}}\right] \tag{14.22a}$$

$$C = \left\{2.16 - 2\log\left[\frac{e}{R_H} + \frac{30}{(\text{Re}_H)\sqrt{f}}\right]\right\}\sqrt{8g} \tag{14.22b}$$

Finally, in the *rough-flow* zone where $e/R_H \gg 30/[(\text{Re}_H)\sqrt{f}]$ in the equation above, we have from the above formulas

$$\frac{1}{\sqrt{f}} = 2.16 - 2\log\left(\frac{e}{R_H}\right) \tag{14.23a}$$

$$C = \left[2.16 - 2\log\left(\frac{e}{R_H}\right)\right]\sqrt{8g} \tag{14.23b}$$

In problems wherein we have *hydraulically smooth* flow or where we have *transition zone flow* you will have to estimate a velocity to get Re_H and thus f. Then find Chézy's C and finally get V from Eq. (14.7). If the solved V differs from assumed V, take the solved V as a starting point for another cycle of computations. You may have to go through three or four cycles to get good accuracy. In solving for f you will have to use trial and error or a programmable calculator.

We now examine two examples, the second of which allows a comparison of the older and newer methods.

Example 14.3. In a planed-wood rectangular channel of width 4 m, water is flowing at a rate of 5 m³/s. The slope of the channel S_0 is 0.0001. The roughness coefficient e is 0.5 mm. What is the height h of the cross section of flow for normal, steady flow? The water has a temperature of 10° C. Go through two cycles of calculation only.

Here as in pipe flow we will guess at a friction factor f and work with the Moody diagram (Fig. 9.16). We assume that $f = 0.02$ and we now go to Eq. (14.16) to determine V. Thus

$$V = \frac{Q}{A} = \left(\frac{8gS_0R_H}{f}\right)^{1/2}$$

$$\therefore \left(\frac{5}{4h}\right)^2 = \frac{(8)(9.81)(0.0001)[4h/(2h+4)]}{0.02} \tag{a}$$

Rewriting the Eq. (a), we have

$$1.991 = \frac{h^3}{h+2} \tag{b}$$

Solving by trial and error, we get

$$h = 1.996 \text{ m} \tag{c}$$

The velocity V of the flow must then be

$$V = \frac{Q}{A} = \frac{5}{(1.996)(4)} = 0.626 \text{ m/s}$$

Let us now check the friction factor. The Reynolds number is

$$\text{Re}_H = \frac{VD_H}{\nu} = \frac{V(4R_H)}{\nu} = \frac{(0.626)(4)\{[(4)(1.996)]/[(2)(1.996) + 4]\}}{1.308 \times 10^{-6}}$$

$$= 1.912 \times 10^6 \tag{d}$$

The relative roughness, $e/D_H = e/4R_H$ for the problem at hand is

$$\frac{e}{4R_H} = \frac{0.5 \times 10^{-3}}{4\{[(4)(1.996)]/[(2)(1.996) + 4]\}} = 0.0001251 \tag{e}$$

From the Moody diagram we get for f the new value of 0.0132. Going back to Eq. (a) we use the new f. We now get for h the value

$$h = 1.693 \text{ m}$$

Example 14.4. Do Example 14.1 using *friction-factor* formulas.

We first determine the flow zone for this case. Accordingly, we compute $V_* e/\nu$ to be able to employ Eq. (14.17) for the flow-zone determination. Noting Eq. (14.18), we have

$$\frac{V_* e}{\nu} = \frac{\left(\sqrt{gR_H S_0}\right)(e)}{\nu} = \frac{\sqrt{(32.2)(5.80)(0.0016)}\,(0.0032)}{1.217 \times 10^{-5}}$$

$$= 143$$

We are thus in the fully rough zone, so we use Eq. (14.23a) for f.

$$\frac{1}{\sqrt{f}} = 2.16 - 2\log\left(\frac{0.0032}{5.80}\right)$$

$$\therefore f = 0.01328$$

Now going to Eq. (14.16) we get

$$V = \left[\frac{8gS_0 R_H}{f}\right]^{1/2} = \left[\frac{(8)(32.2)(0.0016)(5.80)}{0.01328}\right]^{1/2} = 13.42 \text{ ft/s}$$

Hence,

$$Q = VA = (13.42)\left[\tfrac{1}{2}(\pi)(10^2) + (3)(20)\right] = 2913 \text{ ft}^3/\text{s}$$

We get here a result 16 percent less than from the Manning formula. The friction-factor computation above may be considered to be the more accurate.

In a later section, we will consider nonuniform (nonnormal) flow in channels. For the case where the changes of slope, friction, and area are not great (gradually varied flow), we will use at any location in the flow the same f, C, and n as we would have for *normal flow* at the *same depth* and *velocity* at that location. However, before embarking on such a study, we consider in the next section what constitutes the best cross section for a channel.

14.5 BEST HYDRAULIC SECTION

Let us next consider the question—what is the *best hydraulic channel* cross section by which we generally mean the cross section that for a *given Q* requires the *least* cross section area A? With this in mind, let us examine Eq. (14.10). Expressing R_H as A/P we have

$$Q = \frac{\kappa}{n}\left(\frac{A}{P}\right)^{2/3}\sqrt{S_0}\,A = \frac{\kappa}{n}\frac{\sqrt{S_0}}{P^{2/3}}A^{5/3}$$

Solving for A, we get

$$A = \left(\frac{Qn}{\kappa\sqrt{S_0}}\right)^{3/5}P^{2/5} = KP^{2/5} \tag{14.24}$$

For our purposes here, $[Qn/\kappa S_0^{1/2}]^{3/5}$ is a constant K, so we see that if A is *minimized then the wetted perimeter P will also be minimized*. Thus we see that the amount of excavation as determined by A as well as the amount of lining as represented by P will be minimized simultaneously for the best hydraulic cross section. This results in a minimum cost.

In the following example, we will find the best hydraulic cross section for a rectangular channel.

Example 14.5. Consider the rectangular cross section shown in Fig. 14.5. What is the relation between b and y for the best hydraulic cross section? For a flow of 20 m³/s at a speed of 5 m/s, what should the width b be for the best hydraulic section?

For this channel we have

$$A = yb \tag{a}$$

$$P = 2y + b \tag{b}$$

We wish to minimize A and P simultaneously. Solving for b in Eq. (*a*) and

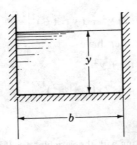

FIGURE 14.5
Rectangular cross section to be optimized.

substituting into Eq. (b) we have

$$P = 2y + \frac{A}{y}$$

Now replace A by $KP^{2/5}$ in accordance with Eq. (14.24) to have an equation with *only P and y as variables*:

$$P = 2y + \frac{KP^{2/5}}{y} \qquad (c)$$

Next, we take the derivative with respect to y and set $dP/dy = 0$ for the extremum.

$$\frac{dP}{dy} = 2 + \frac{K}{y}\frac{2}{5}P^{-3/5}\frac{dP}{dy} + KP^{2/5}(-1)\frac{1}{y^2} = 0$$

$$\therefore 2 - KP^{2/5}\frac{1}{y^2} = 0 \qquad (d)$$

We replace $P^{2/5}$ using Eq. (14.24) by A/K and A/K by yb/K. We then get

$$2 = K\left(\frac{yb}{K}\right)\left(\frac{1}{y^2}\right)$$

$$\therefore y = \frac{b}{2} \qquad (e)$$

Thus, we see that for the best hydraulic section the width b should be twice the height y.[6]

For flow of 20 m³/s at a speed of 5 m/s, we have for the best hydraulic section

$$Q = VA = V(b)(y)$$

$$20 = (5)(b)\left(\frac{b}{2}\right)$$

$$b = 2.83 \text{ m}$$

$$\therefore y = 1.414 \text{ m}$$

Example 14.6. Find the best hydraulic trapezoidal section (Fig. 14.6).
The area A and wetted perimeter P are, respectively,

$$A = ay + y^2 \cot \beta$$

$$P = a + 2\left[y^2 + (y \cot \beta)^2\right]^{1/2}$$

Let $\cot \beta$ be expressed as m. We then have

$$A = ay + y^2 m \qquad (a)$$

$$P = a + 2y(1 + m^2)^{1/2} \qquad (b)$$

---

[6] We could have solved this problem by working with the variable A rather than P.

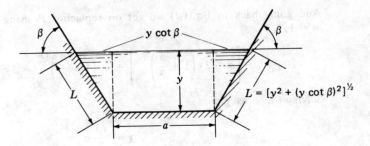

FIGURE 14.6
Trapezoidal section.

Next we solve for a in Eq. (b) and substitute into Eq. (a).

$$A = \left[P - 2y(1 + m^2)^{1/2} \right] y + y^2 m$$

We replace A using Eq. (14.24) so that we will work with P only.

$$KP^{2/5} - \left[P - 2y(1 + m^2)^{1/2} \right] y - y^2 m = 0 \qquad (c)$$

We have two variables in this problem that for a given a govern the perimeter. These variables are y and m. We extremize P by requiring $\partial P/\partial y = 0$ and $\partial P/\partial m = 0$. Thus, differentiating with respect to y, we get

$$K\left(\frac{2}{5}\right) P^{-3/5} \frac{\partial P}{\partial y} - y\frac{\partial P}{\partial y} - P + 4y(1 + m^2)^{1/2} - 2ym = 0$$

Letting $\partial P/\partial y = 0$ we have

$$-P + 4y(1 + m^2)^{1/2} - 2ym = 0 \qquad (d)$$

Now doing the same in Eq. (c), this time with the variable m, we get

$$K\frac{2}{5}P^{-3/5} \frac{\partial P}{\partial m} - y\frac{\partial P}{\partial m} + \frac{2y^2(\frac{1}{2})}{\sqrt{1 + m^2}}(2m) - y^2 = 0$$

Letting $\partial P/\partial m = 0$,

$$y^2\left(\frac{2m}{\sqrt{1 + m^2}} - 1 \right) = 0 \qquad (e)$$

From Eq. (e) we see that

$$\frac{2m}{\sqrt{1 + m^2}} = 1$$

$$\therefore m = \frac{1}{\sqrt{3}} \qquad (f)$$

And going back to Eq. (*d*) we get on replacing *P* using Eq. (*b*) and setting $m = 1/\sqrt{3}$,

$$-\left[a + 2y\left(1 + \frac{1}{3}\right)^{1/2}\right] + 4y\left(1 + \frac{1}{3}\right)^{1/2} - 2y\frac{1}{\sqrt{3}} = 0$$

Solving for *y*, we get

$$y = \frac{\sqrt{3}}{2}a \qquad\qquad (g)$$

For $m = 1/\sqrt{3}$ it follows that $\beta = 60°$. Furthermore, we may solve for *L* (see Fig. 14.6).

$$L = y\left[(1 + m^2)\right]^{1/2} = \frac{\sqrt{3}}{2}a\left[\left(1 + \tfrac{1}{3}\right)\right]^{1/2} = a$$

Thus with $\beta = 60°$ and with the wetted sides of the trapezoidal section equal, it can be concluded that the best hydraulic section is that of *half a hexagon*.

In closing this section, we point out that the semicircular cross section is the best hydraulic cross section of all for open channel flow.

14.6 GRAVITY WAVES

We now use the momentum considerations to study the characteristics of water waves formed by moving an object through the free surface of a liquid at a reasonably high rate.[7] These waves propagate away from the disturbance and have the nature of transverse waves, as studied in elementary physics, where the wave shape moves essentially normal to the motion of the fluid particles in the wave. If surface tension is negligible for such waves, we call them *gravity waves*.

The celerity (speed) and the wave shape of gravity waves have been under study for over 100 years by both hydraulicians and mathematicians. Because of the great complexity of the phenomenon, studies have been restricted generally to shallow-water waves, which means that the waves have a great wavelength λ relative to the depth *d*, as shown in Fig. 14.7, and to the so-called deep-water waves, where *d* is very large compared with any dimension of the wave. Shallow-water theory gives results which may be valid for canals, rivers, and beaches; deep-water theory finds applications in ocean waves.

In this text we can present only the crudest and most elementary examination of water waves. Accordingly, let us consider a solitary shallow-water wave moving with celerity *c* in a channel shown in Fig. 14.8. We are not concerned

---

[7]An object may be lowered through a free surface at a rate slow enough so as not to cause waves, as you may yourself easily verify.

FIGURE 14.7
Shallow-water waves.

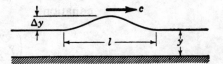

FIGURE 14.8
Wave celerity c.

here with the shape of the wave other than the condition that the distance l be large compared with y and that Δy be small compared with y, as shown in Fig. 14.8. Furthermore, we assume that the wave shape is essentially constant with respect to time. Under these restrictions, it is reasonable to consider that a hydrostatic-pressure distribution exists below the free surface. We can then form a steady flow by giving everything (including the boundary) a velocity $-c$, as explained in Chap. 4. A stationary control volume of unit thickness is then established to include a portion of the wave beginning at its peak and ending at the front of the wave, as we have shown in Fig. 14.9. It is clearly a control volume of finite size, so you will see that we compute certain average values and the detailed knowledge of the wave shape will not be needed.

Assuming uniform flow through the vertical faces of the control volume, we have for the *continuity* equation

$$cy = (c - \Delta V)(y + \Delta y) \tag{14.25}$$

where ΔV is the change in velocity due to the enlargement of the cross section of flow in the control volume. Solving for ΔV, we get

$$\Delta V = \frac{c\,\Delta y}{y + \Delta y} \tag{14.26}$$

Next, using hydrostatic-pressure variations on the vertical sides of the control volume and neglecting friction on the bottom resulting from velocity variation due to the presence of the wave, we then have for the *linear momentum*

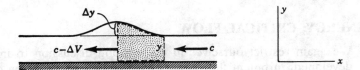

FIGURE 14.9
Stationary gravity wave with control volume.

equation

$$-\left(\gamma\frac{y}{2}\right)y + \left(\gamma\frac{y + \Delta y}{2}\right)(y + \Delta y) = \rho c^2 y - \rho c y(c - \Delta V) \quad (14.27)$$

Carrying out the various products in this equation and simplifying the results gives us

$$\gamma y \, \Delta y + \frac{\gamma(\Delta y)^2}{2} = \rho c y \, \Delta V$$

Dividing through by ρ and using Eq. (14.26) to replace ΔV on the right side of this equation, we get

$$g\left(y \, \Delta y + \frac{(\Delta y)^2}{2}\right) = \frac{c^2 y \, \Delta y}{y + \Delta y}$$

Canceling out Δy and multiplying through by $y + \Delta y$, we get

$$g\left(y + \frac{\Delta y}{2}\right)(y + \Delta y) = c^2 y \quad (14.28)$$

Carrying out the product on the left side of the equation and dropping the term involving $(\Delta y)^2$ as negligible, we then get, on solving for c,

$$c = \sqrt{gy}\left(1 + \frac{3}{2}\frac{\Delta y}{y}\right)^{1/2} \approx \sqrt{gy}\left(1 + \frac{3}{4}\frac{\Delta y}{y}\right) \quad (14.29)$$

wherein we have used the first two terms of a binomial expansion for the approximation on the right. When Δy is very small compared with y, we get the familiar result

$$\boxed{c = \sqrt{gy}} \quad (14.30)$$

which is the celerity of a wave of very small amplitude and of large wavelength compared with the depth. We will soon see that the small-wave concept plays a key role as we continue our study of free-surface flow in succeeding sections.

14.7 SPECIFIC ENERGY; CRITICAL FLOW

We again restrict ourselves in the ensuing discussion to incompressible, fully developed, turbulent flow along a channel where the slope of the bed is small. As before, hydrostatic-pressure distribution is assumed to prevail in the flow. Furthermore, the flow is taken as one-dimensional, where V is essentially parallel to the bed of the channel and is constant over a section normal to the

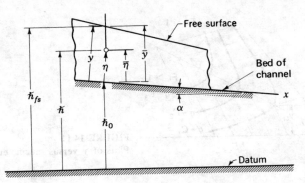

FIGURE 14.10
Channel flow.

bed of the channel. We have shown a portion of such a flow in Fig. 14.10. Note that the elevation *normal to the bed* of the channel to a fluid element is given by η and to the free surface is given by y. Sections normal to the bed are located by the position x along the bed. *Vertical* distance from an element to the bed is given as $\bar{\eta}$ and from free surface to the bed as $\bar{y}$.

Using $\hbar$ as the vertical elevation from a convenient horizontal datum to a fluid element, we have for H_D the total head at this position,

$$H_D = \frac{V^2}{2g} + \frac{p}{\gamma} + \hbar \tag{14.31}$$

From Fig. 14.10 one sees that we can replace $\hbar$ by $\hbar_0 + \eta \cos \alpha$, which for a small slope of the channel bed may be given as $\hbar_0 + \eta$. Furthermore, we may evaluate the pressure p from the hydrostatic variation in the following manner considering $\cos \alpha \approx 1$:

$$p = \gamma(y - \eta) \tag{14.32}$$

We now replace p in Eq. (14.31) using the result above and $\hbar$ by $(\hbar_0 + \eta)$ to get

$$H_D = \frac{V^2}{2g} + (y - \eta) + (\hbar_0 + \eta) \tag{14.33}$$

$$\therefore H_D = \frac{V^2}{2g} + y + \hbar_0$$

We see that the head H_D is *constant for all particles at each section normal to the bed.*

We now define the *specific energy* E_{sp} in the following manner:

$$E_{sp} = H_D - \hbar_0 \tag{14.34}$$

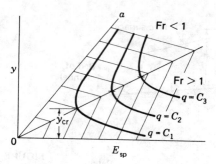

FIGURE 14.11
Plots of y versus specific energy for various flows q.

We see that the specific energy is actually *the head with respect to the bed of the channel as a datum*. Substituting for H_D using Eq. (14.34) into Eq. (14.33) we then have for E_{sp}

$$E_{sp} = \frac{V^2}{2g} + y \qquad (14.35)$$

Like H_D, we can say that E_{sp} is *constant for all fluid elements at any section of flow normal to the bed of the channel*.

Now let us examine the specific energy for the case of a flow in a *rectangular* channel, wherein q is the volume flow per unit width of the channel. Clearly then

$$q = Vy \qquad (14.36)$$

The specific energy then becomes for this flow

$$E_{sp} = \frac{q^2}{2y^2g} + y \qquad (14.37)$$

Let us consider the situation where q is maintained constant and where E_{sp} is a variable. It is then seen from Eq. (14.37) that for any particular value of E_{sp}, there will be a cubic equation in y. One of the roots for y will be negative, so there are either two possible depths of flow y for a given E_{sp} or none, as shown in Fig. 14.11, where we have plotted y versus E_{sp} for different values of q.[8] As $q \to 0$, Eq. (14.37) becomes a straight line, $E_{sp} = y$ and is shown as $0a$ in the diagram. Note that for each q, there is a point of *minimum* specific energy. The depth for this point in the curves is indicated as y_{cr}—the *critical depth* for a particular flow q. We can find this depth easily by taking the partial derivative of

---

[8]Note that for each curve there is one value of E_{sp} for which the two roots become identical.

E_{sp} with respect to y and setting it equal to zero. Thus

$$\frac{\partial E_{sp}}{\partial y} = 0 = -\frac{q^2}{gy_{cr}^3} + 1 \qquad (14.38)$$

Therefore

$$y_{cr} = \left(\frac{q^2}{g}\right)^{1/3} \qquad (14.39)$$

And the velocity for this flow condition, V_{cr}, is then easily determined by substituting for q in the equation above from Eq. (14.36). We then get

$$V_{cr} = \sqrt{gy_{cr}} \qquad (14.40)$$

We now see that the critical speed V_{cr}, is that of the *celerity of a small gravity wave in a shallow liquid*. We may rewrite the equation above as

$$\frac{V_{cr}^2}{gy_{cr}} = 1 \qquad (14.41)$$

You may recall from Chap. 8 that the expression on the left side of Eq. (14.41) was defined as the Froude number (Fr) with the depth y as the length dimension. In channel flow, this expression is taken as the Froude number *squared*. Furthermore, we note from above that at the critical condition the Froude number is *unity*. We will see that the critical depth in channel flow plays the same role as the critical area in a convergent-divergent nozzle (see Fig. 11.10) in compressible flow. The Mach number in the latter is analogous to the Froude number for the former. In the nozzle, the fluid speed equals the celerity of a small pressure wave, thus giving $M = 1$ at the throat area (smallest area). In the channel, the fluid at the critical depth moves at the same speed as the celerity of a small gravity wave, giving a Froude number of unity as per Eq. (14.41). In a nozzle a pressure change downstream of the throat area cannot affect the flow upstream of the throat when $M = 1$ at the throat. This results from the fact that the fluid in the throat moves downstream as fast as a pressure disturbance can move upstream. Similarly, when the critical depth is attained in a channel surface flow, changes downstream of the critical depth cannot be transmitted upstream of the critical depth because the fluid moves downstream as fast as the surface waves move upstream at the critical section.

Also we can see that the specific energy curve for a given value of q is analogous to a Fanno line (see Fig. 11.17). But there are distinctly different effects of friction between the compressible flow as depicted by the Fanno line and the specific energy plot. Because of the second law the entropy must increase as a result of friction in the direction of flow for the adiabatic flow of the Fanno line. Thus the Mach number should always tend toward $M = 1$ as a result of friction. But in *normal* channel flow which must include friction we have a constant specific energy in the direction of flow and hence can remain at one point on the specific energy curve. If *normal* flow does *not* exist in the channel then the flow will tend in the direction of flow toward a normal flow and not necessarily to the critical point of the specific energy curve. Thus the effects of friction differ between constant-area compressible adiabatic flow and channel flow.

For critical conditions, we get from Eq. (14.35)

$$(E_{sp})_{min} = \frac{V_{cr}^2}{2g} + y_{cr} \tag{14.42}$$

Replacing V_{cr} using Eq. (14.40) and solving for y_{cr}, we get

$$y_{cr} = \tfrac{2}{3}(E_{sp})_{min} \tag{14.43}$$

If we have *normal* flow, which is at the same time *critical* flow, then the slope of the channel bed equals the slope of the free surface and, furthermore, the Chézy formula applies. Thus we have

$$V = C\sqrt{R_H S_{cr}} \tag{14.44}$$

where S_{cr} is the channel slope for normal, critical flow. For *wide* rectangular channels we can take $R_H = y_{cr}b/(2y_{cr} + b) \approx y_{cr}$ in Eq. (14.44). Next multiplying Eq. (14.44) by the area of a unit width of the channel, namely, $(1)(y_{cr})$, we get q for the left side of the equation. We can then say

$$q = C\sqrt{y_{cr}S_{cr}}\,(y_{cr}) = Cy_{cr}^{3/2}S_{cr}^{1/2} \tag{14.45}$$

Now we solve for q in Eq. (14.39) for critical flow and substitute into Eq. (14.45).

$$\left[y_{cr}^3 g\right]^{1/2} = Cy_{cr}^{3/2}S_{cr}^{1/2}$$

Solving for S_{cr}, we get

$$S_{cr} = \left(\frac{g}{C^2}\right) \tag{14.46}$$

Next, using Eq. (14.6) to replace C in terms of the friction factor f we get

$$S_{cr} = \frac{f}{8} \tag{14.47}$$

Knowing f from Moody's chart using the hydraulic diameter in the Reynolds number, we can then get the slope for critical, normal flow. Note, finally, that equations following Eq. (14.35) except Eq. (14.44) apply only to rectangular channels.

Example 14.7. A wide rectangular channel has a flow q of 70 ft^3/(s)(ft). What is the critical depth? If the depth is 5 ft, at a section what is the Froude number there? Next, if we were to have normal, critical flow, what must the slope of the channel be? The channel is made from finished cement.

We first find the critical depth for the channel. Thus from Eq. (14.39) we get

$$y_{cr} = \left(\frac{q^2}{g} \right)^{1/3} = \left(\frac{70^2}{32.2} \right)^{1/3} = 5.339 \text{ ft} \qquad (a)$$

Next, we compute the Froude number for a depth of 5 ft and a flow q of 70 ft^3/(s)(ft). Considering Eq. (14.41) we see that

$$\text{Fr} = \frac{V}{\sqrt{gy}} = \frac{(70)/[(1)(5)]}{\sqrt{(g)(5)}} = 1.1034 \qquad (b)$$

Note that Fr is larger than unity, so a wave cannot propagate upstream of this region. Now for normal, critical flow (Fr = 1), we go to Eq. (14.47) for the desired slope S_{cr}. First, we see from Table 14.1 that $e = 0.0032$ ft. The hydraulic diameter is $(4)(R_H) = (4)(5.339) = 21.36$ ft. Finally, the Reynolds number is

$$\text{Re}_H = \frac{(70/5.339)(21.36)}{1.217 \times 10^{-5}} = 2.30 \times 10^7$$

With $e/D_H = 0.0032/21.36 = 0.0001498$, we get from the Moody chart the value $f = 0.0129$. Now using Eq. (14.47) we get

$$S_{cr} = \frac{0.0129}{8} = 0.001612 \qquad (c)$$

Note again that after defining E_{sp} in Eq. (14.35), with the exception of Eq. (14.44) we have been considering only *rectangular* channels in arriving at Eqs. (14.36) to (14.47). We now set forth formulations valid for a *prismatic channel* with *arbitrary* cross sections (see Fig. 14.12). To do this, we start with Eq.

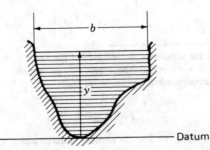

FIGURE 14.12
Channel with arbitrary cross section.

(14.35) and express it as follows:

$$E_{sp} = \frac{V^2}{2g} + y = \frac{Q^2}{2gA^2} + y \qquad (14.48)$$

where Q is the volume flow for the entire section. We now extremize E_{sp} for a given Q to get the critical depth but in so doing we must realize that A will be a function of y. Thus we have

$$\frac{dE_{sp}}{dy} = \left(\frac{Q^2}{2g}\right)(-2)\frac{1}{A^3}\frac{dA}{dy} + 1 = 0$$

Observing Fig. 14.12 we note that dA is an infinitesimal change of the cross-sectional area and is given as $b\,dy$, where b is the width of the free surface. We then get, on replacing dA/dy by b in the preceding equation,

$$\boxed{\frac{bQ^2}{gA_{cr}^3} = 1} \qquad (14.49)$$

where A_{cr} is now the critical area corresponding to Fr = 1.[9] To find the critical depth in a complex cross section we may plot the value bQ^2/gA^3 for a given Q, versus y. When the value becomes equal to 1, we have arrived at the critical depth y_{cr} for that Q. A formula for the minimum specific energy $(E_{sp})_{min}$ can be found by solving for Q^2 in Eq. (14.49), and substituting into Eq. (14.48) we get

$$\boxed{(E_{sp})_{min} = \frac{A_{cr}}{2b} + y_{cr}} \qquad (14.50)$$

To have *normal*, *critical* flow we use the Chézy formula to get Q. Thus from Eq. (14.44) we have

$$Q = C\sqrt{R_H S_{cr}}\,A_{cr}$$

From Eq. (14.49), solve for Q and substitute above:

$$\left[\frac{gA_{cr}^3}{b}\right]^{1/2} = C\sqrt{R_H S_{cr}}\,A_{cr}$$

---

[9]Replacing Q by VA in the expression on the left side of Eq. (14.49) we get the expression V^2b/gA. For rectangular channels this expression becomes V^2/gy, which is the square of the Froude number. In general cross sections, we accordingly say that

$$\frac{V^2b}{gA} = \frac{Q^2b}{gA^3} = \text{Fr}^2$$

Thus in Eq. (13.49) we are setting the Froude number squared equal to unity for critical flow.

Using Eq. (14.6) to replace C, we get on squaring terms

$$\frac{gA_{cr}^3}{b} = \left(\frac{8g}{f}\right)(R_H S_{cr}) A_{cr}^2$$

Solving for S_{cr} and noting that $A_{cr}(R_H) = P_{cr}$, the wetted perimeter, we get

$$S_{cr} = \frac{A_{cr}f}{8bR_H} = \frac{fP_{cr}}{8b} \qquad (14.51)$$

We thus have the formulations needed for a general cross section; we will now illustrate their use.

Example 14.8. A brickwork channel has the shape shown in Fig. 14.4, where the radius is 5 m. For a volume flow of 300 m³/s, what is the critical depth? What is the slope for normal, critical flow? Water is at 5°C.

   We will assume first that Q is large enough so that $b = 10$ m. We go to Eq. (14.49). In this case, we can solve for y_{cr} directly. Thus

$$\frac{bQ^2}{gA_{cr}^3} = \frac{(10)(300^2)}{g\left[(\pi 10^2/8) + (10)(y_{cr} - 5)\right]^3} = 1 \qquad (a)$$

We get

$$y_{cr} = 5.583 \text{ m} \qquad (b)$$

Our assumption about b is thus correct.

   To get the critical slope for normal critical flow, we must find f. From Table 14.1 we get $e = 3.7$ mm. Also we have for R_H

$$R_H = \frac{\left[(\pi)(10^2)\right]/8 + (0.583)(10)}{(\pi(10)/2) + (2)(0.583)} = 2.673 \text{ m} \qquad (c)$$

Hence $e/(4R_H) = 0.000346$. Also for Re_H we get

$$\text{Re}_H = \frac{(300)\left\{\left[(\pi)(10^2)\right]/8 + (0.583)(10)\right\}^{-1}\left[(4)(2.673)\right]}{1.519 \times 10^{-6}}$$

$$= 4.682 \times 10^7 \qquad (d)$$

We get for f from Moody:

$$f = 0.0154 \qquad (e)$$

Going to Eq. (14.51) we get

$$S_{cr} = \frac{(0.0154)\left[(\pi 10/2) + (2)(0.583)\right]}{(8)(10)}$$

$$= 0.00325 \qquad (f)$$

As a final step in this section we return to Eq. (14.37) (valid only for rectangular channels) and solve for q.

$$q = \sqrt{2g}\, y \left(E_{sp} - y \right)^{1/2} \qquad (14.52)$$

We wish to find for a fixed value of E_{sp} the depth y for a *maximum* value of q. Taking the partial derivative with respect to y and setting it equal to zero, we get for Eq. (14.52)

$$\frac{\partial q}{\partial y} = 0 = \sqrt{2g}\left(E_{sp} - y \right)^{1/2} - \sqrt{2g}\, y \frac{1}{2}\left(E_{sp} - y \right)^{-1/2} \qquad (14.53)$$

Solving for y, we get

$$y = \tfrac{2}{3} E_{sp} \qquad (14.54)$$

We thus see from Eq. (14.43) that the depth is the *critical* depth. Thus for any *given specific energy, the maximum flow occurs at the critical depth.*

14.8   VARIED FLOW IN SHORT RECTANGULAR CHANNELS

We now consider *steady* flow over *short* distances in *rectangular* channels where, unlike normal flow, the depth of the flow will be a function of x. The slope of the channel bed, as in the previous section, will be *small* but may vary as one moves along the bed (see Fig. 14.13). We consider a one-dimensional flow, and because we are restricting ourselves to *small distance*, we will *neglect* friction and turbulence effects. Thus, we see that the *total head H_D must be constant*, since there can be no dissipation of mechanical energy. Although the streamlines are no longer straight as in the uniform flow, we will still consider that *hydrostatic* pressure persists normal to the bed in the flow, as we have already done in the previous section.

Let us now consider Figs. 14.13 and 14.14 where we will assume that the critical depth has been reached at position A. What can one expect to the right of A where the value of $\hbar_0$ is decreasing? If we do not change H_D of the flow upstream of A, we will maintain a constant flow rate q per unit width when changes are made downstream. This is due to the critical flow at A which does not allow gravity waves to propagate upstream. Therefore, we will remain on one of the curves in Fig. 14.11 as we move downstream of A. Note that with $\hbar_0$ decreasing, E_{sp} will of necessity increase [see Eq. (14.34)] and we must move to the right of the critical point in our y-versus-E_{sp} curve. For any specific value of

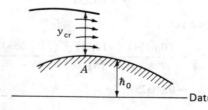

FIGURE 14.13
Flow showing critical depth.

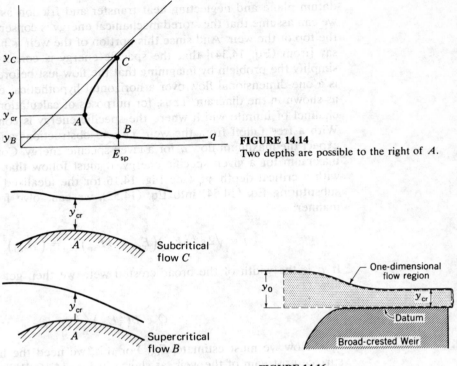

FIGURE 14.14
Two depths are possible to the right of A.

FIGURE 14.15
Two possible flows after critical section A.

FIGURE 14.16
Broad-crested weir (inviscid flow).

h_0 to the right of A, we see from Fig. 14.14 that two depths y_B and y_C are now possible for the given conditions. (This is like the convergent-divergent nozzle, where for given initial conditions there can be either subsonic or supersonic flow downstream of the throat. The kind of flow in the nozzle depends on the conditions downstream in the plenum chamber.) The flow for depth y_B in Fig. 14.14 obviously is faster than at y_{cr} and exceeds the celerity of a gravity wave (Fr > 1). It is called *shooting*, or *supercritical*, flow and obviously corresponds to supersonic flow in the nozzle. The flow at C is slower than that of critical flow (Fr < 1) and we call this flow *tranquil*, or *subcritical*, flow. In Fig. 14.15, we show both possibilities. The particular flow attained depends on the controls downstream of A.

As an example, consider the case of the *broad-crested weir*,[10] which is shown diagrammatically in Fig. 14.16. Using the top surface of the weir as a

---

[10]A weir is a cross section inserted in a channel through which there is fluid flow. The purpose of the weir is to permit calculation of fluid flow in terms of a height measurement of the free surface upstream of the weir. A more detailed discussion of the weir is presented in Appendix A.I.

datum plane and neglecting heat transfer and friction as in the previous flow, we can assume that the stored mechanical energy is conserved for the flow along the top of the weir. And since this portion of the weir is horizontal, we can also say [from (Eq. 14.34)] that the specific energy is conserved. We may further simplify the problem by imagining that the flow just before the crest of the weir is a one-dimensional flow over a horizontal hypothetical extension of the crest, as shown in the diagram. Thus, for purposes of calculation, we have a flow in a channel of infinite width where the specific energy is constant along the flow. With a free falloff from the weir, i.e., no obstructions and no friction, we can expect the maximum flow q for a given specific energy. Consequently, since q is maximum for a given specific energy, it must follow that we have critical flow with a critical depth y_{cr} (see Fig. 14.16 for the idealized frictionless case). By substituting Eq. (14.54) into Eq. (14.52), we can solve for q in the following manner:

$$q = \sqrt{2g}\left(\tfrac{2}{3}E_{sp}\right)\left(E_{sp} - \tfrac{2}{3}E_{sp}\right)^{1/2} = \left(\tfrac{2}{3}E_{sp}\right)^{3/2}\sqrt{g} \qquad (14.55)$$

If b is the width of the broad-crested weir, we then get for the total volume flow Q

$$Q = b\left(\tfrac{2}{3}E_{sp}\right)^{3/2}\sqrt{g} \qquad (14.56)$$

Now we must estimate E_{sp}. For this, we need the height y_0 of the free surface upstream of the weir, as shown in Fig. 14.16. We may then reason that the total head H_D of fluid particles in the channel can be found by considering fluid particles at the free surface far upstream of the weir. We can neglect the velocity head; and if we use gage pressures, as is the customary procedure here, it is clear from Eq. (14.31) that the total head relative to the datum of the problem is y_0. From Eq. (14.34) with $h_0 = 0$, we may use this value as an approximation for our specific energy. Accordingly, substituting into Eq. (14.56), we get for Q

$$Q = b\left(\tfrac{2}{3}y_0\right)^{3/2}\sqrt{g} \qquad (14.57)$$

We have completely neglected friction in arriving at this formulation. Actually friction causes a decrease in the specific energy along the flow. However, for a given q along the crest, the specific energy cannot become less than the specific energy corresponding to the critical depth for that value of q (see Fig. 14.14). The free-surface profile then arranges itself so that at the end of the crest the critical velocity has just been reached (see Fig. 14.17) (much like the one-dimensional compressible flow at the choked condition in constant-area ducts). Equation (14.56) may still be used for the flow, but E_{sp} will be less than y_0, so that Q will be less than the ideal case.

In Fig. 14.18 we show what can happen when the channel bed drops for various flows. You are urged to reason each case out yourself. Note in the

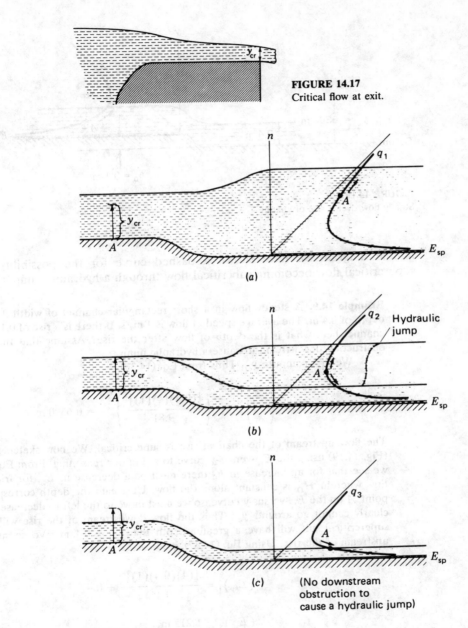

FIGURE 14.17
Critical flow at exit.

FIGURE 14.18
Various flow regimes.

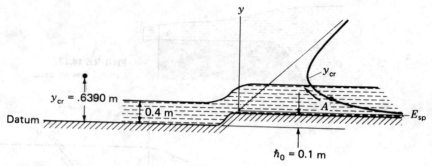

FIGURE 14.19
Steady flow over a rise of channel.

middle diagram we have shown a dashed curve for the possibility of the supercritical flow becoming subcritical flow through a hydraulic jump.

Example 14.9. A steady flow in a short rectangular channel of width 3 m has a depth of 0.4 m. The average speed of flow is 4 m/s. If there is a rise of 0.1 m in the channel bed, what is the depth of flow after the rise? Assume that there is no obstruction downstream to cause a hydraulic jump.

We first compute y_{cr}. Thus from Eq. (14.39),

$$y_{cr} = \left(\frac{q^2}{g}\right)^{1/3} = \left\{\frac{[(0.4)(4)(1)]^2}{9.81}\right\}^{1/3} = 0.6390 \text{ m} \qquad (a)$$

The flow upstream of the channel rise is supercritical. We now sketch this flow (Fig. 14.19) using the E_{sp}-versus-y curve to aid in our reasoning. From Eq. (14.34) we see that for an increase in $\hbar_0$ there must be a decrease in E_{sp} for frictionless flow wherein H_D is constant along the flow. Upstream the depth corresponds to point A on the E_{sp}-versus-y curve, so we must move to the left to decrease E_{sp} but clearly cannot go around y_{cr}. Thus the flow downstream of the rise will still be supercritical but will have a greater depth than $y_1 = 0.4$ m. We compute E_{sp} upstream of the rise. Using Eq. (14.37), we have

$$(E_{sp})_1 = \frac{[(4)(0.4)(1)]^2}{2(0.4^2)g} + 0.4$$

$$\therefore (E_{sp})_1 = 1.215 \text{ m} \qquad (b)$$

Examining Eq. (14.34), using the lower level of the channel as the datum [that is, $(\hbar_0)_1 = 0$] we see that $H_D = (E_{sp})_1 = 1.215$ m. With no friction, H_D is constant, so that from Eq. (14.34) we can compute $(E_{sp})_2$. Thus

$$(E_{sp})_2 = H_D - (\hbar_0)_2 = 1.215 - 0.1 = 1.115 \text{ m} \qquad (c)$$

Now using Eq. (14.37) again, we have

$$1.115 = \frac{[(4)(0.4)(1)]^2}{2 y_2^2 g} + y_2$$

$$\therefore 1.115 = \frac{0.1305}{y_2^2} + y_2$$

(d)

Solving by trial and error, realizing that y_2 must be in the range from 0.4 m to 0.6390 m, we get

$$y_2 = 0.440 \text{ m}$$

(e)

We have thus seen in this section how we may ascertain certain depths in short rectangular channels by using the first law of thermodynamics. We now consider *long* channels, where we must take friction and turbulence into account.

*14.9   GRADUALLY VARIED FLOW OVER LONG CHANNELS

We have thus far considered steady *normal* flow in prismatic channels, where we took friction and turbulence into account. Recall that the depth is constant for such flows. We then considered steady flows in *nonprismatic* rectangular channels over *short* distances. Here we completely neglected friction and turbulence. We consider next steady flow in nonprismatic channels over *long* distances. Because of the long distance we must take friction and turbulence into account, just as we did for flow in long pipes, since friction and turbulence will now definitely affect the flow. We will restrict ourselves to cases where the bottom slope, the roughness, and cross-sectional area change very *slowly along* the channel. For these reasons, we call such flows *gradually varied* flows.

Accordingly, we consider an infinitesimal control volume in a steady, nonuniform flow in Fig. 14.20.[11] We next express the *first law of thermodynamics* for a steady, one-dimensional flow in this control volume. Using Eq. (14.33) for the total head H_D we have, on using gage pressures,

$$\frac{V^2}{2g} + y + \hbar_0 = \left[\frac{V^2}{2g} + d\left(\frac{V^2}{2g}\right)\right] + (y + dy) + (\hbar_0 + d\hbar_0) + d(H_l)$$

(14.58)

---

[11]Note that we are measuring y in the vertical direction here rather than in a direction normal to the bed, as in previous calculations. The velocity is then assumed uniform over a *vertical* section rather than a section normal to the bed. This is acceptable for *small slopes* S_0. We restrict ourselves to this case.

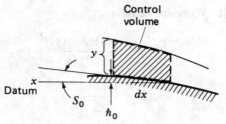

FIGURE 14.20
Infinitesimal control volume for gradually varied flow.

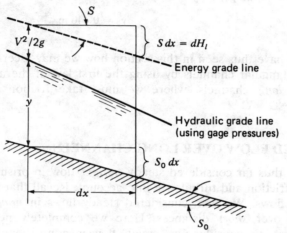

FIGURE 14.21
Gradually varied flow.

where H_l is the *head loss* given as

$$H_l \equiv \left[\frac{1}{g}\left(u_2 - u_1 - \frac{dQ}{dm}\right)\right] \equiv \frac{h_l}{g}$$

Canceling terms in Eq. (14.58) we get

$$d\left(\frac{V^2}{2g}\right) + dy + d\hbar_0 + dH_l = 0 \tag{14.59}$$

Note that $d\hbar_0$ can be expressed as $-S_0\,dx$. Furthermore, the loss in total head H_D is the decrease in elevation of the *energy grade line*[12] (see Fig. 14.21) so that dH_l can be replaced by Sdx, where S is the slope of the energy grade

---

[12]See Sec. 9.11 and particularly Fig. 9.28 to refresh your memory on the energy grade line and to realize that drops in the energy grade line are due to head losses.

line. Replacing $d\hbar_0$ and dH_l as indicated, we get for Eq. (14.59) on dividing by dx

$$\frac{d}{dx}\left(\frac{V^2}{2g}\right) + \frac{dy}{dx} - S_0 + S = 0 \tag{14.60}$$

Now we consider the *continuity* equation for the control volume (Fig. 14.20). Noting that we have steady flow, we can say

$$Q = VA$$

$$\therefore \frac{dQ}{dx} = 0 = V\frac{dA}{dx} + A\frac{dV}{dx} \tag{14.61}$$

The expression dA can be replaced by $b\,dy$, where b, we recall, is the width of the free surface. Solving for dV/dx, we have

$$\frac{dV}{dx} = -\frac{Vb\,dy}{A\,dx} \tag{14.62}$$

Hence, we can say that for the first term in Eq. (14.60)

$$\frac{d}{dx}\left(\frac{V^2}{2g}\right) = \frac{V}{g}\frac{dV}{dx} = -\frac{V^2b}{Ag}\frac{dy}{dx} \tag{14.63}$$

where we have used Eq. (14.62) in the last step. Now using this result in Eq. (14.60) and solving for dy/dx, we get

$$\frac{dy}{dx} = \frac{S_0 - S}{1 - V^2b/Ag} \tag{14.64}$$

The expression V^2b/Ag is dimensionless and is considered in channel flows to be the square of the Froude number Fr, as noted earlier. We can then give dy/dx as

$$\boxed{\frac{dy}{dx} = \left(\frac{S_0 - S}{1 - \text{Fr}^2}\right)} \tag{14.65}$$

This formula is useful for establishing the sign of the slope of the free surface. Clearly it depends on the Froude number (i.e., whether the flow is subcritical or supercritical) and the relative values of the slope of the bed S_0 and the slope of the energy grade line S.

At this point, we will make a key assumption about the slope of the energy grade line S at any position x along the channel. We will say that this slope S equals the *bed slope* for the *same* channel depth and for the *same* volumetric flow Q, but where the flow is *normal* flow. In essence, we are saying that friction effects in our channel correspond to the friction present in the same channel for normal flow at the same Q. Going back to Eq. (14.11), we can replace S in Eq. (14.65) by $(n/\kappa)^2(1/R_H^{4/3})V^2$. Thus we have

$$\frac{dy}{dx} = \left[\frac{S_0 - (n/\kappa)^2(V^2/R_H^{4/3})}{1 - V^2b/gA} \right] \qquad (14.66)$$

We can replace V by Q/A to achieve the following formula:

$$\frac{dy}{dx} = \frac{S_0 - (n/\kappa)^2\left[Q^2/(R_H^{4/3}A^2)\right]}{1 - Q^2b/gA^3} \qquad (14.67)$$

Realizing that b, R_H, and A are functions of y and x while S_0 is a function of x, we can consider that we have an ordinary differential equation of the form

$$\frac{dy}{dx} = f(x, y) \qquad (14.68)$$

(In Chap. 16, we have carefully examined the numerical integration of this type of equation using a computer.)

At this time, we present a simple procedure to estimate the depth y-versus-x along a gradually varied flow. We express Eq. (14.67) in a *finite difference* form as follows[13]:

$$\Delta L = \left\{ \frac{1 - Q^2b/gA^3}{S_0 - (n/\kappa)^2\left[Q^2/(R_H^{4/3}A^2)\right]} \right\} \Delta y \qquad (14.69)$$

where ΔL is the length taken along the channel bed (valid for small S_0) and Δy is the change in elevation of the free surface corresponding to a change in position ΔL along the channel. We can use Eq. (14.69) in a number of ways. Let us *assume that all conditions are known at section 1*. We now discuss various problems to be solved.

1. We wish to know the distance downstream to a position where the depth has a known value y_2. If S_0 does not vary much and if y_2 is close to y_1, then we may employ Eq. (14.69) *once* to determine ΔL by inserting the prescribed value Δy as well as the known values of b, R_H, A, and S_0 corresponding to section 1. A more accurate procedure is to calculate the values of b, R_H, and A at section 2 since we know y_2 and Q and to then get the *linear average* of b, R_H, and A between sections 1 and 2. Using these values and the average value of S_0 we can then go to Eq. (14.69) to determine ΔL. We will ask you to solve such problems as exercises.

---

[13]Note that $Q^2b/gA^3 = \text{Fr}^2$ in Eqs. (14.67) and (14.69).

2. Another problem is to stipulate a short distance ΔL downstream along the bed and to ask for the depth there. This problem can be solved by using successive trial values of y_2 and inserting corresponding values of Δy into Eq. (14.69) with the known values of R_H, A, b, and S_0 at section 1 until the desired distance ΔL is reached on the right side of the equation. A more-accurate procedure is to compute at section 2 the values R_H, b, and A for each trial value y_2 and to then employ the average values of these quantities between sections 1 and 2 as well as the average slope S_0. We continue with successive choices of y_2 until the correct ΔL is found from Eq. (14.69). Again we will not illustrate this simple approach but will leave it to the homework problems.

3. Finally, we come to the case where the *free-surface profile* over a longer distance ΔL is desired, or where a more-accurate calculation of ΔL is desired for a given y_2 (case 1) or finally where a more-accurate calculation of y_2 for a given ΔL (case 2) is desired. In all such cases, choose very *small* increments Δy to work with. The smaller the value of Δy, the more accurate will be your results albeit with considerable increase in work. Now for the first Δy going from section 1 to a section 2 which we will consider terminates the first Δy, proceed as described in case 1, computing $(\Delta L)_{1\text{-}2}$ between sections 1 and 2. Do the same for the next Δy, going from section 2 to 3 finding next $(\Delta L)_{2\text{-}3}$. Proceed until all the increments have been used. We can then plot y versus L using the computed values of Δy and ΔL at each section to form the desired profile. If we are to solve case 1 via this more-detailed approach, we simply use enough small Δy's to reach the desired final depth. The sum of the small ΔL's then gives us the desired overall $(\Delta L)_{\text{total}}$ for the stipulated final depth. As for case 2, we simply carry out the calculations using small successive Δy's until the sum of the ΔL's equals the stipulated overall $(\Delta L)_{\text{total}}$.

We now illustrate such a procedure.

Example 14.10. A flow of 35 m^3/s flows along a trapezoidal concrete channel where (see Fig. 14.6) the base a is 4 m and β is 45°. If at section 1, the depth of the flow is 3 m, what is the water surface profile up to a distance 600 m downstream. The channel is finished concrete and has a constant slope S_0 of 0.001.

We start with $y_1 = 3$ m. At this section we know that

$$A_1 = (3)(4) + \frac{1}{2}(3)(3)(2) = 21 \text{ m}^2$$

$$(R_H)_1 = \frac{A_1}{P_1} = \frac{21}{4 + (2)(3/0.707)} = 1.6818 \text{ m} \qquad (a)$$

$$b_1 = 4 + (2)(3) = 10 \text{ m}$$

We take $n = 0.012$ and $\kappa = 1.00$ and let $y_2 = 3.1$ m. Now we compute A_2, $(R_H)_2$, and b_2.[14]

$$A_2 = (3.1)(4) + \frac{1}{2}(3.1)(3.1)(2) = 22.01 \text{ m}^2$$

$$(R_H)_2 = \frac{22.01}{4 + 2(3.1/0.707)} = 1.7236 \text{ m} \qquad (b)$$

$$b_2 = 4 + (2)(3.1) = 10.20 \text{ m}$$

In the first interval, the average values of A, R_H, and b are

$$(A_{1\text{-}2})_{av} = 21.505 \text{ m}^2$$

$$[(R_H)_{1\text{-}2}]_{av} = 1.7027 \qquad (c)$$

$$(b_{1\text{-}2})_{av} = 10.10 \text{ m}$$

Now we go to Eq. (14.69) to compute $(\Delta L)_{1\text{-}2}$.

$$(\Delta L)_{1\text{-}2} = \left\{ \frac{1 - (35^2)(10.10)\big/[(9.81)(21.505^3)]}{0.001 - (0.012/1)^2(35^2)\big/[(1.7027^{4/3})(21.505^2)]} \right\}(0.1)$$

$$= 107.4 \text{ m}$$

We thus have two points on the free-surface profile. Next we compute A_3, R_{H3}, and b_3 for $y = 3.2$ m. Using Eqs. (b) we now find the average values of these quantities in the interval 2-3. For instance,

$$(A_{2\text{-}3})_{av} = \tfrac{1}{2}\{(22.01) + [(3.2)(4) + (3.2^2)]\} = 22.525 \text{ m}^2$$

We then proceed as indicated for the first interval.

The following table gives the results using six sections so that $L_{total} \approx 600$ m.

Section	y, m	Δy, m	ΔL, m	L_{total}, m
1	3.0	0.1	0	0
2	3.1	0.1	107.4	107.4
3	3.2	0.1	106.9	214.3
4	3.3	0.1	105.6	319.9
5	3.4	0.1	104.7	424.6
6	3.5	0.1	104.1	528.7
7	3.6	0.1	103.8	632.5

We can now plot y versus L, that is, the second and last columns, starting with $y = 3$ for $L = 0$ m and going on to $y = 3.6$ m for $L = 632.5$ m. A smooth curve through these parts gives the approximate desired profile.

Before proceeding further, we note that when we have critical flow we see from Eq. (14.49) that the numerator in Eq. (14.69) is zero. This indicates that

---

[14]We are assuming that the depth y relative to the channel is increasing. If we get a positive result for ΔL, we know that our assumption is correct. If ΔL is negative, then the depth must be decreasing along the channel flow. Use y_2 less than 3 m in that case. In the next section we show diagrams that will permit the quick decision about the general shape of the profile.

$\Delta L = 0$ for finite changes in depth, which has no meaning at this point in our discussion. We can conclude that Eq. (14.69) is not meaningful near critical-flow conditions. The basic assumptions of gradual-flow variation do not apply near critical flow, since approach to or from critical flow is abrupt. The approach to normal flow parenthetically is gradual.

Let us next consider the case of a *very wide* channel where the slope S_0 of the channel is *zero*. The hydraulic radius now becomes y of the channel. Equation (14.67) now can be expressed in the following integral form:

$$x = -\int_{y_1}^{y} \frac{1 - \left[(Q/by)^2/gy\right]}{(n/\kappa)^2(1/y^{4/3})(Q/by)^2} \, dy \tag{14.70}$$

Using $Q/b = q$, where q is the flow per unit width, Eq. (14.70) becomes

$$x = -\int_{y_1}^{y} \frac{1 - (q^2/gy^3)}{(n/\kappa)^2(q^2/y^{10/3})} \, dy$$

This integral may be readily carried out. We get

$$x = -\left(\frac{\kappa}{n}\right)^2\left(\frac{1}{q^2}y^{13/3}\frac{3}{13} - \frac{1}{g}y^{4/3}\frac{3}{4}\right)\Bigg|_{y_1}^{y}$$

$$\therefore \boxed{x = \left(\frac{\kappa}{n}\right)^2\left[\frac{3}{4g}\left(y^{4/3} - y_1^{4/3}\right) - \frac{3}{13q^2}\left(y^{13/3} - y_1^{13/3}\right)\right]} \tag{14.71}$$

We now have from Eq. (14.71) a depth y at any position x, so we have the profile as a continuous function.

★14.10 CLASSIFICATION OF SURFACE PROFILES FOR GRADUALLY VARIED FLOWS

We will consider *wide rectangular channels* again, analyzing the slope of the free surface for a variety of conditions involving (1) what the slope of the channel bed is, (2) whether the flow is supercritical or subcritical, and (3) whether the depth y is larger or smaller than the normal depth y_N considered earlier in Sec. 14.3.

We now return to Eq. (14.67). Replacing R_H by y, Q by qb, and A by yb in the numerator and Q by VA and then A by yb in the denominator, we get

$$\frac{dy}{dx} = \left[\frac{S_0 - (nq/\kappa)^2(1/y^{10/3})}{1 - (V^2/gy)}\right] \tag{14.72}$$

From Sec. 8.8, we note that V^2/gy is the square of the Froude number Fr with depth y as the linear dimension. We then have

$$\frac{dy}{dx} = \left[\frac{S_0 - (nq/\kappa)^2 (1/y^{10/3})}{1 - \text{Fr}^2} \right]$$

(14.73)

Dividing and multiplying by S_0, we then get

$$\frac{dy}{dx} = \frac{S_0 \left[1 - (1/S_0)(nq/\kappa)^2 (1/y^{10/3}) \right]}{1 - \text{Fr}^2}$$

(14.74)

We next replace S_0 inside the brackets using Eq. (14.11), with that of a *hypothetical normal* flow having the same volumetric flow Q of the discussion. We rewrite Eq. (14.11).

$$S_0 = \left(\frac{n}{\kappa} \right)^2 \frac{1}{R_H^{4/3}} V^2$$

For the wide channel under discussion, we can replace R_H by y_N and V by q/y_N, where y_N is the normal depth for the normal flow corresponding to the bed slope S_0. Thus we have

$$S_0 = \left(\frac{n}{\kappa} \right)^2 \frac{1}{y_N^{4/3}} \frac{q^2}{y_N^2} = \left(\frac{n}{\kappa} \right)^2 \frac{q^2}{y_N^{10/3}}$$

(14.75)

Substituting into Eq. (14.74) and canceling terms we get

$$\boxed{\frac{dy}{dx} = \frac{S_0 \left[1 - (y_N/y)^{10/3} \right]}{1 - \text{Fr}^2}}$$

(14.76)

We use Eq. (14.76) to establish certain of the profiles in various flows.

First, we consider a *horizontal* bed slope (see Fig. 14.22a). The normal depth y_N here is infinite[15] for nonzero flow q. Since $S_0 = 0$, we find it best here to consult Eq. (14.73). For subcritical flow, Fr < 1. The slope of the profile, identified in the diagram as H_1, is then negative. For supercritical flow, Fr > 1. The slope of the profile H_2 is then positive. These profiles are shown in the diagram as curves above and below y_{cr}, respectively. At the critical depth, Fr = 1 and the slope of the free surface becomes infinite.

We now consider a slope of the channel bed for which the normal depth y_N exceeds the critical depth y_{cr}. (The horizontal channel for which $y_N \to \infty$ is a special case.) The slope of such a channel must be small and accordingly we say the channel has a *mild* slope (see Fig. 14.22b). If the depth $y > y_N$, the flow of course must be *subcritical* (making Fr < 1), and from Eq. (14.76) we see that the slope is positive and we have curve M_1. Next, if the depth $y < y_N$ but still

---

[15]Note that as $V \to 0$ for normal flow $y \to \infty$ so as to yield a nonzero volume flow.

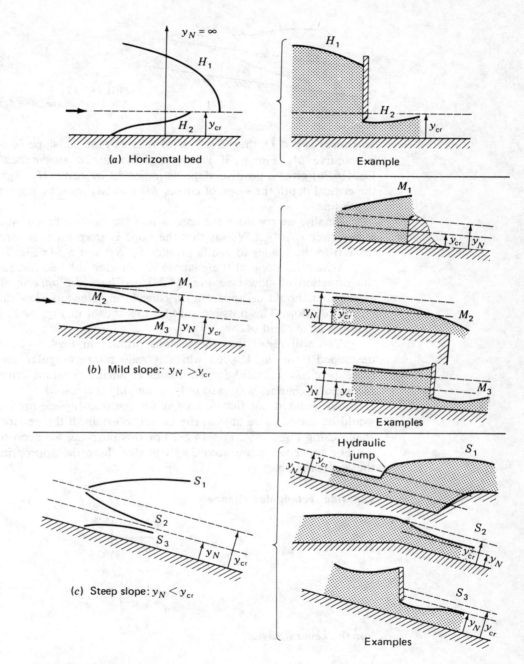

(a) Horizontal bed

Example

(b) Mild slope: $y_N > y_{cr}$

Examples

(c) Steep slope: $y_N < y_{cr}$

Examples

FIGURE 14.22
Flow profiles and examples.

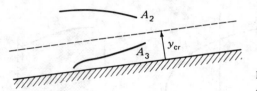

FIGURE 14.23
Adverse slope (negative S_0).

subcritical (Fr < 1), then Eq. (14.76) tells us that the slope is negative and we have curve M_2. Finally, if $y < y_N$ and the flow is *supercritical* Fr > 1), then Eq. (14.76) gives a positive slope as indicated by curve M_3. Again approaching the critical depth the slope of curves M_2 and M_3 must be normal to the critical depth line.

Finally, we consider the case where the normal flow is supercritical. This means that $y_N < y_{cr}$. We say that the slope is *steep* for such circumstances. We leave it to the reader to justify profiles S_1, S_2, and S_3, in Fig. 14.22c.

When the slope of the channel is such that the bed increases elevation in the direction of flow (see Fig. 14.23), we have an *adverse* slope and S_0 is *negative*. It should be apparent on considering Newton's law that there cannot be normal flow. The possible profiles are shown in Fig. 14.22 where they are denoted as A_2 and A_3.

You will recall that the various profiles in Figs. 14.22 and 14.23 were developed from Eq. (14.76), which is valid for rectangular channels. Actually these profiles are valid for all channels having constant cross sections—the rectangular channel was used only to simplify the results.

We pointed out that in solving for the free-surface profile in Sec. 14.8 it would be useful to be able at the outset to establish the general proper profile shape using Figs. 14.22 and 14.23. For this purpose, we need to know y_N and y_{cr}. For convenience we accordingly present here the appropriate formulas for these calculations.

For wide rectangular channels.

$$y_N = \left[\frac{nq}{\kappa\sqrt{S_0}} \right]^{3/5} \tag{14.77}$$

$$y_{cr} = \left(\frac{q^2}{g} \right)^{1/3} \tag{14.78}$$

For the general case.

(a) *Normal flow*

$$Q = \left(\frac{\kappa}{n} \right) R_H^{2/3} \sqrt{S_0} A_N \tag{14.79}$$

Here the value of y_N will enter into the expression for R_H and A_N.

(b) *Critical flow*

$$\frac{bQ^2}{gA_{cr}^3} = 1 \tag{14.80}$$

Here y_{cr} will appear in Q and A.

We now illustrate the procedure stemming from this discussion.

Example 14.11. In Example 14.10 sketch the shape of the free-surface profile in the region of interest.

We must determine y_N and y_{cr} at the outset of the flow for the given $Q = 35$ m^3/s. We start with Eq. (14.79). Thus

$$Q = \left(\frac{\kappa}{n}\right)\left(\frac{A}{P}\right)^{2/3}\sqrt{S_0}\, A = \frac{\kappa}{n}\frac{A^{5/3}}{P^{2/3}}\sqrt{S_0} \tag{a}$$

We now examine A and P.

$$A = (4)(y_N) + 2(\tfrac{1}{2}) y_N^2 = 4y_N + y_N^2 \tag{b}$$

$$P = 4 + 2\frac{y_N}{0.707} = 4 + 2.83 y_N \tag{c}$$

Hence going back to Eq. (*a*),

$$35 = \frac{1}{0.012}\frac{\left(4y_N + y_N^2\right)^{5/3}}{\left(4 + 2.83 y_N\right)^{2/3}}\sqrt{0.001}$$

$$\therefore 2343 = \frac{\left(4y_N + y_N^2\right)^5}{\left(4 + 2.83 y_N\right)^2}$$

We solve by trial and error to get

$$y_N = 1.953 \text{ m} \tag{d}$$

Next, we find y_{cr} using Eq. (14.80). Thus

$$\frac{bQ^2}{gA_{cr}^3} = 1$$

$$\frac{[4 + 2y_{cr}](35^2)}{(9.81)\left(4y_{cr} + y_{cr}^2\right)^3} = 1$$

Solving by trial and error again, we get

$$y_{cr} = 1.708 \text{ m} \tag{e}$$

We thus have at the outset

$$y = 3 \text{ m}$$
$$y_N = 1.953 \text{ m}$$
$$y_{cr} = 1.708 \text{ m}$$

Consulting Fig. 14.22 we see that we have a profile corresponding to M_1. This checks with our calculations in the previous example.

14.11   RAPIDLY VARIED FLOW; THE HYDRAULIC JUMP

We have seen the similarity of certain aspects of free-surface flow and compressible flow in Sec. 14.7 and can therefore expect an action in free-surface flow which is analogous to the shock wave in compressible flow. This action is called the *hydraulic jump*, shown diagrammatically in Fig. 14.24, which illustrates a flow in a horizontal channel of width b. The hydraulic jump may occur when there is supercritical flow in a channel having an obstruction or a rapid change in cross-sectional area. We will later see that when the jump occurs the flow changes from supercritical flow to a subcritical flow having a greater depth.

To study the hydraulic jump, we consider a steady flow wherein the position of the hydraulic jump remains fixed, as shown in Fig. 14.24. A control volume has been drawn to expose the flow conditions upstream (section 1) and downstream (section 2) of the hydraulic jump. At these sections, we consider the flow to be one-dimensional, as in our other analyses in this chapter. By taking sections 1 and 2 reasonably close to the jump, we can neglect the friction at the bed of the channel without serious error.

By using the continuity and momentum equations for this control volume, we can now relate the depths y_1 and y_2 before and after the jump. Thus, the *continuity* equation for incompressible flow for *rectangular channels* gives us

$$by_1V_1 = by_2V_2 = Q \tag{14.81}$$

where Q is the total volume flow, a constant for the problem. Using hydrostatic pressure distributions at sections 1 and 2 of flow, we then have from the *linear momentum* equation in the direction of flow

$$\frac{\gamma y_1}{2}by_1 - \frac{\gamma y_2}{2}by_2 = \rho Q(V_2 - V_1) \tag{14.82}$$

We first divide through by ρ and then on the right side of the equation replace V_2 by Q/by_2 and V_1 by Q/by_1 in accordance with Eq. (14.81). Thus

$$\frac{gby_1^2}{2} - \frac{gby_2^2}{2} = \frac{Q^2}{b}\left(\frac{1}{y_2} - \frac{1}{y_1}\right) \tag{14.83}$$

Collecting terms on the left side of the equation and combining fractions on the

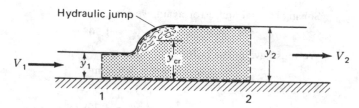

FIGURE 14.24
Control volume attached to stationary hydraulic jump.

right side of the equation, we get

$$\frac{gb}{2}(y_1^2 - y_2^2) = \frac{Q^2}{b}\frac{y_1 - y_2}{y_1 y_2} \tag{14.84}$$

We thus have a relation between y_1 and y_2. Clearly, if $y_1 = y_2$, the equation is satisfied and we have the trivial solution of no hydraulic jump. We therefore cancel $y_1 - y_2$ from the equation to reach a form which will yield a nontrivial answer. Thus

$$\frac{gb}{2}(y_1 + y_2) = \frac{Q^2}{b}\frac{1}{y_1 y_2} \tag{14.85}$$

Multiplying through by $2y_2/gb$ and bringing all terms to the left side of the equation, we then have

$$y_2^2 + y_1 y_2 - \frac{2Q^2}{gb^2}\frac{1}{y_1} = 0 \tag{14.86}$$

Solving for y_2 in terms of y_1 by using the quadratic formula, we get

$$\boxed{y_2 = \frac{-y_1 \pm \sqrt{y_1^2 + (8Q^2/gb^2)(1/y_1)}}{2}} \tag{14.87}$$

It is clear that we must take the positive root if we are to have a positive y_2.

Let us next ascertain what conditions, if any, are necessary if the depth of flow is to *increase* across the hydraulic jump. That is,

$$y_2 > y_1 \tag{14.88}$$

By using Eq. (14.87) for y_2, the inequality above becomes

$$\frac{1}{2}\left(-y_1 + \sqrt{y_1^2 + \frac{8Q^2}{gb^2 y_1}}\right) > y_1 \tag{14.89}$$

Adding $y_1/2$ to both sides of the inequality and then squaring both sides, we get

$$\frac{1}{4}\left(y_1^2 + \frac{8Q^2}{gb^2 y_1}\right) > \left(\frac{9}{4}\right)y_1^2 \tag{14.90}$$

Subtracting $\frac{1}{4}y_1^2$ from both sides of the inequality, we get

$$\frac{2Q^2}{gb^2 y_1} > 2y_1^2 \tag{14.91}$$

Isolating y_1 on the right side of the inequality, we may say that

$$\left(\frac{Q^2}{gb^2}\right)^{1/3} > y_1 \tag{14.92}$$

If in Eq. (14.78) we replace q^2 by Q^2/b^2, we then see from that equation that the left side of the inequality above is the *critical depth*. So we conclude that y_1 must be *less* than the critical depth. *Hence, if the depth y_2 is to exceed y_1, that is, if the flow is to undergo a rise in depth as a result of the hydraulic jump, the flow must be supercritical flow upstream of the hydraulic jump.*

For further information about the hydraulic jump consider the *first law of thermodynamics* for the control volume of Fig. 14.24. Thus[16]

$$\frac{V_1^2}{2g} + y_1 = \frac{V_2^2}{2g} + y_2 + \left[(u_2 - u_1) - \frac{dQ}{dm}\right]\frac{1}{g} \tag{14.93}$$

As in flow through pipes, we can consider the last expression in the equation above as the head loss H_l or, in other words, the loss of useful energy per unit weight from the system. Replacing V_1 by q_T/by_1 and V_2 by q_T/by_2 and rearranging the equation, we get

$$\frac{q_T^2}{2gb^2}\left(\frac{1}{y_1^2} - \frac{1}{y_2^2}\right) + (y_1 - y_2) = H_l \tag{14.94}$$

We next combine fractions in the first expression of the equation above:

$$\frac{q_T^2}{2gb^2 y_1^2}\frac{y_2^2 - y_1^2}{y_2^2} + (y_1 - y_2) = H_l \tag{14.95}$$

From Eq. (14.85), we may solve for $q_T^2/(y_1 b^2)$, getting

$$\frac{q_T^2}{y_1 b^2} = \frac{g}{2}\left(y_2^2 + y_1 y_2\right) \tag{14.96}$$

Now substituting this expression into Eq. (14.95), we get

$$\frac{1}{4y_1}\left(y_2^2 + y_1 y_2\right)\frac{y_2^2 - y_1^2}{y_2^2} + (y_1 - y_2) = H_l \tag{14.97}$$

The terms on the left side of this equation may be combined so that

$$\frac{y_2^3 - y_1^3 + 3y_1 y_2(y_1 - y_2)}{4y_1 y_2} = H_l \tag{14.98}$$

Now a hydraulic jump is a highly irreversible process where there is loss of mechanical energy to heat and internal energy. For hydraulic jumps with an

---

[16]In Eq. (14.93), Q represents heat transfer; in earlier discussion Q represented volume flow. In this immediate discussion, we use q_T for the total volume flow to avoid confusion.

initial Froude number from 20 and upward, loss of mechanical energy can range from 45 up to 85 percent with increase in dissipation accompanying increase in initial-Froude number. Thus, the head loss in the equation above must be *positive*. From the above equation you will be asked to show in Prob. 14.105 that this in turn means that $y_2 > y_1$.

We concluded earlier in the section that for $y_2 > y_1$ the flow upstream of a possible hydraulic jump had to be supercritical. Now we can conclude further that, since $y_2 > y_1$ for *all* hydraulic jumps, as we have just shown, the flow accordingly *upstream of a hydraulic jump has to be supercritical.*

Finally we will show that downstream of the hydraulic jump the flow must be *subcritical*. To do this, we return to Eq. (14.82). We divide through by b; replace Q by bq, where q is the flow per unit width of the channel; and V by q/y from continuity considerations. We get, on rearranging the equation,

$$\frac{\gamma y_1^2}{2} + \frac{\rho q^2}{y_1} = \frac{\gamma y_2^2}{2} + \frac{\rho q^2}{y_2} \qquad (14.99)$$

Thus, we see that the quantity $\gamma y^2/2 + \rho q^2/y$ (which is the sum of the hydrostatic force per unit width at a section plus the linear momentum flow per unit width at the section) remains *constant* along the channel flow. For a given flow q, we can plot $\gamma y^2/2 + \rho q^2/y$ called *specific thrust* versus y. We have shown such a plot in Fig. 14.25. The *minimum* value of $(\gamma y^2/2 + \rho q^2/y)$ occurs at a certain depth y, which can readily be determined by minimizing $\gamma y^2/2 + \rho q^2/y$ with respect to y. We get

$$\frac{\partial}{\partial y}\left[\frac{\gamma y^2}{2} + \frac{\rho q^2}{y}\right] = 0$$

$$\therefore y - \frac{\rho q^2}{y^2} = 0$$

$$\therefore y\left(\gamma - \frac{\rho q^2}{y^3}\right) = 0$$

We get $y = 0$ and $y = (q^2/g)^{1/3}$. From Eq. (14.78) we see that the nonzero value of y found is actually the *critical* depth where the Froude number is *unity*. Hence, that part of the curve above y_{cr} must correspond to subcritical flow, while the lower portion corresponds to supercritical flow. Thus, for an initial supercritical flow A needed for the hydraulic jump, the flow B becomes *subcritical after* the jump to maintain the same value of $(\gamma y^2/2 + \rho q^2/y)$. We thus see that the hydraulic jump is very much like a normal shock wave.

Example 14.12. Figure 14.26 shows a flow over a *spillway* into a rectangular channel, where a hydraulic jump is formed to dissipate mechanical energy. The region downstream of the spillway is called a *stilling basin*. The spillway and stilling basin are 20 m wide. Before the jump the water has a depth of 1 m and a speed of

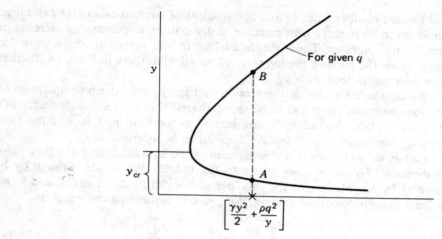

FIGURE 14.25
Hydraulic jump from A to B.

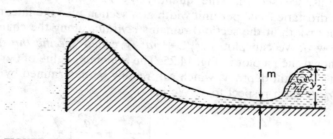

FIGURE 14.26
Flow from spillway into stilling basin.

18 m/s. Determine (*a*) the initial Froude number before the jump, (*b*) the depth of flow after the jump, and (*c*) the head loss H_l in the jump.

(*a*)
$$\text{Fr} = \frac{V_1}{\sqrt{gy_1}} = \frac{18}{\sqrt{(9.81)(1)}} = 5.75$$

(*b*)
$$y_2 = \frac{-y_1 + \sqrt{y_1^2 + (8q_T^2/gb^2)(1/y_1)}}{2}$$

$$= \frac{-1 + \sqrt{1^2 + \frac{(8)[(18)(1)(20)]^2}{(9.81)(20^2)}\left(\frac{1}{1}\right)}}{2}$$

$$= 7.64 \text{ m}$$

(*c*) Using Eq. (14.98) we have

$$H_l = \frac{y_2^3 - y_1^3 + y_1 y_2 (y_1 - y_2)}{4 y_1 y_2} = \frac{7.64^3 - 1^3 + (7.64)(1)(1 - 7.64)}{(4)(7.64)(1)} = 12.90 \text{ m}$$

Thus, there is a loss of mechanical energy of 12.90 N · m per newton weight of flow.

14.12   CLOSURE

We have examined simple free-surface flows in this chapter. The reader should realize that we have only scratched the surface of this vital area of study. Many of the civil engineers will in all likelihood take a separate full course in free-surface or channel flow.

Students who have studied Chap. 11 on compressible flow have seen many analogies between free-surface flow and compressible flow. One can study compressible flow using the shallow rectangular flow in a small wide channel called a *water table*. Here, wave patterns generated in the water table by obstacles in the flow are related to oblique shock waves in two-dimensional supersonic flows.

In the next chapter we shall return to the study of turbomachinery, a topic briefly introduced in Chap. 5, to best illustrate the use of the moment-of-momentum equation. We will have occasion to use what we have learned about external flows and internal flows in this chapter.

PROBLEMS

Problem Categories

Velocity Profile Problems 14.1–14.3
Hydraulic Radius Problems 14.4–14.6
Normal Flow Problems 14.7–14.31
Best Hydraulic Section 14.32–14.35
Gravity Waves 14.36–14.40
Problems Involving Specific Energy 14.41–14.63
Varied Flow in Short Rectangular Channels 14.64–14.73
Gradually Varied Flow over Long Channels 14.74–14.93
Hydraulic Jump 14.94–14.105

Starred Problems

14.70–14.74, 14.86, 14.104

Derivations or Justifications

14.69, 14.104

14.1. In Prob. 10.3, we found that for steady laminar flow in a thin sheet over a flat surface

$$V_z = \frac{\gamma \sin \theta}{\mu} \left[\left(\frac{3q\mu}{\gamma \sin \theta} \right)^{1/3} y - \frac{y^2}{2} \right] \quad (a)$$

where q is the volumetric flow per unit width. What is the thickness t of such a flow of water at 5°C for $\theta = 20°$? The volumetric flow q is 3×10^{-4} m^2/s. *Hint*: Since τ_{yz} is zero at

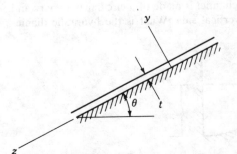

FIGURE P14.1

$y = t$, what can you conclude about dV_z/dy at $y = t$?

14.2. A film of oil of thickness $t = 0.002$ ft moves at uniform speed down at inclined surface having an angle $\theta = 30°$. What is the surface velocity of the film if $\mu = 3 \times 10^{-4}$ lb · s/ft² and γ is 57 lb/ft³? *Hint*: In the previous problem, we found that

$$t = \left(\frac{3q\mu}{\gamma \sin \theta} \right)^{1/3} \qquad (a)$$

Also, see Prob. 14.1. What is the volume flow q per unit width?

14.3. In Prob. 14.1, compute a Reynolds number given as Re $= (V_{at}t)/\nu$. We note that if this Reynolds number is larger than 500, we have turbulent flow rather than laminar flow. Is our laminar flow assumption valid? What is the limiting volumetric flow in Prob. 14.1 wherein the laminar flow assumption is valid? What is the film thickness for this case? Note that $V_{av}t = q$. Also, note result in Prob. 14.2. The thickness t from Prob. 14.1 is 0.7415 × 10^{-3} m.

14.4. What is the hydraulic diameter for a circular sector of radius R and angle 2α degrees?

14.5. What is the hydraulic diameter of a right triangle shown?

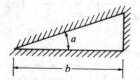

FIGURE P14.5

14.6. A channel is made of a circular boundary and a vertical side. What is the hydraulic diame-

ter? Give the results in terms of h and R only.

14.7. Water at 5°C is flowing in a finished cement rectangular channel of width 10 m with a slope S_0 of 0.001. The height of the water normal to the bed is constant, having the value of 1 m. What is the volume of flow Q for normal flow?

14.8. A wide rectangular channel dug from clean earth is to conduct a flow q of 5 m³/s per meter of width. The slope of the bed is 0.0015. What would be the depth of flow for normal flow?

14.9. A triangular channel is made of corrugated steel and conducts 10 ft³/s at an elevation of 1000 ft to an elevation of 990 ft. What length L should the channel be for normal flow? The depth y is 2 ft.

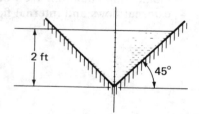

FIGURE P14.9

14.10. What is the depth of normal flow and slope S_0 of a rectangular channel to conduct 5 m³/s of water a distance of 2000 m with a head loss H_l of 15 m? The width of the channel is 2 m. The channel is made of brickwork.

14.11. An asphalt-lined channel has a slope S_0 of 0.0017. The flow of water in the channel is 50 m³/s. What is the normal depth?

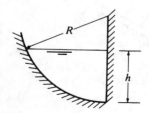

FIGURE P14.6

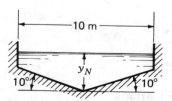

FIGURE P14.11

14.12. Civil engineers frequently encounter flow in pipes wherein the pipe is not full of water. This occurs in sewers, for example, and the flow is then a free-surface flow. Shown is a partially filled pipe discharging 10 ft³/s. If Manning's n is 0.015, what slope is necessary for a normal flow of 50 ft³/s?

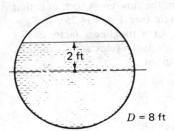

FIGURE P14.12

$D = 8$ ft

2 ft

14.13. What is the flow in Fig. P14.13 when the level of flow has gone above the main channel and extends into the floodways on both sides? The slope of the channel is 0.0007 and the surface is that of clean excavated earth. The slope on all the inclined sides is 45°.

14.14. A rectangular asphalt channel has a slope of 0.00001 and a width of 6 m. Water flows at constant depth to the bed of 1.5 m.
  (a) Find the friction factor f.
  (b) Find the average velocity.
  (c) Find the wall shear stress τ_w.

14.15. Water is flowing uniformly in a trapezoidal channel of unfinished concrete as shown in

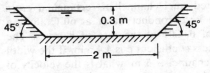

FIGURE P14.15

Fig. P14.15. The slope of the channel is 0.002. What is the volume flow and the wall shear stress?

14.16. Crude oil is flowing in a rectangular channel with a constant depth of 0.5 m. The width of the channel is 3 m. What is the loss in head for 3 m of length of channel? What is the energy dissipated per unit mass of flow in this distance? The channel slope is 0.003. What flow regime are we in if $e = 0.09$ mm and the oil temperature is 60°C?

14.17. A corrugated trapezoidal channel descends 0.2 m/km. The sides are at an angle of 45°. If a flow of 3 m³/s is desired for normal flow, what should be the width b at the bottom for a water depth of 0.8 m?

14.18. What is the ratio of slopes for rectangular channels having the same flow and cross sections for asphalt lining and unfinished concrete? The flow in each case is uniform.

14.19. A concrete channel has a uniform flow of water of 100 ft³/s. The roughness coefficient e is 0.004 ft. The height of the free surface is 3 ft. What should be the slope S_o? The width of the channel is 10 ft. The water is at 60°F.

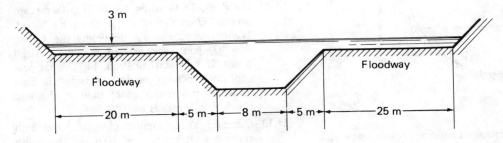

FIGURE P14.13

14.20. A riveted steel channel of rectangular cross section is to conduct water at 60°C along a slope of 0.0001 at the rate of 8 m³/s. The roughness coefficient e is 4.4 mm. If the width of the channel is 5 m, what is the velocity of flow to be expected for uniform flow?

14.21. We are to use a full semicircular channel for uniform flow of water at 60°F at a slope of 0.001. If we wish to have a volume of flow of 100 ft³/s, what radius should the channel be for e = 0.015 ft?

14.22. Compare the ratios of areas and perimeters for triangular cross sections for uniform flow of $Q = 30$ m²/s and $S_0 = 0.001$ for $\alpha = 30°$ and $\alpha = 80°$. The material is asphalt with Manning's $n = 0.016$. Comment on results.

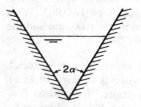

FIGURE P14.22

14.23. A cast iron pipe conducts water. If $R = 3$ m, $\alpha = 60°$, and $S_0 = 0.003$, what is the volume of uniform flow?

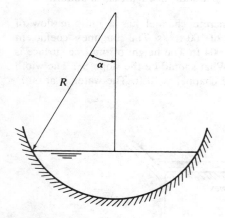

FIGURE P14.23

14.24. Using a friction factor approach employing formulas for f, find the uniform flow in a semicircular channel of radius 4 m flowing full at a slope of 0.0001. The fluid is crude oil at 0°C. The channel is lined with plastic material with $e = 0.28$ mm. Use the simplest formulas for f for an approximate result. *Hint*: Use eqs. (14.20) and (14.7) in an iterative manner.

14.25. Consider channel flow of water at 5°C in a channel wherein the flow cross section is that of a quarter circle (see Fig. P14.25). If $S_o = 0.0025$, find Q for a roughness factor of 0.7 mm. Go through two cycles of calculation. Use Moody diagram.

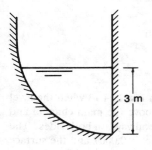

14.26. Do Prob. 14.25 using friction factor formulas.

14.27. The channel in Example 14.1 is to be replaced by a rectangular channel of width 16 ft. What is the ratio of cost of the concrete allowing 2 ft of freeboard (distance above the free surface) for the walls for the channels?

14.28. Do Prob. 14.7 using the friction-factor approach. Assume rough-zone flow and check at the conclusion of the problem whether this is appropriate.

14.29. Do Prob. 14.11 using the friction-factor approach. *Hint*: Guess at y_N and solve for Q. When you have a y_N delivering the known flow Q, you have arrived at your desired result. If $y_N = 2.07$ in Prob. 14.11, what percent difference do you get? Consider the flow to be in the rough zone; then at the end check this. Water is at 10°C.

14.30. Consider a rectangular channel lined with smooth steel where $e = 0.001$ m. The width of the channel is 1 m and the slope S_0 is

0.0001. What is the maximum flow q to have hydraulically smooth steady flow? Water at 5°C is flowing.

14.31. In Prob. 14.30, what is the minimum flow q for rough zone of flow? Take the slope to be 0.01 for this case.

14.32. Suppose in Example 14.6 that the angle β has been specified as 50° to avoid material on the sides of the trapezoidal section from sliding down. For the best hydraulic section for this case, what is the relation between y and a?

14.33. There is a steady flow at 5 m³/s at a slope of $S_0 = 0.002$. What is the proper width b of a rectangular channel if we wish to minimize the wetted perimeter for the sake of economy of construction? The channel is made from unfinished concrete.

14.34. A trapezoidal channel is to conduct 10 m³/s of water at 5°C at a slope $S_0 = 0.003$. For the most efficient design, what is the wetted perimeter? Compare this perimeter with that of a semicircular cross section. The material for the channel is asphalt. Do not consider freeboard (channel wall above the free surface). What is the ratio of the respective perimeters? What is the ratio of the respective cross sections? Comment on the relative efficiency of the two sections.

14.35. A rectangular channel with a slope $S_0 = 0.002$ is to conduct 60 ft³/s of water at $T = 60°F$. What is the perimeter ratio of the most efficient design with that of a trapezoidal section also of the most efficient design for the same Q and S_0? The material is concrete ($n = 0.017$). Use results of Examples 14.5 and 14.6.

14.36. Consider a uniform flow of water of speed V in the x direction with a shallow depth. If $V < \sqrt{gy}$, sketch the circular wave at successive times formed by dropping a pebble into the water. Now consider the case where $V > \sqrt{gy}$. Again show the circular wave at successive time intervals. What is the difference between the patterns? Now explain how if a continuous disturbance is developed at a stationary position in a flow where $V > \sqrt{gy}$, stationary waves having an angle α will be formed as shown. Show that $\sin \alpha = \sqrt{gy}/V$.

14.37. A stream has a speed of about 16 ft/s and is 2 ft deep. If a thin obstruction such as a reed

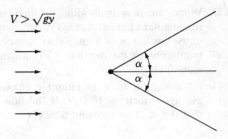

$V > \sqrt{gy}$

FIGURE P14.37

is present, what is the angle of the waves formed relative to the direction of motion of the stream? See Prob. 14.36.

14.38. A small boat is moving in shallow water where the depth is 2 m. A small bow wave is formed so that it makes an angle of 70° with the centerline of the boat. What is the speed of the boat? See Prob. 14.36.

14.39. A stone is thrown into a pond. A wave is formed which has an amplitude of about 1 in and a speed of about 5 ft/s. Estimate the depth of the pond where these measurements are made.

14.40. Consider a uniform flow in a channel of depth 0.4 m with a speed of 3 m/s. A small disturbance is created on the surface, forming a gravity wave. What is the difference in time at which an observer 10 m downstream from the disturbance first feels the wave as compared with an observer 10 m upstream of the disturbance? Observers and center of disturbance are positioned along a straight line.

14.41. A wide rectangular channel excavated clean earth has a flow of 3 m³/(s)(m). W the critical depth and the minimum s energy? What is the slope for critical flow? If $y = 3$ m at a section, wha Froude number at this section for stated above? Water is at 5°C.

14.42. A wide rectangular channel ha depth of 2 m and a critical slope What is the volume of flow for th What is the depth of flow fo with a value of flow $q = 5$ m³/ above slope? Water is at 5°C.

14.43. When the flow in a wide, finished-concrete, rectangular channel has a Froude number of unity the depth is 1 m. What is the Froude number when the depth is 1.5 m for the same mass flow q?

14.44. At a section in a rectangular channel, the average velocity is 10 ft/s. Is the flow a tranquil (Fr < 1) or shooting flow (Fr > 1)?

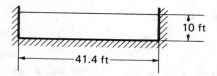

10 ft

41.4 ft

FIGURE P14.44

14.45. In Prob. 14.44 what is the specific energy? What other depth is possible for this energy? The critical depth from Prob. 14.44 is 6.77 ft?

14.46. For 200 ft³/s of water flowing in a rectangular channel 10 ft wide, what is the minimum specific energy possible for this flow? What are the critical depth and the critical velocity?

14.47. What is the critical depth for a rectangular finished-concrete channel of width 3 m. (The channel cannot be considered to be a wide channel.) What is the slope for critical normal flow? The flow Q is 2 m³/s. Water is at 5°C.

14.48. What is the critical depth for a triangular cross section for a flow Q of 5 m³/s? The angle between sides is 60°.

14.49. What is the critical depth of a trapezoidal cross section for a flow Q of 10 m³/s? The width at the base is 3 m and the angle α at the sides is 60°.

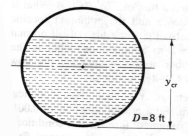

y_{cr}

$D = 8$ ft

GURE P14.50

14.50. Shown in Fig. P14.50 is a partially filled pipe discharging 450 ft³/s. What is the critical depth?

14.51. Oil of S.G. equal to 0.69 and viscosity of 10^{-4} lb · s/ft² is flowing with an average velocity of 10 ft/s at a section of height 1 ft along a rectangular channel with a very small slope.
   (a) What is the total mechanical energy in feet of a flow particle $\frac{1}{2}$ ft from the channel bed relative to the channel bed?
   (b) What is the total mechanical energy per unit mass of a particle at the free surface relative to the bed?

14.52. In the previous problem, determine the critical depth and the critical velocity. What is the actual Froude numbers for the actual flow described and the critical flow? What compressible flow is this flow analogous to?

14.53. Water is flowing in a channel having an equilateral cross section as shown in Fig. P14.53. The volume of flow is 0.707 m³/s. The material is finished concrete. What is the critical depth? What is the minimum specific energy?

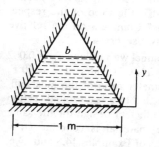

b

y

1 m

FIGURE P14.53

14.54. In a rectangular channel of width 20 ft, what is slope S_0 to have normal and critical flow of 250 ft³/s? The channel is finished concrete and the water flowing is at 60°F.

14.55. Water is to flow at the rate of 10 m³/s at a temperature of 30°C in an asphalt lined semicircular channel. What is the radius of the channel for normal critical flow? The flow must fill the semicircular section. Also, determine the slope S_0.

14.56. A rectangular channel is to have 20 m³/s of 10°C water flowing in normal critical flow. The width of the channel is 5 m and the slope is 0.003. What is the roughness e for such circumstances?

14.57. A channel is shown in Fig. P14.57 to conduct water at 60°C in a normal, critical flow. If the flow is to be 100 m³/s, what is the proper slope? It is an earthen channel with $e = 0.15$ m.

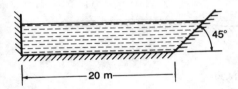

FIGURE P14.57

14.58. What is the critical depth of a parabolic channel when there is a flow of 3 m³/(s)(m)? Position A on the channel has the coordinates shown.

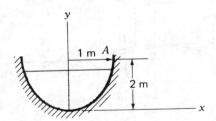

FIGURE P14.58

14.59. In Prob. 14.58, what is the critical slope S_{cr} for normal critical flow? The friction factor $f = 0.015$ and the critical depth and critical area were found to be $y_{cr} = 1.1155$ m and $A_{cr} = (2\sqrt{2}/3)y_{cr}^{3/2}$. *Hint*: Note that $ds \equiv \sqrt{1 + (dy/dx)^2}\, dx$ and that $\int \sqrt{a^2 + x^2}\, dx = \frac{1}{2}[x\sqrt{a^2 + x^2} + a^2 \ln(x + \sqrt{a^2 + x^2})]$.

14.60. At a section in a triangular channel, the average velocity is 10 ft/s. Is the flow tranquil (Fr < 1) or shooting (Fr > 1)? (See Fig. P14.60)

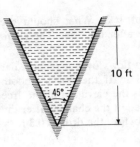

FIGURE P14.60

14.61. Starting with Eq. (14.9) with R_H replaced by A/P, show that

$$\frac{dV}{dy_N} = \frac{\kappa}{n}S_0^{1/2}\left(\frac{2}{3}\right)\left(\frac{A}{P}\right)^{-1/3}\frac{d(A/P)}{dy_N}$$

Show that, if the cross-section area increases faster than the perimeter for increased values of y_N, then V and Q must increase.

14.62. Show for increasing y_N, that V and Q must increase for a given slope S_0 and for a rectangular channel. See Prob. 14.61.

14.63. Do the preceding for a triangular channel.

14.64. Draw sketches similar to those in Fig. P14.18 for the following cases. Indicate the kind of flow to be expected after the small rise. Is y_2 greater than or smaller than y_1?

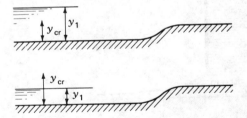

FIGURE P14.64

14.65. Water is moving with a speed of 1 ft/s and a depth of 3 ft. It approaches a smooth rise in

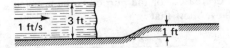

FIGURE P14.65

the channel bed of 1 ft. What should the estimated depth be after the rise? The channel is rectangular.

14.66. A flow of 0.2 ft^3/s flows over a rectangular channel of width 3 ft. If there is a smooth drop of 2 in, what is the elevation of the free surface above the bed of the channel after the drop? The velocity before the drop is 0.3 ft/s.

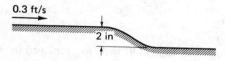

0.3 ft/s

2 in

FIGURE P14.66

14.67. A broad-crested weir has a width of 1 m. The free surface is at a height of 0.3 m above the surface of the weir at a position well upstream of the weir. What is the volume of flow? What is the minimum depth y over the weir and where does it occur?

14.68. Water flows from a reservoir over a runoff, as shown in Fig. P14.68. Estimate the flow q. Comment on the accuracy of your estimate if the opening of the water is increased substantially from 0.6 m given or decreased substantially from this value.

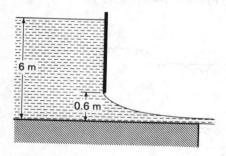

6 m

0.6 m

FIGURE P14.68

14.69. A *Venturi flume* is a region of a rectangular channel where the width has been deliberately decreased for the purpose of measuring the flow. Show that

$$Q^2 = \frac{2g(y_1 - y_2)}{[1/(b_2 y_2)]^2 - [1/(b_1 y_1)]^2}$$

Hint: Use Bernoulli at free surface.

*14.70. Water is flowing in a rectangular channel of width 3 m (see Fig. P14.70). The flow volume is 20 m^3/s. What is the value of y_2 after the flow is made to go over an incline? Neglect friction and use Bernoulli along free surface, and continuity before the incline and after the incline. How would you decide which of the three roots is your valid one?

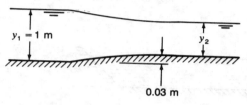

$y_1 = 1$ m

y_2

0.03 m

FIGURE P14.70

*14.71. Shown in Fig. P14.71 is flow in a wide channel over a parabolic bump. The flow q is 5 m^2/s. Estimate $y(x)$ above the bump. We have steady frictionless flow.

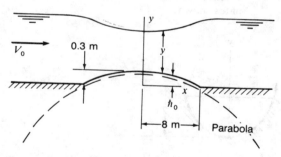

V_0

0.3 m

y

y

x

h_0

8 m

Parabola

FIGURE P14.71

*14.72. Shown in Fig. P14.72 is a flow in a wide channel over a bump described as $h(x)$. We will neglect friction completely. If the bump is part of a circle given as

$$(h - 0.2)^2 + x^2 = 8^2 \qquad (a)$$

and the flow q is 20 m^2/s, form the following equation for y versus x:

$$y + \frac{20.39}{y^2} = \sqrt{64 - x^2} + \text{const.} \qquad (b)$$

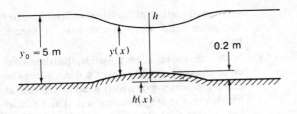

FIGURE P14.72

14.73. In the previous problem determine the constant of integration.

14.74. Water enters a rectangular channel which is 3 ft wide at an average velocity of 0.8 ft/s and a depth of 3 in. The channel has an inclination α of 0.2°. If Manning's n for the surface is 0.012, estimate at what distance L along the channel the elevation will have risen to a depth of 4 in. Use $\Delta y = 0.2$ inches in your numerical calculation.

14.75. In Example 14.10, compute the distance L in one calculation where the free surface has a depth of 3.6 m. Do not average.

14.76. Do Prob. 14.75 using a linear average in the calculations. If in the preceding problem $\Delta L = 649$ m, what is the percent error in not averaging?

14.77. A wide channel is made of finished concrete. It has a slope S_0 equal to 0.0003. The opening from a large reservoir to the channel is a sharp-edged sluice gate. The coefficient of contraction C_c is 0.80 and the coefficient of friction C_f is 0.85. What is the approximate depth at the vena contracta and the flow q of fluid? How far from the vena contracta does the water increase depth by 30 mm? Use one calculation with linear averages.

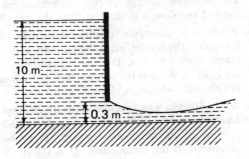

FIGURE P14.77

14.78. In Prob. 14.70, take the channel to be horizontal. For a decrease in depth of $\frac{1}{2}$ in, compute the distance L analytically without the use of numerical methods.

14.79. Water is moving at a speed of 4 m/s in a very wide horizontal channel at a depth of 1 m. If $n = 0.025$ for earth in good condition, at what distance L downstream will the depth increase to 1.1 m? Do problem analytically without numerical methods.

14.80. In a wide horizontal rectangular earth channel with weeds and stones, it is observed that the depth rises 0.2 m from a depth of 1 m in a distance of 6 m. What is the volumetric flow per unit width?

14.81. We have shown a large reservoir of water to which is connected a triangular channel of uniform cross section, as shown in view B-B, Fig. P14.81. We may take n for this channel as 0.012. The water enters the channel by a well-rounded opening of height 3 in, as shown in view B-B in the diagram. If steady flow is established, sketch the free-surface profile as per one of the curves in Fig. 14.22. The angle $\theta = 2°$.

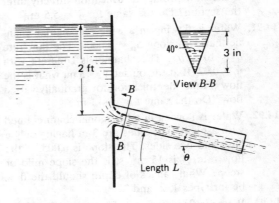

FIGURE P14.81

14.82. In Prob. 14.81, estimate the distance along the channel at which the depth becomes 4 in. Get an estimate in one calculation using only initial data and not averaging.

14.83. In Example 14.1, for a flow of 5000 ft³/s at the geometry shown initially, at what distance downstream is the depth at an elevation of 14 ft? Use initial data and do not average. Do in one calculation.

14.84. Do Prob. 14.83 using a linear average over the interval. Compare results with $\Delta L = 301$ ft from the preceding problem not involving an average. Use one cycle of calculation. What is the error in not averaging in this problem?

14.85. In Example 14.1, find the free-surface profile as the water rises 1 ft from that at the outset. Use five intervals and use average data. The flow Q is 5000 ft^3/s.

*****14.86.** If you have a programmable calculator, do Prob. 14.85 for 20 intervals going from $y = 3$ to $y = 4$.

14.87. In Prob. 14.79 sketch the free-surface profile.

14.88. In Example 14.1, there is a flow of 5000 ft^3/s at the geometry shown initially. Sketch the free-surface profile downstream for steady flow.

14.89. Justify the curves shown in Fig. 14.22c.

14.90. Water is flowing along a riveted triangular channel having a 60° angle at the rate of 100 ft^3/s. The slope of the channel is 0.008. If y at one section in the channel is 4.00 m, what is the flow profile identification directly after the section? Do the same for $y = 5.5$ m.

14.91. Water is flowing in a wide rectangular channel having a slope $S_0 = 0.0008$. The flow rate q is 10 m^2/s. The channel is concrete with $n = 0.02$. If at the outset $y = 3$ m, what is the flow profile identification for gradually varied flow? Do the same for $y = 1$ m.

14.92. Water is flowing in a hexagonal channel lined with asphalt. The sides are 3 m having only a base and two sides. The slope is 0.005. If the flow rate Q is 15 m^3/s, is the slope mild or steep? What ranges of depth should the flow be for types 1, 2, and 3?

14.93. Water is flowing in a brick rectangular channel of width 20 ft. The slope is 0.001 and the volume flow Q is 100 ft^3/s. Is the slope mild or steep? What are the ranges of depth for types 1, 2, and 3 flows?

14.94. A rectangular channel has a width of 10 ft and a flow of 10 ft^3/s. The depth is 3 in. Assume that a hydraulic jump occurs. What

will the elevation of the free surface be after the jump, and what is the loss in kinetic energy?

14.95. Water flows in a rectangular, finished-concrete channel and goes over a dam. After the dam, the water goes into a stilling basin at which there is a hydraulic jump. The channel has a slope of 0.003 and the depth 50 m upstream of the dam is 4 m. What is the depth of flow just before reaching the dam? The volumetric flow Q is 18 m^3/s and the width of the channel is 6 m. Do the problem the simplest way without averaging.

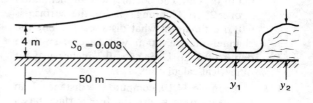

FIGURE P14.95

14.96. Work Prob. 14.95 more accurate using linear averages. In the previous problem, we get $\Delta y = 0.1502$ m using only initial data.

14.97. In Prob. 14.95 we computed the depth before the dam to be 4.1502 m. If we neglect friction over the dam, what is the depth y_2 at the hydraulic jump?

14.98. Water is moving in a rectangular channel is known to have a Froude number of $\sqrt{10}$. The channel is 5 m in width. The depth of flow is 1 m. If the water undergoes a hydraulic jump, what is the Froude number after the jump and what percentage of the mechanical energy of the flow is dissipated from the jump?

14.99. Water in a rectangular channel is seen to go through a hydraulic jump where the depth jumps from 3 to 7 m. The width of the channel is 10 m. What is the volume flow Q? What are the final and initial Froude numbers?

14.100. Water flows in a rectangular concrete channel and undergoes a hydraulic jump such that 60 percent of its mechanical energy is to be dissipated. If the volume flow Q is 100 m^3/s and the width of the channel is 5 m, what must the Froude number be just before the jump? Set up the proper equations but do not actually solve.

14.101. Water is flowing from a spillway into a stilling basin as shown. The elevation y_A ahead of the spillway is 10 m. The width of the rectangular channel is 8 m. If the stilling basin dissipates 60 percent of the mechanical energy, what is the volume flow Q_T? Set up simultaneous equations but do not solve.

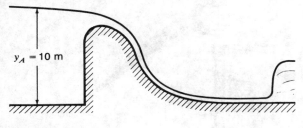

$y_A = 10$ m

FIGURE P14.101

14.102. Water having a steady known volumetric flow Q is moving in a rectangular channel at supercritical speed up an adverse slope. It undergoes a hydraulic jump as shown in Fig. P14.102. If we know y_B and y_A at positions

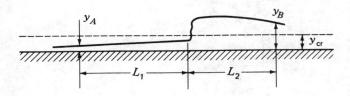

y_A y_B y_{cr}

L_1 L_2

FIGURE P14.102

$L_1 + L_2$ apart, how do we approximately locate the position of the hydraulic jump—i.e., how do we get L_1 and L_2? The channel has a known value of n. Explain the simplest method. The width is b.

14.103. The distance L_1 in Prob. 14.102 is 200 m and y_A is half the critical value. The channel is finished concrete with an adverse slope -0.005. What is y_B at a distance $L_2 = 100$ m from the hydraulic jump? Solve in the simplest manner to get an approximate result. The flow Q is 200 m³/s. The width of the channel is 7 m.

*14.104. As a project find the location of the hydraulic jump in Prob. 14.102 using a digital computer. The following data apply.

$$L_1 + L_2 = 300 \text{ m}$$

$$y_A = 1.500 \text{ m}$$

$$y_B = 5.40 \text{ m}$$

$$n = 0.012$$

$$S_0 = -0.005$$

$$q_T = 200 \text{ m}^3/\text{s}$$

14.105. Formulate Eq. (14.98) from Eq. (14.97). Prove that for $H_l > 0$, it must follow that $y_2 > y_1$.

Final assemblage of the ERDA/NASA wind turbine at Lewis Plum Brook Station. Blades have an overall diameter of 125 ft. Blades will rotate in an 8-mi/h wind and reach maximum power at 19 mi/h. This is the largest wind generator anywhere in the world. The power available is 100 kW under design conditions. Note that the familiar large Dutch windmills develop only about 35 kW.

CHAPTER
15

*TURBOMACHINERY

15.1 INTRODUCTION

As pointed out in Chap. 5, a turbomachine is a device wherein motion of an unconfined[1] fluid is altered in such a way as to transmit power to or from a shaft or is altered to create a propulsive thrust.

Let us consider each of aforementioned types of turbomachines. First, consider machines which transmit power *from* a shaft *to* the fluid. This power is transmitted to the fluid by means of a member on the shaft called a *rotor*, a *runner*, or an *impeller* on which there are blades. The following names are used for such devices:

1. A *pump* is a turbomachine wherein the fluid is a liquid.
2. A *compressor* transmits power to a gas to achieve a high pressure but with little velocity.

---

[1]The fluid is unconfined in that it is at no time trapped as is the case of the reciprocating engine wherein the fluid is trapped for a short time in a cylinder of the machine.

3. A *fan* imparts motion to a gas with small change in pressure.

4. A *blower* imparts substantial velocity and pressure to a gas.

Next, we call devices which transfer power *from* the fluid to the shaft *turbines*. (We have already considered turbines in Chap. 5 when we discussed the Pelton water wheel.) There are two general classifications that are used. In an *impulse* turbine the fluid passes over the blades or buckets of the runner while maintaining a constant static pressure at all times. The Pelton water wheel is such a turbine. In the *reaction* turbine there is a decrease in static pressure as the fluid flows in the region between pairs of blades. There is an expansion taking place in this region and each pair of blades forms as it were a "moving nozzle."

Where the turbomachine gives rise to a thrust, as has been described in Chap. 5 for the jet engine, we have what is called a *propulsion engine*. Generally a fuel is burned in such devices. If the air reacting in the combustion process is taken from the surroundings of the propulsion device through a diffuser we have an *air-breathing machine*. This is clearly the case of the jet engine of Chap.

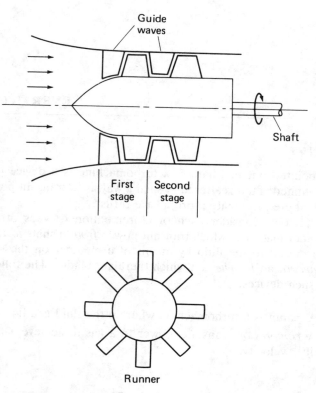

FIGURE 15.1
Two-stage axial-flow compressor.

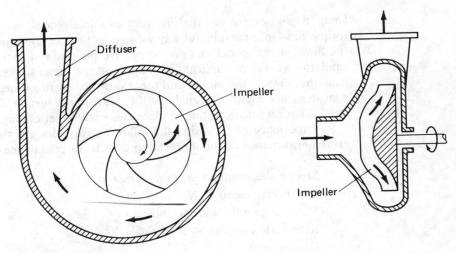

FIGURE 15.2
Centrifugal pump.

5. On the other hand, if the oxidation agent is carried by the propulsive device to enable it to operate in outer space, we have a *rocket*. In the rocket the fuel may be solid or liquid.

  Turbomachines in general are further classified by the direction of the final motion of the fluid relative to the shaft. If the flow through the moving vanes or blades of the turbomachine is essentially *parallel* to the shaft, we consider the turbomachine to be an *axial-flow machine*. The jet engine described in Chap. 5 is an axial-flow machine. Figure 15.1 shows an axial-flow compressor. Note that we have two sets of stationary vanes and two sets of moving vanes. A set of moving and stationary vanes is called a *stage*; we thus show a two-stage compressor. If the motion of the fluid in the rotating part of the machines is now essentially *radial* to the axis of the shaft, we have a *radial-flow turbomachine*. An example of this kind of machine is the centrifugal pump in Fig. 15.2. The water enters axially but on reaching the impeller is then given essentially a radial motion by the blades of the impeller. The water then emerges from the outer periphery of the blades and goes through a diffuser to the outlet. The diffuser changes kinetic head into pressure head in the flow. Finally, we wish to point out that there are *mixed-flow* machines incorporating both radial and axial flow of fluid in the machine. We do not consider such cases in this text.

  We will now discuss similarity relations for families of geometrically similar turbomachines.

15.2   SIMILARITY RELATIONS FOR TURBOMACHINES

It is often desirable to test a model of a turbomachine in order to predict the performance of a larger (or smaller) geometrically similar prototype machine. In

Chap. 8, we pointed out that in order to relate model and prototype force and torque data in a meaningful way we had to achieve *dynamic similarity* between the flows in the model and prototype. We pointed out that to achieve dynamic similarity we had to duplicate all but one of the dimensionless groups characterizing the flows in the model and the prototype. At this time, we will accordingly consider the dimensional analysis of flows through turbomachines which do not have chemical reactions taking place (we exclude jet engines, for instance).

Experience dictates that the following variables are significant for considering performance of the aforestated class of turbomachines:

Size of the machine → diameter D
Rotational speed → N
Volume flow through machine → Q
Kinematic viscosity → ν
Gravity → g
Change in total head through the machine → ΔH_D

where the total head H_D you will recall is

$$H_D = \frac{p}{\gamma} + z + \frac{V^2}{2g} \tag{15.1}$$

There are two basic dimensions L and t involved, so that from Buckingham's π theorem we should have four dimensionless groups characterizing the flow. We will now present a useful set of these dimensionless groups. Their functional relationship to each other characterizes the key aspects of the flows. Thus,

$$f\left(\frac{Q}{ND^3}, \frac{\Delta H_D}{D}, \frac{g}{N^2D}, \frac{ND^2}{\nu} \right) = 0 \tag{15.2}$$

Let us examine the π's given above. The expression ND^2/ν can readily be interpreted as a Reynolds number. This is true since ND is proportional to the *tip speed*, U_t, of the impeller of the machine, so that the grouping ND^2/ν is representable as $k(U_t D/\nu)$, where k is a constant and $U_t D/\nu$ is the so-called *rotational Reynolds number*. As in previous flow considerations, if the rotational Reynolds number is sufficiently large, we can neglect its effect (friction) on the performance of the turbomachine. Solving for $\Delta H_D/D$ for such cases we can then say for Eq. (15.2) that

$$\frac{\Delta H_D}{D} = f_1\left(\frac{Q}{ND^3}, \frac{g}{N^2D} \right) \tag{15.3}$$

We see that to achieve similitude, it is necessary to duplicate the numbers Q/ND^3 and g/N^2D between the flows. The first expression we will see stems directly from the requirement of *kinematic similarity*—i.e., geometric similarity of the streamline patterns between the flows. Thus, we note first that

$$\frac{Q}{ND^3} \equiv \frac{Q/D^2}{ND} \tag{15.4}$$

The numerator on the right side gives a velocity proportional to the exit velocity of the fluid, while the denominator gives a velocity proportional to the tip speed of the impeller as noted earlier. Thus, the expression Q/ND^3 is proportional to the ratio of the fluid exit speed to the tip speed of impeller. For kinematic similarity of the streamlines and thus for dynamic similarity of the flows, this number, called the *flow coefficient*, must be duplicated between the flows. Now we know from experiment that the dimensionless group $\frac{g}{N^2D}$ in Eq. (15.3) actually occurs to the power of minus unity in the relation given and so we can express Eq. (15.3) as

$$\frac{\Delta H_D}{D} = \frac{N^2D}{g} f_2\left(\frac{Q}{ND^3}\right)$$

$$\therefore \frac{g\,\Delta H_D}{(ND)^2} = f_2\left(\frac{Q}{ND^3}\right) \tag{15.5}$$

Let us now consider the dimensionless group on the left side of Eq. (15.5). It is called the *head coefficient*. It is proportional to the ratio of the mechanical energy per unit mass to the square of the tip speed of the impeller.

We are thus left with two π's. If the flow coefficient is duplicated between two geometrically similar machines of different size (model and prototype), then assuming that we have attained dynamic similarity between the flows,[2] we must have the same head coefficients for the flows. Equating these head coefficients we can now determine certain information of the prototype (perhaps ΔH_D) from given data of N and D of the prototype plus the known data from the model.

Also we can plot the head coefficient versus the flow coefficient to form a curve of these two π's using only one machine and testing this machine over a range of speeds and heads to get sufficient points for plotting a curve of the two aforementioned coefficients against each other. This curve can then be used to give information for *any* machines geometrically similar to the test machine provided one can be reasonably sure that at a given flow coefficient the machine being considered from the graph will have a flow dynamically similar to the test machine used to develop the curve.

Next, consider the concept of the *efficiency* of a turbomachine. In the case of a *pump*, the actual total head on discharge will be less than the theoretical head because of losses due to friction and turbulence in the flow as well as

---

[2]Note that we have presented only necessary conditions for similitude; we have not presented sufficiency requirements. Those are determined by experience for various types of turbomachines.

leakage and mechanical losses due to friction in bearings, etc. We then define the efficiency of a pump as

$$e_{pump} = \frac{(\Delta H_D)_{actual}}{(\Delta H_D)_{theoretical}} 100 \qquad (15.6)$$

For a *turbine*, the actual head on discharge will be greater than the theoretical head, so there will be less power transmitted to the turbine. The reasons for this difference in total head are the same as given for the pump. We then say for the efficiency of the turbine

$$e_{turbine} = \frac{(\Delta H_D)_{theoretical}}{(\Delta H_D)_{actual}} 100 \qquad (15.7)$$

Another way of measuring $e_{turbine}$ that we will use later is to divide output power by input power. Thus,

$$e_{turbine} = \frac{\text{output power}}{\text{input power}} 100$$

Hydraulic efficiency up to as high as 90 percent can be achieved by pumps and turbines operating at design conditions.

If the machine is sufficiently large, efficiency e will depend on the size as given by D, the rotative speed N, and the volume flow Q. Dimensional analysis then shows that e is a function of the *flow coefficient* Q/ND^3. Thus,

$$e = f\left(\frac{Q}{ND^3}\right) \qquad (15.8)$$

For a given machine, we can then plot efficiency and head coefficient versus the flow coefficient using a common abscissa. Figure 15.3 shows such curves for a pump. The design condition of the pump corresponds to the highest efficiency.

Example 15.1. A model of mixed-flow water pump has been tested in the laboratory to give performance curves as is shown in Fig. 15.3. What would be the total head ΔH_D delivered by a prototype pump with an impeller size of 1.2 m operating at a speed of 1750 r/min and delivering 1.300 m³/s flow? What is its mechanical efficiency? Consider that dynamic similarity can be achieved between model flow and prototype flow.

We first compute the flow coefficient for the prototype flow.[3]

$$\left(\frac{Q}{ND^3}\right)_p = \frac{1.300}{[(1750)(2\pi)/60](1.2^3)} = 0.00411$$

---

[3] Note that for the flow coefficient to be dimensionless it is necessary that N be given in radians per second.

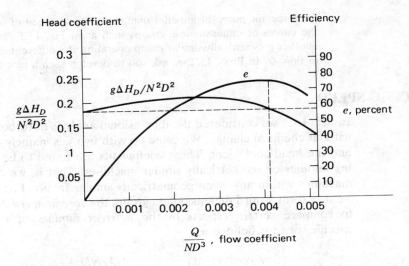

FIGURE 15.3
Head coefficient and efficiency versus flow coefficient.

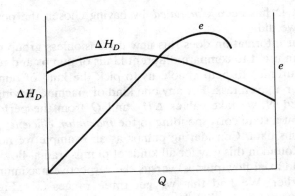

FIGURE 15.4
ΔH_D and e versus Q for a specific pump having an impeller diameter D_0 and a speed N_0.

Examining Fig. 15.3 we see that the efficiency for this operating point is about 75 percent for the prototype. Also note that the head coefficient is about 0.17. Therefore we have for the head ΔH_D of the prototype

$$\frac{g\,\Delta H_D}{N^2D^2} = 0.17$$

$$\therefore \Delta H_D = \frac{0.17}{9.81}\left(1750\frac{2\pi}{60}\right)^2(1.2^2) = 838 \text{ m}$$

Quite often the performance data for a *specific* pump is presented as ΔH_D versus Q, and e versus Q on the same chart (see Fig. 15.4).[4] We can arrive at such a plot

---

[4]This was done in Chap. 9. See Fig. 9.26.

for a specific pump of impeller diameter D_0 and speed of operation N_0 by using the curves of dimensionless groups such as in Fig. 15.3 developed from tests of another geometrically similar pump operating at a different speed and over a range of flow Q. In Prob. 15.2 we ask you to develop such a plot for a proposed pump.

15.3 SPECIFIC SPEED

In Sec. 15.2, we considered the dimensional analysis of turbomachines operating without chemical change. We came up with two π's, namely, the flow coefficient and the head coefficient. These coefficients are found to be useful in considering families of geometrically similar machines. That is, we can compare turbomachines within any given geometrically similar family. From these coefficients we may form a third dimensionless group, the *specific speed* N_s, which permits us to compare certain aspects of the *different families* of turbomachines. The specific speed is defined as

$$N_s = \frac{(\text{flow coefficient})^{1/2}}{(\text{head coefficient})^{3/4}} = \frac{(Q/ND^3)^{1/2}}{(g\,\Delta H_D/N^2D^2)^{3/4}} = \frac{N\sqrt{Q}}{(g\,\Delta H_D)^{3/4}} \quad (15.9)$$

Note that D has been *eliminated* by having chosen the powers for the coefficients as we did.

What information does this new dimensionless group convey? As pointed out, we can use it to compare different kinds of pumps and to compare different kinds of turbines to help enable us to pick the kind of pump or turbine most suitable for a given task. For any one kind of machine, having a given size and a given speed N, we take values ΔH_D and Q from the performance curve (see Fig. 15.4) at a state corresponding to the *maximum efficiency*, and then compute N_s from this data. Considering pumps as an example we now plot the specific speed N_s found in this way for all kinds of pumps (radial-flow pumps, mixed-flow pumps, and axial-flow pumps) versus the respective maximum efficiencies mentioned earlier. We find that we get three ranges (see Fig. 15.5). For small specific speeds, the radial-flow pumps have the greatest efficiency; for high specific speeds the axial-flow machines are most efficient; and the mixed-flow machines are most efficient at the intermediate range of specific speed. A moment's thought will explain this. In radial-flow machines, the inlet region must by necessity be small, limiting one to small Q's for efficient operation. However, because the exit flow occurs at a larger radius than at the inlet, a comparatively large head can then be efficiently developed by such a pump. Examining Eq. (15.9) we see that for efficient operation the small Q and the large ΔH_D make for a small specific speed N_s. For axial-flow motions the inlet area can be large to efficiently allow a large flow of fluid. But each stage of blading can transmit only a limited amount of energy efficiently (to avoid serious separation and even stalling). Accordingly, for efficient operation we can have a large Q but we are restricted to a comparatively small ΔH_D. On inspecting Eq. (15.9) we see that this leads to a large specific speed. And for reasons explained

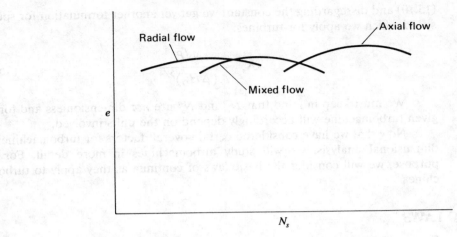

FIGURE 15.5
Maximum efficiency vs. specific speed for various classes of turbomachines.

above, the specific speeds for mixed-flow machines are somewhere in between for most efficient operation. Therefore, we see now how the specific speed helps us choose the kind of pump we desire. Similar arguments can be made for turbines.

In practice, the specific speed is often given the following definition for a *pump*:

$$N_s' = \frac{NQ^{1/2}}{(\Delta H_D)^{3/4}} \tag{15.10}$$

Note that the g is no longer present and consequently this formulation is no longer dimensionless. For such a formulation, note that N_s' is the speed N of the unit corresponding to flow Q of unity and total change in head ΔH_D of unity.

For *turbines*, it is the practice to use another formulation for specific speed N_s. First note that the output power P of a turbine is proportional to the flow Q times the total change in head ΔH_D. That is,

$$P \propto (Q)(\Delta H_D)$$

$$\therefore P = (\text{const})(Q)(\Delta H_D)$$

We can then say for Q that

$$Q = \left(\frac{P}{\Delta H_D}\right)(\text{const}) \tag{15.11}$$

For a given set of units and working fluid, the constant in the equations above is the same value for all turbines. We then substitute for Q given above into Eq.

(15.10) and disregarding the constant we get yet another formulation for specific speed which we apply for turbines:

$$N_s'' = \frac{N\sqrt{P}}{(\Delta H_D)^{5/4}} \tag{15.12}$$

We must keep in mind that N_s' and N_s'' are *not* dimensionless and for any given turbomachine will accordingly depend on the units involved.

Now that we have considered certain overall factors for turbomachines via dimensional analysis, we will study turbomachines in more detail. For this purpose, we will consider the basic laws of continua as they apply to turbomachines.

15.4 THE BASIC LAWS

Figure 15.6*a* shows a generalized turbomachine, including a shaft through which power is transmitted. There is a one-dimensional inlet flow and a one-dimensional outlet flow. The orientation of these flows is arbitrary. We choose a control volume that includes the interior region of the turbomachine. This control volume thus touches the inside surface of the casing of the machine and cuts the shaft in two places. Furthermore, it includes the rotor of the

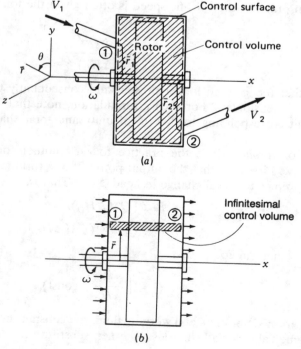

FIGURE 15.6
Control volumes for a generalized turbomachine.

machine and all the fluid inside the casing. We now consider the basic laws for this control volume for the case of *steady flow* with *no heat transfer* and *no chemical reactions*.

1. Continuity. For steady one-dimensional flow at inlet and outlet we have

$$\rho_1 (V_1)_n A_1 = \rho_2 (V_2)_n A_2 \qquad (15.13)$$

where A_1 and A_2 are areas subtended on the control surface by the interior of the inlet and outlet pipes, respectively. For *incompressible* flow we can say that

$$(V_1)_n A_1 = (V_2)_n A_2 = Q \qquad (15.14)$$

2. Moment of momentum. We will find it useful to use the moment-of-momentum equation taken about the axis of the shaft. We accordingly use Eq. (5.29), which for steady flow becomes

$$\oiint_{CS} \bar{r} T_\theta \, dA + \iiint_{CV} \bar{r} B_\theta \rho \, dv = \oiint_{CS} \bar{r} V_\theta (\rho \mathbf{V} \cdot \mathbf{dA}) \qquad (15.15)$$

The body force consists of the weight of the fluid and the weight of the rotor and shafting inside the control volume. This force distribution for axial symmetry gives zero torque about the axis of the shaft. As for the traction forces, we consider the region of the control surface touching fluid. We can say that pressure from the casing onto the fluid at the control surface gives rise to only a very small torque abut the axis of the shaft. This is also true of shear stresses at the control surface touching fluid, since the shear stresses there should be small. This leaves for consideration the region on the control surface cutting the shaft. Here appreciable shear stress occurs and there is a torque M_x that is transmitted through the control surface. Finally, noting the one dimensionality that we have assumed for inlet and outlet flows, we have for Eq. (15.15)

$$M_x = -\bar{r}_1 (V_\theta)_1 \left[\rho_1 (V_1)_n A_1 \right] + \bar{r}_2 (V_\theta)_2 \left[\rho_2 (V_2)_n A_2 \right] \qquad (15.16)$$

For incompressible flow we get[5]

$$\boxed{M_x = \left[\bar{r}_2 (V_\theta)_2 - \bar{r}_1 (V_\theta)_1 \right] \rho Q} \qquad (15.17)$$

Note that M_x is the torque exerted from *outside* the control volume.

3. First law of thermodynamics. For *steady* flow with *one-dimensional* inlet and outlet flows and for *no heat transfer* we may use Eq. (6.15). Thus,

$$\left(\frac{V_1^2}{2} + g y_1 + h_1 \right) - \frac{dW_s}{dm} = \left(\frac{V_2^2}{2} + g y_2 + h_2 \right) \qquad (15.18)$$

---

[5]Equation (15.17) is often called the *Euler turbine equation*.

For *incompressible* flow, we can say that the internal energy u is constant, so that we could replace the enthalpies h_1 and h_2 in Eq. (15.18) by p_1/ρ and p_2/ρ, respectively. Now multiply through by ρQ on the expressions in parentheses and by dm/dt (which equals ρQ) on the shaft-work expression. We then get

$$\left(\frac{V_1^2}{2} + gy_1 + \frac{p_1}{\rho}\right)\rho Q - \frac{dW_s}{dt} = \left(\frac{V_2^2}{2} + gy_2 + \frac{p_2}{\rho}\right)\rho Q \qquad (15.19)$$

If we multiply and divide each expression in parentheses by g we get

$$\left(\frac{V_1^2}{2g} + y_1 + \frac{p_1}{\gamma}\right)\gamma Q - \frac{dW_s}{dt} = \left(\frac{V_2^2}{2g} + y_2 + \frac{p_2}{\gamma}\right)\gamma Q$$

Note now that the terms in parentheses have units of length and are the head terms of H_D. We can now give the first law of thermodynamics for steady incompressible flow as[6]

$$\boxed{\frac{dW_s}{dt} = \gamma Q\left[(H_D)_1 - (H_D)_2\right] = -\gamma Q\,\Delta H_D} \qquad (15.20)$$

where $\Delta H_D = (H_D)_2 - (H_D)_1$.

4. Second law of thermodynamics. When there is compressibility and heat transfer present we must be careful that the second law of thermodynamics is not violated. For such situations, we use it as needed. However, for incompressible flows without heat transfer we will not need to explicitly call upon the second law of thermodynamics.

If we restrict ourselves to incompressible flow, we can make a very useful combination of the moment-of-momentum equation with the first law of thermodynamics. To do this, we multiply Eq. (15.17) by ω, the angular speed of the rotor. The expression $M_x\omega$, you should recall from sophomore mechanics, is the power transmitted from outside the control volume to inside the control volume, through the shaft. That is, it is simply $-dW_s/dt$. Hence we can say for Eq. (15.17) using the subscript t in place of θ that

$$\frac{dW_s}{dt} = \left[\bar{r}_1(V_t)_1 - \bar{r}_2(V_t)_2\right]\rho Q\omega \qquad (15.21)$$

Now using Eq. (15.20) to replace the left side of Eq. (15.21), we get

$$-\gamma Q\,\Delta H_D = \left[\bar{r}_1(V_t)_1 - \bar{r}_2(V_t)_2\right]\rho Q\omega \qquad (15.22)$$

---

[6]Note that when Eq. (15.20) is applied to a turbine $(H_D)_2 < (H_D)_1$ and dW_s/dt is positive, indicating work leaving the control volume. For a pump, $(H_D)_2 > (H_D)_1$, making dW_s/dt negative, indicating work entering the control volume.

Cancel ρQ and note that $\bar{r}\omega$ is the transverse speed of a particle rotating in a circular path at a radial distance $\bar{r}$ from the axis of rotation. In this case, $\bar{r}_1\omega$ and $\bar{r}_2\omega$ may be interpreted as the transverse speed of the rotor at positions $\bar{r}$ corresponding to the inlet and outlet, respectively, of the machine. Using the notation U_t for this quantity, we then get

$$\boxed{g\,\Delta H_D = [(U_t V_t)_2 - (U_t V_t)_1]} \tag{15.23}$$

Note that Eq. (15.23) was developed (see Fig. 15.6a) for a single small inlet at 1 and a single small outlet at 2. We can also apply this equation to a case (see Fig. 15.6b) where the influx and efflux extend over a comparatively large region such as in axial-flow pumps and turbines. Then the control volume could be taken as a small domain, as shown in Fig. 15.6b. In Eq. (15.23) $U_{t_2} = U_{t_1} = U_t$ is then the local speed of the impeller cut by the control surface; $(V_t)_1$ is the entering transverse velocity component of the fluid to the control volume; and $(V_t)_2$ is the exiting transverse velocity component of the fluid from the control volume. The total head change ΔH_D must be the *same* for all such control volumes at any value of $\bar{r}$.

Another useful equation can be directly formed from Eq. (15.21). Noting that $\bar{r}\omega$ can be taken as the speed U_t of the impeller at radial position $\bar{r}$, and, furthermore, noting that ρQ is the mass flow, we can say on dividing Eq. (15.21) by $\rho Q = dm/dt$ that

$$\boxed{\frac{dW_s}{dm} = [(U_t)_1(V_t)_1 - (U_t)_2(V_t)_2]} \tag{15.24}$$

**PART B
TURBINES**

15.5 INTRODUCTORY COMMENTS

A turbine is a turbomachine for transforming kinetic energy from a fluid to shaft work. In this sense, it is opposite to a pump, fan, or compressor in that in these latter machines, shaft work is transferred to higher kinetic energy of the fluid and/or to higher pressure head of the fluid.

We have already pointed out in Chap. 5 that we classify turbines as *impulse* if no expansion takes place in the flow through runner; the static pressure of the fluid remains essentially constant in the runner. The *reaction* turbine, on the other hand, allows expansion to take place in the runner; the static pressure decreases in the runner as the fluid moves through.

In the following section we will consider turbines having compressible and incompressible flows of both impulse type and reaction type.

15.6 IMPULSE TURBINES

We have already studied briefly the Pelton water wheel in Chap. 5. This is an impulse turbine. We now consider an *axial-flow* impulse turbine wherein the blades of the rotor are reasonably short compared with the hub diameter. Because the blades are short, we need only consider an average flow at some given radius. Furthermore, the flow will be incompressible while in the rotor because of the constant static pressure, etc., and includes hot-combustion products for a gas turbine or steam for a steam turbine. We will consider only a single-stage turbine.

Figure 15.7 shows a view of a single-stage impulse turbine. One or more convergent-divergent nozzles expands the working fluid to a supersonic speed V_1

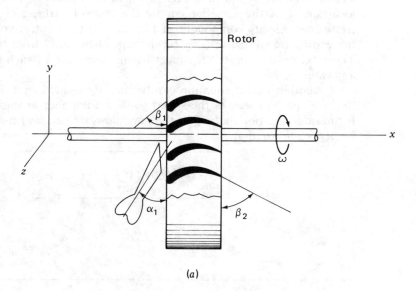

(a)

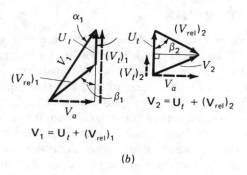

(b)

FIGURE 15.7
Single-stage axial-impulse turbine.

just outside the blades. The angle α_1 is the angle that $\mathbf{V}_1$ makes with the velocity vector $\mathbf{U}_t$ of the blades. The velocity of the fluid relative to the blade $(\mathbf{V}_{\text{rel}})_1$ is then shown in the velocity triangle in Fig. 15.7b at an angle β_1 which for smooth operation should equal the angle β_1 made by the surface of the blade making initial contact with the entering jet (see Fig. 15.7a). Note that $(\mathbf{V}_{\text{rel}})_1 = \mathbf{V}_1 - \mathbf{U}_t$. In the impulse turbine, the speed of the working fluid relative to the blade at exit $(V_{\text{rel}})_2$ is very close in value to the relative speed of the working fluid at entrance $(V_{\text{rel}})_1$—the difference coming from friction. We will neglect friction here thus making $(V_{\text{rel}})_2 = (V_{\text{rel}})_1$.

The power developed by the turbine can be determined from Eq. (15.21). Noting that $\bar{r}_2\omega = \bar{r}_1\omega = U_t$ for the case at hand,

$$\frac{dW_s}{dt} = U_t\big[(V_t)_1 - (V_t)_2\big]\rho Q$$

Note from Fig. 15.7b that

$$(V_t)_1 = V_1 \cos \alpha_1 \tag{15.25a}$$

Also, from Fig. 15.7b we see that

$$(V_t)_1 = (V_{\text{rel}})_1 \cos \beta_1 + U_t \tag{15.25b}$$

and

$$(V_t)_2 = U_t - (V_{\text{rel}})_2 \cos \beta_2 \tag{15.25c}$$

Hence, on using Eqs. (15.25a) and (15.25c) we get for dW_s/dt

$$\frac{dW_s}{dt} = U_t\rho Q\big[V_1(\cos \alpha_1) - U_t + (V_{\text{rel}})_2 \cos \beta_2\big] \tag{15.26}$$

Frequently, the angles β_1 and β_2 are made equal. We then get for Eq. (15.26) on noting that $(V_{\text{rel}})_1 = (V_{\text{rel}})_2$ as pointed out earlier,

$$\frac{dW_s}{dt} = \rho Q U_t\big[V_1(\cos \alpha_1) - U_t + (V_{\text{rel}})_1 \cos \beta_1\big] \tag{15.27}$$

Note from Eq. (15.25b) and then Eq. (15.25a) that

$$(V_{\text{rel}})_1 \cos \beta_1 = (V_t)_1 - U_t = V_1(\cos \alpha_1) - U_t$$

Substituting $(V_{\text{rel}})_1 \cos \beta_1$ as given above into Eq. (15.27), we get

$$\boxed{\frac{dW_s}{dt} = 2\rho Q U_t(V_1 \cos \alpha_1 - U_t)} \tag{15.28}$$

We now ask what the speed U_t of the blade must be to maximize the power output. To find this, take the partial of dW_s/dt with respect to U_t and set it equal to zero. We get

$$\frac{\partial}{\partial U_1}\left(\frac{dW_s}{dt}\right) = 0 = 2\rho Q(V_1 \cos \alpha_1 - 2U_t)$$

$$\boxed{U_t = \tfrac{1}{2}V_1 \cos \alpha_1} \tag{15.29}$$

If α_1 is small, we see that (like the Pelton water wheel) the speed of the jet approaches two times the speed of the blade for maximum power. Also, considering Eq. (15.27), we see that the smaller the approach angle α_1, the more power available.

We will now find a useful formulation for torque. We have assumed that the blades are so shaped that without friction the magnitude of the velocity of the working fluid relative to the blade is constant. That is,

$$(V_{\text{rel}})_1 = (V_{\text{rel}})_2 \tag{15.30}$$

In addition, with equal angles β and β_2, we conclude on considering Fig. 15.7b that the absolute *axial* velocity V_a of the working fluid has the same value at inlet and outlet from the runner. That is,

$$(V_a)_1 = (V_{\text{rel}})_1 \sin \beta_1 = (V_{\text{rel}})_2 \sin \beta_2 = (V_a)_2 \equiv V_a \tag{15.31}$$

Now consider Eq. (15.17). For the case at hand, $\bar{r}_1 = \bar{r}_2 = R$ and $V_\theta = V_t$. We then get for this case

$$M_x = \left[\bar{r}_2(V_\theta)_2 - \bar{r}_1(V_\theta)_1\right]\rho Q = R\rho Q\left[(V_t)_2 - (V_t)_1\right] \tag{15.32}$$

where M_x here is the torque on the shaft *from outside* the control volume and where R is the radial distance out to the center of the jet, impinging on the blades. Next, we employ Eqs. (15.25b) and (15.25c) respectively, for the V_t's in Eq. (15.32) while noting both Eq. (15.30) and the fact that $\beta_1 = \beta_2 = \beta$:

$$M_x = -R\rho Q\left[2(V_{\text{rel}})\cos \beta\right] \tag{15.33}$$

From Eq. (15.31) we have for V_{rel}

$$V_{\text{rel}} = \frac{V_a}{\sin \beta} \tag{15.34}$$

Substituting into Eq. (15.33) we get

$$\boxed{M_x = -2R\rho Q V_a \, \text{ctn}\, \beta} \tag{15.35}$$

We now illustrate the use of the above formulations.

Example 15.2. A single-stage, axial-flow, impulse turbine has a single nozzle delivering 3 kg of steam per second at a speed of 2000 m/s. The angle α for the nozzle is 32° and β for the blades is the same value at both edges. The radius R to the nozzle is 0.8 m. Find the theoretical power and torque for this machine operating at its most efficient speed ω. What should the angle β be?

For the most *efficient speed* ω we make use of Eq. (15.29). Thus

$$U_t = R\omega = \tfrac{1}{2}V_1 \cos \alpha_1$$

$$\therefore \omega = \frac{1}{2R}V_1 \cos \alpha_1 = \frac{1}{(2)(0.8)}2000 \cos 32°$$

$$= 1060 \text{ rad/s} = 10,123 \text{ r/min}$$

Also,
$$U_t = (0.8)(1060) = 848 \text{ m/s}$$

To get the *power output*, we employ Eq. (15.28). Thus

$$\frac{dW_s}{dt} = 2\rho Q U_t(V_1 \cos \alpha_1 - U_t)$$

$$= (2)(3)(848)(2000 \cos 32° - 848)$$

$$= 4.315 \times 10^6 \text{ N} \cdot \text{m/s} = 4.315 \times 10^3 \text{ kW} \qquad (a)$$

The torque M_x on the shaft *from outside* the control volume can be found by noting that

$$M_x\omega = -\frac{dW_s}{dt}$$
$$\qquad\qquad\qquad\qquad\qquad\qquad\qquad\qquad (b)$$
$$M_x = -\frac{1}{\omega}\frac{dW_s}{dt} = -\left(\frac{1}{1060}\right)(4.315 \times 10^6) = -4.071 \times 10^3 \text{ N} \cdot \text{m}$$

We next determine the proper blade *angle* β. Consulting Fig. 15.7b we see that

$$(\mathbf{V}_{\text{rel}})_1 = \mathbf{V}_1 - \mathbf{U}_t = (\mathbf{V}_1)(\cos \alpha_1 \mathbf{j} + \sin \alpha_1 \mathbf{i}) - U_t\mathbf{j}$$

$$= 2000[\cos 32° \mathbf{j} + \sin 32° \mathbf{i}] - 848\mathbf{j}$$

$$= 848\mathbf{j} + 1060\mathbf{i} \quad \text{m/s}$$

The angle β is then determined as follows:

$$\beta = \tan^{-1}\left(\frac{1060}{848}\right) = 51.3° \qquad (c)$$

As a check on our calculations, we now recompute M_x using Eq. (15.35):

$$M_x = -2R\rho Q V_a \operatorname{ctn} \beta$$

$$= -(2)(0.8)(3)[(2000)(\sin 32°)(\operatorname{ctn} 51.3°)] = -4.071 \times 10^3 \text{ N} \cdot \text{m} \quad (d)$$

15.7 RADIAL AND AXIAL-FLOW REACTION TURBINES

We pointed out earlier that for reaction turbines there takes place in the rotor a conversion from pressure head to kinetic energy of the fluid in contrast to an impulse turbine. There are no nozzles in a reaction turbine and the fluid completely fills the vane passages of the rotor.

In the so-called *Francis* turbine, the flow through the runner is in a *radial* direction and accordingly this is a radial-flow machine (see Fig. 15.8). On the other hand, the flow in a runner may also be in the axial direction. We discuss axial-flow reaction turbines in this section as well as in the next. First we deal

FIGURE 15.8
Installing 149-ton turbine runner at Niagara generating plant. The speed of the runner is 120 r/min; the horsepower rating is 70,000 hp; and the diameter is 17 ft.

with hydraulic types of such turbines. Here there is a single row comprising a few blades (4 to 6 blades) which are comparatively large. We call this kind of turbine a *propeller* type of turbine. The flow velocity in such a turbine varies with the radial distance from the axis of the turbine and hence cannot be handled with one-dimensional flow theory. In the following section, we consider turbines having many small, short blades (called a *cascade*) with possibly many stages. Here one-dimensional flow theory can be used.

Accordingly in Fig. 15.9 we show a hydraulic propeller-type turbine, called a *Kaplan* turbine. The guide vanes may have their orientation adjusted to control the inlet flow. The water is given a rotational motion as seen from above by these vanes. Consider a mass dm in the flow after passing the guide vanes. Gravity in the vertical direction acts on this element. We will consider also that

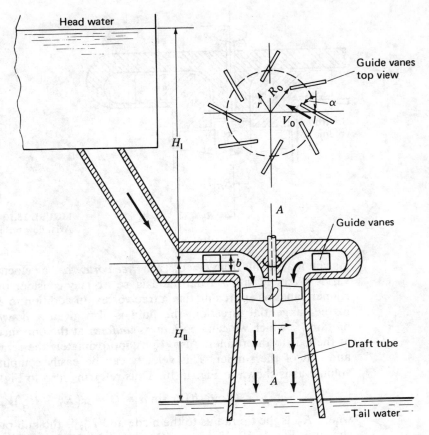

FIGURE 15.9
Axial-flow hydraulic turbine.

the pressure gradient acts *radially inward* toward the center O and hence toward the axis of rotation A-A. We will neglect shear stress in the bulk of the flow. If we consider the moment of linear momentum of dm about axis A-A, we see that the resulting torque from $-\nabla p$ and gravity is zero, so the angular momentum is *conserved* for the mass element about axis A-A. Hence we can say (see Fig. 15.9) that

$$dm(V_0 \cos \alpha) R_0 = dm\, V_t r$$

where V_t is the tangential velocity component of dm at any position r after leaving the guide-vane exit location at $r = R_0$. Solving for V_t we get

$$V_t = \frac{V_0 R_0 \cos \alpha}{r} \tag{15.36}$$

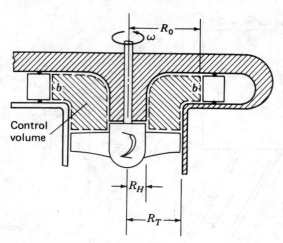

FIGURE 15.10
Axial-flow turbine with control volume.

As you may recall from Chap. 12, a *free vortex* has a velocity about its axis that varies inversely with r from the axis, so we may consider the flow entering the runner vanes as approximating a free vortex. In addition to a free vortex motion in the tangential direction, the fluid is also given a *downward* axial velocity motion V_a which we can consider as *uniform* at the entrance to the runner and at the exit to the runner. Also, V_a is approximately the same value at entrance and exit of the runner. This velocity can be easily computed using a control volume as is shown in Fig. 15.10. Thus referring also to Fig. 15.9 we have

$$(2\pi R_0 b)V_0 \sin \alpha \equiv Q = \pi(R_T^2 - R_H^2)V_a$$

where R_T is the tip radius to the blade and R_H is the hub radius to the blade as shown in Fig. 15.10. Hence, we then have for V_a from this equation

$$V_a = \frac{2\pi R_0 b V_0 \sin \alpha}{\pi(R_T^2 - R_H^2)} \qquad (15.37a)$$

Also, from the same equation

$$V_a = \frac{Q}{\pi(R_T^2 - R_H^2)} \qquad (15.37b)$$

Because V_t is a function of r, we can no longer use a one-dimensional flow through the turbine. Instead, we must consider separately positions of flow at varying radii through the runner with blade angles β_1 and β_2 varying with the radial distance r.

Before proceeding to the examples, we must point out that the water leaving the blades of the turbine will have negative gage pressure. The purpose of the *draft tube* (see Fig. 15.9) is to convert part of this kinetic energy to pressure to eliminate this partial vacuum as we get to the tailwater. The partial vacuum directly after the blade adds to the head across the turbine. To account for this, we take as the *effective head* H_D across the blades the height of the head water above the tailwater, namely, $(H_I + H_{II})$ in Fig. 15.9.

Example 15.3. The following data apply to an axial-flow, hydraulic turbine (see Figs. 15.9 and 15.10):

$$R_0 = 2 \text{ m} \qquad\qquad b = 1 \text{ m}$$

$$R_H = 180 \text{ mm} \qquad\qquad \omega = 350 \text{ r/min}$$

$$R_T = 900 \text{ mm} \qquad H_I + H_{II} = 100 \text{ m}$$

$$\alpha = 13° \qquad\qquad V_0 = 13.74 \text{ m/s}$$

Upon leaving the blades, the water has no angular momentum in the draft tube. Neglecting friction and other losses, determine the ideal power developed by the turbine.

The flow of water leaving the stationary guide vanes (see top of Fig. 15.9) can be equated to Q as follows

$$Q = 2\pi R_0 b V_0 \sin \alpha$$

$$= (2\pi)(2)(1)(V_0 \sin 13°) = 2.83 V_0 \text{ m}^3/\text{s} \qquad (a)$$

We next go to Eq. (15.23), since we know the effective total head H_D to be $H_I + H_{II} = 100$ m. Noting that $(V_t)_2 = 0$ at the exit due to zero angular momentum there, we have at any distance r from the axis of the rotor

$$g \, \Delta H_D = [0 - (U_t V_t)_1] = -r\omega(V_t)_1$$

$$\therefore -(9.81)(100) = -r\left[\frac{(350)(2\pi)}{60}\right](V_t)_1 \qquad (b)$$

$$\therefore (V_t)_1 = \frac{26.77}{r}$$

The velocity V_a is next computed [see Eqs. (15.37) and Eq. (a)].

$$V_a = \frac{Q}{\pi(R_T^2 - R_H^2)}$$

$$= \frac{2\pi R_0 b V_0 \sin \alpha}{\pi(R_T^2 - R_H^2)}$$

$$= \frac{(2.83)(13.74)}{(\pi)(0.9^2 - 0.180^2)}$$

$$= 15.88 \text{ m/s}$$

We will now consider six regions of flow around the blades (see Fig. 15.11). We have denoted the regions, you will note, as a, b, c, d, e, f. Also, the velocity diagram at any position r is shown in Fig. 15.12.

The radius r that we will first be considering is

$$r_a = 900 - 60 = 840 \text{ mm}$$

The tangential velocity ahead of the blades at r_a for the fluid is determined from

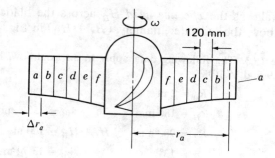

FIGURE 15.11
Blades showing six domains for study.

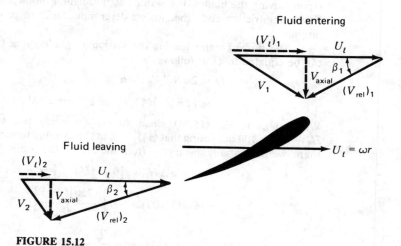

FIGURE 15.12
Velocity triangles for position r.

Eq. (b) depicting the free-vortex motion of the fluid. Thus

$$[(V_t)_1]_{r_a} = \frac{26.77}{r_a} = \frac{26.77}{0.840} = 31.87 \text{ m/s}$$

We now have enough information for region a to compute the power developed. For this purpose, we compute the torque developed on each blade in region a. We accordingly choose a control volume which is a cylindrical region encompassing region a for all the blades as is shown in Fig. 15.13, wherein only two blades are shown.[7] We use the moment-of-momentum equation about axis A-A for steady incompressible flow. Equation (15.17) may be used directly for this control volume. Noting that $r_2 = r_1 = r_a$ in this formula and that $V_\theta \equiv V_t$, we get

---

[7]The control surface accordingly cuts the blades at $r = R_T - 0.120$ m.

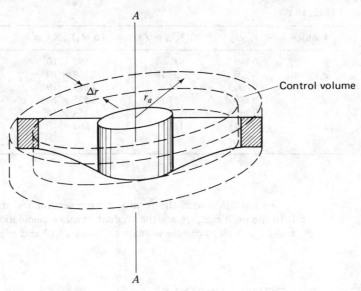

FIGURE 15.13
Control volume of outer ring.

on using Eq. (*b*),

$$(\Delta M_A)_a = [r_a(V_t)_2 - r_a(V_t)_1]\rho Q$$

$$= r_a\left[\frac{0 - 26.77}{r_a}\right]\rho(2\pi r_a)(\Delta r_a)(V_a)$$

$$= -(r_a)\frac{26.77}{r_a}(1000)(2\pi)(r_a)(0.120)(15.88)$$

$$= -3.205 \times 10^5 r_a \qquad (c)$$

The power then is

$$\Delta P_a = -(\Delta M_A)_a(\omega) = (3.205 \times 10^5)(r_a)\left[\frac{(350)(2\pi)}{(60)}\right] = 1.175 \times 10^7 r_a$$

$$= 9867 \text{ kW} \qquad (d)$$

For the power of the *entire* turbine, we have, on using Eq. (*c*) for *any* section at r_i

$$(\Delta P)_{\text{total}} = \sum_i (\Delta M_A)_i(\omega)$$

$$= \sum_i (3.205 \times 10^5 r_i)\left[\frac{(350)(2\pi)}{60}\right]$$

$$= 1.175 \times 10^7(0.840 + 0.720 + 0.600 + 0.480 + 0.360 + 0.240)$$

$$= 3.807 \times 10^7 \text{ W} = 38,070 \text{ kW}$$

TABLE 15.1

Position	r_i, m	$[(V_t)_1]_i$, m / s	$(\Delta M_A)_1$, N $\cdot$ m	$(\Delta P)_i$, kW
a	0.840	31.87	2.692×10^5	9,867
b	0.720	37.18	2.308×10^5	8,460
c	0.600	44.62	1.923×10^5	7,050
d	0.480	55.77	1.538×10^5	5,640
e	0.360	74.4	1.154×10^5	4,230
f	0.240	111.5	7.692×10^4	2,820
Total				38,070

We have tabulated the results at each section and the total above in Table 15.1. In the problems, we ask the student to make calculations involving the angles β_1 and β_2 as well as certain velocities. Figure 15.12 and trigonometry can be used.

15.8 MANY-BLADED REACTION TURBINES (AND COMPRESSORS)

In Sec. 15.7, we discussed the Kaplan hydraulic turbine, where because of the large size and small number of blades we could not carry out a one-dimensional analysis. We will now consider axial-flow reaction turbines where there are many small blades in each stage. Because the theory for axial-flow compressors with many small blades in each section (see Fig. 15.14) is similar, we will (as indicated in the above heading) also consider compressors in this section even though the bulk of the discussion of compressors falls into Part C of this chapter. In such turbines and compressors, the flow does not vary appreciably with the radial distance r per stage and hence may be considered one-dimensional in our study. In our discussion of the turbojet engine in Chap. 5, you will recall there was a compressor section of the type described above and a turbine section. There can be many stages in such devices with 15 not uncommon for compressors. Figure 15.15 shows a row of stationary blades, a single stage following, as well as the associated velocity diagrams for a turbine. Note that the tip diameter of stages will change to accommodate changes in density of the working fluid which might be steam, air, or combustion products. We will consider that density changes and changes in the annular region offset each other so that the axial velocity V_a remains constant.

Let us now consider a single stage of the turbine. Since the density change per stage is small, we can consider *incompressible* flow in each stage. But we should use a different average density in each stage. Hence, using Eq. (15.23),

$$(\Delta H_D)_{1\text{-}2} = \frac{U_t}{g}[(V_t)_2 - (V_t)_1] \tag{15.38}$$

Note from the velocity diagrams in Fig. 15.15 that we can express $(V_t)_2$ and $(V_t)_1$ as follows:

$$(V_t)_1 = V_a \cot \alpha_1$$

$$(V_t)_2 = U_t - V_a \cot \beta_2$$

Hence Eq. (15.38) becomes

$$(\Delta H_D)_{1\text{-}2} = \frac{U_t}{g}[U_t - V_a \cot \beta_2 - V_a \cot \alpha_1]$$

We next rewrite this equation in the following dimensionless form:

$$\frac{g(\Delta H_D)_{1\text{-}2}}{U_t^2} = \left[1 - \frac{V_a}{U_t}(\cot \alpha_1 + \cot \beta_2)\right] \qquad (15.39)$$

Equation (15.39) although derived while considering axial-flow turbines (where ΔH_D is negative) is also valid for axial-flow, many-bladed compressors (where ΔH_D is positive). We consider three domains for these devices, namely,

$$(\cot \alpha_1 + \cot \beta_2) < 0 \qquad (15.40a)$$

$$(\cot \alpha_1 + \cot \beta_2) = 0 \qquad (15.40b)$$

$$(\cot \alpha_1 + \cot \beta_2) > 0 \qquad (15.40c)$$

We next plot the dimensionless group $(\Delta H_D)_{1\text{-}2}g/U_t^2$ versus V_a/U_t for each of the above domains in Fig. 15.16. First you will note that above the abscissa (where $\Delta H_D > 0$) we have compressor operation and below the abscissa (where $\Delta H_D < 0$) we have turbine operation. Note further that for any set of values of α_1 and β_2 satisfying Eq. (15.40a), we must get positive slopes for the ideal performance of that particular machine. Similarly, if we choose α_1 and β_2 to correspond to Eq. (15.40c), we must get negative slopes for the ideal performance of that particular machine, etc. Thus for Eq. (15.40c), which encompasses most compressors, the total head will decrease with increasing value of V_a/U_t. If the speed of the compressor is kept constant as for the case of a compressor driven by a synchronous motor (that is, $U_t = \text{const}$), then the head ΔH_D will decrease with increase in volumetric flow Q.[8] This is shown in Fig. 15.17 as the indicated ideal curve.

---

[8]In Prob. 15.28 you will be asked to explain how compressors with positive performance slope can be unstable in operation.

FIGURE 15.14
Compressor blading. (*Pratt and Whitney Aircraft Group.*)

Let us next consider *actual* performance of synchronous compressors under discussion in this section. The actual value of ΔH_D can be found by subtracting various kinds of head losses from the ideal values shown in Fig. 15.17. The most significant drop in head from the ideal head occurs because the angle β_2 of the blades (see Fig. 15.15) is actually not imposed on the direction of the relative velocity $(V_{\text{rel}})_2$ as has been assumed in the formulation of the

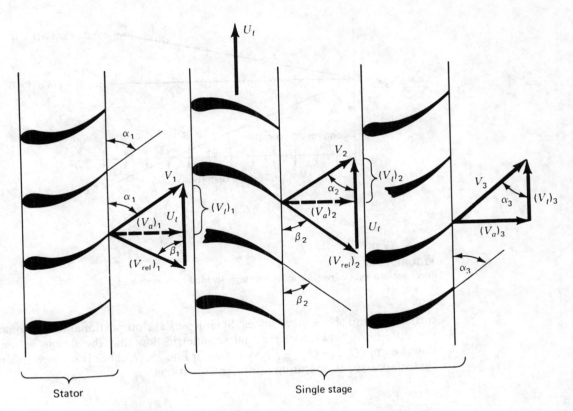

FIGURE 15.15
Velocity diagrams for axial-flow turbines.

equations of this section. Actually, to have the exit-blade angle and the angle of the exit relative velocity at a stage to be identical requires an *infinite* number of blades of zero thickness. In the real case, the actual discharge is *as if* angle β_2 of $(V_{rel})_2$ is *less* than the exit-blade angle. Let us now consider Eq. (15.39). Using a smaller value of β_2 than that of the exit-blade angle in this equation gives us a larger value of $\cot \beta_2$ and hence a smaller ΔH_D than the ideal case. This effect on the head is called the *circulatory flow* effect, shown in Fig. 15.17. Next, there are friction losses ever present in a real flow. This head loss is proportional to Q^2 and is also shown in Fig. 15.17. Now, if the compressor is *not* operating at the design volumetric flow rate, then there will be improper matching of the inlet relative velocity direction of the flow with the orientation of the blades. This results in separation of the flow along the blades and causes

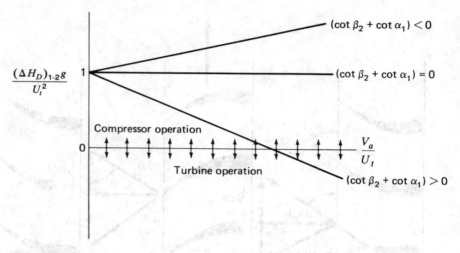

FIGURE 15.16
Ideal-performance curves for axial-flow many-bladed turbomachines.

turbulence and shock-wave losses. Such losses are proportional to the square of the difference between the actual volumetric flow and the design volumetric flow, i.e., to $(Q_{act} - Q_{design})^2$. As shown in Fig. 15.17, there is no such loss when the compressor is operating at design conditions.

PART C
FANS, PUMPS, BLOWERS, AND COMPRESSORS

15.9   INTRODUCTORY REMARKS

Fans, pumps, blowers, and compressors transfer energy from source to the working fluid. In this respect, these turbomachines are directly opposite in function to turbines. Like turbines, pumps and blowers, etc., can be of the radial-flow, axial-flow, or mixed-flow type. For high heads, the radial type of pump (the centrifugal pump) is used; for large flows with small heads, the axial-flow compressor or fan is best. The mixed-flow turbomachines are used in between these extremes for best performance.

15.10   RADIAL-FLOW PUMPS AND BLOWERS

We will now consider radial-flow pumps and blowers. In such machines, the fluid enters the machine while moving axially along the centerline of the impeller (see Fig. 15.18). The fluid is then forced radially into the passageways

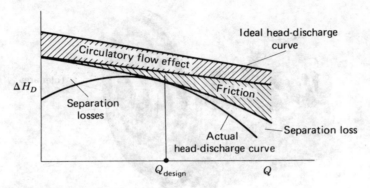

FIGURE 15.17
Head-discharge curves showing various losses for pumps.

of the impeller. In passing through the impeller, the fluid will undergo a change in pressure and a change in velocity relative to the impeller. The flow then emerges from the impeller into the pump casing, which may now act as a diffuser to convert kinetic energy to pressure head.

If the fluid is a liquid such as water or oil, we call the machine described above a centrifugal pump. On the other hand, if the fluid is a gas such as air, we call the machine a centrifugal compressor. Furthermore, the action described in the previous paragraph through a single rotor and then to the outlet is the description of a single-stage machine. There may be a number of stages present

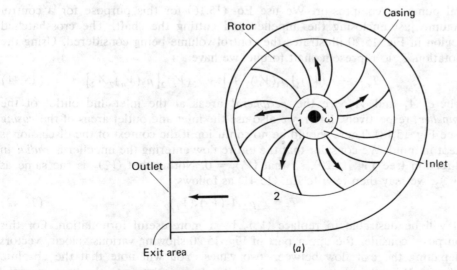

FIGURE 15.18A
Centrifugal blower. (*Courtesy Buffalo Forge Co., Buffalo, N.Y.*)

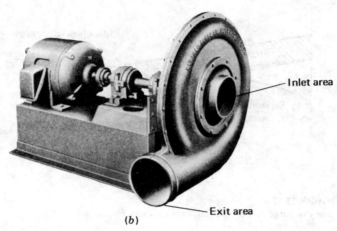

(b)

FIGURE 15.18B

in the centrifugal pump or centrifugal compressor where the output of an impeller is fed axially into the input of a second impeller riding on the same shaft and so forth for yet other impellers mounted on the shaft. Figure 15.19 shows a multistage pump.

We will first evaluate certain common characteristics of the centrifugal pump and the centrifugal compressor and then we will consider individual characteristics separately. We will first note that the flow in the region between the vanes will be considered *one-dimensional* flows relative to the impeller.

Let us now consider the *moment-of-momentum* equation for the centrifugal pump or compressor. We use Eq. (15.16) for this purpose for a control volume just enclosing the impeller and cutting the shaft. The crosshatched region in Fig. 15.20 illustrates the control volume being considered. Using the notation T_s to represent shaft torque, we have

$$T_s = -r_1(V_t)_1[\rho_1(V_n)_1 A_1] + r_2(V_t)_2[\rho_2(V_n)_2 A_2] \qquad (15.41)$$

where A_1 and A_2 are the *peripheral* areas at the inlet and outlet of the *impeller*, respectively. Later, we also use the inlet and outlet areas of the *casing* (see Fig. 15.18). There should be no confusion if the context of the discussion is kept in mind. We consider that the entire flow entering the impeller is *radial* in direction (see Fig. 15.20), so that $(V_t)_1 = 0$. Noting that $(V_n)_2$ is the same as $(V_r)_2$, we may then rewrite Eq. (15.41) as follows:

$$T_s = r_2\rho_2 A_2(V_r)_2(V_t)_2 \qquad (15.42)$$

It will be desirable to replace $(V_t)_2$ by a more-useful formulation. For this purpose, consider the upper part of Fig. 15.20 showing various velocity vectors depicting the exit flow between two vanes. You will note that the absolute velocity $\mathbf{V}_2$ is the vector sum of the exit velocity relative to the vane, namely, $(\mathbf{V}_{rel})_2$, plus the vane velocity $\mathbf{U}_t$. Observing in particular the shaded triangular

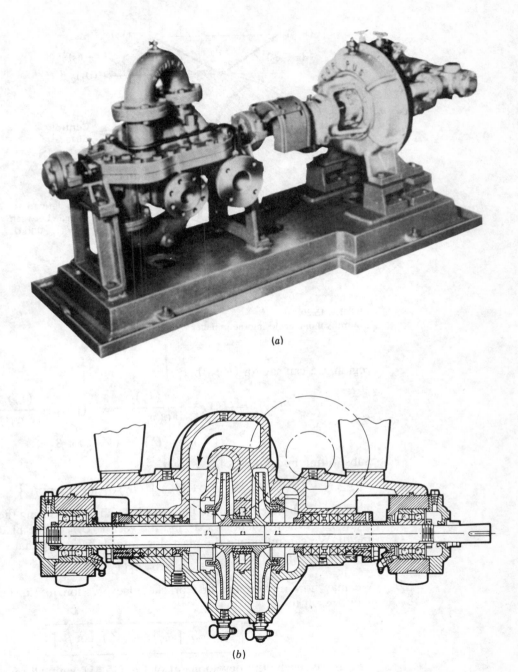

(a)

(b)

FIGURE 15.19
Two-stage pump. (*Courtesy Buffalo Force Co., Buffalo, N.Y.*)

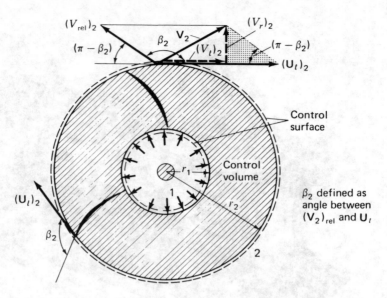

FIGURE 15.20
Control volume enclosing the impeller.

region, we can say for $(V_t)_2$ that:

$$(V_t)_2 = (U_t)_2 - \frac{(V_r)_2}{\tan(\pi - \beta_2)} = (U_t)_2 + \frac{(V_r)_2}{\tan \beta_2}$$

$$\therefore (V_t)_2 = (U_t)_2 + (V_r)_2 \cot \beta_2$$

Substituting into Eq. (15.42), we get

$$T_s = r_2[\rho_2 A_2(V_r)_2][(U_t)_2 + (V_r)_2 \cot \beta_2] \tag{15.43}$$

where you will note that β_2 is the exit-blade angle measuring the angle between the tangent to the blade at exit and the vector $(U_t)_2$ at the blade end (see Fig. 15.20). We can introduce the mass-flow rate $\dot{m}$ by noting that

$$\dot{m} = \rho_2 A_2(V_r)_2 \tag{15.44}$$

We may now replace the first bracketed expression in Eq. (15.43) using the above result to get

$$\boxed{T_s = r_2\dot{m}[(U_t)_2 + (V_r)_2 \cot \beta_2]} \tag{15.45}$$

If we review the development of Eq. (15.41), you will see that friction on the walls of the control volume was entirely neglected. Therefore, using the actual radial velocity $(V_r)_2$ and the actual mass flow in Eq. (15.45), we will not calculate the actual shaft torque T_s needed for the pump but instead a value

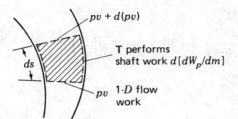

T performs
shaft work $d[dW_p/dm]$

1-*D* flow
work

FIGURE 15.21
Work terms on the fluid system per unit mass.

lower than that which is required. Also, to run the pump there is additional torque needed to overcome bearing and packing friction effects. To account for all such losses, we introduce the *mechanical efficiency* η_m such that

$$\eta_m = \frac{T_s}{(T_s)_{\text{actual}}} \tag{15.46}$$

The actual power $(P)_{\text{input}}$ needed to run the pump or compressor is then

$$(P)_{\text{input}} = (T_s)_{\text{actual}}(\omega) = \frac{T_s}{\eta_m}\omega$$

Hence we have on using Eq. (15.45)

$$(P)_{\text{input}} = \frac{\omega}{\eta_m} r_2 \dot{m} \left[(U_t)_2 + (V_r)_2 \cot \beta_2 \right] \tag{15.47}$$

Let us now look into the details of the flow between the vanes in the impeller. You recall that we consider this to be one-dimensional flow relative to the impeller. We consider an infinitesimal *system* (see Fig. 15.21) in this region and will use the first law of thermodynamics for this system. Note that the work increment at the boundary of the system touching the vanes *from* the rotor is *pump-shaft* work which in the first law of thermodynamics we express as $-(dW_p/dm)\,dm$. The work at the remaining boundary is the *flow work* stemming from pv at the boundary cross sections of the system. If we take W_k to be the *total* work done *on* the infinitesimal system, the first law of thermodynamics states that

$$dE = dQ - dW_k$$

For adiabatic conditions, we may rewrite this equation as

$$dE \equiv d\left[\left(\frac{V^2}{2} dm \right) + u\,dm + gz\,dm \right] = 0 + d\left(\frac{dW_p}{dm} dm \right)$$
$$+ \{ pv - [pv + d(pv)] \}\,dm$$

Dividing through by dm we get

$$d\left(\frac{V^2}{2} \right) + du + g\,dz = d\left(\frac{dW_p}{dm} \right) - d(pv)$$

Rearranging we get replacing v by $1/\rho$

$$d\left(\frac{dW_p}{dm}\right) = V\,dV + du + d\left(\frac{p}{\rho}\right) + g\,dz$$

$$= V\,dV + du + pd\left(\frac{1}{\rho}\right) + \frac{1}{\rho}\,dp + g\,dz \qquad (15.48)$$

Now consider the *combined first and second law* for the system:

$$T\,ds = du + pd\left(\frac{1}{\rho}\right) \qquad (15.49)$$

where s is the entropy. For isentropic conditions $s = $ const, so the left side of Eq. (15.49) is zero. Hence we have

$$du + pd\left(\frac{1}{\rho}\right) = 0$$

Noting this fact in Eq. (15.48) we have:

$$d\left[\frac{dW_p}{dm}\right] = V\,dV + \frac{1}{\rho}\,dp + g\,dz \qquad (15.50)$$

To get a theoretical (and hence maximum) evaluation of the shaft work per unit mass, dW_p/dm, we must integrate Eq. (15.50) for the system undergoing an isentropic process as the system goes from inlet to outlet. In the case of a *liquid* we take ρ as constant and we get:

$$\boxed{\left[\frac{dW_p}{dm}\right]_{\text{theoretical}} = \left[\frac{(V_2)^2_{\text{theoretical}}}{2} - \frac{V_1^2}{2}\right] + \frac{p_2 - p_1}{\rho} + g(z_2 - z_1)} \qquad (15.51)$$

If, on the other hand, we can consider the fluid to be a *perfect gas*, we must consider ρ as a variable. We neglect elevation head and again integrate Eq. (15.50) for an isentropic process. Note for an isentropic process of a perfect gas that

$$pv^k = p\frac{1}{\rho^k} = \frac{p_1}{\rho_1^k}$$

$$\therefore \rho = \left(\frac{p}{p_1}\right)^{1/k} \rho_1 \qquad (15.52)$$

Substituting into Eq. (15.50), we get

$$d\left(\frac{dW_p}{dm}\right) = V\,dV + \left(\frac{p}{p_1}\right)^{-1/k} \frac{1}{\rho_1}\,dp$$

Now integrate from the entrance 1 to the exit 2. We get

$$
\begin{aligned}
\left[\frac{dW_p}{dm}\right]_{\text{theoretical}} &= \left[\frac{(V_2^2)_{\text{theoretical}}}{2} - \frac{V_1^2}{2}\right] + \frac{p_1^{1/k}}{\rho_1}\left(p_2^{1-(1/k)} - p_1^{1-(1/k)}\right)\frac{1}{1 - 1/k} \\
&= \left[\frac{(V_2^2)_{\text{theoretical}}}{2} - \frac{V_1^2}{2}\right] + \frac{k}{k-1}\frac{p_1^{1/k}}{\rho_1}\left(p_2^{(k-1)/k} - p_1^{(k-1)/k}\right) \\
&= \left[\frac{(V_2^2)_{\text{theoretical}}}{2} - \frac{V_1^2}{2}\right] + \frac{k}{k-1}\frac{p_1^{1/k}}{\rho_1}p_1^{(k-1)/k}\left[\left(\frac{p_2}{p_1}\right)^{(k-1)/k} - 1\right]
\end{aligned}
$$

$$(15.53)$$

Consider the coefficient of the last bracketed expression.

$$
\frac{k}{k-1}\frac{p_1^{1/k}}{\rho_1}p_1^{(k-1)/k} = \frac{k}{k-1}\left(\frac{p_1}{\rho_1}\right) = \frac{k}{k-1}\frac{p_1}{\rho_1}\frac{T_1}{T_1}
$$

In the last expression replace $p_1/(\rho_1 T_1)$ by R from the equation of state. We then have

$$
\frac{k}{k-1}\frac{p_1^{1/k}}{\rho_1}p_1^{(k-1)/k} = \frac{k}{k-1}RT_1
$$

Noting that $[k/(k-1)]R$ is simply c_p, the specific heat for constant pressure, we see that the right side of the equation above becomes $c_p T_1$, so we can express Eq. (15.53) in the following manner:

$$
\boxed{\left[\frac{dW_p}{dm}\right]_{\text{theoretical}} = \frac{(V_2)^2_{\text{theoretical}} - V_1^2}{2} + c_p T_1\left[\left(\frac{p_2}{p_1}\right)^{(k-1)/k} - 1\right]} \qquad (15.54)
$$

We will now illustrate the use of the various equations set forth in this section.

Example 15.4. A centrifugal blower (see Fig. 15.18) compresses 0.63 kg/s of air from a pressure of 1 atm and a temperature of 18° C to a pressure of 250 mm water gage. The inner and outer radii of the impeller are 50 mm and 120 mm, respectively. The exit area of the *casing* is 8000 mm² and it is here that the pressure of 250 mm water gage is reached. The inlet area of the *casing*, where we have atmospheric pressure, is 15,000 mm². What power is needed to run the blower under ideal conditions? If $\omega = 5000$ r/min, what should the exit-blade angle β_2 be? Consider that there is no diffuser attached to the blower. The width b of the impeller is 0.3 m. Take $c_p = 0.24$ Btu/(lbm)(°F).

Consider the inlet conditions at the control surface (see Fig. 15.20) to be identified by 1 and the outlet conditions at the control surface to be identified by 2. We will consider that the conditions at 2 are the same as at the outlet of the machine (see Fig. 15.18).

We use Eq. (15.54) to get dW_p/dm. We need to compute $(V_2)_{\text{theoretical}}$ as well as V_1. For this purpose, we first compute p_2.

$$p_2 = (0.250)(9806) = 2451 \text{ Pa gage}$$

$$\therefore \frac{p_2}{p_1} = \frac{101{,}325 + 2451}{101{,}325} = 1.0242$$

We need ρ_1. From the *equation of state* we have

$$p_1 = \rho_1 R T_1$$

$$\therefore \rho_1 = \frac{p_1}{RT_1} = \frac{101{,}325}{(287)(273 + 18)} = 1.213 \text{ kg/m}^3$$

Hence for ρ_2,

$$\frac{\rho_1}{\rho_2} = \left(\frac{p_1}{p_2}\right)^{1/k}$$

$$\rho_2 = \rho_1 \left(\frac{p_2}{p_1}\right)^{1/k} = (1.213)(1.024)^{1/k} = 1.2339 \text{ kg/m}^3$$

We now use the *continuity* equation to calculate the theoretical velocity $(V_2)_{\text{theoretical}}$. Thus

$$\dot{m} = 0.63 = \rho_2 (V_2)_{\text{theoretical}} A_2 = (1.2339)(V_2)_{\text{theoretical}}(8000)(10^{-6})$$

$$(V_2)_{\text{theoretical}} = 63.8 \text{ m/s}$$

Also from *continuity*,

$$\dot{m} = 0.63 = \rho_1 V_1 A_1 = (1.213)(V_1)(15{,}000)(10^{-6})$$

$$\therefore V_1 = 34.62 \text{ m/s}$$

We can now get the theoretical power needed to drive the blower. Employing Eq. (15.54) we get

$$\left(\frac{dW_p}{dt}\right)_{\text{theoretical}} = \frac{dW}{dm}\dot{m} = \left\{\frac{(V_2)^2_{\text{theoretical}} - V_1^2}{2} + c_p T_1\left[\left(\frac{p_2}{p_1}\right)^{(k-1)/k} - 1\right]\right\}\dot{m}$$

$$= \left\{\frac{63.8^2 - 34.6^2}{2} + (0.24)(4187)(273 + 18)\left[(1.0242)^{(1.4-1)/1.4} - 1\right]\right\}(0.63)$$

$$= 2167 \text{N} \cdot \text{m/s} = 2.167 \text{ kW}$$

Actually there will be more power needed to achieve the desired performance of the blower. Thus with a mechanical efficiency of 80 percent we would require $2.167/0.80 = 2.71$-kW power.

Finally, we wish to determine the exit-blade angle β_2. We may go back to Eq. (15.47), which applies to compressible and incompressible flow in the turbomachine. We now rewrite this equation in the following form:

$$\left(\frac{dW_p}{dt}\right)_{\text{actual}}(\eta) = \left(\frac{dW_p}{dt}\right)_{\text{theoretical}} = \omega r_2 \dot{m}\left[(U_t)_2 + (V_r)_2 \cot \beta_2\right]$$

Substituting values, we get

$$2167 = \left[\frac{(5000)(2\pi)}{60} \right](0.120)(0.63)\left[\frac{(5000)(2\pi)}{60}(0.120) + (V_r)_2 \cot \beta_2 \right] \quad (a)$$

To determine $(V_r)_2$, consider the *conservation of mass* for a control volume just enclosing the *impeller* (see Fig. 15.20). We can then say that

$$\dot{m} = 0.63 = \rho_2 (V_r)_2 [(2\pi)(r_2)(b)]$$

$$0.63 = (1.2339)(V_r)_2 [(2\pi)(0.120)(0.300)]$$

$$\therefore (V_r)_2 = 2.26 \text{ m/s}$$

Substituting into Eq. (a), we can solve for the blade-exit angle β_2.

$$\beta_2 = 164.4°$$

15.11 CLOSURE

In this chapter, we have considered certain classes of turbomachinery. In Example 15.3, recall that we carried out calculations for a number of sections of a blade using the hand calculator. Hence, in this chapter and in many of the other chapters, we have seen the need for computer calculations. The advent and the availability of high-speed computers as well as programmable hand calculators have had a profound impact on fluid-mechanics computations. Accordingly, in the final chapter of this text, we examine this topic. In the future, there will inevitably be increasing emphasis on this area of fluid mechanics.

PROBLEMS

Problem Categories

Similarity Problems 15.1–15.3
Specific Speed 15.4
Impulse Turbines 15.5–15.8
Reaction Turbines 15.9–15.23
Many Bladed Reaction Turbines and Compressors 15.24–15.28
Radial Flow Pumps and Blowers 15.29–15.36

Starred Problems

15.27, 15.28

15.1. A centrifugal pump is to deliver a head ΔH_D of 500 m. What flow Q will it have if it is to run at a speed of 2000 r/min and has an impeller diameter of 0.8 m? What efficiency do you expect of this pump? The pump is geometrically similar to that reported in Fig. 15.3.

15.2. Consider a geometrically similar pump to that reported in Fig. 15.3. Plot a performance curve to ΔH_D versus Q for such a pump having an impeller diameter D of 0.7 m and a speed N of 1750 r/min.

15.3. Consider a geometrically similar pump to that reported in Fig. 15.3. For a speed of 1750 r/min, a head 800 m, and a flow of 1 m³/s, what size impeller should be used? Describe first verbally what should be done, then carry out to get approximate result.

15.4. A centrifugal pump is delivering 150 gal/min of water. It runs at a speed of 1750 r/min. The power needed to run the pump is 5 hp and the efficiency is 70 percent. What is the specific speed?

15.5. A Pelton water wheel is to develop 1500 kW. The wheel has a radius to the center of the buckets of 2 m and is to rotate at a speed of

200 r/min. If we have an efficiency of 85 percent at the wheel, what total head is needed if we neglect friction up to the nozzle. What should the cross-sectional area of the nozzle be? The jet speed is $(1/0.48)\,U_t$. The angle β for the buckets is $20°$. *Hint*: See Example 5.14.

15.6. A single-stage in pulse turbine has 5 lbm/s of combustion products at a speed of 6000 ft/s from the nozzles. The radius R to the nozzles is 2 ft. What is the most efficient speed ω of the turbine if $\alpha_1 = 20°$? If the turbine runs at 0.7 of this speed, what is the power developed

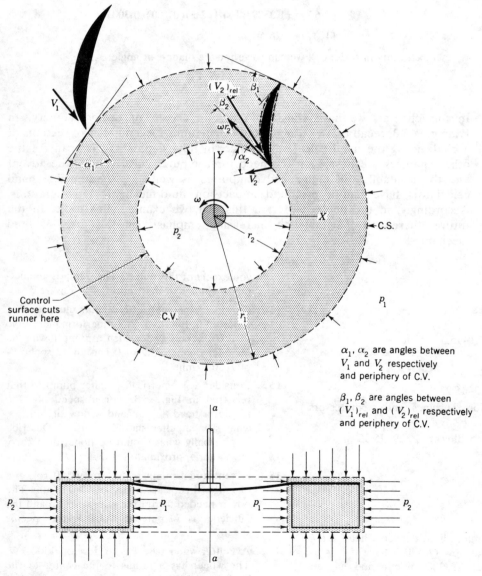

α_1, α_2 are angles between V_1 and V_2 respectively and periphery of C.V.

β_1, β_2 are angles between $(V_1)_{rel}$ and $(V_2)_{rel}$ respectively and periphery of C.V.

FIGURE P15.9

by the turbine? What is the torque transmitted from the turbine? What is β if $\beta_1 = \beta_2 = \beta$?

15.7. Three nozzles expand combustion products to a speed of 1800 m/s. The density of the fluid as it reaches the turbine blades is 0.46 kg/m³. If the turbine is to rotate at its most efficient speed of 12,000 r/min, what should the angle α_1 be for the three nozzles? The nozzles are at a distance R of 0.6 m from the axis of rotation. If the power desired is 6000 hp, what should the exit areas of the nozzles be and what should the blade angles β be? Take $\beta_1 = \beta_2$.

15.8. Consider that there is friction in the axial-single-stage impulse turbine having short blades. Then the relative velocity $(V_{rel})_2$ will be different from $(V_{rel})_1$ and we denote this as $(V_2)_{rel} = K(V_1)_{rel}$, where K is a fraction. Show that for these conditions Eq. (15.29) still holds. Take $\beta_1 = \beta_2$.

15.9. A Francis reaction hydraulic turbine is represented in Fig. P15.9. One outer stationary blade is shown and one inner blade on the runner is shown. Water enters from the stationary blades over the entire outer periphery of the runner and leaves through the inner periphery. Using the control volume shown, show that

Torque $= \rho Q \{ r_1 V_1 (\cos \alpha_1)$

$$+ r_2 [(V_2)_{rel}(\cos \beta_2) - \omega r_2] \}$$

Hint: Note at the exit that $\mathbf{V}_2 = \mathbf{V}_t + (\mathbf{V}_2)_{rel}$ where $\mathbf{U}_t$ is the velocity of the blade at r_2. Draw vector triangle as an aid to replace $V_2 \cos \alpha_2$.

15.10. A reaction water turbine rotates at a speed of 100 r/min and discharges 0.28 m³/s. Radius $r_1 = 0.5$ m, and $r_2 = 0.25$ m. The angle α_1 of the stationary blade ends is 20°. What should the velocity V_1 be for the best operation? Take β_1 as 60°. Do not use the formula from Prob. 15.9 but work from first principles.

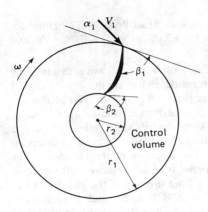

FIGURE P15.10

15.11. If, in Prob. 15.10, we assume incompressible frictionless flow through the runner, determine the torque and power developed for an angle β_2 of 60°. Blades take up 15 percent of the periphery at the inlet and outlet regions of the runner. The velocity V_1 is 7.05 m/s. Work from first principles. Do not use formula from Prob. 15.9.

15.12. In Prob. 15.9, we evaluated the torque developed by a reaction turbine as

$$T = \rho Q \{ r_1 V_1 (\cos \alpha_1)$$
$$+ r_2 [(V_2)_{rel}(\cos \beta_2) - \omega r_2] \} \quad (a)$$

If V_2 is the actual velocity relative to the ground and α_2 is the angle that V_2 makes with the periphery, explain why the equation above can be stated as

$$T = \rho Q (r_1 V_1 \cos \alpha_1 - r_2 V_2 \cos \alpha_2) \quad (b)$$

Now show that the power output of the turbine can be given as

$$\text{Power} = Q \rho N (\Gamma_1 - \Gamma_2)$$

where N is the number of revolutions per unit time and Γ represents *circulation* of the flow given by $\oint \mathbf{V} \cdot \mathbf{dl}$ around a closed path. Thus the power developed is proportional to the change in circulation, a quantity which we have used in Chap. 12.

15.13. In Prob. 15.10, what is the change in circulation of the flow before and after the runner?

Compute power. Read Prob. 15.12 before doing. Note $V_1 = 7.05$ m/s, $V_2 = 7.24$ m/s, and $\alpha_2 = 41.8°$.

15.14. A centrifugal blower is shown schematically where the inside radius of the blades is 3 ft and the outside radius is 4 ft. The blades are 3.5 ft wide. The blower rotates at 36 r/min, and 80,000 ft³/m of air is admitted at the inner radius of the blades in a radial direction as shown. The direction of the air coming out of the impeller is at 30° relative to the periphery as seen from ground. If the density of the air at A is 0.08 lbm/ft³ and at the outer periphery is 0.085 lbm/ft³, what is the value of the velocity leaving the impeller and what are the torque and the power required to run the blower?

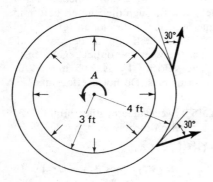

FIGURE P15.14

15.15. In Prob. 15.14, determine β_1 and β_2 for the blades of the impeller. From Prob. 15.14, the absolute exit velocity is 28.5 ft/s.

15.16. In Fig. 15.9, $H_I = 80$ m and $H_{II} = 30$ m. The speed of the rotor is 350 r/min. The value of R_0 at the guide vanes is 1.8 m and the height b is 0.8 m. What should the angle α be for this operation? There is zero angular momentum in the draft tube and the volumetric flow is 25 m³/s.

15.17. A flow of 5 m³/s of water passes through an axial-flow turbine leaving the rotor with no angular momentum. If the hub radius of the rotor blade is 300 mm and the tip radius is 800 mm, what is the tangential velocity $(V_t)_1$ at

$r = 500$ mm? The blade angle β_1 is 30° and $\omega = 200$ r/min.

15.18. If the effective head ΔH_D for an axial-flow hydraulic turbine is 90 m and the volumetric flow Q is 20 m³/s, What should the speed ω of the rotor be? The radius R_0 of the guide vanes is 1.5 m and the height b is 1 m. The exit flow is axial only and the guide vanes are at an angle α of 11.0°.

15.19. An axial flow reaction turbine has a volumetric flow Q of 20 m³/s of water. The following additional data apply.

$$R_0 = 1.8 \text{ m} \qquad b = 1 \text{ m}$$

$$R_H = 200 \text{ mm} \qquad \alpha = 10°$$

$$R_T = 1 \text{ m} \qquad H_I + H_{II} = 80 \text{ m}$$

The water leaves the blade in an axial direction. Determine ω and β_1 at $r = 0.8$ m. Neglect friction.

15.20. In Prob. 15.19, what is the power output of the turbine, neglecting friction and other losses? To decrease this power by 25 percent, what is Q and what should the setting of α be? The angular velocity ω for this operation is 415 r/min.

15.21. In Example 15.3, determine β_1 and β_2 for position c.

15.22. The following data apply to an axial-flow water turbine:

$$R_0 = 6 \text{ ft} \qquad b = 3 \text{ ft}$$

$$R_H = \tfrac{1}{2} \text{ ft} \qquad Q = 6500 \text{ gal/s}$$

$$R_T = 3 \text{ ft} \qquad H_I + H_{II} = 300 \text{ ft}$$

$$\alpha = 15°$$

There is no angular momentum of the water in the draft tube. Find the ideal power developed by the turbine using five sections of the blade.

15.23. In Prob. 15.22, find β_1 and β_2 at the hub and the tip of the blade. The following data were determined from the solution of the previous problem.

$$V_a = 31.61 \text{ ft/s}$$

$$\omega = 56.15 \text{ rad/s}$$

$$V_t = \frac{172.04}{r} \text{ ft/s}$$

15.24. An axial-flow compressor handles 50 kg of air per second at a temperature of 15°C and a pressure of 101,325 Pa. The diameter outside the first stage of the rotor is 1.5 m. Blades are 100 mm in length. The angle α_1 of the stationary blades ahead of the first stage is 35° and the angle β_2 of the rotor blades is 52°. If the rotor is turning at 3000 r/min, what theoretical head is being imparted to the air by the first stage? What horsepower theoretically is needed to drive the first stage?

15.25. The first stage of an axial-flow gas turbine has a hub-to-tip diameter ratio of 0.85 with the tip having a diameter of 4 ft. A flow of 50 lbm/s of combustion products flows through the machine. At the first stage, the pressure is 5 atm and the temperature is 1800° R. If $\alpha_1 = 38°$ and $\beta_2 = 55°$ for the first stage and it is to develop 25 hp, at what speed should the rotor turn? Consider the gas constant R to be that of air for the working fluid.

15.26. A volumetric flow Q of 15 m³/s of water goes through an axial-flow turbine delivering a head H_D to the turbine rotor of 100 m. Neglecting losses, how much horsepower is produced? If the turbine rotor turns at 360 r/min, what torque is being developed on the rotor? At a position $r = 0.8$ m, what should the tangential velocity V_t be going into the rotor? The flow leaving the rotor has no rotation about the rotor axis.

* **15.27.** An axial-flow reaction turbine has a mean radius out to a stage of 2 ft and rotates at 10,000 r/min. The blades are 3 in in length. The working fluid is steam, which expands from a temperature of 900° F and a pressure of 100 lb/in² absolute to a pressure of 80 lb/in² absolute. What is the change in kinetic energy per pound-mass of the steam going through the stage in question? The flow is 500,000

lbm/h. The angle α_1 for the stator preceding the stage is 48° and α_2 is 50° for the next stator (see Fig. 15.15). *Hint:* Use a *Mollier* diagram and *steam tables*. Work from first principles of angular momentum and energy.

15.28. A centrifugal blower compresses 1.5 lbm/s of air at a temperature of 60° F and a pressure of 14.00 lb/in² absolute to a pressure of 15.50 lb/in² absolute. The impeller runs at a speed of 3550 r/min. The exit area of the blower casing is 0.15 ft² and the inlet areas is 0.5 ft². What power is needed to run this blower if we have an efficiency of 80 percent for the blower? Neglect diffusion in the casing and consider that we do not have a diffuser. What torque is required?

15.29. In Prob. 15.28, we found that it took 10.54 hp to do the job. If the impeller has an outside radius of 8 in, what should the blade angle β_2 be at exit? The width of the impeller is 1.5 in. The density ρ_2 found in the preceding solution is 0.002429 slug/ft³.

15.30. An air compressor has a mass flow of 0.6 kg/s, taking the fluid from a pressure of 100 kPa at the inlet to a static pressure corresponding to 35 mm water in a U tube. The inlet temperature is 15° C. The exit angle β_2 is 160°. The inlet area is 7000 mm² and the exit area is 6000 mm². Neglect diffusion in the casing and consider that we do not have a diffuser. What is the speed of the compressor if the diameter of the rotor is 175 mm? The efficiency is 80 percent. The width of the impeller is 80 mm.

15.31. A performance chart is shown in Fig. P15.31 for a 3550 r/min blower from a bulletin of the Buffalo Forge Co. For a 21-in wheel having a flow of 3500 ft³/min, what is the efficiency and the required torque? The exit diameter is 8 in and the entrance diameter is 16 in. At the entrance $p = 14.7$ lb/in² and $T = 70°$ F. The capacity is given at the exit pressure. *Note:* The upper curves apply to the left ordinate and the lower curves apply to the right ordinate.

15.32. A centrifugal pump takes 200 gal/min of water from a pressure of 13 lb/in² to a pressure p_2. The inside radius of the impeller is 3 in and

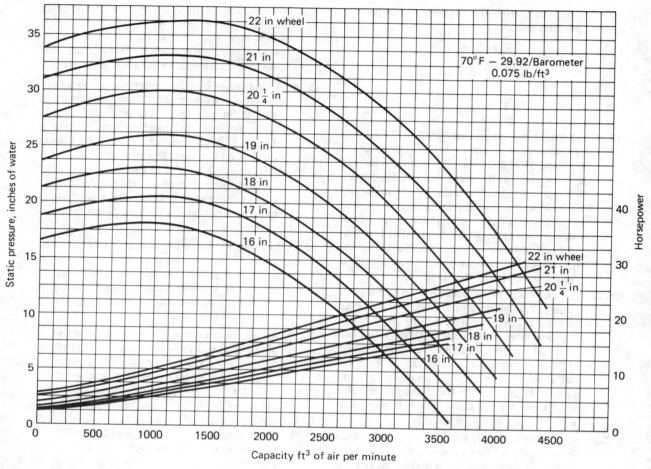

FIGURE P15.31

the outside radius of the impeller is 0.8 ft. If the output of the pump is 26.5 hp running at a speed of 1750 r/min, what should the blade angle β_2 be? The efficiency is 80 percent. The width b of the vanes is 2 in. If the exiting fluid is at the same elevation as the incoming fluid, what is p_2? Neglect diffusion in casing. *Hint:* From (V_r), get $(V_{rel})_2$. Then use the law of cosines to get V_2.

15.33. A centrifugal pump moves 10.0 L of water per second from a pressure of 100 kPa to a pressure of 550 kPa in one stage. The section inlet

has a diameter of 50 mm and the discharge has a diameter of 40 mm. The impeller diameter is 0.4 m and β_2 is 150°. The width b of the impeller is 30 mm. What power is required if there is an efficiency of 65 percent? What speed should the pump run at for the above condition? The discharge pipe is 0.3 m above the inlet pipe. Use theoretical torque in computing ω.

15.34. Shown in Fig. P15.34 are performance curves as found in a Buffalo Forge Co. bulletin for boiler feed pumps. It describes the operating

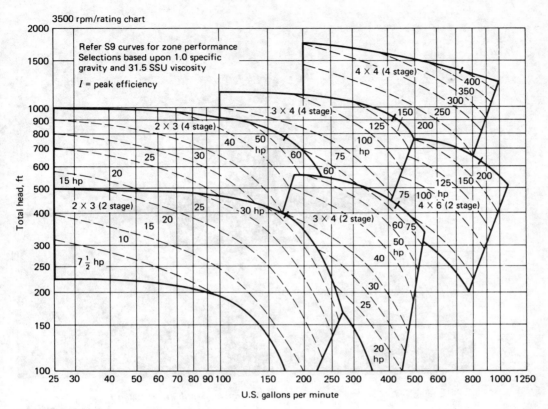

3500 rpm/rating chart

Refer S9 curves for zone performance
Selections based upon 1.0 specific
gravity and 31.5 SSU viscosity

I = peak efficiency

U.S. gallons per minute

Total head, ft

FIGURE P15.34

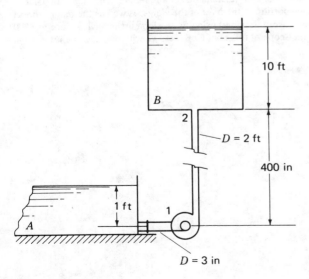

FIGURE P15.35

characteristics of various multistage centrifugal
pumps. The number designations $2 \times 3, 4 \times 4$,
etc., give the exit and the inlet diameters, respectively. For a 2×3 (four-stage) pump operating at 30 hp with a flow Q of 90 gal/min of
water, compute the efficiency. What is the total
required torque on the four rotors?

15.35. Using the Buffalo Forge rating chart in Prob.
15.34, what centrifugal pump would you choose
to move 100 gal/min of water from tank A to
tank B? The inlet and outlet pipes are steel
pipes. Neglect head loss for the short inlet
piping. Water is at 60° F. See Fig. P15.35.

Boeing 747. (*Courtesy Boeing Aircraft Corporation, Seattle, Washington.*)

With the advent of the high-speed computer, numerical methods have become a key part of design with such techniques as finite difference, finite elements, boundary elements, and others. The Boeing 747 illustrates this. The fuselage portion at the base of the wings, as well as the main support structure connecting wings and fuselage, were designed using finite elements involving close to 8000 unknowns. This technique is beyond the scope of this text. However, we will look at some interesting finite difference calculations.

CHAPTER
16

*COMPUTATIONAL FLUID MECHANICS

By DALE TAULBEE
Professor of Mechanical and Aerospace Engineering,
State University of New York at Buffalo

16.1 INTRODUCTION

At the outset of the book, we talked about theoretical approaches and experimental approaches. With the advent of the computer, a third regime has become discernible, namely, *computational fluid mechanics*. We engage in numerical simulation of fluid dynamics, using the computer while changing at will various parameters of interest built into the program. New phenomena have been discovered via this route before experimentation has uncovered them. Computational fluid mechanics can thus be considered a separate discipline supplementing theoretical and experimental fluid dynamics.

Industry routinely uses computers to help solve the fluid-flow problems needed for the design of such devices as pumps, compressors, and engines Aircraft engineers simulate three-dimensional flows about entire aircraft on the computer to predict flight characteristics. In fact, a significant portion of a development budget is often allocated to computational fluid-dynamic studies. This can result in significant savings when one can replace, by these studies, the expensive and time-consuming experimentation that would otherwise be needed.

This chapter thus brings to bear the power of the digital computer in dealing with problems in fluid mechanics. Numerical methods, involving the discretization of operations on continuous mathematical functions, will be discussed and applied to some of the flow phenomena dealt with earlier in this book. The procedure for facilitating the numerical methods and some aspects of the computer programming will be described.

PART A
NUMERICAL METHODS I

16.2 NUMERICAL OPERATIONS FOR DIFFERENTIATION AND INTEGRATION

In this section, we show how a continuously varying function can be represented by functional values at *discrete* points and how the basic operations of the calculus, namely differentiation and integration, can be performed on a discretized function. Primarily, we set forth three different so-called interpolation formulas—namely, the *linear*, *quadratic*, and *central difference interpolation formulas*.

We start with a Taylor series representation of $f(x)$ developed about some point x_0.

$$f(x) = \sum_{n=0}^{N} \frac{f^{(n)}(x_0)}{n!} (x - x_0)^n + R(x) \tag{16.1}$$

where $f^{(n)}(x_0)$ represents $(d^n f/dx^n)_{x=x_0}$ and $R(x)$ is the *remainder*, which you may recall from your calculus course is[1]

$$R(x) = \frac{f^{(n+1)}(\zeta)}{(n+1)!} (x - x_0)^{n+1} \qquad x_0 \leq \zeta \leq x \tag{16.2}$$

Let us next presume that we *know* the values of a function f at *equally* spaced positions x_i. That is

$$f(x_i) \equiv f_i \text{ (known)} \qquad \text{where } x_{i+1} - x_i = \Delta x$$

Now consider the simplest approximation of $f(x)$ in Eq. (16.1) using a two-term Taylor series expansion and noting that $[O(\ )]$ represents order of magnitude:

$$f(x) = f(x_0) + [f'(x_0)](x - x_0) + R \tag{16.3}$$

---

[1]Also note that $f' \equiv df/dx$, $f'' \equiv d^2f/dx^2$, and so forth.

where from Eq. (16.2) with $n = 1$ we can readily see that

$$R \propto (x - x_0)^2$$

$$\therefore R \approx \left[O(x - x_0)^2\right] \tag{16.4}$$

We may conclude that the formulation

$$f(x) = f(x_0) + f'(x_0)(x - x_0) \tag{16.5}$$

is in error by the remainder term $[O(x - x_0)^2]$. This kind of error is called a *truncation* error, i.e., an error resulting from terminating the series to obtain a given approximation. We will in general not try to evaluate the truncation error precisely; instead, we will be able to give only its approximate size. For this purpose, we express this error as $[O(\Delta x^n)]$, where clearly the larger the value of n and the smaller the interval Δx, the more accurate the approximation. The approximation of Eq. (16.5) gives the correct value of f at x_0, as you readily see, by setting $x = x_0$. We now denote f at x_0 to be f_0 and at x_n to be f_n. Accordingly,

$$f(x_1) \equiv f_1 = f_0 + f_0'\Delta x + \left[O(\Delta x^2)\right] \tag{16.6}$$

where f_0' is $f'(x_0)$. This in turn means that

$$f_0' = \frac{f_1 - f_0}{\Delta x} + \left[O(\Delta x)\right] \tag{16.7}$$

Now substitute for $f'(x_0) \equiv (f_0')$ in Eq. (16.3). Using Eq. (16.7), we get

$$f(x) = f_0 + \left\{\frac{f_1 - f_0}{\Delta x} + [O(\Delta x)]\right\}(x - x_0) + \left[O(x - x_0)^2\right]$$

$$\therefore \boxed{f(x) = f_0 + (f_1 - f_0)\left(\frac{x - x_0}{\Delta x}\right) + \left[O(\Delta x^2)\right]} \tag{16.8}$$

where we have approximated $(x - x_0) \approx \Delta x$ in evaluating the order of magnitude of the remainder error. By this process, we have found a function $f(x)$ as given by Eq. (16.8) which equals the correct values of f at x_0 and has, as we say, second-order accuracy in the interval x_0, x_1. Equation (16.8) is a *linear-interpolation* formula. Note that $f(x)$ in Eq. (16.8) is given in terms of the discretized values of f at $x = x_0$ and $x = x_1$.

Now let us produce the *quadratic-interpolation* formula by considering the Taylor series expansion for three terms and a remainder. That is,

$$f(x) = f_0 + (f_0')(x - x_0) + \frac{1}{2!}(f_0'')(x - x_0)^2 + \left[O(x - x_0)^3\right] \tag{16.9}$$

We can formulate this representation of f in terms of values of f at x_0, x_1, and x_2.

$$f_1 = f_0 + f_0' \, \Delta x + \tfrac{1}{2} f_0'' (\Delta x)^2 + \left[O(\Delta x)^3 \right] \tag{16.10}$$

$$f_2 = f_0 + 2 f_0' \, \Delta x + 2 f_0'' (\Delta x)^2 + \left[O(\Delta x)^3 \right] \tag{16.11}$$

Multiply Eq. (16.10) by 4 and from it subtract Eq. (16.11):

$$4 f_1 - f_2 = 3 f_0 + 2 f_0' \, \Delta x + \left[O(\Delta x)^3 \right]$$

Solving for f_0' we get:

$$\boxed{ f_0' = \frac{-3 f_0 + 4 f_1 - f_2}{2 \, \Delta x} + \left[O(\Delta x)^2 \right] } \tag{16.12}$$

Next multiply Eq. (16.10) by 2 and subtract Eq. (16.11).

$$2 f_1 - f_2 = f_0 - f_0'' (\Delta x)^2 - \left[O(\Delta x)^3 \right]$$

Solving for f_0'', we get

$$\boxed{ f_0'' = \frac{f_0 - 2 f_1 + f_2}{(\Delta x)^2} + \left[O(\Delta x) \right] } \tag{16.13}$$

With these values for f_0' and f_0'', we now go back to Eq. (16.9) and get, on insertion of these values,

$$f(x) = f_0 + \left\{ \frac{-3 f_0 + 4 f_1 - f_2}{2 \, \Delta x} + \left[O(\Delta x)^2 \right] \right\} (x - x_0)$$

$$+ \frac{1}{2} \left\{ \frac{f_0 - 2 f_1 - f_2}{(\Delta x)^2} + \left[O(\Delta x) \right] \right\} (x - x_0)^2 + \left[O(\Delta x)^3 \right]$$

Now rearranging the terms, we get

$$\boxed{ \begin{aligned} f(x) = f_0 &+ (f_1 - f_0) \frac{x - x_0}{\Delta x} + \frac{1}{2} (f_0 - 2 f_1 + f_2) \left(\frac{x - x_0}{\Delta x} \right) \\ &\times \left(\frac{x - x_0}{\Delta x} - 1 \right) + \left[O(\Delta x^3) \right] \end{aligned} } \tag{16.14}$$

This is a *quadratic-interpolation* formula in the interval x_0 to x_2. We now introduce the following notation:

$$u = \left[\frac{x - x_0}{\Delta x} \right]$$

$$\Delta f_i = f_{i+1} - f_i \qquad \text{First difference}$$

$$\Delta^2 f_i = \Delta f_{i+1} - \Delta f_i \qquad \text{Second difference}^2 \qquad (16.15)$$

$$\Delta^3 f_i = \Delta^2 f_{i+1} - \Delta^2 f_i \qquad \text{Third difference}$$

$$\vdots$$

$$\Delta^p f_i = \Delta^{p-1} f_{i+1} - \Delta^{p-1} f_i \qquad p\text{th difference}$$

Thus, consider the expression $(f_0 - 2f_1 + f_2)$ in Eq. (16.14). This can be written as $(f_2 - f_1) - (f_1 - f_0)$ as you may readily demonstrate. Using the notation in Eq. (16.15), we can say that

$$(f_0 - 2f_1 + f_2) = (\Delta f_1 - \Delta f_0) = \Delta^2 f_0$$

Thus you see that Δ is a difference, Δ^2 is a difference of a difference, and so forth. Now the quadratic-interpolation formula can be given as follows using $u = (x - x_0)/\Delta x$:

$$f(x) = f_0 + (\Delta f_0)u + \frac{1}{2!}(\Delta^2 f_0)(u)(u - 1) + \left[O(\Delta x)^3 \right]$$

If we continue the process up through more discretized values of f in the sequence, namely, $f_4, \ldots, f_n, \ldots$, we can extrapolate the formula above to get

$$f(x) = f_0 + (\Delta f_0)u + \frac{1}{2!}(\Delta^2 f_0)(u)(u - 1) + \frac{1}{3!}(\Delta^3 f_0)(u)(u - 1)(u - 2)$$

$$+ \frac{1}{4!}(\Delta^4 f_0)(u)(u - 1)(u - 2)(u - 3) + \cdots$$

$$(16.16)$$

A different formula can be obtained if we use points to the *left* of x_0, which we denote as x_{-1}, x_{-2}, and so forth. For this purpose, we go back to Eq. (16.9) wherein we will require that $f(x)$ take on a prescribed value f_{-1} at

---

[2] As an aid in understanding the various differences, note that

$$\Delta^2 f_i = \Delta f_{i+1} - \Delta f_i = (f_{i+2} - f_{i+1}) - (f_{i+1} - f_i)$$

$$\Delta^3 f_i = \Delta^2 f_{i+1} - \Delta^2 f_i = [(f_{i+3} - f_{i+2}) - (f_{i+2} - f_{i+1})] - [(f_{i+2} - f_{i+1}) - (f_{i+1} - f_i)]$$

and so on.

position x_{-1} to the left of x_0 by the amount Δx. Thus,

$$f_{-1} = f_0 + f_0'(x_{-1} - x_0) + \frac{1}{2!}f_0''(x_{-1} - x_0)^2 + \left[O(\Delta x)^3\right] \quad (16.17a)$$

$$\therefore f_{-1} = f_0 - f_0'(\Delta x) + \frac{1}{2!}f_0''(\Delta x)^2 - \left[O(\Delta x)^3\right] \quad (16.17b)$$

Next subtract Eq. (16.17b) from Eq. (16.10):

$$f_1 - f_{-1} = 2f_0'(\Delta x) + \left[O(\Delta x)^3\right] \quad (16.18)$$

Solving for f_0', we get

$$\boxed{f_0' = \frac{f_1 - f_{-1}}{2\,\Delta x} + \left[O(\Delta x)^2\right]} \quad (16.19)$$

Now add Eq. (16.18) to two times Eq. (16.17b). Note that the third-order terms exactly cancel out. We then get:

$$2f_{-1} + f_1 - f_{-1} = 2f_0 + f_0''(\Delta x)^2 + \left[O(\Delta x)^4\right]$$

Solving for f_0'' we get

$$\boxed{f_0'' = \frac{f_1 - 2f_0 + f_{-1}}{(\Delta x)^2} + \left[O(\Delta x)^2\right]} \quad (16.20)$$

Note that Eq. (16.20) for f_0'', having an order of magnitude error of $[O(\Delta x)^2]$, is more accurate than the corresponding formulation for f_0'' given by Eq. (16.13) having a larger order of magnitude error of $[O(\Delta x)]$. Now we go back to Eq. (16.9). Using Eqs. (16.19) and (16.20) for f_0' and f_0'' we get

$$\boxed{f(x) = f_0 + \frac{f_1 - f_{-1}}{2\,\Delta x}(x - x_0) + \frac{1}{2}\frac{f_1 - 2f_0 + f_{-1}}{(\Delta x)^2}(x - x_0)^2 + \left[O(\Delta x)^3\right]}$$

$$(16.21)$$

This interpolation formula is in terms of f_{-1}, f_0, and f_1 is called a *central difference interpolation formula*. It has third-order accuracy. A general expression analogous to Eq. (16.16) could be written for the central difference formula.

The interpolation formulas that we have developed allow us to express the approximate values of the function in between positions where the function is known. You will note that in the process of deriving these formulas, we have produced formulas for the approximation of first and second derivatives at the points where f is known.

Often in the numerical solution of problems, we must differentiate a tabulated function. That is, we have $f(x_i)$ for a range of i from 1 to I (we express this range as $i = 1, I$) and we would like to have $df(x_i)/dx$ at the points x_i in the range. Furthermore, we want the same accuracy $[O(\Delta x)^2]$ at every point in the range. To do this, we use Eq. (16.12) at the *first* point (recall it starts from x_0) and an analogous formula for the *last* point in the range. The reason for this choice is that Eq. (16.12) for the first point requires no information to the left of this point, and the equation analogous to (16.12) requires no information of f to the right of the last point in the range. For the points interior to the end points, we can use Eq. (16.19), where we have information to the left and to the right of every point x_c. The following subroutine using these formulas accomplishes this task. The call statement in the main program gives the tabulated function F, the number of values $I_{\max}$, and the interval DX. The subroutine gives the differentiated values FD.

SUBROUTINE 1. DIFFERENTIATION

```
SUBROUTINE DERIV (F, IMAX, DX, DF)
DIMENSION F(101), DF(101)
FD(1) = (-3.*F(1) + 4.*F(2) - F(3)) / (2.*DX)
IMAX1 = IMAX - 1
DO 10 I = 2, IMAX1
10   DF(I) = (F(I + 1) - F(I - 1)) / (2.*DX)
DF(IMAX) = (F(IMAX - 2) - 4.*F(IMAX - 1) + 3.*F(IMAX) / (2.*DX)
RETURN
END
```

Similarly, it is often necessary to integrate a tabulated function. To obtain the appropriate formula, we simply integrate the interpolation formula as given by Eq. (16.8). That is, for linear interpolation,

$$\int_{x_0}^{x_1} f(x)\, dx = f_0\, \Delta x + \frac{(f_1 - f_0)}{\Delta x} \int_{x_0}^{x_1} (x - x_0)\, dx + \left[O(\Delta x^2) \right]$$

$$= \frac{f_1 + f_0}{2}\, \Delta x + \left[O(\Delta x)^2 \right] \tag{16.22}$$

Greater accuracy could be obtained were we to use Eq. (16.14)—i.e., use the quadratic-interpolation formula instead of the linear-interpolation formula. However, the simpler formulation above has the same accuracy as the formula used in the derivative subroutine and hence will suffice for our purposes. Clearly, to integrate a tabulated function, we simply apply Eq. (16.22) to successive intervals. The following subroutine carries out the required operations with the same call statements as in the differentiation subroutine.

```
SUBROUTINE 2. INTEGRATION
SUBROUTINE INTEG (F, IMAX, DX FI)
DIMENSION F(101), FI(101)
FI(1) = 0.0
DO 10 I = 2, IMAX
10  FI(I) = FI(I - 1) + (F(I - 1) + F(I))*DX / 2.
RETURN
END
```

We are now ready to consider flow problems satisfying ordinary differential equations.

PART B
FLUID-FLOW PROBLEMS REPRESENTED BY
ORDINARY DIFFERENTIAL EQUATIONS

16.3 A COMMENT

Many problems in fluid mechanics involve one or more ordinary differential equations wherein the dependent variables such as pressure, temperature, or velocity are functions of a single independent variable such as time or a space variable. We list the following examples of such flows:

1. One-dimensional, compressible flows involving simultaneously two or more of the effects such as friction, heat transfer, etc., that were examined individually in Chap. 11.
2. Solution of the boundary-layer thickness as a function of position arising from use of the von Kármán integral momentum equation of Chap. 13, particularly when there is a nonzero pressure gradient in the free stream
3. Solution of the nonlinear Blasius equation (Sec. 13.3) for determining the velocity profile

Accordingly, we now consider the solution of certain ordinary differential equations.

16.4 INTRODUCTION TO NUMERICAL INTEGRATION OF ORDINARY DIFFERENTIAL EQUATIONS

Let us begin by considering a simple differential equation given as

$$\boxed{y' = f(x, y)} \tag{16.23}$$

where f is some function of the dependent variable y and the independent variable x. Let us suppose, furthermore, that at a position x_{i-1} we know the value of y. That is, we assume that $y_{i-1} = y(x_{i-1})$ is known. It is desired to find y satisfying Eq. (16.23) at a succeeding position x_i as well as at other positions beyond x_i. For this purpose, consider the Taylor series representation of y_i in terms of y_{i-1} a distance Δx away. That is,

$$y_i = y_{i-1} + y'_{i-1} \Delta x + \frac{1}{2!} y''_{i-1} \Delta x^2 + \cdots \tag{16.24}$$

From Eq. (16.23) (assuming it is of simple form), we could formulate higher-order derivatives of y. For instance, for y'' we have

$$y'' = \frac{dy'}{dx} = \frac{\partial f}{\partial x} + \frac{\partial f}{\partial y} \left(\frac{dy}{dx} \right) = \frac{\partial f}{\partial x} + f \frac{\partial f}{\partial y} \tag{16.25}$$

Higher-order derivatives could be similarly formulated; in this way we could express Eq. (16.24) in terms of f and ever higher-order derivatives of f taken at $i - 1$. However, we will limit the expansion given by Eq. (16.24) to two terms so that we have

$$\boxed{y_i = y_{i-1} + f(x_{i-1}, y_{i-1}) \Delta x} \tag{16.26}$$

This relationship is called *Euler's formula*. We can use this formula for the numerical solution of Eq. (16.23). Thus, say that we know y_1 at x_1. We can immediately calculate $f(x_1, y_1)$ so that with Euler's formula we get y_2. This procedure is repeated at succeeding points x_i so that we can give y at discrete positions x_i to cover the domain x_1 to x_l. Note that the reason for the simplicity of approach here is that the right side of Euler's formula does not contain y_i, so the formulation is *explicit* in y_i.

What price do we pay for such simplicity? Clearly, there is a local truncation error of order $[O(\Delta x^2)]$. Furthermore, the error developed at each point *accumulates* as one moves to successive points. Indeed, it can be shown that because of this, the overall error from truncation is actually $[O(\Delta x)]$ and not the smaller error of $[O(\Delta x^2)]$. Finally, we must point out that depending on the size of the steps Δx and the particular differential equation, the resulting formulation of y's may become *unstable* and thereby become meaningless to the investigator. What then happens in the computer is that the numbers become so big or so small that they exceed the capacity of the computer.

Accordingly, to decrease the error associated with use of Euler's formula and also to improve the stability characteristics of the process, we seek another methodology. This time, we set forth a Taylor series going *backward*, as it were, from point x_i. Thus,

$$y_{i-1} = y_i + y_i'(-\Delta x) + \frac{1}{2!}y_i''(-\Delta x)^2 + \frac{1}{3!}y_i'''(-\Delta x)^3 + \cdots$$

$$= y_i - y_i'\Delta x + \tfrac{1}{2}y_i''\Delta x^2 - \tfrac{1}{6}y_i'''\Delta x^3 + \cdots$$

Solving for y_i, we get

$$y_i = y_{i-1} + y_i'\Delta x - \tfrac{1}{2}y_i''\Delta x^2 + \left[O(\Delta x^3)\right] \tag{16.27}$$

Now add Eqs. (16.27) and Eq. (16.24) and divide by 2.

$$y_i = y_{i-1} + \tfrac{1}{2}(y_{i-1}' + y_i')\Delta x + \tfrac{1}{4}(y_{i-1}'' - y_i'')\Delta x^2 + \left[O(\Delta x^3)\right] \tag{16.28}$$

Consider y_i'' in the second parentheses. We can expand this term in a Taylor series as follows:

$$y_i'' = y_{i-1}'' + y_{i-1}'''\Delta x + \cdots$$

Substitute the formulation above for y_i'' limited to two terms into the second parentheses of Eq. (16.28).

$$y_i = y_{i-1} + \tfrac{1}{2}(y_{i-1}' + y_i')\Delta x + \tfrac{1}{4}(y_{i-1}'' - y_{i-1}'' - y_{i-1}'''\Delta x)(\Delta x^2) + \cdots$$

We then may express the result above on using Eq. (16.23) in the second expression on the right side of the equation above, and canceling terms in the next expression.

$$\boxed{y_i = y_{i-1} + \tfrac{1}{2}\left[f(x_{i-1}, y_{i-1}) + f(x_i, y_i)\right]\Delta x + \left[O(\Delta x^3)\right]} \tag{16.29}$$

Note now that the local truncation error has been reduced to $O(\Delta x^3)$ and is thus more accurate than the Euler formula [Eq. (16.26)]. Actually, the formula above is equivalent to the central difference approximation. However, it has the disadvantage that y_i appears on *both sides* of the equation and, as a result, the equation may be *implicit* in y_i. If indeed y_i cannot be solved explicitly in terms of x_i for a particular differential equation, then a simple iteration process can be used to determine y_i as well as succeeding y's.

We may generalize Eq. (16.29) to be used when there is more than one ordinary differential equation with one independent variable. Thus, consider two simultaneous differential equations with two dependent variables such as

$$y' = f(x, y, z)$$

$$z' = g(x, y, z)$$

where z and y are the dependent variables and x is the independent variable. We may formulate equations for y_i and z_i similar to Eq. (16.29). Omitting the remainder expressions, we have

$$y_i = y_{i-1} + \tfrac{1}{2}\big[f(x_{i-1}, y_{i-1}, z_{i-1}) + f(x_i, y_i, z_i)\big]\,\Delta x \qquad (16.30a)$$

$$z_i = z_{i-1} + \tfrac{1}{2}\big[g(x_{i-1}, y_{i-1}, z_{i-1}) + g(x_i, y_i, z_i)\big]\,\Delta x \qquad (16.30b)$$

The extension to more than two simultaneous ordinary differential equations should be obvious.

16.5 PROGRAMMING NOTES

The programming procedure for the methodology of the previous section is relatively simple. We will discuss now the salient features for the programming of the solution of two simultaneous equations as per Eqs. (16.30).

We start by taking as dimensioned variables for x_i, y_i, z_i, the expressions X(I), Y(I), and Z(I). The key steps then are as follows:

1. Specify the total number of points at which the solution is to be found as IT. Hence I goes from 1 to IT.
2. Specify the domain for which the solution is desired by giving X(1) and X(IT).
3. The length of the grid increment DX must then be

$$DX = (X(IT) - X(1))/FLOAT(IT - 1)$$

   Clearly, IT − 1 represents the number of incremental segments in the domain of the sought-for solution.
4. The grid point locations can be determined using a DO loop over the range (I = 1, IT) given as

$$X(I) = X(1) + FLOAT(I - 1)*DX$$

5. Specify the initial values of the dependent variables Y(1) and Z(1).
6. Calculate the solution Y(I) and Z(I) in another DO loop for the interval (I = 2, IT).

We leave the detailed working out of these steps to the student for particular problems and projects.

16.6 PROBLEMS

We will now discuss several fluid problems which are governed by ordinary differential equations and which are difficult to solve analytically. In each such problem, the equations will be put into proper form for numerical solution and some details relative to the numerical procedures will be discussed. In certain instances, numerical results from a computer program will be presented. However, the detailed computer program will be left for the student to develop as part of possible projects. We start by first considering incompressible, viscous flow in pipes which was first presented in Chap. 9.

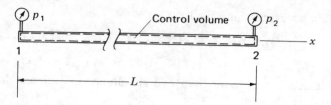

FIGURE 16.1
Unsteady flow in a pipe.

Case I. Pipe flow. We now consider the *unsteady* flow in a constant-area pipe of length L (Fig. 16.1) when the fluid is subject to a *time-varying difference of pressures between the ends* or when the *pressure difference between the ends is changed suddenly from one constant value to another constant value*. The *linear momentum* equation in the x direction for the control volume shown is given as

$$p_1 A - p_2 A - \tau_w \pi DL = -\iint_{A_1} V_x(\rho V_x \, dA) + \iint_{A_2} V_x(\rho V_x \, dA)$$

$$+ \frac{\partial}{\partial t} \iiint_{CV} V_x(\rho \, dv) \tag{16.31}$$

Since V_x is the same at any time throughout the pipe (the flow is incompressible), we see that the surface integrals cancel. (Note that if the pipe is inclined, we would have to include a component of the gravity body force.) Let us consider the volume integral from the equation above. If we consider V to be the *average* fluid velocity in the pipe at any time, we can say

$$\frac{\partial}{\partial t} \iiint_{CV} V_x(\rho \, dv) = \frac{\partial}{\partial t} V \iiint_{CV} \rho \, dv = \frac{\partial V}{\partial t}(\rho AL)$$

Employing this result in Eq. (16.31) and dividing through by ρAL, we get

$$\frac{\partial V}{\partial t} = \frac{1}{\rho} \frac{p_1 - p_2}{L} - \frac{\tau_w}{\rho} \frac{4}{D} \tag{16.32}$$

Note next that τ_w will depend only on V for incompressible flow, so there is only one dependent variable t making this equation an ordinary differential equation. Using Eq. (9.17), $\tau_w = (f/4)(\rho V^2 / 2)$, we next replace τ_w in terms of the friction factor f. Thus, we have our working equation[3]

$$\boxed{\frac{dV}{dt} = \frac{1}{\rho} \frac{p_1 - p_2}{L} - \frac{f}{2D} V^2} \tag{16.33}$$

---

[3] Note that the working equation is of the form $dV/dt = f(V, t)$ and thus is of the same form as Eq. (16.23).

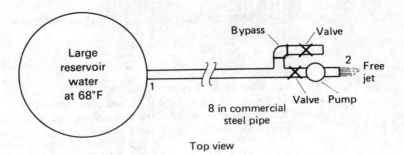

FIGURE 16.2
Flow problem from reservoir.

For pipe flow in the *transition* zone or *rough-pipe* zone, we use Eq. (9.20) for f. That is,

$$f = \frac{0.25}{\left\{\log\left[(e/3.7D) + 5.74/(\text{Re})^{0.9}\right]\right\}^2} \quad \begin{array}{l} 5 \times 10^3 \leq \text{Re} \leq 10^8 \\ 10^{-6} \leq \dfrac{e}{D} \leq 10^{-2} \end{array} \quad (16.34)$$

while for the *hydraulically smooth* zone, we have for f Eq. (9.21)

$$f = \frac{0.3164}{\text{Re}^{1/4}} \qquad \text{Re} \leq 100{,}000 \qquad (16.35)$$

We now illustrate the procedure in the following example.

Example 16.1. A 200-ft horizontal 8-in-diameter pipe is fed by a large reservoir in which the free surface is 50 ft above the pipe entrance. At the end of the pipe is a pump and a bypass around the pump (see Fig. 16.2). With the pump in operation and the bypass closed, the volumetric steady flow rate is 19.81 ft^3/s. A set of quick operating valves closes off the pump and opens the bypass. What is the flow rate as a function of time until it reaches a new steady state? The pipe is commercial steel pipe. Neglect minor losses.

Consider the flow in the large reservoir. We may use *Bernoulli's* equation between the free surface and position 1 at the entrance to the pipe. Thus, neglecting kinetic energy at the free surface, we have

$$\frac{p_{\text{atm}}}{\rho} + gh = \frac{p_i}{\rho} + \frac{V^2}{2} \qquad (a)$$

$$\therefore \frac{p_1 - p_{\text{atm}}}{\rho} = gh - \frac{V^2}{2}$$

where V is the average velocity (q/A) in the pipe. Now go to Eq. (16.33) for the pipe and substitute from above for $(p_1 - p_2)/\rho$, noting that $p_2 = p_{atm}$. We get

$$\frac{dV}{dt} = -\frac{V^2}{2L} + \frac{gh}{L} - \frac{f}{2D}V^2 \qquad (b)$$

We compute the initial Reynolds number to be 3.48×10^6, so with $e/D = 0.0002$, we are at the outer fringe of the transition zone (see Fig. 9.16). We can expect that as the flow reaches its second steady state, we will still remain in the transition zone. And so we use Eq. (16.34) for f, which for the data of the problem becomes

$$f = \frac{0.25}{\left\{\log\left[(5.41 \times 10^{-5}) + (2.822 \times 10^{-4}/V^{0.9})\right]\right\}^2} \qquad (c)$$

We thus have a differential equation of the form

$$\frac{dV}{dt} = f(V)$$

and we use the *Euler method* for solving it. In Fortran variables, we calculate velocity $V(I)$ for each time $T(I)$ where the independent variable has the range $I = 1, IT$. The initial values are

$$V(1) = \frac{q}{A} = \frac{19.81}{\dfrac{\pi}{4}\left(\dfrac{8^2}{144}\right)} = 56.75 \quad \text{for } T(1) = 0$$

We take IT to be 50 and we take a large enough total time interval of 18 s to ensure steady state.[4] This means that $\Delta t \approx 0.36$ s. Some of the results for this calculation are given in Table 16.1 A calculation for steady-state operation with the pump off and bypass open by methods presented in Chap. 9 will verify the fact that we have arrived at the correct steady state in the numerical solution. A more-accurate transient description can be arrived at by taking $\Delta t = 0.180$ s—i.e., half the previous value. We give in Table 16.2 the results corresponding to those given in Table 16.1 for comparison.

For a more-accurate transient, even smaller time intervals can be used or one could use a more-accurate procedure such as that given by Eq. (16.29), which is equivalent to the central difference method.

Case II. Trajectory problem. We next consider the motion of a particle such as a small sphere moving through the atmosphere from which is developed a drag force collinear with the relative velocity vector between the sphere and the atmosphere. This velocity vector we denote as $\mathbf{V} - \mathbf{V}_p$ where $\mathbf{V}$ is the velocity of the atmosphere and $\mathbf{V}_p$ is the velocity of the projectile both taken relative to the

---

[4]Also, you can program your calculation to stop when $(V_{i+1} - V_i)$ is less than a prescribed small number, to signal the arrival of steady state.

TABLE 16.1
Transient solution for $\Delta t = 0.360$ s

t, s	V, ft / s	Re	f
0.36	47.09	2.88×10^6	0.01417
0.72	40.90	2.50×10^6	0.01422
1.44	33.72	2.06×10^6	0.01430
2.88	27.82	1.50×10^6	0.01440
5.76	25.03	1.53×10^6	0.01446
11.52	24.56	1.50×10^6	0.01488

TABLE 16.2
Transient solution for $\Delta t = 0.180$ s

t, s	V, ft / s	Re	f
0.36	47.11	2.88×10^6	0.01417
0.72	40.89	2.50×10^6	0.01421
1.44	33.69	2.06×10^6	0.01430
2.88	27.76	1.70×10^6	0.01441
5.76	24.96	1.53×10^6	0.01447
11.52	24.47	1.50×10^6	0.01448

ground. *Newton's* law for a sphere of diameter D and mass m is

$$m\frac{d\mathbf{V}_p}{dt} = \frac{1}{2}C_D\rho|\mathbf{V} - \mathbf{V}_p|^2\left(\frac{\pi D^2}{4}\right)\left\{\frac{\mathbf{V} - \mathbf{V}_p}{|\mathbf{V} - \mathbf{V}_p|}\right\} - mg\mathbf{k} \qquad (16.36)$$

where $(\mathbf{V} - \mathbf{V}_p)/|\mathbf{V} - \mathbf{V}_p|$ is the unit vector in the direction of the relative velocity and $\mathbf{k}$ is in the opposite direction of gravity. Note that we have neglected buoyancy in Eq. (16.36). If we divide by m and express $m/3\pi D\mu$ having dimension of time as τ, and if we express the Reynolds number as $\rho|\mathbf{V} - \mathbf{V}_p|D/\mu$, we get for Eq. (16.36)

$$\frac{d\mathbf{V}_p}{dt} = \frac{C_D \, \text{Re}}{24}\frac{\mathbf{V} - \mathbf{V}_p}{\tau} - g\mathbf{k} \qquad (16.37)$$

The scalar equations for this equation are

$$\frac{du_p}{dt} = \frac{C_D \, \text{Re}}{24}\frac{u - u_p}{\tau} \qquad (16.38a)$$

$$\frac{dv_p}{dt} = \frac{C_D \, \text{Re}}{24}\frac{v - v_p}{\tau} \qquad (16.38b)$$

$$\frac{dw_p}{dt} = \frac{C_D \, \text{Re}}{24}\frac{w - w_p}{\tau} - g \qquad (16.38c)$$

wherein $u_p = dx/dt$, $v_p = dy/dt$, and $w_p = dz/dt$. We thus have a set of simultaneous, ordinary, differential equations of the form discussed which will be solved numerically from some initial position and initial velocity.

Example 16.2. We are to find the trajectory and time of flight at discrete points on the trajectory for a golf ball having an initial speed of 120 ft/s in the xy plane at an angle of 30° with the x axis. The ball weighs 1.5 oz and is 1.75 in in diameter. Assume that the golf ball is not spinning. Take $\mu = 0.375 \times 10^{-6}$ lb · s/ft².

We idealize the drag coefficient for simplicity, as has been shown in Fig. 16.3 Note that Re $= 9 \times 10^4$ is the critical Reynolds number discussed in Chap. 13, wherein separation suddenly occurs from a turbulent boundary layer rather than a laminar boundary layer. The parameter τ (called the relaxation time) becomes

$$\tau = \frac{m}{3\pi D\mu} = \frac{1.5/[(16)(g)]}{(3\pi)(1.75/12)(0.375 \times 10^{-6})} = 5650 \text{ s} \qquad (a)$$

In this problem, we use the method of *central differences* to afford a more-accurate, stable solution. That is, taking Eqs. (16.38), we have, remembering that C_D and Re depend on $\mathbf{V}_p = u_p \mathbf{i} + v_p \mathbf{j} + w_p \mathbf{k}$,

$$\dot{u}_p = f(u_p, v_p, w_p) = \frac{C_D \text{ Re}}{24} \frac{u - u_p}{\tau}$$

$$\dot{v}_p = g(u_p, v_p, w_p) = \frac{C_D \text{ Re}}{24} \frac{v - v_p}{\tau} \qquad (b)$$

$$\dot{w}_p = h(u_p, v_p, w_p) = \frac{C_D \text{ Re}}{24} \frac{w - w_p}{\tau} - g$$

where u, v, w, are the velocities of the wind and will be taken as constant. We will now assume for the time intervals that C_D and Re are constant in each interval. Hence, using Eq. (16.29), we can say for the equations above:

$$u_p(t) = u_p(t - \Delta t) + \frac{C_D \text{ Re}}{24\tau}\left[u - \frac{u_p(t) + u_p(t - \Delta t)}{2}\right]\Delta t$$

$$v_p(t) = v_p(t - \Delta t) + \frac{C_D \text{ Re}}{24\tau}\left[v - \frac{v_p(t) + v_p(t - \Delta t)}{2}\right]\Delta t \qquad (c)$$

$$w_p(t) = w_p(t - \Delta t) + \frac{C_D \text{ Re}}{24\tau}\left[w - \frac{w_p(t) + w_p(t - \Delta t)}{2}\right]\Delta t - g \Delta t$$

Now let

$$\alpha = \frac{C_D \text{ Re}}{24\tau}\Delta t$$

where

$$\text{Re} = \frac{D\sqrt{(u - u_p)^2 + (v - v_p)^2 + (w - w_p)^2}}{\nu} \qquad (d)$$

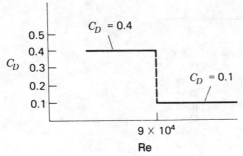

FIGURE 16.3
Idealized plot of C_D versus Re for a golf ball.

Using vectors, the equations above then become

$$\mathbf{V}_p(t) = \mathbf{V}_p(t - \Delta t) + \alpha \mathbf{V} - \frac{\alpha}{2}\mathbf{V}_p t - \frac{\alpha}{2}\mathbf{V}_p(t - \Delta t) - g\mathbf{k}\,\Delta t$$

This becomes

$$\mathbf{V}_p(t) = \frac{\alpha \mathbf{V} - g\,\Delta t\,\mathbf{k} + (1 - \alpha/2)\mathbf{V}_p(t - \Delta t)}{1 + \alpha/2} \qquad (e)$$

Furthermore, the position of the ball at time t can be given as follows:

$$\mathbf{r}(t) = \mathbf{r}(t - \Delta t) + \frac{\Delta t}{2}\left[\mathbf{V}_p(t - \Delta t) + \mathbf{V}_p(t)\right] \qquad (f)$$

To obtain an estimate of the time of flight and thereby determine a time interval Δt for calculation, we leave it to you to determine that for no friction the maximum time of flight is 3.73 s. We then plan for a flight of roughly 4 s and make 100 time increments each of value 0.04 s. We arrange the calculation to end when $z = 0$—i.e., when the ball hits the ground.

The procedure is as follows. For an initial time $(t - \Delta t)$ using known data at this time, compute α and C_D using Fig. 16.3 and Eq. (d). Next from Eq. (e) compute $\mathbf{V}$ at time t and from Eq. (f) compute $\mathbf{r}$ at time t. Now consider these computed data as the starting point for a new initial time again denoted as $t - \Delta t$, and once again find α and C_D for the next interval, so that from Eqs. (e) and (f) we get $\mathbf{V}_p$ and $\mathbf{r}$ at the end of the second interval, etc. The procedure is repeated for each time increment until the ball is again at $z = 0$. In Table 16.3, we show the results for this problem with no wind given for selected grid points.

Case III. Emptying a pressurized tank. Consider a liquid which partially fills a tank (Fig. 16.4). A pressurized gas of volume $\mathscr{V}$ is shown above the liquid. We wish to determine the rate of flow from the tank. As the key equation in the liquid, we use the *Bernoulli* equation—thereby adopting a *quasi-steady*, inviscid flow analysis. For the gas, we will consider an *isentropic* expansion. Thus we can

TABLE 16.3

t, s	x, ft	z, ft
0.04	4.149	2.370
0.20	20.62	11.26
0.40	40.91	21.08
0.60	60.91	29.47
0.80	80.62	36.45
1.00	100.1	42.07
1.20	119.1	46.30
1.40	137.3	49.08
1.60	154.6	50.47
1.80	171.1	50.53
2.00	186.9	49.34
2.20	202.1	46.93
2.40	216.6	43.35
2.60	230.6	38.66
2.80	244.1	32.88
3.00	257.0	26.06
3.20	269.4	18.24
3.40	281.4	9.467
3.60	292.8	−0.228

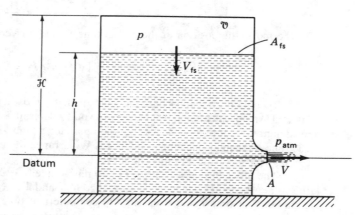

FIGURE 16.4
Emptying of a pressurized tank.

say for the free surface and the free jet of the liquid that

$$\frac{p}{\rho} + \frac{V_{fs}^2}{2} + gh = \frac{p_{atm}}{\rho} + \frac{V^2}{2} \tag{16.39}$$

From *continuity*, we require for incompressible flow of the liquid that

$$Q = V_{fs} A_{fs} = VA \tag{16.40}$$

Furthermore, we note that (see Fig. 16.4)

$$V_{fs} = -\frac{dh}{dt} \tag{16.41}$$

Meanwhile, for the gas we have on using subscript i for initial conditions

$$\frac{p}{(p)_i} = \left[\frac{\rho}{(\rho)_i}\right]^k \tag{16.42}$$

where $\rho = M/\mathcal{V}$ with M as the mass of the gas and with $\mathcal{V}$ as its volume. We can accordingly express $\rho/(\rho)_i$ as follows:

$$\frac{\rho}{(\rho)_i} = \frac{M/\mathcal{V}}{M/(\mathcal{V})_i} = \frac{\mathcal{V}_i}{\mathcal{V}} \tag{16.43}$$

Since the value $\mathcal{V}$ of the gas is $(\mathcal{H} - h)A_{fs}$, the equation above becomes

$$\frac{\rho}{(\rho)_i} = \frac{\mathcal{H} - h_i}{\mathcal{H} - h} \tag{16.44}$$

so that Eq. (16.42) becomes

$$\frac{p}{(p)_i} = \left(\frac{\mathcal{H} - h_i}{\mathcal{H} - h}\right)^k \tag{16.45}$$

Now let us return to the Bernoulli equation [Eq. (16.39)] and replace V_{fs} on the left side by $-dh/dt$ in accordance with Eq. (16.41). Next, replace V by $V_{fs}(A_{fs}/A)$ in accordance with Eq. (16.40) so that on using Eq. (16.41) again to replace V_{fs} we have $V = -(dh/dt)(A_{fs}/A)$. Using the latter result for V on the right side of Eq. (16.39), we then have

$$\frac{p}{\rho} + \frac{1}{2}\left(\frac{dh}{dt}\right)^2 + gh = \frac{p_{atm}}{\rho} + \frac{1}{2}\left(\frac{A_{fs}}{A}\right)^2\left(\frac{dh}{dt}\right)^2$$

Solving for dh/dt we get

$$\frac{dh}{dt} = -\left\{\frac{[2(p - p_{atm})/\rho] + 2gh}{(A_{fs}/A)^2 - 1}\right\}^{1/2}$$

Substituting for p using Eq. (16.45) and neglecting unity in the above denominator compared with $(A_{fs}/A)^2$ we get

$$\frac{dh}{dt} = -\sqrt{2}\left(\frac{A}{A_{fs}}\right)\left[\frac{(p)_i}{\rho}\left(\frac{\mathcal{H} - h_i}{\mathcal{H} - h}\right)^k - \frac{p_{atm}}{\rho} + gh\right]^{1/2}$$

Now multiply and divide the first expression in the last brackets, which is raised to $\frac{1}{2}$ power, by p_{atm}. We then get

$$\boxed{\frac{dh}{dt} = -\sqrt{2}\left(\frac{A}{A_{fs}}\right)\left\{\frac{p_{atm}}{\rho}\left[\frac{(p)_i}{p_{atm}}\right]\left(\frac{\mathscr{H} - h_i}{\mathscr{H} - h}\right)^k - \frac{p_{atm}}{\rho} + gh\right\}^{1/2}}$$

(16.46)

We have here a nonlinear ordinary differential equation to be solved for $h(t)$ starting from initial conditions h_i and $(p)_i$. Note that a closed-form solution cannot be found, so we use a numerical solution.

> **Example 16.3.** Consider a tank (Fig. 16.4) having the following data:
>
> $$\frac{A}{A_{fs}} = 0.01 \qquad \mathscr{H} = 4 \text{ ft}$$
>
> The initial conditions are
>
> $$h_i = 3 \text{ ft}$$
> $$p_i = 5p_{atm}$$
>
> As a first step, we estimate the time required to empty the tank so that we can choose a good time interval Δt for the numerical procedure. To do this, we take the case where the pressure p is constant and simply is the atmosphere pressure p_{atm}. In that case, using the result $V = \sqrt{2gh}$ from Bernoulli [Eq. (16.39), neglecting $V_{fs}^2/2$], we note from *continuity* that
>
> $$A_{fs}\frac{dh}{dt} = -AV = -A\sqrt{2gh} \qquad (a)$$
>
> Separating variables h and t, we can integrate this equation to get
>
> $$\frac{dh}{h^{1/2}} = -\left(\frac{A}{A_{fs}}\right)\sqrt{2g}\, dt$$
>
> (b)
>
> $$\therefore h^{1/2}(2) = -\frac{A}{A_{fs}}\sqrt{2g}\, t + C_1$$
>
> When $t = 0$, then $h = h_0$—hence $C_1 = 2\sqrt{h_0}$. Solving for h, we get
>
> $$h = \left(\sqrt{h_0} - \frac{1}{2}\frac{A}{A_{fs}}\sqrt{2g}\, t\right)^2 \qquad (c)$$
>
> When the tank stops, $h = 0$ and we can solve for the time t_f. We get
>
> $$t_f = \left(\frac{2h_0}{g}\right)^{1/2}\left(\frac{A_{fs}}{A}\right) \qquad (d)$$
>
> Inserting the known data for this example, we get
>
> $$t_f = \left[\frac{(2)(3)}{32.2}\right]^{1/2}\left(\frac{1}{0.01414}\right) = 30.53 \text{ s} \qquad (e)$$
>
> For 300 time steps in this period, the time increment would be 0.10 s.

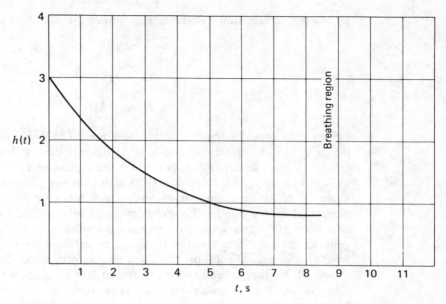

FIGURE 16.5
Plot of water level in tank versus time.

We solve Eq. (16.46) using the simple Euler method given by Eq. (16.26). A plot is shown in Fig. 16.5 giving h versus t. Note that the calculation reaches a velocity $\dot{h} = 0$ *before* the tank is empty. What has happened is that the pressure p of the gas in the tank has decreased below atmospheric pressure to a value such that the total head H_D at the exit of the tank $(p/\gamma + h)$ just equals the pressure head from the atmosphere (p_{atm}/γ) at the exit of the jet. What happens now? Actually, we would expect the tank to "breathe," in that bubbles of air would come into the exit. This would increase the pressure p so that more water could get out. There would then be this pulsating flow out of the tank with a bubble coming in for each pulse. If the exit were small enough, there could also be surface tension effects that would have to be considered.

Case IV. Solution of Blasius' equation for flat-plate boundary layer. In Sec. 13.3, we developed the Blasius equation for flat-plate boundary-layer flow [Eq. (13.34)]. We now restate this equation.

$$f''' + \tfrac{1}{2}ff'' = 0 \tag{16.47}$$

Recall that f is a function of $\eta = y/g(x)$.

The boundary conditions [see Eqs. (13.28)] are

$$f(0) = 0 \qquad f'(0) = 0 \qquad f'(\infty) = 1 \tag{16.48}$$

The differential equation is different from those we have thus far undertaken in that it is a third-order equation and has boundary conditions (at 0 and at ∞) rather than simply initial conditions. We can rewrite Eq. (16.47) as three first-order, differential equations as follows[5]:

$$f' = G \qquad (16.49a)$$

$$G' = H \qquad (16.49b)$$

$$H' = -\tfrac{1}{2}fH \qquad (16.49c)$$

Let us next consider the boundary conditions. The condition $f(0) = 0$ remains [see Eq. (16.48)]. Next the condition $f'(0) = 0$ means that $G(0) = 0$. Finally $f'(\infty) = 1$ gives us $G(\infty) = 1$. Note that the equations for f and G have initial values. However, the value for $H(0)$ is *not known*. Because of the unknown value of $H(0)$, we do not have an initial-value problem as indicated earlier. Nevertheless, we handle this problem as an initial-value problem by choosing values of $H(0)$ and solving by numerical methods $f(\eta)$, $G(\eta)$, and $H(\eta)$. In general, the condition $G(\infty) = 1$ will not be satisfied for the function G arising from the numerical solution. We then choose other initial values of H so that eventually we find an $H(0)$ which results in $G(\infty) = 1$. This method is called the *shooting* method—crudely put, we keep shooting until we hit the correct boundary condition $G(\infty) = 1$.

Instead of just repeatedly guessing $H(0)$, we can set up a more systematic, efficient procedure. For this purpose, consider Fig. 16.6a where we plotted two solutions of G versus η for two different values of $H(0)$ which are, respectively, the slopes of the G curves at the origin. The values $G(\infty)$ are inferred from the G curves and are plotted in Fig. 16.6b versus the corresponding values of $H(0)$. A straight line is drawn between these plotted points. The value of $H(0)$ desired can be estimated by finding the value $\bar{H}(0)$ at which the line 1-2 crosses the line $G(\infty) = 1$. By using similar triangles shown shaded we can say

$$\frac{\bar{H}(0) - H(0)_1}{1 - G(\infty)_1} = \frac{H(0)_2 - H(0)_1}{G(\infty)_2 - G(\infty)_1} \qquad (16.50)$$

permitting us to solve for $\bar{H}(0)$. We make the same kind of calculation as above by using $\bar{H}(0)$ and the better of the two initial values $H(0)$. In this way, we can find another improved value of $H(0)$ which we denote as $\bar{\bar{H}}(0)$. This process may be continued. That is, we would use $\bar{\bar{H}}(0)$ and $\bar{H}(0)$ as a pair of values to find $\bar{\bar{\bar{H}}}(0)$. Then we would use $\bar{\bar{\bar{H}}}(0)$ and $\bar{\bar{H}}(0)$, and so forth. Of course, keep in mind that for each value of $H(0)$, the curve $G(\eta)$ versus η would have to be examined in some way to get the proper values of $G(\infty)$. The process is carried out until the change of $H(0)$ at successive computations is less than some small

---

[5] You can verify this by starting with Eq. (16.49c). Replace H by G' and H' by G''. Finally replace G by f'. You should get Eq. (16.47) back.

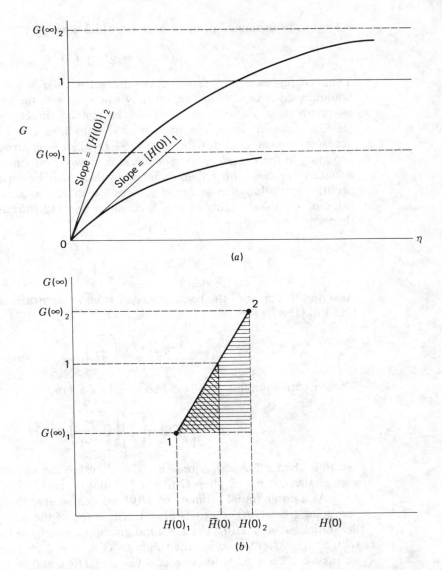

FIGURE 16.6
Procedure to estimate $H(0)$.

prescribed value ϵ. That is, using a subscript instead of an overbar,

$$\frac{[H(0)]_{k+1} - [H(0)]_k}{[H(0)]_k} < \epsilon \tag{16.51}$$

The question now arises about how we find the value of $G(\infty)$ for any specific value of $H(0)$. To answer this, note that from Eq. (13.20a) and the fact that

$h/g = U$, as explained directly after Eq. (13.24), we can say that

$$u(x, y) = Uf' = UG \qquad (16.52)$$

Thus G represents in terms of η the self-similar velocity *profiles* for the boundary-layer flow. Now theoretically when $\eta \to \infty$, then $u \to U$, which gives us the boundary condition $G(\infty) = 1$ as can be directly seen above in Eq. (16.52). However, we do not have to go to ∞ to have $u \to U$. With little loss in accuracy, we can consider the $u(x, y) \to U$ when η approaches a value corresponding to the *boundary-layer thickness* δ at *any* position x. Thus, we can use a finite value of η rather than an infinite value. Also, because of the self-similar profiles, one value of η so found suffices for all positions x. To determine this value of η corresponding to $y = \delta$, consider Eqs. (13.16b) and (13.32). We can then say

$$\eta = \frac{y}{g(x)} = y\sqrt{\frac{U}{\nu x}} \qquad (16.53)$$

Also, the thickness of the boundary layer from an approximate integral solution [see Eq. (13.48)] is

$$\delta = \sqrt{30\left(\frac{\nu x}{U}\right)}$$

Then η corresponding to $y = \delta$ becomes in Eq. (16.53)

$$\eta_\delta = \left[\sqrt{30\left(\frac{\nu x}{U}\right)}\right]\left(\sqrt{\frac{U}{\nu x}}\right) = \sqrt{30} \qquad (16.54)$$

We thus choose for η in estimating $G(\infty)$ a value somewhat greater than $\sqrt{30}$ such as the number 6. Then $G(6) \approx G(\infty)$ that we seek.

As a crude initial estimate of $H(0)$, we can assume that $G(\eta)$ is a straight line going from $G(0) = 0$ to $G(6) = 1$. The slope of the line, namely $\frac{1}{6}$, is then the initial guess at $H(0)$. For a second guess, we simply add 0.1 to the value of $H(0)$ above. After we obtain the solutions G at $\eta = 6$ for these values of $H(0)$, we then use Eq. (16.51) to get a new value of $H(0)$, and so forth.

Now we set forth the finite-difference versions of Eqs. (16.49). Considering Eq. (16.29) (central difference), we can say for f that

$$f(\eta) = f(\eta - \Delta\eta) + \tfrac{1}{2}[G(\eta - \Delta\eta) + G(\eta)]\,\Delta\eta \qquad (16.55)$$

Similarly, for G we have

$$G(\eta) = G(\eta - \Delta\eta) + \tfrac{1}{2}[H(\eta - \Delta\eta) + H(\eta)]\,\Delta\eta \qquad (16.56)$$

Finally, we have for H using Eq. (16.26) (Euler's formula) and later using Eq.

TABLE 16.4

Grid index	η	f	G	H
1	0	0	0	0
11	0.30	0.014928	0.099513	0.331549
21	0.60	0.059685	0.198788	0.329930
41	1.20	0.237803	0.393651	0.316862
61	1.80	0.529374	0.574950	0.283715
81	2.40	0.922420	0.729741	0.229133
101	3.00	1.397569	0.847381	0.162165
121	3.60	1.931196	0.924986	0.098317
141	4.20	2.500709	0.968592	0.050256
161	4.80	3.088908	0.989203	0.021444
181	5.40	3.685280	0.997334	0.007594
201	6.00	4.284639	1.000000	0.002224

(16.49c)

$$H(\eta) = H(\eta - \Delta\eta) + \left[H'(\eta - \Delta\eta) \right] \Delta\eta$$

$$\therefore H(\eta) = H(\eta - \Delta\eta) - \tfrac{1}{2} f(\eta - \Delta\eta) H(\eta - \Delta\eta) \Delta\eta \qquad (16.57)$$

We have used the central difference approximations in Eqs. (16.55) and (16.56) to improve accuracy since H is determined from Eq. (16.57) to be used in Eq. (16.56) for $G(\eta)$, and $G(\eta)$ is determined by Eq. (16.56) to be used in Eq. (16.55).

Example 16.4. Solve the Blasius equation to an accuracy where successive values of $G(\infty)$ have differences such that $\epsilon < 1 \times 10^{-5}$ in Eq. (16.51). Use 200 grid points. Determine $u(y)$ at a position $x = 3$ ft. The free-stream speed $U = 20$ ft/s and the fluid is air at 60° F.

The results for G, H, and f are shown in Table 16.4 for every tenth grid point. These results stem from seven solutions. To get more accurate solutions, the finite value η used in the expression $G(\infty)$ should be made larger and we should use more grid points. We thus know G for 200 values of η going from zero to 6.0. From Eq. (13.16b) we can relate η to y as follows:

$$y = g(x)\eta \qquad (a)$$

And from Eq. (13.32), we have for g at $x = 3$

$$g = \sqrt{\frac{\nu x}{U}} = \sqrt{\frac{(1.7 \times 10^{-4})(3)}{20}} = 0.00505 \qquad (b)$$

Therefore, for each value of η we can easily determine y from Eqs. (a) and (b). We therefore can determine u versus y using Eq. (16.52) and the next-to-the-last column of Table 16.4. We can accordingly find the velocity profile u over the thickness of the boundary layer at $x = 3$ ft. We leave this simple task to the reader.

16.7   STEADY-FLOW BOUNDARY-VALUE PROBLEMS—AN INTRODUCTION

We now return to basics of numerical methods to set the stage for solving partial differential equations. We restrict ourselves to steady-state, boundary-value problems at this time where conditions are specified on the boundary of the domain of interest. A common procedure in such problems is to discretize the partial differential equation using finite differences. The result is a system of algebraic equations for the dependent variable at the grid points inside the boundary. Some or all of these equations involve the known values along the grid points on the boundary.

For this purpose, consider a system of five grid points in Fig. 16.7 where grid point P is at the center surrounded by grid points N (north), S (south), W (west), and E (east). We can give partial derivatives in the x and y directions at point P in terms of the other grid points by using Eqs. (16.19) and (16.20). Thus we get

$$\left(\frac{\partial f}{\partial x}\right)_P = \frac{f_E - f_W}{2\,\Delta x} + \left[O(\Delta x^2)\right] \tag{16.58a}$$

$$\left(\frac{\partial f}{\partial y}\right)_P = \frac{f_N - f_S}{2\,\Delta y} + \left[O(\Delta y^2)\right] \tag{16.58b}$$

$$\left(\frac{\partial^2 f}{\partial x^2}\right)_P = \frac{f_E - 2f_P + f_W}{(\Delta x)^2} + \left[O(\Delta x^2)\right] \tag{16.58c}$$

$$\left(\frac{\partial^2 f}{\partial y^2}\right)_P = \frac{f_N - 2f_P + f_S}{(\Delta y)^2} + \left[O(\Delta y^2)\right] \tag{16.58d}$$

If the partial differential equation involves the preceding partial derivatives, the procedure would be to express the partial differential equation using

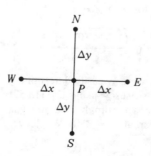

FIGURE 16.7
Five grid points using compass designations.

Eqs. (16.58) at each internal grid point. (Clearly we would need other formulas such as given above for equations with higher-order or mixed partial derivatives.) By expressing the partial differential equation thus at each internal grid point, we would arrive at a system of n simultaneous algebraic equations for the dependent variable f at each of the n internal grid points. For the boundary conditions, we remind you of several factors to keep in mind in properly assigning the values of f at the boundary grid points. If f is the stream function ψ in an inviscid flow, then along an impermeable boundary the values of f must be a specified *constant*. And, in a viscous flow the f chosen along an impermeable boundary must be such as to ensure *zero velocity* at grid points along such a boundary.

To illustrate these procedures and formulations in the most simple way, consider ideal, two-dimensional, steady flow. Such flows are governed by Laplace's equation, $\nabla^2\psi = 0$, where ψ is the stream function. At any internal grid point, we can say for such a flow that

$$\nabla^2\psi = \frac{\partial^2\psi}{\partial x^2} + \frac{\partial^2\psi}{\partial y^2} \approx \frac{\psi_E - 2\psi_P + \psi_W}{(\Delta x)^2} + \frac{\psi_N - 2\psi_P + \psi_S}{(\Delta y)^2} = 0 \quad (16.59)$$

Rearranging we can say on multiplying through by $(\Delta x)^2$ and using β to represent $\Delta x / \Delta y$ that

$$\psi_E + \psi_W + \beta^2(\psi_S + \psi_N) - 2(1 + \beta^2)\psi_P = 0 \quad (16.60)$$

We note now that the grid spacing for the x direction can be different from that in the y direction. This algebraic equation is the discretized algorithm for the numerical solution of the partial differential equation.

To illustrate this further, consider a two-dimensional, inviscid flow (see Fig. 16.8) having a uniform flow U_1 at the left end and a smaller uniform flow at the right end. The bottom of the passage is impermeable and the top of the passage is porous. What is the stream function inside the aforestated rectangular boundary of the flow if there is a uniform velocity through the porous wall in the y direction? Employ the following data.

$$L = 3 \text{ ft} \qquad h = 1.5 \text{ ft} \qquad U_1 = 30 \text{ ft/s} \qquad U_2 = 10 \text{ ft/s}$$

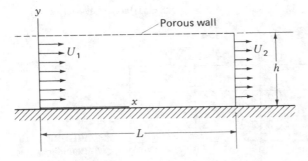

FIGURE 16.8
Boundary for a 2-dimensional inviscid flow.

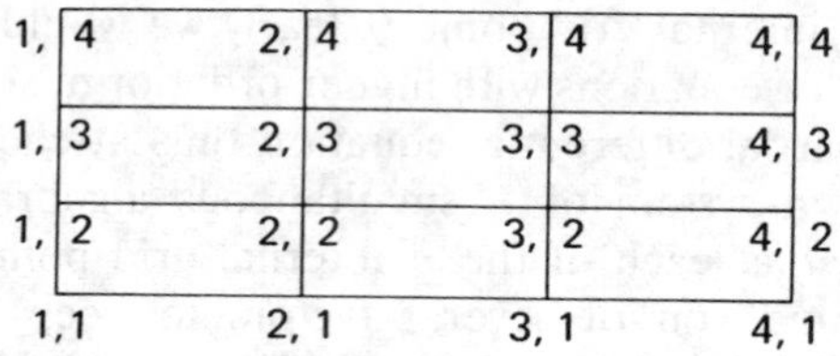

1, 4
2, 4
3, 4
4, 4
1, 3
2, 3
3, 3
4, 3
1, 2
2, 2
3, 2
4, 2
1,1
2, 1
3, 1
4, 1

FIGURE 16.9
Grid numbering system.

The uniform velocity through the porous top wall must then be 10 ft/s. We use, for illustrative purpose, a very course grid system $\Delta x = 1.0$ ft and $\Delta y = 0.5$ ft giving $\beta = 2.0$. In Fig. 16.9 we have identified the grid points using the usual matrix scheme for identifying elements.

Consider the boundaries of the domain. Noting that $V_x = \partial\psi/\partial y$, we can say that along the left vertical inlet section

$$\psi = U_1 y \tag{16.61a}$$

where we set the arbitrary constant equal to zero. Similarly for the right exit we have

$$\psi = U_2 y \tag{16.61b}$$

Along the bottom wall, clearly $\psi = 0$. We may consider that the flow through the upper surface is *uniform* so that we have for ψ along $y = h$ on considering both the definition of ψ and continuity

$$\psi(x, h) = U_1 h - (U_1 h - U_2 h)\left(\frac{x}{L}\right) \tag{16.62}$$

We can now determine ψ along the boundary grid points. Note that with no flow across the impermeable boundary, we have set $\psi = 0$ at the four grid points in Fig. 16.10. The values along the other boundary grid points follow from Eqs. (16.61*a*), (16.61*b*), and (16.62). The internal grid points have been identified with our matrix identification and have been circled to further identify them.

Our next step is to use Eq. (16.60) for each of the inside grid points. Let us look at grid point (2, 2). We can say that

$$15 + \psi_{32} + 4(0 + \psi_{23}) - 2(1 + 4)\psi_{22} = 0$$

$$\therefore -10\psi_{22} + \psi_{32} + 4\psi_{23} = -15$$

-35
-25
-45
-15
-30
-10
-15
-5
2, 3
3, 3
2, 2
3, 2
0
0
0
0

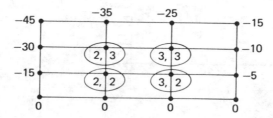

FIGURE 16.10
Boundary values for ψ.

We do the same for the other three grid points leading us to four simultaneous equations which we now set forth.

$$-10\psi_{22} + \psi_{32} + 4\psi_{23} \qquad = -15$$

$$\psi_{22} - 10\psi_{32} \qquad + 4\psi_{33} = -5$$

$$4\psi_{22} \qquad - 10\psi_{23} + \psi_{33} = -170$$

$$4\psi_{32} + \psi_{23} - 10\psi_{33} = -110$$

We can solve for the four unknowns here. In using a finer grid for greater accuracy we would have to use a computer to solve many simultaneous equations. Most computer facilities have library subroutines for solving simultaneous equations. If not, it is not too difficult to program this process.[6]

16.8 POTENTIAL FLOW

We have presented some general concepts for the solution of partial differential equations and then illustrated some of these ideas by considering a very simple, two-dimensional, potential flow. At this time we will continue the discussion of potential flow in greater depth.

We consider an inside grid point located by indices i, j, where the stream function ψ is described as $\psi_{i,j}$. Going back to Eq. (16.60) and considering ψ_p to correspond to $\psi_{i,j}$, we now express this equation, stemming from the discretized Laplace's equation by solving for $\psi_p \equiv \psi_{i,j}$ as follows:

$$\psi_{i,j} = \frac{1}{2(1 + \beta^2)} \left(\psi_{i+1,j} + \psi_{i-1,j} \right) + \frac{\beta^2}{2(1 + \beta^2)} \left(\psi_{i,j+1} + \psi_{i,j-1} \right) \quad (16.63)$$

where you will recall $\beta = \Delta y / \Delta x$ of the grid system. Letting i and j cover all internal grid points, we then have a system of simultaneous algebraic equations as before in Sec. 16.7. When such a system of equations is large or if the equations were nonlinear (they were not for potential flow), then it is usually better to use iteration (successive substitution) methods rather than the Gauss elimination method described earlier (see footnote 6). This is especially true if there are many unknowns but where in any one equation there are only a few nonzero coefficients of the unknowns. Clearly, in Eq. (16.63) no matter how many unknowns are present in the problem, there are only five unknowns in any one equation, so we should use the iteration procedure. Parenthetically, these iteration methods are not hard to program.

---

[6] In this regard, when the algebraic equations are linear and their number is not large, we suggest that you use the Gauss elimination method. Thus, in the first equation, solve for one unknown, say ψ_{22}. Substitute for ψ_{22} using this result in all the other equations. Go to the second equation now and solve for a different unknown, say ψ_{32} and substitute for ψ_{32} in all remaining equations. Continue in this way until you end up with one equation and one unknown.

The basic iteration procedure is to guess first at the values of the unknowns. Most simply we could choose them to be zero. The system of guessed values we denote is $\psi_{i,j}^{(1)}$. We point out now that the rate of convergence does not depend too much on the initial guesses. Next, going back to Eqs. (16.63) and using $\psi^{(1)}$ values on the right side of each equation we calculate from these equations a new set of values for $\psi_{i,j}$ on the left sides of the equations which we denote as $\psi_{i,j}^{(2)}$—that is, a second approximation.[7] With $\psi_{i,j}^{(2)}$ replacing our initial guess $\psi_{i,j}^{(1)}$, we again return to Eqs. (16.63) and using $\psi^{(2)}$ values on the right sides of the equation, we come up with $\psi_{i,j}^{(3)}$, on the left sides. We repeat this process until the *change* of $\psi_{i,j}$ from one iteration to the next has a magnitude less than a prescribed value ϵ at all internal grid points. That is to say

$$\frac{|\psi_{i,j}^{(k)} - \psi_{i,j}^{(k-1)}|}{|\psi_{i,j}^{k-1}|} \le \epsilon$$

The method we have described is the *Richardson* iteration method.

In the Richardson method, note that all the new data $\psi_{i,j}^{(k)}$ are calculated by using old data only $\psi_{i,j}^{(k-1)}$. That is to say, we replaced $\psi_{i,k}^{(k-1)}$ by $\psi_{i,k}^{(k)}$ only after all the calculations had been made. To speed up convergence we should replace the old data $\psi_{i,k}^{(k-1)}$ at a point *immediately* by the new data $\psi_{i,k}^{(k)}$ as it is calculated and *before* moving on to the next point. Thus the calculation of all but one of the $\psi^{(k)}$ values will involve not only old $\psi^{(k-1)}$ values but also $\psi^{(k)}$ values. Thus, if we are "covering" a field from left to right (i increasing) and from bottom to top (j increasing), Eq. (16.63) then becomes for this method

$$\psi_{i,j}^{(k)} = \frac{1}{2(1+\beta^2)}\left[\psi_{i+1,j}^{(k-1)} + \psi_{i-1,j}^{(k)}\right] + \frac{\beta^2}{2(1+\beta^2)}\left[\psi_{i,j+1}^{(k-1)} + \psi_{i,j-1}^{(k)}\right] \quad (16.64)$$

Note in this equation that we have already covered the $i-1, j$ and the $i, j-1$ points for which new values $\psi^{(k)}$ are used. This method is called the *Gauss-Seidel iterative method*.[8]

To speed up the convergence even further, we alter the calculated value $\psi_{i,j}^{(k)}$ by modifying it by a factor r_f, called a *relaxation factor*. Thus, let $\psi_{i,j}^{(k)}*$ be the computed value from Gauss-Seidel at the kth iteration. The value of $\psi_{i,j}^{(k)}$ to

---

[7]Note that if $\Delta x = \Delta y$, then $\beta = 1$ and $\psi_{i,j}^{(2)}$ equals $\frac{1}{4}$ the sum of the $\psi^{(1)}$ at the four surrounding points.

[8]One can demonstrate that the Gauss-Seidel method will converge for linear algebraic equations if the sum of the absolute values of the coefficients on each of the right sides of Eqs. (16.64) is *less than* or *equal to* the magnitude of the coefficient of the term on the left side of the equation and, in addition, the sum of the magnitudes of the coefficients on the right side of Eqs. (16.64) is *less than* the magnitude of the coefficient on the left side of the equation for *at least one* equation. Since for Eq. (16.64), $(2)\{(1/[2(1+\beta^2)])\} + 2\{(\beta^2/[2(1+\beta^2)])\}$ sum to 1 for all β's for the four expressions on the right side, the Gauss-Seidel method converges for two-dimensional, potential flow problems.

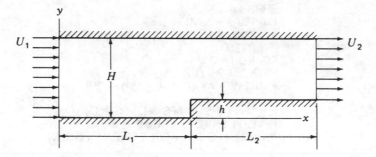

FIGURE 16.11
Two-dimensional channel with a step.

be used as the *latest* value at the point i, j is given as

$$\psi_{i,j}^{(k)} = \psi_{i,j}^{(k-1)} + r_f\left[\psi_{i,j}^{(k)}{}^* - \psi_{i,j}^{(k-1)}\right] \quad \text{where } 1 \leq r_f \leq 2 \quad (16.65)$$

Hence on considering Eq. (16.65) we see that we are adding more than 100 percent of the change of ψ at a grid point generated by the Gauss-Seidel method. Optimum values of r_f are around 1.6, resulting in a significant decrease in work involved.

Example 16.5. Determine the potential flow field of a two-dimensional channel with a step on one wall (see Fig. 16.11). The velocity is uniform at the entrance and at the exit.
From *continuity*, we see that

$$U_2 = \frac{H}{H - h}U_1 \quad (a)$$

Furthermore, the stream function is taken as zero on the bottom wall. It follows from the definition ψ that $\psi = U_1 H$ on the top wall. At the entrance we have $\psi = U_1 y$ and at the exit we have $\psi = U_2 y$.
For this example, use the following data.

$$U_1 = 1 \text{ m/s} \qquad \Delta y = \tfrac{3}{18} \text{ m}$$
$$H = 3 \text{ m} \qquad \Delta x = \tfrac{9}{18} \text{ m}$$
$$h = 1 \text{ m} \qquad \Delta x / \Delta y = \beta = 3 \qquad (b)$$
$$L_1 = 20 \text{ m}$$
$$L_2 = 20 \text{ m}$$

As an aid to help you program around the step, we present part of the program, using as a reference Fig. 16.12. Note that JM1 is the grid ordinate just before the top grid ordinate, namely, JM. Also, the grid abscissa IN1 is just before IN at the step, and grid abscissa IP1 is just before the end IP. The program portion to get around the step is shown below. These statements calculate one sweep of the field, giving a new set of approximations as well as the largest error for all grid points.

```
        ERRØR = 0.0
        DØ 100 J = 2,JM1
        II = IN1
        IF (J.GT.JD)II = IP1
        DØ 100 I = 2,II
        PSIØLD = PSI (I,J)
        PSINEW = A* (PSI (I + 1,J) + PSI (I − 1,J))
                                         +B* (PSI (I,J + 1) + PSI (I,J − 1))
        E = ABS ( (PSINEW-PSIØLD)/PSIØLD)
        IF (E.GT.ERRØR) ERRØR = E
  100   PSI (I,J) = PSIØLD + RF* (PSINEW − PSIØLD)
```

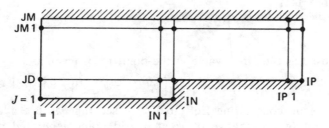

FIGURE 16.12
Domain showing certain grid lines.

We would next compare the ERROR to our tolerance, say EPSILON = 0.001. If this error exceeds the value 0.001, using a GO TO we would repeat the sweep given above until we get the proper desired accuracy.

Figure 16.13 is the computer output for grid points I from 1 to 41 and J from 1 to 16.

16.9 VISCOUS LAMINAR INCOMPRESSIBLE FLOW IN A DUCT

In Sec. 7.8 we developed the velocity profiles for incompressible, laminar flow in a circular pipe. Only one spatial variable had to be used for steady flow leading to simple analytic solutions. However, if the duct cross section does not have the axial symmetry of that of a circular pipe, then the problem becomes much more complex, making the use of numerical methods very desirable.

Consider a duct of constant cross section in which there is laminar steady flow. We take the x direction to be parallel to the axis of the duct. Then the Navier-Stokes equation [see Eq. (10.13)] in the x direction becomes

$$0 = -\frac{\partial p}{\partial x} + \mu\left(\frac{\partial^2 u}{\partial y^2} + \frac{\partial^2 u}{\partial z^2}\right) \tag{16.66}$$

where $\partial p/\partial x$ is constant. This is the well-known Poisson equation, which we wish to solve to get the desired velocity profile. We will consider now that the cross section is that of a rectangle (see Fig. 16.14).

The X-Component of the Velocity Field

1.000	1.000	1.000	1.000	1.000	1.000	1.000	1.000	1.000	1.000	1.000	1.000
1.000	1.000	1.000	1.000	1.000	1.000	1.000	1.000	1.000	1.000	1.000	1.000
1.000	1.000	1.000	1.000	1.000	1.000	1.000	1.000	1.000	1.000	1.000	1.000
1.000	1.000	1.000	1.000	1.000	1.000	1.000	1.000	1.000	1.000	1.000	1.000
1.000	1.000	1.000	1.000	1.000	1.000	1.000	1.000	1.000	1.000	1.000	1.000
1.000	1.000	1.000	1.000	1.000	1.000	1.000	1.000	1.000	1.000	1.000	1.000
.999	.999	1.000	1.000	1.000	1.000	1.000	1.000	1.000	1.000	1.000	1.000
.999	.999	.999	.999	.999	1.000	1.000	1.000	1.000	1.000	1.000	1.000
.999	.999	.999	.999	.999	.999	1.000	1.000	1.000	1.000	1.001	1.001
.998	.998	.998	.998	.998	.999	.999	1.000	1.000	1.000	1.001	1.001
.996	.996	.996	.997	.997	.998	.999	.999	1.000	1.001	1.001	1.002
.993	.993	.994	.995	.996	.997	.998	.999	1.000	1.001	1.002	1.003
.988	.989	.990	.991	.993	.994	.996	.998	1.000	1.002	1.004	1.006
.981	.981	.983	.985	.988	.991	.994	.997	1.000	1.004	1.007	1.010
.967	.969	.971	.975	.979	.984	.990	.996	1.001	1.007	1.012	1.017
.943	.946	.951	.958	.966	.975	.984	.994	1.003	1.013	1.021	1.029
.901	.907	.916	.928	.943	.959	.976	.993	1.010	1.025	1.038	1.050
.826	.837	.854	.878	.906	.937	.969	.999	1.026	1.050	1.071	1.087
.686	.705	.738	.785	.843	.909	.974	1.030	1.075	1.109	1.135	1.155
.427	.454	.504	.584	.706	.892	1.065	1.161	1.214	1.243	1.259	1.268
0.000	0.000	0.000	0.000	0.000	.983	1.857	1.678	1.567	1.496	1.449	1.417
					1.639	1.625	1.594	1.559	1.527	1.499	1.477
					1.554	1.551	1.542	1.531	1.518	1.506	1.494
					1.523	1.522	1.519	1.515	1.510	1.504	1.499
					1.510	1.510	1.509	1.507	1.505	1.502	1.500
					1.505	1.505	1.504	1.503	1.502	1.501	1.500
					1.502	1.502	1.502	1.502	1.501	1.501	1.500
					1.501	1.501	1.501	1.501	1.500	1.500	1.500
					1.500	1.500	1.500	1.500	1.500	1.500	1.500
					1.500	1.500	1.500	1.500	1.500	1.500	1.500
					1.500	1.500	1.500	1.500	1.500	1.500	1.500
					1.500	1.500	1.500	1.500	1.500	1.500	1.500
					1.500	1.500	1.500	1.500	1.500	1.500	1.500
					1.500	1.500	1.500	1.500	1.500	1.500	1.500
					1.500	1.500	1.500	1.500	1.500	1.500	1.500
					1.500	1.500	1.500	1.500	1.500	1.500	1.500
					1.500	1.500	1.500	1.500	1.500	1.500	1.500
					1.500	1.500	1.500	1.500	1.500	1.500	1.500
					1.500	1.500	1.500	1.500	1.500	1.500	1.500
					1.500	1.500	1.500	1.500	1.500	1.500	1.500
					1.500	1.500	1.500	1.500	1.500	1.500	1.500

Flow

Step

Flow

FIGURE 16.13
X component of velocity for grid points $I = 1$ to $I = 41$, $J = 1$ to $J = 16$.

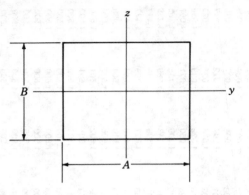

FIGURE 16.14
Rectangular duct.

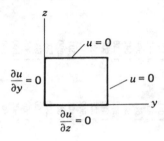

FIGURE 16.15
First quadrant of duct with boundary conditions.

Owing to the symmetry about the y and z axes, we need consider only one quadrant. Using the first quadrant and inserting known boundary conditions, we have shown the domain of interest to us in Fig. 16.15. Note that along the lines of symmetry the derivative normal to the lines of symmetry must be zero.

If we nondimensionalize the equation, our solution will apply to all geometrically similar cross sections (i.e., same aspect ratio A/B). Accordingly, we introduce the following nondimensional variables:

$$y^* = \frac{y}{A/2} \qquad\qquad (16.67a)$$

$$z^* = \frac{z}{A/2} \qquad\qquad (16.67b)$$

$$u^* = \left[\frac{\mu}{(A^2/4)(-dp/dx)} \right] u \qquad\qquad (16.67c)$$

Then on substituting y, z, and u as given above into Eq. (16.66) we get

$$\frac{\partial^2 u^*}{\partial y^{*2}} + \frac{\partial^2 u^*}{\partial z^{*2}} = -1 \qquad\qquad (16.68)$$

and the domain of interest for the variables with asterisks becomes

$$0 \leq y^* \leq 1$$
$$0 \leq z^* \leq \frac{B}{A} \tag{16.69}$$

Now, as explained in Sec. 16.7 for Laplace's equation, we employ finite differences [see Eqs. (16.59), (16.60), and (16.63) as applied to Poisson's equation] to Eq. (16.68) for a set of grid points to get

$$u^*_{i,j} = \tfrac{1}{4}\left(u^*_{i+1,j} + u^*_{i,j+1} + u^*_{i-1,j} + u^*_{i,j-1} + \Delta y^{*2}\right) \tag{16.70}$$

wherein $\Delta z^* = \Delta y^*$ and $\beta = 1$. The procedure here is almost identical to that employed for Laplace's equation as discussed in Sec. 16.7. Thus choose a grid system and for the interior grid points proceed as before with the successive substitution scheme. From Fig. 16.15 and Eq. (16.67c) we see that $u^* = 0$ on the top- and right-hand boundaries because of no-slipping conditions. Also, the normal derivative of u^* on the other two boundaries must be zero because of the symmetry conditions. To ensure the last pair of conditions, we go back to Eq. (16.7), which is the forward difference formula for a derivative. The subscript zero in Eq. (16.7) will correspond to unity for our numbering system here, while the subscript 1 corresponds to 2 here. Replacing the variable x by y in Eq. (16.7), we conclude that $\partial u^*/\partial y = 0$ for the left vertical boundary (Fig. 16.14) corresponds to

$$\frac{u^*_{i,2} - u^*_{i,1}}{\Delta y} = 0 \tag{16.71}$$

$$\therefore u^*_{i,1} = u^*_{i,2}$$

Similarly, for the bottom horizontal boundary, the use of Eq. (16.7) with z as the independent variable leads us to conclude that

$$u^*_{1,j} = u^*_{2,j} \tag{16.72}$$

It is evident that the values of u^* on the symmetric boundaries are unknown; they must be handled by incorporating the previous two equations into the iteration process. Note however that the symmetric boundaries do not change the total number of unknowns and the basic procedure is not changed in any significant way.

Once the solution for $u^*_{i,j}$ has been iterated to an acceptable degree of convergence, we will want to determine $u_{i,j}$. We now need the value of dp/dx. For this purpose, go back to Eq. (16.67c) and average both sides of the equation over a cross section. The average of u is Q/area, where Q, the volumetric flow, is usually known. Denoting averages with overbars we then have

$$\bar{u} = -\frac{dp}{dx}\frac{A^2}{4\mu}\bar{u}^* \tag{16.73}$$

We can get $\bar{u}^*$ by calculating the average u^* over the cross section as follows:

$$\bar{u}^* = \frac{\int_0^{B/A} \int_0^1 u^* \, dy^* \, dz^*}{(B/A)(1)}$$

(Note that we are using nondimensional terms for the rectangular quadrant.) This integration can be accomplished using subroutine INTEG given in Sec. 16.2. The procedure is to integrate $\int_0^1 u^* \, dy^*$ over y^* for each z_j^* by repeatedly applying the subroutine. We then integrate the computed quantities $[\int_0^1 u^* \, dy^*]$ for each z_j^* over z^* between the limits of zero and B/A again using the aforementioned subroutine.

Finally, knowing $\bar{u}$ and $\bar{u}^*$ we can get dp/dx from Eq. (16.73) and thus we can easily get the head loss. And from Eq. (16.67c) we can get $u_{i,j}$ knowing $u_{i,j}^*$ as well as dp/dx.

Example 16.6. Compute the nondimensional flow field for a rectangular duct where $A = B = 1$ ft. Use a tolerance of 0.001, which is the maximum fractional change at a point.

We use grid spacings of $\Delta y^* = 0.05$ thus giving 10 intervals in each direction in the first quadrant. It took 223 iterations to achieve the desired tolerance. The calculation of $u_{i,j}^*$ took about 7 s on a CYBER 175. The results are shown in Fig. 16.16 for the first quadrant using every fourth grid point.

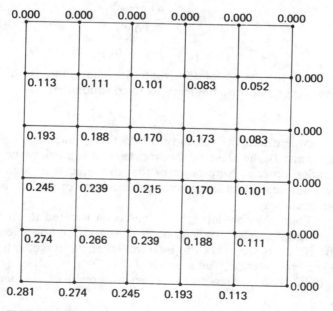

FIGURE 16.16
$u_{i,j}^*$ for every fourth grid point.

16.10   PROJECTS

Project 1. Work out the details of the time-dependent pipe flow of Example 16.1 and check the results.

Project 2. Work out the details of the trajectory problem of the golf ball presented in Example 16.2. Verify the given results.

Project 3. Work out the details of the transient tank flow problem given in Example 16.3. Verify the given results.

Project 4. Work out the details of the solution of the Blasius equation as presented in Example 16.4. Verify the results.

Project 5. Work out the details of the potential flow over a step as presented in Example 16.5. Verify the results presented.

Project 6. Assuming quasi-steady flow, formulate an equation for determining the time behavior of the height h of the fluid in the conical shaped tank. Numerically solve the equation for $h(t)$ for the following data:

$$\text{Fluid} = \text{water}$$
$$h_0 = 12 \text{ in}$$
$$L = 12 \text{ in}$$
$$D_i = \tfrac{1}{8} \text{ in}$$

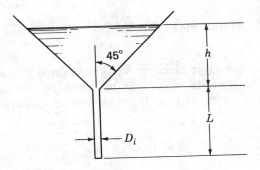

45°

h

L

D_i

FIGURE P16.6

Project 7. The golf ball in the example given in case II is hit when there is a 25 ft/s cross wind. Calculate the trajectory and determine the lateral drift due to the cross wind.

Project 8. Solve for the ideal, two-dimensional flow through the 90° bend. Assume the velocity profiles at the entrance and exit are uniform. Take the

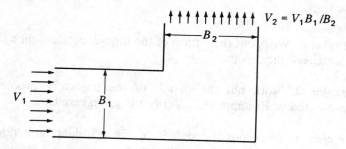

FIGURE P16.8

following data:

$$V_1 = 10 \text{ ft/s}$$
$$B_1 = 5 \text{ ft}$$
$$B_2 = 10 \text{ ft}$$

Project 9. Determine the fully developed velocity profile in a duct with right-triangular cross section. With this result, determine the relation between the flow rate in the duct and the pressure gradient.

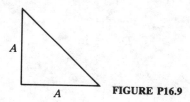

FIGURE P16.9

Project 10. Boundary layer bleed or suction through a porous surface can be used to delay transition from laminar to turbulent flow and thus reduce the friction of the flow on the surface. If the suction velocity at the porous surface is given by

$$v(x,0) = -\frac{1}{2}\alpha\sqrt{\frac{\nu U_\infty}{x}}$$

the boundary layer problem (see Sec. 13.3 in Chapter 13) can be formulated as

$$f''' + \tfrac{1}{2}ff'' = 0$$
$$f(0) = \alpha \qquad f'(0) = 0 \qquad f'(\infty) = 1$$

Note that for the suction parameter $\alpha = 0$, this problem is the same as the plate boundary layer. Numerically solve for the velocity profiles for different α and determine the corresponding shear stress at the surface.

ANSWERS TO SELECTED PROBLEMS

CHAPTER 1

1.4. $1 \text{ ft/s}^2 \equiv 0.305 \text{ m/s}^2$

1.8. H is dimensionless

1.10. $1 \text{ P} \equiv 2.09 \times 10^{-3} \text{ (lbf)(s)/(ft)}^2$

1.12. $\tau_w = \beta D/4; \ \tau = -\beta D/8; \ \text{drag} = \beta D^2 \pi L/4$

1.14. 6.11 m/s

1.16. 314 lbf

1.18. $1.988 \times 10^{-5} \text{ N} \cdot \text{m}$

1.20. $0.0254 \text{ in} \cdot \text{lb}$

1.22. 16.55 m/s

1.24. $(R) \equiv (FL/(M \circ R)); \ v = 4.15 \text{ ft}^3/\text{lbm}$

1.26. $385K = 112.2°\text{C}$

1.28. 408.9 ft^3

1.30. 3138 ft/s

1.32. $.1585 \text{ m}^3$

1.34. 0.6295 mm

1.38. 29.33 lbm

1.40. 14.89 mm

1.42. 11.81 mm

1.44. $h \approx 5.14 \text{ mm}$

1.46. 4.16 Pa gage

CHAPTER 2

2.2. $\mathbf{V} = 85\mathbf{i} + 47\mathbf{j} + 20\mathbf{k}; \ |\mathbf{V}| = 99.2 \text{ m/s}$

2.4. $\mathbf{F} = 108\mathbf{i} + 94.5\mathbf{j} + 72\mathbf{k} \text{ lb}$

2.12. $A_i B_j C_j = \begin{bmatrix} 210 \\ 105 \\ 70 \end{bmatrix}; \qquad A_k B_n C_k = \begin{bmatrix} 190 \\ 38 \\ -57 \end{bmatrix}$

2.16. $\bar{\sigma}(0, 10, 2) = 336 \text{ MPa}$

2.18. $|\nabla\phi| = 37.0$

2.20. $\nabla = \dfrac{\partial}{\partial r}\boldsymbol{\epsilon}_r + \dfrac{\partial}{r\,\partial\theta}\boldsymbol{\epsilon}_\theta + \dfrac{\partial}{\partial z}\boldsymbol{\epsilon}_z$

CHAPTER 3

3.2. $\rho = \gamma z^2/K$

3.6. $3214.7 \text{ N} \cdot \text{m}^2$

3.8. $77,262 \text{ Pa}$

3.10. 0.895 ft

3.12. $2520 \text{ Pa}; \ -2520 \text{ Pa}$

3.14. $S = 0.500$

3.16. $86,224 \text{ Pa}$

3.18. $p_E - p_B \approx (\gamma_2 - \gamma_1)d$

3.20. 65.77 psi abs.

3.22. $p = 13.93$ psi

3.24. $n = 1.2351$

3.28. $z = 7778$ m

3.30. 5.43×10^{-2} ft $\cdot$ lb/ft^3

3.32. 27,340 ft

3.36. $r = 0.8171$ m; $M_{AB} = 1.048 \times 10^5$ N $\cdot$ m

3.38. $F = 4.381 \times 10^4$ lb; $d = 5.716$ ft from B

3.40. $F_1 = 706$ kN; $F_2 = 235$ kN $y' - y_c = 0.067$;
$F_3 = 71.7$ kN $y' - y_c = 0.0577$ m

3.42. 7200 N $\cdot$ m

3.44. $F = 4.969 \times 10^6$ N; $y' - y_c = 0.00961$ m

3.46. $F = 14,195$ lb; $d = 2.0332$ ft

3.48. $p = \left[\dfrac{1}{2} \dfrac{\gamma_0}{\sqrt{p_{\text{atm}}}} z + \sqrt{p_{\text{atm}}} \right]^2$

$F = b \left[\dfrac{\gamma_0^2 h^3}{12 \, p_{\text{atm}}} + \dfrac{\gamma_0 h^2}{2} + p_{\text{atm}} h \right]$

3.50. $\mathbf{F} = -4464\mathbf{j} - 596.9\mathbf{k}$ N

3.52. $F_x = -1630.3$ N

3.54. $\mathbf{F} = 17,643\mathbf{i} + 3529\mathbf{j}$ lb

3.56. 176,828 lb

3.58. $\mathbf{F} = 7,800\mathbf{i} - 8,653\mathbf{j}$ lb

3.60. $\mathbf{F} = 198.6\mathbf{i} + 312\mathbf{j}$ kN; $\theta = 57.5°$

3.62. 2607 lb downward

3.64. $\tau_{xx} = 3.14 \times 10^8$ Pa; $\tau_{yy} = 1.423 \times 10^8$ Pa
$F = 89.3$ kN per bolt

3.66. 7775 N

3.68. $-39,426$ N

3.70. 1199 lb downward

3.72. $F_V = 1241.6$ lb; $F_H = 1695$ lb

3.74. 2.03 MN

3.76. 0.0454 m^3

3.80. 25.2×10^3 m^3

3.82. $h = 1.220$ m; $h' = 1.2407$ m

3.84. 34.8°

3.86. $P = 159.2 - 8.4x$ lb

3.90. Load $= 0.6588$ N; $z = 283.9$ m

3.92. $\mathbf{F}_{\text{inside}} = 3023\mathbf{i} + 153.0\mathbf{j}$ N
$\mathbf{F}_{\text{outside}} = -2597.7\mathbf{i} + 10,743\mathbf{j}$ N

3.94. 148.5 m

3.96. 31.8 m; 27,147 kN $\cdot$ m

3.98. $(\overline{MC})_x = 0.648$ m; $(\overline{MC})_y = 27.88$ m
$(\overline{MC})_{AA} = 7.455$ m

3.100. $S = 0.3832$

CHAPTER 4

4.2. $dy/dx = -0.0559$

4.4. $\mathbf{a} = 360\mathbf{i} + 216\mathbf{j} - 7\mathbf{k}$ m/s^2

4.6. $\mathbf{a} = -58\mathbf{i} - 10\mathbf{j}$ m/s^2

4.8. $D\mathbf{F}/Dt = 10^{-5}\{[(18 + zy)[(z^2 + y^2)z$
$+ (18 + zy)(2y)]\mathbf{i} - [(20)(z^2 + y^2)(20x + t^2)$
$+ (18 + zy)(2y)(20x + t^2) + (2t)(z^2 + y^2)]\mathbf{j}$
$- [(18 + zy)(2t) + (z)(18 + zy)(10 + t^2)]\mathbf{k}\}$ N/s

4.14. $\mathbf{a}_{xyz} = 300\mathbf{i} + 4077\mathbf{k}$ m/s^2
$\mathbf{a}_{XYZ} = 300\mathbf{i} - 1805\mathbf{j} + 3793\mathbf{k}$ m/s^2

4.16. $\mathbf{a} = \dfrac{4V_0^2}{a} \sin \theta_0 \cos \theta_0 \boldsymbol{\epsilon}_\theta - \dfrac{4V_0^2 \sin^2 \theta}{a} \boldsymbol{\epsilon}_r$

4.18. $\boldsymbol{\omega} = -40\mathbf{i} + 5\mathbf{k}$ r/s; $\omega = 40.31$ r/s

4.20. $\boldsymbol{\omega} = -(3/2)\mathbf{i}$ r/s

4.22. $\mathbf{V} = (-6xy + 3)\mathbf{i} - (3x^2 + 6y)\mathbf{j} - 12t\,\mathbf{k}$

4.30. 1.333

4.32. $\dot{m} = (\rho \Delta l / 2\pi)\ln(R/a)$

CHAPTER 5

5.2. $y = .211$ m

5.4. 5.29 m/s

5.6. $V = 1.839$ m/s; $\rho = 953.7$ kg/m^3

5.8. 21.37 ft/s

5.10. 32.33 ft^3/s

5.12. $Q_2 = 12$ ft^3/s; $dh_1/dt = 0.04$ ft/s

5.14. $dh/dt = 0.00268$ m/s; $\Delta t = 82.1$ s

5.16. $(\Delta N) = (\pi a^2 V n)(\Delta t)$

5.18. 1.818 ft/s

5.20. $K_x = 13.06$ N; 1.451 N

5.22. $K_x = 7864$ N; $K_y = -980.6$ N

5.24. $K_x = 721$ lb

5.26. $K_x = 3707$ N

5.28. $K_x = -24.54$ N

5.30. $K_x = -9359$ N; $K_y = -7720$N

5.32. (a) 119.4 N $\cdot$ m
    (b) $M - 60.68$ N $\cdot$ m
    (c) 58.72 N $\cdot$ m

5.34. $K_y = -453.4$ kN

5.36. $(K_y)_{\text{oil}} = 3734$ N; $(K_y)_{\text{water}} = -3478$ N
    $(K_y)_{\text{total}} = 256$ N

5.38. $F_x = F_y = 0$; $F_z = 7469$ N

5.40. $K_n = -355$ N

5.42. $K_x = V_0^2 nhm\{\cos[\alpha + \tan^{-1}(\epsilon \tan \alpha)] - 1\}$

5.44. 10.194 kN

5.46. $h_2 = 9.40$ ft

5.48. 490 ft/s = 334 mph

5.50. $Q = 1.111$ m^3/s; $K_x = -49.38$ kN

5.52. 84.9 N $\cdot$ m

5.54. -10.13 N

5.56. $K_x = 67.4$ N; $M = 40.4$ N $\cdot$ m

5.58. 6837 N (including atmosphere)

5.60. $T_x = -(p_e)_g A_2 + \rho_1 V_1 A_1 (V_1 - V_2) -$
    $(w_f/g_0)V_2$

5.62. 3280 lb

5.64. 13,759 N

5.66. $\iint V_x \rho \mathbf{V} \cdot \mathbf{dA} = (V_c^2 \rho D^2 \pi/4 \tan^2 \alpha)\ln(\sec^2 \alpha)$

5.68. 6203 kN

5.70. $F = 306t^2 + 1.5$ N

5.74. -5.33 ft/s^2

5.76. $\mathbf{K} = -12.53\mathbf{i} + 108.6\mathbf{j} + 44.1\mathbf{k}$ N

5.78. $\mathbf{M}_K = -1780\mathbf{i} + 782\mathbf{j} + 304\mathbf{k}$ ft $\cdot$ lb

5.80. $35.01\mathbf{k}$ N $\cdot$ m

5.82. $\mathbf{M} = -1.209 \times 10^4 \mathbf{i} - 5.093 \times 10^3 \mathbf{j} + 1.380 \times$
    $10^5 \mathbf{k}$ N $\cdot$ m
    $\mathbf{K} = -2.865 \times 10^3 \mathbf{i} - 2.018 \times 10^4 \mathbf{j} + 1.273 \times$
    $10^3 \mathbf{k}$ N

5.84. 2.69 ft-lb

5.86. $\omega = 26.13$ r/s

5.88. $\omega = 1.389$ r/s

5.90. $\mathbf{K}_s = -9499\mathbf{k}$ N
    $\mathbf{C}_s = -7.77\mathbf{i} + 3893\mathbf{k}$ N $\cdot$ m

5.92. $\omega = -0.353$ r/s

5.94. $T_s = r_0^2 \kappa (w/g)l + I\kappa$

5.98. $\omega = -1.442$ r/s

5.100. $\mathbf{T} = -(0.951\mathbf{j} + 1.570\mathbf{k})$ N $\cdot$ m

5.102. $\mathbf{T}_{\text{total}} = 5.16\mathbf{i} - 17.75\mathbf{j} + 35.6\mathbf{k}$ N $\cdot$ m

CHAPTER 6

6.2. 0.463 kW

6.4. 0.463 kW = 0.6209 hp

6.6. $Q = 4.8$ ft^3/s

6.10. 8000 Btu/h

6.12. $\dot{Q} = -1.489 \times 10^5$ Btu/h

6.14. -204.1 Btu/s

6.16. $V = 9.50$ m/s; $Q = 0.1679$ m^3/s

6.18. 0.1197 m^3/s

6.20. $\dot{m} = 4.70$ kg/s

6.24. $q = \left[-2gd \Big/ \left(\dfrac{1}{h^2} - \left(\dfrac{1}{k - d - \delta} \right)^2 \right) \right]^{1/2}$

6.26. $h = 10.16$ m

6.30. 6972 N

6.32. 33.96 kg/s

6.34. 3.85 hp

6.36. $\alpha_1 = 69.3°$; $\alpha_2 = 75.7°$

6.38. 878 lb

6.40. 46.9 hp

6.42. $T_x = -206$ lb

6.44. $A_1 = 1.24$ m^2

6.46. 484.6 hp

6.48. $V_e = 110.4$ s; $\mathbf{M} = 46,931\mathbf{i} + 44,968\mathbf{k}$ ft-lb

6.50. $T = 1077$ ft-lb; $F = 613$ lb

CHAPTER 7

7.6. $\partial \rho/\partial t = 6.16\rho_0$

7.8. $\mathbf{a} = -0.072\mathbf{i} - 0.001\mathbf{j} - 9.83\mathbf{k}$ m/s^2

7.10. $\partial p/\partial x = -10\rho$ Pa/m at $(1, 1, 0)$
    $\partial p/\partial z = -\rho[16t^2 + 32t + 11.81]$ Pa/m at
    $(1, 0, 2)$

7.12. $a_y = 1.962$ m/s^2

7.14. $F = 76.9$ lb

7.16. $F = 76.9$ lb; 0.4777 ft above B

7.18. $F = 776$ N

7.20. $h = 29.7$ mm

7.22. $\omega = 7.92$ r/s

7.24. $\mathbf{B} = (1/\rho)\{-(1500x^2 + 20z)\mathbf{i}$
    $+(1600yx^2 + 1000y^2)\mathbf{j} + (2000zy)\mathbf{k}\}$
    N/kg

7.26. $\mathbf{a} = (1/\rho)[1500\mathbf{i} - 7200\mathbf{j} - \rho g\,\mathbf{k}]$ m/s^2

7.32. $r\dfrac{dp}{dr} + \rho\left[\dfrac{\omega a}{\ln(a/b)}\ln\left(\dfrac{r}{b}\right)\right]^2 = -2p$

7.34. Error = 2.48%

CHAPTER 8

8.48. $F_{\text{oil}}/F_{\text{water}} = 19.76$
8.50. $q = 157.2$ L/s
8.52. $V_M = 258$ kn
8.54. $Q = 35.7$ m^3/s; $\Delta H_D = 30$ m
8.56. $D_P/D_M = 2.7 \times 10^4$
8.58. $V_M = 2.236$ kn; $\nu_M = 3.762 \times 10^{-8}$ ft^2/s; $D_M/D_P = 1.250 \times 10^{-4}$
8.60. $F_M/F_P = 1.214 \times 10^{-4}$
8.62. $V_M = 8.150$ m/s; $M_P/M_M = 36.07$
8.64. $V_M = 5.14$ mph; $D_M/D_P = 3.065 \times 10^{-3}$
8.66. $V_M = 46.3$ km/h; $F_P/F_M = 1.259$
8.68. Scale factor = 2.58; $\Delta H_D = 5.15$ m; $Q_P = 37.94$ L/s

CHAPTER 9

9.2. Re = 1.41×10^6
9.4. $(q_B)_{\text{max}} = 5.16 \times 10^{-3}$ L/s
9.8. 3.11 Btu leaving pipe
9.10. $V = 0.8973$ ft/s
9.12. $r = 3.873$ in
9.14. $q = 3.72 \times 10^{-4}$ L/s
9.16. 2.06 N
9.18. $D = 0.359$ mm
9.20. $h = 0.27165$ m; $h = 0.1832$ m
9.22. $p = 7.947 \times 10^5$ Pa gage
9.24. $V = 1.05 \times 10^{-2}$ m/s
9.26. $q = 0.04475$ L/s; $q = 0.1092$ L/s
9.28. 133.4 diameters; 15.99 diameters
9.30. 0.1522 hp
9.32. 770.5 kW
9.34. $p = 28.777$ lb/ft^2 gage
9.36. $p = 11.60$ lb/in^2 gage
9.38. $q = 113.6$ L/s

9.40. 2.02%
9.42. $q = 1.453$ L/s; $-17{,}950$ Pa
9.44. 1.6215×10^4 N
9.46. $q = 3.44$ ft^3/s
9.48. $D = 1.223$ ft
9.52. $0.1475D$
9.54. $h_l = 0.5009$ ft · lb/slug
9.56. 64.96 hp
9.58. $D = 7.81$ in
9.60. $D = 311$ mm
9.62. $q = 2545$ L/s
9.64. $\alpha = 0.9192°$
9.66. $\Delta H_D = 20.39$ m
9.68. 2.305 km; $\Delta H_D = 11.85$ m; 100 kW
9.70. $D = 2.54$ ft
9.72. 93.52 hp; 246.7 lb
9.74. $h = 22.8$ m
9.76. 1.620 L/s
9.78. 30.89 kW; 65.03 kW
9.82. $L = 0.2309$ m
9.84. $a = 0.2285$ ft = 2.743 in
9.86. 15.34 m/s
9.88. $\tau_w = 0.0774$ Pa
9.90. $a = 2.74$ in
9.92. 5106 L/s
9.96. Blasius, 0.01779; Prandtl, 0.01799; Moody, 0.1080
9.98. 24.67 kW
9.100. $\Delta H_D = 18.06$ m; $D = 141$ mm
9.102. $\lambda = 1.245 \times 10^{-4}$ ft
9.104. 9.92 m/s
9.106. 94.6 lb/in^2 gage
9.108. 435 kPa gage
9.110. 13.52 kW
9.112. $q_1 = 227$ ft^3/s; $q_2 = 243$ ft^3/s; $q_3 = 467$ ft^3/s

CHAPTER 10

10.2. $\tau_{xx} = -10$ lb/ft^2; $\tau_{yy} = -5.56$ lb/ft^2; $\tau_{zz} = -10.96$ lb/ft^2

10.12. $V_z = \dfrac{\gamma}{4\mu}\left\{r^2 - r_2^2 + \dfrac{r_2^2 - r_1^2}{\ln(r_1/r_2)}\ln\dfrac{r}{r_2}\right\}$

CHAPTER 11

11.4. $v_2 = 0.340$ m^3/kg; $W_k = -86.130$ N $\cdot$ m/kg

11.6. $V = 890$ ft/s; $V = 778$ ft/s

11.8. $k = 1.40$

11.10. $t = 0.777$ s

11.14. $V = 116.8$ m/s; $T_2 = 548$ K

11.16. 610 K

11.18. $A^* = 0.364$ ft^2; $A_e = 1.540$ ft^2; 12,190 lb

11.20. 5.5% error

11.22. $A_e = 0.1488$ ft^2; $A^* = 0.0767$ ft^2

11.24. $A^* = 0.365$ ft^2; $A_e = 0.431$ ft^2; $p_2 = 100.6$ lb/in^2 abs

11.26. $M_e = 2.20$; $T = 4.097 \times 10^5$ N

11.28. $w = 4.317$ kg/s

11.30. $T_e = 5.9°C$

11.32. (a) 47.420 Pa
(b) 29,420 Pa
(c) 18,000 Pa
(d) 29,420 Pa

11.36. 0.0236 m^2

11.38. $c_0 = 342.2$ m/s; $c^* = 312.3$ m/s

11.40. M $= 0.732$; $V = 798.3$ ft/s

11.46. M $= 2.287$

11.48. 710 m/s

11.54. 1167 N

11.56. $p_e = 270$ kPa; $T_e = 475$ K; $c_e = 437$ m/s; $V_e = 321$ m/s

11.58. M $= 0.664$; M$_e = 0.360$; $p_2 = 65.2$ lb/in^2 abs; $p_0 = 86$ lb/in^2 abs; $T_0 = 200°F$

11.60. $w = 0.382$ kg/s; $V_e = 223$ m/s

11.62. $p_e = 93,300$ Pa; $p^* = 291,500$ Pa

11.64. 1.047×10^5 N

11.66. 18 in from outlet

11.68. $\alpha = 36.93°$; M$_2 = 1.875$

11.70. M$_2 = 0.814$

11.72. M $= 0.61$

11.74. M $= 0.838$; Re $= 8 \times 10^5$

11.76. M $= 0.3044$; $L = 92.86$ m

11.78. $T = 299$ K; $(T_0)_1 = (T_0)_2 = 328.9$ K

11.82. M$_2 = 0.62$

11.84. M $= 0.653$; $p_0 = 2.005 \times 10^4$ Pa; $A^* = 0.351$ m^2

11.86. $\Delta p_0 = -2.768 \times 10^4$ Pa abs

11.88. $D = 1361$ N/m

11.90. $q = -75,730$ N-m/kg

11.92. M$_2 = 0.136$; $T_2 = 498°C$; $p_2 = 122,000$ Pa; $V_2 = 75.2$ m/s

CHAPTER 12

12.4. $4t^2$

12.6. $\mathbf{V} = -4\mathbf{i} + 10\mathbf{j}$

12.8. $\phi = y^2 - x^2 + C$

12.12. $\psi = -2Axy$; $\psi = -2Ar^2 \cos\theta \sin\theta$

12.18. $\psi = x^2 y + 2x^2$

12.20. $\mathbf{a} = (100xy^2 + 50x^3 + 100z^3x)\mathbf{i}$
$\quad + (100x^2y + 90t^2z^2 + 900yz^4)\mathbf{j}$
$\quad + (150x^2z^2 + 300z^5 + 180yt^2z$
$\quad + 1800y^2z^3)\mathbf{k}$

12.28. $\psi = U\left\{ \dfrac{y^2}{2h} - \left[\dfrac{h^2}{2\mu U}\dfrac{\partial p}{\partial x} \right]\left[\dfrac{y^2}{2h} - \dfrac{y^3}{3h^2} \right] \right\} + C$

12.32. $V_r = 6.997$ ft/s; $V_\theta = -7.19$ ft/s

12.34. For $0 < \alpha < \pi$, $r = 0$, $V_r = V_\theta = 0$
For $\pi < \alpha < 2\pi$, $r = 0$, $V_r = V_\theta = \infty$

12.42. $\mathbf{V} = 0.3753\mathbf{i} + 0.701\,\mathbf{j}$ m/s

12.46. $p = -241.42$ Pa gage

12.48. $V_0 = 12.5$ m/s

12.52. $h = 0.063$ m

12.54. $D = 70.7$ N

12.56. $\mathbf{V} = -0.03707\mathbf{i} + 5.939\mathbf{j}$ m/s

12.58. 676 lb/ft

12.60. 127.4 kN

12.62. 2.14×10^5 N $\cdot$ m

12.66. $\mathbf{V} = 2\mathbf{i} + 6\mathbf{j}$ m/s; $\mathbf{V} = 1.3815\mathbf{i} + 0.3261\mathbf{j}$ m/s

12.68. $\psi = -(m/4\pi)\cos\beta$

12.78. $\psi = -\frac{1}{2}V_0 r^2$; $\phi = -V_0 z$

12.80. $D \approx \frac{9}{16}\pi\rho V_0^2 b^2 \sin^4\alpha$

12.84. 4.17 ft^3/s

12.86. $L = 0.61147$ m at $B = 45°$, $r = 36.1$ mm

CHAPTER 13

13.4. $u = U\sin(\pi y/2\delta)$; $\delta^* = 0.363\delta$

13.10. $x = 0.1156$ m

13.16. $x_0 = 2.155$ m; $V_0 = 0.467$ m/s

13.18. 5.166 m/s; 5.139 m/s

13.24. $D = 0.529$ N

13.26. 0.096 kW

13.28. $\delta = 0.623$ in; $\tau_w = 0.01200$ lb/ft^2

13.30. $\delta = 1.992$ m

13.32. $D_{max} = 7.76$ lb; $D_{min} = 7.06$ lb

13.36. 11.35 ft

13.38. $D = 526$ N

13.40. 61.3 kW

13.42. 170.0 N

13.44. $D = 3052$ N

13.46. 9.16 N · m

13.48. 1462 N · m

13.50. 0.4496 N

13.52. $a = 19.70$ ft/s^2

13.54. $D = 1192$ N; 47.68 kW

13.58. 0.00170; $\tau_0 = 0.0593$ lb/ft^2

13.60. $D = 8,827$ N

13.62. $e = 0.1376$ mm

13.64. 6908 N

13.68. $C_D = 0.370$

13.70. Torque = 0.366 N · m

13.72. M = 4374 N · m

13.74. $A = 48.76$ m^2

13.76. 856 N · m

13.78. 34.7 mpg

13.80. $(V_{air})_1 = 121.4$ m/s; $V_{H_2O} = 4.43$ m/s; $(V_{air})_2 = 61.17$ m/s

13.82. $x = 279$ m

13.84. $D = 60.8$ MN

13.86. 1.424 lb

13.88. $\eta = 0.669$

13.90. 8041 lb

13.92. 1169 m

13.94. 40.8 kW; 19,846 kN

CHAPTER 14

14.2. $V = 0.1900$ ft/s; $q = 2.53 \times 10^4$ ft^2/s

14.4. $D_H = (2\pi\alpha R)/(\pi\alpha + 180)$

14.6. $D_H = 2\left\{ R^2 \cos^{-1}\left(\dfrac{R - h}{R} \right) - (R - h) \right.$

$$\left. \times \left[R^2 - (R - h)^2 \right]^{1/2} \right\} \Big/ \left[h + R \cos^{-1}\left(\dfrac{R - h}{R} \right) \right]$$

14.8. $y_N = 2.02$ m

14.10. $y_N = 0.795$ m

14.12. $S_0 = 0.0000481$

14.14. $f = 0.02009$; $V = 12.35$ m/s; $\tau_w = 383$ Pa

14.16. $\Delta H_D = 0.009$; $h_l = 0.08829$ N · m/kg
Hydraulically smooth zone

14.18. 1.1378

14.20. $V = 0.6957$ m/s; $h = 2.30$ m

14.22. Area ratio = 1.262; Perimeter ratio = 1.787

14.24. $Q = 28.82$ m^3/s

14.26. $Q = 25.06$ m^3/s

14.28. $Q = 20.5$ m^3/s

14.30. 0.00391 m^3/s

14.32. $b = 0.933y$

14.34. $P_1/P_2 = 1.0631$; $\dfrac{A_1}{A_2} = 1.0257$

14.38. $V = 4.714$ m/s

14.40. $\Delta t = 7.805$ s

14.42. $y_n = 1.366$ m;
$q = 8.86$ m^2/s

14.46. $y_{cr} = 2.316$; $(E_{sp})_{min} = 3.474$ ft;
$V_{cr} = 8.64$ ft/s

14.48. $y_{cr} = 1.725$ m

14.50. $y_{cr} = 5.368$ ft

14.52. $y_{cr} = 1.459$ ft; $V_{cr} = 6.854$ ft/s;
Fr = 1.762

14.54. $S_{cr} = 0.002485$

14.56. $e = 1.28$ mm

14.58. $y_{cr} = 1.1155$

14.66. $y_2 = 0.3896$

14.68. 25.1 m^3/s/m

14.70. $y_2 = 2.983$

14.74. 23.76 ft

14.76. 630 m; 2.93%

14.78. 4.77 m

14.80. $q = 4.50$ m^2/s

14.82. $\Delta L = 6.38$ ft
14.84. $\Delta L = 239.5$ ft
14.92. For S_1: $y > 1.255$ m
For S_2: $1.026 < y < 1.255$
for S_3: $y < 1.027$
14.94. $y_2 = 4.67$ in; 4.70 ft-lb/slug
14.96. $\Delta y = 0.1502$ m
14.98. $(Fr)_2 = 0.3953$; $\Delta H_l = 3.1875$ m

CHAPTER 15

15.4. 0.2632
15.6. $\omega = 13,460$ r/min; 4084 hp; 2272 ft · lb
$\beta = 29.25°$

15.10. $V_1 = 7.05$ m/s
15.14. $V_2 = 28.5$ ft/s; $T_{shaft} = 327$ ft · lb; 2.24 hp
15.16. $\alpha = 9.59°$
15.18. $\omega = 515$ r/min
15.20. 21,040 hp; $Q = 15$ m^3/s; $\alpha = 7.53°$
15.22. 29,477 hp
15.24. 341 m; 224 hp
15.26. 19,725 hp; 390,000 N · m;
$V_t = 32.5$ m/s
15.28. $T = 15.60$ ft-lb; 10.54 hp
15.30. $\omega = 7631$ r/min
15.32. $\beta_2 = 168.9°$; $p_2 = 157.6$ lb/in^2 abs
15.34. $T = 45$ ft-lb

SELECTIVE LIST OF ADVANCED OR SPECIALIZED BOOKS ON FLUID MECHANICS

Batchelor G. K.: *An Introduction to Fluid Dynamics*, Cambridge University Press, 1967.

Bird, R. B., W. E. Stewart, and E. N. Lightfoot: *Transport Phenomena*, John Wiley & Sons, Inc., New York, 1960.

Cambel, A. B., and B. H. Jennings: *Gas Dynamics*, McGraw-Hill Book Company, Inc., New York, 1958.

Chow, V. T.: *Open Channel Hydraulics*, McGraw-Hill Book Co., New York, 1959.

Corcoran, W. H., J. B. Opfell, and B. H. Sage: *Momentum Transfer in Fluids*, Academic Press, Inc., New York, 1956.

Dwinnell, J. H.: *Principles of Aerodynamics*, McGraw-Hill Book Company, Inc., New York, 1949.

Goldstein, S.: *Modern Developments in Fluid Dynamics*, 2 vols., Oxford University Press, New York, 1938.

Hayes, W. D., and R. F. Probstein: *Hypersonic Flow Theory*, Academic Press, Inc., New York, 1959.

Henderson, F. M.: *Open Channel Flow*, Macmillan Publishing Co., New York, 1966.

Hinze, J. O.: *Turbulence*, McGraw-Hill Book Company, Inc., New York, 1959.

Hughes, W. H.: *An Introduction to Viscous Flows*, Hemisphere Publishing Corp., 1979.

Knudsen, J. G., and D. L. Katz: *Fluid Dynamics and Heat Transfer*, McGraw-Hill Book Company, Inc., New York, 1958.

Kochin, N. E., I. A. Kibel, and N. V. Roze: *Theoretical Hydrodynamics*, Interscience Publishers, 1964.

Kuethe, A. M., and J. D. Schetzer: *Foundations of Aerodynamics*, John Wiley & Sons, Inc., New York, 1950.

Lamb, H.: *Hydrodynamics*, Dover Publications, New York, 1932.

Landau, L. D., and E. M. Lifshitz: *Mechanics of Continuous Media—Hydrodynamics*, Addison-Wesley Publishing Company, Reading, Mass., 1958.

Liepmann, H. W., and A. Roshko: *Elements of Gasdynamics*, John Wiley & Sons, Inc., New York, 1957.

McCormack P. D., and L. Crane,: *Physical Fluid Dynamics*, Academic Press, New York, 1973.

Milne-Thomson L. M.: *Theoretical Hydrodynamics*, The Macmillan Company, New York, 1950.

Loitsyanski, L. G.: *Mechanics of Liquids and Gases*, Pergamon Press, 1966.

Oswatitsch, K.: *Gas Dynamics*, Academic Press, Inc., New York, 1956.

Pai, S. I.: *Fluid Dynamics of Jets*, D. Van Nostrand Company, Princeton, N.J., 1957.

———: *Viscous Flow Theory*, 2 vols., D. Van Nostrand Company, Princeton, N.J., 1958.

———: *Introduction to the Theory of Compressible Flow*, D. Van Nostrand Company, Princeton, N.J., 1959.

Patterson, G. H.: *Molecular Flow of Gases*, John Wiley & Sons, New York, 1956.

Prandtl, L.: *Essentials of Fluid Dynamics*, Hafner Publishing Company, New York, 1952.

——— and O. G. Tietjens: *Applied Hydro- and Aeromechanics*, McGraw-Hill Book Company, Inc., New York, 1934.

——— and ———: *Fundamentals of Hydro- and Aeromechanics*, McGraw-Hill Book Company, Inc., New York, 1934.

Rauscher, M.: *Introduction to Aeronautical Dynamics*, John Wiley & Sons, Inc., New York, 1953.

Rouse, H.: *Fluid Mechanics for Hydraulic Engineers*, McGraw-Hill Book Company, Inc., New York, 1938.

———(ed.): *Advanced Mechanics of Fluids*, John Wiley & Sons, Inc., New York, 1959.

Sauer, R.: *Introduction to Theoretical Gas Dynamics*, J. W. Edwards, Publisher, Inc., Ann Arbor, Mich., 1947.

Schlichting, H.: *Boundary Layer Theory*, McGraw-Hill Book Company, Inc., New York, 1955.

Shapiro, A. H.: *The Dynamics and Thermodynamics of Compressible Fluid Flow*, 2 vols., The Ronald Press Company, New York, 1953.

Stoker, J. J.: *Water Waves*, Interscience Publishers, Inc., New York, 1957.

Streeter, V. L.: *Fluid Dynamics*, McGraw-Hill Book Company, Inc., New York, 1948.

———(ed.): *Handbook of Fluid Dynamics*, McGraw-Hill Book Company, Inc., New York, 1961.

Tennekes, H., and J. L. Lumley: *A First Course in Turbulence*, MIT Press, Cambridge, 1972.

Vavra, M. H.: *Aero-Thermodynamics and Flow in Turbomachines*, John Wiley & Sons, New York, 1960.

Von Mises, R.: *Theory of Flight*, McGraw-Hill Book Company, Inc., New York, 1945.

———: *Mathematical Theory of Compressible Flow*, Academic Press, Inc., New York, 1958.

White, F. M.: *Viscous Fluid Flow*, McGraw-Hill Book Co., New York, 1974.

Yuan, S. W.: *Foundations of Fluid Mechanics*, Prentice-Hall Inc., 1967.

APPENDIX
A.I

MEASUREMENT
METHODS

A.I.1 INTRODUCTION

We will now discuss techniques used to measure certain properties in fluids as well as flow parameters. We have already discussed some techniques at various times earlier in the text for purposes of illustrating the theory. We also wanted to use these measurement techniques in the text discussions and in the problems and examples. Here we only remind the reader of these techniques. Furthermore, we do not discuss all possible methods in this appendix but will choose those that are important as well as those whose explanations are primarily in terms of fluid-flow theory as presented in this text. Those devices not covered we believe can be understood from manufacturers' instructions and specifications.

A.I.2 PRESSURE MEASUREMENTS

In the case of static fluids, we have discussed the technique of *manometry* in Chap. 3 and have had opportunity to calculate pressures from manometry measurements. To measure the static pressure in a fluid moving through a *smooth* pipe using manometry (see Fig. A.I.1), we make a *piezometer* opening at right angles to the pipe. The piezometer opening should have a length l at least twice as large as the diameter of the opening. Also, there should be no burrs at the inside edge of the opening—otherwise there will be local eddies that will lead to an incorrect measurement. As you will recall from Chap. 9, the pressure in the pipe will vary hydrostatically in the vertical direction, since we have parallel flow in the horizontal direction.

To measure the *stagnation pressure* or *total pressure* at a point A, we pointed out in Chap. 11 that we could use the *pitot tube*, which we show again in Fig. A.I.2. Also in Fig. A.I.3, we show the *pitot-static tube* for measuring at point A both the static pressure

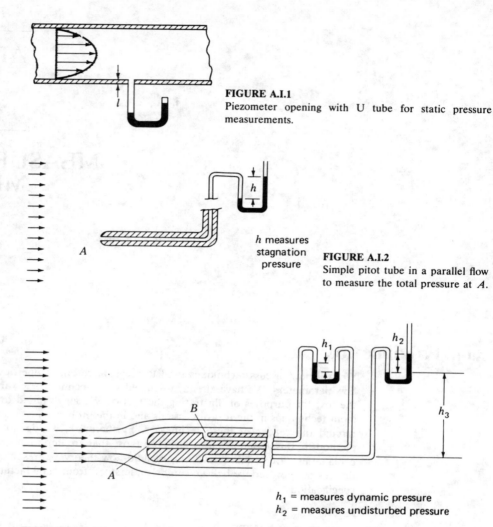

FIGURE A.I.1
Piezometer opening with U tube for static pressure measurements.

h measures stagnation pressure

FIGURE A.I.2
Simple pitot tube in a parallel flow to measure the total pressure at *A*.

h_1 = measures dynamic pressure
h_2 = measures undisturbed pressure

FIGURE A.I.3
Pitot-static tube to measure dynamic and static pressure at *A*.

and the *dynamic pressure*, which is the difference between the total pressure and the static pressure. This device can be used to find the velocity profile in a parallel flow or the static pressure near the wall if there is a rough surface present precluding the use of the piezometer hole discussed earlier.

We wish now to point out the practical fact that the pitot tube is not greatly sensitive to misalignment with the parallel flow. An error of only a few percent is incurred if the alignment of the tube is up to 15° off from the flow.

If large pressures are to be measured, a pressure gage such as a Bourdon pressure gage is used instead of a manometer. Inside the Bourdon gage is a tube closed at one

FIGURE A.I.4
Bourdon pressure gage. (*Crosby Steam Gage and Valve Co.*)

end and open to the fluid in contact with the pressure to be measured at the other end. This tube is C shaped with the closed end free to move (see Fig. A.I.4). The pressure inside the tube above atmosphere will cause the tube to straighten somewhat with a degree dependent on the inside gage pressure. The resulting movement of the free end is picked up by a mechanism to turn the pointer on the front face of the instrument. The measure of this rotation of the pointer is calibrated to read the gage pressure.

If the pressure being measured is changing fairly rapidly, one may use as a detector a *piezoelectric crystal* which generates a voltage when under pressure. Also, one can use as a detector a thin diaphragm on which there is a strain gage. Under a net pressure on one side, the diaphragm deforms directly with the amount of net pressure. The strain-gage resistance is thereby affected and this change is noted by suitable electronic instrumentation calibrated to read the gage pressure on the diaphragm.

A.I.3   VELOCITY MEASUREMENTS

The pressure measurements from the pitot-static tube or the pitot tube permit us to ascertain velocity V at the point A, corresponding to the tip of the pitot tube before its insertion. For small pressures, one need only use knowledge of ideal incompressible flow to compute V from pressure data.

Next, we note that small turbinelike devices called *anemometers* are in use to measure the velocity of fluids. The device has either cups or vanes which the moving fluid drives when the device (see Fig. A.I.5) is aligned so that the axis of rotation is along the direction of flow. The vanes or cups generally drive a little generator, which causes a pointer to rotate around a graduated scale calibrated to measure the velocity of the fluid.

FIGURE A.I.5
Air anemometer. (*Taylor Instruments.*)

When there are rapid velocity fluctuations to be measured, such as in turbulent flow, and also when we are interested in velocity measurements at a point in a small region such as in a thin boundary layer, we find that the *hot-wire anemometer* is of great use. Such a device has a very short response time, permitting it to pick up rapid fluctuations in velocity. Furthermore, the probe of the device is very small, so rather than getting average values of velocity over a comparatively large region as in the case of the pitot tube and the diaphragm probes, we get instead an average over a much smaller region and can consider the measurements for all practical purposes as valid for a point in the flow. For this reason, the hot-wire anemometer is valuable in the study of thin films.

The hot-wire probe consists of a thin (diameter is 5×10^{-6} m), short, platinum or tungsten wire through which a current of electricity is passed. This current will cause the wire to heat up due to resistance. However, the flow of fluid around the exposed wire will tend to cool the wire with a degree dependent on the local fluid velocity. There are two kinds of hot-wire anemometers that can now be set forth. In one, the *current* is to be kept constant. Now, depending on the local fluid velocity at the probe, the temperature and hence the resistance of the wire changes. We can then calibrate the *voltage* needed to maintain constant current with the local velocity of fluid. If the wire *temperature* is now to be kept constant and hence its resistance, we must change the *current*. The required current is related to the local fluid velocity whose changes are tending to change the temperature. We can then calibrate the current with the local fluid velocity at the probe.

Hot-wire anemometers have been used mainly with gas flows. In liquids there are difficulties associated with corrosion. However, there are techniques using a thin-coated film rather than a wire to get around these difficulties.

A.I.4   FLOW RATE MEASUREMENTS FOR
INCOMPRESSIBLE FLOW IN PIPES

In Example 6.5 we describe the measurement of flow from an orifice through which we have efflux from a reservoir. In that case, we could calculate q, the volume flow, with the aid of the coefficient of discharge C_d.

We will now consider *flow nozzles* which are inserted into a pipe, as shown in Fig. A.I.6*a* and *square-edged orifices*, as shown in Fig. A.I.6*b*. Note that the vena contracta in the square-edged orifice flow is very much like the one discussed for the sharp-edged reservoir opening of Example 6.5. Note also that the pressure p_2 at the manometer leg is positioned at the vena contracta where we can consider that we have parallel flow.

We will first consider *incompressible flows*. Knowing the geometry of the pipe and nozzle or orifice and h of the manometer, we can determine a theoretical value of q, the volume flow. Thus, consider *Bernoulli's* equation between points 1 and 2 in Fig. A.I.6*a* and A.I.6*b*.

$$\frac{p_1}{\rho} + \frac{V_1^2}{2} = \frac{p_2}{\rho} + \frac{V_2^2}{2} \qquad (A.I.1)$$

Using *continuity* for 1-D flow, we have

$$\rho V_1 A_1 = \rho V_2 A_2$$

$$\therefore V_1 = V_2 \frac{A_2}{A_1} \qquad (A.I.2)$$

Substituting from Eq. (A.I.2) into Eq. (A.I.1), we have

$$\frac{p_1}{\rho} + \frac{V_2^2}{2}\left(\frac{A_2}{A_1}\right)^2 = \frac{p_2}{\rho} + \frac{V_2^2}{2}$$

$$\therefore V_2 = \sqrt{\frac{2[(p_1 - p_2)/\rho]}{\left[1 - (A_2/A_1)^2\right]}}$$

The *ideal* volume flow for the flow nozzle and for square-edged orifices is then

$$q_{\text{theoretical}} = V_2 A_2 = A_2 \left\{ \frac{[2(p_1 - p_2)]/\rho}{\left[1 - (A_2/A_1)^2\right]} \right\}^{1/2} \qquad (A.I.3)$$

In the case of the flow nozzle of Fig. A.I.6*a*, we know A_2, and to account for frictional effects, we include a coefficient of discharge $(C_d)_{\text{noz}}$ which must be determined experimentally. That is,

$$q_{\text{act}} = (C_d)_{\text{noz}} A_2 \left\{ \frac{2[(p_1 - p_2)/\rho]}{\left[1 - (A_2/A_1)^2\right]} \right\}^{1/2} \qquad (A.I.4)$$

Now $(C_d)_{\text{noz}}$ depends on the Reynolds number of the flow in the pipe for any given sized pipe and nozzle opening. Clearly $(C_d)_{\text{noz}} = q_{\text{act}}/q_{\text{theoretical}}$ and is accordingly dimensionless. From a review of many tests on ASME long-radius-type flow nozzles that are 2 in or longer, an empirical equation has been developed for C_d for the range of Reynolds

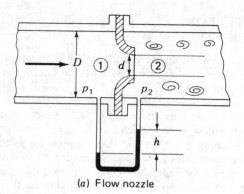

(a) Flow nozzle

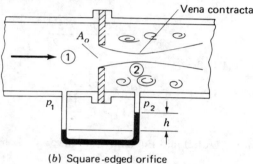

(b) Square-edged orifice

FIGURE A.I.6
Pipe flow nozzle and orifice.

number $10^4 \leq \mathrm{Re}_d \leq 10^6$ where the Reynolds number is based on the nozzle diameter. Taking β to be the ratio of nozzle diameter to the pipe diameter, we have for the range $0.30 \leq \beta \leq 0.825$ the following formula for C_d[1]:

$$C_d = 0.99622 + 0.00059D - \left(6.36 + 0.13D - 0.24\beta^2\right)\frac{1}{\sqrt{\mathrm{Re}_d}} \qquad \text{(A.I.5)}$$

where D, the inside diameter of the pipe, is in inches and Re_d is based on the nozzle diameter d.

In the case of the square-edged orifice, we do not know A_2 at the vena contracta. Accordingly, we replace A_2 by $C_c A_o$, where C_c is the *coefficient of contraction* and A_o is the area of the orifice opening. Including the coefficient of contraction as well as the correction for friction in Eq. (A.I.3), we arrive at the following formulation with a new *coefficient of discharge* $(C_d)_o$ for square-edged orifices.

$$q_{\mathrm{act}} = (C_d)_o A_o \left\{\frac{2[(p_1 - p_2)/\rho]}{\left[1 - (A_o/A_1)^2\right]}\right\}^{1/2} \qquad \text{(A.I.6)}$$

where A_1, recall, is the inside cross-sectional area of the pipe.

---

[1]See *Fluid Meters—Their Theory and Applications*, 1971, ASME, pp. 64–65.

The data for square-edged orifices is not well enough established to cover a wide and comprehensive range of conditions. However, data in the form of tables for certain geometries have been presented in *Fluid Meters*, 6th ed., ASME, 1971, pp. 202–207. For specific geometries not reported, one must interpolate between charts.

Example A.I.1. Find q for flow of water in a pipe of inside diameter $D = 100$ mm using a long-radius-type flow nozzle [see Fig. A.I.6(a)]. The pipe is horizontal for this problem and the value of h for the manometer is 140 mm. What is the mass flow of water? The throat diameter of the nozzle is 60 mm. Take $\rho = 999$ kg/m^3 and $\nu = 1.12 \times 10^{-3}$ m^2/s.

First we can compute $p_1 - p_2$ as follows:

$$p_1 - p_2 = h(\gamma_{\text{Hg}} - \gamma_{\text{H}_2\text{O}}) = 0.140(13.6 - 1)(999)(9.81) = 17.300 \text{ kPa} \quad (a)$$

Next we go to Eq. (A.I.5) to determine the discharge coefficient as a function of Re$_d$. Thus, assuming for now that Re is in the allowed range,

$$(C_d)_{\text{noz}} = \left(0.99622 + 0.00059\frac{10.0}{2.54}\right) - \left[6.36 + 0.13\frac{10.0}{2.54} - (0.24)(0.6^2)\right]\frac{1}{\sqrt{\text{Re}_d}}$$

$$= 0.998 - 6.785\frac{1}{\sqrt{\text{Re}_d}} \quad (b)$$

Now go to Eq. (A.I.4) and substitute values:

$$q_{\text{act}} = (C_d)_{\text{noz}} A_2 \left\{ \frac{2[(p_1 - p_2)/\rho]}{1 - \beta^4} \right\}^2$$

$$\therefore q_{\text{act}} \equiv A_2 V_2 = \left[0.998 - 6.785\frac{1}{\sqrt{\text{Re}_d}}\right](A_2)\left[\frac{2(17,300/999)}{1 - 0.6^4}\right]^{1/2}$$

This equation becomes

$$V_2 = \left(0.998 - \frac{6.785}{\sqrt{\rho V_1 d/\mu}}\right)(6.31)$$

$$= \left[0.998 - \frac{6.785}{\sqrt{[(999)(V_1)(0.060)]/(1.12 \times 10^{-3})}}\right](6.31) \quad (c)$$

From *continuity*,

$$V_2 = \left(\frac{100}{60}\right)^2 V_1 \quad (d)$$

Therefore, substituting for V_2 in Eq. (c) using Eq. (d),

$$V_1 = \left(0.998 - \frac{0.0293}{\sqrt{V_1}}\right)(2.27)$$

Solving by trial and error we get

$$V_1 = 2.22 \text{ m/s}$$

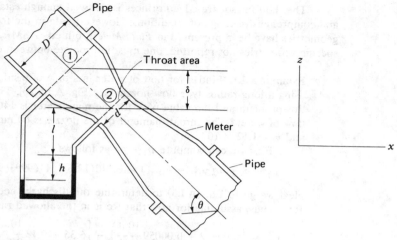

FIGURE A.I.7
Venturi meter

The Reynolds number is then

$$\mathrm{Re}_d = \left[\frac{(999)(0.060)(2.22)}{1.12 \times 10^{-3}}\right] = 1.189 \times 10^5$$

We are well within the range of Eq. (A.I.5), so we can get q as

$$q = \left[\frac{(\pi)(0.100^2)}{4}\right](2.22) = 0.01744 \text{ m}^3/\text{s}$$

We next consider the *Venturi meter* for measuring flow rate in a pipe (Fig. A.I.7). The Venturi meter is a section of piping inserted into the pipe line after cutting away a portion of piping to make room. The Venturi meter resembles a nozzle in that it has a converging portion from the inside pipe diameter to a throat, followed by a diverging section from the throat back to the inside diameter of the pipe. A differential manometer is used to enable one to determine the pressure difference between positions 1 and 2 as shown in the diagram.

We can evaluate $q_{\text{theoretical}}$ from this instrument for incompressible flow by first using Bernoulli's equation between positions 1 and 2 as well as the one-dimensional continuity equation exactly as was done for the orifice and nozzle leading to Eq. (A.I.3). The only difference now is the presence of the elevation terms z_1 and z_2 in Bernoulli's equation. We leave it to you to show that for ideal conditions

$$q_{\text{theoretical}} = A_2 \left\{2\frac{[(p_1 - p_2)/\rho] + (z_1 - z_2)g}{1 - (A_2/A_1)^2}\right\}^{1/2} \tag{A.I.7}$$

To get the actual flow, we again use an experimentally determined coefficient $(C_d)_{\text{ven}}$. Thus we get

$$q_{\text{act}} = (C_d)_{\text{ven}} A_2 \left\{ 2 \frac{[(p_1 - p_2)/\rho] + (z_1 - z_2)g}{1 - (A_2/A_1)^2} \right\}^{1/2} \qquad \text{(A.I.8)}$$

For evaluating $(C_d)_{\text{ven}}$, we present the results set forth by the International Organization for Standardization. Noting that $\beta = d/D$ we have

1. *Rough-cast convergent inlet:*

$$101.6 \text{ mm} \leq D \leq 813 \text{ mm}$$
$$0.3 \leq \beta \leq 0.75$$
$$2 \times 10^5 \leq \text{Re}_D \leq 2 \times 10^6$$
$$(C_d)_{\text{ven}} = 0.984 \pm 0.70\%$$

2. *Machined convergent inlet section:*

$$50.8 \text{ mm} \leq D \leq 813 \text{ mm}$$
$$0.4 \leq \beta \leq 0.75$$
$$2 \times 10^5 \leq \text{Re}_D \leq 1 \times 10^6$$
$$(C_d)_{\text{ven}} = 0.995 \pm 1.0\%$$

3. *Rough-welded sheet-iron convergent inlet:*

$$203 \text{ mm} \leq D \leq 1219 \text{ mm}$$
$$0.4 \leq \beta \leq 0.7$$
$$2 \times 10^5 \leq \text{Re}_D \leq 2 \times 10^6$$
$$(C_d)_{\text{ven}} = 0.985 \pm 1.5\%$$

The Re_D is based on the inside diameter D of the pipe. We now consider an example for a flow-rate meter.

Example A.I.2. Find q for flow of water in a pipe of inside diameter D of 100 mm using a Venturi meter with a machined convergent section. The throat diameter d of the Venturi meter is 60 mm. The Venturi meter is in a pipe section having an inclination θ (see Fig. A.I.7) of $45°$. The distance from 1 to 2 in the meter is 120 mm and the value of h for the manometer is 140 mm. Take $\rho = 999 \text{ kg/m}^3$ and $\mu = 1.12 \times 10^{-3} \text{ kg/m} \cdot \text{s}$.

We consider now that $(C_d)_{\text{ven}}$ for this case is 0.995. We need next the value of $p_1 - p_2$. This is easily determined as follows from *manometry*:

$$p_1 + \gamma_{H_2O}(\delta + l + h) = p_2 + l\gamma_{H_2O} + h\gamma_{Hg}$$

$$\therefore p_1 - p_2 = h(\gamma_{Hg} - \gamma_{H_2O}) - \gamma_{H_2O}\delta$$

$$= (0.140)(13.6 - 1)(999)(9.81)$$

$$- (999)(9.81)(0.120)(0.707)$$

$$= 16{,}456 \text{ Pa} \qquad (a)$$

We now compute V_2 in Eq. (A.I.8) by deleting A_2 in the formula. Thus,

$$V_2 = 0.995 \left\{ 2 \frac{(16{,}456/999) + (0.120)(0.707)(9.81)}{1 - (60/100)^4} \right\}^{1/2} \qquad (b)$$

where we used diameters squared in place of areas. We get from Eq. (b) and continuity

$$V_2 = 6.27 \text{ m/s} \qquad V_1 = (0.6^2)(6.27) = 2.26 \text{ m/s}$$

We must next compute Re_D for the pipe.

$$\mathrm{Re}_D = \frac{\rho V_1 D}{\mu} = \left[\frac{(999)(2.26)(0.100)}{1.12 \times 10^{-3}} \right]$$

$$= 2.02 \times 10^5$$

We see that the Reynolds number is within the proper range for the $(C_d)_{\text{ven}}$ we have used and so we have for q

$$q = (V_2)(A_2) = (6.27) \frac{(\pi)(0.060^2)}{4}$$

$$= 0.01773 \text{ m}^3/\text{s}$$

It should be kept in mind in closing that if the situation at hand for determining C_d is such that one is out of range of the reported formulations, then one must determine C_d for the range of interest by carrying out one's own test. Simply weigh the fluid discharge over a period of time for a chosen set of conditions. Then compute the weight of the fluid ($\rho q \, \Delta t$) from the ideal theory for the time Δt. The ratio of the actual weight to the theoretical weight gives the desired C_d for the conditions tested. Do this test over the range of conditions for which the flow device is to be used and plot C_d versus Reynolds number. In problems for computing q where such a plot is available, *guess* at C_d and *solve* for Reynolds number. Then check to see if your guessed C_d is close to the one demanded by your curve. If not, repeat the process using the Reynolds number just found from the curve. Repeat again as necessary. This is very much like the friction-factor problems of Chap. 9.

A.I.5 FLOW RATE MEASUREMENTS FOR COMPRESSIBLE FLOW IN PIPES

Let us first consider isentropic flow of a perfect gas through a pipe having inserted in it either a flow nozzle as in Fig. A.I.6a or a Venturi meter as in Fig. A.I.7. We first write the one-dimensional first *law of thermodynamics* for a control volume between sections 1 and 2 for these meters. Neglecting potential energy of gravity, we have for steady, subsonic, isentropic flow of a perfect gas

$$c_p T_1 + \frac{V_1^2}{2} = c_p T_2 + \frac{V_2^2}{2} \qquad (A.I.9)$$

and for *continuity* for this control volume, we have

$$\rho_1 V_1 A_1 = \rho_2 V_2 A_2$$

Solving for V_1 in the preceding equation,

$$V_1 = \frac{\rho_2}{\rho_1} \frac{A_2}{A_1} V_2 \tag{A.I.10}$$

Next, replace ρ_2/ρ_1 by $(p_2/p_1)^{1/k}$ in Eq. (A.I.10) for isentropic pressure change of a perfect gas. Thus,

$$V_1 = \left(\frac{p_2}{p_1}\right)^{1/k} \left(\frac{A_2}{A_1}\right) V_2 \tag{A.I.11}$$

Substituting Eq. (A.I.11) into Eq. (A.I.9), we have

$$c_p T_1 + \left(\frac{p_2}{p_1}\right)^{2/k} \left(\frac{A_2}{A_1}\right)^2 \frac{V_2^2}{2} = c_p T_2 + \frac{V_2^2}{2}$$

Solving for V_2 we get

$$V_2 = \left[\frac{2 c_p (T_1 - T_2)}{1 - (p_2/p_1)^{2/k} (A_2/A_1)^2}\right]^{1/2}$$

$$= \left[\frac{2 c_p T_1 (1 - T_2/T_1)}{1 - (p_2/p_1)^{2/k} (A_2/A_1)^2}\right]^{1/2}$$

Replace c_p by $[k/(k-1)](R)$ from Eq. (11.6a) and T_2/T_1 by $(p_2/p_1)^{(k-1)/k}$ from Eq. (11.9b):

$$V_2 = \left\{\frac{[2k/(k-1)](RT_1)\left[1 - (p_2/p_1)^{(k-1)/k}\right]}{1 - (p_2/p_1)^{2/k}(A_2/A_1)^2}\right\}^{1/2} \tag{A.I.12}$$

Replace RT_1 by p_1/ρ_1 and get $\dot{m}$, the mass flow, by multiplying both sides by $A_2 \rho_2$. Thus,

$$\dot{m}_{\text{isen}} = A_2 \rho_2 \left\{\frac{[2k/(k-1)](p_1/\rho_1)\left[1 - (p_2/p_1)^{(k-1)/k}\right]}{1 - (p_2/p_1)^{2/k}(A_2/A_1)^2}\right\}^{1/2}$$

Put ρ_2 inside the bracket and multiply it by ρ_1/ρ_1.

$$\dot{m}_{\text{isen}} = A_2 \left\{\frac{[2k/(k-1)]p_1(\rho_2^2/\rho_1)(\rho_1/\rho_1)\left[1 - (p_2/p_1)^{(k-1/k)}\right]}{1 - (p_2/p_1)^{2/k}(A_2/A_1)^2}\right\}^{1/2}$$

Hence, replacing $(\rho_2/\rho_1)^2$ by $(p_2/p_1)^{2/k}$, we get for nozzle orifice and Venturi meters the result

$$\dot{m}_{\text{isen}} = A_2 \left\{\frac{[2k/(k-1)]p_1\rho_1(p_2/p_1)^{2/k}\left[1 - (p_2/p_1)^{(k-1)/k}\right]}{1 - (p_2/p_1)^{2/k}(A_2/A_1)^2}\right\}^{1/2} \tag{A.I.13}$$

The actual flow must take friction into account, so we must multiply by a discharge

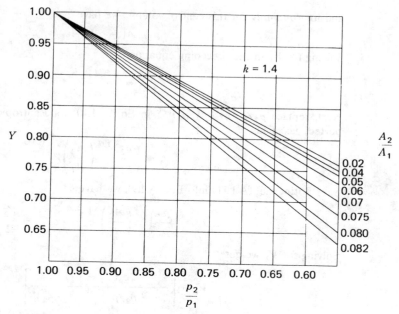

FIGURE A.I.8
Compressibility factor Y for flow nozzles and Venturi meters. $k = 1.4$.

coefficient, $(C_d)_o$ for the nozzle orifice and $(C_d)_{ven}$ for the Venturi meter. $(C_d)_o$ and $(C_d)_{ven}$ are functions of Reynolds numbers Re_d and Re_D, respectively, and are evaluated as in the incompressible flow discussed in Sec. A.I.4.

It is the usual practice to rewrite Eq. (A.I.13) in a form similar to the incompressible case as given by Eq. (A.I.4) with a factor called the *compressibility factor*, Y, included. Thus,

$$\dot{m} = \rho_1 C_d A_2 \left\{ \frac{2[(p_1 - p_2)/\rho_1]}{1 - (A_2/A_1)^2} \right\}^{1/2} Y \qquad (A.I.14)$$

where

$$Y = \left\{ \frac{[k/(k-1)](p_2/p_1)^{2/k} \left[1 - (p_2/p_1)^{(k-1)/k}\right] \left[1 - (A_2/A_1)^2\right]}{\left[1 - (A_2/A_1)^2 (p_2/p_1)^{2/k}\right] \left[1 - p_2/p_1\right]} \right\}^{1/2} \qquad (A.I.15)$$

The compressibility factor Y has been plotted in Fig. A.I.8 for $k = 1.4$ for various area ratios A_2/A_1 as a function of p_2/p_1 for flow nozzles as well as Venturi meters.

Example A.I.3. Consider in Example A.I.2 that we have a flow of air at a temperature of 40°C and a pressure of 200 kPa absolute. The pipe as before is 100 mm in inside diameter, but the diameter in the Venturi is now 20 mm. The height h is 140 mm of mercury. What is the mass flow per hour?

We can use the same C_d as in incompressible flow, which is 0.995. We next determine Y, so we can use Eq. (A.I.14). We need p_2 for this purpose. Thus,

$$p_2 = p_1 - \Delta p = 200{,}000 - (0.140)(9806)(13.6)$$

$$= 181.3 \text{ kPa}$$

The ratio p_2/p_1 is then

$$\frac{p_2}{p_1} = \frac{181.3}{200} = 0.907$$

Noting that $A_2/A_1 = (d/D)^2 = (0.020/0.100)^2 = 0.04$, we go to Fig. A.I.8 to get Y.

$$Y = 0.95$$

Finally, we need ρ_1. Using the equation of state of a perfect gas,

$$\rho_1 = \frac{p_1}{RT_1} = \frac{200{,}000}{(287)(313)} = 2.2264 \text{ kg/m}^3$$

We are now ready to go to Eq. (A.I.14).

$$\dot{m} = \rho_1 C_d A_2 \left\{ \frac{2[(p_1 - p_2)/\rho_1]}{1 - (A_2/A_1)^2} \right\}^{1/2} Y$$

$$= (2.2264)(0.995) \left[\frac{(\pi)(0.020^2)}{4} \right] \left\{ \frac{2[(200{,}000 - 181{,}300)/2.2264]}{1 - (0.04^2)} \right\}^{1/2} (0.95) \quad (0.95)$$

$$= 0.0858 \text{ kg/s} = 308.7 \text{ kg/h}$$

In the case of *flow nozzles*, we would use Fig. A.I.8 to get the compressibility factor Y knowing p_2/p_1 and β. Then, we proceed as we did for the incompressible case to solve Eqs. (A.I.5) and (A.I.14) simultaneously after replacing $\dot{m}$ by $\rho_1 V_1 A_1$.

We have thus far considered the flow nozzle and the Venturi meter. We now consider the square-edged thin-plate orifice meter of area A_o, where we encounter as in the incompressible case a vena contracta. The size of the vena contracta A_2 will depend on Reynolds number as before but will now depend in addition on the Mach number.

We can once again express the mass flow in the same form as Eq. (A.I.4). Thus,

$$\dot{m}_{\text{act}} = \rho_1 (C_d)_o A_o \left\{ \frac{2[(p_1 - p_2)/\rho_1]}{1 - (A_o/A_1)^2} \right\}^{1/2} Y \quad (A.I.16)$$

We use the same $(C_d)_o$ as in the incompressible case. However, because of the vena contracta, we cannot evaluate Y as we did for the flow nozzle and the Venturi meter [Eq. (A.I.15)] but must find its value as a function of k, A_o/A_1, and p_2/p_1 experimentally. We have shown a plot of Y in terms of β, p_2/p_1, and k in Fig. A.I.9. The empirical formula for these curves is

$$Y = 1 - \frac{(0.41 + 0.35\beta^4)(1 - p_2/p_1)}{k} \quad (A.I.17)$$

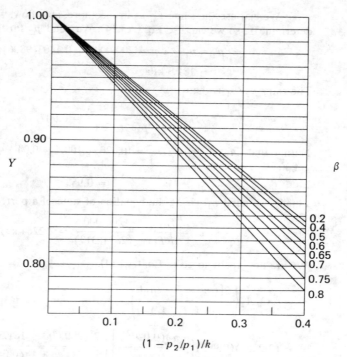

FIGURE A.I.9
Compressibility factors for thin square-edged orifices.

A.I.6 FREE-SURFACE FLOW MEASUREMENTS; THE WEIR

A weir may be thought of as a dam over which the liquid must pass. By making certain measurements of the flow near the weir, one can determine the flow q with the aid of Bernoulli's equation plus some empirical data. The weir is generally a comparatively thin plate with a sharp edge on the upstream side (see Fig. A.I.10). This is called a *sharp-crested, rectangular weir*. The height h of the flow upstream from the weir above

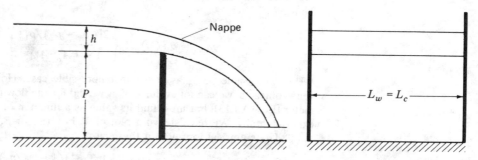

FIGURE A.I.10
Sharp-crested rectangular weir.

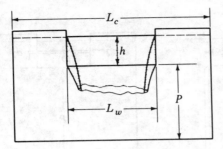

FIGURE A.I.11
Weir with side contractions of the cross section of the nappe.

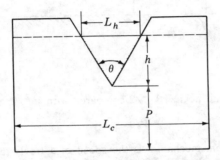

FIGURE A.I.12
Triangular-notch weir.

the edge of the weir is called the *head*, which is the key measurement for evaluating the flow. The sheet of fluid moving over the weir is called the *nappe*. Like a free jet, we can consider the pressure in the nappe to be close to atmospheric in our upcoming calculations. The top surface of the weir over which the liquid flows is called the *crest* of the weir; in Fig. A.I.10 the crest has a length L_w that extends over the entire width L_c of the channel. If the crest is shorter than the width of the channel, then we have a weir "with contraction" (see Fig. A.I.11). The weir crest may have shapes other than a straight edge (Fig. A.I.10) or rectangle (Fig. A.I.11). The V, or triangular, notch (see Fig. A.I.12) may be used for low rates of flow. Other special notches include hyperbolic and parabolic notches. They are intended to have a constant discharge coefficient or to have the head h directly proportional to the rate of flow q.

We will now consider the *contracted* rectangular weir in detail. Notice in Figs. A.I.10 and A.I.11 that there is a contraction of the nappe as it goes by the weir. We have seen this effect before in sharp-edged orifices. It is a result primarily of the radial velocity needed by the liquid to get by the sharp edges of the weir crest and the sides of the weir. Also, surface tension makes a contribution to this effect. We will now delete both these effects and assume for simplicity that we have parallel flow in the nappe, as shown in Fig. A.I.13. Next, we omit viscosity and use *Bernoulli* between a point 1 on the free surface upstream of the weir and any point in the nappe above the crest of the weir. Thus, we have using the crest of the weir as the datum for potential energy and measuring y downward from the free surface:

$$\frac{V_1^2}{2} + gh = \frac{V_2^2}{2} + g(h - y)$$

From *continuity* considerations, we can consider V_1, the so-called *velocity of approach*,

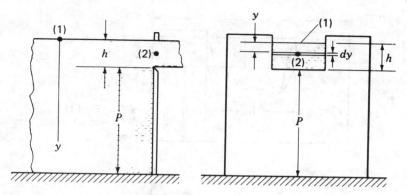

FIGURE A.I.13
Highly idealized weir nappe.

much smaller than V_2, and neglect $V_1^2/2$ compared with $V_2^2/2$. Solving for V_2 we get

$$V_2 = \sqrt{2gy} \qquad (A.I.18)$$

which you may recall from mechanics is the speed of a freely falling particle. To get $q_{\text{theoretical}}$ (the total volume flow rate) we integrate over the area of the nappe. We thus have

$$q_{\text{theoretical}} = \int_0^h (L_w \, dy)V_2 = \int_0^h L_w \sqrt{2gy} \; dy$$

$$= L_w \sqrt{2g} \; y^{3/2} \left(\tfrac{2}{3}\right)\Big|_0^h = \tfrac{2}{3}\sqrt{2gh^3} \qquad (A.I.19)$$

Because of friction, surface tension, and contraction effects, the actual flow q_{act} will be less than $q_{\text{theoretical}}$ as given above. Again, to take this into account we introduce through experiment the coefficient of discharge C_d defined once again as

$$C_d = \frac{q_{\text{act}}}{q_{\text{theoretical}}} \qquad (A.I.20)$$

It should be pointed out additionally that C_d depends also on the volume of flow q_{act}. We present here the procedures set forth by the ASME 1971 publication of *Fluid Meters*, chap. I-8, based on work done at the Georgia Institute of Technology. By these procedures Eq. (A.I.19) becomes

$$q_{\text{act}} = \tfrac{2}{3} C_d' L_w' \sqrt{2g(h')^3} \qquad (A.I.21)$$

where $C_d' = f(L_w/L_c, h/P)$
$L_w' = L_w + b$
$h' = h + 0.900$ mm (or $h + 0.003$ ft)

The term b is given in Fig. A.I.14 as a function of L_w/L_c, and C_d' is given in Fig. A.I.15. Note that the case where $L_w = L_c$ [i.e., the uncontracted, sharp-crested weir (Fig. A.I.10)] is included.

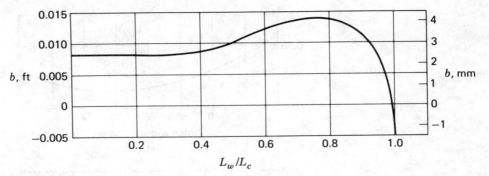

FIGURE A.I.14
Evaluation of correction factor b.

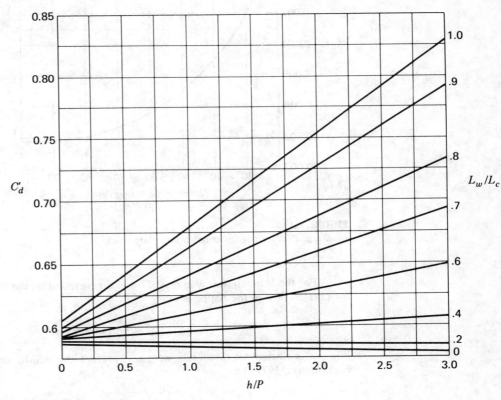

FIGURE A.I.15
C_d' versus h/P and L_w/L_c for Eq. (A.I.21).

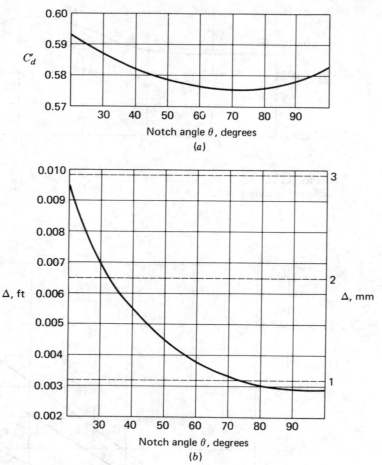

FIGURE A.I.16
Discharge data for triangular notch thin-plate weirs.

For the V notch weir, you may show as an exercise that the theoretical volume flow for parallel flow in the nappe is

$$q_{\text{theoretical}} = \frac{8}{15} \tan \frac{\theta}{2} \sqrt{2gh^5} \qquad (A.I.22)$$

For the coefficient of discharge, we again present the ASME recommendations from *Fluid Meters*, 1971. Thus,

$$q_{\text{act}} = \frac{8}{15} C_d' \tan \frac{\theta}{2} \sqrt{2g(h')^5} \qquad \text{where } h' = h + \Delta \qquad (A.I.23)$$

A plot of C_d' versus θ is shown in Fig. A.I.16a and a plot of Δ versus θ is shown in Fig. A.I.16b. These data take into account the primary dependence of C_d' on the notch angle.

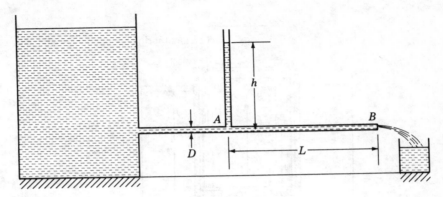

FIGURE A.I.17
Capillary tube for measuring viscosity.

The use of h' takes into account the secondary affect of the ratio h/P on the coefficient of discharge. Other effects such as kinematic viscosity and surface tension have been deleted since data to incorporate these effects are as yet inadequate.

A.I.7   MEASUREMENT OF VISCOSITY

There are a number of ways to measure viscosity. We will consider several of these methods, namely, the flow through a capillary tube, the rotating viscosimeter, and the small sphere falling slowly in a fluid. In all such cases, laminar-flow theory permits us to relate the coefficient of viscosity μ with the simple respective flows.

We first consider the flow through a capillary tube such as is shown in Fig. A.I.17, where the fluid under consideration flows from a reservoir through a capillary tube of internal diameter D. At a position A where fully developed laminar flow is established, we measure the gage pressure by noting the height h in the vertical capillary tube. We have shown in Chap. 9 that the pressure drop between position A and the exit B, a distance L apart, is given as

$$p_A - p_B = \frac{128qL\mu}{\pi D^4}$$

But $p_A - p_B = \rho gh$, so we have

$$\rho gh = \frac{128qL\mu}{\pi D^4}$$

$$\therefore \mu = \frac{\pi D^4 \rho g (h/L)}{128q} \tag{A.I.24}$$

The volume flow q is measured by collecting the fluid for a given time interval.[2]

---

[2]An adaptation of the capillary tube for industrial purposes is the Saybolt viscosimeter.

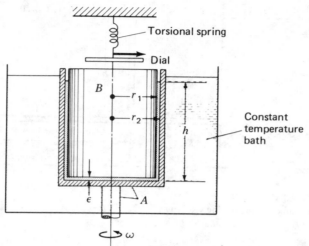

FIGURE A.I.18
Concentric cylinder viscosimeter.

The second of our devices is the concentric cylinder, rotating viscosimeter shown in Fig. A.I.18. A cylindrical container A surrounded on the outside by a constant-temperature bath is made to rotate at a constant angular speed of ω radians per second. Inside container A is a cylinder B with an outside radius r_1 only slightly smaller than inside radius r_2 of container A. The cylinder B is separated at the bottom from container A by a distance ϵ. Between A and B is the fluid whose viscosity is being measured. Due to viscous action, container A will cause cylinder B to rotate with A. However, a torsional spring at the top of B resists this rotation so that depending on ω, the viscosity μ of the fluid, and the geometry, cylinder B will rotate a fixed amount as indicated by the dial (see Fig. A.I.18). This dial gives the torque T developed on cylinder B from container A.

We can relate T with μ and other known data by assuming a linear velocity profile of the fluid at the side of the cylinder and at bottom of the cylinder. We first compute the torque T_1 from the side of B using *Newton's viscosity* law.

$$
\begin{aligned}
T_1 &= (\tau_{\text{shear}})(\text{area})(r_2) \\[2mm]
&= \left[\mu \left(\frac{dV_r}{dr} \right)_B \right] [(2\pi r_1)(h)] r_1 \\[2mm]
&= \mu \left(\frac{\omega r_1 - 0}{r_2 - r_1} \right) [(2\pi r_1)(h)] r_1 \\[2mm]
&= \frac{2\mu \pi r_1^{3} \omega h}{r_2 - r_1}
\end{aligned}
\tag{A.I.25}
$$

Integrating over the bottom surface we get

$$
T_2 = \int_0^{2\pi} \int_0^{r_1} \frac{\mu \omega}{\epsilon} r^3 \, dr \, d\theta = \frac{\mu \omega r_1^{4}}{4\epsilon} 2\pi
\tag{A.I.26}
$$

The total torque that the pointer registers on the dial is then

$$T = T_1 + T_2 = 2\pi\mu\omega r_1^3\left(\frac{h}{r_2 - r_1} + \frac{r_1}{4\epsilon}\right) \tag{A.I.27}$$

Denoting $r_2 - r_1$ as α we have on solving for μ

$$\mu = \frac{T}{2\pi\omega r_1^3(h/\alpha + r_1/4\epsilon)} \tag{A.I.28}$$

The final method to be discussed involves the motion of a small sphere in a glass container of the liquid being tested. The inside diameter of the container should be large compared with the diameter of the sphere. If the specific weight of the sphere is close to that of the liquid, the sphere will approach a small constant speed downward after being released in the liquid. In that case we can make use of *Stoke's law* for steady creeping flow around a sphere where the drag F on the sphere of radius R is

$$F = 6\pi\mu VR \tag{A.I.29}$$

With the sphere moving at constant speed (i.e., it has reached its terminal velocity V_T) we have equilibrium, so from Newton's law the sum of the weight, the buoyant force, and the drag is zero. Thus we can say that

$$\underset{\text{Weight}}{\tfrac{4}{3}\pi R^3[\rho_S g]} - \underset{\text{Buoyant force}}{\tfrac{4}{3}\pi R^3[\rho_L g]} - \underset{\text{Drag}}{6\pi\mu V_T R} = 0$$

where ρ_S is the density of the sphere and ρ_L is the density of the liquid. Solving for μ we get

$$\mu = \frac{2}{9}\frac{gR^2}{V_T}[\rho_S - \rho_L] \tag{A.I.30}$$

The terminal velocity V_T can be measured easily by observing the time for the sphere to pass two points a known distance apart once acceleration has ceased.

A.I.8   CLOSURE

In this introductory study of measuring techniques in fluid mechanics, we have considered devices and methods you will most likely encounter in your laboratory courses in your junior and senior years. Your author hopes that this material will be of help to you in these laboratory experiments and their subsequent writeups.

APPENDIX
B

CURVES AND TABLES

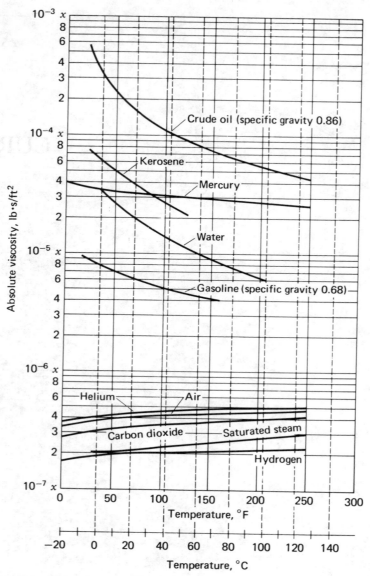

FIGURE B.1
Absolute viscosity curves. For μ in S.I. units multiply by 47.9.

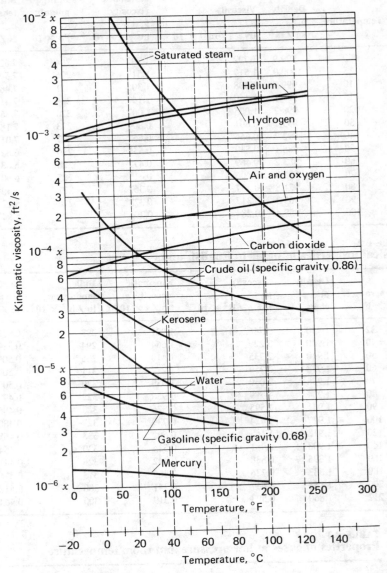

FIGURE B.2
Kinematic viscosity at atmospheric pressure. For S.I. units multiply ν by 0.0929.

The following tables have been adapted from these sources:

Ames Research Staff, *Equations, Tables, and Charts for Compressible Flow, NACA Rept*. 1135, 1953.

J. H. Keenan and J. Kaye, *Gas Tables*, Wiley, New York, 1948.

NACA. TN 1428.

TABLE B.1
Physical properties of water in S.I. units

Temperature, °C	Density ρ, kg/m^3	Viscosity μ, (N·s/m^2) × 10^{-3}	Kinematic viscosity ν, m^2/s × 10^{-6}	Bulk modulus κ, Pa × 10^7	Surface tension σ, N/m × 10^{-2}	Vapor pressure, Pa
0	999.9	1.792	1.792	204	7.62	588
5	1000.0	1.519	1.519	206	7.54	882
10	999.7	1.308	1.308	211	7.48	1,176
15	999.1	1.140	1.141	214	7.41	1,666
20	998.2	1.005	1.007	220	7.36	2,447
30	995.7	0.801	0.804	223	7.18	4,297
40	992.2	0.656	0.661	227	7.01	7,400
50	988.1	0.549	0.556	230	6.82	12,220
60	983.2	0.469	0.477	228	6.68	19,600
70	977.8	0.406	0.415	225	6.50	30,700
80	971.8	0.357	0.367	221	6.30	46,400
90	965.3	0.317	0.328	216	6.12	68,200
100	958.4	0.284	0.296	207	5.94	97,500

TABLE B.2
Physical properties of water in English units

Temperature, °F	Density ρ, slugs/ft^3	Viscosity μ, lb·s/ft^2 × 10^{-5}	Kinematic viscosity ν, ft^2/s × 10^{-5}	Bulk modulus κ, lb/in^2 × 10^3	Surface tension σ, lb/ft × 10^{-2}	Vapor pressure, lb/ft^2
32	1.940	3.746	1.931	293	0.518	12.5
40	1.940	3.229	1.664	294	0.514	17.5
50	1.940	2.735	1.410	305	0.509	25.6
60	1.938	2.359	1.217	311	0.504	36.9
70	1.936	2.050	1.059	320	0.500	52.3
80	1.934	1.799	0.930	322	0.492	72.8
90	1.931	1.595	0.826	323	0.486	100
100	1.927	1.424	0.739	327	0.480	135
120	1.918	1.168	0.609	333	0.465	241
140	1.908	0.981	0.514	330	0.454	409
160	1.896	0.838	0.442	326	0.441	668
180	1.883	0.726	0.385	313	0.426	1050
200	1.868	0.637	0.341	308	0.412	1599
212	1.860	0.593	0.319	300	0.404	2028

TABLE B.3
Properties of gases at low pressure and room temperature

Gas	k	Molecular weight	R N·m/(kg)(K)	R ft·lb/(lbm)(°R)	Specific heat, Btu/(lbm)(°R) c_p	Specific heat, Btu/(lbm)(°R) c_v	Specific heat, J/kg·K c_p	Specific heat, J/kg·K c_v
Air	1.40	29	287	53.3	0.240	0.171	1,004	717.4
CO	1.40	28	297	55.2	0.249	0.178	1,039	742.1
He	1.67	4.00	2077	386	1.25	0.753	5,225	3,147
H$_2$	1.40	2.02	4121	766	3.43	2.44	14,180	10,060
N$_2$	1.40	28.0	297	55.2	0.248	0.177	1,039	742.0

TABLE B.4
A standard-atmosphere table

Altitude, ft	Temperature, °F	Pressure in Hg	Pressure lb / ft^2	ρ / ρ_0†	c, ft / s
0	59.0	29.92	2116	1.000	1117
1,000	55.4	28.86	2041	0.9711	1113
2,000	51.9	27.82	1968	0.9428	1109
3,000	48.3	26.82	1897	0.9151	1105
4,000	44.7	25.84	1828	0.8881	1101
5,000	41.2	24.90	1761	0.8617	1098
6,000	37.6	23.98	1696	0.8359	1094
7,000	34.0	23.09	1633	0.8106	1090
8,000	30.5	22.22	1572	0.7860	1086
9,000	26.9	21.39	1513	0.7620	1082
10,000	23.3	20.58	1455	0.7385	1078
11,000	19.8	19.79	1400	0.7156	1074
12,000	16.2	19.03	1346	0.6932	1070
13,000	12.6	18.29	1294	0.6713	1066
14,000	9.1	17.58	1243	0.6500	1062
15,000	5.5	16.89	1194	0.6292	1058
16,000	1.9	16.22	1147	0.6090	1054
17,000	−1.6	15.57	1101	0.5892	1050
18,000	−5.2	14.94	1057	0.5699	1045
19,000	−8.8	14.34	1014	0.5511	1041
20,000	−12.3	13.75	972.5	0.5328	1037
21,000	−15.9	13.18	932.4	0.5150	1033
22,000	−19.5	12.64	893.7	0.4976	1029
23,000	−23.0	12.11	856.3	0.4806	1025
24,000	−26.6	11.60	820.2	0.4642	1021
25,000	−30.2	11.10	785.3	0.4481	1016
26,000	−33.7	10.63	751.6	0.4325	1012
27,000	−37.3	10.17	719.1	0.4173	1008
28,000	−40.9	9.725	687.8	0.4025	1004
29,000	−44.4	9.297	657.6	0.3881	999
30,000	−48.0	8.885	628.4	0.3741	995
31,000	−51.6	8.488	600.3	0.3605	991
32,000	−55.1	8.106	573.3	0.3473	986
33,000	−58.7	7.737	547.2	0.3345	982
34,000	−62.2	7.382	522.1	0.3220	978
35,000	−65.8	7.041	498.0	0.3099	973
36,000	−67.6	6.702	474.8	0.2971	971
37,000	−67.6	6.397	452.5	0.2844	971

†$\rho_0 = 0.002378$ slug/ft^3.

TABLE B.5
One-dimensional isentropic relations†

M	A/A^*	p/p_0	ρ/ρ_0	T/T_0	M	A/A^*	p/p_0	ρ/ρ_0	T/T_0
0.00	···	1.000	1.000	1.000	0.86	1.02	0.617	0.708	0.871
0.01	57.87	0.9999	0.9999	0.9999	0.88	1.01	0.604	0.698	0.865
0.02	28.94	0.9997	0.9999	0.9999	0.90	1.01	0.591	0.687	0.860
0.04	14.48	0.999	0.999	0.9996	0.92	1.01	0.578	0.676	0.855
0.06	9.67	0.997	0.998	0.999	0.94	1.00	0.566	0.666	0.850
0.08	7.26	0.996	0.997	0.999	0.96	1.00	0.553	0.655	0.844
0.10	5.82	0.993	0.995	0.998	0.98	1.00	0.541	0.645	0.839
0.12	4.86	0.990	0.993	0.997	1.00	1.00	0.528	0.632	0.833
0.14	4.18	0.986	0.990	0.996	1.02	1.00	0.516	0.623	0.828
0.16	3.67	0.982	0.987	0.995	1.04	1.00	0.504	0.613	0.822
0.18	3.28	0.978	0.984	0.994	1.06	1.00	0.492	0.602	0.817
0.20	2.96	0.973	0.980	0.992	1.08	1.01	0.480	0.592	0.810
0.22	2.71	0.967	0.976	0.990	1.10	1.01	0.468	0.582	0.805
0.24	2.50	0.961	0.972	0.989	1.12	1.01	0.457	0.571	0.799
0.26	2.32	0.954	0.967	0.987	1.14	1.02	0.445	0.561	0.794
0.28	2.17	0.947	0.962	0.985	1.16	1.02	0.434	0.551	0.788
0.30	2.04	0.939	0.956	0.982	1.18	1.02	0.423	0.541	0.782
0.32	1.92	0.932	0.951	0.980	1.20	1.03	0.412	0.531	0.776
0.34	1.82	0.923	0.944	0.977	1.22	1.04	0.402	0.521	0.771
0.36	1.74	0.914	0.938	0.975	1.24	1.04	0.391	0.512	0.765
0.38	1.66	0.905	0.931	0.972	1.26	1.05	0.381	0.502	0.759
0.40	1.59	0.896	0.924	0.969	1.28	1.06	0.371	0.492	0.753
0.42	1.53	0.886	0.917	0.966	1.30	1.07	0.361	0.483	0.747
0.44	1.47	0.876	0.909	0.963	1.32	1.08	0.351	0.474	0.742
0.46	1.42	0.865	0.902	0.959	1.34	1.08	0.342	0.464	0.736
0.48	1.38	0.854	0.893	0.956	1.36	1.09	0.332	0.455	0.730
0.50	1.34	0.843	0.885	0.952	1.38	1.10	0.323	0.446	0.724
0.52	1.30	0.832	0.877	0.949	1.40	1.11	0.314	0.437	0.718
0.54	1.27	0.820	0.868	0.945	1.42	1.13	0.305	0.429	0.713
0.56	1.24	0.808	0.859	0.941	1.44	1.14	0.297	0.420	0.707
0.58	1.21	0.796	0.850	0.937	1.46	1.15	0.289	0.412	0.701
0.60	1.19	0.784	0.840	0.933	1.48	1.16	0.280	0.403	0.695
0.62	1.17	0.772	0.831	0.929	1.50	1.18	0.272	0.395	0.690
0.64	1.16	0.759	0.821	0.924	1.52	1.19	0.265	0.387	0.684
0.66	1.13	0.747	0.812	0.920	1.54	1.20	0.257	0.379	0.678
0.68	1.12	0.734	0.802	0.915	1.56	1.22	0.250	0.371	0.672
0.70	1.09	0.721	0.792	0.911	1.58	1.23	0.242	0.363	0.667
0.72	1.08	0.708	0.781	0.906	1.60	1.25	0.235	0.356	0.661
0.74	1.07	0.695	0.771	0.901	1.62	1.27	0.228	0.348	0.656
0.76	1.06	0.682	0.761	0.896	1.64	1.28	0.222	0.341	0.650
0.78	1.05	0.669	0.750	0.891	1.66	1.30	0.215	0.334	0.645
0.80	1.04	0.656	0.740	0.886	1.68	1.32	0.209	0.327	0.639
0.82	1.03	0.643	0.729	0.881	1.70	1.34	0.203	0.320	0.634
0.84	1.02	0.630	0.719	0.876	1.72	1.36	0.197	0.313	0.628

TABLE B5 *Continued*

M	A/A^*	p/p_0	ρ/ρ_0	T/T_0	M	A/A^*	p/p_0	ρ/ρ_0	T/T_0
1.74	1.38	0.191	0.306	0.623	2.50	2.64	0.059	0.132	0.444
1.76	1.40	0.185	0.300	0.617	2.52	2.69	0.057	0.129	0.441
1.78	1.42	0.179	0.293	0.612	2.54	2.74	0.055	0.126	0.437
1.80	1.44	0.174	0.287	0.607	2.56	2.79	0.053	0.123	0.433
1.82	1.46	0.169	0.281	0.602	2.58	2.84	0.052	0.121	0.429
1.84	1.48	0.164	0.275	0.596	2.60	2.90	0.050	0.118	0.425
1.86	1.51	0.159	0.269	0.591	2.62	2.95	0.049	0.115	0.421
1.88	1.53	0.154	0.263	0.586	2.64	3.01	0.047	0.113	0.418
1.90	1.56	0.149	0.257	0.581	2.66	3.06	0.046	0.110	0.414
1.92	1.58	0.145	0.251	0.576	2.68	3.12	0.044	0.108	0.410
1.94	1.61	0.140	0.246	0.571	2.70	3.18	0.043	0.106	0.407
1.96	1.63	0.136	0.240	0.566	2.72	3.24	0.042	0.103	0.403
1.98	1.66	0.132	0.235	0.561	2.74	3.31	0.040	0.101	0.400
2.00	1.69	0.128	0.230	0.556	2.76	3.37	0.039	0.099	0.396
2.02	1.72	0.124	0.225	0.551	2.78	3.43	0.038	0.097	0.393
2.04	1.75	0.120	0.220	0.546	2.80	3.50	0.037	0.095	0.389
2.06	1.78	0.116	0.215	0.541	2.82	3.57	0.036	0.093	0.386
2.08	1.81	0.113	0.210	0.536	2.84	3.64	0.035	0.091	0.383
2.10	1.84	0.109	0.206	0.531	2.86	3.71	0.034	0.089	0.379
2.12	1.87	0.106	0.201	0.526	2.88	3.78	0.033	0.087	0.376
2.14	1.90	0.103	0.197	0.522	2.90	3.85	0.032	0.085	0.373
2.16	1.94	0.100	0.192	0.517	2.92	3.92	0.031	0.083	0.370
2.18	1.97	0.097	0.188	0.513	2.94	4.00	0.030	0.081	0.366
2.20	2.01	0.094	0.184	0.508	2.96	4.08	0.029	0.080	0.363
2.22	2.04	0.091	0.180	0.504	2.98	4.15	0.028	0.078	0.360
2.24	2.08	0.088	0.176	0.499	3.00	4.23	0.027	0.076	0.357
2.26	2.12	0.085	0.172	0.495	3.10	4.66	0.023	0.0685	0.342
2.28	2.15	0.083	0.168	0.490	3.20	5.12	0.020	0.062	0.328
2.30	2.19	0.080	0.165	0.486	3.3	5.63	0.0175	0.0555	0.315
2.32	2.23	0.078	0.161	0.482	3.4	6.18	0.015	0.050	0.302
2.34	2.27	0.075	0.157	0.477	3.5	6.79	0.013	0.045	0.290
2.36	2.32	0.073	0.154	0.473	3.6	7.45	0.0114	0.041	0.278
2.38	2.36	0.071	0.150	0.469	3.7	8.17	0.0099	0.037	0.2675
2.40	2.40	0.068	0.147	0.465	3.8	8.95	0.0086	0.0335	0.257
2.42	2.45	0.066	0.144	0.461	3.9	9.80	0.0075	0.030	0.247
2.44	2.49	0.064	0.141	0.456	4.0	10.72	0.0066	0.028	0.238
2.46	2.54	0.062	0.138	0.452					
2.48	2.59	0.060	0.135	0.448					

†For a perfect gas with constant specific heat, $k = 1.4$

TABLE B.6
One-dimensional normal-shock relations†

M_1	M_2	$\dfrac{p_2}{p_1}$	$\dfrac{T_2}{T_1}$	$\dfrac{(p_0)_2}{(p_0)_1}$	$\dfrac{(A*)_2}{(A*)_1}$	M_1	M_2	$\dfrac{p_2}{p_1}$	$\dfrac{T_2}{T_1}$	$\dfrac{(p_0)_2}{(p_0)_1}$	$\dfrac{(A*)_2}{(A*)_1}$
1.00	1.000	1.000	1.000	1.000	1.000	2.02	0.574	4.594	1.704	0.711	1.411
1.02	0.980	1.047	1.013	1.000	1.000	2.04	0.571	4.689	1.720	0.702	1.430
1.04	0.962	1.095	1.026	1.000	1.000	2.06	0.567	4.784	1.737	0.693	1.447
1.06	0.944	1.144	1.039	1.000	1.000	2.08	0.564	4.881	1.754	0.683	1.467
1.08	0.928	1.194	1.052	0.999	1.000	2.10	0.561	4.978	1.770	0.674	1.485
1.10	0.912	1.245	1.065	0.999	1.000	2.12	0.558	5.077	1.787	0.665	1.504
1.12	0.896	1.297	1.078	0.998	1.000	2.14	0.555	5.176	1.805	0.656	1.522
1.14	0.882	1.350	1.090	0.997	1.010	2.16	0.553	5.277	1.822	0.646	1.551
1.16	0.868	1.403	1.103	0.996	1.000	2.18	0.550	5.378	1.839	0.637	1.570
1.18	0.855	1.458	1.115	0.995	1.000	2.20	0.547	5.480	1.857	0.628	1.667
1.20	0.842	1.513	1.128	0.993	1.010	2.22	0.544	5.583	1.875	0.619	1.614
1.22	0.830	1.570	1.140	0.991	1.015	2.24	0.542	5.687	1.892	0.610	1.642
1.24	0.818	1.627	1.153	0.988	1.010	2.26	0.539	5.792	1.910	0.601	1.667
1.26	0.807	1.686	1.166	0.986	1.013	2.28	0.537	5.898	1.929	0.592	1.686
1.28	0.796	1.745	1.178	0.983	1.017	2.30	0.534	6.005	1.947	0.583	1.712
1.30	0.786	1.805	1.191	0.979	1.022	2.32	0.532	6.113	1.965	0.575	1.739
1.32	0.776	1.866	1.204	0.976	1.027	2.34	0.530	6.222	1.984	0.566	1.767
1.34	0.766	1.928	1.216	0.972	1.022	2.36	0.527	6.331	2.003	0.557	1.798
1.36	0.757	1.991	1.229	0.968	1.026	2.38	0.525	6.442	2.021	0.549	1.825
1.38	0.748	2.055	1.242	0.963	1.032	2.40	0.523	6.553	2.040	0.540	1.852
1.40	0.740	2.120	1.255	0.958	1.037	2.42	0.521	6.666	2.060	0.532	1.886
1.42	0.731	2.186	1.268	0.953	1.051	2.44	0.519	6.779	2.079	0.523	1.912
1.44	0.723	2.253	1.281	0.948	1.057	2.46	0.517	6.894	2.098	0.515	1.945
1.46	0.716	2.320	1.294	0.942	1.063	2.48	0.515	7.009	2.118	0.507	1.977
1.48	0.708	2.389	1.307	0.936	1.068	2.50	0.513	7.125	2.138	0.499	2.009
1.50	0.701	2.458	1.320	0.930	1.083	2.52	0.511	7.242	2.157	0.491	2.041
1.52	0.694	2.529	1.334	0.923	1.077	2.54	0.509	7.360	2.177	0.483	2.073
1.54	0.687	2.600	1.347	0.917	1.081	2.56	0.507	7.479	2.198	0.475	2.104
1.56	0.681	2.673	1.361	0.910	1.090	2.58	0.506	7.599	2.218	0.468	2.139
1.58	0.675	2.746	1.374	0.903	1.095	2.60	0.504	7.720	2.238	0.460	2.177
1.60	0.668	2.820	1.388	0.895	1.110	2.62	0.502	7.842	2.260	0.453	2.208
1.62	0.663	2.895	1.402	0.888	1.125	2.64	0.500	7.965	2.280	0.445	2.246
1.64	0.657	2.971	1.416	0.880	1.128	2.66	0.499	8.088	2.301	0.438	2.280
1.66	0.651	3.048	1.430	0.872	1.136	2.68	0.497	8.213	2.322	0.431	2.318
1.68	0.646	3.126	1.444	0.864	1.147	2.70	0.496	8.338	2.343	0.424	2.359
1.70	0.641	3.205	1.458	0.856	1.156	2.72	0.494	8.465	2.364	0.417	2.396
1.72	0.635	3.285	1.473	0.847	1.169	2.74	0.493	9.592	2.396	0.410	2.445
1.74	0.631	3.366	1.487	0.839	1.185	2.76	0.491	8.721	2.407	0.403	2.482
1.76	0.626	3.447	1.502	0.830	1.20	2.78	0.490	8.850	2.429	0.396	2.522
1.78	0.621	3.530	1.517	0.821	1.214	2.80	0.488	8.980	2.451	0.389	2.566
1.80	0.617	3.613	1.532	0.813	1.228	2.82	0.487	9.111	2.473	0.383	2.613
1.82	0.612	3.698	1.547	0.804	1.239	2.84	0.485	9.243	2.496	0.376	2.657
1.84	0.608	3.783	1.562	0.795	1.252	2.86	0.484	9.376	2.518	0.370	2.704
1.86	0.604	3.869	1.577	0.786	1.273	2.88	0.483	9.510	2.541	0.364	2.751
1.88	0.600	3.957	1.592	0.777	1.286	2.90	0.481	9.645	2.563	0.358	2.794
1.90	0.596	4.045	1.608	0.767	1.307	2.92	0.480	9.781	2.586	0.352	2.841
1.92	0.592	4.134	1.624	0.758	1.319	2.94	0.479	9.918	2.609	0.346	2.894
1.94	0.588	4.224	1.639	0.749	1.339	2.96	0.478	10.055	2.632	0.340	2.948
1.96	0.584	4.315	1.655	0.740	1.352	2.98	0.476	10.194	2.656	0.334	2.990
1.98	0.581	4.407	1.671	0.730	1.373	3.00	0.475	10.333	2.679	0.328	3.043
2.00	0.577	4.500	1.688	0.721	1.391						

†For a perfect gas with $k = 1.4$.

TABLE B.7
Fanno line†

M	$\dfrac{T}{T^*}$	$\dfrac{p}{p^*}$	$\dfrac{p_0}{p_0^*}$	$\dfrac{V}{V^*}$	M	$\dfrac{T}{T^*}$	$\dfrac{p}{p^*}$	$\dfrac{p_0}{p_0^*}$	$\dfrac{V}{V^*}$
0	1.200	∞	∞	0	0.78	1.070	1.33	1.05	0.807
0.01	1.200	109.54	57.87	0.011	0.80	1.064	1.29	1.04	0.825
0.02	1.200	57.77	28.94	0.022	0.82	1.058	1.25	1.03	0.843
0.04	1.200	27.38	14.48	0.044	0.84	1.052	1.22	1.02	0.861
0.06	1.199	18.25	9.67	0.066	0.86	1.045	1.19	1.02	0.879
0.08	1.199	13.68	7.26	0.088	0.88	1.039	1.16	1.01	0.897
0.10	1.198	10.94	5.82	0.109	0.90	1.033	1.13	1.01	0.914
0.12	1.197	9.12	4.86	0.131	0.92	1.026	1.10	1.01	0.932
0.14	1.195	7.81	4.18	0.153	0.94	1.020	1.07	1.00	0.949
0.16	1.194	6.83	3.67	0.175	0.96	1.013	1.05	1.00	0.966
0.18	1.192	6.07	3.28	0.197	0.98	1.007	1.02	1.00	0.983
0.20	1.191	5.46	2.96	0.218	1.00	1.000	1.00	1.00	1.000
0.22	1.189	4.96	2.71	0.240	1.02	0.993	0.98	1.00	1.016
0.24	1.186	4.54	2.50	0.261	1.04	0.986	0.96	1.00	1.033
0.26	1.184	4.19	2.32	0.283	1.06	0.980	0.93	1.00	1.049
0.28	1.182	3.88	2.17	0.304	1.08	0.973	0.91	1.01	1.065
0.30	1.179	3.62	2.04	0.326	1.10	0.966	0.89	1.01	1.081
0.32	1.176	3.39	1.92	0.347	1.12	0.959	0.87	1.01	1.097
0.34	1.173	3.19	1.82	0.368	1.14	0.952	0.86	1.02	1.113
0.36	1.170	3.00	1.74	0.389	1.16	0.946	0.84	1.02	1.128
0.38	1.166	2.84	1.66	0.410	1.18	0.939	0.82	1.02	1.143
0.40	1.163	2.70	1.59	0.431	1.20	0.932	0.80	1.03	1.158
0.42	1.159	2.56	1.53	0.452	1.22	0.925	0.79	1.04	1.173
0.44	1.155	2.44	1.47	0.473	1.24	0.918	0.77	1.04	1.188
0.46	1.151	2.33	1.42	0.494	1.26	0.911	0.76	1.05	1.203
0.48	1.147	2.23	1.38	0.514	1.28	0.904	0.74	1.06	1.217
0.50	1.143	2.14	1.34	0.535	1.30	0.897	0.73	1.07	1.231
0.52	1.138	2.05	1.30	0.555	1.32	0.890	0.71	1.08	1.245
0.54	1.134	1.97	1.27	0.575	1.34	0.883	0.70	1.08	1.259
0.56	1.129	1.90	1.24	0.595	1.36	0.876	0.69	1.09	1.273
0.58	1.124	1.83	1.21	0.615	1.38	0.869	0.68	1.10	1.286
0.60	1.119	1.76	1.19	0.635	1.40	0.862	0.66	1.11	1.300
0.62	1.114	1.70	1.17	0.654	1.42	0.855	0.65	1.13	1.313
0.64	1.109	1.65	1.15	0.674	1.44	0.848	0.64	1.14	1.326
0.66	1.104	1.59	1.13	0.693	1.46	0.841	0.63	1.15	1.339
0.68	1.098	1.54	1.11	0.713	1.48	0.834	0.62	1.16	1.352
0.70	1.093	1.49	1.09	0.732	1.50	0.828	0.61	1.18	1.365
0.72	1.087	1.45	1.08	0.751	1.52	0.821	0.60	1.19	1.377
0.74	1.082	1.41	1.07	0.770	1.54	0.814	0.59	1.20	1.389
0.76	1.076	1.36	1.06	0.788	1.56	0.807	0.58	1.22	1.402

TABLE B.7 *Continued*

M	$\dfrac{T}{T^*}$	$\dfrac{p}{p^*}$	$\dfrac{p_0}{p_0^*}$	$\dfrac{V}{V^*}$	M	$\dfrac{T}{T^*}$	$\dfrac{p}{p^*}$	$\dfrac{p_0}{p_0^*}$	$\dfrac{V}{V^*}$
1.58	0.800	0.57	1.23	1.414	2.30	0.583	0.33	2.19	1.756
1.60	0.794	0.56	1.25	1.425	2.32	0.578	0.33	2.23	1.764
1.62	0.787	0.55	1.27	1.437	2.34	0.573	0.32	2.27	1.771
1.64	0.780	0.54	1.28	1.449	2.36	0.568	0.32	2.32	1.778
1.66	0.774	0.53	1.30	1.460	2.38	0.563	0.32	2.36	1.785
1.68	0.767	0.52	1.32	1.471	2.40	0.558	0.31	2.40	1.792
1.70	0.760	0.51	1.34	1.483	2.42	0.553	0.31	2.45	1.799
1.72	0.754	0.50	1.36	1.494	2.44	0.548	0.30	2.49	1.806
1.74	0.747	0.50	1.38	1.504	2.46	0.543	0.30	2.54	1.813
1.76	0.741	0.49	1.40	1.515	2.48	0.538	0.30	2.59	1.819
1.78	0.735	0.48	1.42	1.526	2.50	0.533	0.29	2.64	1.826
1.80	0.728	0.47	1.44	1.536	2.52	0.529	0.29	2.69	1.832
1.82	0.722	0.47	1.46	1.546	2.54	0.524	0.28	2.74	1.839
1.84	0.716	0.46	1.48	1.556	2.56	0.519	0.28	2.79	1.845
1.86	0.709	0.45	1.51	1.566	2.58	0.515	0.28	2.84	1.851
1.88	0.703	0.45	1.53	1.576	2.60	0.510	0.27	2.90	1.857
1.90	0.697	0.44	1.56	1.586	2.62	0.506	0.27	2.95	1.863
1.92	0.691	0.43	1.58	1.596	2.64	0.501	0.27	3.01	1.869
1.94	0.685	0.43	1.61	1.605	2.66	0.497	0.26	3.06	1.875
1.96	0.679	0.42	1.63	1.615	2.68	0.492	0.26	3.12	1.881
1.98	0.673	0.41	1.66	1.624	2.70	0.488	0.26	3.18	1.886
2.00	0.667	0.41	1.69	1.633	2.72	0.484	0.26	3.24	1.892
2.02	0.661	0.40	1.72	1.642	2.74	0.480	0.25	3.31	1.898
2.04	0.655	0.40	1.75	1.651	2.76	0.476	0.25	3.37	1.903
2.06	0.649	0.39	1.78	1.660	2.78	0.471	0.25	3.43	1.909
2.08	0.643	0.38	1.81	1.668	2.80	0.467	0.24	3.50	1.914
2.10	0.638	0.38	1.84	1.677	2.82	0.463	0.24	3.57	1.919
2.12	0.632	0.37	1.87	1.685	2.84	0.459	0.24	3.64	1.925
2.14	0.626	0.37	1.90	1.694	2.86	0.455	0.24	3.71	1.930
2.16	0.621	0.36	1.94	1.702	2.88	0.451	0.23	3.78	1.935
2.18	0.615	0.36	1.97	1.710	2.90	0.447	0.23	3.85	1.940
2.20	0.610	0.35	2.00	1.718	2.92	0.444	0.23	3.92	1.945
2.22	0.604	0.35	2.04	1.726	2.94	0.440	0.22	4.00	1.950
2.24	0.599	0.34	2.08	1.734	2.96	0.436	0.22	4.08	1.954
2.26	0.594	0.34	2.12	1.741	2.98	0.432	0.22	4.15	1.959
2.28	0.588	0.34	2.15	1.749	3.00	0.428	0.22	4.23	1.964

†For a perfect gas with $k = 1.4$.

TABLE B.8
Rayleigh line†

M	$\dfrac{T_0}{T_0^*}$	$\dfrac{T}{T^*}$	$\dfrac{p}{p^*}$	$\dfrac{p_0}{p_0^*}$	$\dfrac{V}{V^*}$	M	$\dfrac{T_0}{T_0^*}$	$\dfrac{T}{T^*}$	$\dfrac{p}{p^*}$	$\dfrac{p_0}{p_0^*}$	$\dfrac{V}{V^*}$
0	0	0	2.40	1.27	0	0.86	0.984	1.028	1.179	1.010	0.872
0.01	0.000	0.000	2.40	1.27	0.000	0.88	0.988	1.027	1.152	1.007	0.892
0.02	0.002	0.002	2.40	1.27	0.001	0.90	0.992	1.024	1.125	1.005	0.911
0.04	0.008	0.009	2.39	1.27	0.004	0.92	0.995	1.021	1.098	1.003	0.930
0.06	0.017	0.020	2.39	1.26	0.009	0.94	0.997	1.017	1.073	1.002	0.948
0.08	0.030	0.036	2.38	1.26	0.015	0.96	0.999	1.012	1.048	1.001	0.966
0.10	0.047	0.056	2.37	1.26	0.024	0.98	1.000	1.006	1.024	1.000	0.983
0.12	0.067	0.080	2.35	1.26	0.034	1.00	1.000	1.000	1.000	1.000	1.000
0.14	0.089	0.107	2.34	1.25	0.046	1.02	1.000	0.993	0.977	1.000	1.016
0.16	0.115	0.137	2.32	1.25	0.059	1.04	0.999	0.986	0.954	1.001	1.032
0.18	0.143	0.171	2.30	1.24	0.074	1.06	0.998	0.978	0.933	1.002	1.048
0.20	0.174	0.207	2.27	1.23	0.091	1.08	0.996	0.969	0.911	1.003	1.063
0.22	0.206	0.244	2.25	1.23	0.109	1.10	0.994	0.960	0.891	1.005	1.078
0.24	0.239	0.284	2.22	1.22	0.128	1.12	0.991	0.951	0.871	1.007	1.092
0.26	0.274	0.325	2.19	1.21	0.148	1.14	0.989	0.942	0.851	1.010	1.106
0.28	0.310	0.367	2.16	1.21	0.170	1.16	0.986	0.932	0.832	1.012	1.120
0.30	0.347	0.409	2.13	1.20	0.192	1.18	0.982	0.922	0.814	1.016	1.133
0.32	0.384	0.451	2.10	1.19	0.215	1.20	0.979	0.912	0.796	1.019	1.146
0.34	0.421	0.493	2.06	1.18	0.239	1.22	0.975	0.902	0.778	1.023	1.158
0.36	0.457	0.535	2.03	1.17	0.263	1.24	0.971	0.891	0.761	1.028	1.171
0.38	0.493	0.576	2.00	1.16	0.288	1.26	0.967	0.881	0.745	1.033	1.182
0.40	0.529	0.615	1.96	1.16	0.314	1.28	0.962	0.870	0.729	1.038	1.194
0.42	0.564	0.653	1.92	1.15	0.340	1.30	0.958	0.859	0.713	1.044	1.205
0.44	0.597	0.690	1.89	1.14	0.366	1.32	0.953	0.848	0.698	1.050	1.216
0.46	0.630	0.725	1.85	1.13	0.392	1.34	0.949	0.838	0.683	1.056	1.226
0.48	0.661	0.759	1.81	1.12	0.418	1.36	0.944	0.827	0.669	1.063	1.237
0.50	0.691	0.790	1.78	1.11	0.444	1.38	0.939	0.816	0.655	1.070	1.247
0.52	0.720	0.820	1.74	1.10	0.471	1.40	0.934	0.805	0.641	1.078	1.256
0.54	0.747	0.847	1.70	1.10	0.497	1.42	0.929	0.795	0.628	1.086	1.266
0.56	0.772	0.872	1.67	1.09	0.523	1.44	0.924	0.784	0.615	1.094	1.275
0.58	0.796	0.896	1.63	1.08	0.549	1.46	0.919	0.773	0.602	1.103	1.284
0.60	0.819	0.917	1.60	1.08	0.574	1.48	0.914	0.763	0.590	1.112	1.293
0.62	0.840	0.936	1.56	1.07	0.600	1.50	0.909	0.752	0.578	1.122	1.301
0.64	0.859	0.953	1.52	1.06	0.625	1.52	0.904	0.742	0.567	1.132	1.309
0.66	0.877	0.968	1.49	1.06	0.649	1.54	0.899	0.732	0.556	1.142	1.318
0.68	0.894	0.981	1.46	1.05	0.674	1.56	0.894	0.722	0.544	1.153	1.325
0.70	0.908	0.993	1.423	1.043	0.698	1.58	0.889	0.712	0.534	1.164	1.333
0.72	0.922	1.003	1.391	1.038	0.721	1.60	0.884	0.702	0.524	1.176	1.340
0.74	0.934	1.011	1.358	1.032	0.744	1.62	0.879	0.692	0.513	1.188	1.348
0.76	0.945	1.017	1.327	1.028	0.766	1.64	0.874	0.682	0.504	1.200	1.355
0.78	0.955	1.022	1.296	1.023	0.788	1.66	0.869	0.672	0.494	1.213	1.361
0.80	0.964	1.025	1.266	1.019	0.810	1.68	0.864	0.663	0.485	1.226	1.368
0.82	0.972	1.028	1.236	1.016	0.831	1.70	0.860	0.654	0.476	1.240	1.374
0.84	0.978	1.028	1.207	1.012	0.852	1.72	0.855	0.644	0.467	1.254	1.381

TABLE B.8 *Continued*

M	$\dfrac{T_0}{T_0^*}$	$\dfrac{T}{T^*}$	$\dfrac{p}{p^*}$	$\dfrac{p_0}{p_0^*}$	$\dfrac{V}{V^*}$	M	$\dfrac{T_0}{T_0^*}$	$\dfrac{T}{T^*}$	$\dfrac{p}{p^*}$	$\dfrac{p_0}{p_0^*}$	$\dfrac{V}{V^*}$
1.74	0.850	0.635	0.458	1.269	1.387	2.38	0.727	0.409	0.269	2.012	1.522
1.76	0.846	0.626	0.450	1.284	1.393	2.40	0.724	0.404	0.265	2.045	1.525
1.78	0.841	0.618	0.442	1.300	1.399	2.42	0.721	0.399	0.261	2.079	1.528
1.80	0.836	0.609	0.434	1.316	1.405	2.44	0.718	0.384	0.257	2.114	1.531
1.82	0.832	0.600	0.426	1.332	1.410	2.46	0.716	0.388	0.253	2.149	1.533
1.84	0.827	0.592	0.418	1.349	1.416	2.48	0.713	0.384	0.250	2.185	1.536
1.86	1.823	0.584	0.411	1.367	1.421	2.50	0.710	0.379	0.246	2.222	1.538
1.88	0.818	0.575	0.403	1.385	1.426	2.52	0.707	0.374	0.243	2.259	1.541
1.90	0.814	0.567	0.396	1.403	1.431	2.54	0.705	0.369	0.239	2.298	1.543
1.92	0.810	0.559	0.390	1.422	1.436	2.56	0.702	0.365	0.236	2.337	1.546
1.94	0.806	0.552	0.383	1.442	1.441	2.58	0.700	0.360	0.232	2.377	1.548
1.96	0.802	0.544	0.376	1.462	1.446	2.60	0.697	0.356	0.229	2.418	1.551
1.98	0.797	0.536	0.370	1.482	1.450	2.62	0.694	0.351	0.226	2.459	1.553
2.00	0.793	0.529	0.364	1.503	1.454	2.64	0.692	0.347	0.223	2.502	1.555
2.02	0.789	0.522	0.357	1.525	1.459	2.66	0.690	0.343	0.220	2.545	1.557
2.04	0.785	0.514	0.352	1.547	1.463	2.68	0.687	0.338	0.217	2.589	1.559
2.06	0.782	0.507	0.346	1.569	1.467	2.70	0.685	0.334	0.214	2.634	1.561
2.08	0.778	0.500	0.340	1.592	1.471	2.72	0.683	0.330	0.211	2.680	1.563
2.10	0.774	0.494	0.334	1.616	1.475	2.74	0.680	0.326	0.208	2.727	1.565
2.12	0.770	0.487	0.329	1.640	1.479	2.76	0.678	0.322	0.206	2.775	1.567
2.14	0.767	0.480	0.324	1.665	1.483	2.78	0.676	0.319	0.203	2.824	1.569
2.16	0.763	0.474	0.319	1.691	1.487	2.80	0.674	0.315	0.200	2.873	1.571
2.18	0.760	0.467	0.314	1.717	1.490	2.82	0.672	0.311	0.198	2.924	1.573
2.20	0.756	0.461	0.309	1.743	1.494	2.84	0.670	0.307	0.195	2.975	1.575
2.22	0.753	0.455	0.304	1.771	1.497	2.86	0.668	0.304	0.193	3.028	1.577
2.24	0.749	0.449	0.299	1.799	1.501	2.88	0.665	0.300	0.190	3.081	1.578
2.26	0.746	0.443	0.294	1.827	1.504	2.90	0.664	0.297	0.188	3.136	1.580
2.28	0.743	0.437	0.290	1.856	1.507	2.92	0.662	0.293	0.186	3.191	1.582
2.30	0.740	0.431	0.286	1.886	1.510	2.94	0.660	0.290	0.183	3.248	1.583
2.32	0.736	0.426	0.281	1.916	1.513	2.96	0.658	0.287	0.181	3.306	1.585
2.34	0.733	0.420	0.277	1.948	1.516	2.98	0.656	0.283	0.179	3.365	1.587
2.36	0.730	0.414	0.273	1.979	1.520	3.00	0.654	0.280	0.176	3.424	1.588

†Perfect gas, $k = 1.4$.

INDEX